FOETAL AND NEONATAL PHYSIOLOGY

Published by the Syndics of the Cambridge University Press
Bentley House, 200 Euston Road, London NW1 2DB
American Branch: 32 East 57th Street, New York, N.Y.10022

Library of Congress Catalogue Card Number: 72-93673

ISBN: 0 521 20178 0

Printed in Great Britain
at the University Printing House, Cambridge
(Brooke Crutchley, University Printer)

Joseph Barcroft.
3 Sept. 1939

FOETAL AND NEONATAL PHYSIOLOGY

PROCEEDINGS OF THE SIR JOSEPH BARCROFT CENTENARY SYMPOSIUM
HELD AT
THE PHYSIOLOGICAL LABORATORY
CAMBRIDGE
25 TO 27 JULY 1972

CAMBRIDGE
AT THE UNIVERSITY PRES
1973

CONTENTS

SESSION 3: PLACENTAL FUNCTION

Chairman: PROFESSOR J. METCALFE

SESSION 4: METABOLISM

Chairman: PROFESSOR R. A. McCANCE FRS

FOREWORD

When, in 1969, the Committee of the Physiological Society invited me to Chair a Sub-Committee to celebrate the Centenary of the birth of Sir Joseph Barcroft (26 July 1872) by organising a Symposium on foetal and neonatal physiology there were some matters which required no thought at all and others which presented mammoth problems.

It was obvious that the Symposium must be in his old Laboratory in Cambridge. King's was his College and the Commemoration Dinner should be held there, attended by his family, his old colleagues, and those of us who consciously or unconsciously celebrate his work in our present pursuits. The great fund of good will which he had generated needed to be drawn upon so that 'The Chancellor, the Scholars and the Members of this University' all co-operated to achieve these simple but splendid ends.

When one came to the details of the Symposium one could no longer drift so effortlessly. Barcroft's work on haemoglobin and ruminant digestion had already been considered at other Symposia and his breadth of interest indicated that we should take no narrow view of his work on the foetus. How wide could we afford to be? In these days of specialisation could we in any way be said to be celebrating Barcroft if we had the maximum number crowded on some narrow point – however much this could be called a *growing* point?

We decided that we would not limit ourselves to the particular aspects of the foetus in which he had been interested (and he was interested in a great deal) but would look at 'the present state of the art' where the position was lively and changing and seemed likely to remain in that stimulating state. There had to be ruthless exclusions. The immunology of the foetus and maternal tolerance to this unruly antigen was excluded. This was not because the Committee felt the subject to be irrelevant or unimportant but because it would have required all and more of the time available.

Three years ago we set about choosing contributors who had clear-cut features; these were men who, in a further three years, were still to be active in the field, with something fresh to say and an ability to say it. It is a great source of satisfaction to me to know that there are 68 members of the World Scientific Community who each unreservedly accepted one of our decisions.

In editing this volume we have decided, contrary to present trends, not to present the discussion. The papers appear to us to be sufficiently comprehensive to present an up-to-date view over a considerable range, and my colleagues and I on the editorial board felt that our main task was to accelerate publication by all means at our disposal. Probably the most valuable discussion occurred around the courts of St Catharine's, King's and Corpus well into each night.

The Symposium ended on 27 July 1972 and is sent to press as I date this, with some pride, 11 August 1972.

K. W. CROSS

The Members of the Barcroft Sub-Committee of the Physiological Society: H. Barcroft FRS, D. H. Barron, R. H. Beard,* R. S. Comline,* K. W. Cross,* G. S. Dawes FRS,* R. J. Fitzpatrick, Sir Bryan Matthews FRS, P. W. Nathanielsz*

* Organising Committee.

ACKNOWLEDGEMENTS

The Physiological Society expresses its gratitude for support from:

The Wellcome Trust
The National Institutes of Health, USA
The Commonwealth Foundation
The Royal Society
The International Union of Physiological Sciences
The National Foundation, USA
The Neonatal Society
The Blair Bell Society
Glaxo Laboratories
Smith, Kline & French Research Institute
ICI
An undisclosed American benefactor who gave repeated donations 'for the Barcroft Dinner'.

The Society is also grateful to the many Universities and Research Institutes whose support has been indispensable to the success of this Symposium.

We would most particularly mention crucial support from the Wellcome Trustees who guaranteed even further funds if all else failed in a financially disturbed period. The National Institutes of Health gave most understanding and considerable help – particularly for an expatriate symposium. Lastly, an earmarked grant from Smith, Kline & French Research Institute for the pre-publication of the participants' illustrations made ready comprehension and recollection of the useful material a great aid to discussion.

All our secretaries undertook much extra work to make communication possible, but we would particularly like to mention Miss M. Sutherland who was one of Sir Joseph's last secretaries, who organised the Secretariat and made available to us her great experience in this difficult field. We asked much of the technical and administrative staff of the Physiological Laboratory in Cambridge and their co-operation helped enormously to make the Symposium a success.

SIR JOSEPH BARCROFT

Few men of science endeared themselves to their students, their colleagues and their friends as Joseph Barcroft; few contributed so much to the advancement of physiology in the first half of this century. His pupils have carried his teachings to leading positions in science and in medicine in Britain and abroad; his researches won him a place in that brilliant succession of British students of respiration which includes Boyle, Hales, Priestley and Haldane. His monograph on *The Respiratory Function of the Blood* and his treatise on *Features in the Architecture of Physiological Function* are classics in the literature of physiology, and the growing interest in the physiology of intra-uterine life has its roots in the last book he wrote, *Researches on Pre-natal Life.*

Concerned as he was with the affairs of science, active in research and exploring new lines until the final moments of his life on 21 March 1947, Barcroft's death was felt as a personal loss by men in all parts of the world. To honour Barcroft and to pay tribute to his memory, plans were soon made by a group of his friends for a conference on his major research interest, haemoglobin, to be held in Cambridge (England), where he spent the whole of his career in research and teaching.

The opening session of the conference, held in June of the following year, was appropriately devoted to personal reminiscences of Barcroft – 'J.B.' as he was known to his friends – and to tributes to his many fine personal qualities by colleagues in science who had been associated with him and knew him intimately. Each tribute bore testimony to the wise and good man Joseph Barcroft was, to his optimism and to his faith in the goodness of man, to his courage and to his lively sense of humour, which, next to courage, is the most precious of human qualities. These tributes, published in the informal style in which they were presented, reveal, perhaps better than anything that has been written about Barcroft, how he endeared himself to so many and why he continues to live on in the hearts of those who knew him.

Joseph Barcroft was born on 26 July 1872 to Henry and Anna – née Malcolmson – Barcroft at Newry, County Down, Northern Ireland. His parents were of Quaker stock and from them he appears to have inherited that emotional balance for which so many strive but so few achieve. It never deserted him. Equanimity and self-mastery came early to him, for he once remarked in his late sixties that he could recall the place in the family garden where, at the age of four, he resolved never to cry again. His heritage appears to have included, as well, a belief that the only guide to a man's conduct must be his own 'inner light', that truth is to be sought, that life is to be lived, not experienced. These guides to life appear to have been inborn, for those who knew him at school, and enjoyed his close friendship afterwards, mention that Barcroft seemed to change with the years far less than most men. As a lad he had the emotional balance of maturity, as a man he retained the child's wonder, a love of learning and a boyish enthusiasm for his enterprises.

The family environment was a comfortable one in which broad interests were encouraged – interests in the out-of-doors, in recreational sports, in reading and in painting. Curiosity was nurtured by his mother. Questions were asked in the family circle; answers were sought. The availability of a well-equipped workshop invited expression of an inventive interest in things mechanical which Barcroft appears to have inherited from his father.

Barcroft's formal education, which began at the age of seven with a tutor, was continued at the Friends' School in Bootham, York, and then at the Leys School in Cambridge where he entered in the summer of 1888. The Leys was chosen because it was one of the few nonconformist schools in Britain in the 1880s where science, his area of interest, was encouraged. For reasons which are not clear, the year after he entered the Leys, in 1889, Barcroft matriculated at the University of London, then only an examining and degree-granting body. Two years later he was awarded the B.Sc., an unusual accomplishment for someone still at school.

But the preparation for the examination undoubtedly required a concentration of effort which limited his associations at school and it appears to have taxed his health to the degree that he was advised, after he left the Leys in the summer of 1892, to spend a year resting and recuperating at home with his family. Parenthetically, in the final chapter of the *Features in the Architecture* . . . he refers to this illness in his discussion of 'The Significance of Pain'.

When he went up to King's College, Cambridge, in October 1893 it was Barcroft's intention to read sufficiently in the physical sciences to provide a background on which to base a career in the law of patents. He was interested in science, but saw it as a supplement to a career and not as the basis for one. Not until some time after he began to try his hand at research in physiology and at teaching did he abandon the plan to read law, for he was admitted into Lincoln's Inn in January 1898, two years after he obtained the B.A. with first class honours in Part I of the Natural Sciences Tripos; in 1897 he was awarded a first class in Part II.

Barcroft was introduced, as he put it later, to 'the fascination of physiology' by C. W. Cummins, the senior science master at the Leys, and his interest was furthered by his supervisor in physiology, Hugh K. Anderson, a fellow and later Master of Caius. Anderson, known to physiologists outside Britain through his collaboration with Langley, in studies on the autonomic nervous system, was one of the best teachers of the day. All he taught was presented with an original outlook. Barcroft's fascination was further increased by W. H. Gaskell's Part II lectures. As Gaskell's two books demonstrate, the one on *The Involuntary Nervous System* and the other on *The Origin of the Vertebrates*, he was a man of ideas with an interest in broad generalizations. Gaskell used both to excite the minds of the Part II students to whom he conveyed his own enthusiasm for research, presenting it as an intellectual adventure.

In both of these men Barcroft found much to admire personally, in their 'way of life' and in their teaching. Anderson became one of his closest friends. Encouraged by them he decided to test his aptitudes for a career in research and in the teaching of physiology before entering upon the study of law. In later years when speaking of his undergraduate days and his early years of research the names most often on his lips were those of Anderson and Gaskell.

When Barcroft began to try his hand at research in the autumn of 1897, the subject he chose was one suggested by Langley, 'The Metabolism of the sub-Maxillary Gland'. Because of the accessibility of its duct, the veins that drain it, and its nerve supply, the gland was then a favourite object for studies on the role of the nervous system in the regulation of glandular secretion. It appeared to be equally suitable for a study of organ metabolism. At the outset, as he appeared to do in later years, Barcroft chose a field for research that was relatively uncultivated, one in which he was very much on his own. His approach to physiology was to be through chemistry – he appears to have decided as an undergraduate that the more important advances were to be made along chemical lines – and he knew he could expect little in the way of guidance in the technical aspects of his research from his teachers, Langley, Gaskell and Anderson, for their studies on the sympathetic nervous system were carried out with the techniques of morphology and histology.

Interest in the relation between the quantity of oxygen utilized by an organ at rest and whilst active was a natural outgrowth of the progress which had been made on the estimation of the gaseous exchange of animals and man in a variety of circumstances. A few attempts had been made earlier to study the relation in organs by using the artificial perfusion techniques introduced by the Ludwig School. The results were quite unsatisfactory. To obtain reliable data, methods were required for the estimations of the rate of blood-flow through the organ *in situ* and of the change in the oxygen content of the blood wrought by its passage through it. Barcroft developed the necessary techniques and applied them.

Barcroft's immediate objective in research was always a limited one; he was content to increase his understanding of phenomena by a series of small gains. His experiments were designed to provide an unambiguous answer to a simple question. In his initial research the question was, 'Is the oxygen consumption of the sub-maxillary gland increased over the resting level when it is actively secreting?'

To extract the oxygen and carbon dioxide from small samples taken from the blood entering and leaving the gland, he modified the Toepler Pump. It was a cumbersome apparatus and one which required patience, such as Barcroft had, to yield satisfactory results. Those which he obtained in his first year of research apparently impressed interested members of the International Congress of Physiology held in Cambridge in August 1898, and the next year they won for him a Fellowship at King's. When published, the observations established Barcroft as an experimental physiologist of the first rank.

Although he preferred to work in fields which were uncultivated, Barcroft unhesitatingly sought the counsel and the aid of those who could help him in the tilling. Early in the planning stage of the research he visited those experienced in the technical aspects of the work and discussed with them the details involved. In the development of methods for the estimation of the gas content of small blood samples he consulted J. S. Haldane and together they developed for that purpose what has come to be known as the Barcroft/Haldane apparatus. Barcroft continued to modify the apparatus to increase its sensitivity and later devised the differential manometer which enabled him to deal with as little as 0.1 ml of blood. He devised as well the tonometer known by his name and developed methods for the preparation of the oxygen dissociation curves of blood and haemoglobin. His last paper on apparatus, which appeared in 1934, described tonometers especially designed for the equilibration of small samples of blood.

Out of his initial observations on the salivary gland, Barcroft's research interests developed in the natural way of a dinner conversation, observation giving rise to question, the search for an answer, to another observation, etc., without any apparent break in the continuity though to the reader some of his published papers may appear to deal with unrelated subjects. They were related to his experience and born of it.

His observations on the oxygen consumption of the salivary gland at rest and during activity led to similar studies on other organs which responded to different stimuli and performed different types of work, the pancreas, the kidney, the liver and the heart. In each study he collaborated with a worker experienced with the organ involved. The reports of this work reveal a growing interest in the determination of the quantitative relation between the increased work done by an organ in response to a natural stimulus and the added oxygen consumed.

He saw in the heart an organ especially suited to such a study, but the work was delayed by research along other allied lines; first, on the nature of the oxygen dissociation curve, and then later in the service of the Government in World War I. When he felt free to return to the problem much of the ground had been cleared by the work of Lovatt Evans and Starling. In later years when he expressed his admiration of their work he occasionally added the comment that it was work he would have enjoyed doing himself as the natural sequel to the studies he began on the salivary gland.

The interest in the nature of the oxygen dissociation curve of blood arose from one in an explanation for the unexpectedly high oxygen content of saliva formed during its active secretion and so the desire to calculate the oxygen tension in the gland and in the blood perfusing its capillaries.

As he wrote later, Barcroft was thus brought 'face to face with the problem of whether it was possible from the percentage saturation of a given sample of blood, to calculate from the dissociation curve of blood, the oxygen pressure to which the blood was exposed'. With the aid of the new techniques which he had developed for blood gas analysis, the problem appeared of easy solution; but it proved otherwise. The results of the initial attempts to prepare a reproducible dissociation curve led to confusion which would have discouraged most investigators. But again Barcroft's patience, and his conviction that the facts could be put into order, won the day. Order began to appear when Barcroft and Camis, who worked with him, discovered that the affinity of haemoglobin for oxygen is profoundly influenced by the nature of the electrolytes in the solution. Then followed a series of studies of the effect of acid, of exercise, and of high altitude on the position of the dissociation curve.

It was during the expedition to Tenerife in 1910 to study the effects of a reduction in the alveolar carbon dioxide tension on the position of the oxygen dissociation curve that Barcroft appears to have experienced for the first time the physical and mental effects of a reduced oxygen pressure in the atmosphere. That experience appears to have given birth to his interest in 'what happens when the organs of the body cannot get the oxygen they require', and 'in the degree to which a person's mentality is affected by trifling alterations in the composition of his blood'.

In 1915, not long after gas warfare was introduced by the Germans,

Barcroft was invited to undertake studies of the effects on the respiratory and circulatory systems of the gases used, and those proposed for use. He accepted the invitation, but the decision to do so obliged him to take a course which many in the Society of Friends could not approve. Here he appears to have been guided by 'his own inner light' and from an understanding of his character one may assume that his decision was reached only after a careful weighing of the service he felt he could render his fellow man against principles of Faith. Once the decision was made concern about it was undoubtedly at an end for he often remarked, in substance, that there are two categories of things one should never worry about: those that you can do nothing about, and those about which you can do something.

There are a number of stories of his activities in the war years which have been handed down to illustrate his imperturbability in the face of danger, and the courage with which he supported his convictions. There is one told of him, whilst on a visit to France, standing calmly wearing a bowler hat at a crossroads 'of dubious reputation since the range was accurately known to the Germans', making enquiries of his guide about points of interest in the area. But the best known is the one of his demonstration that the dog and man differ in their susceptibility to the inhalation of hydrocyanic acid. It was for that purpose Barcroft went into a respiration chamber accompanied by a dog and whilst unprotected by a respirator released sufficient hydrocyanic acid to give the air a concentration of about one part in two thousand. The dog was unconscious in about a minute and appeared near death thirty seconds later when Barcroft left the chamber without ever feeling breathlessness or any other symptom.

During the war years Barcroft saw at first hand the respiratory problems of patients who were victims of poison gas attacks and the results of their treatment with oxygen therapy in wards constructed of plate glass. The former appears to have quickened his interest in the role, if any, of the pulmonary epithelium in the exchange of the respiratory gases between the alveolar air and the blood; the latter appears to have suggested a means of settling the question 'Is the movement of oxygen across the pulmonary epithelium a consequence of the difference of its tension on the two sides or is the gas actively secreted by it?'

Despite the evidence for the 'diffusion theory' provided earlier by Krogh and others, the question remained unanswered, for Haldane and Douglas interpreted their observations made on Pike's Peak in 1912 to support the theory of oxygen secretion. When Barcroft returned to Cambridge after the War a glass chamber was built in the laboratory. In it he lived at a simulated altitude of 18,000 feet for six days. At the end of that time, his arterial blood, drawn directly from his left radial artery which had been exposed by a surgeon, contained less oxygen than samples of the same blood exposed to his alveolar air. The experiment provided no evidence for oxygen secretion. Again, as he had earlier at Tenerife, Barcroft experienced at first hand the effects of low oxygen pressure on his own nervous system – a distressing headache and the inability to see an object clearly without concentrating on it.

But Haldane objected that Barcroft's physical condition at the end of the experiment was evidence that he was not 'acclimatized', the inference being that his failure to do so was evidence of a failure of the pulmonary epithelium to perform its function of oxygen secretion. The expedition to Cerro de Pasco in the Peruvian Andes was undertaken to meet that objection and to study the mechanisms of acclimatization. There a number of observations and tests were made of the effects of altitude on the mental attitude and mental efficiency of the members of the party. From a study of his lectures about the work of the expedition one senses the growth of Barcroft's interest in the relation between the activity of the nervous system – mental activity in particular – and the stability of the internal environment as regulated through the blood.

The apparent change in the circulating blood volume which was observed in the members of the party of the expedition when estimated in a warm climate and again in a cold one attracted Barcroft's attention and on his return he began a series of observations to identify the blood stores in the body. Particular attention was devoted to the spleen, and to satisfy his wish always 'to study things as they are' Barcroft exteriorized the spleen so the changes in its size could be seen. Permanent records of the changes were made by tracing the outline of the spleen on a piece of cellophane laid over it.

The concept of the spleen as a blood store was not entirely new but Barcroft's studies revitalized it and gave it meaning by demonstrating the variety of stressful circumstances – exercise, haemorrhage, emotion, etc. – in which it delivered its store of blood into the general circulation.

When the exteriorized spleen of one of the dogs 'suddenly became very small, quite pale in colour and responded poorly to exercise', the dog was killed for autopsy which revealed that 'it was in an advanced state of pregnancy' and the distended state of the uterine veins. Those observations raised the question 'is the blood in the uterine vein during pregnancy stored there and if so to what purpose?' The answer to that question lay in the rate of its 'turnover' – in the rate of the blood flow through the uterus.

It was whilst engaged in estimating that rate in rabbits that Barcroft was struck by the fall – with advancing gestation – of the oxygen saturation of the uterine venous blood, and moved to speculate about the oxygen pressure in which the foetus lives *in utero*; to compare the oxygen pressure to which the foetal blood is exposed during its arterialization in the placenta with that in the alveolar air of a mountaineer on Mount Everest; and to consider the possibility that the adaptive mechanisms which made survival possible were similar in the two situations.

From remarks which Barcroft made later it is clear that although he wrote of the 'environment of the foetus', he was thinking more specifically about the environment of the foetal brain, of the adaptive mechanisms which enable the brain to grow and develop in an environment in which the functions of an adult brain would be seriously impaired. He had written earlier 'Man lives primarily by his intellect', – 'his intellectual development to my thinking is conditioned by his capacity for the exact regulation of the properties of his "milieu intérieur"'.

Here was a new relatively uncultivated field for research – the physiology of the foetus – in which all his earlier interests were joined, the metabolism of an organ, the respiratory function of the blood, the influence of hypoxaemia on the central nervous system, and adaptation to hypoxaemia. Barcroft entered it with that enthusiasm for new adventures in learning which he never lost, and as he did so began drawing younger colleagues to opportunities in it. He was then in his early sixties. Though he developed other interests, among them animal nutrition, the role of the environment in the regulation of the development and the function of the central nervous system held a prime place in his thinking until the end of his life.

The direction of Barcroft's research was not, as a superficial survey might suggest, that of a craft carried by the winds and currents on a random course, but one determined by a sure hand on a tiller that took advantage of both on a voyage of discovery. Like Claude Bernard, Barcroft saw in the familiar – the dilated veins of the pregnant uterus is an example – questions which opened the way to the disclosure of new mechanisms and a broader understanding of their role in the economy of the whole animal. His interests were in architecture and only secondarily in the building materials. His capacity for discovery appears to have been based on a belief 'that accidents happen in nature as elsewhere', but he wrote 'I range myself on the side of those who regard a phenomenon as more likely to have a significance than not. Those who think with me must shoulder the burden of discovering what the significance may be, but on our opponents rests the much heavier burden of proving the phenomenon to be an accident, if indeed it be such.'

And like Bernard, Barcroft had a talent for recognizing when to turn from one line of research to another, 'To divine when thus to turn aside or when not to turn aside, but to go straight on regardless of side issues, however tempting, is perhaps the chiefest sign of genius in enquiry.' 'Instinct guided him to leave the road at the right turning and to follow a bye path which brought him to great results.' These lines written by Michael Foster about Claude Bernard are no less descriptive of Barcroft.

As so many who are successful in the art of research, Barcroft appears to have had from the outset a very broad question in mind which was never lost to sight. His experiments were carefully planned with an objective easy to obtain and a broader more remote goal. This scheme avoids failure from the attempt to achieve too much. He often put aside an experiment proposed by a colleague on the grounds that a definite answer could not be obtained to the question involved; that with more thought the question could be reframed and answered unequivocally. Thus he seemed always to have a place in his scheme of things for the answers to the questions he put to nature by experiment or to his colleagues in conversation. In analysis he seemed never to lose the totality. Like the architect who has in mind the design of his building he seemed to select from nature's offering those which would give it form. Undoubtedly he would have credited some of his success in this respect to his interest in the physiology of the whole animal and occasionally in his later years he remarked with a touch of regret that he was the last of those in Britain who were.

Barcroft's research methods are reflected too in the way in which he published his observations and both may be fairly said to be representative of his 'way of life'. The observations were usually published with a minimum of delay in a short paper in which the research objective was stated in a few words, on occasion in a single sentence. There was always a clear description of the techniques involved. Discussion was kept to the minimum; he made no attempt to have it comprehensive or inclusive.

These papers were usually written at a sitting, often without pause, when time planned for another purpose could not be so used. Apparently the material had been 'worked over' and wholly arranged in his mind before he began to write. He needed only the opportunity to put it on paper. After the draft had been typed up and Mrs Thacker had added the requisite 'three or four references' it was carried about in his briefcase for a short time for rereading, before it was sent off for publication. The original draft needed little or no revision. The broader question to which the observations in the short papers were related is not apparent in many of them; it was dealt with in a review, a lecture or a monograph.

Both in his writing and in his lectures Barcroft's style was informal and conversational; he was as gifted with his pen as he was engaging as a relating structural and functional characteristics of the various corticopetal systems.

The structural development of the neocortex has been thoroughly studied since the classical investigations by Vignal (1888), Kölliker (1890), Cajal (1929) and Lorente de Nó (1933). During the last decades electron microscopic as well as autoradiographic studies have added much to our present knowledge on the developing neocortex. Data on cytoarchitecture, synaptology, myelination, and histochemistry constitute a basis for the analysis of functional ontogenesis. There is, however, a considerable discrepancy between the amount of structural data and the sparse observations on the functional properties of the neocortex, particularly during the early phases of development.

The building up of the organization of recipient functions within the cortex during ontogenesis can be studied by the analysis of gross and unitary cortical responses following activation of various sensory pathways. Recipient functions of the cortex may also be studied by using activation of other corticopetal systems, such as the transcallosal system. Furthermore, the characteristics of the cortical activity as expressed by the electro-corticogram may be used for such an analysis since it depends upon a ubcortical drive. The changes during development of these electrocortical ctivities have been studied in a number of mammalian species (for refer-nces see Himwich, 1970). However, with a few exceptions such studies have een confined to the postnatal phase of ontogenesis and the early formative riod during prenatal life has often been neglected. In many species several cipient cortical functions are already present by the time of birth, even animals born in an altricial condition such as the rat, cat and dog.

Sir Joseph Barcroft's classical studies on prenatal circulation and aviour performed on the sheep (see Barcroft & Barron, 1939) suggested t this animal would be appropriate for use in electrophysiological estigations on the development of neocortical functions. The sheep's ational period comprises 145 days and the weight of the foetus at full is about 3000 g. Like other herd animals the sheep is relatively mature irth from a neurophysiological and behavioural point of view (cf. ebusch, 1971). As shown by a morphological study of the developing x in the foetal sheep, each phase of the cortical epigenesis seems to be protracted than in animals with a shorter gestation period (Åström, . The fact, furthermore, that the process of maturation is almost leted before birth implies that various nervous functions can be niently studied under the same physiological and technical conditions the entire cycle of development before the transition takes place ntrauterine to extrauterine life.

lecturer – accomplished in the use of analogy and anecdote to join the experience of his audience with the material he was presenting. With them he captured the interest of his audience, then led them, often in quite devious ways, into subjects which but for his art in presentation would have been dull and difficult. As an example in that classic *The Respiratory Function of the Blood* which opens with 'The specific oxygen capacity of the blood', a subject in itself not likely to win a large audience, he began by speculating about the role of minor events in shaping the course of history and then presented the details against a matrix of the fun and excitement in their collecting. His lectures were an unusual blend of the familiar and the new leavened with flashes of Irish wit and humour.

As Barcroft lectured without notes his audience was never burdened by unnecessary details, a feature especially appreciated by the undergraduates. As he said, 'I look things up sometime before the lecture. If a student is expected to remember what he hears for all time, he who tries to teach ought to be able to retain in his head for a few hours that which he intends to say.' In these days in which the wealth of factual material with which the students are showered often obscures the purpose of its presentation, one would hope for a return to Barcroft's style. None complained that the material he presented was not relevant to his goal.

Another of Barcroft's talents was his ability to attract younger colleagues to his enterprises and to aid them in the discovery and the development of their potentialities for research. Most of his work was done in collaboration. He often remarked, in substance, 'better than doing something in a mediocre fashion yourself, interest some young man in the problem who has the ability to do it well'.

As he was content with limited gains from experiment those who joined him seldom experienced failure. His enthusiasm for the work was contagious – irresistible for some; his pleasure in success was inspiring, his patience an example.

Where he saw latent talent – and he had a very good eye for it – he encouraged it and opened the way to areas where it could be fruitfully applied. For many his encouragement and guidance removed the natural doubts they entertained about their capacities for original research. He understood them, for he had had his own. In reporting on any work which was a joint effort he appeared to take particular pleasure in assigning the major role in its successful outcome to the efforts and skills of his associates whilst claiming for himself in a humorous way any shortcomings it might have.

Equanimity is an essential quality for sound judgment and Barcroft's natural endowment enabled him, and helped others, to 'see things as they are'. He could be relied upon to keep a confidence. These virtues made him a trusted and a valued counsellor of his many friends.

Barcroft's ability as a teacher and an administrator were recognized by his election to the Chair of Physiology in 1925 as Langley's successor; his War and other Government services by a knighthood in 1935; and for his achievements in research he was awarded the Copley Medal by the Royal Society in 1943. Other honours came to him at home and abroad. He enjoyed them all but none gave him greater pleasure than his place in the hearts of men, for of Joseph Barcroft it can be said:

None knew thee but to love thee,
Nor named thee but to praise.

26 July 1972 D. H. BARRON

SESSION 1

CENTRAL NERVOUS SYSTEM

MORPHOLOGICAL AND PHYSIOLOGI ASPECTS OF THE DEVELOPMENT RECIPIENT FUNCTIONS IN THE CEREBRAL CORTEX

By C. G. BERNHARD and B. A. MEYERS
Department of Physiology II, Karolinska Institut
104 01 Stockholm 60, Sweden

An analysis of the development of behaviour demands da neuronal background of perception is established during study of the functional development of the different affer provides a basis for discussing the possible influence wh inflow may have on the rate of maturation. In view of the plexity of behaviour during development the involvement in afferent functions deserves special attention. However the functional ontogenesis of an afferent system, compr from the receptor to the specific area of the cerebral c kept in mind that different parts of such a system reach during different developmental periods. For instance responses both in the visual and auditory areas can be to tract stimulation long before the receptors in the excitable (Marty, 1962; Marty & Thomas, 1963). stimulation may activate the somaesthetic system at still no differentiation of the specialized cutaneous n Bryant, 1969; Kasprzak *et al.* 1970). There is also activation of an afferent system from the peripher stage when the central pathway has not yet reache the cortex. These facts have to be considered wher mental hierarchy of afferent systems in various sp 1971).

Studies of how the processing of sensory inforn in the developing individual are based on a descri

* Present address: Department of Neurosurgery, Kar

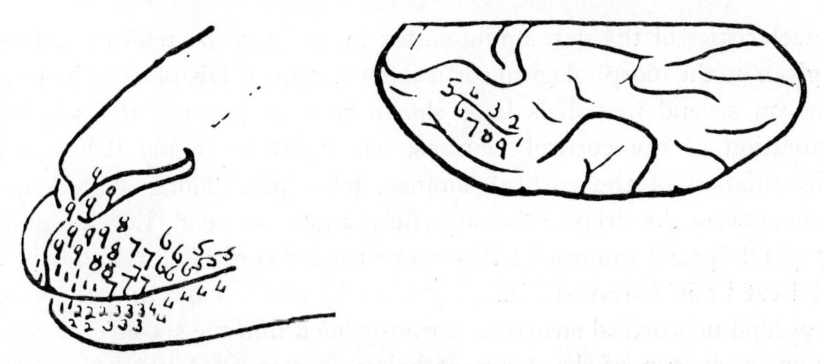

Fig. 1. Somaesthetic cortical representation of the ipsilateral lips in the sheep. (From Adrian, 1943).

Since, apart from developmental studies, the sheep has been relatively seldom used as an experimental animal, data on the morphology and physiology of the adult sheep brain are relatively sparse. Plate 1 illustrates that the sheep's brain has a comparatively advanced pattern of convolutions and fissures. The motor area as well as the specific somaesthetic, visual, and auditory projection areas are marked. As to the motor side, the pyramidal part of the cortico-spinal system is relatively poorly developed in the sheep. On the other hand the motor cortex in sheep seems to be well supplied by interhemispheric neocommissural connections, a fact which is of importance in our further discussion. The somaesthetic region was first mapped by Adrian (1943), who found that the perioral representation in the cortex is relatively extensive and that the upper lip occupies a large part of the projection area (Fig. 1). This was supposed to have a functional significance for the feeding behaviour of a grazing animal like the sheep. Adrian also found that there is a dominance of the ipsilateral projection from the trigeminal nose area and this finding has been confirmed later by several investigators (for references see Bernhard *et al.* 1972). It has also been shown that the projection of the ipsilateral perioral region occupies an exceptionally large part of the thalamus. As a matter of fact it has not been possible to find any distinct subdivision of the thalamus devoted to projections from other parts of the body (Cabral & Johnson, 1971). Also, it has been shown that in the cuneate–gracile complex projections from the rear half of the body occupy a smaller area than projections from the front half and that the spinal trigeminal nucleus is relatively large (Woudenberg, 1970).

The microstructure of the adult sheep's cerebral cortex has been studied by Rose (1942), who concluded that in essence the cytoarchitecture and lamination do not differ from that found in other animals, such as the cat and rabbit.

As a background for the forthcoming description of the functional characteristics of the developing cortex in sheep, some relevant data will be given on the morphological cortical development. Histological investigations on several mammals have shown that, in general, the sequential maturation of the cortical neurons, which occurs in parallel with the differentiation of the cortical laminae, takes place along a corticopetal gradient from the deep to the superficial strata (see, e.g., Lorente de Nó, 1933). Of special interest in this context are recent Golgi studies in the foetal cat brain (Marin-Padilla, 1970, 1971) which show how the simple primordial neocortical structure is transformed into the six-layered organization, each step of development being characterized by the arrival of afferent fibres at a specific cortical level and by the formation of neurons at that level. Following the arrival of corticopetal fibres at the marginal zone a primitive layer develops, characterized by a few immature neurons situated between the fibres. This primordial plexiform layer, visible between the 20th and 25th day of foetal age, is assumed to lack functional activity at this early stage. The following subdivision of this layer into an outer and inner zone represents the formation of layers I and VI which remain as such throughout development. During this stage – between the 25th and 45th day – three basic types of neurons appear: the horizontal cells of layer I with descending axons, terminating in layer VI, Marinotti neurons of layer VI with ascending neurons terminating in layer I and stellate neurons in layer VI sending recurrent collaterals to layer I. The next step begins around the 45th day when new types of afferent fibres arrive at the lower region of the cortical plate followed by maturation of pyramidal neurons. The sequence of development proceeds in proximodistal direction from layer V to layer IV and later on to the deep part of layer III. Towards the end of the foetal life of the cat, fibrillar and neuronal development proceeds to the upper region of layer III and layer II.

A similar developmental gradient has also been demonstrated for the maturation of the cortical neurons with regard to the elaboration of the basal dendrites, as described by Cajal (1911, see Fig. 2). The drawing in Fig. 2 was made from a Golgi preparation of the cortex of a newborn mouse, and demonstrates that neurons in the superficial strata are of more primitive bipolar shape than those in the depth which display comparatively elaborate dendrites and axon collaterals. Data obtained with the aid of the electron microscope indicate that developmental changes of cellular organelles also follow a similar vertical gradient (Caley & Maxwell, 1968). The mechanism behind the sequential histogenesis of the cortical neurons has been revealed by autoradiographic methods. It has been demonstrated that the cells migrate outward through the cortex and that those formed early in development are displaced towards the depth by the younger cells,

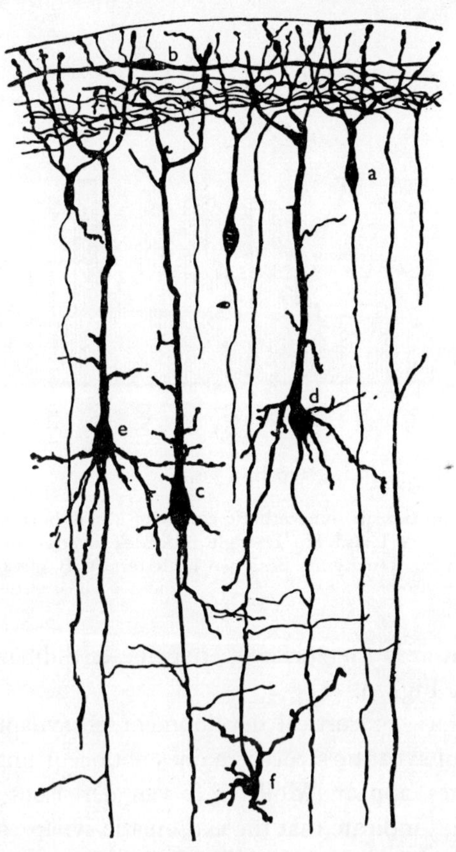

Fig. 2. Golgi-preparation of the cerebral cortex of a 4-day mouse. (*a*) Small pyramidal cell in the bipolar stage. (*b*) Horizontal cell of the plexiform layer. (*c, d, e*) Pyramidal cells. (*f*) Martinotti cell. (From Cajal, 1911.)

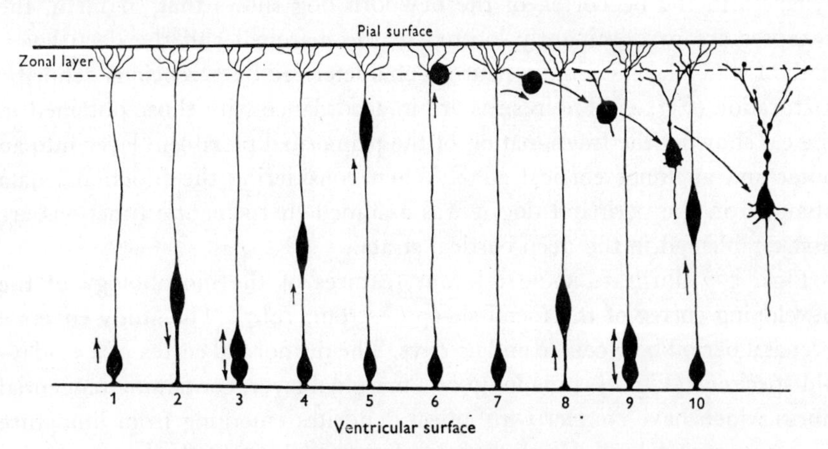

Fig. 3. Schematic representation of the suggested manner of migration and differentiation of neuroblasts. (From Berry & Rogers, 1965.)

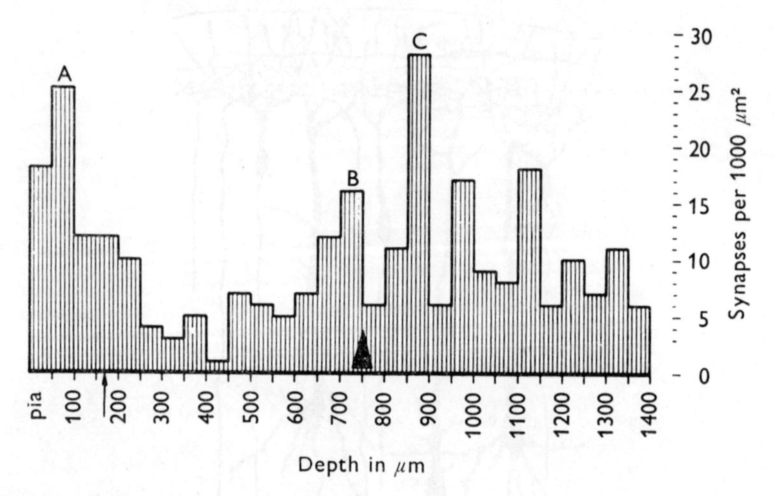

Fig. 4. Synapse distribution in somaesthetic cortex of a new-born dog. Arrow points to interphase between layers I and II. Triangle indicates location of layer of large cells. Peaks of highest synaptic density are designed by letters A, B and C. (From Molliver & Van der Loos, 1970.)

which migrate towards the surface (Angevine & Sidman, 1961; Berry & Rogers, 1965; see Fig. 3).

In the early phase of cortical development no synaptic structures are seen. Formation of synaptic structures does not occur until the stage when elaborate dendrites appear (Molliver & van der Loos, 1970). Electron microscopic studies indicate that the axosomatic synapses are formed later than the axodendritic synapses (e.g. Voeller *et al.* 1963; Gruner & Zahnd, 1967). The study by Molliver & van der Loos (1970) of the distribution of synapses in the neocortex of the newborn dog shows that, at birth, the synapses are predominantly located in the marginal and the deep layer, both of which layers contain neurons characterized by precocious dendritic maturation (Fig. 4). The results are in accordance with those obtained in the cat showing the fractionation of the primordial plexiform layer into an outer and an inner cortical zone. When considering the functional data obtained in the perinatal dog it was assumed that synaptic functions are first established in the deep cortical strata.

Figs. 5–7 illustrate some relevant features of the morphology of the developing cortex of the foetal sheep (Åström, 1967). The study covers a prenatal period between 40 and 90 days. The primordial cortex of a 50-day-old specimen (Fig. 5) is made up of a marginal layer, containing tangential fibres which have contact with apical dendrites emerging from immature bipolar pyramidal cells. In the intermediate zone beneath the cortex proper, stellate cells are seen. At this stage no afferent connections of the thalamic

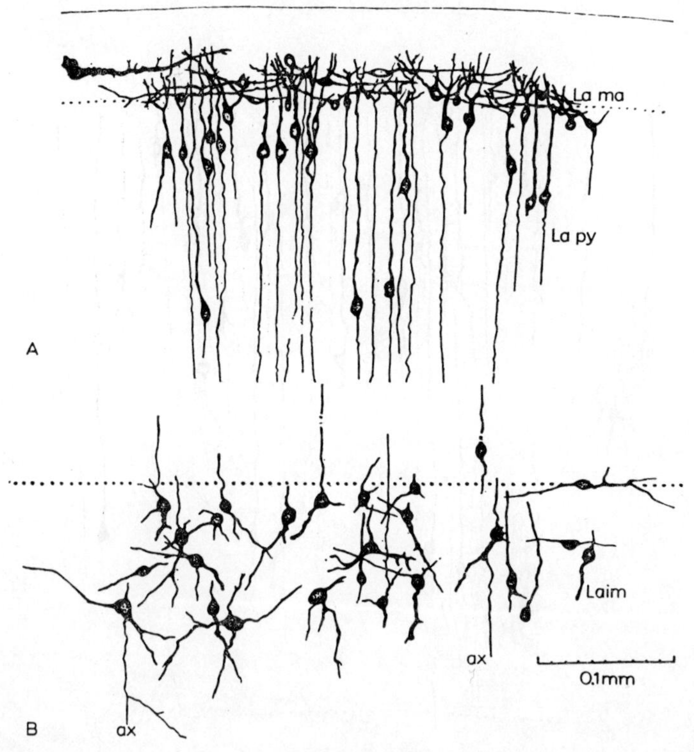

Fig. 5. Golgi-preparation of the neocortex of a 50-day sheep foetus. Composite picture of marginal and pyramidal (A) and subpyramidal (B) strata. Most cells in superficial part of pyramidal layer are of a primitive bipolar type. Stellate cells in (B) are located in the intermediate zone (Laim). Marginal zone (La ma) contains a Cajal–Retzius cell (upper left). (From Åström, 1967.)

or callosal type were seen to penetrate the cortex. At about 60 days of foetal life (Fig. 6) the pyramidal cells in the upper stratum are still undifferentiated, whereas those in the middle and lower part are more mature, displaying short basal dendrites. Pyramidal cells with ascending axons (Martinotti cells, CM) may be seen. During the period of 65–70 days (Fig. 7) there is apparently an acceleration of cortical development. Axon collaterals of the pyramidal cells are elaborated and distributed between the cells within the same stratum. The pyramidal cells of the middle stratum have numerous basal dendrites with indication of spines. Stellate cells are incorporated within the pyramidal zone and presumably constitute the primordium of the 4th layer of the mature cortex. At a somewhat later stage, there is a more distinct 4-layer stratification of the pyramidal layer. At the age of about 80–90 days the cortex displays a comparatively mature picture and by then the superficial pyramidal cells have also lost their primitive bipolar shape.

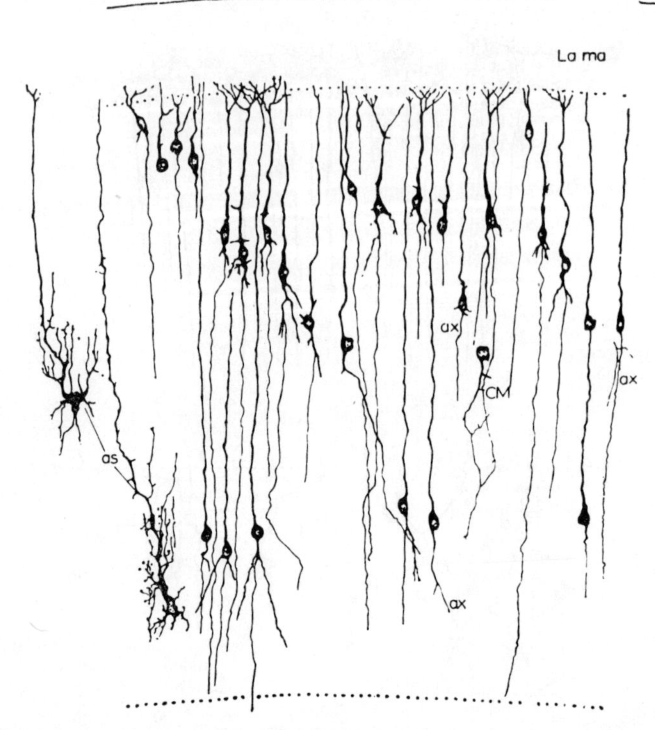

Fig. 6. Golgi-preparation of the neocortex of a 60-day sheep foetus. Cells in superficial stratum of pyramidal layer have immature bipolar appearance. Cells in middle part are more highly developed than those in deeper stratum of the layer. Martinotti cell with ascending axon (CM); Astrocytes (as). (From Åström, 1967.)

However, details on this latter developmental period are unfortunately still lacking. In his description of the morphology of the developing cortex of the sheep, Åström adopts the concept of a vertical gradient of maturation through the cortex. However, this author claims that there are exceptions to this principle as shown by the fact that pyramidal cells in the deepest stratum may have a somewhat retarded maturation in comparison with those in the middle third of the pyramidal zone (cf. Marty & Pujol, 1966).

The following discussion of the early ontogeny of the functional capacities of the neocortex is based mainly on data obtained in electrophysiological experiments on sheep foetuses in which the characteristics of the spontaneous cortical activity and the cortical responses to various types of corticopetal influences were analysed. The non-anaesthetized exteriorized foetus was kept in umbilical contact with the placenta *in situ* of the decerebrate ewe, kept on artificial respiration (for details see Meyerson, 1968).

When working with sheep foetuses younger than 60 days, one is con-

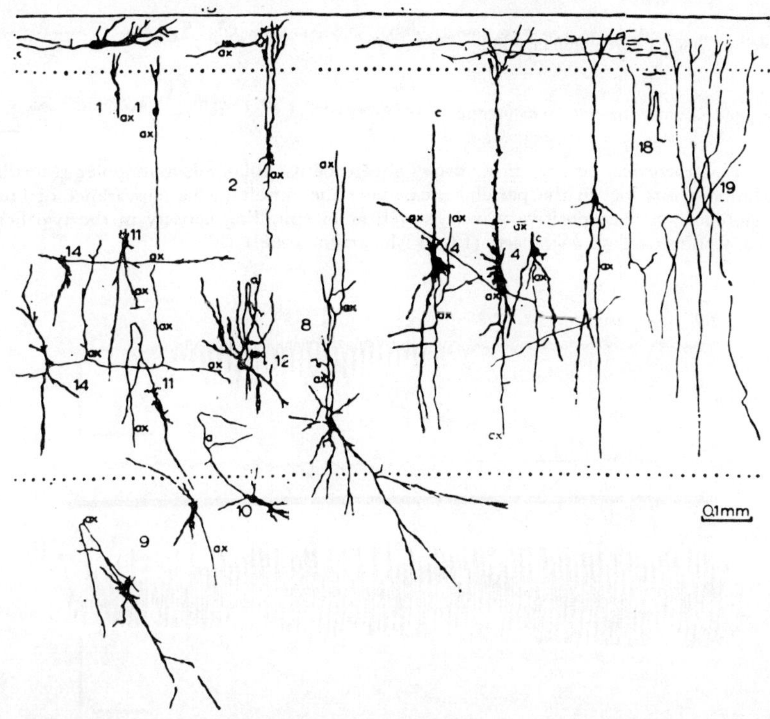

Fig. 7. Golgi-preparation of the neocortex of a 67-day sheep foetus. Note the elaboration of stellate cells (11, 12, 14) within the pyramidal layer, (9, 10) subpyramidal stellate cells of the type seen in younger specimens, (8) Martinotti cells with ascending axons. The superficial pyramidal cells (1, 2) are less mature than those in deeper strata (4) which have axon collaterals. Ascending afferent fibres (19) reach the superficial zone. (Modified from Åström, 1967.)

fronted with a lissencephalic cortex which is electrically silent as judged by the electrocorticogram led off from the surface of the exposed brain. Typical for an area posterior and medial to the anlage of the somatosensory cortex, is the appearance, around the 65th day, of short epochs of regular spindle-like waves with a frequency of about 8–10 per sec. Somewhat later, around the 70th day, spontaneous bursts of regular monophasic surface positive waves may also appear and very often these bursts of arcade-shaped positive waves are preceded or followed by bursts of spindles (Fig. 8). These types of spontaneous bursts of activity, which may be regarded as prenatal precursors of the spindle activity of the EEG in the mature animal, are characteristic for the foetal period up to about the 90th day (see Bernhard *et al.* 1959, 1967 and 1972; Meyerson, 1968). Similar intermittent spontaneous activity has been observed to represent an early type of EEG activity in other animals as well as in man (for references see Bernhard & Meyerson, 1968).

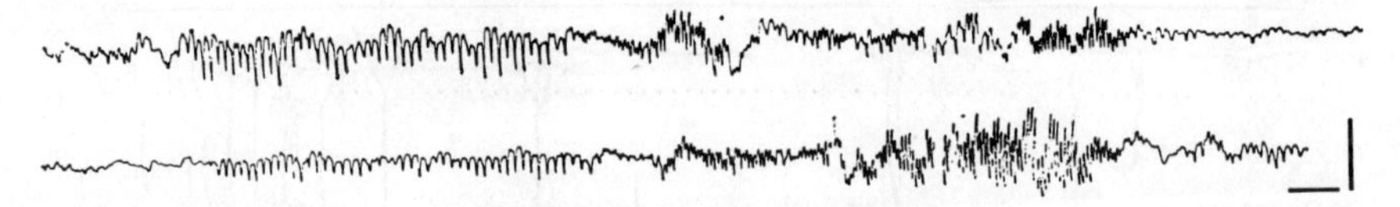

Fig. 8. Electrocortical activity in a 70-day sheep foetus. Bilateral, monopolar recordings from homologous loci in the parietal area. Note the simultaneous appearance of bursts of monophasic surface positive waves as well as of spindling activity in the two hemispheres. Calibration, 50 μV; 1 sec. (From Meyerson, 1968.)

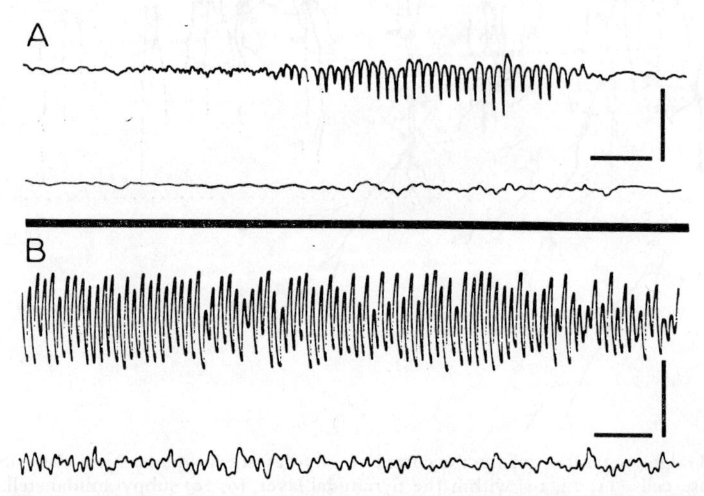

Fig. 9. Electrocortical activity, monopolar recordings from the same hemisphere. A, records from a 70-day sheep foetus, with separation of recording points 1.5 mm. B, records from an 85-day foetus with separation of recording points 3 mm. Calibration, 50 μV; 1 sec. (From Meyerson, 1968.)

The spindle bursts are always confined to very limited cortical areas. The activity thus has an 'insular' appearance and the location of the active cortical 'islands' shifts from one place to another, the rest of the cortex being more or less silent. This is in contrast to the complex multiform and sustained cortical activity in more mature foetuses and in the adult animal. The discreteness of such spindling areas in young foetuses is illustrated by the records in Fig. 9. As shown by the upper tracings a lively activity appears in one region which is separated by only a few mm from a nearby region which does not display any such activity as illustrated by the lower tracings.

Of great interest is that these spatially very restricted spontaneous bursts have a bilateral simultaneous appearance within homologous cortical areas (Fig. 10; see also Fig. 8). When either of the focal electrodes is moved from the homologous area the temporal correlation is lost. Cross-correlation analysis shows that the individual waves are bilaterally coupled (Fig. 11).

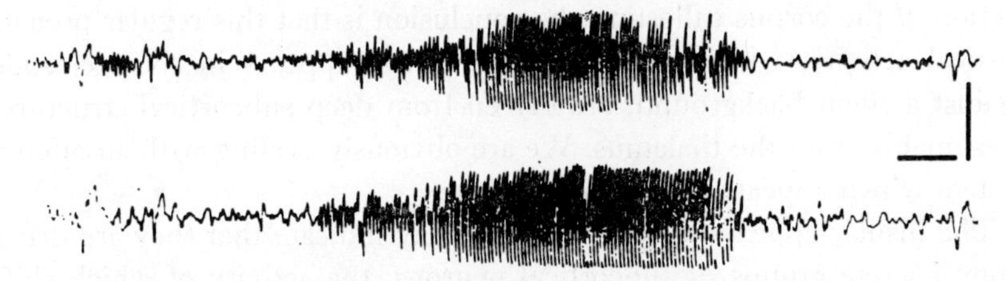

Fig. 10. Electrocortical activity in an 80-day sheep foetus. Bilateral, monopolar recordings from *g. suprasylvius*. Note the well synchronized appearance of spindling activity of 14–16 Hz. Calibration, 50 μV; 1 sec. (From Meyerson, 1968.)

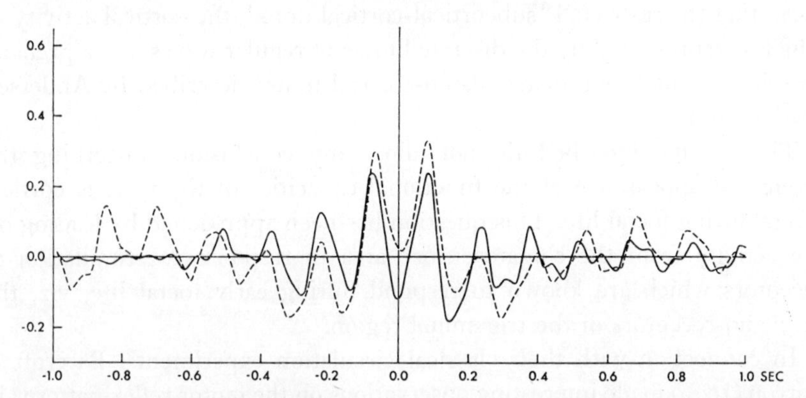

Fig. 11. Cross-correlograms based on two consecutive periods of recording of electrocortical activity in a 70-day sheep foetus. The activity was recorded from bilateral homologous points. Note that the correlograms exhibit periodicity even at long delays, indicating a considerable stability of the frequency of the correlated components. The dip at zero delay in the correlograms is caused by the opposite phasing of the waves in the two hemispheres. (From Meyerson, 1968.)

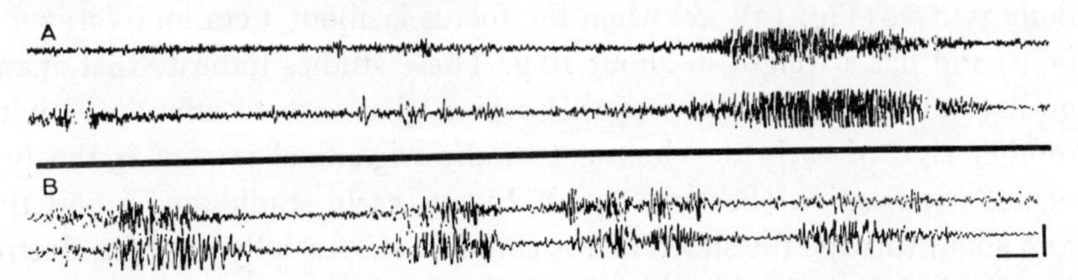

Fig. 12. Electrocortical activity in a 95-day sheep foetus after transection of the corpus callosum. Bilateral, monopolar recordings from the middle (A) and the anterior (B) portions of *g. suprasylvius.* Note the preservation of the simultaneous appearance of bursting activity in the two hemispheres. Calibration, 50 μV; 2 sec. (From Meyerson, 1968.)

The question then arises of whether or not the coupling of the insular activity, which appears simultaneously in the two hemispheres, is mediated by the callosal connections. An answer to this question was obtained in experiments in which the corpus callosum was transected. It was found that after transection the bilateral episodes of activity still had a synchronous appearance (Fig. 12). Since the coordination remains after transection of the corpus callosum, the conclusion is that this regular prenatal cortical activity, which at an early foetal age appears in short episodes against a silent background, are driven from deep subcortical structures, presumably from the thalamus. We are obviously dealing with an afferent system which appears early in ontogeny.

The insular appearance of the bursts may indicate that they are driven from discrete groups of subcortical neurons, the activity of which shifts from one group to another while intermediate groups are temporarily silent. One may assume that the drive has a thalamic origin which would mean that the restricted 'subcortical-cortical units', the cortical activity of which is represented by the discrete bursts of regular waves, are a prenatal correlate to the 'columnar thalamo-cortical units' described by Andersen *et al.* (1967).

The results described do not allow any conclusion concerning the sequential appearance of the functional capacities of the various cortical layers during foetal life. This question has been approached by leading off the activity from the somatosensory area in response to stimulation of receptors which are known to respond during early foetal life, viz. the mechano-receptors of the trigeminal region.

In connection with their classical circulation experiments, Barcroft & Barron (1939) made interesting observations on the motor reflex patterns in response to tactile stimulation and presented basic data on the development of motor reactions to tactile stimulation within the trigeminal region. Barcroft & Barron found that the earliest motor reflex response to afferent stimulation could be elicited from the trigeminal region at a foetal age of about 35 days (Fig. 13), i.e. when the foetus is about 3 cm long (crown to rump) and has a weight of about 10 g. These studies indicate that at this age natural skin stimulation of this region gives rise to afferent activity. Another sign of early development of the trigeminal system is the precocious myelination (Barlow, 1969). In our early studies we found that from about this age the sheep foetus could be successfully used for electrophysiological analysis. At this stage the brain of the foetal sheep is dominated by the diencephalon and mesencephalon and the lissencephalic cerebral hemispheres appear as two small thin-walled vesicles.

As described above the structural data on the epigenesis of the neocortex indicate a vertical gradient of maturation with regard to cortical neurons and lamination and suggest that in the early phase of development corticopetal activation takes place primarily in the depth of the cortex. It is therefore to be expected that a volley of corticopetal impulses activating the precocious neuronal elements located subcortically or in the deepest cortical strata would appear as a positive transient when recorded from the cortical surface.

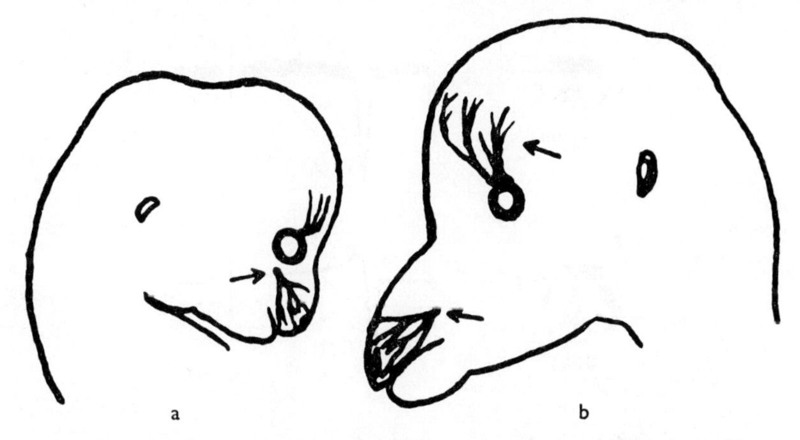

Fig. 13. Approximate extension of the peripheral trigeminal nerve in sheep foetuses of 36 and 40 days of age. (From Barcroft & Barron, 1939.)

The first investigation of the development of the *cortical surface response* to tactile lip stimulation in prenatal sheep was made by Molliver (1967). He found that an evoked cortical response could be obtained already at a foetal age of about 50 days. At its earliest appearance the response was found to be purely positive. During the period between 70–90 days it successively changed into a diphasic positive–negative potential sequence. Around the 90th day the response turned into a predominantly negative potential.

The sequence of the developmental changes of the evoked cortical response during the period between the 60th to the 90th day is illustrated in Fig. 14 from a recent experimental series by Meyerson & Persson (1969), whose results confirm Molliver's conclusions. *A* is from a 60-day foetus and represents the initial developmental phase, characterized by a pure positive surface deflection. The youngest foetuses examined in this series had a weight of 10–15 g, corresponding to 45 days, and in this foetus too a significant positive surface response was obtained. Since for several reasons successful experiments of this type could not be performed on smaller foetuses, we are not able to say when the positive response first appears during the prenatal period prior to the 45th day. The records *B–D* show the changes in configuration of the response during the following period, when the positivity successively changes via a positive–negative response into a predominant negativity. In the following discussion we shall concentrate on the events represented by the initial phase of the surface response.

The localization in depth of the activity, signalled by the surface potential at various ages, has been studied by Persson (1973) in special investigations the details of which are published in this volume. Both focal potentials and

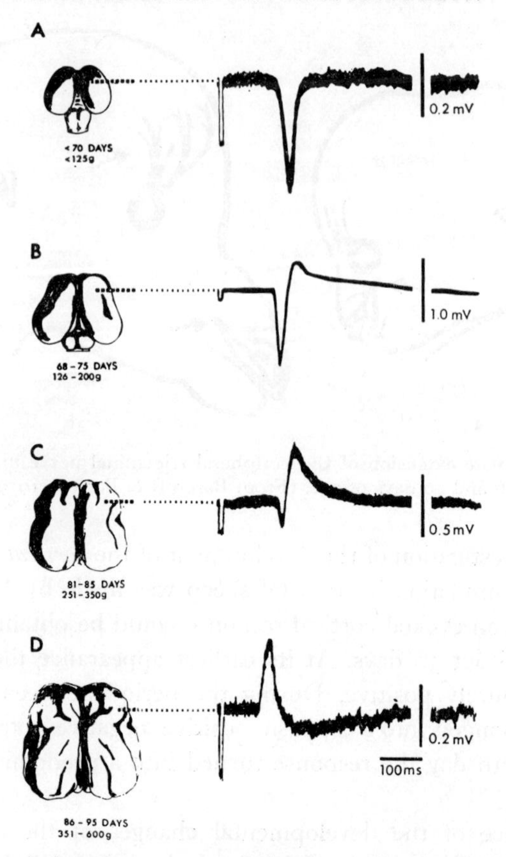

Fig. 14. Evoked cortical surface responses following tactile stimulation of the ipsilateral upper lip obtained in sheep foetuses of different ages. To the left is shown the area from which maximal responses were obtained (indicated by broken line). In this as in all following figures negativity is upwards. (From Bernhard *et al.* 1972.)

unit activity from single cortical neurons were recorded at various cortical depths in response to mechanical stimulation of the upper lip. A comparative analysis of the depth potential profile and the vertical intracortical distribution of the unit discharges revealed that, during development, there is a progressive activation of cortical neurons from the deeper to the more superficial layers. Thus, generally speaking, the early phase of the intracortical functional development proceeds along a corticopetal gradient. In a somewhat later stage of cortical epigenesis, when activation occurs throughout the whole cortex, the picture becomes more complex as the formative processes involve rebalancing of excitatory and inhibitory influences (see Meyerson & Persson, 1973).

In this context it should be mentioned that several investigations on the evoked cortical surface response in sensory areas have been made on various animals during the *postnatal* period. Such experiments were made on cats and rabbits and in all these studies an evoked response was found to be present at birth and to consist of a predominantly *negative* deflection (see, e.g., Scherrer & Oeconomos, 1954; Purpura, 1961; Marty, 1962; Laget & Delhaye, 1962; Rose & Lindsley, 1968). The neonatal negative evoked response has been interpreted as a sign of EPSPs located in the superficial layers of the cortex and initiated by corticopetal activation of apical dendrites of perikarya in the pyramidal layer II and III (Purpura *et al.* 1964). It has also been suggested that inhibitory postsynaptic events play a role in the genesis of the surface negative response and that the inhibitory mechanisms dominate during early neurogenesis (Purpura *et al.* 1965). This view does not harmonize with the fact that there is a delayed maturation of the axosomatic synapses which suggests instead that inhibitory synaptic activation appears comparatively late (cf. Huttenlocher, 1967). As judged by the results obtained on the prenatal sheep material – which includes the very early developmental phases – the studies on the postnatal animals cover a period corresponding to the stage of the sheep foetus which is characterized by the predominating negativity, i.e. a stage when the cortex has passed the developmental period characterized by the pure positive response. This is further supported by the fact that the postnatal changes in the configuration of the evoked response in cats and rabbits are similar to those which occur during the prenatal period in sheep, subsequent to the stage characterized by the predominating negativity. It should be noted that the 90-day-old foetal sheep in which the evoked somaesthetic response consists of predominating negativity has reached a comparatively mature stage of cortical morphogenesis in comparison with that of the newborn kitten. This developmental phase of the sheep has not been studied in detail and ultrastructural observations are still missing. However, the structural and functional data available indicate the difficulties which one encounters when comparing developmental stages in different species.

One may ask if, during development, there appear similar changes of the surface response to activation of other afferent pathways entering the cortex, for instance the callosal system which projects into cortical areas outside the somatosensory cortex as well. It has been mentioned that the callosal system is well represented in the sheep. In this animal the formation of the corpus callosum starts towards the end of the first trimester (see Meyerson, 1968).

Whereas the trigeminal somaesthetic system shows evidence of function at a comparatively early stage of ontogenesis, the neocommissural callosal system matures late. The callosal fibres display a delayed maturation both

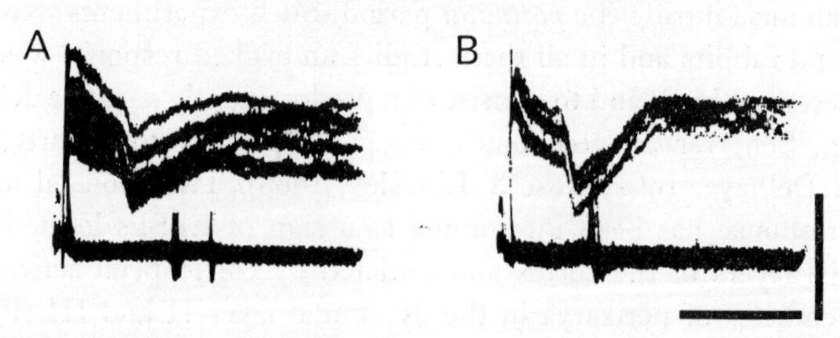

Fig. 15. Transcallosal gross and unitary responses obtained from an 82-day sheep foetus. Gross responses were recorded from the cortical surface. Unitary activity was recorded in the same penetration, in A at a depth of 500 μm, in B at 1500 μm. Calibration, 100 μV; 100 msec. (From Meyerson & Persson, 1973.)

in terms of growth and myelination and the latter is actually not completed until a late postnatal stage (Plate 2; cf. Romanes, 1947).

It has been shown that cortical stimulation does not evoke any response within contralateral homologous cortical areas before the 68th day. The reason is that the cortical structures are not electrically excitable until that day as shown by the fact that one does not obtain any direct cortical response (DCR) before that time (Eidelberg *et al.* 1965). When the inter-hemispheric surface response appears on the 68th day, it is purely positive. Preliminary experiments with depth recordings of focal and unitary responses indicate that this surface positive response to transcallosal stimulation represents activity located in the deeper cortical strata (Meyerson & Persson, 1973; Fig. 15). The callosal origin of the response is shown by the fact that it is abolished by transection of the corpus callosum.

Up to an age of about 85 days this so-called transcallosal response (TCR) generally consists of a pure positive deflection, sometimes followed by a low amplitude slow negativity. In the following period, the trans-callosal response is altered in form by the appearance of a predominating negative transient. In foetuses over 95–100 days the response may be pure negative (Meyerson, 1968). These phasic changes during development are shown in Fig. 16. The transcallosal response does not develop synchronously in the different cortical areas from which it can be obtained in the adult animal. Thus, for instance, the specific somatosensory area does not display any response to contralateral cortical stimulation until the age of about 80 days and in some areas it does not appear until the 95th day. However, the sequence of the phasic changes of the response during development is always the same irrespective of its timing during maturation.

The changing form of the transcallosal response during development is not as regular as that of the evoked somaesthetic response. The fact that the transition from positivity to predominating negativity of these responses

PLATE I

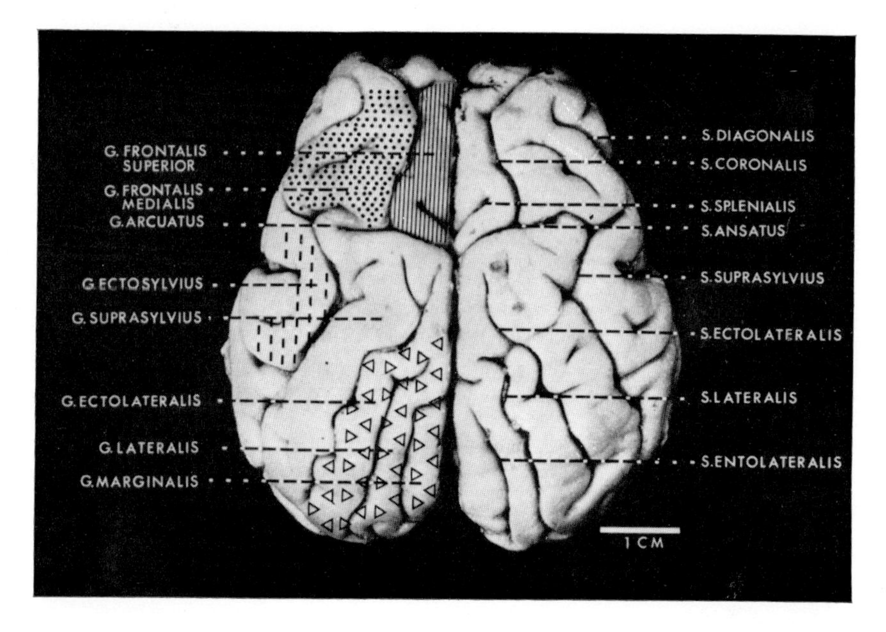

Dorsal view of adult sheep brain with identification of fissures and convolutions. On the left hemisphere is marked the approximate extension of the motor (vertical lines), somatosensory (dotted), auditory (broken lines) and visual (triangles) areas.

(*Facing p.* 16)

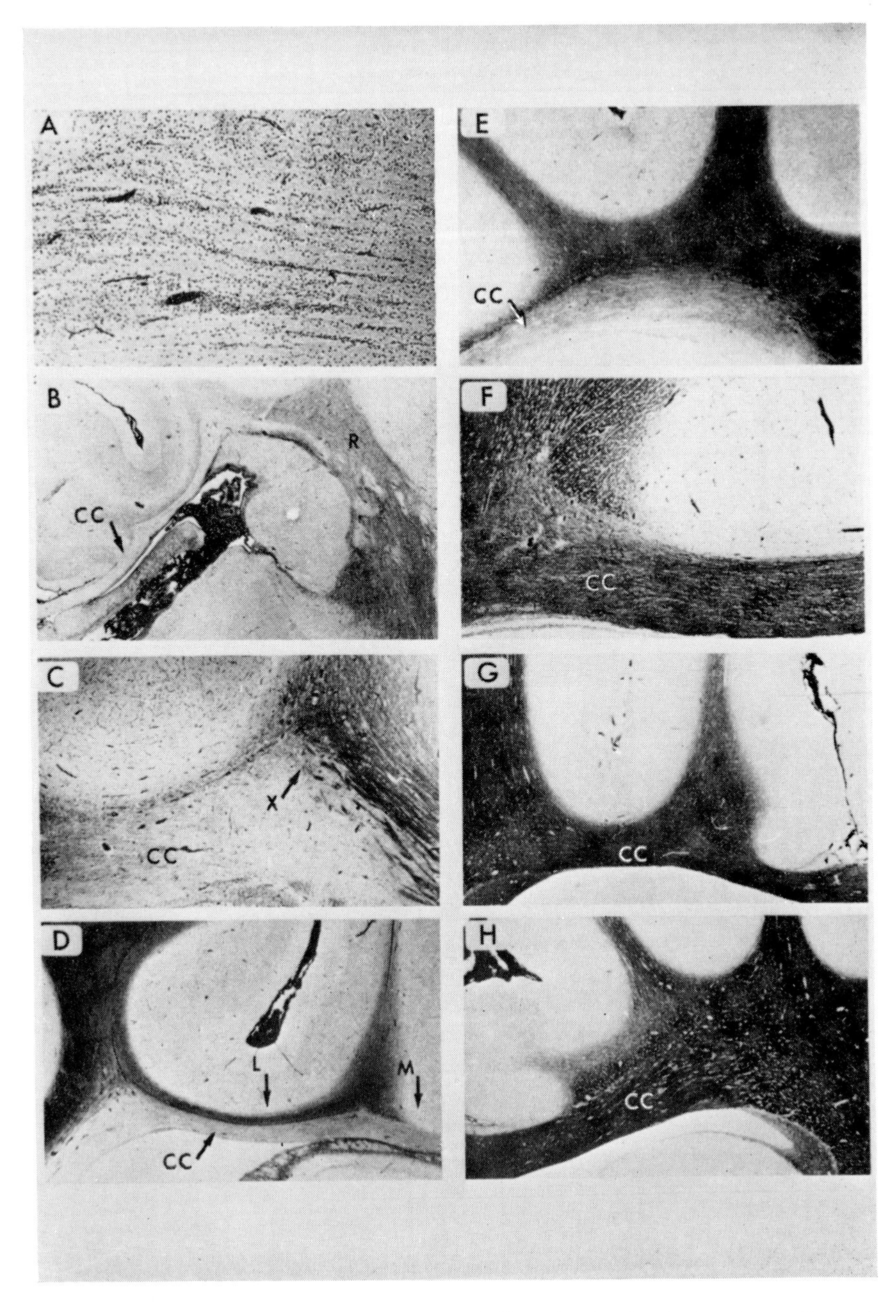

Myelination of callosal (CC) and radiating (R) fibres in the developing sheep. Coronal sections of the brain. A, gliosis as a sign of incipient myelination of radiating fibres in a 97-day foetus. Foetuses of 110 (B), 137 (C) and lambs of 2 (D), 15 (E), 24 (F), 70 (G) days of age and adult sheep (H). In C, letter X denotes intersection between callosal and stalk fibres. In D, letters L and M denote lateral and paramedial portions respectively of the callosal system. Loyez stain. (From Meyerson, 1968.)

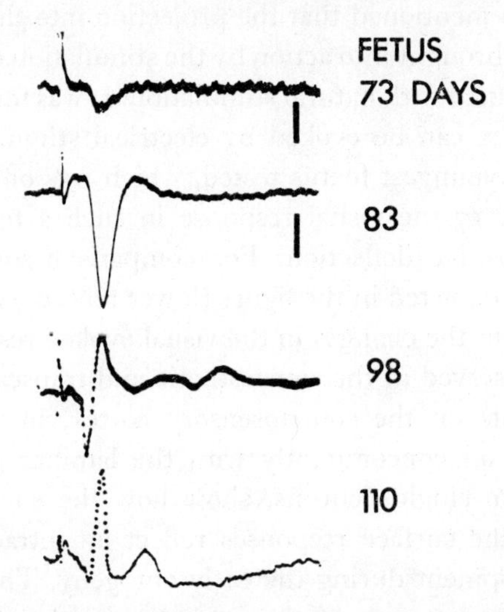

Fig. 16. Transcallosal responses from *g. frontalis superior* evoked by stimulation of homologous contralateral loci in sheep foetuses of different ages. Calibration, 50 μV in the first record and 100 μV in the following records. Horizontal bar, 100 msec. (From Meyerson, 1968.)

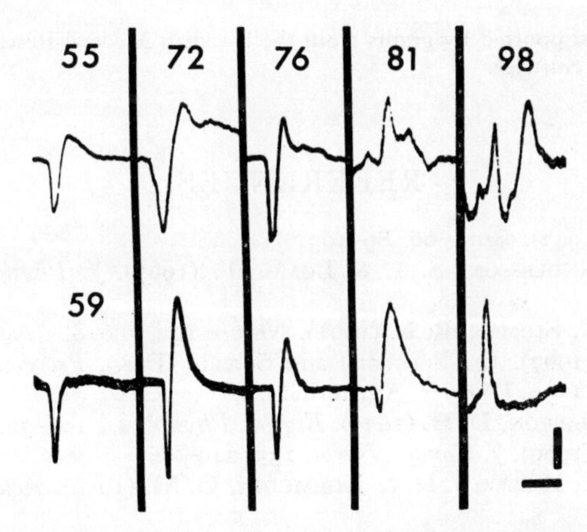

Fig. 17. Cortical surface responses elicited by electrical stimulation of the optic tract (upper row) and somaesthetic responses to trigeminal stimulation (lower row) in sheep foetuses of different ages. Both visual and somaesthetic responses in each age group are from the same animal except the first two pairs of responses. Numbers indicate gestational age in days. Calibration, 100 μV; 100 msec. (From Persson & Stenberg, 1972.)

occurs at different foetal ages illustrates the heterochronous development of these two corticopetal systems.

To pursue the discussion on cortical recipient functions during early ontogenesis, it is of interest to refer to some of the salient findings in a recent study of the developing visual evoked response (Persson & Stenberg,

1972). It has been mentioned that the projection into the visual and auditory cortex can be brought into action by the stimulation of the tracts before responses are obtained with natural stimulation. It was found that responses in the visual cortex can be evoked by electrical stimulation of the optic nerve even in the youngest foetus tested, which was only 55 days old. As illustrated in Fig. 17 the visual response in such a foetus consists of a predominantly positive deflection. For comparison somaesthetic evoked responses are also depicted in the figure (lower records) and it appears that during development the changes in the visual evoked response are remarkably like those observed in the somaesthetic and transcallosal response.

The experiments on the somatosensory cortex, in which the surface potential was led off concomitantly with the laminar potentials and the spike activity from single neurons, show how the successive changes in configuration of the surface responses reflect an intracortical functional gradient of development during the early ontogeny. The similarity in the phase shift during ontogeny of the responses to the two other types of afferent activation described, may be taken as indication of a general principle for the development of corticopetal activation.

This work was supported by grants from the Swedish Medical Research Council and Magnus Bergvalls Stiftelse.

REFERENCES

ADRIAN, E. D. (1943). *Brain* **66**, 89–103.

ANDERSEN, P., ANDERSSON, S. A. & LØMO, T. (1967). *J. Physiol.* (*Lond.*) **192**, 283–307.

ANGEVINE, J. B. & SIDMAN, R. L. (1961). *Nature* **192**, 766–8.

ÅSTRÖM, K.-E. (1967). In: Bernhard and Schadé (Eds.), *Progress in Brain Res.*, Vol. 26, pp. 1–59. Elsevier, Amsterdam.

BARCROFT, J. & BARRON, D. H. (1939). *Ergebn. Physiol.* **42**, 107–52.

BARLOW, R. M. (1969). *J. Comp. Neurol.* **135**, 249–62.

BERNHARD, C. G., KAISER, I. H. & KOLMODIN, G. M. (1959). *Acta physiol. scand.* **47**, 333–49.

BERNHARD, C. G., KOLMODIN, G. M. & MEYERSON, B. A. (1967). In: Bernhard and Schadé (Eds.), *Progress in Brain Res.*, Vol. 26, pp. 60–77. Elsevier, Amsterdam.

BERNHARD, C. G. & MEYERSON, B. A. (1968). In: Kellaway and Petersén (Eds.), *Clinical Electroencephalography of Children*, pp. 1–29. Almqvist and Wiksell and Grune and Stratton, Stockholm and New York.

BERNHARD, C. G., MEYERSON, B. A. & PERSSON, H. E. (1972). *Actual. Neurophysiol.* (*Paris*). **9**, 119–144.

BERRY, M. & ROGERS, A. W. (1965). *J. Anat.* (*Lond.*) **99**, 691–709.

CABRAL, R. J. & JOHNSON, J. I. (1971). *J. comp. Neurol.* **141**, 17–36.

CAJAL, S. RAMÓN Y. (1911). In: *Histologie du Système Nerveux de l'Homme et des Vertébrés*, Vol. 2. Maloine, Paris.

CAJAL, S. RAMON Y. (1929). In: *Etudes sur la Neurogenèse de Quelques Vertébrés.* Madrid.
CALEY, D. W. & MAXWELL, D. S. (1968). *J. Comp. Neurol.* **133**, 17–44.
EIDELBERG, E., KOLMODIN, G. M. & MEYERSON, B. A. (1965). *Exp. Neurol.* **12**, 198–214.
GOTTLIEB, G. (1971). In: Tobach, Aronson and Shaw (Eds.), *The Biopsychology of Development*, pp. 67–128. Academic Press, New York.
GRUNER, J. E. & ZAHND, J. P. (1967). In: Minkowski (Ed.), *Regional Development of the brain in Early Life*, pp. 125–33. Blackwell, Oxford.
HIMWICH, W. A. (1970). *Developmental Neurobiology.* Charles C. Thomas, Springfield, Illinois.
HOGG, I. D. & BRYANT, J. W. (1969). *J. Neurol.* **136**, 33–56.
HUTTENLOCHER, P. R. (1967). *Exp. Neurol.* **17**, 247–62.
KASPRZAK, H., TAPPER, D. N. & CRAIG, P. H. (1970). *Exp. Neurol.* **26**, 439–46.
KÖLLIKER, A. (1890). *Zeit. f. Wissen. Zool.* **51**, 1–54.
LAGET, P. & DELHAYE, N. (1962). *Actual. Neurophysiol.* (*Paris*) **4**, 259–84.
LORENTE DE NÓ, R. (1933). *J. Psychol. Neurol.* (*Lpz.*) **45**, 381–438.
MARIN-PADILLA, M. (1970). *Brain Research* **23**, 167–83.
MARIN-PADILLA, M. (1971). *Z. Anat. Entwickl.-Gesch.* **134**, 117–45.
MARTY, R. (1962). *Arch. d'anat. microscop. morphol. expériment.* **51**, 129–264.
MARTY, R. & PUJOL, R. (1966). In: Hassler and Stephan (Eds.), *Evolution of the Forebrain*, pp. 405–18. Georg Thieme Verlag, Stuttgart.
MARTY, R. & THOMAS, J. (1963). *J. Physiol.* (*Paris*) **55**, 165–6.
MEYERSON, B. A. (1968). *Acta physiol. scand.* Suppl. 312.
MEYERSON, B. A. & PERSSON, H. E. (1969). *Nature* **221**, 1248–9.
MEYERSON, B. A. & PERSSON, H. E. (1973). In: Gottlieb (Ed.), *Advances in Ontogeny of Behavior and Nervous System.* Academic Press, New York.
MOLLIVER, M. (1967). In: Bernhard and Schadé (Eds.), *Progress in Brain Research*, Vol. 26, pp. 78–90. Elsevier, Amsterdam.
MOLLIVER, M. E. & VAN DER LOOS, H. (1970). *Ergebn. Anat. Entwickl.-Gesch.* **42**, 7–53.
PERSSON, H. E. (1973). This volume, pp. 20–7.
PERSSON, H. E. & STENBERG, D. (1972). *Exp. Neurol.* **37**, 199–202.
PURPURA, D. P. (1961). *Ann. N.Y. Acad. Sci.* **92**, 840–59.
PURPURA, D. P., SHOFER, R. J., HOUSEPIAN, E. M. & NOBACK, C. R. (1964). In: Purpura and Schadé (Eds.), *Progress in Brain Research*, Vol. 4, pp. 187–221. Elsevier, Amsterdam.
PURPURA, D. P., SHOFER, R. J. & SCARFF, T. (1965). *J. Neurophysiol.* **28**, 925–42.
ROMANES, G. J. (1947). *J. Anat.* (*Lond.*) **81**, 64–81.
ROSE, J. E. (1942). *J. Comp. Neurol.* **77**, 469–523.
ROSE, G. H. & LINDSLEY, D. B. (1968). *J. Neurophysiol.* **31**, 607.
RUCKEBUSCH, Y. (1971). *Rev. méd. Vét.* **34**, 483–510.
SCHERRER, J. & OECONOMOS, D. (1954). *Etud. néonatal* **3**, 199.
VIGNAL, W. (1888). *Arch. Physiol. Norm. Path.* **2**, ser. 4, 311–38.
VOELLER, K., PAPPAS, G. D. & PURPURA, D. P. (1963). *Exp. Neurol.* **7**, 107–30.
WOUDENBERG, R. A. (1970). *Brain Research* **17**, 417–37.

FUNCTIONAL DEVELOPMENT IN THE SOMATOSENSORY CORTEX OF FOETAL SHEEP

By H. E. PERSSON

Department of Physiology II, Karolinska Institutet,
104 01 Stockholm 60, Sweden

The functional development in a sensory cortical area is the result of a delicate interplay between the maturation of the cortical afferent pathways and of the neurones in the neocortex. The present investigation was undertaken to study the functional development in the somatosensory cortex of foetal sheep by using sensory evoked cortical responses.

The experiments were performed on sheep foetuses (from 42 to 135 days of age; full term 145 days) kept in contact with the placenta by the intact umbilical cord. Decerebration of the mother ewe made it possible to avoid the use of general anaesthesia. Tactile stimulation of the trigeminal nose region was used to elicit somatic sensory cortical activity. Recordings were made of both evoked gross potentials and evoked single-unit activity from the surface and from varying depths in the neocortex.

During development the somatosensory evoked cortical surface response underwent a series of characteristic changes in cortical distribution, configuration, latency and fatiguability (cf. Molliver, 1967). In the foetuses less than 75 days of age the surface cortical response was distributed over the entire anterior third of the lissencephalic brain (Fig. 1A). It is possible to trace the convolutional pattern of the adult brain in foetal brains from the age of about 85 days (Meyerson, 1968). At this developmental stage the cortical response to tactile stimulation was confined to those areas in the foetal brain (Fig. 1B) which correspond to the specific somaesthetic areas in the adult sheep (Adrian, 1943; Woolsey & Fairman, 1946). A somatosensory evoked response in the parietal association area was not obtained until late prenatal stages. It has been shown that cortical arousal reactions to cutaneous stimulation are only obtained in older foetuses near term (Bernhard, Kaiser & Kolmodin, 1959).

All these observations are consistent with the view that the somatosensory cortex is activated by the specific set of afferent fibres during the early prenatal stages and not until later stages is there a generalized activation of the neocortex through unspecific somaesthetic afferent pathways (cf. Marty & Scherrer, 1964).

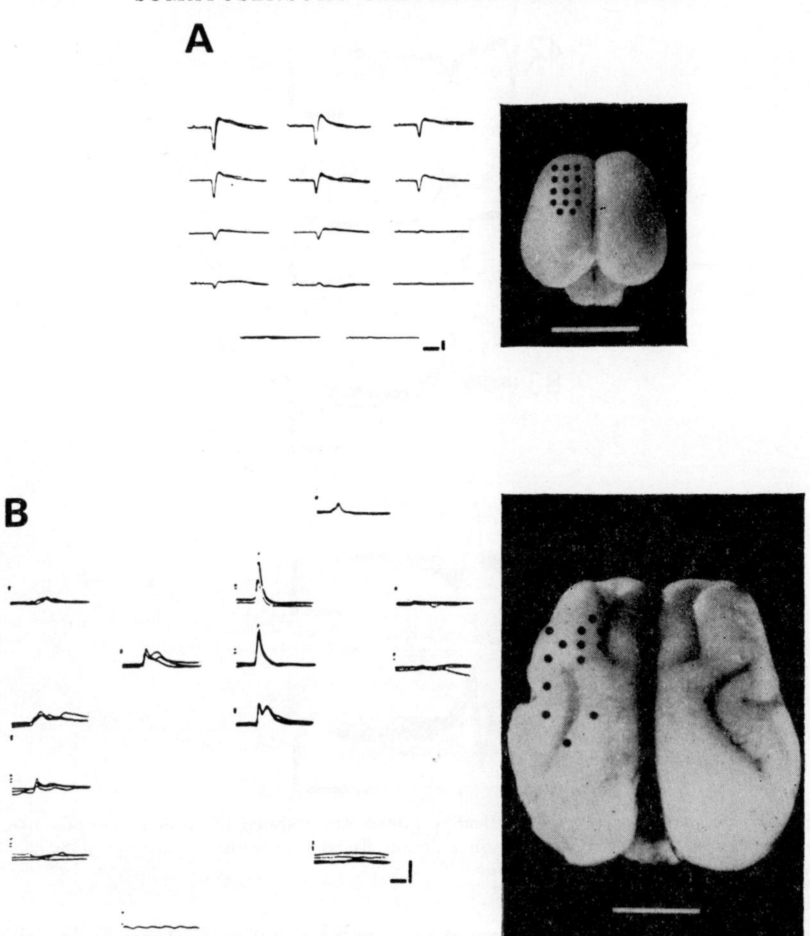

Fig. 1. Distribution of somaesthetic cortical responses evoked by tactile stimulation of the ipsilateral upper lip. A, from a 69-day-old sheep foetus; B, from a 91-day-old foetus. The records of cortical responses correspond to the relative positions of dots on the brains. Horizontal bars below the brains represent 1 cm. Calibration: vertical bars: 200 μV; horizontal bars: 100 ms. In this figure, as in all of the following figures, negativity is an upward deflection.

A remarkable developmental feature of the somaesthetic cortical response is its gradual alteration in configuration, which is thought to reflect the functional maturation of the cerebral cortex. In very young foetuses the evoked cortical response consisted of a long-latency surface positive wave (Fig. 2; 42, cf. Molliver, 1967). At a foetal age of about 70 days a small negativity appeared after the initial positive deflection (Fig. 2; 72). During the following developmental period this negativity increased in amplitude and, at about 90 days, it dominated the response (Fig. 2; 83 and 97). In

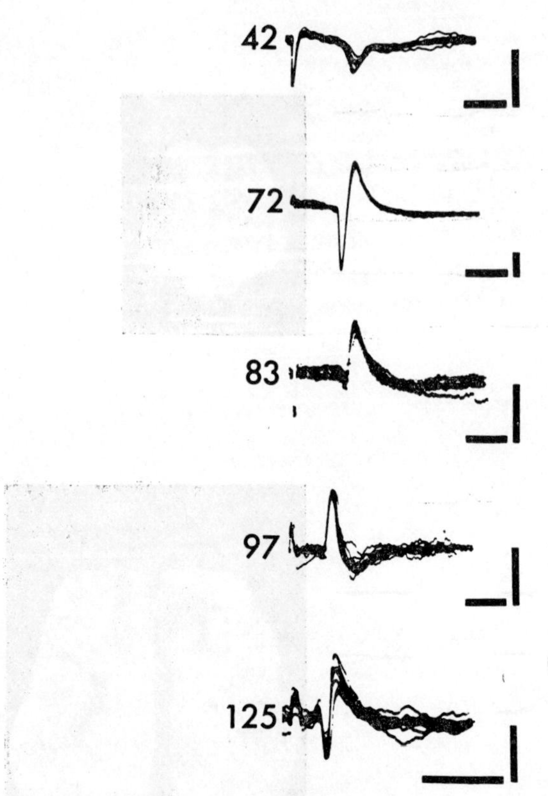

Fig. 2. Development of evoked cortical responses from sheep foetuses of various ages. The numbers to the left of each response denote foetal ages in days. Calibration: vertical bars: 200 μV; horizontal bars: 100 ms.

later stages (around 100 days) a small positive component preceding the negativity reappeared in the response. This positivity increased in amplitude and, at the age of about 125 days, the somatosensory cortical response displayed a positive–negative configuration similar to that of the adult animal (Fig. 2; 125).

To determine the sequence in which neurones in various cortical layers become activated by the afferent inflow at the different stages of the prenatal development, recordings were made of evoked field potentials and evoked single-unit activity at varying cortical depths. The voltage-values of the field potentials, recorded at various depths, were measured corresponding to the peaks of the positivity and negativity of the surface evoked responses. Depth-potential profiles were constructed from these measurements. On the basis of the depth-potential profiles vertical current gradients were calculated (see Humphrey, 1968). A positive value of the vertical current gradient is considered to indicate a 'source' and a negative

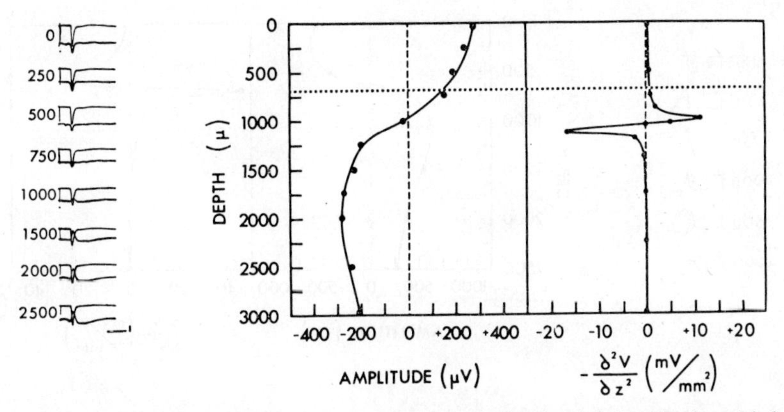

Fig. 3. Laminar analysis of somatosensory evoked cortical potentials in a 60-day-old sheep foetus. The upper traces to the left are surface potentials; lower are potentials recorded with a microelectrode at indicated depths. Calibration: vertical bar: 200 μV; horizontal bar: 100 ms. Depth-potential profile and estimate of the vertical current gradient corresponding to the peak of the surface positivity ares hown in the diagram to the right. A positive value of the vertical current gradient indicates a net current 'source' and a negative value indicates a relative 'sink'. The horizontal dotted line represents approximate cortical depth.

value to indicate a relative 'sink'. These current gradient estimations showed that the evoked somaesthetic cortical response originated from neuronal activity located at various cortical depths at different developmental stages.

The depth-potential profile and the estimate of vertical current gradient in the neocortex of a 60-day old foetus, which was characterized by a surface positive somatosensory response, are shown in Fig. 3. As can be seen the site of the 'sink' was located at a depth of 1200 μm (cf. Meyerson & Persson, 1969). The evoked neuronal discharges were repetitive (Fig. 6; 65) and confined to the same depths below the cortical surface. The 'sink' and the evoked unitary discharges arose from the *subcortical* strata which contained the developing afferent fibre projection. This observation was ascertained by histological examination of brain sections, in which the depths of penetration of the microelectrode tips had been marked with Prussian blue. Neurones within the primitive cortex were not activated by the afferent trigeminal inflow.

The next developmental stage was characterized by a somaesthetic response displaying a biphasic positive–negative form in which the positivity predominated. The estimate of the vertical current gradient in the neocortex during the evoked cortical response from this stage of development revealed that the net maximal 'sink' corresponding to both the surface positivity and the surface negativity was located at various

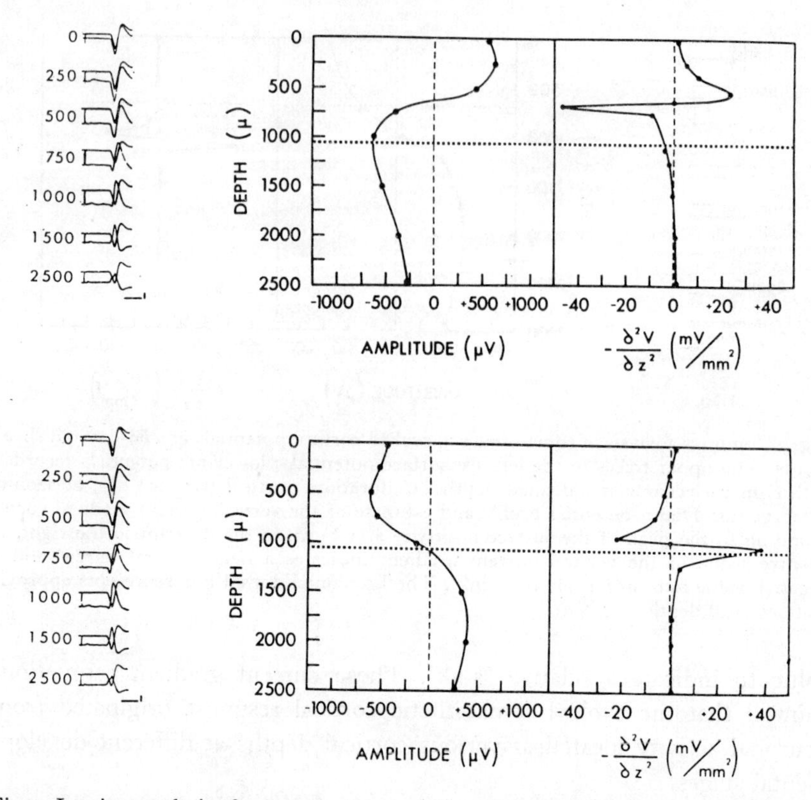

Fig. 4. Laminar analysis of somatosensory evoked cortical potentials in a 71-day-old sheep foetus. The upper traces are surface potentials; lower are field potentials recorded at indicated depths. Calibration: vertical bars: 200 μV, horizontal bars: 100 ms. Depth-potential profile and estimate of the vertical current gradient during the peak of the surface positivity are shown in the upper diagram and the same parameters corresponding to the peak of the surface negativity are shown in the lower diagram. The horizontal dotted lines represent the approximate cortical depth.

depths within the lower layers of the cortex (Fig. 4). This finding indicated that tactile stimulation resulted in excitatory neuronal activation within deep cortical layers, a conclusion which was confirmed by the fact that sensory evoked unit activity was obtained from the same strata (Fig. 7A). The evoked neuronal activity was always a single discharge (Fig. 6; 72) and repetitive spikes were never seen. It has been shown in experiments on newborn cats (Purpura, Shofer & Scarff, 1965) that synaptic activation of immature cortical neurones produces a single discharge. Thus, at this developmental stage the cortical neurones in the deeper layers of the neocortex for the first time become synaptically activated by the afferent trigeminal inflow.

During the next developmental period the negativity of the somato-

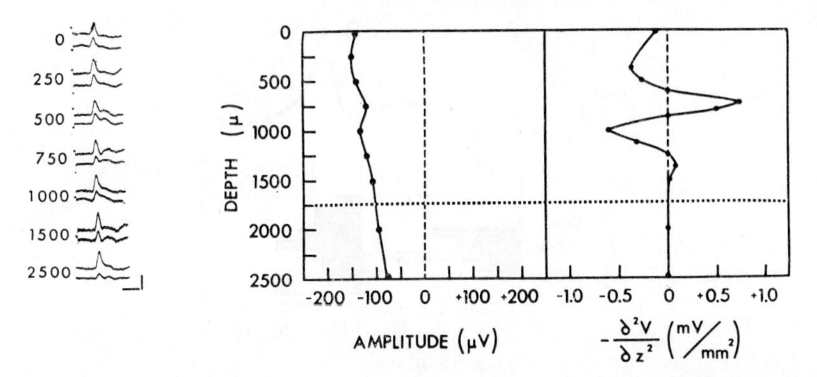

Fig. 5. Laminar analysis of somatosensory evoked cortical potentials in a 95-day-old sheep foetus. The upper traces to the left are surface potentials; lower are field potentials recorded at indicated depths. Calibration: vertical bar: 200 μV; horizontal bar: 100 ms. Depth-potential profile and estimate of vertical current gradient corresponding to the peak of the surface negativity are shown in the diagram. The horizontal dotted line represents the approximate cortical depth.

sensory response increased in amplitude and became the dominating component. The vertical spatial distribution of estimated 'sources and sinks' (Fig. 5) and of single neuronal activity (Fig. 7B) evoked by cutaneous stimulation showed that cortical neurones in more superficial layers were now also synaptically driven. The activated neurones usually discharged with a single spike (Fig. 6; 83 and 96) and the repetitive pattern of dis-

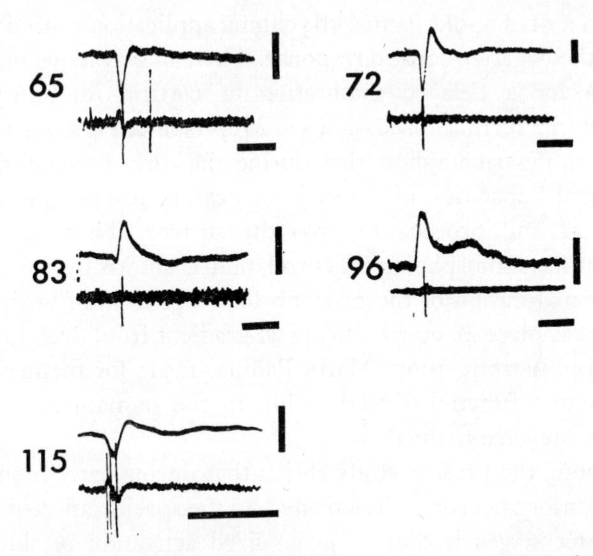

Fig. 6. Cortical activity in response to tactile stimulation of the trigeminal nose region in sheep foetuses of various ages. The numbers to the left of each response represent the gestational ages. Calibration: vertical bars: 200 μV; horizontal bars: 100 ms.

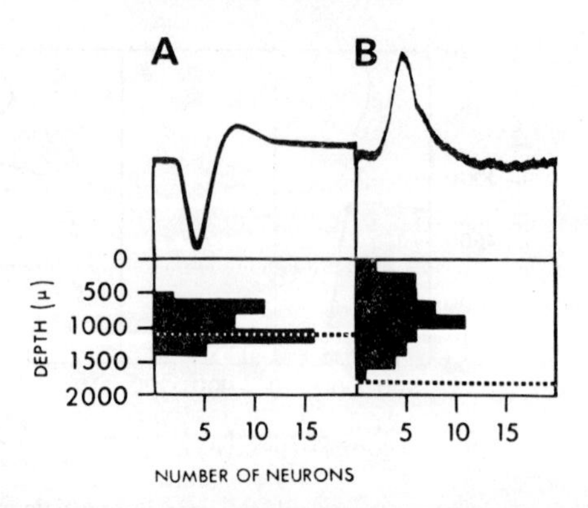

Fig. 7. Depth distribution in the somaesthetic cortex of evoked single-unit activity during the developmental stage characterized by a biphasic positive–negative somatosensory response (A) and the stage with a negative response (B). The dotted horizontal lines represent the approximate depths of the cortex in the two developmental stages.

charge, characteristic of mature somatic sensory cortical neurones (Mountcastle, Davies & Berman, 1957), was not obtained until later stages of development (Fig. 6; 115). It is of prime interest to find out if inhibitory synaptic activities are generated in the immature neocortex by sensory stimulation. In the present study it was found that from a foetal age of 75 days signs of cortical inhibition were obtained as shown by sensory induced arrest of single-unit activity. At that age, topical cortical application of strychnine had minor effects on the somaesthetic cortical surface response. However, in a 91-day-old foetus, strychnine application caused a dramatic increase of the negative cortical response. These observations may be taken as evidence for a delayed maturation of cortical inhibitory synaptic mechanisms (for further discussion see Meyerson & Persson, 1972).

The results described show that during the early prenatal ontogenesis the functional capacities of sensory cortical neurones appear in deep cortical layers and proceed towards the surface. That this functional development has a morphological correlation is shown by the fact that, in general, the maturation of the neurofibrillar organization in the cortex of mammals takes place along a corticopetal gradient from deep to superficial strata (e.g. see Åström, 1967; Marin-Padilla, 1971; for further discussion on the structuro-functional relationship in the immature neocortex see Bernhard & Meyerson, 1972).

To conclude, the present study shows that during early prenatal stages the somatosensory neocortex is activated by the specific afferent inflow and not until later stages is there a generalized activation of the neocortex through nonspecific somaesthetic afferent pathways. In the early period of development the somatosensory system operates without functional contacts with the somaesthetic cortex. The functional development within the somatic sensory cortex, as reflected by the sensory activation of neurones, starts in deep neocortical layers and proceeds towards the surface and is expressed in the gradual alteration in the configuration of the cortical response. This gradient of functional development in the neocortex corresponds to a similar gradient of morphological maturation of the cortical neurofibrillar organization.

This work has been supported by grants from the Swedish Medical Research Council and by personal grants from the Karolinska Institute (Reservationsanslaget), Svenska Sällskapet för Medicinsk Forskning, and the Hierta-Retzius Scientific Funds.

This presentation is preliminary and a detailed account of the results obtained is awaiting publication.

REFERENCES

ADRIAN, E. D. (1943). *Brain* **66**, 89–103.
ÅSTRÖM, K. E. (1967). *Progress in Brain Research* **26**, 1–59.
BERNHARD, C. G., KAISER, I. H. & KOLMODIN, G. M. (1959). *Acta physiol. scand.* **47**, 333–49.
BERNHARD, C. G. & MEYERSON, B. A. (1972). This volume.
HUMPHREY, D. R. (1968). *Electroenceph. clin. Neurophysiol.* **24**, 116–29.
MARIN-PADILLA, M. D. (1971). *Z. Anat. Entwickl.-Gesch.* **134**, 117–45.
MARTY, R. & SCHERRER, J. (1964). *Progress in Brain Research* **4**, 222–36.
MEYERSON, B. A. (1968). *Acta physiol. scand.*, Suppl. 312.
MEYERSON, B. A. & PERSSON, H. E. (1969). *Nature* **221**, 1248–9.
MEYERSON, B. A. & PERSSON, H. E. (1973). *Advances in Ontogeny of Behavior and Nervous System*. Vol. 2. Academic Press, New York.
MOLLIVER, M. E. (1967). *Progress in Brain Research* **26**, 78–91.
MOUNTCASTLE, V. B., DAVIES, P. W. & BERMAN, A. L. (1957). *J. Neurophysiol.* **20**, 374–407.
PURPURA, D. P., SHOFER, R. J. & SCARFF, T. (1965). *J. Neurophysiol.* **28**, 925–42.
WOOLSEY, C. N. & FAIRMAN, D. (1946). *Surgery* **19**, 684–702.

FUNCTIONAL INTEGRITY OF NEONATAL RAT CEREBRAL CORTEX

By M. ARMSTRONG-JAMES
Department of Physiology,
The London Hospital Medical College, London, E1 2AD

The cerebral cortex of the neonatal rat has been shown, in previous investigations, to undergo a remarkable expansion of synaptic growth between birth and three months postnatal (Armstrong-James & Johnson, 1970; Johnson & Armstrong-James, 1970). From histological investigations using electron-microscopy, synaptic function at birth is calculated to be at about 0.1 % of the adult level, reaching 4 % at 7 days and 11 % at 14 days. Since the greatest rate of increase of synaptic development is during the first fortnight of postnatal life, it was considered to be of interest how well primary somato-sensory cortex is capable of handling sensory information at this stage. Preliminary experiments were performed to this end, and it was found that the earliest stage at which neurones responded regularly to tactile stimulation was at 7 days postnatal. A series of experiments was therefore carried out on some 48 7-day-old animals under urethane anaesthesia. The results were compared with equivalent experiments on adult animals.

Glass micro-electrodes were advanced automatically through the somato-sensory cortex and at 25-micron intervals, light general tactile stimulation was carried out to determine the presence or absence of responsive single cells. On finding a responsive unit, the responses were analysed in terms of mean numbers of action potentials in response to at least 10 standard mechanical stimuli per site, and secondly spike interval histograms were constructed for all responsive and spontaneously active cells.

Analysis of all unitary activity was carried out in 16 experiments at 7 days of age. Of the 309 cells encountered, 80 % were activated by tactile stimulation but showed no spontaneous activity, a further 14 % showed spontaneous activity and were also activated by tactile stimulation. The remaining 6 % were spontaneously active, but were not affected by peripheral tactile stimulation, or joint manipulation where the latter was carried out. A very large majority of cells was, therefore, silent without intentional stimulation at 7 days of age. On the other hand, in the adult cortex the great majority of responsive cells was also spontaneously active. Seven-day-old spontaneously active cells fired at a wide range of overall frequencies; some fired only once every 30 seconds or so, others at rates up to 20 times a second. Those cells which fired rapidly showed very little tendency to fire in short bursts. Comparative adult cells, on the other hand, predominantly fired in short bursts of activity with intervals as short as 1.5 msec. In summary, spontaneous activity at 7 days of age could be termed fundamentally tonic, whereas adult cells normally fired phasically.

Phasic activity of cortical cells may indicate the presence of feedback inhibition, the action potential train being cut off in this manner; as indeed it is in the adult. If this is the case then such inhibition might be poorly developed at 7 days of age.

The depth at which cells were found in the cortex was logged, and it was found that peak numbers of responsive and spontaneously active cells were located within layer 4, the inner stellate cell layer. This is the layer in which primary afferents relayed from the thalamus terminate. All penetrations of the cortex which gave responsive cells always contained cells within the inner stellate cell layer.

At 7 days of age strong evidence was found for the functional organisation of somato-sensory cortex into vertical columns of cells. Within a particular vertical column all responsive cells encountered were driven by similar receptive fields. Responsive columns were found in all peripheral sites including the vibrissae, and these columns were arranged in an orderly homunculus over the somato-sensory cortex.

The vertical columnar organisation was confirmed by using slightly oblique electrode penetrations designed to pass across two or more adjacent columns.

Fig. 1 shows the receptive fields of cells from an oblique trajectory through the forearm area of somato-sensory cortex. The first cells encountered in superficial cortex are predominantly driven by plantar palmar stimulation, whereas the remaining six cells in lower cortex are predominantly driven by little finger stimulation. The electrode had clearly transgressed two columns, one dominantly responding to palmar stimulation, the other predominantly to little finger stimulation. This type of transformation occurs with oblique penetrations for all other areas of somato-sensory cortex, examples also being found for flank, vibrissa and leg columns.

Receptive fields for vibrissae were found to involve several or all vibrissae at 7 days of age. It has been reported previously by Welker (1971) that individual vertical columns of cells exist for each and every vibrissae in the adult. Such an organisation was not found at 7 days of age, since several or all of the vibrissae afferents at this stage converge on single units within the vertical penetration. It was therefore decided to make comparisons between

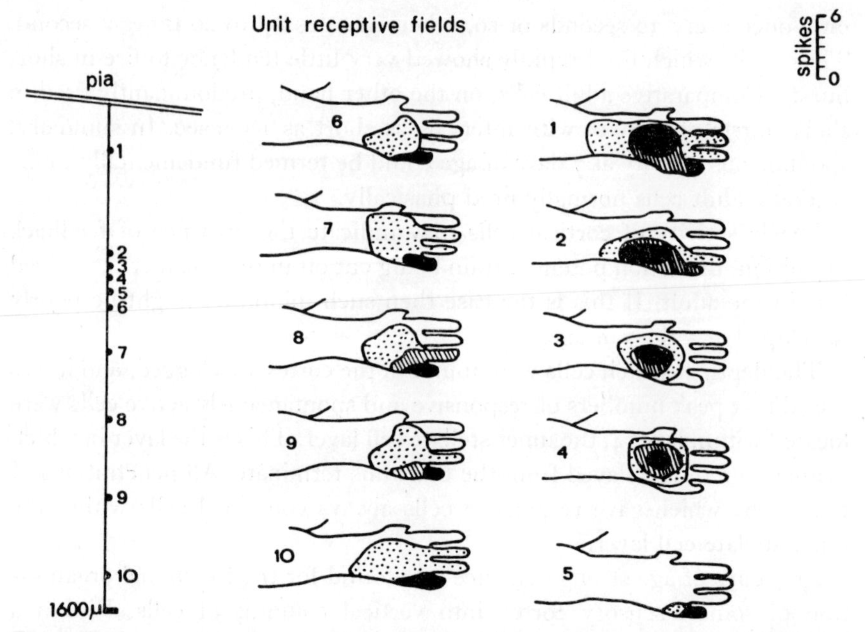

Fig. 1. Black areas show the extent of the maximal focus of the responsive field. The total extent of the responsive field where stimulation caused liminal activity in cortical neurones is indicated by the stippled area. Numbers on the figurines correspond to units encountered in the electrode trajectory, shown to the left of the diagram.

sizes of receptive fields of single cells between 7-day-old and adult animals.

Fig. 2 shows a comparison between receptive fields of cells in vertical columns within the distal hind limb zone at 7 days and maturity. Whereas cells in the adult column were driven by tactile stimulation of one or two digits only, those found in a typical equivalent 7-day-old column were driven by a far greater number of discrete anatomical sites. Distal hind limb fields involving one or two digits only at 7 days of age, were indeed never found. Findings were similar for all other tactile sites, namely that, equivalent receptive fields were far greater in size at 7 days of age than at maturity. In other words the columnar organisation at 7 days is considerably less discrete than at maturity.

Now, in the adult cerebral cortex the means by which parcellation into columns occurs may be due to profoundly organised inhibitory mechanisms, such as surround or adjacent inhibition. In adult experiments such inhibition was fairly common. In contrast only two cells have been encountered at 7 days of age which showed a decrease in their spontaneous firing rate when tactile stimulation was carried out. On a small number of occasions the receptive fields of silent cells have exhibited a highly irregular

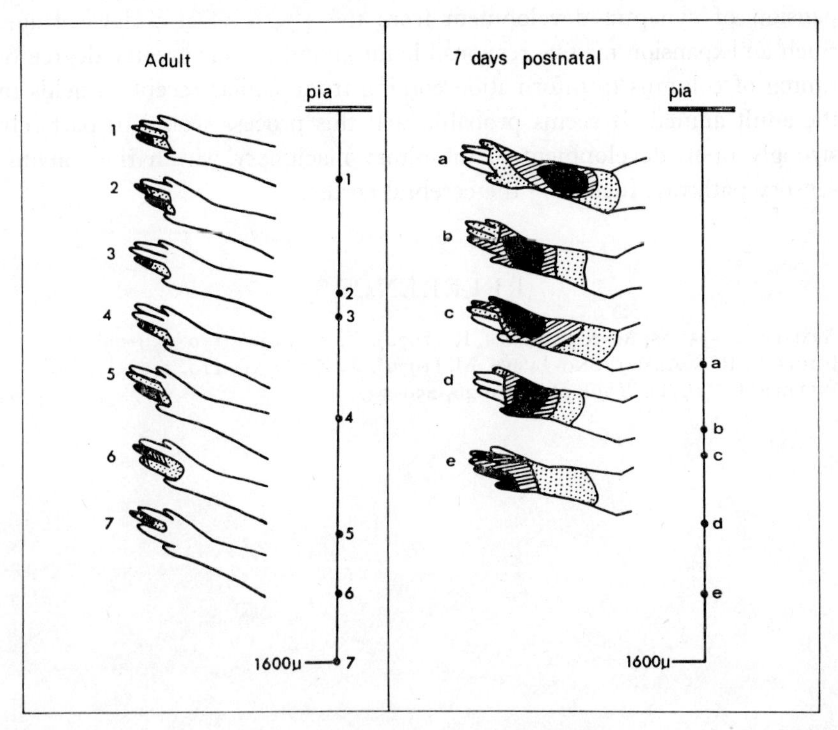

Fig. 2. Distal hind limb units at 7 days and maturity. Maximal focus of receptive fields is shown by the black areas on the figurines. The numbers and letters on the figurines correspond to units shown in the equivalent electrode trajectory to the right of the figurines. Stippled areas show the maximum extent of the receptive fields.

shape which may be indicative of surround or adjacent inhibition. For example, a vertical trajectory showed cells which were maximally responsive to either lateral palm stimulation or lateral elbow stimulation; other cells within the same column showed maximal receptive fields for both lateral elbow and lateral palm. Such a situation has not been found in adult primary somato-sensory cortex. It is possible that the intervening area between palm and elbow was yielding information of a mixed excitatory and inhibitory influence, thus dividing the receptive field into two zones.

In summary, it appears that 7 days after the commencement of postnatal life represents a critical period in postnatal development of somato-sensory pathways in the rat. It is at this stage that thalamic neurones rather suddenly first become regularly capable of driving cortical cells in response to tactile stimulation. Secondly, a remarkable degree of functional organisation is already present at this age, and one is led to conclude that the main anatomical substrate for such an organisation has already been laid. Further development would clearly in part rely on a general expansion of synaptic development from the 4% level at 7 days of age. Such an expansion may be responsible for giving the far greater degree of tuning of columns to information coming from similar receptive fields in the adult animal. It seems probable that this process would in part rely strongly upon development of inhibitory machinery within the somatosensory pathway, including the cerebral cortex.

REFERENCES

Armstrong-James, M. & Johnson, R. (1970). *Z. Zellforsch.* **110**, 559–68.
Johnson, R. & Armstrong-James, M. (1970). *Z. Zellforsch.* **110**, 540–58.
Welker, C. (1971). *Brain Research* **26**, 259–75.

DEVELOPMENT OF BLOOD–BRAIN BARRIER MECHANISMS IN THE FOETAL SHEEP

By J. M. REYNOLDS, M. L. REYNOLDS
and N. R. SAUNDERS

University College London, Gower Street, London, W.C.1

INTRODUCTION

The importance of the stability of the brain's environment for normal central nervous system neurone function has long been recognised and Sir Joseph Barcroft discussed it in his Terry Lectures (1938). He did not distinguish between the local environment of the central nervous system and the more general internal environment of the whole animal. However, it is now clear that the local environment of the brain is kept even more stable in composition than the rest of the internal environment of the whole animal, and that there are several different mechanisms which regulate the composition of brain extracellular fluid (ECF) and cerebrospinal fluid (CSF) rather than the single impermeable barrier implied by the term 'blood–brain barrier'.

It has quite frequently been suggested that the blood–brain barrier is immature in the late foetal and newborn period (Lee, 1971); the evidence for this was based on studies of kernicterus (deposition of unconjugated bilirubin in the brain) or dye penetration from blood into brain. Both of these materials bind to protein, and the degree of binding can be affected by factors such as blood pH which were usually not measured, let alone controlled. The picture is further confused by the fact that the different species used develop *in utero* at different rates, and are born at different stages of maturity.

The experiments described in this paper have attempted to overcome some of these problems by choosing a species (the sheep) which is large enough for the foetus to be accessible to a range of physiological methods over a long period of its intra-uterine development, and by using substances, e.g. sucrose, which are not metabolised or specifically transported into brain, and which are not protein-bound or affected by pH changes.

However, there remain three important factors (Fig. 1) to be taken into account when considering the penetration of substances such as sucrose from blood into brain and CSF:

(1) *The permeability barriers which the sucrose has to penetrate.* These are probably in the tight junctions (Brightman & Reese, 1969) between cells

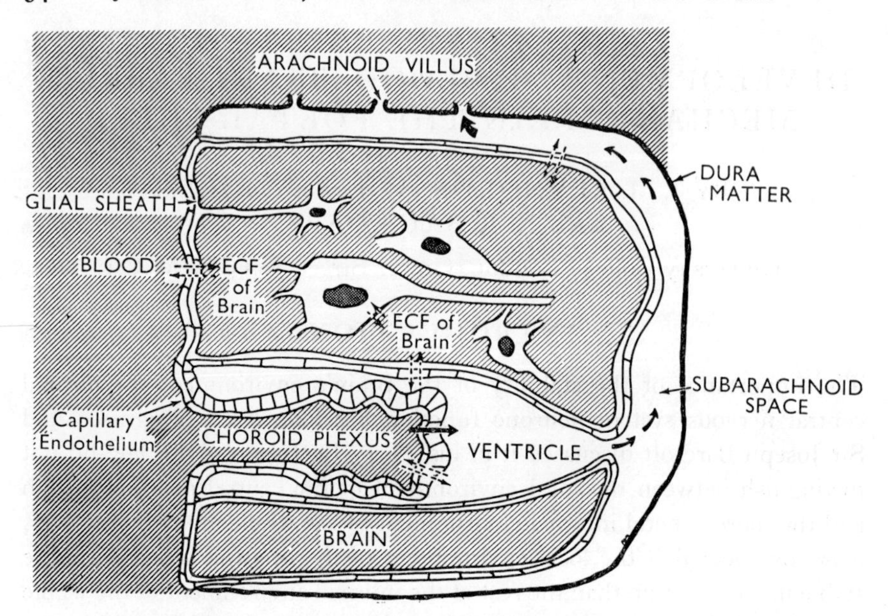

Fig. 1. Diagram to illustrate some of the factors which affect penetration of substances from blood into brain and CSF.

Barriers: Blood–brain barrier across capillary endothelium. Blood–CSF barrier across choroid plexus epithelium. At both these sites tight-junctions between adjacent cells place a restraint upon the movement of small molecular weight lipid-insoluble molecules, e.g. sucrose.

Spaces: Brain extracellular fluid (ECF). Cerebrospinal fluid. Brain intracellular space. There is relatively free exchange between brain ECF and CSF across the ependyma lining the cerebral ventricles.

Sink effect of CSF: The flow of CSF is through the ventricular system to the subarachnoid space and back into the circulation via the arachnoid villi. This has the effect of keeping at a low concentration substances which penetrate only slowly into brain ECS and CSF.

From *Clinical Disorders of Fluid and Electrolyte Metabolism*, 2nd ed., pp. 1023–42 (eds. M. H. Maxwell & C. R. Kleeman), © McGraw-Hill, 1972. Reproduced by permission of Dr M. W. B. Bradbury and McGraw-Hill Book Company.

in the cerebral capillary endothelium (blood–brain barrier) and the choroid plexus epithelium (blood–CSF barrier). These place a restraint on the penetration of small molecular weight lipid-insoluble substances from blood into brain and CSF.

(2) *The volume of extracellular space in the foetal brain.* This has been suggested to be larger in the foetus than in the adult. If it is, this would be expected to increase the sucrose accumulation in foetal brain.

(3) *The formation of CSF by the choroid plexus* and its flow from the ventricular system to the subarachnoid space and back into the circulation via the subarachnoid villi acts as a sink (Davson, 1967).* A decrease in this

* Substances which penetrate the brain and CSF only slowly are kept at a low concentration by this sink effect.

effect, for example because of a lower rate of CSF secretion in the foetus, would be expected to increase the rate of accumulation and, more particularly, the final steady state level of sucrose in the brain.

METHODS

The methods used were based on those described by Davson (1967) and Dawes (1968) and have been demonstrated to the Physiological Society (Evans, Reynolds, Reynolds, Saunders & Segal, 1972).

Ewes of known gestational age were anaesthetised with thiopentone and chloralose. Foetuses older than 90 days were delivered on to a heated side table; a femoral artery and vein were cannulated and the kidneys tied off. Foetuses younger than 90 days were left *in utero* and placental vessels were cannulated (Meschia, Makowski & Battaglia, 1969). Foetuses were kept at a temperature of 39 ± 1 °C by external heating. The condition of the ewe was monitored throughout the experiment (arterial blood pressure, heart rate, end tidal CO_2, arterial pH and blood gases). Similar measurements were made in the foetuses, although in the smallest there was only sufficient circulating blood volume for a single determination of blood gases at the end of the experiment.

Stable blood concentrations of ^{14}C and/or ^{3}H sucrose were maintained by i.v. infusion. After times between 25 min and $4\frac{1}{2}$ h, CSF was obtained by cisternal puncture, blood was drained from the head by cutting the carotid arteries and jugular veins, and the brain was removed.

Samples of plasma, CSF, and homogenised brain were solubilised in 'Soluene' (Packard) and activities were estimated by liquid scintillation counting (using Packard 'Instagel' mixed with PPO/POP as scintillant and a Packard Tricarb Spectrometer). Blood contamination of brain and CSF samples was estimated by injecting ^{113}Indium mixed with plasma 5 min before the end of the experiment (Sisson, Oldendorf & Cassen, 1970). Results are expressed as the apparent sucrose space or volume of distribution of sucrose, i.e.

$$\frac{\text{DPM/g brain or CSF}}{\text{DPM/g plasma}} \%.$$

RESULTS

The penetration of sucrose from blood into brain and CSF at different foetal ages is shown in Fig. 2. In these experiments the blood level of sucrose was maintained for 90 min. At 125 days' gestation and in the newborn period the level of brain sucrose is that which would be expected

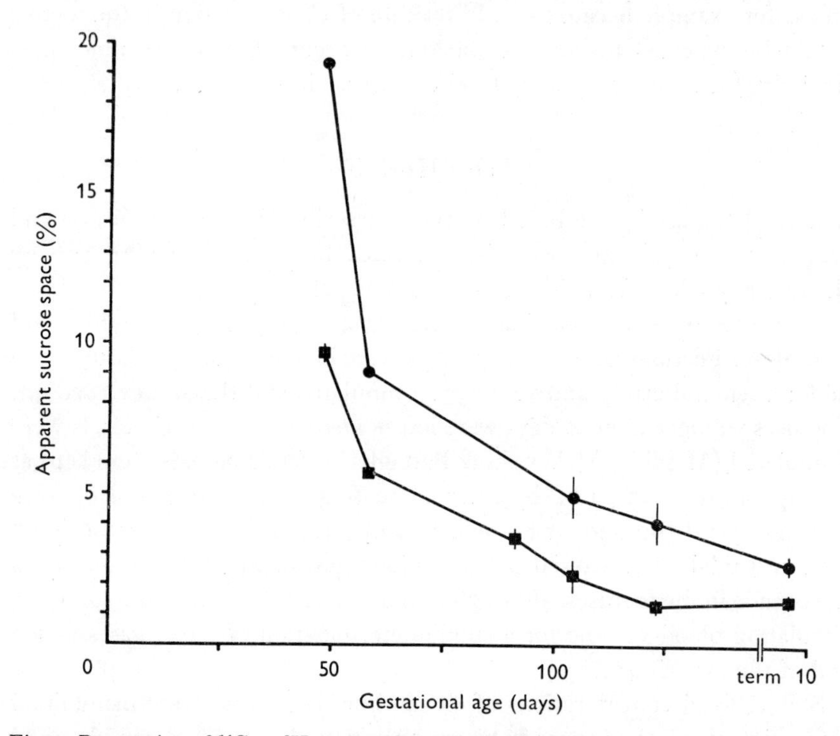

Fig. 2. Penetration of ^{14}C or ^{3}H sucrose from blood into foetal and newborn sheep brain (■—■) and CSF (●—●) at different ages. 90 min. i.v. infusion of isotope. In this and Fig. 3 the apparent sucrose space $= \frac{\text{DPM/g brain or CSF}}{\text{DPM/g plasma}}\%$.
Bars indicate ±SEM (some fall within the symbols); $n = 3$–7, except for the 50-day CSF point ($n = 2$).

in an adult animal (Oldendorf & Davson, 1967) and this has been confirmed in a few ewes, using chemical estimation of sucrose (Reynolds, unpublished). There is a small but significant increase in the brain sucrose space around 100 days' gestation, but a marked increase in brain sucrose accumulation only occurs at a much earlier stage of development, around 50 to 60 days' gestation. The level of sucrose in CSF shows a similar pattern with decreasing age.

In order to obtain information about the rate of sucrose penetration into brain at different ages, the blood concentration was maintained for different periods of time before removing the CSF and brain. These experiments are illustrated in Fig. 3 which shows the course of penetration of sucrose into brain during 3 h in 48-day foetuses and during $4\frac{1}{2}$ h in 59-day foetuses. The curves for the more mature foetuses are incomplete, but it can be predicted from work on adult animals that the level of brain sucrose at, for

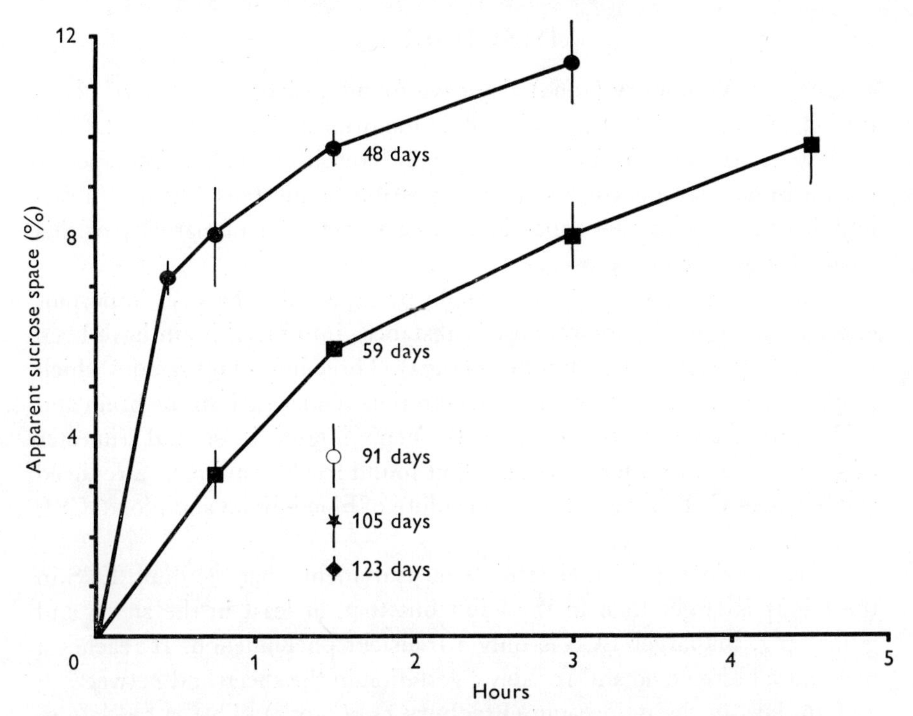

Fig. 3. Rate of penetration of ^{14}C or ^{3}H sucrose from blood into foetal sheep brain at different ages; $n = 3$–7.

example, 123 days would be unlikely to rise above 2% even after 4–6 h i.v. infusion.

In order to assess the sink effect of CSF secretion in the foetus, ventriculo-cisternal perfusions (Oldendorf & Davson, 1967) were carried out in foetuses aged 60 to 140 days, and the CSF secretion rate was measured by dye (Blue Dextran) dilution. The absolute rate of CSF secretion increases with age, being 11.1 ± 2.4 SEM μl/min ($n = 5$) at 60 days and 59.6 ± 10.0 μl/min ($n = 9$) at term. This is hardly surprising since the brain size increases from about 2 to 40 g between these ages. In order to have some estimate of 'sink-effect' in relation to brain size the results have also been calculated as a secretion rate per g brain. On this basis the rate is similar between the ages of 90 days *in utero* and adult (1.6 μl/min·g) but is markedly greater in the younger foetuses (6.0 μl/min·g at 60 days). Thus the washing out or sink effect would be greater in the younger foetuses, i.e. it would be expected that the sucrose space would be smaller rather than larger, as was found to be the case.

DISCUSSION

Ferguson & Woodbury (1969) also have found an increased rate of penetration of sucrose into the brains of immature rats (in the late foetal and early newborn period). As Barcroft (1938) himself points out, interspecies comparisons may be confusing, but possibly a newborn rat and a 60-day sheep foetus may be equated, at least in terms of immaturity of this particular barrier to sucrose.

Factors which in general terms might be expected to have an important effect on the rate of penetration of substances into foetal brain have been discussed by Saunders & Bradbury (1972). Three important factors which might contribute to the increased penetration of sucrose into the brain and CSF of immature sheep foetuses at the beginning of the second trimester of gestation are: a larger ECS than that found in older animals, a reduced sink effect of CSF and greater permeability of blood–brain and blood–CSF barriers.

There is evidence from electrolyte measurements that the brain ECS in the foetus is larger than in the adult but that, at least in the sheep and guinea-pig, the larger ECS is only a transient phenomenon. It reaches a maximum between 90 and 105 days' gestation in the sheep and between 39 and 46 days in the guinea-pig (Bradbury *et al.* 1972). This is too late in gestation to account for the main increase in sucrose penetration in the sheep which occurs between 50 and 60 days. But this increase in brain ECS may contribute to the slightly larger brain sucrose space around 90 to 100 days (Fig. 2).

A smaller sink-effect does not seem to contribute to the large sucrose space in the immature foetus: the CSF secretion rate per unit weight of brain is substantially *higher* in the very small foetuses. A more appropriate comparison of the sink-effect at different ages may be the turnover rate of CSF since the volume of CSF in the smaller foetuses seems to be larger in relation to brain size than in the older foetuses. However, so far we do not have any satisfactory measurements of CSF volume.

Other evidence for the very early development of blood–brain, blood–CSF barrier mechanisms comes from the experiments of Bradbury *et al.* (1972) on the distribution of electrolytes between CSF and plasma at different foetal ages. In the sheep, gradients are established for magnesium before 48 days, sodium and chloride by 65 days, potassium by 90 days and calcium after birth. Bito & Myers (1970) have made similar findings in the monkey. In the guinea-pig (Bradbury *et al.* 1972) the order in which these gradients develop is the same but presumably because of the shorter period of gestation and more rapid rate of development they are compressed into an even earlier and shorter period of gestation. The development of a barrier mechanism which limits the inward penetration of sucrose in the period 50 to 60 days' gestation corresponds well with the development of CSF: plasma magnesium, sodium and chloride gradients at about the same time since this also implies the development of a barrier mechanism which prevents rapid outward movement of these ions down the concentration gradients set up by the onset of specific transport processes in the choroid plexuses and/or cerebral capillary endothelium.

This work was supported by grants from the Medical Research Council.

REFERENCES

Barcroft, J. (1938). *The Brain and its Environment*. New Haven: Yale Univ. Press.

Bito, L. T. & Myers, R. E. (1970). *J. Physiol. Lond.* **208**, 153–70.

Bradbury, M. W. B., Crowder, J., Desai, S., Reynolds, J. M., Reynolds, M., & Saunders, N. R. *J. Physiol. Lond.* **227**, 1–20.

Brightman, M. W. & Reese, T. S. (1969). *J. cell. Biol.* **40**, 648–77.

Davson, H. (1967). *Physiology of the Cerebrospinal Fluid*. London: Churchill.

Dawes, G. S. (1968). *Foetal and Neonatal Physiology*. Chicago: Year Book Medical Publishers.

Evans, C. A. N., Reynolds, J. M., Reynolds, M., Saunders, N. R. & Segal, M. B. (1972). *J. Physiol. Lond.* **224**, 15–16*P*.

Ferguson, R. K. & Woodbury, D. M. (1969). *Expl. Brain Res.* **7**, 181–94.

Lee, J. C. (1971). *Progress in Neuropathology* **1**, 84–145.

Meschia, G., Makowski, E. L. & Battaglia, F. C. (1969). *Yale J. Biol. Med.* **42**, 154–65.

Oldendorf, W. M. & Davson, H. (1967). *Archs. Neurol.* **17**, 196–205.

Saunders, N. R. & Bradbury, M. W. B. (1972). In *Fetal Pharmacology*. Ed. L. Boreus. New York: Raven. (In press.)

Sisson, W. B., Oldendorf, W. H. & Cassen, B. (1970). *J. Nuclear Med.* **11**, 749–52.

ONTOGENESIS OF BLOOD–BRAIN-BARRIER FUNCTION IN PRIMATE: CSF CATION REGULATION

BY RONALD E. MYERS* AND LASZLO Z. BITO†

The 'blood–brain-barrier' was first recognized and continues to be studied primarily in its property of excluding intravascularly injected dyes or other markers from brain tissue. However, the 'blood–brain-barrier' also actively regulates the ionic composition of the various nervous system fluids.

In considering the barrier systems of the brain, two major anatomical subdivisions can be identified: the one lying between the circulating blood and the ventricular fluid, the other interposed between the nervous system capillary blood and the fluid in the extracellular space. The former is chiefly (but perhaps not exclusively) related to the activity of the cuboidal epithelium of the choroid plexus while the latter lies in intimate relation to the capillaries of the brain.

The present paper focuses on the dynamic functions of the barrier systems in regulating the ionic composition of the various brain fluids. The timetable for the maturation of these regulatory functions in the several nervous system areas are examined during primate ontogeny during the latter part of gestation through to adult life. In adults, the cerebrospinal fluid (CSF) samples were obtained from sites in the lateral ventricles, the cisterna magna, the lumbar subarachnoid space, and the subarachnoid space located over the hemispheral convexity (Bito & Davson, 1966). CSF sampling in foetal and juvenile monkeys was carried out only at the cisterna magna and the cortical subarachnoid space. Arterial blood samples withdrawn simultaneously served as sources of plasma samples. All animals were anaesthetized with pentobarbital.

All plasma and CSF samples were analysed for potassium, magnesium and calcium concentrations using a Perkin–Elmer atomic absorption spectrophotometer. The dialysis ratios for these cations in adult plasma were also determined by placing 1.2 ml plasma samples in dialysis sacs and stirring for 6–12 hours in 1.2 ml samples of monkey CSF or protein-free tissue culture media. The results of all determinations carried out on the adult samples appear in Table 1.

* Laboratory of Perinatal Physiology, National Institute of Neurological Diseases and Stroke, National Institutes of Health, United States Public Health Service, Department of Health, Education, and Welfare, Bethesda, Maryland 20014; † Department of Ophthalmology, Columbia University School of Medicine, New York, New York 10027.

Table 1. *The concentrations of K, Mg and Ca in blood plasma and cerebrospinal fluids, and the concentration gradients of these cations between blood and CSF and within CSF system of adult rhesus monkeys*

		Absolute concns (m-equiv/kg H_2O)						Relative concns (distribution ratios)				
		Plasma*		Cerebrospinal fluids								
		$P_{-\frac{1}{2}}$	P_0	Ventricular	Cisternal	Cortical	Spinal	Dialysis ratio	Cisternal / P_0	Ventricular / Cisternal	Cortical / Cisternal	Spinal / Cisternal
K	Mean	4.49	4.41	2.57	2.55	1.99	2.56	0.92	0.60	1.02	0.81	1.02
	SE	±0.51	±0.21	±0.12	±0.13	±0.12	±0.13	±0.01	±0.05	±0.06	±0.03	±0.05
	(*n*)	(5)	(9)	(6)	(9)	(8)	(8)	(7)	(8)	(6)	(8)	(8)
	$P <$	—	—	—	—	—	—	—	0.001	n.s.	0.01	n.s.
Mg	Mean	1.18	1.18	1.99	1.86	2.00	1.78	0.75	1.66	1.07	1.07	0.96
	SE	±0.04	±0.07	±0.03	±0.05	±0.08	±0.05	±0.003	±0.10	±0.02	±0.02	±0.01
	(*n*)	(5)	(9)	(6)	(9)	(9)	(8)	(7)	(8)	(6)	(9)	(8)
	$P <$	—	—	—	—	—	—	—	0.001	0.005	0.01	0.01
Ca	Mean	4.65	4.81	2.12	2.28	2.37	2.34	0.58	0.47	0.97	1.04	1.06
	SE	±0.25	±0.15	±0.02	±0.09	±0.09	±0.07	±0.02	±0.02	±0.01	±0.02	±0.05
	(*n*)	(5)	(9)	(6)	(9)	(9)	(8)	(7)	(8)	(6)	(9)	(8)
	$P <$	—	—	—	—	—	—	—	0.001	0.05	0.02	0.002

* Cation concentrations in plasma of blood samples taken half an hour before ($P_{-\frac{1}{2}}$) and directly before (P_0) induction of anaesthesia.

The adult potassium concentrations show a sharp downward regulation of this ion from the plasma to the ventricular fluid. No additional potassium concentration changes appear with the further passage of the CSF through the cisterna magna and the spinal subarachnoid spaces. However, CSF flow through the cortical subarachnoid space is associated with significant further decreases in the potassium content.

In contrast to the situation of potassium, magnesium is significantly increased in its concentration in the ventricular fluid compared to the plasma. This distribution is particularly significant considering the plasma magnesium dialysis ratio of 0.75. Significant slight decreases then appear in the CSF magnesium content with its passage from the lateral ventricles to the cisterna magna and from the cisterna magna to the lumbar subarachnoid space. However, as the CSF passes over the cortical convexity in the cortical subarachnoid space, a significant rise in the CSF magnesium content again appears.

Calcium, like potassium, decreases in concentration from the adult blood plasma to the ventricular fluid. The calcium content then increases as the CSF passes from the ventricles to the cisterna magna and from the cisterna magna to the cortical and lumbar subarachnoid spaces.

The present data show that in the adult, while the potassium and calcium concentrations are normally regulated sharply downwards, that of magnesium greatly increases from plasma to ventricular fluid. Thus, close to the site of CSF formation major differential effects already appear relating to the handling of the different cations. Further significant alterations in cation concentration profiles in fluid samples taken at different sites along the channels of CSF flow provide evidence for a continuing regulation of fluid composition within the CSF system. That these concentration changes are due to active regulation rather than to ion 'leakage' between blood and CSF is evident from the fact that the directions of cation movement are frequently opposite in direction to existent diffusion gradients and that the migration of two different ionic species may be opposite in their direction despite similarities in electrical charge.

The presumed patterns of ion migration accounting for the shifts in cation concentrations in the CSF as it passes through the cortical subarachnoid spaces appear in Fig. 1. An unimpeded movement of ions and of dye particles occurs across the pia–glia limiting membrane (Bito, Bradbury & Davson, 1966; Davson, 1967). Thus, a free exchange of constituents occurs between the fluid perfusing the cortical subarachnoid spaces and that permeating the cortical interstitial spaces (the brain extracellular fluid). Under these circumstances compositional changes occurring in one of these two fluids are transmitted with a time lag to the other.

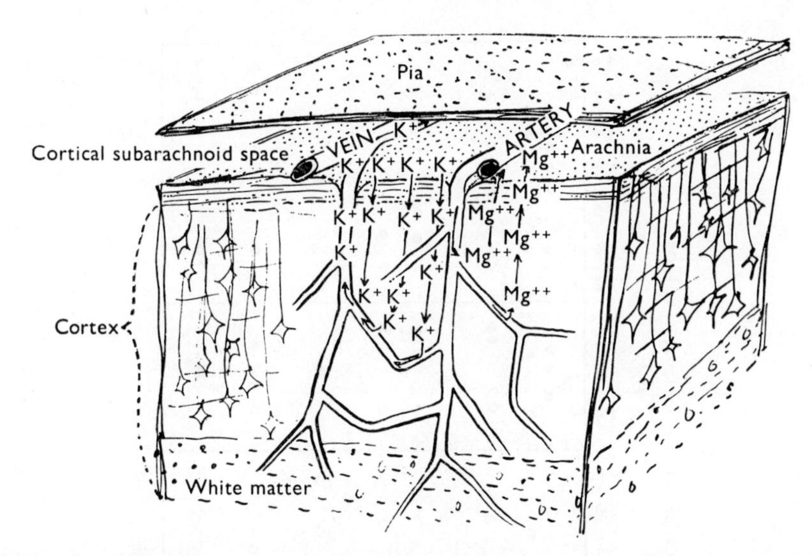

Fig. 1. Presumed patterns of potassium and magnesium ion migration in the cortex. The potassium ions are believed to be actively transported from the extracellular fluid to the blood by the blood–brain-barrier mechanism. Potassium ions then move from the cortical subarachnoid to the extracellular fluid compartments along the diffusion gradient thus created. Magnesium ions are actively moved in the opposite direction by the blood–brain-barrier mechanism causing increases in their concentration in the extracellular and cortical subarachnoid fluids.

The present discussion assumes an active transport of cations by membrane structures associated with the cerebral blood vessels: the capillary endothelium, the astrocytic perivascular membrane, or the two acting together. It is further assumed that potassium is actively extruded into the blood from the brain extracellular fluid by the blood–brain-barrier mechanism to account for the further downward regulation of the potassium concentration in the CSF on its passage through the cortical subarachnoid space. This extrusion leads to a lowering of the potassium concentration in the extracellular fluid and a net movement of potassium from the cortical subarachnoid into the brain extracellular fluid along a diffusion gradient. In a similar but opposite fashion, the magnesium ion is transported from the blood to the brain extracellular fluid. With magnesium accumulation in this site, a diffusion gradient develops leading to magnesium migration into the contiguous cortical subarachnoid fluid. Similar but less dramatic ion movements occur in relation to calcium.

Thus, the composition of the cerebrospinal fluid constantly alters as the fluid is conducted along its channels of movement from its site of formation at the choroid plexus. The specific compositional changes exhibited reflect the activities of the various components of the blood–brain-barrier

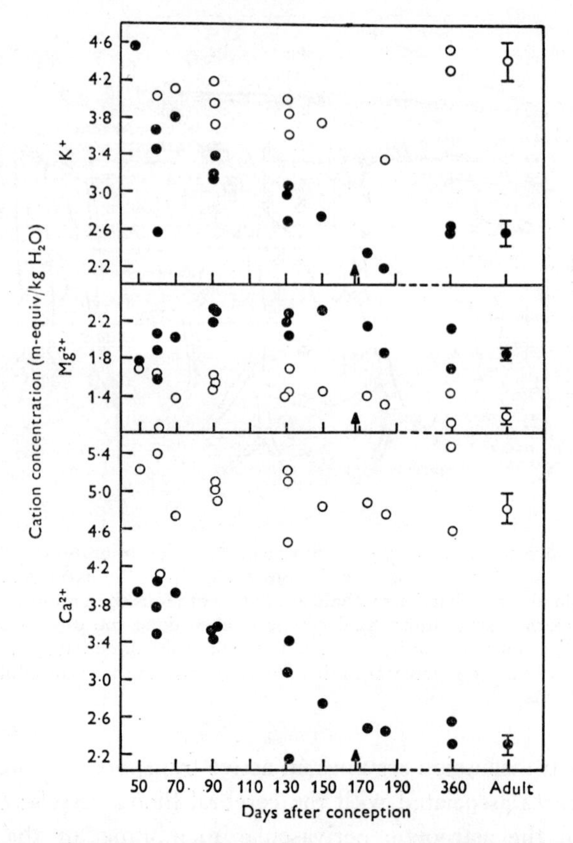

Fig. 2. The concentrations of K^+, Mg^{++} and Ca^{++} in the blood plasma (○) and cisternal cerebrospinal fluid (●) of foetal and postnatal monkeys as a function of development. In this and in all following figures the abscissa denotes days after conception linearly up to approximately 22 days of postnatal life. The normal time of delivery (168 days) is indicated by the arrow. The values for the 6-month-old (1 year after conception) and adult monkeys are placed arbitrarily at the end of the scale.

system as they regulate the fluid composition at the various sites (Bito & Davson, 1966; Bito, 1969).

Considerable differences appear in the concentrations of the various cations in the fluid sampled at the same anatomical sites but at different stages in development. Fig. 2 depicts the potassium, magnesium, and calcium concentrations in the plasma and the cisternal CSF from early gestation through to adult life. In the plasma of the foetus, the potassium concentration is slightly depressed while the magnesium and calcium concentrations are elevated compared to the adult.

The cisterna magna CSF potassium concentration when examined early in gestation shows a value only slightly lower than that of plasma.

Thus, the sharp downward regulation of the potassium concentration characteristic of the cisternal fluid in the adult appears only slightly developed at earlier gestational ages. However, with further in-utero development the cisterna magna CSF potassium concentration gradually decreases until, close to the time of birth, its concentration is the same as, or, as in the early days following delivery, slightly lower than are the adult values.

In contrast to the behaviour of potassium, magnesium is upwardly regulated in the cisternal fluid in the earliest samples taken during gestation. It remains elevated throughout later gestation at levels equal to or greater than the concentrations observed in the adult cisternal fluid. Calcium behaves in a fashion similar to potassium. Its cisterna magna fluid concentrations early in gestation are considerably greater than those found in the adult and close to that of the plasma. During in-utero development, its concentration at the cisterna magna gradually diminishes to attain finally adult levels near the time of birth. Thus, during foetal development, the mechanisms involved in the active downward (K^+ and Ca^{++}) or upward (Mg^{++}) regulation of cation concentrations at this site show a gradual but ultimately full functional development.

The distribution ratios of these three cations between the cisterna magna fluid and the plasma appear in Fig. 3. The regression line of the potassium distribution ratio shows a gradual evolution from 0.9 at 70 days of gestation to 0.65 at birth. This latter value already falls within the normal adult range. Calcium exhibits a similar evolution in its cisternal fluid to plasma distribution ratio from 0.82 at 70 days of gestation to 0.50 at delivery. The regression line of the cisterna magna to plasma magnesium distribution ratio remains relatively flat throughout in-utero development varying only between 1.45 and 1.49.

Thus, the blood–brain-barrier mechanisms which regulate the potassium and calcium concentrations in the cisternal fluid mature during the latter part of in-utero development to the extent that, by birth, these ions are regulated close to their adult levels. On the other hand, the close-to-adult cisterna to plasma magnesium distribution ratio observed already at 70 days of gestation indicates an earlier maturation of the magnesium regulatory mechanism.

The developmental data presented thus far relate to the cation concentrations in the CSF only at the cisterna magna. However, as already described, the adult alters further the cation concentrations of the CSF as it passes from the cisterna magna to the cortical subarachnoid spaces. The concentration gradients along the cortical CSF system are depicted for the three cations in Fig. 4 as the ratios of their concentrations in the

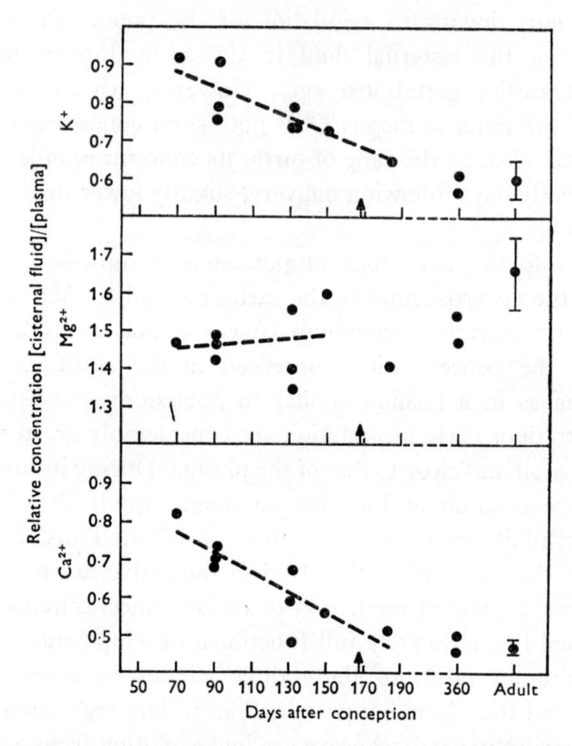

Fig. 3. The development of an adult type of cisternal-CSF/plasma distribution ratio. In this and the following figure the regression lines (dashed lines) were calculated by the method of least squares using all values within the range of the indicated regression line. The correlations between the distribution ratios and gestational age are significant for K and Ca ($P < 0.01$), but not for Mg. Abscissa: see legend to Fig. 2.

cortical subarachnoid as compared to the cisternal fluids. The regression lines depicting the changes in the ratios of their concentrations in these two sites at different developmental stages also appear.

The regression line of the ratios of the potassium concentrations observed in the cortical subarachnoid and the cisternal fluids remains at about 1.05 throughout in-utero development and during the newborn period. It is only by the sixth month after birth that the adult concentration ratio between these two sites is achieved. Thus, the cortical blood–brain-barrier system which regulates the potassium concentrations of the cortical extracellular and, secondarily, of the subarachnoid fluids begins to function at adult levels only during the first postnatal year. In contrast to this, the magnesium concentrations in the cortical CSF are diminished below those in the cisternal fluid early during gestation (as early as 60 days). Thereafter, however, the ratios of these two concentrations gradually increase until, by birth, the magnesium concentrations in the cortical

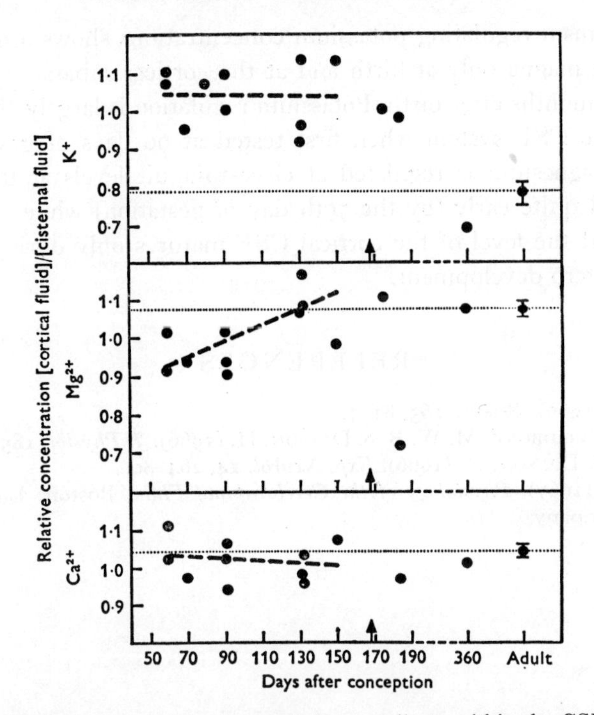

Fig. 4. The development of cation concentration gradients within the CSF system. The adult types of cortical-subarachnoid-fluid/cisternal-fluid ratios are indicated by the dotted lines. Only the Mg regression line is significant ($P < 0.01$). Abscissa: see legend to Fig. 2. Regression lines: see legend to Fig. 3.

subarachnoid fluid are greater than in the cisternal fluid. At this time and when sampled during the newborn period, the relative magnesium concentrations in these two sites remain unaltered and compare closely to those found in the adult. Thus, the cortical magnesium handling system undergoes its maturation during in-utero rather than during postnatal development.

The regulation of calcium in the cortical CSF system appears relatively fixed throughout in-utero and postnatal development. The regression line characterizing the cortical subarachnoid to cisternal fluid calcium concentration ratios remains at 1.00–1.05 throughout the later part of gestation.

The present data reveal significant differences in the concentrations of several different cations within the CSF system not only close to the site of CSF formation but also as it passes through the ventricles and the cortical and spinal subarachnoid spaces. At each of these sites the adult CSF exhibits a characteristic composition. During ontogenetic development the various morphological subdivisions of the regulatory or barrier mechanisms mature at different times for the different cations studied.

The mechanism regulating potassium concentrations shows a maturity at the cisterna magna only at birth and at the cortical subarachnoid spaces only by six months after birth. Potassium regulation is largely absent at all levels of the CSF system when first tested at 60 days of gestation. By contrast, magnesium is regulated at close-to-adult levels at the cisterna magna level quite early (by the 70th day of gestation) while magnesium regulation at the level of the cortical CSF matures only during the later part of in-utero development.

REFERENCES

BITO, L. Z. (1969). *Science* **165**, 81–3.

BITO, L. Z., BRADBURY, M. W. B. & DAVSON, H. (1966). *J. Physiol.* **185**, 323–54.

BITO, L. Z. & DAVSON, H. (1966). *Exp. Neurol.* **14**, 264–80.

DAVSON, H. (1967). *Physiology of the Cerebrospinal Fluid*. Boston: Little, Brown and Company.

BREATHING AND RAPID-EYE-MOVEMENT SLEEP BEFORE BIRTH

By G. S. DAWES

Nuffield Institute for Medical Research,
University of Oxford

There have been many anecdotal reports of foetal breathing *in utero*, in both man and animals. But these reports have been discounted because of doubts as to whether the movements observed were spontaneous or were caused as a result of the conditions under which the observations were made. Thus Windle (1941) suggested that they might result from tactile, thermal or other sensory stimuli or as a result of partial asphyxia. In recent years opinion moved towards the conclusion that spontaneous breathing movements were not normally present *in utero* for three reasons. First, Barcroft & Barron (1937*a*, *b*) and Barcroft (1946) came to the conclusion, from observations on foetal lambs observed under maternal spinal or general anaesthesia, that though movements of the respiratory musculature could easily be elicited by tactile stimulation of the foetus between 40 and 60 days' gestational age, such movements were not present thereafter. Secondly, injection of radiopaque contrast medium into the amniotic fluid of women (McLain, 1964) or animals (King & Becker, 1964) towards the end of gestation led to its appearance soon afterwards in the gut, but not the lungs. So if respiratory movements were present *in utero*, they were not associated with sufficient movement of the tracheal fluid to clear the dead space of the tracheo-bronchial tree. Thirdly, many observers within the past 15 years have delivered foetal lambs towards the end of gestation, under maternal spinal or local anaesthesia, onto a warm table alongside the mother, under reasonable physiological conditions as ascertained by measurements of the foetal arterial blood gases; yet movements of the respiratory musculature in such preparations are very uncommon.

On the other hand if respiratory movements did not normally occur *in utero* it was hard to understand the remarkable ability of the respiratory musculature to maintain vigorous movements after birth, especially in premature infants suffering from the respiratory distress syndrome. It seemed improbable that the newborn could be capable of sustaining prolonged hyperpnoea without previous exercise *in utero*. Movements of the limbs from time to time were well attested. It seemed that a search for such movements might be profitable (Dawes, 1968).

The first observations were made on foetal lambs of 40–98 days' gestational age delivered into a warm saline bath under maternal epidural or spinal anaesthesia in the autumn of 1969. It was found that all such animals made spontaneous irregular breathing movements. The arterial blood gas values were judged to be reasonable, by the criteria then available (mean P_{O_2} 22 mm Hg; mean pH 7.36) and it was noted that the usual response to tactile stimulation of the nose or face was an *expiratory* effort. The breathing movements were irregular in both rate and depth, and were episodic, lasting for a few minutes at a time.

In January 1970 Professor G. C. Liggins came to spend three months' sabbatical leave in Oxford, and among the experimental proposals which were discussed was the possibility of recording breathing movements *in utero* by implanting a catheter in the trachea. The first experiments showed breathing movements *in utero* as evidenced by irregular falls of tracheal pressure similar to those observed in animals delivered into a warm saline bath. There were no comparable changes in amniotic fluid pressure. The account of foetal breathing *in utero* which follows is based on both this first series of 27 chronic foetal lamb experiments (Dawes, Fox, Leduc, Liggins & Richards, 1970, 1972), supplemented by some 80 further experiments carried out by K. Boddy & J. S. Robinson together with S. Pinter, R. Fischer & C. Gibbs.

Changes in foetal tracheal and oesophageal pressures similar to those described as rapid irregular breathing were reported at the same time by Merlet, Hoerter, Devilleneuve & Tchobroutsky (1970) in 8 lambs of 113–133 days' gestational age, using the same methods.

RAPID IRREGULAR BREATHING AND SLEEP IN UTERO

Intrauterine foetal breathing movements, as evidenced by falls of tracheal or oesophageal pressures, are irregular both in rate and depth. Both increase from 80 days' gestational age to term, when they reach frequencies of 3–4 Hz and a depth of more than 30 mm Hg (Fig. 1). The movements cannot be detected by palpation of the maternal abdomen. The evidence that the movements which cause these falls of pressure involve the respiratory musculature depends upon several observations. First, they are accompanied by backward-and-forward fluid flow within the trachea as measured by an electromagnetic flowmeter. The movement normally is small, rarely exceeding ±1 ml, commensurate with the relatively high viscosity and density of the tracheal fluid as compared with air (Fig. 2). This movement is insufficient to clear the dead space of the trachea

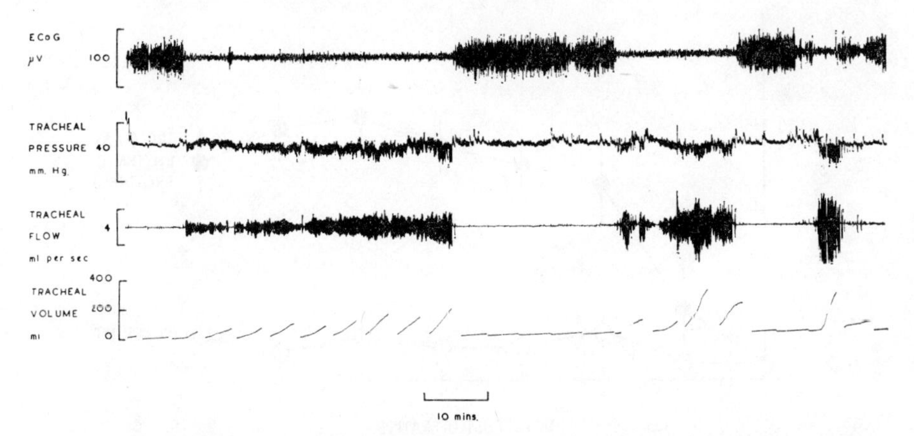

Fig. 1. Foetal lamb at 142 days' gestation *in utero*. Records from above downwards of the biparietal electrocorticogram, tracheal pressure, tracheal flow (from an electromagnetic flowmeter implanted just below the trachea) and gross tracheal flow integrated over 5-min periods (analogous to minute volume of breathing after birth).

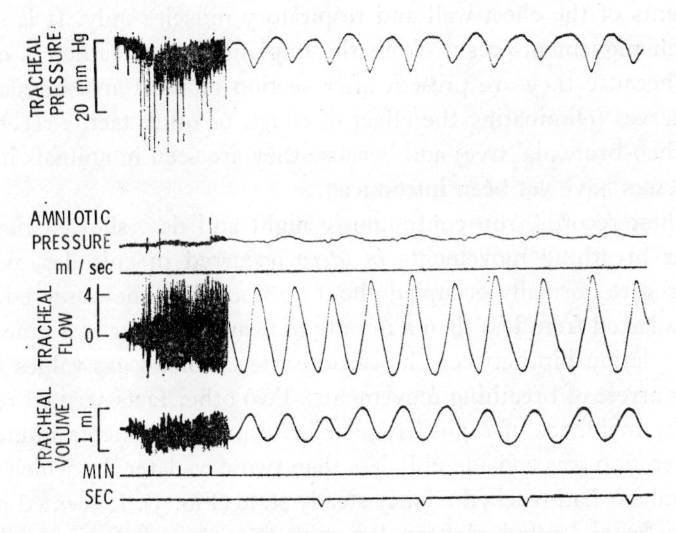

Fig. 2. Records from a lamb at 135 days *in utero* of tracheal and amniotic fluid pressures (recorded at the same manometer sensitivity) and of tracheal flow and fluid displacement from a flowmeter implanted below the larynx (Dawes *et al.* 1972).

(8–10 ml in a lamb near term), an observation which is consistent with the fact that radiopaque material does not normally enter the tracheo-bronchial tree after introduction into the amniotic fluid. Secondly, when a foetus in which such catheters have been implanted is delivered some days later by Caesarean section under maternal epidural anaesthesia into a warm saline bath, the characteristic pressure changes are seen to be accompanied by

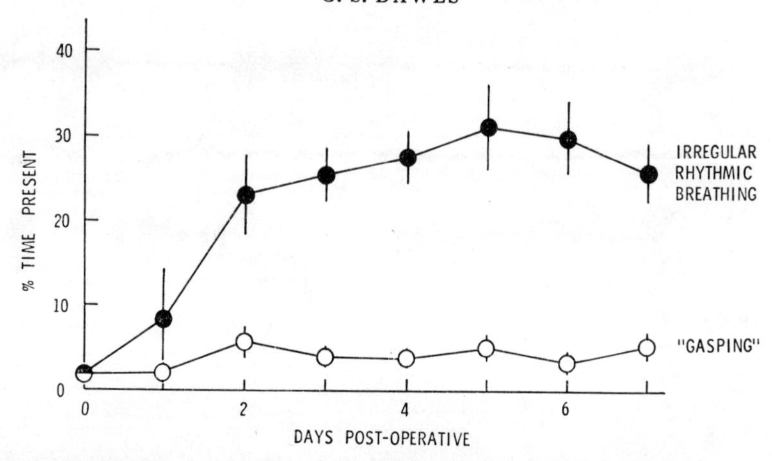

Fig. 3. The incidence of gasping and rapid irregular breathing in 13 foetal lambs, of 108 days' gestation to term, in each 24-hour period over the first 7 days after operation (means ± SE; Dawes *et al.* 1972).

movements of the chest wall and respiratory muscles only. It is unlikely that such movements result from the implantation of catheters or flowmeters because they are present after section or local anaesthesia of the vagus nerves (eliminating the effect of cough or other tactile receptors in the tracheo-bronchial tree) and because they are seen in animals in which no catheters have yet been introduced.

The first records, run continuously night and day, showed that rapid irregular breathing movements *in utero* occurred in episodes, which in the aggregate normally occupied about 40% of the time observed. These episodes lasted from less than a minute to nearly an hour at a time. There was no relationship between the carotid arterial blood gas values and the onset or arrest of breathing movements. Two other facts seemed relevant. First, the incidence of rapid irregular breathing movements immediately after operation was considerably less than two days later, by which time in most lambs it had reached a fairly steady state (Fig. 3). It seemed possible that the foetal lamb had been left with a sore neck from the incision required to insert the catheters. Secondly it had been observed that breathing movements did not normally occur in foetuses which were restrained, whether they were operated on or not. This directed attention to the possibility that breathing movements might be associated with a change in the state of consciousness. Mann (1969) had reported changes in the EEG of a foetal lamb during the last 10 days of gestation which suggested an alteration from light to deep sleep or arousal. And Jost made similar electroencephalographic observations on foetal lambs (see Jost *et al.* 1972).

In order to obtain direct evidence for changes in behaviour we made observations on foetal lambs delivered into a warm saline bath under maternal epidural anaesthesia. The umbilical cord was kept intact and the lambs were not restrained but floated freely in the warm saline bath. Observation over many hours at a time showed that there were indeed changes in behaviour, from arousal or wakefulness (comparatively infrequent), to quiet sleep (about 55% of the time) or rapid-eye-movement sleep (about 40% of the time). The criteria used in coming to these conclusions were the same as those used in making observations on newborn infants, i.e. records of spontaneous changes of movement and in response to tactile stimuli. It was noticed that rapid irregular breathing movements coincided with episodes of rapid-eye-movement sleep. These observations were repeated with similar results, recording breathing movements and eye movements in chronic foetal lamb preparations *in utero*.

The observation of Jost & Mann on the variable character of the sheep foetal electrocorticogram was confirmed. Ruckebusch (1971*a*, *b*) also has made extensive observations on the correlation between the changes in the character of the foetal lambs' electrocorticogram and the state of arousal *in utero*. In our experiments it was found that episodes of rapid irregular breathing only occurred during periods when the electrocorticogram was predominantly fast and of low voltage, as in rapid-eye-movement sleep (Fig. 1). This observation has been substantiated by many days of continous recordings in different lambs. Incidentally, in human newborn infants rapid-eye-movement sleep is associated with an irregularity of breathing, perhaps as a result of the activation of the same system.

It was concluded that the rapid irregular breathing movements characteristic of foetal life, which are present about 40% of the time, are associated with rapid-eye-movement sleep. The electrocortical characteristics of rapid-eye-movement sleep are only well developed by 125–130 days' gestation. There is some reason to believe from casual observations that rapid-eye-movement sleep may be detectable before this time. No systematic analysis has been made at an earlier gestational age. On the other hand rapid irregular breathing movements are present from 40 days' gestation, when the cortex is in a very early stage of development and there is no spontaneous electrocortical activity according to Bernhard & Meyerson (1968). At this early age it would appear that spontaneous respiratory activity must exist in the absence of the electrocortical signs associated with rapid-eye-movement sleep.

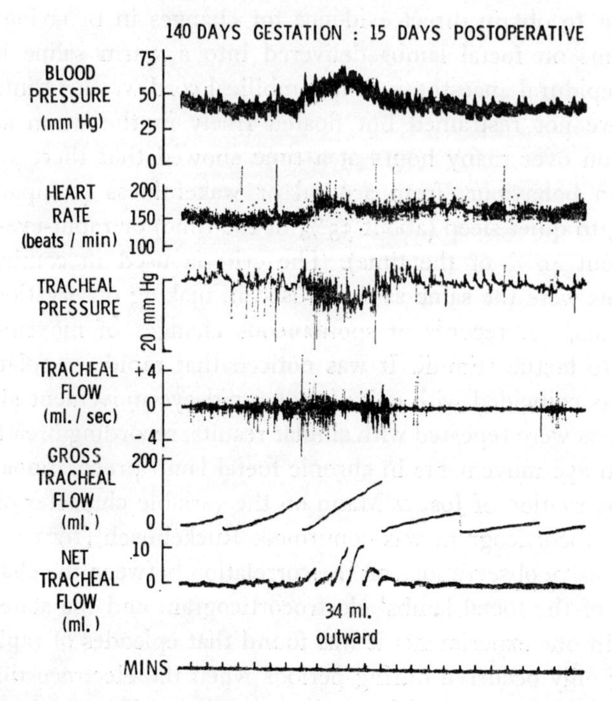

Fig. 4. Observations on a foetal lamb *in utero* to show one episode of large outward net tracheal fluid flow. Such episodes occur 6–12 times a day and are accompanied by large circulatory changes.

'GASPING' OR 'SIGHING' MOVEMENTS

As well as rapid irregular breathing movements there is another form of respiratory activity seen from time to time in foetal lambs. This consists of large single inspiratory efforts (20 mm Hg or more) repeated at a frequency of 1–4 per minute and usually lasting for a few minutes. These gasping or sighing movements may occur in association with or independently of rapid irregular breathing movements; they are present about 5% of the time (Fig. 3). They are not associated with any particular type of electrocortical activity, nor with any significant changes in blood gas values. During severe asphyxia, shortly before the death of a foetus *in utero*, there is a series of regular large gasping movements in the absence of rapid irregular breathing. These asphyxial gasps are regular, more prolonged and deeper than those described above.

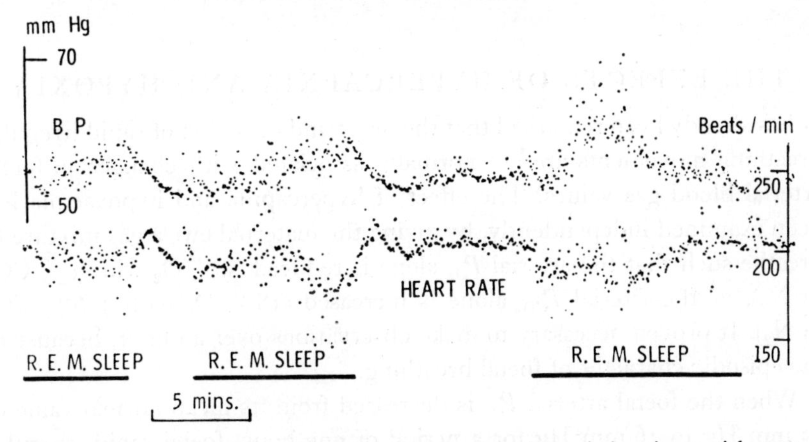

Fig. 5. Same lamb as in Fig. 1, to illustrate the large episodic variations in arterial pressure (carotid–amniotic fluid pressures) and heart rate in association with foetal breathing movements and rapid-eye-movement sleep (as inferred from the electrocorticogram).

PANTING

On exposure to excessive warmth adult sheep pant at frequencies up to 4 Hz. This raised the question as to whether the phenomenon described as rapid irregular foetal breathing *in utero* could be equated with panting, especially as intra-uterine temperature is high (about 0.5 °C above maternal rectal temperature, or 39.5–40 °C *in utero* in sheep).

J. S. Robinson (unpublished) delivered foetal lambs of 75–135 days' gestational age into a warm saline bath, with intact umbilical cords, under maternal epidural anaesthesia. Raising the bath temperature from 39.5 °C by 1 °C caused the onset of rapid foetal breathing movements accompanied by protrusion of the tongue. But these panting movements were not usually maintained for more than 5–10 minutes. They were accompanied by expiratory efforts, as shown by rises of intrathoracic pressure above the basal level. In this they differed from foetal rapid irregular breathing movements.

Panting may occur *in utero*. For instance disturbance of a ewe, e.g. by washing its cage, can induce a rise of maternal temperature (and hence foetal temperature) by nearly 1 °C in 10 minutes. In these circumstances there is an episode of vigorous foetal breathing, with strong expiratory efforts, lasting upwards of 10 minutes and indistinguishable from records made on warm exposure in a saline bath.

THE EFFECTS OF HYPERCAPNIA AND HYPOXIA

It has already been remarked that the onset and cessation of rapid irregular breathing movements are not normally associated with changes in foetal arterial blood gas values. The effect of hypercapnia and hypoxaemia has been examined independently, by giving the maternal ewe a gas mixture to breathe such that the arterial P_{O_2} alone is reduced (9% O_2 with 3% CO_2 in N_2), or the arterial P_{CO_2} alone is increased (18% O_2 with 4–6% CO_2 in N_2). It proved necessary to make observations over an hour, because of the episodic character of foetal breathing.

When the foetal arterial P_{O_2} is decreased from its mean normal value of 25 mm Hg to 16 mm Hg for a period of one hour, foetal rapid irregular breathing is abolished for the whole hour. In a few lambs this period of hypoxaemia was not accompanied by a fall in arterial pH; foetal breathing reappeared within a few minutes after the end of the hour. In other lambs hypoxaemia was accompanied by a small fall of arterial pH and in these, though rapid irregular breathing was quickly restored, the incidence of breathing during the next six hours was reduced below that which had been observed in the six hours preceding hypoxaemia. In these lambs the arterial pH took several hours to return to its initial value. Where the foetal electrocorticogram also was recorded, in no instance was there evidence of any gross reduction of voltage during the period of hypoxaemia. On the contrary, hypoxaemia of this degree was associated with a higher incidence of electrocortical activity characteristic of quiet sleep, i.e. of predominantly slow large-voltage.

The question then arose as to whether oxygen supply at the normal average P_{O_2} of 23 mm Hg limits rapid irregular breathing. This was tested by administering 50% oxygen in nitrogen to the ewe. The consequent rise in foetal arterial P_{O_2}, of a few mm Hg, was not accompanied by an increase in the incidence of vigour of foetal breathing. Also in some foetal lambs much higher arterial P_{O_2} values were recorded, up to 35 mm Hg. In these the incidence and vigour of foetal rapid irregular breathing was similar to those at the average arterial P_{O_2} of 23 mm Hg. So far as these experiments go, therefore, foetal breathing *in utero* is not normally limited by oxygen supply.

There are two classes of foetuses which call for further comment. First, in some chronic experiments there was a progressive deterioration in foetal oxygen supply some days after implantation of catheters, as shown by a fall in foetal carotid arterial P_{O_2} (measured daily) and a fall in arterial pH and rise in plasma lactate concentration. In such preparations administration of a high oxygen mixture to the ewe increased the incidence of rapid irregular breathing towards the 40% previously established as characteristic of the normal.

Secondly, we have from time to time come across a few lambs with exceptionally low arterial P_{O_2} values, of the order of 10 mm Hg, as measured from the time of implantation of catheters, i.e. there was no evidence of progressive deterioration. In these lambs, in spite of the exceptionally low P_{O_2} values, rapid irregular breathing was present over many days' observation, though usually occupying less than 40% of the time. In all such foetuses the haematocrit was very high, exceeding 50 as compared with a normal foetal arterial value of about 35. The evidence is interpreted to suggest that these animals had a well compensated hypoxaemia of long standing.

The effect of hypercapnia was to cause a sustained increase in foetal breathing. The minute volume of breathing was calculated by integrating the full-wave-rectified output of the tracheal flowmeter over 5-minute internals. An increase of foetal arterial P_{CO_2} by 8–10 mm Hg was associated with 2–3 times the normal minute volume of foetal breathing (averaged over an hour). The incidence of rapid irregular breathing movements was somewhat increased by hypercapnia; in addition the rate and depth of breathing movements were both increased during episodes of predominantly rapid small-voltage electrocortical activity ascribed to rapid-eye-movement sleep.

Thus the effects of changes in the blood gas values on foetal breathing *in utero* are substantially different from those observed after birth. Foetal hypercapnia does cause hyperpnoea (if it is permissible to use this term of foetal liquid breathing movements), but the system is less sensitive than postnatally. The effect of a moderate degree of hypoxaemia in the foetus is to arrest breathing movements; there was no evidence of hyperpnoea. It is pertinent that other observations have suggested that the carotid bodies are normally inert during foetal life (Purves & Biscoe, 1966; Dawes, Duncan, Lewis, Merlet, Owen-Thomas & Reeves, 1969).

PULMONARY FLUID FLOW

The effect of foetal breathing movements in causing a small fluid movement into and out of the lungs, as measured by an electromagnetic flowmeter implanted in the trachea, has already been described. The *net* outward fluid flow was measured by integrating the (unrectified) output of the flowmeter over a period of one minute. Such records readily demonstrate a change in inward or outward flow ($>$ 1 ml/min) over a short period of time.

In some lambs large rapid irregular breathing movements are regularly associated with a net *inward* tracheal fluid flow of several ml over 2–3 minutes. *Outward* fluid flows are more common. A small outward flow of 1–2 ml is often seen from time to time, especially accompanying a brief expiratory effort as indicated by a rise of intrathoracic pressure. Less commonly, 6–12 times a day near term, there is an outward gush of tracheal fluid of the order of 20–40 ml over a period of 3–5 minutes, often associated with a rise in arterial pressure of up to 20 mm Hg (Fig. 4). The physiological description of these episodes is still incomplete; it is possible that they may be associated with swallowing movements. They are not necessarily connected with any particular form of foetal breathing.

In four foetal lambs of 110–145 days' gestation, net tracheal flow was analysed over 61 complete days' recording, starting at least two days after operation. The net outward flow increased with age, to reach a volume up to 100 ml Kg^{-1} day^{-1} near term. The volume contributed by episodic outward gushes was more than half the total. There was no evidence of a diurnal rhythm.

The volume of fluid in the alveoli and tracheo-bronchial tree depends on many factors, on the tone of the thoracic musculature, on the volume of the heart and great vessels, on the tone of the bronchial and tracheal muscles, on the patency of the glottis and on the rates of formation of alveolar and bronchial fluids. We have comparatively little direct evidence on many of these factors. It is evident that the glottis must be open much of the time, when there is vigorous to-and-fro movement of tracheal fluid with foetal breathing (the flowmeter was implanted just below the larynx). There are also, from time to time, small rises of intrathoracic pressure of a few mm Hg, with no change in amniotic fluid pressure or evidence of net outward tracheal fluid flow. More direct evidence of glottal patency is required. The evidence of Adams, Delilets & Towers (1967) on the passage of radiopaque contrast medium through the larynx of foetal lambs is difficult to interpret because of the experimental conditions, and the absence of pressure measurements.

EFFECTS UPON THE CIRCULATION

The true arterial pressure of the foetus *in utero* is recorded as the difference between arterial and amniotic fluid pressures, the zeros of the two manometer systems being identical.

Records of carotid arterial pressure and heart rate taken continuously at the same time as those of foetal breathing and electrocortical activity show a wide episodic variation. The falls of intrathoracic pressure associated with respiratory movement are transmitted to the great vessels and are detected as gross irregularities in the foetal pulse recorded from the carotid artery. The flow pulsations in the descending aorta (recorded from an electromagnetic flowmeter implanted below the origin of the renal arteries) also become irregular with the onset of foetal breathing. When breathing movements are vigorous, especially in lambs near term, there is often a rise of arterial pressure which can be quite substantial. The heart rate usually decreases with change in the electrocorticogram from predominantly slow large-voltage to rapid low-voltage activity. There may be a subsequent further change when foetal breathing is vigorous. Thus in some foetal lambs there is a cyclical variation in blood pressure and heart rate with the onset and arrest of breathing movements (Fig. 5).

There is also the possibility to be considered that, as in the adult, the reflex control of the circulation may alter with changes in the state of sleep or arousal, independently of any mechanical effect from changes in intrathoracic pressure or volume. It is clear that investigations by transient methods, such as cineangiography or the injection of isotope-labelled microspheres, must be related to the phases of sleep, breathing and other activities *in utero*.

EXPERIMENTAL CONDITIONS

It has already been noted that breathing movements *in utero* are reduced or abolished by handling or operating upon the foetus, sometimes for many hours. Goodlin & Rudolph (1970) observed neither breathing movements nor continued tracheal fluid efflux in foetal lambs on the day of operation. Experiments on a chronic foetal lamb preparation are best not begun until at least two days have elapsed, and until there is evidence of a steady breathing state over 24 hours (bearing in mind the diurnal rhythm; Boddy, Dawes & Robinson, 1973), and foetal arterial blood gas values have stabilized. Even this is no guarantee that foetal blood or amniotic fluid concentrations of ions or metabolites will have stabilized (Mellor & Slater, 1971; Shelley, 1972) and early observations (before five days from operation) must be treated with caution. The effects of removing a pregnant ewe into an unfamiliar environment, of unusual noise or new personnel are unpredictable and can produce large metabolic, thermal and cardiovascular changes in the mother and consequential effects on the foetus.

Limited observations on pregnant sheep in metabolic cages within the laboratory suggest that the foetus does not readily respond directly to changes in the maternal environment, i.e. to noise or light. Foetal breathing is unaffected by the mother standing up or sitting down, or by the movements of a twin, even one in the same uterine horn.

GENERAL ANAESTHESIA

Small doses of general anaesthetics such as chloralose or pentobarbitone, administered to the maternal ewe, cause an arrest of foetal breathing. Pentobarbitone has been studied most thoroughly from 80 days' gestation age. A dose of 4 mg/kg i.v. to the ewe, given while the foetus is breathing during a period of rapid small-voltage electrocortical activity, causes arrest of breathing within 2–3 minutes and a reversion to predominantly slow large-voltage electrocortical activity, both for about 25–30 minutes. Variations in foetal arterial pressure and heart rate are much reduced. This dose of pentobarbitone has little overt effect on the ewe who, if standing, remains so with little or no head drop. The dose of pentobarbitone required to cause surgical anaesthesia in adult sheep is about 30 mg/kg i.v.; much smaller doses given to the mother make it possible to operate on the foetus without evidence of painful response. We must also take into account the much lower concentrations of anaesthetic to be expected in the foetal bloodstream after maternal administration (e.g. Dawes, 1972). These observations suggest that the foetal brain is more sensitive to low blood concentrations of pentobarbitone than is the adult. This greater sensitivity may account in part for the prolonged effects of such anaesthetics administered after birth (as well as less rapid detoxification or elimination).

FOETAL BREATHING IN DIFFERENT SPECIES

So far, most experiments have been done on foetal sheep, in which it is comparatively easy to insert catheters. A few observations were made in foetal rabbits during the last two days of gestation (Merlet, Hoerter & Tchobroutsky, 1970), and Leduc & Robinson (unpublished) independently observed breathing movement from intrathoracic catheters implanted for 1–2 days. Their irregularity and episodic character were similar to the rapid irregular breathing movements observed in sheep.

During the past two years, on a number of occasions, it was observed that not only hypoxaemia, but also hypoglycaemia and maternal peritoneal infection were associated with the disappearance of foetal breathing in chronic sheep preparations. It seemed that the presence of foetal breathing might be a useful guide to health and wellbeing. There was therefore a strong incentive to discover whether it was possible to detect foetal breathing in the human infant *in utero*. Boddy & Robinson (1971) used an ultrasonic A-scan method to detect the movement of the chest wall during foetal breathing in sheep, and were able to use it successfully in man.

Further investigations with this method are in progress (Boddy & Mantell). Foetal breathing is normally present in human infants *in utero*. Its episodic character resembles that observed in the sheep, but it is present for a greater proportion of the time. The presence of human foetal breathing is normally imperceptible to the mother or by abdominal palpation.

CONCLUSION

The observations described impinge on several different physiological systems. They suggest that irregular spontaneous discharges may normally occur in the respiratory centre at least as early as muscular movement can be detected. The fact that they are later associated with the electrocortical signs of rapid-eye-movement sleep may have no functional physiological significance (i.e. it may be a result of casual irradiation in an otherwise comparatively inactive nervous system) or might prove to be a necessary association in normal development.

The episodic expulsion of foetal tracheo–bronchial fluid may perhaps explain how pulmonary lecithin enters the amnion late in gestation (if the lung does indeed prove to be its main source). Also, it is interesting that in foetal lambs the volume of tracheo–bronchial fluid effluent each day is not very different from that which is swallowed (Bradley & Mistretta, 1972). Yet the contribution of foetal urine to amniotic fluid is thought to be large. We need simultaneous measurements of the volumes of pulmonary effluent, of swallowing and of micturition in the same lamb to determine whether there are other large fluid movements across the foetal membranes, skin or umbilical cord.

The large variations from time to time in the foetal circulation and in sleep or arousal, and the presence of diurnal rhythms in breathing and the electrocorticogram (Boddy, Dawes & Robinson, 1972) are factors which must be considered in relation to many other aspects of foetal physiology, and especially in relation to the endocrine systems.

Finally, we do not yet know whether episodic rapid irregular foetal breathing *in utero* can persist throughout labour; in a few instances it has ceased before birth. Its presence *in utero* does not explain the onset of regular continuous rhythmic breathing after birth. It does, however, provide the necessary background information about respiratory control mechanisms in the preceding period from which we may, in the future, hope to construct a more reasonable account of the change which occurs at birth.

This paper summarizes the work of many colleagues over the past two years. It was supported by grants from the Medical Research Council. In addition to those mentioned I must thank Dr W. M. Harris, and also K. F. Bolton, N. P. J. Green, H. C. Holt and A. D. S. Stevens for their technical assistance.

REFERENCES

ADAMS, F. H., DELILETS, D. T. & TOWERS, B. (1967). *Resp. Physiol.* **2**, 302–10.

BARCROFT, J. (1946). *Researches on Prenatal Life.* Oxford: Blackwells Scientific Publications.

BARCROFT, J. & BARRON, D. H. (1937*a*). *J. Physiol.* **88**, 56–61.

BARCROFT, J. & BARRON, D. H. (1937*b*). *J. Physiol.* **91**, 329–51.

BERNHARD, C. G. & MEYERSON, B. A. (1968). In *Clinical Electroencephalography of Children*, ed. P. Kellaway & I. Petersen. Stockholm: Almqvist & Wiksell. Pp. 11–29.

BODDY, K. & ROBINSON, J. S. (1971). *Lancet* **2**, 1231–3.

BODDY, K., DAWES, G. S. & ROBINSON, J. S. (1973). This book.

BRADLEY, R. & MISTRETTA, C. (1973). This book.

DAWES, G. S. (1968). *Fetal and Neonatal Physiology.* Chicago: Year Book Medical Publishers.

DAWES, G. S. (1972). In *International Symposium on Fetal Pharmacology*, ed. L. O. Boreus. New York: Raven Press.

DAWES, G. S., DUNCAN, S. L. B., LEWIS, B. V., MERLET, C. V., OWEN-THOMAS, J. B. & REEVES, J. T. (1969). *J. Physiol.* **201**, 105–16.

DAWES, G. S., FOX, H. E., LEDUC, B. M., LIGGINS, G. C. & RICHARDS, R. T. (1970). *J. Physiol.* **210**, 47–8*P*.

DAWES, G. S., FOX, H. E., LEDUC, B. M., LIGGINS, G. C. & RICHARDS, R. T. (1972). *J. Physiol.* **220**, 119–43.

GOODLIN, R. C. & RUDOLPH, A. M. (1970). *Amer. J. Obstet. Gynec.* **106**, 597–606.

JOST, R. G. (1969). M.D. Thesis, Yale University, Medical School.

JOST, R. G., QUILLIGAN, E. J., YEH, A. & ANDERSON, A. (1972). *Amer. J. Obstet. Gynec.* **114**, 535–39.

KING, J. E. & BECKER, R. F. (1964). *Am. J. Obstet. Gynec.* **90**, 257–63.

MCLAIN, C. R. (1964). *Obstet. Gynec.* **23**, 45–50.

MANN, L. I. (1969). In *Perinatal Factors Affecting Human Development*. Washington, D.C.: Pan American Health Organization Scientific Publication 15, pp. 149–57.

MELLOR, D. J. & SLATER, J. S. (1971). *J. Physiol.* **217**, 573–604.

MERLET, C., HOERTER, J., DEVILLENEUVE, C. & TCHOBROUTSKY, C. (1970). *C. r. hebd. Séanc. Acad. Sci., Paris* **270**, 2462–4.

PURVES, M. J. & BISCOE, T. J. (1966). *Br. med. Bull.* **22**, 56–60.

RUCKEBUSCH, Y. (1971*a*). *Revue med. vet.* **122**, 483–510.

RUCKEBUSCH, Y. (1971*b*). *Electroenceph. clin. Neurophysiol.* **32**, 119–28.

SHELLEY, H. J. (1972). This book.

WINDLE, W. F. (1941). *Physiology of the Fetus.* Philadelphia: Saunders.

A 24-HOUR RHYTHM IN THE FOETUS

By K. BODDY, G. S. DAWES and
J. S. ROBINSON
Nuffield Institute for Medical Research,
University of Oxford

Episodic breathing movements were described in foetal lambs by Dawes *et al.* (1972) and Merlet *et al.* (1970). Rapid irregular breathing movements are a normal phenomenon of intra-uterine life and are associated with rapid-eye-movement sleep. During subsequent investigations of foetal breathing it was noted that the most vigorous periods occurred most often in the late evening, and a search was made for a circadian rhythm.

METHODS

In 23 foetal lambs (between 97 and 137 days' gestation) tracheal and carotid catheters, biparietal cortical electrodes and a tracheal electromagnetic flowmeter were implanted as described by Dawes *et al.* (1972). Observations were made between 100 days and full term (about 147 days), on one foetus only if there were twins. Records were made continuously from soon after the operation, but to allow for recovery from surgery only those at least 57 hours from operation were included in the present analysis. Foetal and maternal arterial blood samples were taken daily. All foetuses considered had normal blood gas values; there were no signs of deterioration.

Foetal breathing was measured in three ways. First, the to-and-fro tracheal flow signal was full-wave rectified and integrated over 5 minutes, to give a measure of the minute volume of fluid breathing. Secondly, the number of minutes during which rapid breathing movements occurred in each hour was calculated. Thirdly, the rate of breathing was measured by integrating the number of tracheal flow pulses over one minute periods.

RESULTS

In 12 foetal lambs the total volume of tracheal fluid flow (to and fro) in each 24-hour period increased with gestational age from near zero at 100 days' gestation to about 12 litres at term (2.5 ml/kg. min). There was considerable variation from lamb to lamb at the same gestational age.

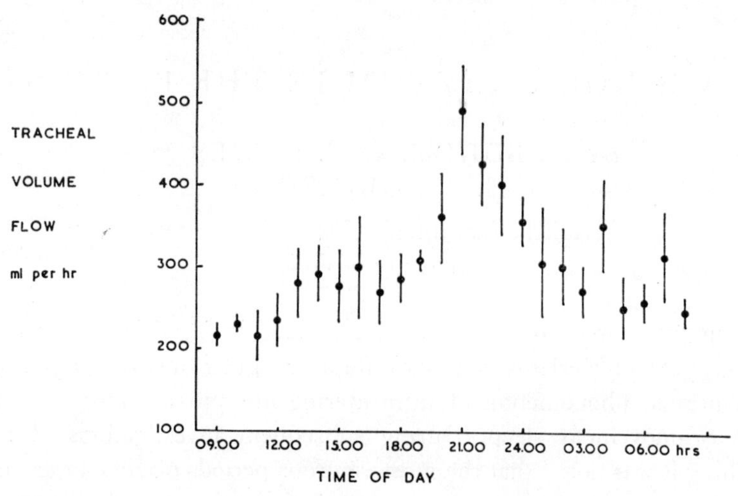

Fig. 1. Tracheal volume flow (ml) in a foetal lamb over 5 consecutive days (gestational age 112–117 days; mean values (●) ± S.E.)

In 7 foetal lambs of 110–142 days' gestation the total tracheal volume flow per hour was plotted against the time of day over 5–12 consecutive days (45 days total). There was clear evidence of a circadian rhythm. In every lamb there was a peak in the hourly tracheal volume flow in the late evening, and a trough in the early hours of the morning (Fig. 1). In lambs observed in the late spring and early autumn the peaks tended to occur later in the evening than those observed in mid-winter. The combination of the results from all 7 lambs gave a broad flat peak in the late evening (133–134% at 19.00–22.00 hours, the tracheal volume flow per hour being expressed as a percentage of the mean hourly volume averaged over the day). The difference between the peak and the trough was highly significant ($P < 0.001$). In individual lambs the peak hourly tracheal volume flow was 2.0–2.8 times that observed in the trough, and the peak was of shorter duration, usually less than two hours.

The incidence of rapid breathing movements (per hour), assessed from records of tracheal pressure, was calculated in 6 lambs of 115–133 days' gestation and showed a highly significant variation ($P < 0.001$) from 15.8 ± 1.8 min (SE) at the trough (08.00–09.00 hours) to 30.3 ± 2.3 min at the peak (20.00–21.00 hours).

Observations of individual records suggested that the peak periods of foetal breathing were associated with an increase in both rate and depth (measured from either tracheal flow or pressure) as compared with the trough.

The electrocorticogram was recorded in 7 foetal lambs of 124–137 days' gestation over a total period of 22 days. Rapid foetal breathing movements are always associated with low-voltage, rapid-frequency electrocortical activity. The number of minutes per hour of this electrocortical activity was calculated. The mean incidence increased to reach a peak in the late evening (40.2 ± 1.3, SE, min/hr at 19.00–20.00 hours), falling to a trough in the early hours of the morning (28.2 ± 1.0 min/hr at 04.00–05.00 hours). The peaks in tracheal volume flow (ml/hr), in tracheal pressure changes with rapid breathing movements (min/h), and in low-voltage electrocortical activity all occurred in the late evening. The troughs all were flat and lasted 4–5 hours.

DISCUSSION

The fact that there is little or no foetal tracheal flow before 100 days' gestation agrees with the observation that the alveoli are solid before this age (Fauré-Fremiet & Dragoiu, 1923) and the lungs inexpansible (Born, Dawes & Mott, 1955). Thereafter to-and-fro flow increases rapidly with gestational age and foetal weight. In the present experiments the minute volume of tracheal flow near term reached a value of 2·5 ml/kg min, using a 3 mm i.d. cannulated electromagnetic flowmeter (Clark & Wyatt, 1969) implanted into the trachea just below the larynx. Preliminary observations suggest that a larger bore flowmeter may be desirable in foetal lambs near term to ensure that the flow of viscous fluid is not limited by the bore.

Our experiments show that in the foetal lambs there is a circadian rhythm of both foetal breathing and rapid low-voltage electrocortical activity (probably mainly associated with rapid-eye-movement sleep). This circadian rhythm of foetal breathing is certainly present by 110 days' gestation and might be due to exogenous causes (e.g. light and sound affecting the foetus directly), maternal causes (e.g. circadian variations in maternal placental blood flow, maternal blood concentrations of glucose or other blood constituents, maternal activity or maternal temperature) or to endogenous causes (e.g. an inherent foetal neural rhythm). Preliminary observations show that there is a circadian variation in foetal temperature (passively following that of the mother, which it exceeds by about 0.5 °C).

Buddingh *et al.* (1971) also have reported evidence of a 24-hour variation in renal excretion in four foetal lambs from 127 days' gestation onwards. The peak renal water clearance at 10.00 hours was followed by a trough at 15.00 hours. It is clear that evidence must be sought for circadian variations in other neuro-endocrine systems. The effect of keeping the ewe in an environment where temperature, light and sound are controlled is being studied.

This work was carried out with the help of grants from the Medical Research Council. We received able technical assistance from A. Stevens, K. Bolton, Mrs J. McCairns and N. Green.

REFERENCES

BORN, G. V. R., DAWES, G. S. & MOTT, J. C. (1955). The viability of premature lambs. *J. Physiol.* (*Lond.*) **130**, 191–212.

BUDDINGH, F., PARKER, H. R., ISHIZAKI, G. & TYLER, W. S. (1971). Long term studies of the functional development of the fetal kidney in sheep. *Amer. J. Vet. Res.* **32**, 1993–8.

CLARK, D. M. & WYATT, D. G. (1969). An improved perivascular electromagnetic flowmeter. *Med. & Biol. Engng* **1**, 185–90.

DAWES, G. S., FOX, H. E., LEDUC, B. M., LIGGINS, G. C. & RICHARDS, R. T. (1972). Respiratory movements and rapid eye movement sleep in the foetal lamb. *J. Physiol.* (*Lond.*) **220**, 119–43.

FAURÉ-FREMIET, E. & DRAGOIU, J. (1923). Le développement du poumon foetal chez le mouton. *Arch. Anat. micr.* **19**, 411–74.

MERLET, C., HOERTER, J., DEVILLENEUVE, C. & TCHOBROUTSKY, C. (1970). Mise en evidence de mouvements respiratoires chez le foetus d'agneau *in utero* au cours du demier mois de la gestation. *C. r. hebd. Séanc. Acad. Sci., Paris* **270**, 2462–4.

VENTILATION IN NEW-BORN INFANTS IN DIFFERENT SLEEP STATES

By M. K. S. HATHORN
Department of Physiology,
The London Hospital Medical College, London E1 2AD

Numerous studies have shown that during sleep characterised by rapid eye movements (REM), respiratory rate is higher and shows greater irregularity than during periods of regular sleep when eye movements are absent (NREM sleep). Prechtl & Lenard (1967) put forward the hypothesis that REM and the irregular fluctuations in many other parameters may be regarded as the result of random noise in the CNS. On this basis one would expect to find evidence of a large random component in ventilation in REM sleep. It was therefore decided to investigate some of the characteristics of ventilation in new-born infants: this might provide evidence of differences in supramedullary influences on respiratory centre activity in the two sleep states.

Ten female and four male full-term infants with Apgar scores of 9 or 10 at 15 min were studied during the first week after delivery, within 2 hours of the mid-morning feed. Ventilation was measured in the trunk plethysmograph (Cross, 1949), using a pressure transducer and pen-recorder. Intrathoracic pressure was monitored by means of an oesophageal balloon. The sleep state was determined using the criteria of Prechtl & Beintema (1964). For each infant, the duration and amplitude of each respiratory cycle during periods of REM and NREM sleep were measured, together with calibrations for volume and time. From these measurements, instantaneous respiratory rate, tidal volume and pulmonary ventilation for each breath were calculated.

It was found that mean instantaneous pulmonary ventilation and respiratory rate were significantly greater in all the infants in REM than in NREM sleep, while mean tidal volume was somewhat reduced. There was significantly greater variation in instantaneous pulmonary ventilation, tidal volume and respiratory rate in REM sleep than in NREM sleep. These findings confirm those recently described by S. Herman (personal communication), who measured ventilation, averaged over 20-second intervals, in the two sleep states in new-born infants.

The random noise hypothesis would imply that each successive breath was statistically independent of the previous breath. It was therefore

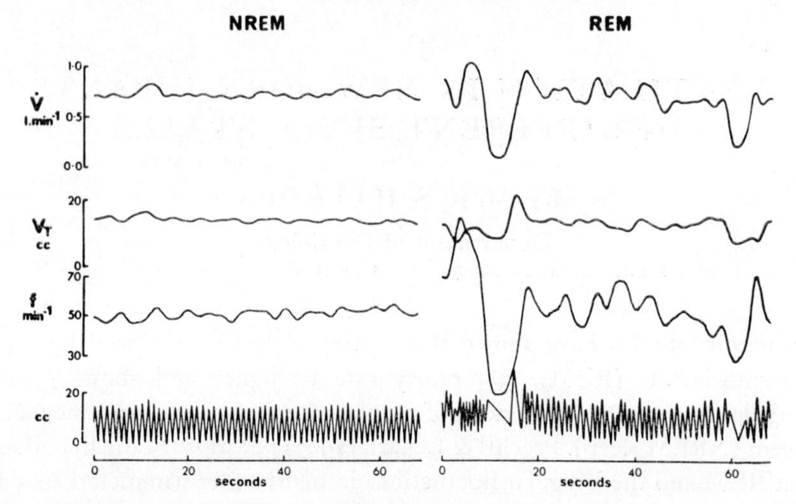

Fig. 1. Ventilation in NREM and REM sleep in an infant. From above downwards: smoothed instantaneous pulmonary ventilation, tidal volume and respiratory rate, and lines joining the end-expiratory and end-inspiratory points, providing a facsimile of the original tracing.

necessary to establish whether there was serial dependence in time between successive respiratory cycles. This was done by sampling instantaneous pulmonary ventilation, tidal volume and respiratory rate at equi-spaced time intervals, followed by repeated smoothing with binomial coefficients and decimation as described by Blackman & Tukey (1959). The smoothed data showed the presence of oscillations in the rate and depth of breathing, as well as total ventilation (Fig. 1). Autocovariance and spectral analysis (Bendat & Piersol, 1966) confirmed the presence of these periodicities, and showed the main frequencies to lie in the region of 0.07 to 0.25 Hz, i.e. a cycle length of 4 to 14 seconds.

In order to demonstrate these oscillations more clearly, the original time-sampled data was passed through a digital Tukey filter designed to pass only those signals lying within this frequency range. Graphs produced on a computer graph-plotter showed the virtual identity between the oscillations in the smoothed data and those in the filtered data (Fig. 2).

Comparison of the mean cycle length of these oscillations in respiratory rate and tidal volume showed no significant differences in the two sleep states; they fell within the range of 5.8 to 8.8 sec in the different infants studied. There was a marked difference, however, in the amplitude of the oscillations: all infants studied showed a larger amplitude in REM than in NREM sleep, for both respiratory rate (Fig. 2) and tidal volume. The mean ratio of these amplitudes in REM and NREM sleep was 2.46 for tidal volume and 3.61 for respiratory rate ($P < 0.001$). These high-

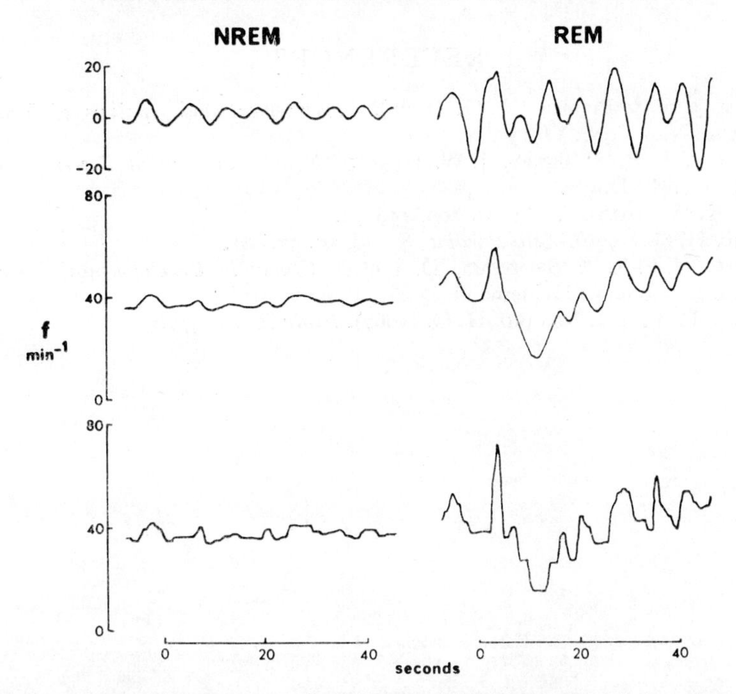

Fig. 2. Instantaneous respiratory rate in NREM and REM sleep in an infant. From below upwards (all plotted on the same scale): original time-sampled data, smoothed data and filtered data (variation around mean level).

amplitude oscillations largely account for the increased variability of ventilation in REM sleep.

A number of neurological differences have been described in the two sleep states, including a reduction in amplitude of the EEG and inhibition of various afferent pathways in REM sleep (reviewed by Lenard, 1970). It was the latter which led Prechtl & Lenard (1967) to suggest that the increased variability in a number of parameters in REM sleep was due to random noise in the CNS consequent upon a reduction in afferent patterning in this sleep state. There is, however, no evidence of inhibition of chemoreceptor afferent activity in REM sleep (S. Herman, personal communication).

During REM sleep, therefore, there is an organised rhythmicity in respiration. The effective gain of the neural oscillatory mechanisms controlling respiration appears to be considerably increased in this sleep state. A possible explanation is that there is a reduction in damping in respiratory centre activity in REM sleep, leading to an increased amplitude of the oscillations already present in NREM sleep.

REFERENCES

BENDAT, J. S. & PIERSOL, A. G. (1966). *Measurement and Analysis of Random Data*. New York: Wiley.

BLACKMAN, R. B. & TUKEY, J. W. (1959). *The Measurement of Power Spectra*. New York: Dover.

CROSS, K. W. (1949). *J. Physiol.* **109**, 459–74.

LENARD, H. G. (1970). *Acta Paediat. Scand.* **59**, 572–81.

PRECHTL, H. F. R. & BEINTEMA, D. (1964). *Clinics in Developmental Medicine* No. 12. London: Heinemann.

PRECHTL, H. F. R. & LENARD, H. G. (1967). *Brain Res.* **5**, 477.

ON THE FOETAL EEG DURING PARTURITION

By MORTIMER G. ROSEN AND JOSEPH J. SCIBETTA

Department of Obstetrics–Gynecology,
The University of Rochester School of Medicine and Dentistry,
Rochester, New York 14620

To evaluate the brain during parturition a suction electrode was devised (Rosen & Scibetta, 1969; Rosen *et al.* 1970) which isolated the recording silver point from the electrically conductive in-utero environment surrounding the foetal head. Two sensors were used for bipolar foetal electroencephalography (FEEG). The information obtained was correlated with the foetal heart rate (FHR) which was obtained from a single scalp sensor and the in-utero pressure (IUP) which was obtained from a water-filled plastic tube connected to a strain gauge pressure transducer. With the use of this experimental design, the FEEG will be described specifically with respect to the results of birth, myometrial contractions, and heart rate decelerations. The descriptive parameters will be the FEEG, IUP, and FHR.

RESULTS

The FEEG patterns are similar to those observed after birth and display patterns described in wave frequencies and amplitudes which are directly related to foetal maturity (Rosen & Scibetta, 1969; Rosen *et al.* 1970). FEEG does not vary with thc transition from the in-utero environment to the extra-uterine atmosphere.

Abrupt changes in wave amplitude and frequency are easily apparent when using a continuous monitoring technique. Therefore, using the FEEG prior to that acute event as its own control for baseline data, variations in wave amplitudes, frequencies, and overall patterns can be related to those changes perceived during that period of time.

The visual appearance of the FEEG between uterine contractions does not change during contractions (Fig. 1). This is so even though the IUP recorded may reach 90 mm Hg as seen in the second stage of labour when maternal abdominal 'pushing' (Fig. 2) assists the expulsive uterine forces in delivery of the foetus. At times, most commonly near the end of labour, the onset of a contraction is associated with FHR deceleration described

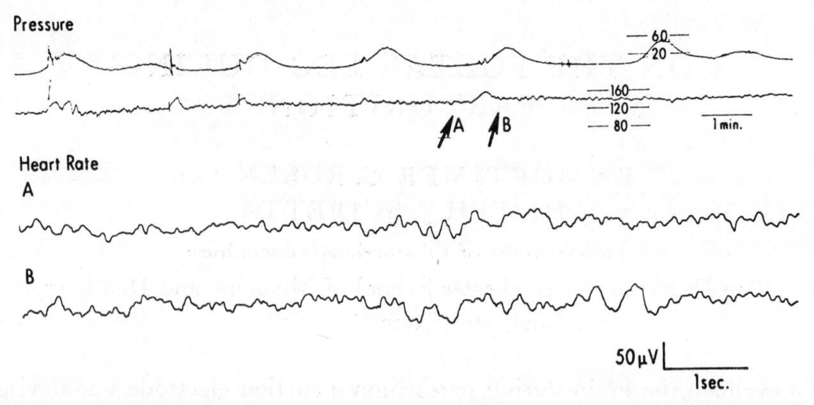

Fig. 1. Upper two lines are slow speed (30 mm/min) recordings describing IUP and FHR. Points A and B are shown in lower 2 lines at faster FEEG recording speed (30 mm/sec). No change in FEEG is seen.

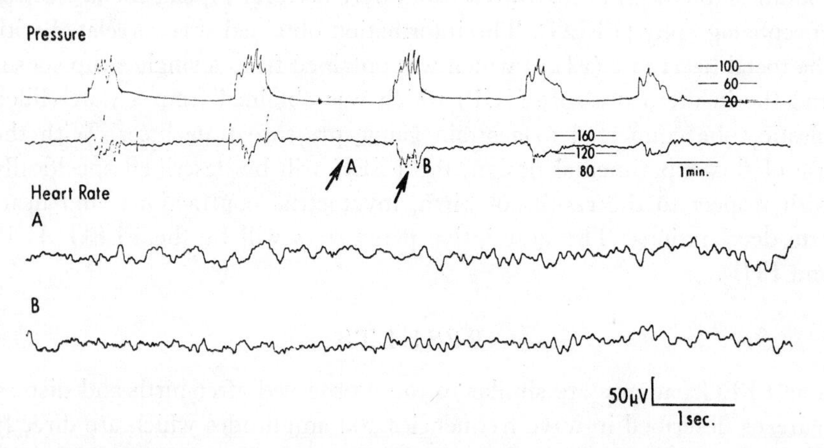

Fig. 2. IUP and FHR seen in upper two lines (30 mm/min). Sharp margins at height of IUP curve indicates maternal 'bearing down' efforts during second stage of labour. Points A and B are shown in lower two lines seen at the time of the high IUP and depressed FHR of early deceleration. The EEG appears unchanged.

as early deceleration. During this situation the FEEG generally remains unchanged (Fig. 2).

Two other heart rate decelerative patterns classified as variable (Fig. 3) and as delayed deceleration (Fig. 4) have been studied in relation to the continuously recorded FEEG. In both situations the FEEG does change in parallel with the heart rate alterations. As an example of this in Fig. 5 the EEG shows a continuum of changes from pre-existing amplitudes and frequencies to an almost isoelectric baseline. The small baseline undulations still seen are difficult to differentiate from movement artifacts which produce electrical deflections with amplitudes less than 5 μV and fre-

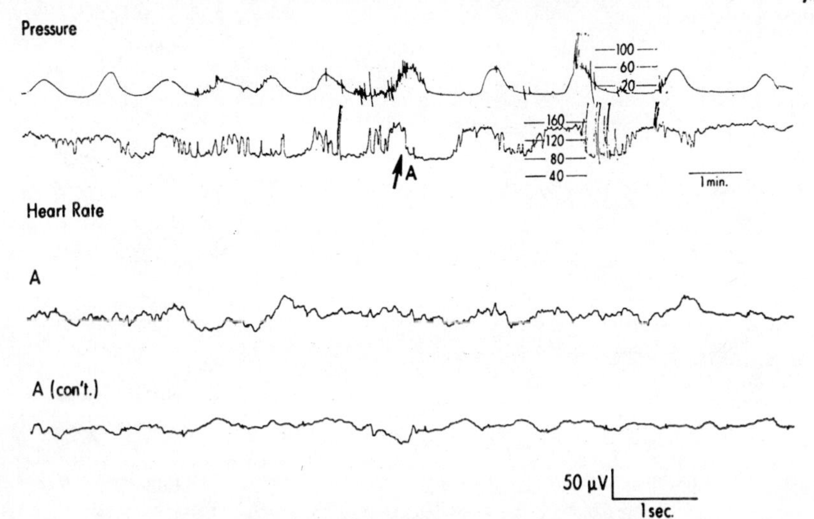

Fig. 3. The slow speed (30 mm/min) recordings in upper two lines display a variable decelerative FHR pattern. Point A marks the beginning of two consecutive 9-second strips of FEEG seen in lower two lines. The EEG seen at first shows slow and fast frequencies, but in lower trace a clear change to 'isoelectricity' is apparent.

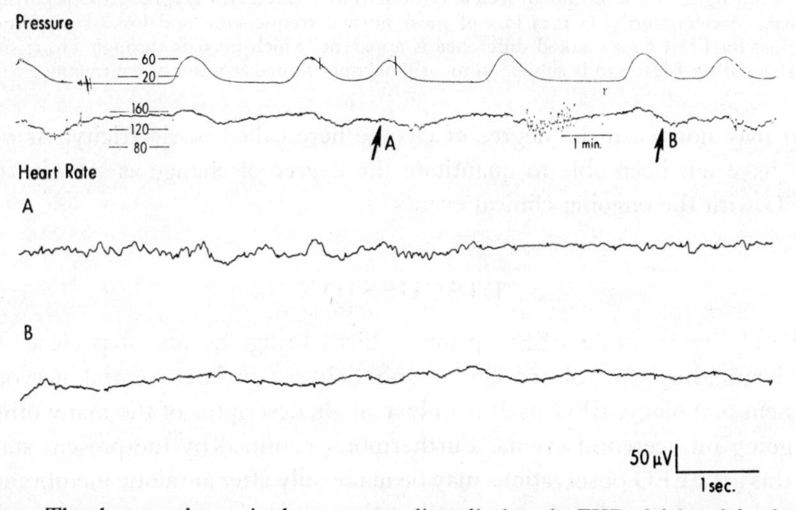

Fig. 4. The slow speed trace in the upper two lines displays the FHR of delayed deceleration. Point A and line A indicate FEEG taken at the onset of the deceleration when FEEG is beginning to change. Point B and line B indicate FEEG later in another delayed deceleration with most apparent FEEG voltage suppression.

quencies of one cycle or greater per second. As the contraction ends the FHR returns to its previous levels, and the FEEG quickly returns to the visual patterns seen earlier. At times it would appear that the FEEG does first change with loss of faster frequencies and diminution of amplitude

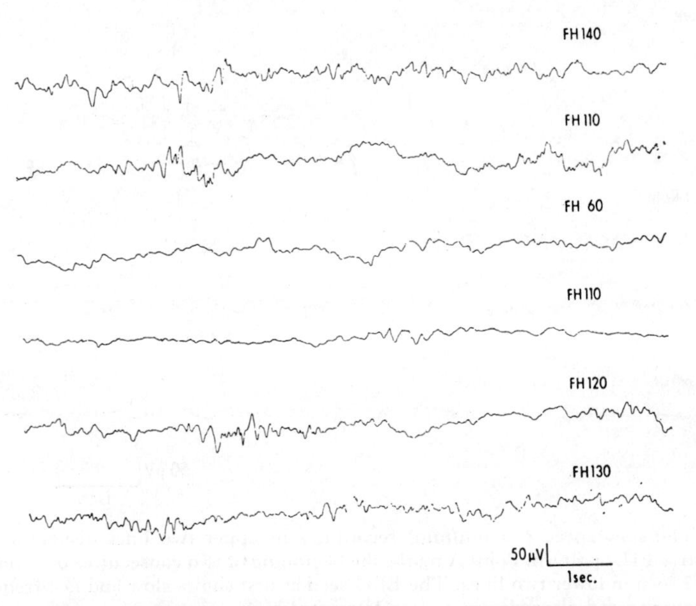

Fig. 5. Example of FEEG seen in the course of one pathological decelerative pattern. FHR during each 9-second strip seen at right margin of trace. At FHR 140 EEG is normal. During deceleration (FH 110) loss of most normal frequencies, and loss of amplitude begins. By FHR 60 a marked difference is apparent, which persists through 110 (rising FHR) and by FHR 130 is similar to its original appearance at onset of contraction.

but may not reach the degree of change here called isoelectricity. As yet we have not been able to quantitate the degree of change as seen in the EEG with the ongoing clinical events.

DISCUSSION

The ability to study FEEG prior to birth brings us one step closer to understanding the role of labour and delivery in later central nervous system pathology. EEG itself is only a single descriptor of the many other ongoing intracerebral events. Furthermore, confined by the present state of this art FEEG observations may be made only after amniotic membranes have been ruptured and represent but the application of two electrodes in bipolar fashion recording from large areas of the brain.

Nevertheless, despite these limitations new information is being derived and may be described in relation to the previously mentioned IUP and FHR parameters. This may be important since IUP and FHR are being used for clinical treatment and FEEG may provide more understanding and intelligence in their use.

Several statements may be made at this time when comparing the foetal heart and brain using heart rate and EEG as parameters for study. In certain cases described above as early deceleration, the FEEG is unaltered and at least suggests that the electrical activity of the brain in the cortex nearest the recording points is not affected by or correlated with the FHR alterations. In other cases described above as delayed and variable deceleration, the two systems, heart and brain, do seem to alter at the same time and are associated with foetal events which would appear to regulate both systems.

Thus it becomes apparent that the systems being studied are both linked closely in some clinical events, and operate independently in others. Stated differently, FHR alone may not represent all of the information vital to understanding central nervous system morbidity.

IUP increases during a contraction, or during spontaneous birth, do not appear to alter the FEEG. A report is being prepared (Rosen *et al.* in press) on the relation of IUP, spontaneous birth and forceps delivery, but is not available at this time. However, during normal spontaneous vertex birth the FEEG continues to maintain pre-existing patterns.

In our earlier studies (Rosen *et al.* 1970) we have noted that patterns present in the foetus appear to be the electrical analogues of those events described in the neonate as the deep sleep state and the alert state. These electrical patterns are apparently unaffected by the normal labour process itself. However, these patterns are interrupted by the previously described events as well as the use of maternal medications such as meperidine (Rosen *et al.* 1970; Rosen *et al.*, in press).

The transition of FEEG from pre-existing wave amplitudes and frequencies to low voltage 'isoelectric' states describes intracerebral electrical events not seen any time later in the neonatal period or in adult life. 'Isoelectricity' itself is an EEG phenomenon well known to electroencephalographers and usually associated with a stressful state, loss of consciousness for short periods of time such as syncope or longer periods of time such as coma secondary to drugs or trauma. It is also known that these electrical phenomena may be reversible and may be unassociated with permanent brain damage.

As noted earlier these short-lived FEEG changes which occur in association with the described decelerations are repetitive in frequency and may be seen often during some labours. There is no adequate model with which to compare these events in the human foetus. At this time we may describe such events but we cannot ascribe morbidity to their occurrence. The developmental status of these infants is now part of a five-year follow-up programme.

CONCLUSION

FEEG provides a more direct potential for the study of the brain during birth. The normal birth process and in-utero contractions do not alter the FEEG from pre-existing patterns and from the same patterns seen after birth. At times FHR decelerations (variable and delayed) correlate closely with FEEG amplitudes and frequency changes. During other FHR decelerations (early) the FEEG appears to operate independently.

Supported in part by The Grant Foundation, Inc., and by the Department of Health, Education, and Welfare, Public Health Service Grant No. 1 POI HD 05566-01A1.

REFERENCES

Rosen, M. G. & Scibetta, J. J. (1969). *Am. J. Obs. Gyn.* **104**, 1057.
Rosen, M. G., Scibetta, J. J., Chik, L. & Borgstedt, A. D. (1973). In press.
Rosen, M. G., Scibetta, J. J. & Hochberg, C. J. (1970). *Obs. Gyn.* **36**, 132.

THE SENSE OF TASTE AND SWALLOWING ACTIVITY IN FOETAL SHEEP

By ROBERT M. BRADLEY AND CHARLOTTE M. MISTRETTA
Nuffield Institute for Medical Research, University of Oxford

It is through the senses that we receive and first process information about our surroundings. Since sensory receptors are present in foetal life, the foetus may be able to monitor its environment. The foetal environment is liquid, and therefore the sense of taste may be particularly important in intra-uterine life. Taste buds are present in the human foetal tongue epithelium by 14 weeks of gestation (Bradley & Stern, 1967) and swallowing begins at about this age (Davis & Potter, 1946). Possibly the taste buds are stimulated by amniotic fluid and provide information on chemical composition of the fluid to the foetus.

As in the human, taste buds appear on the tongue of the foetal lamb during the first third of gestation (Bradley & Mistretta, 1972). At 50 days localized collections of cells (the presumptive taste buds) are present in the epithelium overlying fungiform papillae. Taste pores have developed by 90–100 days. To determine whether the foetal taste buds respond to stimuli, recordings were made of neural activity in the foetal chorda tympani nerve while applying chemicals on the tongue (Bradley & Mistretta, 1971, 1972). These experiments demonstrated that the peripheral taste system of the foetal lamb is functional over the last third of gestation. It is therefore pertinent to ask if the foetal lamb can respond to, or perceive, taste stimuli *in utero* during this period.

Experiments were undertaken in which foetal swallowing was used as an index of taste perception. If the foetus perceives taste stimuli *in utero*, then possibly it will swallow more when the amniotic fluid is artificially flavoured to taste pleasant, and less when it tastes unpleasant. Swallowing activity was measured with an electromagnetic flow transducer chronically implanted in the foetal oesophagus. Preliminary data on daily swallowing are reported here.

METHODS

Under maternal epidural anaesthesia the foetal head is delivered and a 2–3 cm midline incision is made in the neck. A cannulated electromagnetic flow transducer (Clark & Wyatt, 1969) is inserted into the oesophageal lumen and the oesophageal incision is closed. Catheters are inserted into the foetal carotid artery and trachea. An amniotic fluid catheter, for delivery of chemical solutions, is sutured to the foetal head and snout, with the end over the nose. Alternatively, a curved catheter is placed with the end inside the foetal mouth, held in place by sutures on the outside of the cheek. A second catheter, for sampling amniotic fluid, is sutured to the back of the foetal neck.

The ewe is placed in a metabolism cage after surgery. The foetal carotid and tracheal catheters are connected to pressure transducers. The output of the flowmeter is recorded continuously, providing a record of fluid flow in the oesophagus 24 hours per day. The flow output is also integrated electronically, and the integrated flow (volume swallowed) is recorded. Maternal and foetal arterial blood samples are taken daily and analysed for blood gases and pH.

RESULTS

Swallowing results in net fluid flow down the oesophagus (Fig. 1). The recorded deflections can readily be distinguished from the biphasic pen deflections which sometimes occur during foetal respiratory and other movements. Interpretation of foetal swallowing records has been corroborated by direct observations on a foetus, with an implanted oesophageal flow transducer, delivered into a saline bath.

Records from 8 foetuses of 101–143 days' gestational age have shown that the foetus swallows large volumes of amniotic fluid in 2–7 'bouts' per day. A bout consists of a period of swallowing during which 20–200 ml of amniotic fluid are consumed. The bouts are 1–9 min in duration and are accompanied by a rise of arterial pressure of up to 20 mm Hg. In between these episodes of relatively large volume swallowing the foetus frequently swallows small volumes (1–10 ml). Fig. 2 illustrates the integrated recordings of swallowing bouts over 24 hours from a foetus on the third postoperative day; the bouts were randomly spaced throughout the day. No obvious pattern of swallowing episodes is established from day to day. The average daily volume swallowed by 4 foetuses ranged from 247 to 477 ml (Table 1).

In deteriorating foetal preparations swallowing stops, sometimes days before foetal death occurs. In Fig. 3 the daily swallowing volumes for

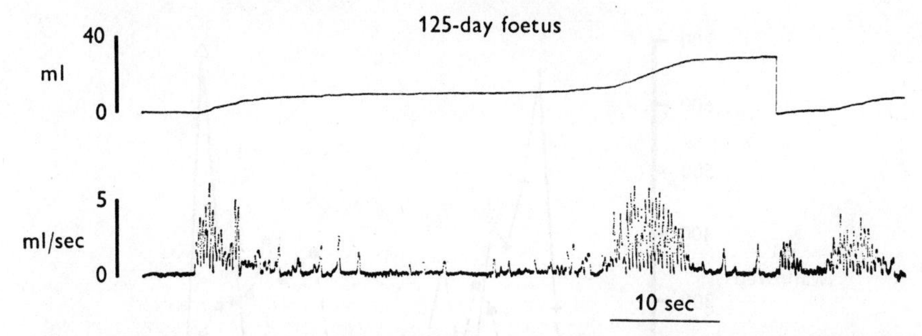

Fig. 1. High speed record of swallowing activity from Foctus No. 85 on thc fourth postoperative day. The lower trace is the direct output of the electromagnetic flowmeter; positive deflections result from swallowing. The upper trace is the integrated flow and indicates volume swallowed. The integrator automatically resets to zero every 60 sec.

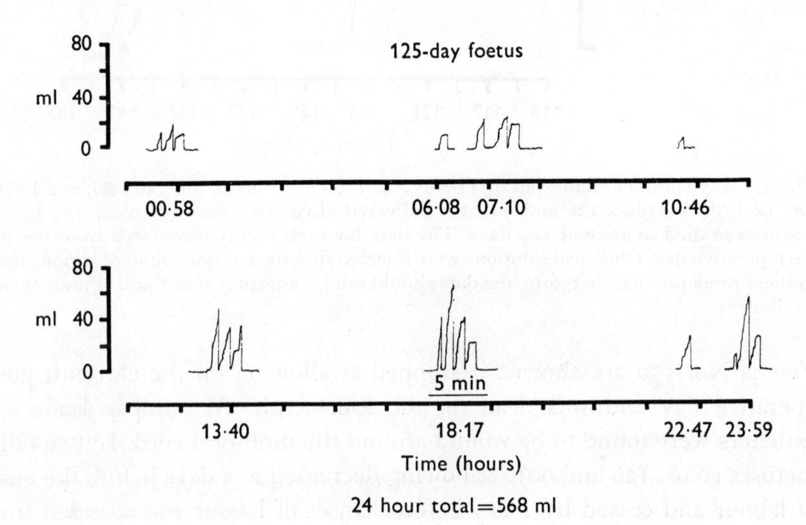

Fig. 2. Integrated records of swallowing bouts over a 24-hour period from Foetus No. 85 on the third postoperative day. The time at which swallowing occurred is indicated under each integrated recording.

Table 1. *Average daily volumes swallowed by foetuses with flowmeter heads chronically implanted in the oesophagus*

Foetus no.	Age (days)	Vol. swallowed per day Mean ± SD (ml)	No. of bouts per day Mean ± SD	Wt of foetus at delivery (kg)
170	115–129	359 ± 147 (9)*	5 ± 2	2.88
126	118–141	247 ± 86 (19)	4 ± 1	3.92
85	122–128	477 ± 203 (3)	6 ± 2	3.10
90	135–143	344 ± 222 (5)	4 ± 2	3.15

* The number of days on which the average is based is given in parentheses.

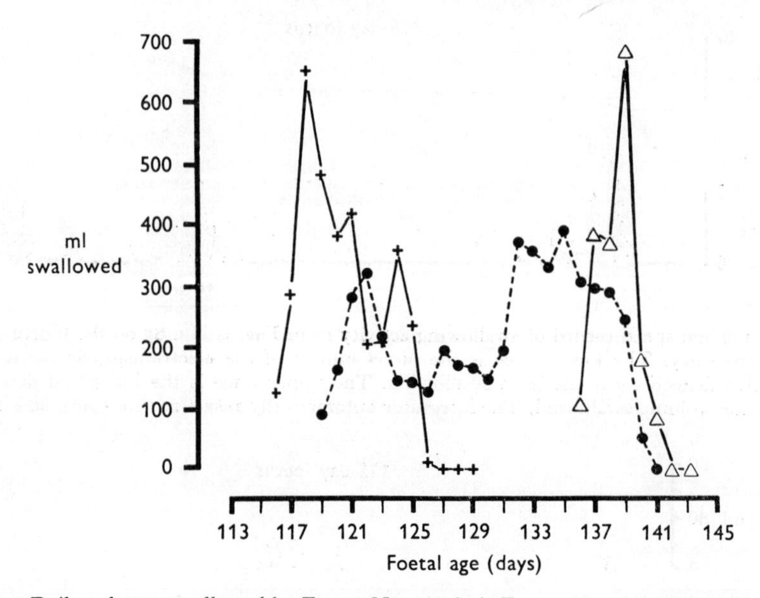

Fig. 3. Daily volumes swallowed by Foetus No. 170 (+), Foetus No. 126 (●), and Foetus No. 90 (△). Foetuses 126 and 90 were delivered alive after the ewes went into labour. Foetus 170 died *in utero* at 129 days. The data for each foetus are plotted from the first postoperative day. Chemical solutions were injected into the amniotic fluid at various times in these preparations; therefore, the data should not be interpreted as those of undisturbed swallowing.

Foetus No. 170 are shown: it stopped swallowing on the eleventh post-operative day and was dead on the fourteenth. At autopsy leads and catheters were found to be wound around the umbilical cord. In two other foetuses (Nos. 126 and 90) swallowing decreased 2–3 days before the onset of labour and ceased by 1 day before. Onset of labour was assessed from evidence of uterine contractions on pressure recordings.

DISCUSSION

It is not yet possible to determine what initiates foetal swallowing bouts. Foetal swallowing stops in deteriorating preparations and before the ewe goes into labour; it therefore seems that swallowing activity is influenced to some extent by the general condition of the foetus. Swallowing begins at approximately 80 days of gestation in sheep (Duncan & Phillipson, 1951), the age at which Mellor & Slater (1971) first detected entry of foetal urine into the amniotic sac. It is interesting to speculate that foetal urine, by altering the composition and taste of the amniotic fluid, may stimulate the taste buds and thereby induce swallowing episodes.

Induction of swallowing by changing the taste of amniotic fluid might have various applications, e.g. 'feeding' undernourished foetuses by periodic injections of a nutritious, palatable fluid. Hamilton & Behrman (1972) have described treatment of foetal acidosis by intra-amniotic infusion of bicarbonate. Since large volumes of amniotic fluid (up to 200 ml) may be swallowed during a single, brief episode, it is possible to administer substantial quantities of solution via this route. We have induced swallowing bouts repeatedly in one foetus by injecting 25 ml 50% (w/v) glucose into the amniotic fluid, but in other foetuses results have not been reproducible. Further experiments are needed to demonstrate foetal taste perception and any relation between the gustatory sense and swallowing activity *in utero*.

This work was carried out with the aid of a grant from the Medical Research Council. We are grateful to Dr G. S. Dawes for his advice and criticisms throughout the course of this study. We also wish to thank Drs K. Boddy, C. Gibbs and J. Robinson for their invaluable surgical assistance.

REFERENCES

BRADLEY, R. M. & MISTRETTA, C. M. (1971). *J. Physiol.* **218**, 104*P*.

BRADLEY, R. M. & MISTRETTA, C. M. (1972). In *Oral Physiology*, ed. N. Emmelin & Y. Zotterman, pp. 239–53. Oxford: Pergamon Press.

BRADLEY, R. M. & STERN, I. B. (1967). *J. Anat.* **101**, 743–52.

CLARK, D. M. & WYATT, D. G. (1969). *Med. Electron. Biol. Engng* **7**, 185–90.

DAVIS, M. E. & POTTER, E. L. (1946). *J. Am. Med. Ass.* **131**, 1194–201.

DUNCAN, D. L. & PHILLIPSON, A. T. (1951). *J. Exp. Biol.* **28**, 32–40.

HAMILTON, L. A. & BEHRMAN, R. E. (1972). *Am. J. Obstet. Gynec.* **112**, 834–47.

MELLOR, D. J. & SLATER, J. S. (1971). *J. Physiol.* **217**, 573–604.

MEMBRANE POTENTIAL IN MUSCLE CELLS OF DEVELOPING CHICKS AND RATS

By J. BOËTHIUS

The Department of Physiology,
Karolinska Institutet, Stockholm, Sweden

The postnatal development of the muscle membrane potential has been investigated in rats (Fudel-Osipova & Martynenko, 1962; Novikova, 1964; Hazlewood & Nichols, 1967, 1969), mice (Harris & Luff, 1970) and in newly-hatched chicks (Karzel, 1968), but no data are available on muscle membrane potentials during earlier developmental stages. At birth, the mean membrane potential was reported to be as low as 23–27 mV in rats and chicks, thereafter to rise gradually, and to reach adult values at some time between the 15th day (Fudel-Osipova & Martynenko, 1962) and the 40th day (Hazlewood & Nichols, 1969) in the rat, and on the 16th day in the chick (Karzel, 1968).

The present report concerns the development of the membrane potential in embryonic chicks and postnatal rats. The study indicates that the membrane potential of the chick thigh musculature is low and constant at early developmental stages and rises to adult values during a well-defined period of time. A similar developmental pattern was found in the comparative studies which were made on the neck musculature and on the sartorius and gastrocnemius muscles of neonatal rats. Finally, some preliminary experiments are reported which deal with the effect of denervation on the development of the rat membrane potential.

The membrane potential was measured by pushing a conventional glass micro-pipette through the muscle tissue. The measure of the potential was the amplitude of the initial voltage deflection recorded when the electrode penetrated a cell. Both in chicks and rats the measurements were made *in vivo*. In chicks no anaesthetics were used. The rats were anaesthetized with urethane. (For further details see Boëthius & Knutsson, 1970; Boëthius, 1971.)

The development of the membrane potential in the chick is shown in Fig. 1. In the 3-day embryo the mean resting potential was about 25 mV and the largest resting potentials about 40 mV. During the subsequent stages, from the 6th to the 15th day of incubation, the mean value of the resting potential was approximately the same as in the 3-day chick, but the highest potential values were 60–65 mV. After the 15th day there was a

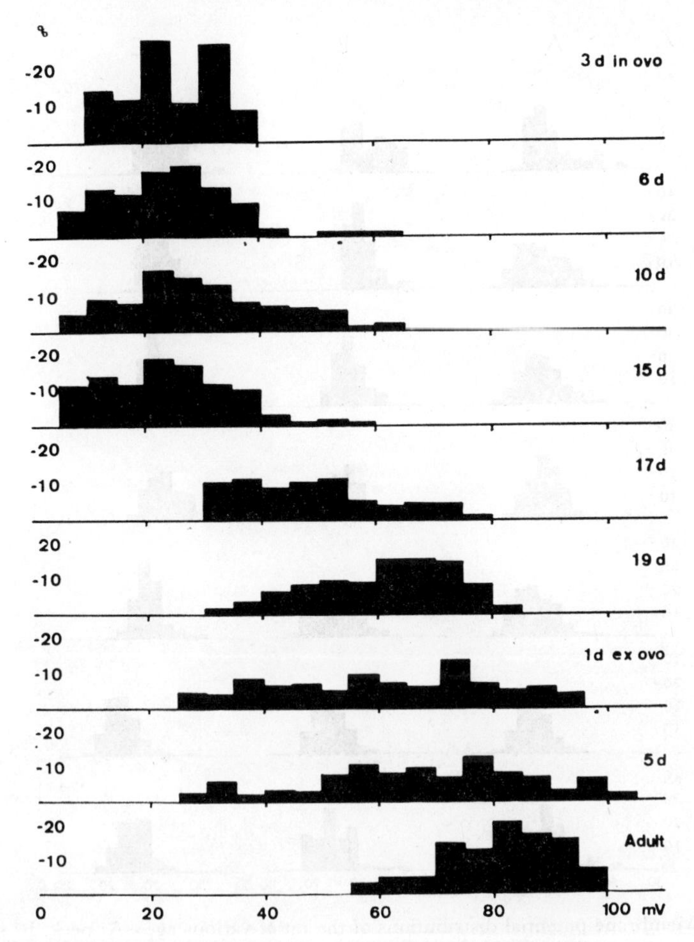

Fig. 1. Membrane potential distribution of the chick thigh musculature at different developmental stages *in ovo* and after hatching. Membrane potential on the abscissae and number of observations in percentages on the ordinates.

marked increase in the membrane potential, the mean values by the 17th day being 47 mV, and the highest potentials about 80 mV. On the 19th day the corresponding values were 62 and 86 mV respectively. At hatching, the mean membrane potential was 65 mV, and the highest potentials were 95–100 mV. By the 5th day *ex ovo* no further appreciable increase was observed. In the adult chick the membrane potentials ranged between 55 and 105 mV, the bulk of the observations being concentrated between 75 and 95 mV. The membrane potential of the chick thus follows a developmental sequence characterized by a first phase of low and constant potential followed by a second phase of rapidly increasing potential. At the end of the second phase the highest potentials have become of adult magnitude.

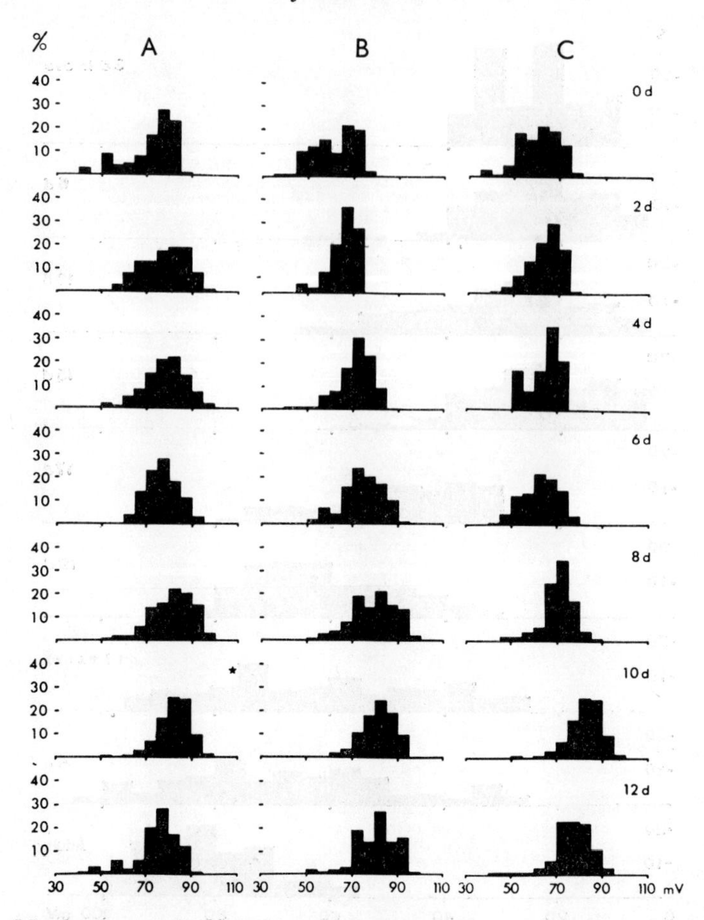

Fig. 2. Membrane potential distributions of the rat at various ages. A: neck, B: sartorius and C: gastrocnemius membrane potentials. At every age the potentials were obtained from the same animal, with the exception of the neck potentials in the 10-day row (asterisk), which were recorded in a 9-day animal.

In the rat, membrane potential measurements were made in the neck muscles and the sartorius and gastrocnemius muscles. The results (Fig. 2) demonstrated a general increase in mean membrane potential values between the first and the 10th postnatal day in all three muscles. This increase was earliest in the neck muscle, the potential of which rose from a value of about 70 mV at birth up to an adult value of 80 mV on the 2nd and 3rd day of life. The sartorius mean membrane potential was about 60 mV at the time of birth, and at 8 days it had reached 80 mV. Finally, at the time of birth, the mean membrane potential of the gastrocnemius muscle was roughly the same as the potential of the sartorius muscle, i.e. about 60 mV. Between birth and the 4th day there was no

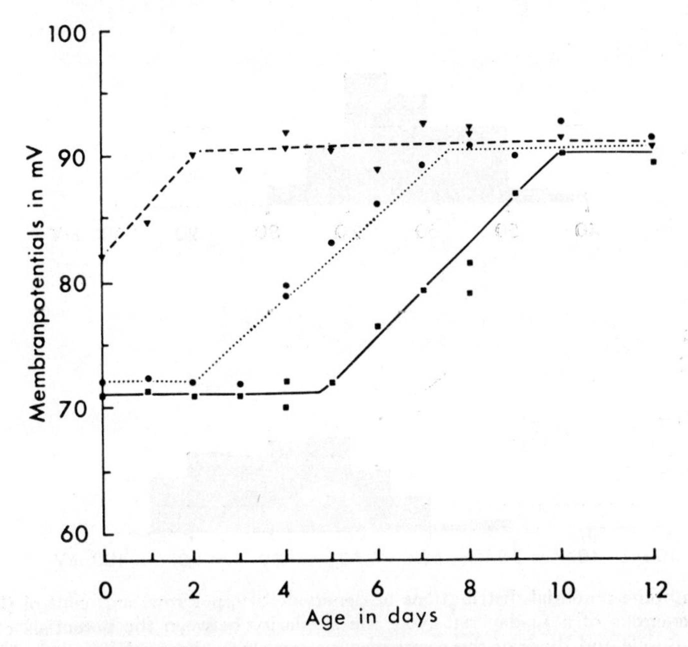

Fig. 3. Developmental patterns of the 'maximum potential'. Each point represents the mean of the highest potential from 10 to 18 electrode tracks. Symbols: neck (broken line, triangles), sartorius (dotted line, circles) and gastrocnemius (solid line, squares).

significant increase in potential. On the 5th day, however, the potential started to increase, and on the 10th day it had reached the adult level. The distribution of the membrane potential values around the mean value was similar in all ages and muscles. The standard deviation varied between 4 and 9 mV, i.e. values corresponding to those of adult rats (Li *et al.* 1957; Zierler, 1959).

It could be argued that the high potentials afford the best measure of the 'true' membrane potential since the low potentials are more likely to be influenced by errors caused by imperfect electrode penetrations. In the rat, the membrane potential was therefore also estimated by using the mean of the highest potential in each electrode track (cf. Boëthius, 1971). The developmental change of the 'maximum potential' was essentially similar to that of the mean membrane potential (Fig. 3). At birth, the maximum potential of the neck muscle was higher than that of the leg muscles, the mean value being 82 mV. During the next two days the potentials rose and already reached the highest level (90 mV) on the 2nd day. Initially, the mean maximum potential of both the sartorius and the gastrocnemius muscle was about 70 mV. After the 2nd day the sartorius maximum potential started to rise and reached the adult level of about

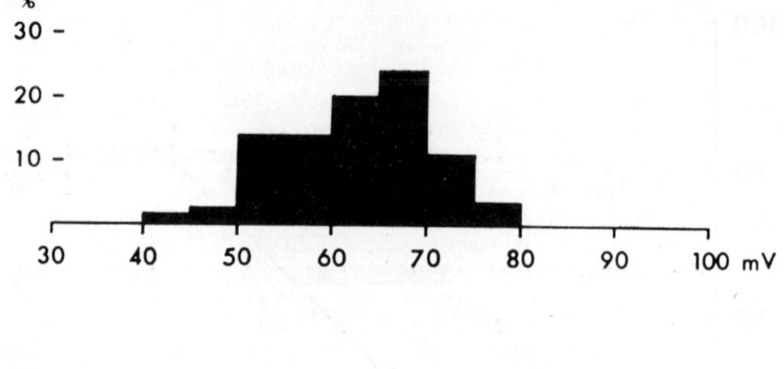

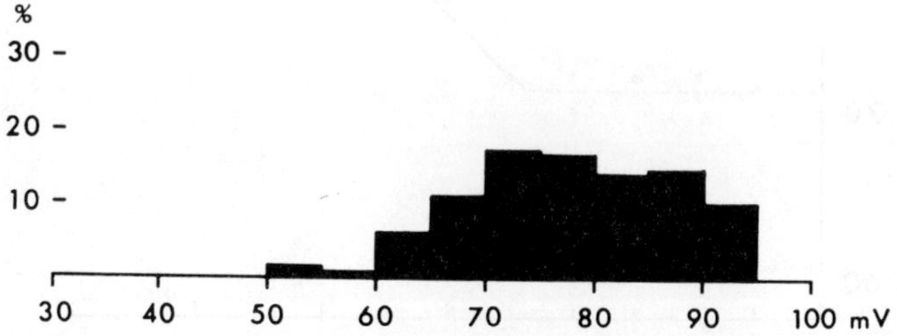

Fig. 4. Membrane potential distributions of denervated (upper row) and control (lower row) gastrocnemius of a 14-day rat. Note the similarity between the potentials of the denervated muscle and those of the gastrocnemius muscle in the newborn rat in Fig. 1.

90 mV on the 7th day. The gastrocnemius maximum potential started to increase on the 5th day and reached the top level on the 10th day. The distribution of the maximum potential values was approximately normal. The standard deviation was 2–4 mV. The scatter did not increase at the time when there was an increase in the membrane potential. This finding may indicate that the increase in potential of the individual muscle cell took about 5 days, since one would otherwise expect the scatter to increase during the rising phase.

The present results seem to indicate that the development of the membrane potential in general proceeds with an initial phase of low potential followed by a fairly short period of increasing membrane potential whereafter the adult level is reached. This pattern prompts the question of the mechanism of this sudden developmental rise in potential. It is interesting to note that denervation causes a decrease in the membrane potential of adult rats (Lüllmann & Pracht, 1957; Thesleff, 1963). This finding suggests the possibility of a neuronal influence on the development of the muscle membrane potential. Some preliminary experiments were therefore done on denervated developing muscles in rats.

Newborn rats were denervated by removing the sciatic nerve in the thigh. In the acute experiment the EMG of the denervated muscle was first recorded in order to determine whether or not re-innervation had occurred. When the membrane potential was measured in the denervated gastrocnemius of a 14-day-old rat the membrane potential level remained the same as in the newborn gastrocnemius (Fig. 4). After re-innervation, on the other hand, the level of the membrane potential had increased and was the same as that in the contralateral control muscle, viz. 80–90 mV. These preliminary experiments suggest that among the manifold trophic functions of the motor nerve should be included also the regulation of the rise in muscle membrane potential during development.

This study was supported by grants to the Department of Physiology II from the Association for the Aid of Crippled Children, the Swedish Research Council, Magnus Bergvalls Stiftelse and by personal grants from Svenska Sällskapet för Medicinsk Forskning and Karolinska Institutet.

REFERENCES

Boëthius, J. (1971). *Acta Physiol. Scand.* **81**, 492–507.

Boëthius, J. & Knutsson, E. (1970). *J. Exp. Zool.* **174**, 281–6.

Fudel-Osipova, S. K. & Martynenko, O. A. (1962). *Transl. in Fed. Proc., Transl. Suppl.* 1964 **23**, 28–30.

Harris, J. B. & Luff, A. R. (1970). *Comp. Biochem. Physiol.* **33**, 923–31.

Hazlewood, C. F. & Nichols, B. L. (1967). *Nature* **213**, 935–6.

Hazlewood, C. F. & Nichols, B. L. (1969). *Hopkins Med. J.* **125**, 119–33.

Karzel, K. (1968). *J. Physiol.* (*Lond.*) **196**, 86–7*P*.

Li, C.-L., Shy, G. M. & Wells, J. (1957). *J. Physiol.* (*Lond.*) **135**, 522–35.

Lüllmann, H. & Pracht, W. (1957). *Experientia* **13**, 288–9.

Novikova, A. I. (1964). *Fiziol. Zh.* (*Mosk.*) **50**, 626–30.

Thesleff, S. (1963). In *The Effect of Use and Disuse on Neuromuscular Functions*, eds. E. Gutmann and P. Hnik, pp. 41–51. Publishing House of the Czechoslovak Academy of Sciences, Prague.

Zierler, K. L. (1959). *Am. J. Physiol.* **197**, 515–23.

SESSION 2

CIRCULATION AND BREATHING

CONTROL OF THE FOETAL CIRCULATION

By ABRAHAM M. RUDOLPH and
MICHAEL A. HEYMANN*
The Cardiovascular Research Institute and Departments of
Pediatrics and Physiology, University of California,
San Francisco, California 94122, U.S.A.

Inherent in any attempt to study biological function is some modification of those processes being examined. Early physiological observations of the circulation of the mammalian foetus were performed on exteriorized foetuses. The ovine species has been studied rather extensively in this manner, as the uterus does not undergo forceful contraction and, thereby, separate the placental attachment after the foetus has been removed. The foetus, prevented from establishing pulmonary respiration, may continue to survive on the umbilical–placental circulation for several hours. There is, however, accumulating evidence that exteriorization exerts a profound influence on the foetal circulation.

EFFECTS OF FOETAL EXTERIORIZATION

We have previously demonstrated that umbilical blood flow falls after exteriorization of the lamb foetus (Heymann & Rudolph, 1967). In the exteriorized previable human foetus, the umbilical blood flow drops even more rapidly and may almost cease within 30 minutes (Rudolph *et al.* 1971). Since umbilical blood flow represents about 40% of the combined output of the left and right ventricles in the late gestation foetus, an increase in umbilical–placental vascular resistance could markedly alter the distribution of cardiac output, flow patterns and, if severe enough, foetal blood gases. It is of interest that umbilical venous blood gases are most unreliable in assessing foetal metabolic status, as the reduced umbilical blood flow in exteriorized foetuses is associated with well-maintained

* Recipient Research Career Development Award HD 35398 of the National Institute of Child Health and Human Development.

H^+ concentration and P_{CO_2} levels, and a normal, or often increased, P_{O_2} level, while foetal arterial blood shows a fall in P_{O_2} and a rise in P_{CO_2} and H^+ concentration (Rudolph *et al.* 1971).

The decreased umbilical blood flow results in a decrease in blood returning to the heart through the inferior vena cava with, therefore, a relative increase in superior vena caval flow. This disturbance is probably responsible for the increase in superior vena caval blood which passes through the foramen ovale. We have noted an increase of superior vena caval flow across the foramen ovale from the usual 2–3% to more than 50% when foetal arterial H^+ concentration rises and P_{O_2} falls markedly (Rudolph & Heymann, 1967). An additional factor that has not been studied is the possible effect of increased umbilical–placental vascular resistance in causing an elevated arterial pressure with baroreceptor stimulation and altered circulatory distribution patterns.

ACUTE AND CHRONIC PREPARATIONS *IN UTERO*

In order to overcome these effects of exteriorization on the circulation, attempts have been made to observe the foetus *in utero*. Although limited studies have been made in sub-primate models, most investigators have focused on the sheep.

Various monitoring devices have been used, such as placement of plastic catheters in foetal arteries and veins and in the trachea, attachment of electrocardiographic and electroencephalographic electrodes, and application of electromagnetic flow transducers. The uterine and abdominal incisions were closed and the catheters and leads made available on the ewe's flank. Cardiac output and its distribution and blood flow to foetal organs have been studied using radio-nuclide labelled microspheres (Rudolph & Heymann, 1967, 1970).

Initially, observations were made acutely, within the first 24 hours after surgery. Evidence is now accumulating that the circulation, as well as other foetal functions, may show changes for 2–5 days after a surgical insult. Foetal surgery results in a fall of arterial P_{O_2} and rise of H^+ and P_{CO_2}; this is most evident if the umbilical vessels are catheterized. Heart rate is increased, probably as a result of catecholamine release, and there is also evidence of increased adrenal cortical activity (Bassett & Thorburn, 1969). There is often an interference with normal intrauterine respiratory movements and EEG activity (Dawes *et al.* 1972) and foetal metabolism is also influenced (Gresham *et al.* 1972). The recovery period is very much related to the type of the procedure and its duration. The effects of surgery also seem to be more marked in lamb foetuses in later gestation.

It is thus evident that, ideally, observations on cardiovascular function should be made in chronic in-utero preparations only after an adequate period of adjustment from surgical manipulations has been allowed. Even under these circumstances, the foetus may not be in a normal physiological state, as there is a high incidence of spontaneous parturition 5–7 days after surgery in later gestation.

DEVELOPMENT OF AUTONOMIC REGULATION OF THE FOETAL CIRCULATION

Effective autonomic nervous control of the circulation is dependent on growth of nerves into the heart and blood vessels, ability to release the active neurotransmitter, development of the effector organ (cardiac or smooth muscle cells) and of the receptor sites in the organ. The development of sympathetic innervation of the heart has been examined in several species by histochemical means. Using the monoamine fluorescence technique, Friedman (Friedman *et al.* 1968) found that sympathetic innervation to the rabbit heart developed largely postnatally. Recent studies (Lipp & Rudolph, 1972) have shown that there is a marked species difference in cardiac sympathetic innervation. The rat, relatively immature at birth, has no cardiac sympathetic nerves demonstrable until about a week after birth. The guinea pig, quite mature and self-sufficient at birth, has well-developed sympathetic nerves in the heart almost equivalent to those found in the adult. Even at 0.66 gestation, cardiac sympathetic nerves are already evident in the guinea pig. Innervation in the lamb's heart is similar to that of the guinea pig. At 75 days' (0.5) gestation, no sympathetic nerves are yet evident, but there are large numbers of dopamine-containing cells scattered throughout the myocardium (Lebowitz, Novick & Rudolph, 1972). The role of these cells is obscure. Since dopamine has myocardial inotropic properties, it is possible that they have a role in cardiac regulation in early gestation. It is also possible that they may provide a source for norepinephrine release, as a single enzyme, dopamine beta-oxidase, can convert dopamine to norepinephrine. The pattern of innervation of the heart is similar in all species studied. Nerve endings are first seen in the right atrium and then, in sequence, the left atrium, right ventricle and left ventricle, and finally the apical area. The coronary circulation receives its innervation earlier than the myocardium and it has a much more extensive network at all stages of development.

Detailed histochemical observations are not available for other foetal vascular structures; it would be important to extend these studies to determine when blood vessels in different organs are innervated. On the basis of these observations it is evident that the heart is not subject to reflex control mediated through the sympathetic nervous system in early gestation and, in some species, not till after birth. It has been intimated that the foetal myocardial response to norepinephrine is relatively greater than that of the adult, suggesting that it is demonstrating supersensitivity as found in adult structures subjected to denervation (Friedman *et al.* 1968).

Limited studies of the lamb ductus arteriosus have shown the presence of sympathetic nerve fibres in the latter third of gestation. Histochemical demonstration of acetylcholinesterase, supporting the presence of parasympathetic nerve endings in the ductus arteriosus of late gestation lambs, has also been reported (Silva & Ikeda, 1971), but the precise developmental pattern has not yet been examined.

The ability of the heart and blood vessels to respond to autonomic neurotransmitter substances has been studied both in intact foetal preparations, and in some isolated segments in tissue baths. The chick embryo heart shows an increase of heart rate as early as 4 days when exposed to beta-receptor stimulation with isoproterenol; this effect is also noted in the rabbit embryo of 5–6 days' development. Infusion of isoproterenol into the hind limb vein of 60-day gestation lamb foetuses (0.4 gestation) produced a marked increase in heart rate (Barrett, Heymann & Rudolph, 1972). It has also been shown that the amount of isoproterenol, based on foetal weight, required to produce an increase in heart rate, is similar in foetal lambs from 85 to 150 days' (0.55–1.0) gestation. The effects of beta-adrenergic stimulation on myocardial contractility have not been examined as extensively, but studies of isolated myocardial strips from late gestation lamb hearts show that there is an increased contractility in response to isoproterenol, similar to that observed in the adult sheep (Friedman *et al.* 1969). Beta-adrenergic stimulation has also been shown to produce coronary vasodilation from as early as 60 days' gestation in foetal lambs (Barrett, Heymann & Rudolph, 1972) and pulmonary vasodilation as early as 75–90 days (Dawes, 1968). Alpha-adrenergic receptor activity has been demonstrated by 75 days' (0.5) gestation in foetal lambs. Infusion of methoxamine, a specific alpha-adrenergic stimulant, produced a rise of arterial pressure, and vasoconstriction in the kidney and peripheral circulation (Barrett, Heymann & Rudolph, 1972). It is also known that norepinephrine can produce pulmonary vasoconstriction in 75–90 days' gestation foetal lambs (Dawes, 1968). No obvious difference in the pattern of response of the circulation at varying gestational ages could be demonstrated.

The response of foetal vascular structures to the parasympathetic neurotransmitter, acetylcholine, has been examined in isolated heart muscle, and isolated ductus arteriosus preparations. Contractility of ventricular muscle from the heart of late gestation lamb foetuses is decreased by acetylcholine. Ductus arteriosus rings obtained from lambs of 80–90 days' (0.6) gestation (McMurphy, Heymann, Rudolph & Melmon, 1972) and from human foetuses of 10–20 weeks (McMurphy & Boréus, 1971), are constricted by acetylcholine. The response of the lamb ductus increases with advancing gestational age, but this could be related to the increase in muscular development, rather than an increase in sensitivity to acetylcholine. Acetylcholine injected into the hindlimb vein of intact foetal lambs produces a fall in heart rate as early as 80–85 days' gestation and there does not appear to be any increased responsiveness with advancing gestation (Vapaavuori *et al.* 1972). The pulmonary vessels may be dilated by infusion of acetylcholine as early as 75 days' (0.5) gestation in lambs (Dawes, 1968).

It is thus evident that beta- and alpha-adrenergic receptor binding sites and parasympathetic receptors are exhibited early in foetal development both in heart muscle and vascular smooth muscle. Whether there is any change in the actual number of sites on individual cells during gestation and in different organs remains, however, to be examined.

Although the existence of end-organ receptor sites can be demonstrated in early gestation, the actual role of the autonomic nervous system in influencing resting vasomotor tone, heart rate and myocardial contractility in the foetus has not been examined extensively. After 100 days' (0.66) gestation, hexamethonium, a ganglion blocker, produced a fall in heart rate and blood pressure, suggesting there is some sympathetic influence by this age in the lamb (Dawes, 1968). This study was performed on exteriorized foetuses and it is possible that there was abnormal stimulation, not normally present in the foetus. The role of the autonomic nervous system in regulation of the circulation has been studied in chronic foetal lamb preparations from 85 to 150 days' gestation by selective blockade of specific autonomic functions (Vapaavuori *et al.* 1972). Parasympathetic blockade, produced by atropine in doses of 0.2 mg/kg foetal weight, resulted in an increase in heart rate in even the youngest foetus, indicating there was some vagal effect on the heart early in development. The effect was, however, quite small in foetuses under 100 days' (0.66) gestation, and increased strikingly between 100 and 120 days. There was no significant difference in response with advancing gestation beyond 120 days, and, in fact, the behaviour of the newly-born lamb was comparable to that of the late gestation foetus.

The effects of alpha-sympathetic blockade, achieved by administering phenoxybenzamine or phentolamine (5 mg/kg), were similar in pattern, in terms of chronological development, to those of parasympathetic blockade. Only minimal decrease in blood pressure occurred in foetuses under 100 days' (0.66) gestation. Beyond this period, responses were similar in foetal as well as neonatal lambs.

The pattern of response to beta-sympathetic blockade, produced by propranolol (1 mg/kg) or by practalol (2 mg/kg) was quite different. Major reduction in heart rate was noted only after about 120 days' (0.8) gestation, and there was also a greater effect in neonatal as compared with foetal lambs.

Although no statement can be made regarding the earliest onset of sympathetic and parasympathetic influences on the foetal circulation, it appears that there is a more rapid development of resting parasympathetic and alpha-sympathetic regulations. Not only is the maturation of beta-sympathetic activity on the heart relatively delayed, but there is continuing development after birth. This is in accord with the observation that sympathetic innervation to the heart is incomplete in the foetus and continues to develop after birth.

It should be pointed out that in studies with sympathetic blocking agents one cannot conclude that the results are entirely related to blockade of autonomic nerve endings. It is possible that the agents interfere with the local effects of circulating catecholamines. Although Comline & Silver (1966) have indicated that secretory activity of the adrenal medulla is low except under conditions of severe hypoxaemia in the foetal lamb, it is possible that chromaffin tissue in other sites may secrete catecholamines.

CENTRAL NERVOUS INFLUENCE ON THE CIRCULATION

Cardiovascular function in the adult is influenced in a complex manner by cerebral cortical, hypothalamic and medullary activity. The potential of the central nervous system to moderate heart rate and blood pressure in the foetus at various stages of development has not been examined. These observations have been hampered by the fact that foetal studies were usually done on anaesthetized animals, and anaesthesia markedly alters central nervous system function.

Recent studies have demonstrated that electrical stimulation of the hypothalamus in lambs as young as 115 days' (0.75) gestation produces

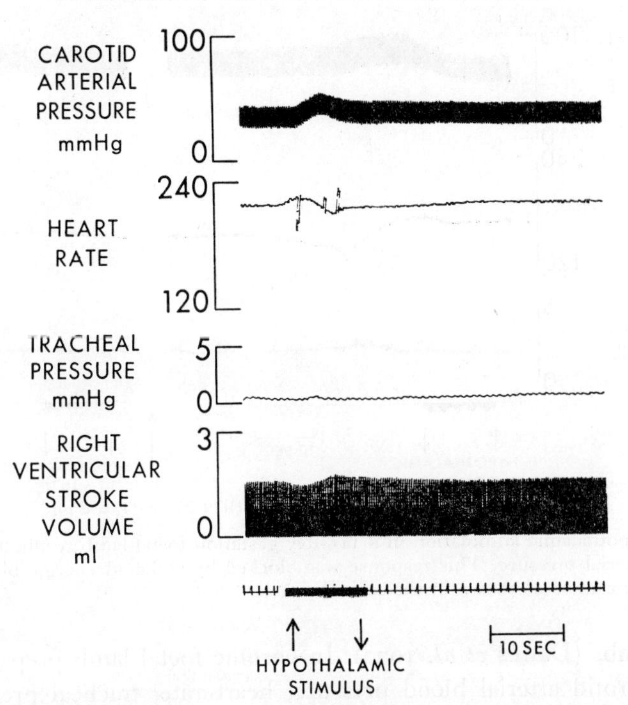

Fig. 1. Hypothalamic stimulation in a 115-day gestation foetal lamb produced a rise in carotid arterial pressure with an initial rise in heart rate followed by slowing, probably due to baroreceptor stimulation. Right ventricular stroke volume showed an initial decrease associated with the increase in heart rate and then a small increase. No respiratory effort was induced.

changes in heart rate and blood pressure (Williams, Hof, Heymann & Rudolph, 1972). Polyvinyl catheters were placed in carotid and femoral arteries to record foetal blood pressure and heart rate, in the trachea to monitor respiratory movements, and in the amniotic cavity, to assess any uterine or maternal abdominal pressure changes. A monopolar steel electrode was inserted into the foetal hypothalamus stereoactically using predetermined coordinates. Electrical stimuli were applied at the time of surgery to produce changes in heart rate or blood pressure. After recovery for several days, the lambs were studied while the ewe stood quietly. In some instances, muscular paralysis was induced pharmacologically to eliminate any respiratory movements while hypothalamic stimulation was studied. Increases in blood pressure and heart rate could be elicited in the youngest lambs thus far examined (115 days' gestation) (Fig. 1). The pressure response was prevented by alpha-sympathetic blockade (Fig. 2) and the heart rate increase was blocked by propranolol.

Cerebral cortical activity has also been shown to affect the circulation in

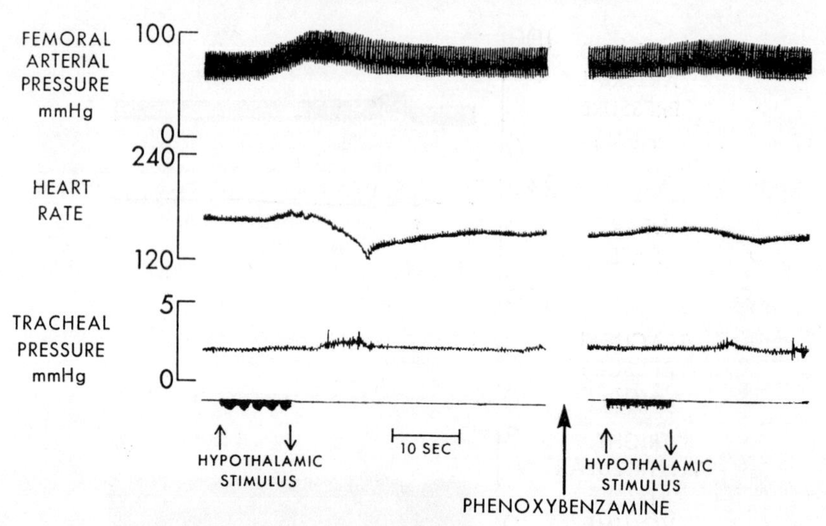

Fig. 2. Hypothalamic stimulation in a 115-day gestation foetal lamb resulted in a rise in femoral arterial pressure. This response was blocked by alpha-adrenergic blockade with phenoxybenzamine as shown in the second panel.

foetal lambs (Dawes *et al.* 1972). In chronic foetal lamb preparations in which carotid arterial blood pressure, heart rate, tracheal pressure and biparietal electrocorticogram were continuously monitored, spontaneous variations in heart rate and blood pressure were frequently noted. A change in electrocorticogram from slow high voltage to rapid low voltage activity which appeared to represent rapid-eye-movement sleep, was associated with irregular respiratory activity, and a rise in heart rate and blood pressure.

We have made similar observations in foetal lambs and noted that the spontaneous changes in heart rate, blood pressure and right ventricular output frequently occur in relation to respiratory activity. Circulatory changes are more likely to occur at the onset of the respiratory activity and especially if there are large swings in tracheal pressure; they are usually also evanescent and a return to resting levels tends to occur in spite of continued chest movement. Larger, but brief, responses have been observed with occasional major respiratory movements. Spontaneous increases in intratracheal pressure lasting 2–10 seconds produce a fall in heart rate, blood pressure and right ventricular output, and deep respiratory movements lasting 1–3 seconds are followed by an evanescent increase in these parameters. It has been difficult to separate the mechanical effects of chest motion from possible cerebral activity in promoting these cardiovascular changes. We have also noted that spontaneous changes may occur unrelated to respiratory activity.

Additional evidence that cortical impulses may influence the circulation in the foetus has been obtained in chronic studies on lidocaine toxicity (Teramo, Rudolph & Heymann, 1972). Infusion of lidocaine into late gestation (>0.8) foetal lambs *in utero* produces intermittent marked increases in blood pressure. These have been associated with bursts of irregular very high voltage electrocortical activity, probably representing convulsive episodes. Administration of muscle relaxants did not prevent these effects but barbiturates (pentobarbital) eliminated the abnormal cortical activity and the pressor responses.

The evidence that the central nervous system regulates the foetal circulation is, as yet, scanty. Recent work has shown that by 0.75 gestation, the lamb exhibits the ability to modify some circulatory functions by central nervous activity. However, there is little indication as to the degree to which it affects the circulation under normal in-utero conditions, nor to the extent of development of its control.

DEVELOPMENT OF CARDIOVASCULAR REFLEX FUNCTION

In all species thus far examined, newly-born animals demonstrate baroreceptor and chemoreceptor activity which is similar to that observed in adults. Studies in the foetus have usually been performed acutely in exteriorized preparations and frequently under general anaesthesia. Since these experimental conditions may markedly depress or modify reflex activity, little information is available regarding normal reflex function in the foetus.

BAROREFLEX ACTIVITY

Barcroft & Barron (1945) showed that the acute bradycardia resulting from epinephrine or norepinephrine injection into foetal lambs could be abolished by cutting the vagi, and it was later shown that similar responses could be elicited in foetuses of about 0.6 gestation (Dawes, Mott & Rennick, 1956). The recording of impulses consistent with the heart rate from the carotid sinus nerve in late gestation foetal lambs, confirms the fact that the carotid sinus baroreceptor is operative (Biscoe, Purves & Sampson, 1969).

Recently, studies have been performed in foetal lambs at gestational periods of 85 days to term (0.6–1.0) and in newborn lambs, to assess the sensitivity of the baroreflex (Shinebourne *et al.* 1972). In addition to placing a catheter in a foetal carotid artery, a small balloon catheter was passed into the descending aorta from a femoral artery and electro-

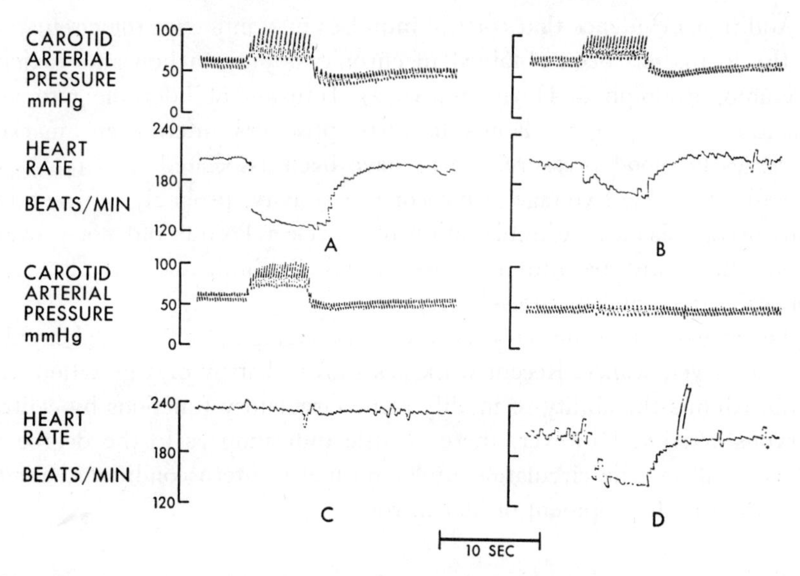

Fig. 3. Baroreceptor responses in a 130-day gestation foetal lamb. The fall in heart rate associated with an increase in systemic arterial pressure produced by inflating a balloon in the descending aorta is seen in panel A. Panel B shows that this response is decreased after section of both carotid sinus nerves. Panel C shows complete ablation of the response after bilateral carotid sinus nerve section and aortic stripping. The integrity of the vagus nerve is shown in Panel D in which heart rate fell after direct stimulation of the cervical vagus.

cardiographic leads were applied. After recovery from surgery, studies were conducted with the ewe standing quietly. Carotid and upper aortic pressures were suddenly increased by inflating the balloon while pressure and heart rate were continuously recorded on magnetic tape. In one acute experiment, the fall in heart rate associated with the pressure increase was abolished by cutting the carotid sinus nerves and stripping the aorta confirming that the response was baroreflex in nature (Fig. 3). The sensitivity of the reflex was examined in a similar manner to the method described by Smyth (Smyth, Sleight & Pickering, 1969). These studies indicated that baroreflex activity was present in the youngest lambs, but a significant response could not always be elicited. With advancing gestation, there was an increased sensitivity of the reflex, and the proportion of significant positive responses increased. It was of interest that in several lambs studied just before birth and after spontaneous delivery, there was no change in baroreflex sensitivity in the neonatal period. Whether the baroreflex serves any function in regulating the circulation in the foetus is not known; its possible role in the foetus will be discussed below.

CHEMORECEPTOR REFLEX ACTIVITY

It has been suggested that, since the carotid chemoreceptors are primarily concerned with control of respiration, and aortic chemoreceptors largely involved in regulation of circulation, only the aortic chemoreceptors should be operative in the foetus (Dawes, 1969). Currently available evidence indicates that there is, indeed, very little tonic activity of the carotid chemoreceptor even in the immature foetal lamb, but there is a dramatic increase in activity immediately after ventilation is established and the umbilical cord is clamped (Purves & Biscoe, 1966). Observations on aortic chemoreceptor stimulation suggest that these bodies are very sensitive to modest changes in foetal P_{O_2}, and that aortic chemoreflexes are most important in promoting foetal survival during hypoxemia (Dawes, 1969). Foetal hypoxaemia was induced by administering low oxygen gas mixtures to the ewes under chloralose anaesthesia. The increase in arterial pressure, tachycardia and hindlimb vasoconstriction induced by this hypoxemia, was prevented after section of the aortic nerves or the cervical vagi. It was thus suggested that foetal aortic chemoreceptors are stimulated by hypoxaemia and modify cardiovascular function. In acute studies in exteriorized late gestation lambs, in which a long tube was placed in the carotid artery to produce a delay between the aortic and carotid chemoreceptors, evidence was obtained that both receptors are capable of responding (Goodlin & Rudolph, 1972) and producing cardiovascular responses. Aortic and carotid stimulation was induced by injections of sodium cyanide (200 μg) or $NaHCO_3$ (1 ml, 1 M). Aortic chemoreceptor stimulation most frequently caused bradycardia with either mild hypotension or hypertension, whereas carotid chemoreceptor stimulation usually caused hypertension and either no change or an increase in heart rate. The results of these studies were quite variable and inconclusive, probably due to their acute nature and the variable physiological status of the foetus. In preliminary studies in chronic foetal lamb preparations we have found that stimulation of either aortic or carotid chemoreceptors with sodium cyanide consistently produced bradycardia and hypotension.

CARDIAC OUTPUT AND ITS REGULATION

Since the foetal circulation is not, as in the adult, a series circuit, it is convenient to consider cardiac output as the combined output of the left and right ventricles. Cardiac output in the foetus has been measured by several techniques in the exteriorized foetal lamb in late gestation. Using the Fick method, a combined ventricular output of 315 ml/kg foetal weight·min was reported (Dawes, Mott & Widdicombe, 1954); an output of 362 ml/kg·min was recorded using indicator dilution methods (Mahon, Goodwin & Paul, 1966); and with use of electromagnetic flowmeters the combined output was 235 ml/kg·min excluding coronary flow (Assali, Morris & Beck, 1965). In acute studies of foetal lambs *in utero*, umbilical blood flow was estimated by the steady state diffusion of antipyrine using the Fick method, and distribution of cardiac output was measured with radionuclide-labelled plastic microspheres. Cardiac output was similar in relation to foetal weight in foetuses from about 60 days' (0.4) gestation to term, and averaged about 525 ml/kg·min (Rudolph & Heymann, 1970). More recently we have measured combined ventricular output in chronic foetal lamb preparations using radionuclide-labelled microspheres. These studies show that the cardiac output was slightly lower than in the acute preparations, averaging 500 ml/kg·min (Heymann & Rudolph, 1972). Although the microsphere method has been shown to be most reliable in measuring blood flow, it suffers from the disadvantage that only limited observations can be made at individual points in time.

We have developed a different model for studying ventricular output continuously for long periods *in utero*. Polyvinyl catheters were inserted into a hindlimb artery and vein, a carotid artery and jugular vein and the trachea, in foetal lambs of 105–128 days' gestation. A uterine incision was made over the left thorax, and a thoracotomy performed in the third intercostal space. The main pulmonary trunk just beyond the valve was isolated and a cuff-type precalibrated electromagnetic flow transducer (Statham SP 2202) placed around it. A polyvinyl catheter was also inserted into the right ventricle, and two platinum electrodes were passed through the pericardium onto the left atrium for electrical pacing. In some lambs, platinum wires were also positioned on the left vagus nerve in the neck. The chest was closed, and a catheter left in the thoracic and amniotic cavities. All catheters, the wires, and the flow transducer leads were passed through the abdominal wall at the ewe's flank, and protected by a teflon cloth patch. The uterine and abdominal incisions were sutured and the sheep allowed to recover. Repeated observations were made for periods of 5–21 days after surgery.

Foetal arterial pressures and blood gases were in what was considered to be the normal range on the day after surgery. However, right ventricular output increased over the first 2–5 days by 20 to 100% of the level recorded at the time of surgery. After the rapid postoperative rise, there was a more gradual increase which presumably was related to foetal growth. This was not associated with any increase in heart rate, but with

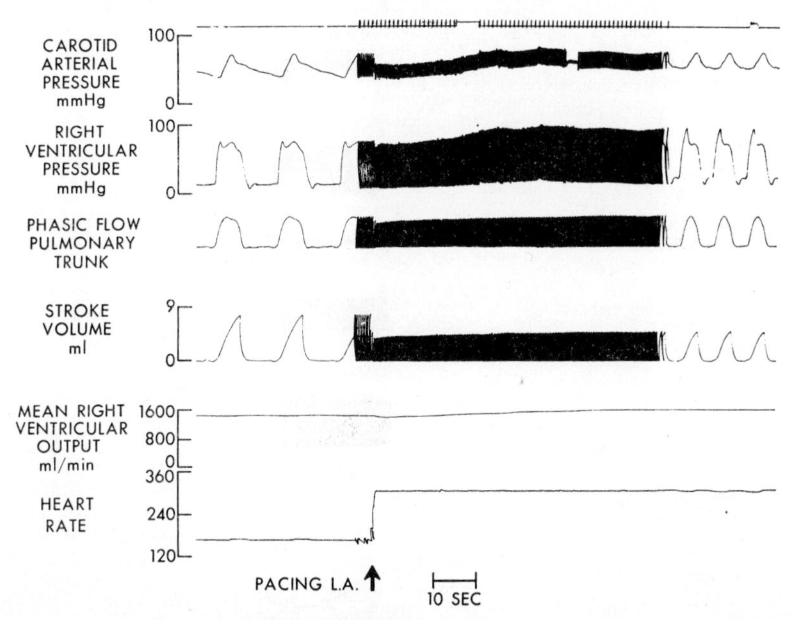

Fig. 4. Response of the circulation to left atrial pacing in a 125-day gestation foetal lamb. When heart rate is increased to 300/min there is an increase in right ventricular output in spite of a marked fall in stroke volume.

an increase in stroke volume. Right ventricular output showed sudden spontaneous variations. In many instances, these changes were associated with onset of respiratory movements. There was usually a small decrease, followed by a small increase, but even though respiratory movements continued, flow returned to control levels within 15–30 seconds. Most of the spontaneous changes in flow were associated with changes in heart rate, increasing with tachycardia and decreasing with bradycardia. In view of these close relationships between cardiac rate and right ventricular output noted, we examined the effects of left atrial pacing.

EFFECT OF HEART RATE ON VENTRICULAR OUTPUT AND STROKE VOLUME

The resting heart rate *in utero* usually ranged from 150 to 180 per minute. The left atrium was stimulated with pulses of 2 msec and voltages of 4–8 V at rates of 180–330/min. In all lambs studied, right ventricular output rose as rate was increased to levels of 240–260/min. No further increase occurred to rates of 285–315/min but levels of output were maintained (Fig. 4). With rates above 300–325/min, there was a decrease in ventricular output.

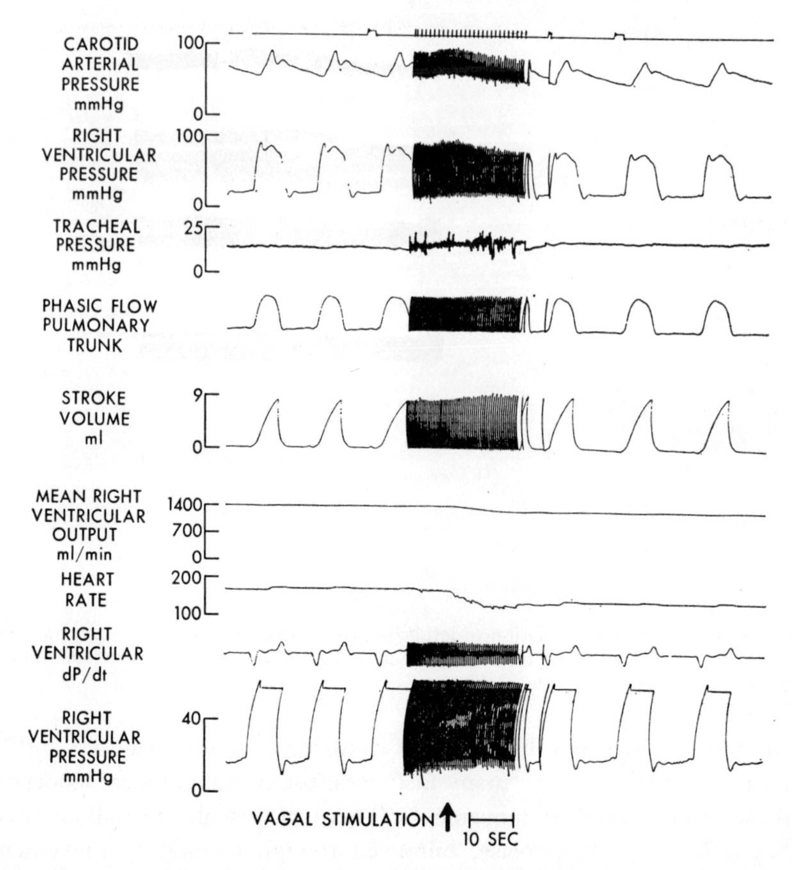

Fig. 5. Response of the circulation to cervical vagal stimulation in a 125-day gestation foetal lamb. Stimulation of the left vagus resulted in a fall in heart rate. Although stroke volume increased slightly, it was not adequate to compensate for the slowing and right ventricular output fell.

The heart rate was slowed by stimulating the left vagus at a frequency of 15/sec with pulses of 2–3 msec, 5–25 V. Rate could be decreased to and maintained at levels of 115–125/min at maximal stimulation. The right ventricular output decreased markedly in association with the fall in heart rate (Fig. 5). A typical pattern of the relationship between varying heart rates and right ventricular output is shown in Fig. 6.

Right ventricular stroke volume was measured by electronically integrating the flowmeter recording. As heart rate was increased above 180/min, stroke volume fell and at rates above 300/min, it fell precipitously. With reduction of heart rate below 160–180/min, there was only a small increase in stroke volume. The effects of heart rate on stroke volume are shown in Fig. 7 depicting a single study which is typical of all experiments.

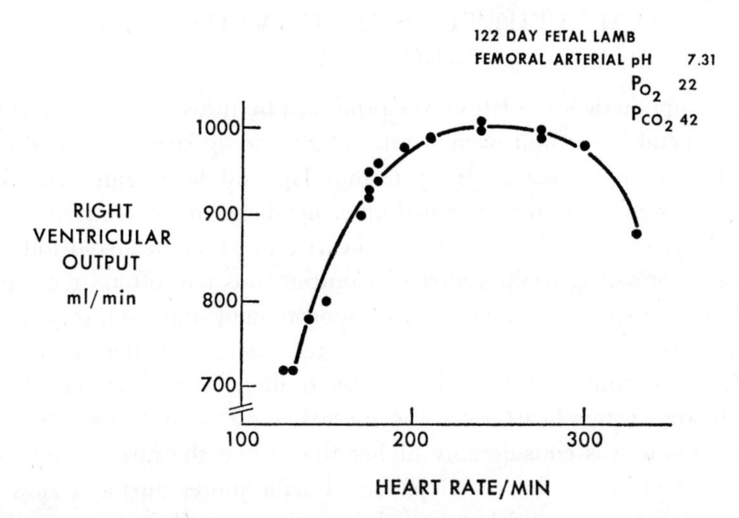

Fig. 6. The relationship between right ventricular output and heart rate is shown, demonstrating that output increases markedly as rate is increased from 125 to 200/min. A rapid fall in right ventricular output occurs at rates above 300/min.

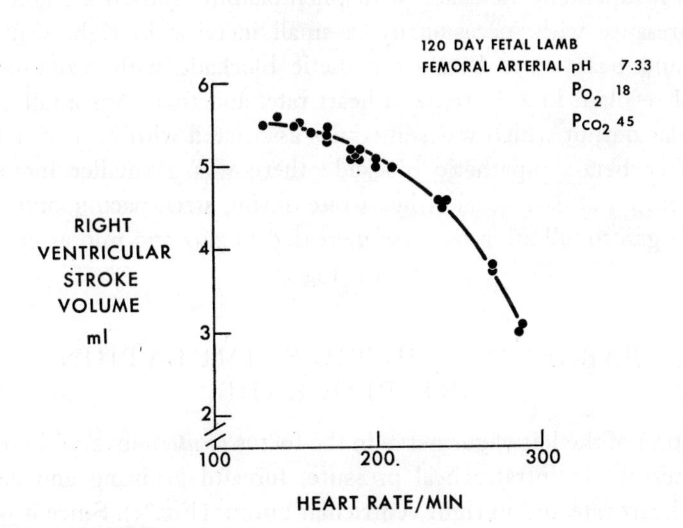

Fig. 7. The relationship between right ventricular stroke volume and heart rate is shown in a 120-day gestation foetal lamb. In the range between 125 and 180/min the stroke volume is relatively constant. With a further increase in heart rate it falls rapidly.

EFFECTS OF ALPHA- AND BETA-SYMPATHETIC STIMULATION AND BLOCKADE

Alpha-sympathetic stimulation was produced by infusion of methoxamine into the foetal lower limb vein at rates of 25–100 μg/kg·min. Foetal mean arterial pressure increased by 5–10 mm Hg and heart rate fell. Right ventricular stroke volume remained unchanged and in some instances even fell slightly, in spite of the decreased heart rate which occurred following the rise in pressure. Right ventricular output thus fell, often by 25–30%.

Beta-sympathetic stimulation by isoproterenol infusion (0.5–2.0 μg/kg·min) accelerated the heart rate and resulted in an increase in right ventricular output. Stroke volume was quite well maintained. In the same foetus, when heart rate was increased by isoproterenol infusion, stroke volume was considerably higher than when the rate was increased to the same level by left atrial pacing. Furthermore, during a slow isoproterenol infusion, which resulted in only a small increase in heart rate, left atrial pacing resulted in a larger cardiac output with higher stroke volume than occurred with left atrial pacing alone, at any level of heart rate.

Alpha-sympathetic blockade with phentolamine caused a slight fall in blood pressure with, occasionally, a small increase in right ventricular output and heart rate. Beta-sympathetic blockade with propranolol or practalol resulted in a decrease in heart rate, and there was a fall of right ventricular output which was sometimes associated with heart rate reduction. After beta-sympathetic blockade there was a smaller increase in cardiac output at each level of heart rate during atrial pacing, and cardiac output began to fall after rate was increased to 240–270/min as compared with the usual level of 305–325/min (Fig. 8).

PARASYMPATHETIC STIMULATION AND BLOCKADE

Stimulation of the left vagus nerve in the foetus *in utero* invariably resulted in an increase in intratracheal pressure, forceful grunting and gasping, a fall in heart rate and in right ventricular output (Fig. 5). Since it was not possible to assess the direct effects of vagal stimulation separately from those due to the vigorous respiratory changes, the foetus was paralysed temporarily with gallamine (2 mg/kg). Vagal stimulation produced a reduction in heart rate which at maximal stimulation fell to about 120–125/min from resting levels of 150–180/min. In an attempt to determine

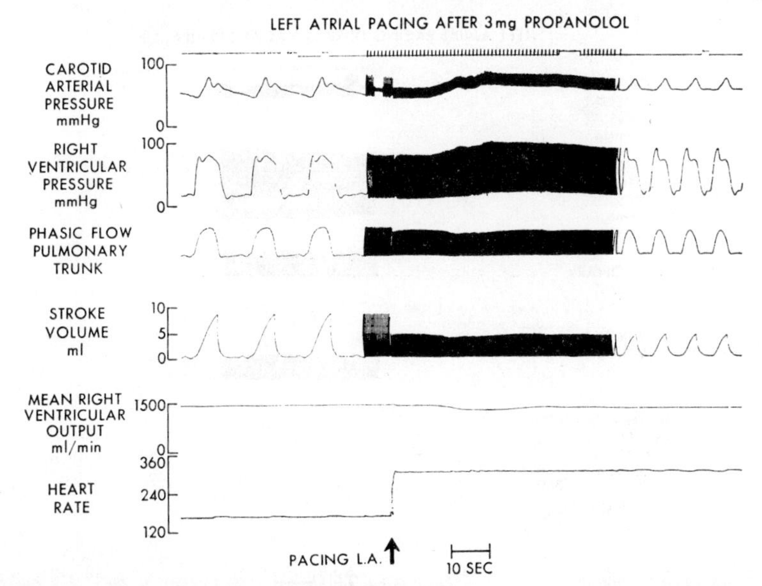

Fig. 8. The effects of beta-adrenergic blockade with propranolol during left atrial pacing are shown in the same foetus as depicted in Fig. 4. After blockade left atrial pacing at 300/min results in a decrease in right ventricular output.

the effect of vagal stimulation on myocardial contractility, right ventricular output and stroke volume were measured at the same heart rates produced by atrial pacing, both with and without vagal stimulation. Stroke volumes and ventricular output were consistently lower at each heart rate during vagal stimulation, suggesting that vagal stimulation had a direct negative inotropic effect on the ventricular myocardium (Fig. 9).

Parasympathetic blockade with atropine resulted in a variable small increase in heart rate and ventricular output.

EFFECTS OF INCREASED OUTFLOW IMPEDANCE

In some foetal lambs, a balloon catheter was positioned in the descending aorta after insertion through a femoral artery. Inflation of the balloon adequate to produce even small increases in carotid arterial pressure invariably resulted in a decrease in heart rate and right ventricular output and stroke volume. When the decreased heart rate was obliterated by blocking the baroreflex with atropine, a decrease in stroke volume was noted as long as the obstruction was maintained.

In one lamb in which the flow transducer produced a modest constriction of the pulmonary trunk at the time of surgery, the flow increased

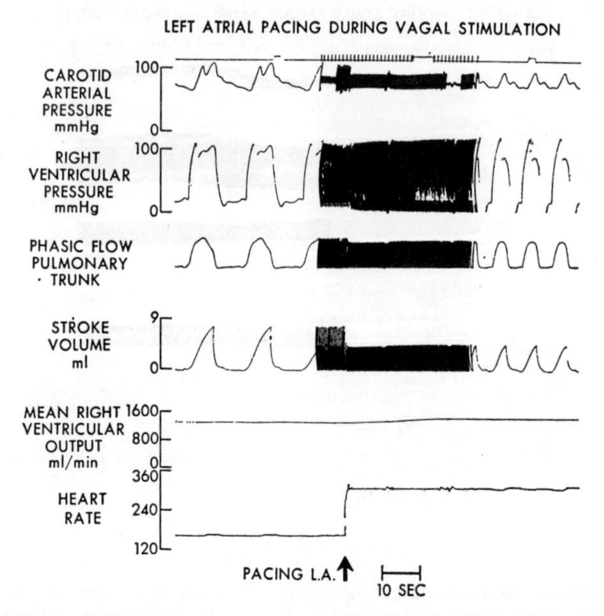

Fig. 9. The effects of vagal stimulation during left atrial pacing are shown in the same foetus as in Figs. 4 and 8. Resting heart rate and right ventricular output are lower and during pacing there is a smaller increase in right ventricular output as compared to the control as seen in Fig. 4.

during the first 7 days from 750 to 1000 ml and then gradually fell to 360 ml over the next 10 days, when the lamb was delivered by Caesarian section. Heart rate did not change significantly over this period, but right ventricular stroke volume first increased from 4.2 to 6.0 ml and then fell to 2.4 ml. The pulmonary trunk was found to be constricted by about 30% by the flow transducer in this lamb.

These findings suggest that both acutely and chronically, an increase in impedance to the ventricle results in a decreased stroke volume and ventricular output.

TRANSMURAL PRESSURE RELATIONSHIPS IN THE FOETAL AND NEONATAL HEART

We examined the pressure relationships between the right ventricular cavity, the intrapleural space, the trachea and the amniotic cavity in several chronic preparations in standing sheep. In one lamb, we also had the opportunity to study these relationships both before and after birth. There is a dramatic change in transmural pressure across the ventricle at birth. *In utero* the intrapleural pressure is only 1–2 mm Hg lower than right ventricular pressure and 2–3 mm Hg higher than amniotic pressure. Immediately after birth, with spontaneous breathing, right ventricular end diastolic pressure is 5–8 mm Hg higher than intrapleural pressure which in turn is 2–3 mm Hg lower than atmospheric pressure at end expiration.

The effects of inspiration were also observed *in utero* in one foetus that made deep inspiratory efforts occasionally, while all these parameters were being monitored (Fig. 10). A decrease of intrapleural pressure of 28 mm Hg was not associated with any drop of right ventricular end diastolic pressure. Immediately following this, right ventricular and aortic systolic pressure rose markedly, and right ventricular stroke volume increased slightly, indicating a large increase in right ventricular stroke work, presumably as a result of the Frank–Starling mechanism. These variations in transmural pressures only occur intermittently *in utero* and are probably not functionally significant. However, the dramatic changes which occur at birth may be important in postnatal circulatory adaptation.

CIRCULATORY RESPONSES TO HYPOXAEMIA AND ACIDAEMIA

Although the effects of asphyxia on the foetal circulation have been examined by several investigators, most reports deal with studies on exteriorized foetuses. In view of the criticisms mentioned above, we have assessed the effects of hypoxaemia alone and of hypoxaemia and acidaemia in chronic preparations in foetal lambs at 120–138 days' gestation (Cohn, Sacks, Heymann & Rudolph, 1972). Administration of low oxygen gas mixtures to the ewe produced hypoxaemia alone in some lambs and hypoxaemia and acidaemia in others. Foetal arterial P_{O_2} fell to 12–13 mm Hg, and pH in the group in which it changed fell to 7.27. Heart rate fell from the onset, and no period of tachycardia, as has previously been reported, was noted. Cardiac output and its distribution were measured with radionuclide-labelled microspheres. There was a consistent rise in arterial pressure and fall in combined ventricular output, greater in those animals with acidaemia.

The reduction of cardiac output was related to the fall in heart rate and calculated stroke volume did not change. The umbilical blood flow was maintained at control levels, the reduction in cardiac output occurring entirely in blood flow to the foetal body. We found, as has been previously reported in acute exteriorized preparations (Campbell, Dawes, Fishman & Hyman, 1967), that there was marked peripheral vasoconstriction, with increase of flow to the myocardium and maintenance of brain flow, and

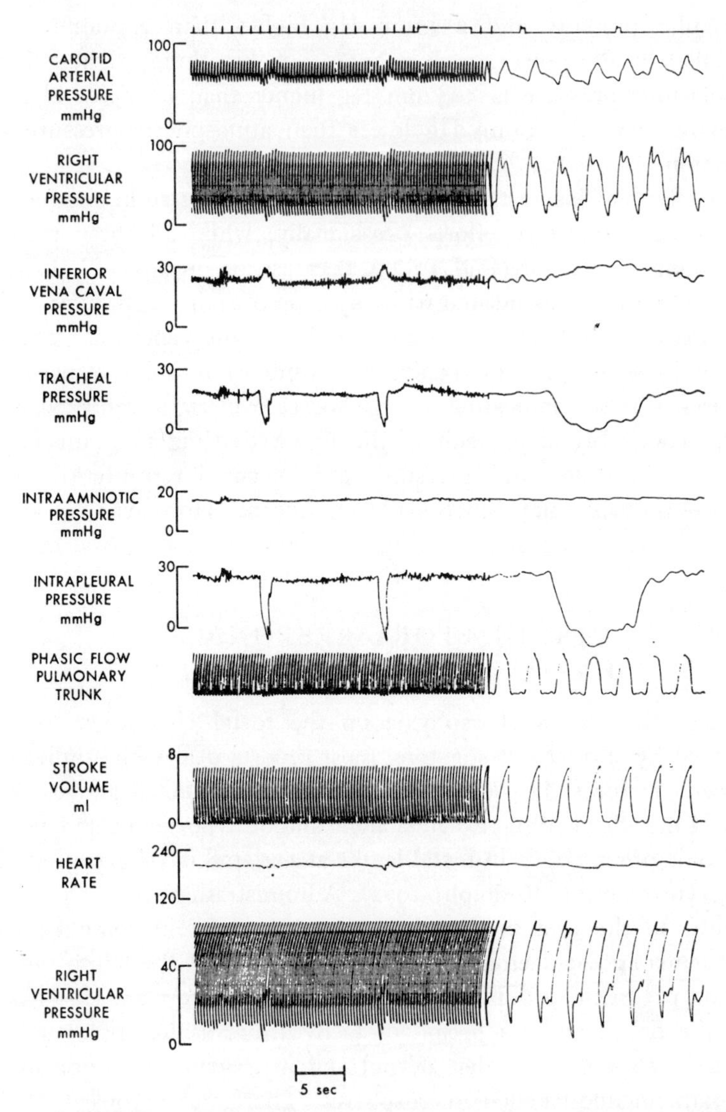

Fig. 10. Effects of intrauterine inspiratory effort on circulatory dynamics. With deep inspiration the intrapleural pressure fell from a resting level of 23 mm Hg to −5 mm Hg. It is evident that right ventricular end-diastolic pressure does not drop accordingly so that transmural pressure across the right ventricle is markedly increased. During the expiratory phase there is an increase in right ventricular end-diastolic and carotid arterial and right ventricular systolic pressures.

marked pulmonary vasoconstriction. We also noted a marked increase in blood flow to the adrenal gland. The circulatory response to hypoxaemia is thus largely a redistribution of blood flow in order to maintain respiratory exchange, and vital organ function.

CONCLUSIONS

Analysis of the available data indicates that cardiac output in the foetus is altered almost exclusively by changes in heart rate. The foetal ventricles show little capability of increasing stroke volume when heart rate is decreased, so that cardiac output falls. This limitation on increase in stroke volume also results in a decreased cardiac output when vascular impedance is increased. The reason for these differences in the foetal heart as compared to the adult heart are not as yet clearly defined. We have demonstrated that stroke volume may be increased by rapid volume loading, and also that infusion of isoproterenol produces an increased stroke volume at all heart rates. It is possible that the poor response of the foetal myocardium could be related to immaturity of sympathetic innervation to the ventricles.

Whereas stroke volume tends to fall in the adult when heart rate is increased above resting levels, the foetus maintains stroke volume, so that cardiac output is increased; cardiac output falls in the adult sheep when heart rate exceeds about 160–180/min, but it does not fall in the foetus until rates of 300–325/min are reached.

In the adult animal, the Frank–Starling mechanism serves the primary function of providing a balance between the outputs of the left and right ventricles to maintain respective circulatory volumes. Since the two ventricles tend to function in parallel in the foetus, this mechanism is not particularly important. Foetal ventricular myocardium does demonstrate a similar qualitative length–tension relationship as in the adult, but absolute quantitative correlations have not yet been established. The changes in cardiac transmural pressure which occur with lung expansion may be important in maintaining stroke volume in the face of the sudden increased impedance associated with elimination of the placental circulation.

Hypoxaemia and acidaemia result in a rise in arterial pressure and a redistribution of the cardiac output which favours maintenance of umbilical placental flow and of myocardial, adrenal and cerebral blood flow. This response is partly related to local effects, but is probably mainly mediated through aortic chemoreceptor reflexes. The combined ventricular output falls during hypoxaemia, largely due to the decreased heart rate observed in late gestation foetal lambs. Since the decrease in heart rate in moderate hypoxaemia can be overcome with atropine, it appears that it is associated with a baroreflex response. This raises the interesting question as to the role of mechano-receptors in the foetal circulation. The resultant decrease in heart rate and cardiac output and reflex peripheral vasodilation could be deleterious to the foetus. The baroreflex matures in late gestation and it may be a mechanism which is developing for postnatal circulatory adjustments, but which does not have a useful function during foetal life. Further studies are needed to clarify the role of the mechanoreceptors in the foetal circulation.

This research was supported by PHS Grant HL 06285 from the National Heart and Lung Institute.

REFERENCES

Assali, N. S., Morris, J. A. & Beck, R. (1965). *Am. J. Physiol.* **208**, 122.
Barcroft, J. & Barron, D. H. (1945). *J. Exp. Biol.* **22**, 63.
Barrett, C. T., Heymann, M. A. & Rudolph, A. M. (1972). *Amer. J. Obstet. Gynec.* **112**, 114.
Bassett, J. M. & Thorburn, G. D. (1969). *J. Endocr.* **44**, 285.
Biscoe, T. J., Purves, M. J. & Sampson, S. R. (1969). *J. Physiol.* (*London*) **202**, 1.
Campbell, A. G. M., Dawes, G. S., Fishman, A. P. & Hyman, A. L. (1967). *Circulation Res.* **21**, 229.
Cohn, H. E., Sacks, E. J., Heymann, M. A. & Rudolph, A. M. (1972). To be published.
Comline, R. S. & Silver, M. (1966). *Brit. Med. Bull.* **22**, 16.
Dawes, G. S. (1968). *Foetal and Neonatal Physiology*. Chicago: Year Book.
Dawes, G. S. (1969). In *Foetal Autonomy*, Ciba Foundation Symposium, ed. Wolstenholme, G. E. W. and O'Connor, M. London: Churchill.
Dawes, G. S., Fox, H. E., Leduc, B. M., Liggins, G. C. & Richards, R. T. (1972). *J. Physiol.* **220**, 119.
Dawes, G. S., Mott, J. C. & Rennick, B. R. (1956). *J. Physiol.* (*London*) **134**, 139.
Dawes, G. S., Mott, J. C. & Widdicombe, J. G. (1954). *J. Physiol.* (*London*) **126**, 563.
Friedman, W. F., Cooper, C., Pool, P., Jacobowitz, D. & Braunwald, E. (1969). *Proceedings Am. Pediatric Society*, **79**, 1.
Friedman, W. F., Pool, P. E., Jacobowitz, D., Seagren, S. C. & Braunwald, E. (1968). *Circulation Res.* **23**, 25.
Goodlin, R. C. & Rudolph, A. M. (1972). In *Proceedings of the International Symposium on Physiological Biochemistry of the Fetus*, ed. Hodari, A. A. and Mariona, F. G. Springfield: Thomas.
Gresham, E. L., Rankin, J. H. G., Makowski, E. L., Meschia, G. & Battaglia, F. C. (1972). *J. Clin. Invest.* **51**, 149.
Heymann, M. A. & Rudolph, A. M. (1967). *Circulation Res.* **21**, 741.
Heymann, M. A. & Rudolph, A. M. (1972). To be published.
Lebowitz, E. A., Novick, J. S. & Rudolph, A. M. (1972). To be published.
Lipp, J. A. M. & Rudolph, A. M. (1972). *Biology of the Neonate*. In press.
McMurphy, D. M. & Boréus, L. O. (1971). *Amer. J. Obstet. Gynec.* **109**, 937.
McMurphy, D. M., Heymann, M. A., Rudolph, A. M. & Melson, K. L. (1972). *Pediatric Res.* **6**, 231.
Mahon, W. A., Goodwin, J. W. & Paul, W. M. (1966). *Circulation Res.* **19**, 191.
Purves, M. J. & Biscoe, T. J. (1966). *Brit. Med. Bull.* **22**, 56.
Rudolph, A. M. & Heymann, M. A. (1967). *Circulation Res.* **21**, 163.
Rudolph, A. M. & Heymann, M. A. (1970). *Circulation Res.* **26**, 289.
Rudolph, A. M., Heymann, M. A., Teramo, K. A. W., Barrett, C. T. & Räihä, N. C. R. (1971). *Pediatric Res.* **5**, 452.

SHINEBOURNE, E. A., VAPAAVUORI, E. K., WILLIAMS, R. L., HEYMANN, M. A. & RUDOLPH, A. M. (1972). To be published.
SILVA, D. G. & IKEDA, M. (1971). *J. Ultrastructure Res.* **34**, 358.
SMYTH, H. S., SLEIGHT, P. & PICKERING, G. W. (1969). *Circulation Res.* **24**, 109.
TERAMO, K A. W., RUDOLPH, A. M. & HEYMANN, M. A. (1972). To be published.
VAPAAVUORI, E. K., SHINEBOURNE, E. A., WILLIAMS, R. L., HEYMANN, M. A. & RUDOLPH, A. M. (1972). To be published.
WILLIAMS, R. L., HOF, R., HEYMANN, M. A. & RUDOLPH, A. M. (1972). To be published.

THE STARLING RESISTOR MODEL IN THE FOETAL AND NEONATAL PULMONARY CIRCULATION

By S. CASSIN

Department of Physiology, College of Medicine,
University of Florida, Gainesville, Florida 32601, U.S.A.

The fact that hypoxia or asphyxia may cause a rise in pulmonary arterial pressure has been known for a long time (Daly & Hebb, 1966). However, until 1946, it was not clear whether this rise was due to an increase in cardiac output or to a local pulmonary vasoconstriction. Von Euler & Liljestrand, in 1946, published critical experiments on anaesthetized adult cats which demonstrated that hypoxia produced a significant rise in pulmonary arterial pressure and attributed this to pulmonary vasoconstriction. Since then, many papers have been published which have established with reasonable assurance that the vasculature of the lung constricts in response to a local reduction of oxygen in intact as well as isolated preparations. There is also evidence that the response may be mediated (Daly & Hebb, 1966) through peripheral chemoreceptors. The response of foetal pulmonary vessels to reduced oxygen tension is very similar to that of the adult (Dawes, 1968; Rudolph & Heymann, 1968). Expansion of foetal lungs with any gas causes a reduction in pulmonary artery pressure and an increase in pulmonary artery flow even though there may be no changes in blood gas composition (Cassin *et al.* 1964). It is interesting, however, that expansion of foetal lungs with liquid (saline or dextran) containing a reduced oxygen tension does not produce a decrease in pressure or an increase in flow as seen after expansion with air, but results in no change or a decrease in pulmonary blood flow (Dawes *et al.* 1953; Lauer *et al.* 1965). The reduction of foetal pulmonary vascular resistance upon expansion of the lungs with gas has been shown to be due to: (*a*) mechanical events which accompany expansion of the lung, and (*b*) local effects on the pulmonary vasculature of increased oxygen or decreased carbon dioxide tension. In addition, Campbell, Cockburn, Dawes & Milligan (1967) have shown that alterations in gas tensions of the blood perfusing unventilated lungs will produce changes in pulmonary vascular resistance.

Very little has been reported concerning efforts to localize sites of changes in vascular resistance (Gilbert *et al.* 1972) in foetal and neonatal lungs in response to ventilation, changes in blood–gas tensions and drugs. However, studies on pulmonary circulation of adult animals have been described which were concerned with distribution of vascular resistance as well as the localization of sites sensitive to hypoxia (Bergofsky *et al.* 1968). Similar determinations have not been made for foetal and newborn lungs.

The action of drugs on the pulmonary circulation and their relationship to the pulmonary vascular response to hypoxia have been studied in both foetal and adult animals. Single injections of 0.05–0.2 μg of norepinephrine or epinephrine directly into the pulmonary artery of immature foetal lambs produced a marked pulmonary vasoconstriction (Cassin, Dawes & Ross, 1964). In adult lungs epinephrine also causes vascular constriction which has been localized to the venous segment (Gilbert *et al.* 1958). In both mature and immature foetal lambs, vasodilator agents, such as acetylcholine, histamine and bradykinin, cause a pronounced increase in pulmonary blood flow. In mature lambs the increase in flow is as great as the increase subsequent to expansion of the lungs with air (Dawes, 1968). Yet when these drugs are injected into the pulmonary arteries of foetal lambs and goats immediately after they have been ventilated with air, and in which pulmonary blood flow has already increased, they have little or no action. In short, these drugs are powerful vasodilators in unventilated vasoconstricted lungs, but have little or no effect on ventilated dilated lungs. In adult intact dogs acetylcholine has in fact been shown to constrict pulmonary vasculature (Hyman, 1969).

Bradykinin decreases pulmonary vascular resistance in isolated as well as in intact adult dog lungs and normal human adult lungs (De Freitas *et al.* 1964; De Pasquale *et al.* 1969; Waaler, 1961). The unventilated foetal lamb lung has been shown to be exquisitely sensitive to minute doses (nanogram quantities) of bradykinin injected directly into the pulmonary artery (Dawes, 1968). Heymann *et al.* (1969) have suggested that bradykinin is the mediator for the profound fall in pulmonary vascular resistance consequent to ventilation of foetal animals. On the other hand, Hyman (1968) has studied the effects of bradykinin on the pulmonary veins of adult dogs; he has suggested that bradykinin causes an active constriction of pulmonary veins and that vessels upstream to the veins, presumably the arteries, are passively distended in response to an increase in cardiac output. No one has attempted to localize the site of action of these drugs on the foetal and neonatal pulmonary circulation.

Over the years various models have been designed to describe flow through adult pulmonary vasculature. Of special interest to us has been a model described by Bannister & Torrance (1960), modified by Lopez-Muniz *et al.* (1968) and Permutt *et al.* (1962), which suggests that blood flow to adult lungs behaves as if it were controlled by vessels acting as Starling resistors. A Starling resistor is an easily collapsible tube through which flow is related to the pressure drop between inflow and the surrounding pressure rather than the pressure drop between inflow and outflow pressure, as long as the surrounding pressure is greater than outflow pressure. In 1967, MacDonald & Butler devised a model involving Starling resistors in order to analyse the distribution of vascular resistance in the adult pulmonary circulation. The present studies were designed to characterize, by utilizing the concept of Starling resistor, sites of change in foetal pulmonary vascular resistance in response to changes in pulmonary blood gases, gaseous expansion of the lungs and injected drugs. We used a modification of the MacDonald & Butler (1967) model which permits us to calculate resistances in two segments of the pulmonary vasculature. Resistance was calculated in a segment proximal to the presumed Starling resistors as well as in the segment distal to them. The calculated resistances in these two segments were then used to determine sites of change in pulmonary vascular resistance in each of the experimental conditions mentioned. The work presented is the result of a collaborative effort by S. Cassin, R. Gilbert, J. Hessler & D. Eitzman over the past few years.

Complete details of the physical model which we used to set up our analyses have been described elsewhere (Gilbert *et al.* 1972).

ANIMAL MODEL

Pregnant goats close to term were anaesthetized with chloralose and foetuses were delivered by Caesarean section as described previously (Gilbert *et al.* 1972). Care was taken to keep the foetuses from breathing air and to prevent heat loss. The left carotid artery was connected by means of a cannula to the pulmonary artery of the lower lobe of the left lung (Fig. 1). Flow from the carotid to the lung was controlled by way of a finger pump. Distal to the finger pump sites were provided for measurement of pressure and flow. The pulmonary vein from the lower lobe of the left lung was cannulated through the left atrium. Blood from the lower left lung was collected in a reservoir. The upper and middle lobes were essentially isolated from the circulation and the vessels to these segments tied off. Provision was made to measure flow and pressure in the outflow cannula from the lower left lobe. The reservoir into which the flow from the lower left lobe passed was raised or lowered by a pulley. Blood from the reservoir passed through a heat exchanger and then by way of a finger pump to the femoral artery of the animal. Pulmonary artery pressure

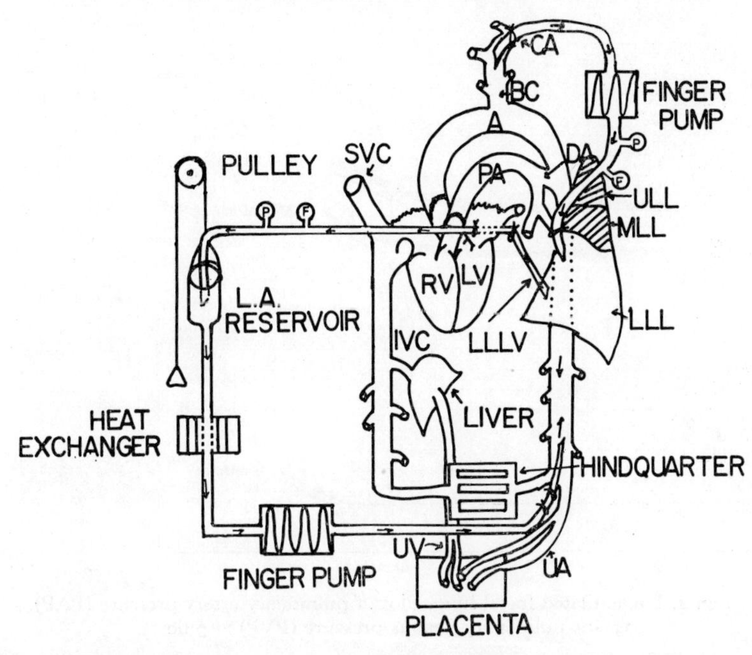

Fig. 1. Isolated perfused lower lobe of left lung in foetal goat (CA, carotid artery; ULL, MLL, LLL, upper, middle and lower left lobe respectively).

was plotted against pulmonary venous pressure on a Grass polygraph as well as on a Houston XY recorder. A curve was generated by slowly raising the pulmonary venous reservoir from a position below P_S (the pressure which tends to close the Starling resistor) while pulmonary artery flow was held constant. An example of such a plot is presented in Fig. 2. Throughout the procedure blood gases and pH were analysed on a Radiometer blood gas analyser. A set of curves was drawn first for the unventilated foetal lung with tracheal pressure held at atmospheric pressure. The curves were drawn at zero flow, then at approximately 37 ml/min and at 55 ml/min flow. The animals were subsequently ventilated with specific gas mixtures or given drugs and another set of curves drawn after the experimental procedure. Resistances proximal and distal to the vessels acting as Starling resistors were calculated as for the model (Gilbert *et al.* 1972). In the ventilated animals measurements were made of resistance at zero end expiratory pressures. Curves obtained from the animals, in contrast to those of the model, did not exhibit a very sharply defined break point. Thus, the break point for each of the curves obtained from the animals was arbitrarily defined as the point on each curve which was raised 1 mm above the base line for that curve. We believe that the curves obtained from the animals did not demonstrate a sharp break point

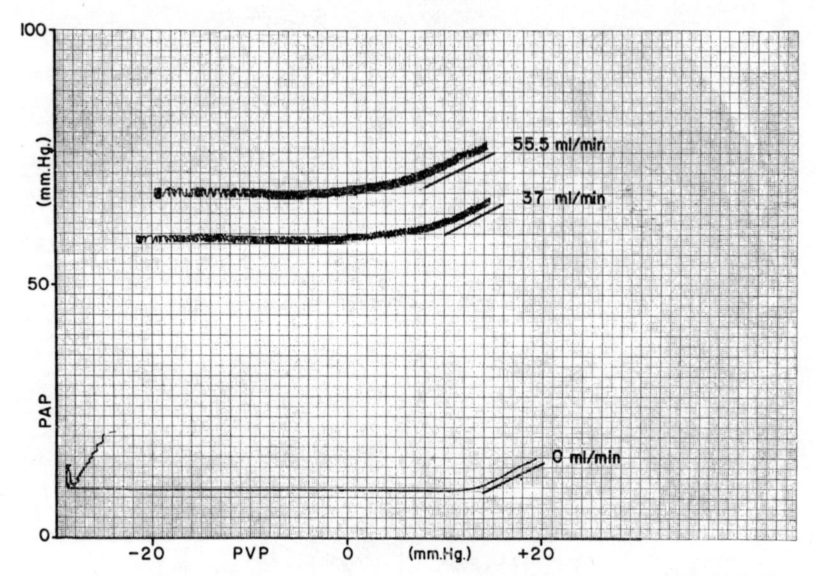

Fig. 2. Unventilated foetal lung. Plot of pulmonary artery pressure (PAP) versus pulmonary venous pressure (PVP) at 3 flows.

because of the existence of multiple P_S values in the pulmonary circulation. The method which we used for determining the break point in the curves obtained from animal studies provides a value for P_S which is not the lowest, but is in the low range of spectrum of all the P_S values in the pulmonary circulation.

THE EFFECT OF VENTILATION ON RESISTANCE PROXIMAL AND DISTAL TO THE STARLING RESISTOR

In these experiments a set of curves was drawn first for the unventilated foetal lung. Another set of curves was obtained when the lungs were expanded with a gas mixture intended not to change the blood gas of the foetus (4% oxygen and 6% carbon dioxide in nitrogen), the foetal gas mixture. The animals were ventilated with positive pressure in such a fashion that the peak inspiratory pressure was never greater than 30 cm of water. Expiratory pressure was equal to atmospheric pressure. Finally, a third set of curves was obtained when the lungs were expanded with air. In each of the three states described proximal resistance was larger than distal resistance (Fig. 3).

Upon ventilation with the foetal gas mixture, proximal resistance decreased significantly from 3.78 to 2.95 mm Hg·kg·min·ml^{-1}. Distal

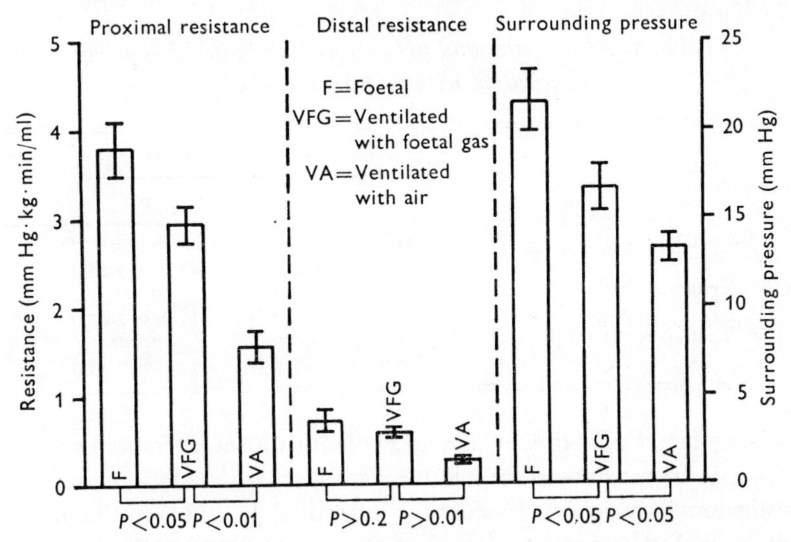

Fig. 3. Calculated proximal and distal resistances in foetal lungs in foetal state, ventilated with foetal gas and ventilated with air.

resistance decreased but not significantly from 0.714 to 0.569 mm Hg·kg·min·ml^{-1}. Upon substitution of air for the foetal gas mixture both proximal and distal resistances fell significantly. Proximal resistance decreased to 1.57 mm Hg·kg·min·ml^{-1}, and distal resistance decreased to 0.262 mm Hg·kg·min·ml^{-1}. The surrounding pressure of P_S was rather high in the unventilated foetal lung, averaging about 21 mm Hg. However, upon ventilation with foetal gas this dropped to about 16 mm Hg. Further ventilation with air decreased P_S significantly to about 13 mm Hg. The mean and the range of values for P_{O_2}, P_{CO_2} and pH in the pulmonary artery for each of the conditions are shown in Table 1. Ventilation with the foetal gas decreased the P_{O_2} and the pH slightly and increased P_{CO_2}. All of these changes should theoretically increase resistance to flow; however, the effects shown in the figure are that of a reduction in resistance. Therefore, the reduction in resistance must be due to mechanical expansion of the lungs and not to changes in blood gases or pH. Further ventilation with air increased the pulmonary arterial P_{O_2} and decreased P_{CO_2}. This, of course, was associated with a marked decrease in resistance in the proximal and distal segments.

At this point, one might question the significance of P_S. In the physical model P_S is clearly a surrounding pressure due to the hydrostatic pressure of water around the Starling resistor. However, any force which would tend to close the Starling resistor could contribute to P_S. Thus, in the animal P_S is probably not a simple hydrostatic pressure surrounding the

Table 1. *Blood gases and pH in foetal state and following ventilation with foetal gas and air*

	Pulmonary arterial blood		
	P_{O_2}	P_{CO_2}	pH
Foetal	25 (20–33)	53 (34–65)	7.202 (7.093–7.285)
Ventilated with foetal gas	21 (15–28)	60 (36–81)	7.121 (7.010–7.271)
Ventilated with air	60 (32–96)	45 (27–67)	7.220 (7.008–7.435)

vessels. In fact, P_S is probably due to a combination of interstitial pressure, alveolar pressure, smooth muscle tone and perhaps other forces. Several investigators have reported values for interstitial pressures in the lungs of adult animals (Levine *et al.* 1967; Meyer, Meyer & Guyton, 1968). Adult animals with closed chest seem to have interstitial pressures of about −9 mm Hg. All animals used in the present study, however, had their chests open. Since interstitial pressure seems to be related to pleural pressure as suggested by Mellins *et al.* (1969), it appears that interstitial pressure in an animal with an open chest might not be as negative as that in an animal with a closed chest. Furthermore, foetal animals are different from adults in that their lungs are filled with liquid rather than air. Again, Mellins *et al.* (1969) have indicated that interstitial pressure in a fluid-filled lung is not as negative as an air-filled lung. Lung interstitial pressure is not greater than atmospheric except during edema formation (Meyer, Meyer & Guyton, 1968). In the fluid-filled lung of open chest foetal goats used in our preparation, P_S averaged about +22 mm Hg. Obviously, P_S does not reflect directly interstitial pressure since this is so much higher than the reported values. However, changes in interstitial pressure probably influence P_S. Previous work has indicated (Lopez-Muniz, 1968) that alveolar pressure may be an important factor in determining P_S. In isolated adult dog lungs if alveolar pressure is less than 5 mm Hg, P_S is independent of the alveolar pressure. However, if the alveolar pressure is greater than 5 mm Hg, P_S is approximately equal to the alveolar pressure. In the experiments reported here all measurements including P_S were made with tracheal pressures at atmospheric or zero in the foetal animals. In animals that were ventilated, measurements of P_S were made at an end expiratory pressure of zero. Clearly in our experiments alveolar pressures did not contribute in any way to P_S. We have suggested above that smooth muscle tone may contribute to P_S. In several

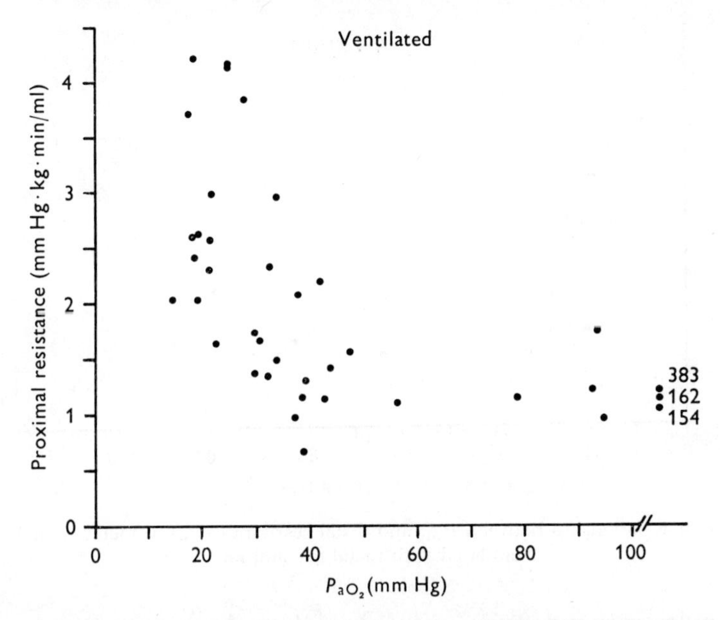

Fig. 4. Relationship between P_{aO_2} and proximal resistance (P_R) in foetal animals ventilated with foetal gas and air.

preliminary experiments injection of 15 mg papaverine hydrochloride directly into the pulmonary artery in unventilated animals produced a significant decrease in P_S. Although P_S diminished considerably, in no instance did the P_S decrease to zero. The lowest value for P_S that we were able to measure after papaverine was +6 mm Hg. We must assume, therefore, that some factor or factors other than interstitial pressure, alveolar pressure or smooth muscle tone must also influence P_S.

The relationship between proximal resistance and pulmonary arterial oxygen tension for animals ventilated with foetal gas and ventilated with air is shown in Fig. 4. Proximal resistance decreases until a critical P_{O_2} of approximately 40 mm Hg is reached and then remains constant. The relationship between proximal or distal resistance and P_{O_2} in unventilated lungs could not be made because of a small range of values for the pulmonary arterial P_{O_2}. Distal resistance, however, is related to venous P_{O_2} in ventilated animals as shown in Fig. 5. The distal resistance tends to decrease as P_{O_2} increases to 40 mm Hg, and then remains constant. We were not able to demonstrate a relationship between proximal or distal resistance and P_{CO_2} or pH.

Total resistance in ventilated foetal animals was calculated as the difference between pulmonary artery and pulmonary venous pressures divided by flow, at a point on a curve of pulmonary arterial pressure

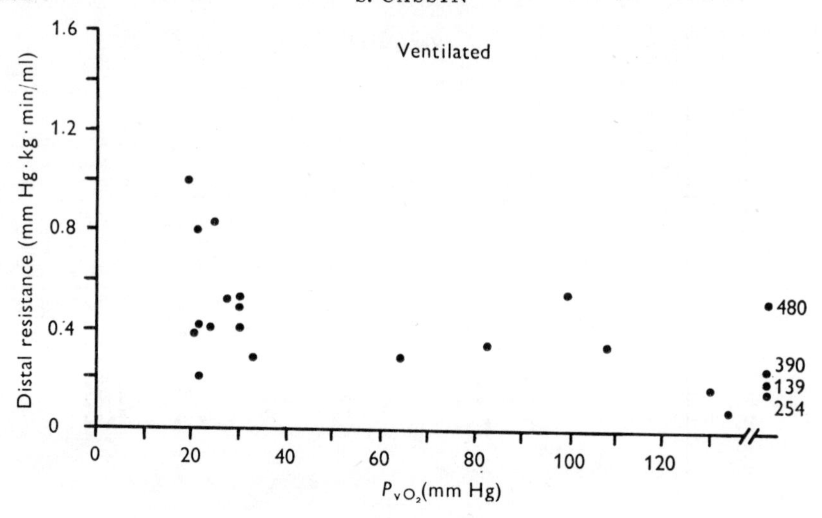

Fig. 5. Relationship between P_{vO_2} and distal resistance (R_D) in foetal animals ventilated with foetal gas and air.

versus pulmonary venous pressure, where the pulmonary venous pressure influenced pulmonary artery pressure in a one-to-one relationship (Gilbert *et al.* 1972). This total resistance should equal the sum of the calculated proximal and distal resistances; they were very close (Fig. 6). The bars labelled T are total and those labelled P+D are the values obtained for proximal and distal resistances in each of the three states. The difference between calculated total resistance and that obtained by adding proximal and distal resistances is not statistically significant. Note that after ventilation the drop in pulmonary artery pressure is due not only to a change in proximal resistance, but to a change in proximal resistance and surrounding pressure. Upon ventilation of the lungs the pressure inside the Starling resistor or P_S dropped to 13.3 mm Hg. Because of this decrease, one might anticipate that the pulmonary artery pressure would decrease also. However, the pulmonary artery pressure dropped to a larger extent than can be accounted for by the decrease in P_S, indicating that something else had to decrease. If one considers the equation relating flow to proximal resistance and P_S, it is clear PAP $= R_P\dot{Q}+P_S$ (where R_P is proximal resistance). Therefore, to account for the greater diminution in pulmonary artery pressure than can be accounted for by the fall in P_S, one must assume that both flow ($\dot{Q}$) and proximal resistance must have decreased.

Of particular interest is the actual location of vessels which behave as if they are Starling resistors. At this time we cannot say where they are in the pulmonary circulation, although we can make some guesses as to their approximate locations. When alveolar pressure in the adult animal is

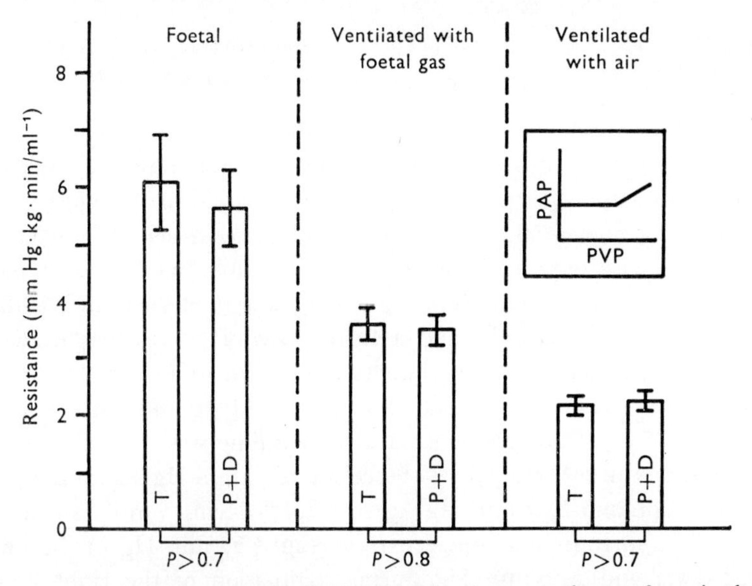

Fig. 6. Comparison of calculated total resistance and sum of proximal and distal resistances in various states.

higher than 5 mm Hg, as shown by Lopez-Muniz *et al.* (1968), the vessels acting as Starling resistors are probably those exposed to alveolar pressure (presumably the capillaries and small arterioles and venules). Since in our experiment measurements were made when alveolar pressure was atmospheric the vessels acting as Starling resistors may not necessarily be the same ones as those when alveolar pressure is greater than 5 mm Hg. There is some evidence, however, which suggests that Starling resistors may be located in vessels distal to the capillaries in both fluid-filled and ventilated lung (Ross, 1963). Fluid in foetal lungs is produced primarily by filtration at the capillary level. Capillary pressure high enough for filtration would be possible if Starling resistors were distal to the capillaries. Filtration would be difficult to explain if the Starling resistors were proximal to the capillaries. In comparing the percentage contribution of proximal resistance to the total resistance in foetal lungs ventilated with air in our experiments, we found that about 84% of the total resistance was accounted for by proximal resistance and only some 16% by distal resistance. If the distribution of resistance in this group of ventilated animals is the same as has been described for adult dogs by Brody *et al.* (1968), this would suggest that R_P included resistance offered by arteries and capillaries. It might indicate that the Starling resistors are least at the end of the capillaries. However, further investigation is necessary in order to locate more precisely the site of the vessels acting as resistors.

EFFECT OF ALTERATION OF BLOOD GASES ON PROXIMAL AND DISTAL RESISTANCES

In another series of animals the bronchus to the left lung was ligated and cut distal to the tie in order to permit fluid to drain from the lung at atmospheric pressure. The procedure allows ventilation of the right lung while measuring flow to the left unventilated lung. A set of pressure–pressure curves was obtained from the left lower lobe while both lungs were in the unventilated state. The right lung was then ventilated with: (*a*) foetal gas mixture, (*b*) air, and a further set of curves was drawn. The results of this series of experiments are shown in Fig. 7. In seven animals proximal resistance, distal resistance and P_S did not change significantly in the left lower lobe when the right lung was ventilated with the foetal gas mixture. Proximal resistance was 2.59 mm Hg·kg·min·ml^{-1} before ventilation and 2.85 mm Hg·kg·min·ml^{-1} during ventilation with foetal gas. The surrounding pressure P_S averaged 22 mm Hg in the unventilated state and 21.7 mm Hg during ventilation of the right lung with foetal gas. In contrast, when the right lung was ventilated with air, proximal resistance in the left lung decreased significantly ($P < 0.01$) to 1.96 mm Hg·kg·min·ml^{-1}. The distal resistance decreased significantly also ($P < 0.05$) to 0.50 mm Hg·kg·min·ml^{-1}. P_S likewise was reduced significantly to 16.9 mm Hg ($P < 0.01$). It should be noted that ventilation of the right lung with air resulted in an increase in P_{O_2}, and a decrease in P_{CO_2} in the left pulmonary arterial blood in each animal. These data are in agreement with those reported previously (Cassin *et al.* 1964) and are presumably due to a direct effect of the change in gas tension locally. We were not able to distinguish whether the response was due to an increased P_{O_2} or a decreased P_{CO_2} since ventilation of the right lung caused both of these changes to occur. Clearly without a change in blood gas tensions and without mechanical insufflation of the lungs the resistance across the lung does not change.

These experiments also indicate that probably no vasodilator material is released when the lung is ventilated with a gas containing oxygen at the same oxygen tension as is found in foetal arterial blood. This release has been suggested for bradykinin by Heymann *et al.* (1969) on ventilation with 100% oxygen. Since the blood perfusing the left lung is obtained directly from the carotid artery, which was supplied with blood coming from the right lung, it might be anticipated that if there were a dilator substance, such as bradykinin, released by the lung, the left lung would have been dilated by this substance. When only the right lung was ventilated with foetal gas there was no reduction of resistance in the left lung. It is

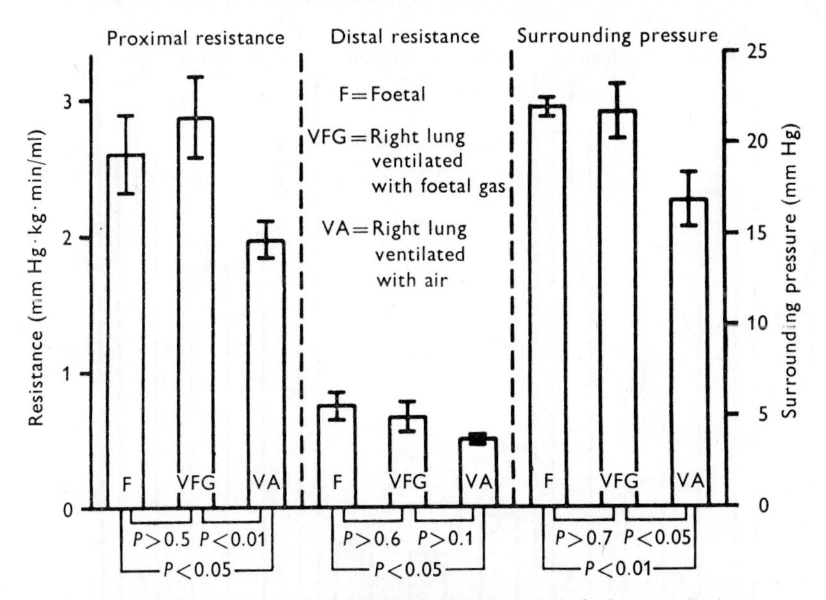

Fig. 7. R_P, R_D and P_S in left lung in foetal state and after ventilation of right lung with foetal gas or air.

possible that with a higher P_{O_2} bradykinin is released and, in fact, it may be possible that part of the response that we saw in terms of reduction in pulmonary vascular resistance in relation to ventilation with air is due to bradykinin. Although bradykinin is an extremely powerful vasodilator, it probably is not entirely responsible for the decreased resistance upon ventilation; we, as well as others, have shown that there are mechanical effects accompanying ventilation which played a part in reducing pulmonary vascular resistance.

THE EFFECT OF BRADYKININ ON PULMONARY VASCULAR RESISTANCE IN FOETAL ANIMALS

The effect of direct arterial infusion of bradykinin in the pulmonary circulation was studied in six foetal animals before and after ventilation. Resistances proximal and distal to the Starling resistor were calculated for the foetal unventilated lung. The animals were then infused with 60–240 ng/min bradykinin directly into the pulmonary artery and another set of pressure–pressure curves as generated for calculations. The animals were then ventilated with air and resistances were calculated before, during and after infusion of bradykinin. Left pulmonary arterial and venous blood were analyzed for P_{O_2}, P_{CO_2} and pH.

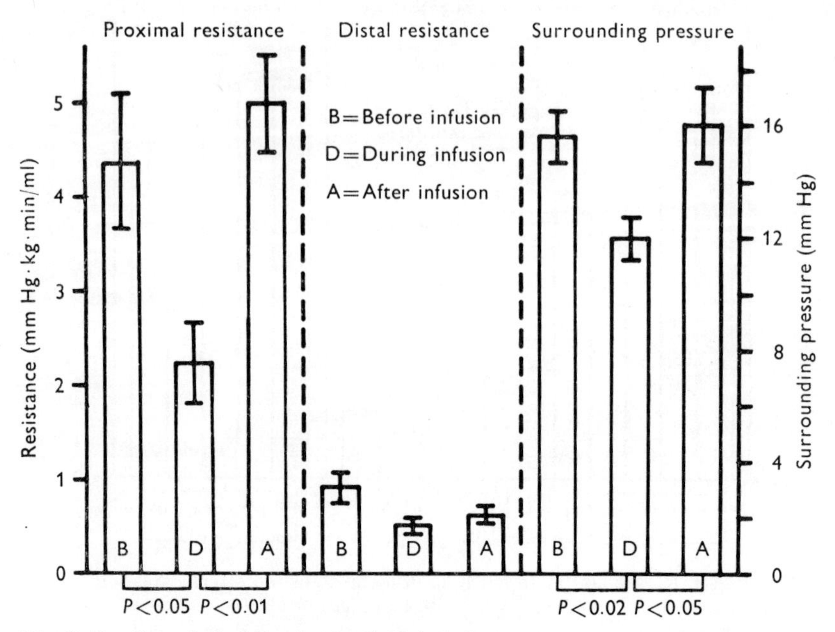

Fig. 8. R_P, R_D and P_S following bradykinin infusion directly into pulmonary artery prior to ventilation.

Infusion of bradykinin decreased both R_P ($P < 0.05$) and P_S significantly (Fig. 8). Proximal resistance decreased from a value of 4.38 to 2.25 mm Hg·kg·min·ml^{-1}. The surrounding pressure fell from 15.6 to 12 mm Hg. After the infusion of bradykinin was terminated both R_P and P_S returned to the pre-infusion levels; R_P increased to 5.05 mm Hg·kg·min·ml^{-1} and P_S increased to 16 mm Hg. Although the distal resistance decreased from 0.93 to 0.53 mm Hg·kg·min·ml^{-1} during infusion of bradykinin, the decrease was not statistically significant. Fig. 9 shows that during infusion of bradykinin directly into the pulmonary artery of ventilated lungs there was a decrease only in proximal resistance. R_P fell from 1.81 to 1.15 mm Hg·kg·min·ml^{-1} ($P < 0.02$). P_S decreased during the infusion of bradykinin but not significantly. R_D did not change with the infusion of bradykinin. The changes in resistance in both ventilated and unventilated foetal lungs are similar to reports previously described in the literature for the action of bradykinin in the foetal animal (Dawes, 1968).

Our results are in contradistinction to those of Hyman (1968) who showed that bradykinin caused active pulmonary venoconstriction. In our experiments bradykinin did not increase R_D significantly in either ventilated or unventilated animals. Most likely, the resistances offered by the pulmonary veins are reflected in the measurement of R_D. Also, our data

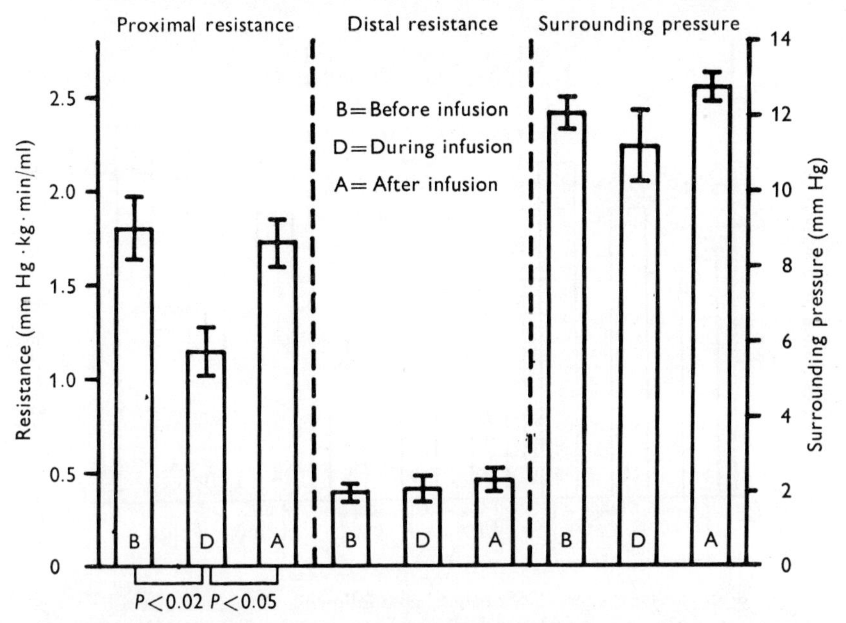

Fig. 9. R_P, R_D and P_S following bradykinin infusion directly into pulmonary artery after ventilation of foetal lung.

showed a decrease in proximal resistance which was not passive since R_P was calculated at the same flow rate as before, during and after infusion of bradykinin.

In analysing the data obtained for the bradykinin experiments and those obtained with changes in blood gases, one is struck with an interesting observation. Changing blood gas tension significantly reduced R_D, the distal resistance, whereas the infusion of bradykinin resulted in an insignificant reduction in R_D. Even though the increase in oxygen tension produced by changing the blood gases may affect the release of bradykinin, it is conceivable that changes in blood gases *per se* have a distinct effect on the vessels which is quite different from that of the action of bradykinin. It is possible that the decrease in P_{CO_2} with ventilation and change of blood gases affects the vessels directly while the response to P_{O_2} is mediated through the release of bradykinin. This difference in response to oxygen and bradykinin requires further examination to clarify the mechanism of action.

NEONATES

Finally, the effects of hypoxia on the pulmonary circulation of neonatal kids was investigated in order to determine whether the newborn lung behaved just as the foetal lung. Kids were anaesthetized with chloralose

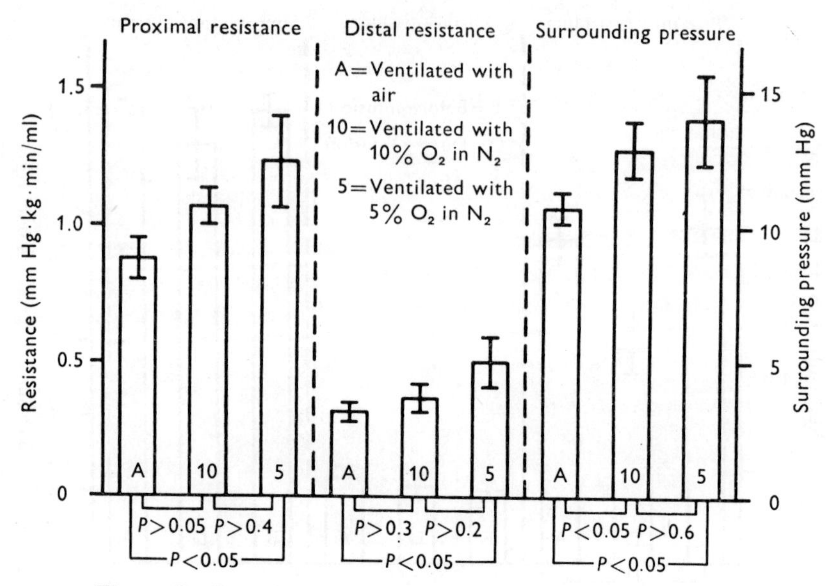

Fig. 10. R_P, R_D and P_S of neonatal goat following ventilation with air, or with 10 % or 5 % O_2 in N_2.

and surgically prepared in the same fashion as the foetal animals. After opening the chest they were maintained on positive pressure ventilation with air. A control series of observations was made. The animals were then ventilated with a mixture of 10% oxygen in nitrogen and another set of curves obtained. Finally, the gas was changed to 5% oxygen in nitrogen and a final set of curves obtained. Fig. 10 clearly indicates that the response of the pulmonary vasculature of newborn goats to hypoxia follows a similar pattern to that of foetal animals. The reduction in inspired P_{O_2} to 10% and then 5% oxygen produced a significant increase in proximal resistances in newborn animals. Similarly, there was an increase in distal resistance and in surrounding pressure. Although the change from room air to 10% oxygen was not significant, the change in every instance from room air to 5% oxygen was significant. The effect of reduction of oxygen tension probably is best explained by constriction of vascular smooth muscle in response to hypoxia. The effects of oxygen or lack of it on the pulmonary circulation may also be due in part to reflex nervous control initiated by chemoreceptors. However, there is some evidence from the work of Aherne & Dawkins (1964) that the water content of the lungs increases in response to hypoxia. If in fact the water is lost from the pulmonary vessels into the perivascular spaces, this may increase interstitial pressure and in turn cause a rise in R_P, R_D and P_S in response to hypoxia.

SUMMARY AND CONCLUSIONS

We have proposed a model which can be used to calculate resistances in two segments of pulmonary circulation based on the concept that the pulmonary vascular bed behaves as if it contains Starling resistors. This model may be used to explain the relationship seen in the foetal and neonatal pulmonary vascular bed between pulmonary artery pressure and pulmonary venous pressure, and also between pulmonary artery pressure and flow. Some of our studies were designed to describe the changes that occurred in the resistance proximal to and distal to the postulated Starling resistors as well as in the surrounding pressure around the Starling resistors in response to ventilation, changes in blood gas tension and the drug bradykinin in mature foetal goats. We also described the effects of hypoxia on the changes of proximal and distal resistances in newborn goats. Ventilation of both lungs in the foetus with a gas that did not change blood gases resulted in a decrease in both proximal resistance and surrounding pressure. Ventilation with air caused an even further decrease in proximal resistance, distal resistance and surrounding pressure. Ventilation of one lung, the right lung, while measuring changes in flow to the contralateral lung, indicated that there was no change in R_P, R_D or P_S when foetal gas was utilized, but all three values showed a marked decrease when the right lung was ventilated with air. In newborn goats proximal resistance, distal resistance and surrounding pressure all increased in response to hypoxia (5% oxygen in nitrogen). The drug bradykinin when infused into the pulmonary circulation caused a significant decrease in proximal resistance and surrounding pressure in the unventilated foetus, but resulted in a decrease only in the proximal resistance in foetuses ventilated with air. Although the exact site of the Starling resistor in the pulmonary bed is not known, we suggest that it is probably distal to the capillaries in both unventilated and ventilated animals.

REFERENCES

AHERNE, W. & DAWKINS, M. J. R. (1964). *Biol. Neonat.* **7**, 214–29.
BANNISTER, J. & TORRANCE, R. W. (1960). *Quart. J. Exp. Physiol.* **45**, 352–76.
BERGOFSKY, E. H., HAAS, F. & PORCELLI, R. (1968). *Fed. Proc.* **27**, 1420–5.
BRODY, J. S., STEMMLER, E. J. & DUBOIS, A. B. (1968). *J. Clin. Invest.* **47**, 783–99.
CAMPBELL, A. G. M., COCKBURN, F., DAWES, G. S. & MILLIGAN, J. E. (1967). *J. Physiol.* **192**, 111–21.
CASSIN, S., DAWES, G. S., MOTT, J. C., ROSS, B. B. & STRANG, L. B. (1964). *J. Physiol.* **171**, 61–79.
CASSIN, S., DAWES, G. S. & ROSS, B. B. (1964). *J. Physiol.* **171**, 80–9, 1964.

DALY, I. DEB. & HEBB, C. (1966). *Pulmonary and Bronchial Vascular Systems.* Baltimore, Williams and Wilkins Co., 1966, p. 250.

DAWES, G. S. (1968). *Fetal and Neonatal Physiology.* Chicago: Year Book Medical Publishers, Inc.

DAWES, G. S., MOTT, J. C., WIDDICOMBE, J. G. & WYATT, D. G. (1953). *J. Physiol.* **121**, 141–62.

DEFREITAS, F. M., FARACO, E. Z. & DEAZUEDO, D. F. (1964). *Circulation* **29**, 66–70.

DE PASQUALE, N. P., SANCHEZ, G., BURCH, G. E. & QUIROS, A. C. (1969). *Am. Heart J.* **78**, 802–6.

EULER, U. S. VON & LILJESTRAND, G. (1946). *Acta Physiol. Scandinav.* **12**, 301–20.

GILBERT, R. D., HESSLER, J. R., EITZMAN, D. V. & CASSIN, S. (1972). *J. Appl. Physiol.* **32**, 47–53.

GILBERT, R. P., HINSHAW, L. B., KUIDA, H. & VISSCHER, M. B. (1958). *Am. J. Physiol.* **194**, 165–70.

HEYMANN, M. A., RUDOLPH, A. M., NEIS, A. S. & MELMAN, K. L. (1969). *Arc. Res.* **25**, 521–34.

HYMAN, A. L. (1968). *J. Pharmacol. Exp. Therap.* **161**, 78–87.

HYMAN, A. L. (1969). *J. Pharmacol. Exp. Therap.* **168**, 96–105.

LAUER, R. M., EVANS, J. A., AOKI, H. & KITTLE, C. F. (1965). *J. Pediat.* **67**, 568–77.

LEVINE, O. R., MELLINS, R. B., SENIOR, R. M. & FISHMAN, A. P. (1967). *J. Clin. Invest.* **46**, 934–44.

LOPEZ-MUNIZ, R., STEPHENS, M. L., BROMBERGER-BARNEA, B., PERMUTT, S. & RILEY, R. L. (1968). *J. Appl. Physiol.* **24**, 625–35.

MCDONALD, F. G. & BUTLER, J. (1967). *J. Appl. Physiol.* **23**, 463–74.

MELLINS, R. B., LEVINE, O. R., SKALAK, R. & FISHMAN, A. P. (1969). *Circ. Res.* **24**, 197–212.

MEYER, B. J., MEYER, A. & GUYTON, A. C. (1968). *Circ. Res.* **22**, 263–71.

PERMUTT, S., BROMBERGER-BARNEA, B. & BANE, H. N. (1962). *Med. Thoracalis.* **19**, 239–60.

RUDOLPH, A. M. & HEYMANN, M. A. (1968). *Ann. Rev. Med.* **19**, 195–206.

ROSS, B. B. (1963). *Nature* **199**, 1100.

WAALER, B. A. (1961). *J. Physiol.* **157**, 475–83.

QUANTITATION OF BLOOD FLOW PATTERNS IN THE FOETAL LAMB *IN UTERO*

By MICHAEL A. HEYMANN,*
ROBERT K. CREASY AND ABRAHAM M. RUDOLPH

The Cardiovascular Research Institute and
Departments of Pediatrics, Physiology and
Obstetrics and Gynaecology,
University of California, San Francisco, California 94122, U.S.A.

Few results are so certain as to free the observer of the responsibility of checking them by some quite independent method. (Sir Joseph Barcroft, 1946)

We have previously reported on measurements of cardiac output and distribution of blood flow in foetal lambs (Rudolph & Heymann, 1967*a*, 1970). These studies were performed acutely with the ewe under low spinal anaesthesia. Catheters were placed in various foetal vessels, and the uterine and abdominal incisions then closed. After 2–3 hours of recovery, cardiac output and its distribution were measured by injecting 50 μm radionuclide-labelled plastic microspheres into the foetus; one nuclide into an upper limb vein and another one into a lower limb vein (Rudolph & Heymann, 1967*a*). Quantitation was permitted by measuring blood flow to one reference organ, the placenta, by an independent method, the steady state diffusion Fick method using an infusion of antipyrine into the foetus (Rudolph & Heymann, 1967*b*). Several problems and potential errors were inherent in the methods as originally described. Since pulmonary venous, azygous venous and coronary venous returns were not included in the calculations, a 5–10% error in flows to certain organs could have resulted. In addition, the antipyrine technique required a prolonged period of at least 30–45 min of infusion before equilibration was reached. If there were any changes in umbilical blood flow a further long period had to be allowed to provide for steady state measurements. We have devised newer methods to overcome these problems and to allow more accurate measurements of cardiac output and organ blood flows in chronic foetal lamb preparations. We have also corroborated the accuracy of the microsphere technique in measuring flows in the foetal lamb by a 'quite independent method', the electromagnetic flow transducer.

* Recipient Research Career Development Award HD 35398 of the National Institute of Child Health and Human Development.

METHODS

In order to obviate the use of the antipyrine method to obtain an independent lower body organ flow we have substituted the use of a femoral arterial reference sample. Blood is withdrawn at a constant rate into a syringe. In addition, we no longer derive the coronary and upper body organ blood flows from an assumed distribution of microspheres in the inferior vena caval return. We now measure these flows more accurately using a second reference sample withdrawn from a carotid or brachial artery and obtained simultaneously with that from the femoral artery. Fifteen-micron microspheres are injected over a 10- to 15-second period and the reference samples are started 5 seconds before the microsphere injection is commenced. The reference samples are then collected for a total of one minute at a withdrawal rate for each of about 5–8 ml/min. This technique has provided adequate numbers of microspheres in the reference samples to allow accurate flow measurements. (Buckberg *et al.* 1971). The use of two reference samples also obviates the problems of streaming or inadequate mixing of microspheres injected into the inferior vena caval stream.

Using this approach it is possible to accurately measure blood flow to all organs excluding the lungs by a single injection of microspheres into the inferior vena caval stream from a lower limb vein.

Pulmonary blood flow can still be calculated as originally described (Rudolph & Heymann, 1967*a*) using the distribution to the lungs of both superior and inferior vena caval returns. This method requires injection of microspheres into both an upper and lower limb vein. However, it is now also possible to measure pulmonary blood flow more accurately by obtaining a third reference sample directly from the pulmonary artery during the single injection of microspheres into a lower limb vein.

RESULTS

We compared cardiac output, expressed as combined left and right ventricular outputs, as measured using the calculations originally described with only a lower body independent reference sample and as measured using both a carotid arterial and femoral arterial reference sample in 24 foetal lambs of 110 to 145 days gestation. These observations were made under varying conditions of acid–base status. As shown in Fig. 1 the older method consistently underestimated cardiac output by about 5–10%. It was of interest that the six animals in which the underestimate was greater than 10% all had large pulmonary blood flows, in excess of

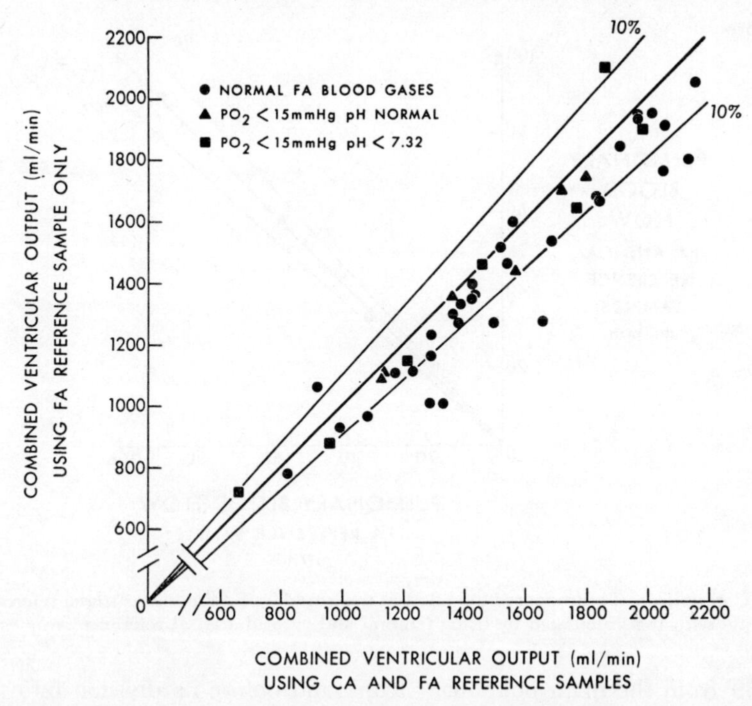

Fig. 1. Comparison of combined ventricular output as measured by using both a carotid and femoral arterial reference sample with that calculated by using a femoral arterial reference sample alone.

200 ml/min. Thus, failure to take pulmonary venous return into account may produce considerable error in calculating coronary and upper body organ blood flows, particularly when pulmonary blood flow is large. This is overcome by using the new approach of obtaining both carotid and femoral arterial reference samples.

We also compared pulmonary blood flow as measured directly using a pulmonary arterial reference sample and a lower limb vein microsphere injection with that calculated from the distribution of microspheres injected into both an upper and lower limb vein using both carotid and femoral arterial reference samples. As shown in Fig. 2 the correlation is excellent. One of the major problems involved in the pulmonary arterial reference sample technique is adequate mixing and the catheter should be as distal as possible. We have found that when the reference sample is obtained in the right ventricle there is a poorer correlation.

We also compared the measurement of pulmonary blood flow by both microsphere techniques with an independent method on nine occasions in five foetal lambs. A precalibrated electromagnetic flow transducer (Statham SP 2202) was placed on the main pulmonary artery immediately after its

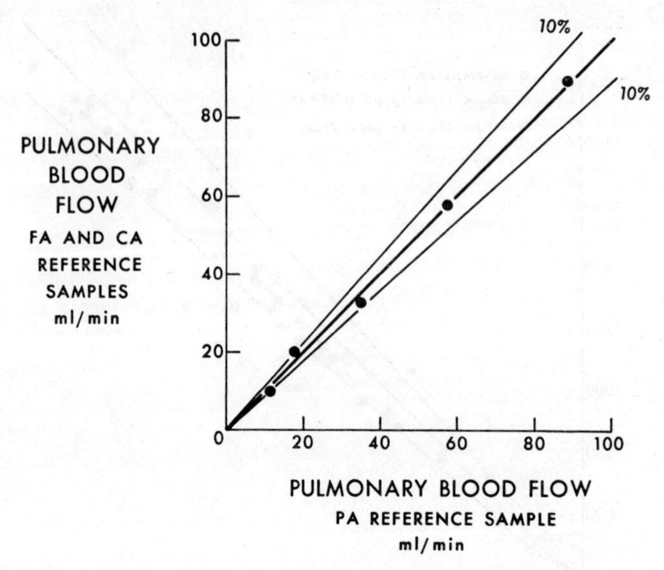

Fig. 2. Comparison of pulmonary blood flow as measured by a pulmonary arterial reference sample with that calculated by using femoral and carotid arterial reference samples.

take-off from the main pulmonary trunk and before its division into the left and right pulmonary arteries. Measurements of flow were then made simultaneously with the flow transducer and by injection of microspheres. In two instances pulmonary blood flow was increased above the resting level by the continuous infusion of acetylcholine into an external jugular vein. All studies were performed three to five days after implantation of the transducer and the catheters, with the exception of one animal in which the observations were made the day after surgery. This observation is indicated (+) in the figure. With this exception an extremely good correlation between the two different methods is shown.

In addition to improving the accuracy of measurements of organ blood flows made with the microsphere method, the use of carotid, femoral and pulmonary arterial reference samples affords the opportunity to calculate flow through the major arteries as well as to determine the relative contributions of the left and right ventricles to cardiac output. The equations used to calculate these values are shown in the appendix. Again one of the major problems involved in these calculations relates to the difficulty in obtaining adequate reference samples in which the microspheres have been well mixed within the blood stream. Current use of 15 μm microspheres instead of larger ones may help to overcome some of the mixing problems as they more closely follow the distribution patterns of red cells (Phibbs & Dong, 1970). Since considerable numbers of microspheres

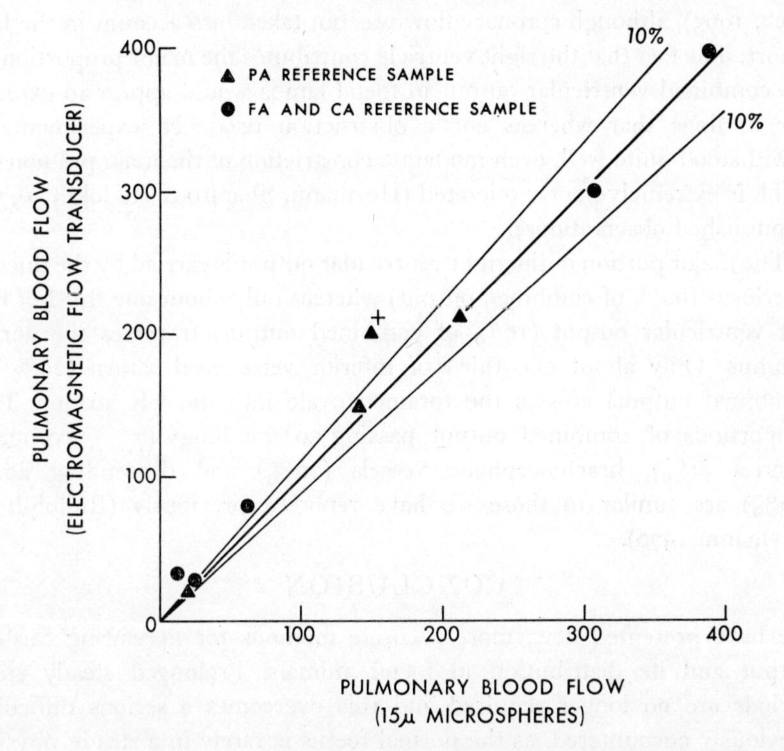

Fig. 3. Comparison of pulmonary blood flow as measured by an electromagnetic flow transducer with that measured using 15 μm microspheres.

need to be injected to overcome the statistical problems of random distribution (Buckberg *et al.* 1971) an added advantage of using the smaller microspheres is that they will produce less disturbance of the circulation than would the same numbers of larger sized microspheres.

Based on data obtained from these calculations as well as from the combined use of these calculations and measurements of right ventricular output by means of an electromagnetic flow transducer placed on the main pulmonary trunk (Rudolph & Heymann, 1972), we have derived preliminary data on the major flow patterns within the heart and great vessels in foetal lambs of 125–145 days' gestation. These values were obtained in chronic in-utero preparations at least 3–5 days following surgery, and are expressed as percentages of combined ventricular output. Right ventricular output was about 67% and left ventricular 33% of combined ventricular output. Previous estimates, all performed in anaesthetized, exteriorized foetuses, showed right ventricular output as 45% of combined ventricular output (Dawes, Mott & Widdicombe, 1954), 50% (Mahon, Goodwin & Paul, 1966) and about 58% (Assali, Morris &

Beck, 1965), although coronary flow was not taken into account in the last report. The fact that the right ventricle contributes the major proportion of the combined ventricular output in foetal lambs would appear to explain our findings that whereas aortic obstruction produced experimentally is withstood quite well, even moderate constriction of the main pulmonary trunk is extremely poorly tolerated (Heymann, Shapiro & Rudolph, 1972; unpublished observations).

The major portion of the right ventricular output is carried by the ductus arteriosus (60% of combined output) whereas only about one third of the left ventricular output (10% of combined output) traverses the aortic isthmus. Only about one third of inferior vena caval return (26% of combined output) crosses the foramen ovale into the left atrium. The proportions of combined output passing to the lungs (7%), coronary arteries (3%), brachiocephalic vessels (20%) and descending aorta (70%) are similar to those we have reported previously (Rudolph & Heymann, 1970).

CONCLUSION

We have presented new, more accurate methods for measuring cardiac output and its distribution in foetal animals. Prolonged steady state periods are no longer required and this overcomes a serious difficulty previously encountered, as the normal foetus is rarely in a steady physiological state for more than a few minutes. The methods employed require collecting arterial reference samples simultaneously from the femoral and carotid arteries with injection of 15 μm microspheres into both an upper and lower limb vein. If a pulmonary arterial reference sample is also obtained only a single lower limb vein injection is required. Comparisons of different approaches for measuring pulmonary blood flow are presented as well as an independent validation of these with an electromagnetic flow transducer. Previous observations on exteriorized foetal lambs have shown that right and left ventricular outputs are similar; however, our studies indicate that right ventricular output represents about two thirds of the combined ventricular output *in utero*.

This research was supported by PHS Grant HL06285 from the National Heart and Lung Institute.

APPENDIX

Terms used

DAo Descending aorta
IS Aortic isthmus
DA Ductus arteriosus

$\dot{Q}$ Flow (ml/min)

CPA Concentration pulmonary arterial reference sample $\left(\frac{\text{Cts/min}}{\dot{Q}}\right)$

CCA Concentration carotid arterial reference sample $\left(\frac{\text{Cts/min}}{\dot{Q}}\right)$

CFA Concentration femoral arterial reference sample $\left(\frac{\text{Cts/min}}{\dot{Q}}\right)$

$$\dot{Q}\text{DAo} = \dot{Q}\text{DA} + \dot{Q}\text{IS} \tag{1}$$

$$\text{therefore } \dot{Q}\text{IS} = \dot{Q}\text{DAo} - \dot{Q}\text{DA} \tag{2}$$

$$(\dot{Q}\text{DA})(\text{CPA}) + (\dot{Q}\text{IS})(\text{CCA}) = (\dot{Q}\text{DAo})(\text{CFA}) \tag{3}$$

$$\text{from (3) } \dot{Q}\text{DA} = \frac{(\dot{Q}\text{DAo})(\text{CFA}) - (\dot{Q}\text{IS})(\text{CCA})}{\text{CPA}} \tag{4}$$

substituting (4) in (2)

$$\dot{Q}\text{IS} = \dot{Q}\text{DAo} - \left[\frac{(\dot{Q}\text{DAo})(\text{CFA}) - (\dot{Q}\text{IS})(\text{CCA})}{\text{CPA}}\right] \tag{5}$$

$$(\dot{Q}\text{IS})(\text{CPA}) = (\dot{Q}\text{DAo})(\text{CPA}) - (\dot{Q}\text{DAo})(\text{CFA}) + (\dot{Q}\text{IS})(\text{CCA}) \tag{6}$$

$$(\dot{Q}\text{IS})(\text{CPA}) - (\dot{Q}\text{IS})(\text{CCA}) = (\dot{Q}\text{DAo})(\text{CPA}) - (\dot{Q}\text{DAo})(\text{CFA}) \tag{7}$$

$$\dot{Q}\text{IS}(\text{CPA} - \text{CCA}) = \dot{Q}\text{DAo}(\text{CPA} - \text{CFA}) \tag{8}$$

$$\dot{Q}\text{IS} = \frac{\dot{Q}\text{DAo}(\text{CPA} - \text{CFA})}{\text{CPA} - \text{CCA}} \tag{9}$$

$\dot{Q}$DAo is known, therefore $\dot{Q}$DA can be calculated

$$\dot{Q}\text{DA} = \dot{Q}\text{DAo} - \dot{Q}\text{IS} \tag{10}$$

$\dot{Q}$ Lungs is known therefore right ventricular output ($\dot{Q}$RV) can be calculated

$$\dot{Q}\text{RV} = \dot{Q}\text{ Lungs} + \dot{Q}\text{DA} \tag{11}$$

REFERENCES

ASSALI, N. S., MORRIS, J. A. & BECK, R. (1965). *Am. J. Physiol.* **208**, 122.

BARCROFT, J. (1946). *Researches on Prenatal Life*. Oxford: Blackwell.

BUCKBERG, G. D., LUCK, J. C., PAYNE, B. D., HOFFMAN, J. I. E., ARCHIE, J. P. & FIXLER, D. E. (1971). *J. appl. Physiol.* **31**, 598.

DAWES, G. S., MOTT, J. C. & WIDDICOMBE, J. G. (1954). *J. Physiol. (London)* **126**, 563.

MAHON, W. A., GOODWIN, J. W. & PAUL, W. M. (1966). *Circulation Res.* **19**, 191.

PHIBBS, R. H. & DONG, L. (1970). *Can. J. Physiol. Pharm.* **48**, 415.

RUDOLPH, A. M. & HEYMANN, M. A. (1967*a*). *Circulation Res.* **21**, 163.

RUDOLPH, A. M. & HEYMANN, M. A. (1967*b*). *Circulation Res.* **21**, 185.

RUDOLPH, A. M. & HEYMANN, M. A. (1970). *Circulation Res.* **26**, 289.

RUDOLPH, A. M. & HEYMANN, M. A. (1972). *Proceedings of the Barcroft Centenary Symposium.*

BIOCHEMICAL BASIS FOR RESPONSE OF DUCTUS ARTERIOSUS TO OXYGEN

By FREDRIC S. FAY

Department of Physiology,
University of Massachusetts Medical School,
419 Belmont Street, Worcester, Massachusetts 01604, U.S.A.

Numerous clinical (Record & McKeown, 1953; Penaloza *et al.* 1964) and experimental (Kennedy & Clark, 1942; Born *et al.* 1956; Kovalcik, 1963; Fay, 1971) observations have drawn attention to the importance of the postnatal increase in arterial oxygen pressure (P_{O_2}) for muscular closure of the ductus arteriosus. Interest has recently centred about the mechanism whereby increased oxygen pressure triggers contraction of the smooth muscle in the wall of the ductus. The effect of oxygen appears to be exerted on the smooth muscle cells themselves rather than some intermediate cell type (Kovalcik, 1963; Fay, 1971). What then makes the muscle cells in the wall of the ductus so uniquely sensitive to changes in oxygen pressure? Studies to be reported represent an attempt to trace the events whereby an increase in P_{O_2} initiates contraction. All experiments were performed using rings of ductus arteriosus obtained from neonatal guinea pigs and studied in an organ bath under isometric conditions. Previous studies (Kovalcik *et al.* 1968; Fay, 1971) indicated that agents which inhibit or uncouple oxidative phosphorylation also inhibit the contractile response to oxygen although they have little or no effect on the contractile response to other stimuli. This led to the suggestion that oxygen initiates contraction of ductal smooth muscle by interacting with the terminal oxidase of the cytochrome chain.

To test this possibility, the effect of CO on the response to oxygen was determined (Fay, 1971; Fay & Jöbsis, 1972). CO selectively depressed the contractile response of the ductus to O_2 but its inhibitory effect was almost completely reversed by illumination of the ductus preparation, presumably due to the photodissociation of CO from a molecular species which was involved in the response to oxygen. The effectiveness of light of equal intensity but different wavelengths in reversing the CO-inhibition of the oxygen response was determined in an effort to pinpoint the site of CO-inhibition. The results of a typical experiment are shown in Fig. 1. In this experiment the action spectrum contained a maximum between 425 and 430 nm. The averaged action spectrum based on seven prepara-

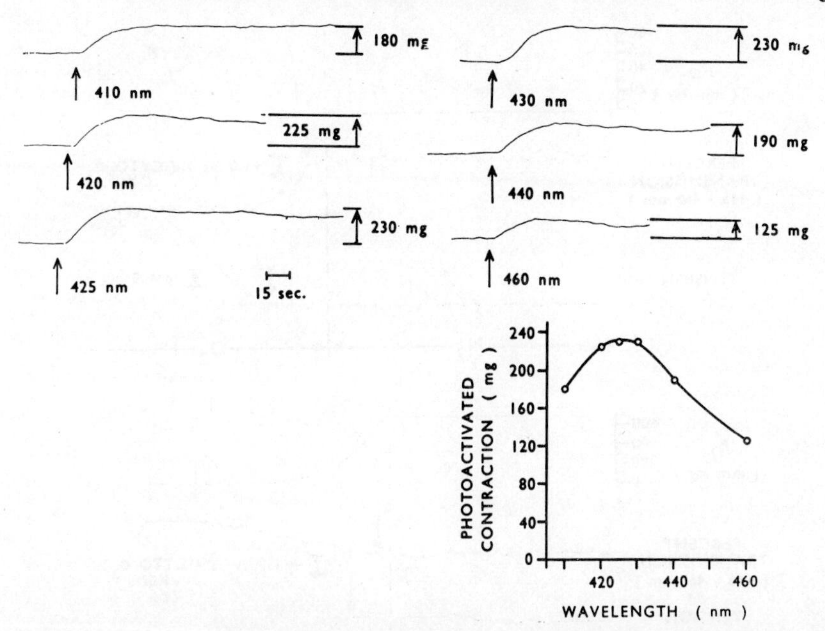

Fig. 1. The effect on a ductus arteriosus of illumination with monochromatic light in the presence of 85.5 % CO and 9.5 % O_2. The upper part of the figure shows actual records of the increase in tension upon illumination with light of different wavelengths but equal intensities. The responses were elicited in order going from 410 to 460 nm. The numbers at the right of each record indicate the amplitude of the light-induced tension increment at its peak. The graph in the lower right shows the amplitude of these photo-induced contractions plotted as a function of wavelength.

tions exhibited a peak between 420 and 425 nm, rather close to the 430 nm peak for the photodissociation of the cytochrome a_3–CO complex (Castor & Chance, 1955). The small discrepancy may either be a consequence of errors introduced by working with an intact tissue or possibly may indicate that the response of the ductus to oxygen is subserved by a unique cytochrome pigment.

In order to check for the latter possibility, the cytochrome complement of the ductus was assessed by spectrophotometric measurements made on the intact tissue. The difference in light absorption of two ducts, one exposed to nitrogen, the other to oxygen, was determined as a function of wavelength using a split-beam spectrophotometer (Fay & Jöbsis, 1972). The spectrum contained maxima at 430 and 445 nm characteristic of cytochromes *b* and a_3 respectively. No evidence of an atypical cytochrome component was noted; the one unusual feature of the spectra, when compared to that for striated muscle (Jöbsis, 1963), was the relative preponderance of cytochrome *b* over a_3, but this was also seen in samples of aortic tissue.

If in fact oxygen triggers contraction as a result of a primary interaction

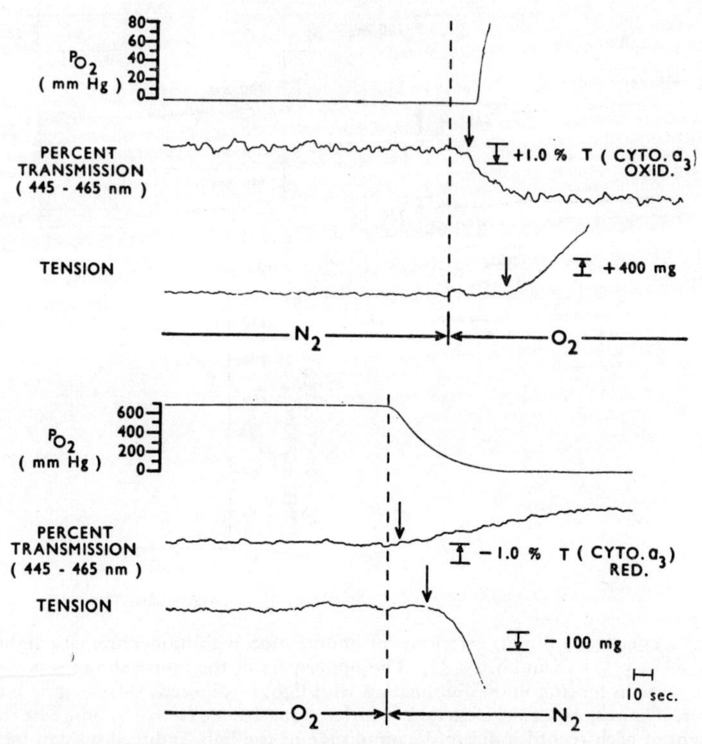

Fig. 2. Simultaneous measurement of bath P_{O_2}, and cytochrome a_3 oxidation and tension of a ductus arteriosus preparation. The upper and lower groups of three polygraph records were obtained from the same preparation. The uppermost trace in each group registers bath P_{O_2} as monitored with an oxygen electrode. The optical traces indicate cytochrome a_3 oxidation state as measured at 445–465 nm; oxidation of cytochrome a_3 results in a decrease in transmission at 445–465 nm. Upper: At the dotted line 95 % O_2, 5 % CO_2 was substituted for 95 % N_2, 5 % CO_2 bubbling through the tissue chamber. The arrows above the optical and mechanical records indicate the first noticeable change in these traces after switching gases. The output of the P_{O_2} electrode went off-scale above 80 mm Hg and the record is therefore discontinued above 80 mm Hg. Lower: At the dotted line 95 % N_2, 5 % CO_2 was substituted for 95 % O_2, 5 % CO_2 bubbling through the bath; coincident with the switch in gases, the chamber was rapidly rinsed with 3 × its volume of Krebs–Ringer bicarbonate saline which had been pre-equilibrated with 95 % N_2, 5 % CO_2 at 37 °C. The arrows above the optical and mechanical records indicate the first noticeable change in these traces after switching to anaerobic conditions.

with cytochrome a_3 then one might expect: (1) that upon changes in P_{O_2}, changes in cytochrome oxidation should *always* precede any change in force; and (2) that in the entire range over which ductal tension is sensitive to changes in P_{O_2}, cytochrome a_3 should exhibit changes in redox state. The proposed kinetic and steady state relationships were investigated with the aid of a dual-wavelength spectrophotometer. The redox state of cytochrome a_3 was monitored by comparing the absorption at 445 and 465 nm of a ductus arteriosus while tension exerted by the same preparation was continuously recorded (Fay & Jöbsis, 1972). The results of a typical experiment are shown in Fig. 2. As was found in all preparations studied, cytochrome a_3 oxidation always preceded any increase in force in response to oxygen; upon rapidly decreasing the P_{O_2} in the organ bath reduction of cytochrome a_3 always preceded any decrease in force. The time between the first noticeable change in a_3 oxidation state and tension averaged 81 ± 13 seconds in the former case and 15 seconds in the latter. Simultaneous measurement of tension and cytochrome a_3 oxidation as P_{O_2} was increased in small increments from 0 to 680 mm Hg indicated that tension and cytochrome a_3 oxidation are related by a hyperbolic curve such that changes in tension are always associated with changes in a_3 oxidation. Both requirements stipulated by the hypothesis that cytochrome a_3 is the primary site of oxygen interaction within ductal smooth muscle are thus fulfilled.

The proposed role of cytochrome a_3 in the response of the ductus to oxygen, however, leaves an apparent contradiction. The ductus arteriosus of the newborn guinea pig exhibits its maximum sensitivity to oxygen when the P_{O_2} is between 0 and 140 mm Hg (Fay, 1971), yet the cytochrome a_3–cytochrome oxidase system is sensitive in a range two orders of magnitude below this (Jöbsis, 1964). This apparent contradiction might be resolved if a steep gradient for oxygen existed within the walls of the ductus such that the smooth muscle cells deep within the walls are exposed to a much lower range of P_{O_2} when the outside of the vessel is exposed to P_{O_2}'s between 0 and 140 mm Hg. Direct measurements of P_{O_2} within the wall of the ductus, using an ultramicro oxygen cathode, indicate that steep oxygen gradients do in fact exist (Fay, Nair & Whalen, unpublished observations), exactly in keeping with the proposed mechanism.

The sequence of events postulated to underly the contractile response of the ductus arteriosus to oxygen is summarized in Fig. 3. As indicated, increased turnover of the cytochrome chain initiates contraction as a consequence of the increased synthesis of high-energy phosphate compounds; this follows from the selective inhibition of the response to oxygen produced by 2,4-dinitrophenol and oligomycin (Fay, 1971), which are known to interfere with oxidative phosphorylation. Increased synthesis of ATP somehow stimulates the ductus to contract. It seems unlikely that this is a direct result of increased availability of ATP to the contractile apparatus, since the ductus is fully capable of a sustained contraction under anaerobic conditions in response to K^+ or acetylcholine (Fay, 1971). The increase in force in going from anaerobic to aerobic conditions is therefore probably not a consequence of a transition from a state in which activity of the contractile apparatus is limited by the availability of high-energy phosphate

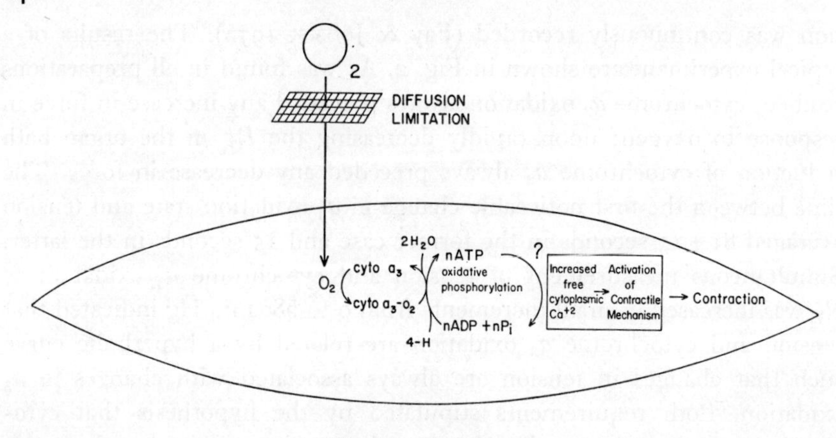

Fig. 3. Proposed mechanism for O_2-induced constriction of the ductus arteriosus. Only the terminal member of the cytochrome chain is shown explicitly. Details of the proposal are discussed in the text.

compounds to one in which this limitation is relieved. Rather, it is proposed that increased oxidative synthesis of ATP in some manner leads to an increase in free cytoplasmic calcium which in turn triggers the contractile mechanism. It may be therefore that the unique oxygen sensitivity of the ductus resides in the link between oxidative synthesis of high-energy phosphate compounds and levels of cytoplasmic calcium.

This work was supported in part by grants from the National Institutes of Health (HE 14523-01) and the Massachusetts Heart Association (1034).

REFERENCES

Born, G. V. R., Dawes, G. S., Mott, J. C. & Rennick, B. R. (1956). *J. Physiol.* **132**, 304–42.

Castor, L. N. & Chance, B. (1955). *J. Biol. Chem.* **217**, 453–65.

Fay, F. S. (1971). *Am. J. Physiol.* **221**, 470–9.

Fay, F. S. & Jöbsis, F. F. (1972). *Am. J. Physiol.*, **223**, 588–95.

Jöbsis, F. E. (1963). *J. Gen. Physiol.* **46**, 905–28.

Jöbsis, F. F. (1964). *Handbook of Physiology, Respiration.* Sect. 3, Vol. I, pp. 63–124.

Kennedy, J. A. & Clark, S. L. (1942). *Am. J. Physiol.* **136**, 140–7.

Kovalcik, V. (1963). *J. Physiol.* **165**, 185–97.

Kovalcik, V., Kriska, M., Slamova, J. & Dolezel, S. (1968). *Proceedings of the International Union of Physiological Sciences, XXIV International Congress*, Volume VII, p. 246.

Penaloza, D., Arias-Stella, J., Sime, F., Recavarren, S. & Marticorena, E. (1964). *Pediatrics* **34**, 568–82.

Record, R. G. & McKeown, T. (1953). *Brit. Heart J.* **15**, 376–86.

A BIOPHYSICAL LOOK AT THE RELATIONSHIP OF STRUCTURE AND FUNCTION IN THE UMBILICAL ARTERY

By MARGOT R. ROACH

Departments of Biophysics and Medicine,
University of Western Ontario, London 72, Ontario, Canada

The umbilical arteries are unique in a number of ways, including:

(1) Their length, which goes up to about one hundred centimetres in man and some of the larger animals, usually with few or no branches.

(2) Their lack of innervation in most if not all species.

(3) Their relatively large proportion of muscle compared to elastin and collagen.

(4) The fact they must remain patent *in utero*, but close completely immediately after birth.

(5) Their relative lack of vasa vasorum.

Considering the ease with which these arteries can be obtained, it is surprising that so few physiologists have studied them. Barclay, Franklin & Prichard (1944) have reviewed the literature on the umbilical vessels up to 1945, and Dawes (1968) has summarized some of it since. Most of the literature seems devoted either to a histological search for nerves, or else to a study of reactivity of the vessels in response to a variety of chemical stimuli.

INNERVATION

Spivack (1943) has probably done the most thorough anatomical study of guinea pig and human cords in the search for nerves. She concluded that there was a rich innervation of the intra-abdominal vessels, but that the extra-abdominal arteries were non-innervated, except perhaps at the skin edge. The controversy still rages. In 1969, Fox & Jacobson, using a modified methylene blue immersion technique, demonstrated nerves in human cords of both term babies and those from first trimester therapeutic abortions. They claimed that the only fibrillary structures stained by this technique are nerves. They found nerve trunks in Wharton's jelly that gave off branches to form a plexus in the walls of the umbilical vessels. In 1970, Nadkarni did both light and electron microscope studies on human umbilical cords. He found no nerve fibres with the light microscope after silver staining, but did see structures he felt were myelinated nerves within the smooth muscle cells of the extrafoetal segments of the umbilical artery. While his evidence seems fairly convincing he concluded 'it is felt that more studies of these structures by other investigators are necesary to confirm our findings'.

Hülsemann, in 1971, did similar electron microscopic studies on guinea pig cords. He found no innervation of the extrafoetal parts of the umbilical vessels, but did find a rich intrafoetal innervation. Large nerve bundles accompanied the artery, running in the adventitia, and giving off smaller branches to the media. The density of innervation increased as the artery passed proximally to join the iliacs. At times several axons went to individual muscle cells. In the veins, by contrast, the nerve fibres tended to run with the vasa vasorum. The density of innervation increased as the liver was approached. He concluded that Nadkarni (1970) had misinterpreted his electron micrographs.

Fluorescence studies (Ehinger *et al.* 1968) have also failed to demonstrate evidence of innervation extra-abdominally, yet they found a rich intra-abdominal innervation.

To my knowledge, no one has attempted to study these nerves physiologically. Since I personally am not competent to assess the histological results of others, I must conclude that the question of innervation is still unsettled. However, the weight of evidence seems to favour the presence of nerves intra-abdominally, but the absence of nerves extra-abdominally. Since we have unpublished evidence that the intra-abdominal part of the umbilical artery is still patent, and pulsating, seven to ten days after delivery in sheep, cow, and pig, I would suspect that the two parts of the cord may respond differently. The extra-abdominal vessels, where closure occurs, seem likely to contract in response to non-neural stimuli.

Reactivity

A number of groups (Lewis, 1968; Eltherington, Stoff, Hughes & Melmon, 1968; Somlyo, Woo & Somlyo, 1965; Altura, Malaviya, Reich & Orkin, 1972; Davignon, Lorenz & Shepherd, 1965; Gokhale, Gulati, Kelkar & Kelkar, 1966) have studied the reactivity of the umbilical arteries, using a variety of preparations. There is so little agreement between groups, and even within one set of experiments, that one must question if the artery is as uniform structurally as it appears.

Biophysically, two questions are intriguing. First, how can a hole the size of this be eliminated? Second, is the muscle in the artery unique, since it appears to be the only natural source of non-innervated vascular smooth muscle?

CLOSURE OF THE UMBILICAL ARTERY

In a reasonably thorough survey of the literature, I found no discussion of how a hole 6–8 mm in diameter could be eliminated. This is a major mechanical feat, equalled only by the closure of the ductus arteriosus which occurs over several days instead of instantaneously. Yet, it is obvious the hole must close. These arteries are attached almost directly to the aorta, and hence have systemic pressure available to force blood through their cut ends. How is this closure achieved?

Since most authors have studied helical strips (which test mainly circular muscle), one must assume that the consensus of opinion is that contraction of circular muscle is responsible for closure. There are many fallacies in this assumption.

First, contraction of circular muscle, particularly on the intimal side, can produce a smaller lumen, but it is still circular. This process can continue but requires much more muscle than is available (Roach, 1970). By contrast, contraction of longitudinal muscle will decrease the lumen more readily. If we assume that the arterial wall has a constant volume, then the wall will become thicker due to an increase in cross-sectional area in proportion to the degree of shortening (volume = area × length). By contrast, contraction of circular muscles decreases the lumen proportional to the degree of shortening of the fibre – i.e. to the radius instead of the square of the radius of the wall. This is illustrated in Fig. 1. The details have been discussed elsewhere (Roach, 1970).

What evidence is there that longitudinal muscles are involved in closure of umbilical arteries? Our evidence is three-fold. First, we have histological evidence that there appears to be much more longitudinal muscle in the contracted umbilical artery of both sheep (Plate 1) and man (Plate 2) than there is in the relaxed arteries from the same species. Sections from contracted arteries of cow and pig also show prominent bellies of longitudinal muscle protruding into the lumen, although we do not have relaxed arteries from these species for comparison.

Second, we have evidence from sheep umbilical arteries which contract uniformly (in contrast to human ones that contract segmentally) that the degree of shortening is comparable (within 2%) to the increase in cross-sectional area of the wall. This evidence was obtained at the Nuffield Institute for Medical Research in Oxford, where cords from lambs delivered by Caesarian section were cross-clamped, the arteries dissected out, their length measured, and a section fixed for histological study while in the relaxed or dilated state. The rest of the artery was then unclamped, and the shortening measured with a ruler and/or a travelling microscope

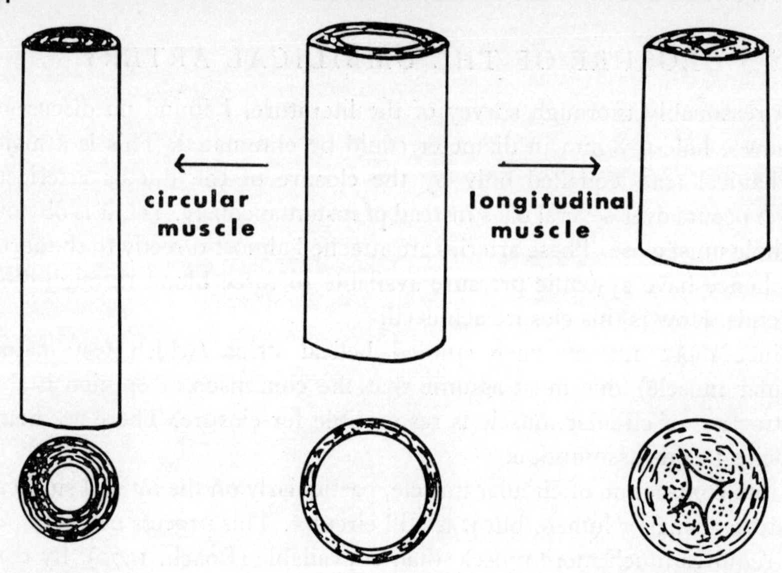

Fig. 1. Schematic diagram of the difference in circular and longitudinal muscle contraction in a cylindrical vessel. With contraction of circular fibres the vessel elongates, and the external diameter decreases, but the wall area changes very little. With longitudinal muscle contraction the vessel shortens, the lumen becomes non-circular, and the cross-sectional wall area increases.

as contraction occurred. In all segments (average 6–8 cm long), the degree of shortening ranged from 23 to 32%. The contracted segment was also studied histologically. The wall areas were studied by projecting the slides with an enlarger and then measuring the wall and lumen areas either with a planimeter, or by weighing paper.

Fig. 2*a* shows a set of three sections from an artery that contracted partially but did not close. The lumen is still circular, although smaller, and there is no change in the area of the wall (Table 1). Histologically the longitudinal muscles were not prominent.

Fig. 2*b* shows a partially, and an almost completely closed artery. Note that the external diameter (area) is unchanged, but the lumen is stellate, and the wall area has increased by 31%. This was associated with 29% shortening. (Table 2).

Fig. 2*c* shows three stages of narrowing in a third artery. In the top two, the lumen is circular, and the wall area comparable, although the outside diameter has decreased. By contrast the third section has the same outside area as the first section, but is almost completely closed with a crescentic lumen, and a 29% increase in wall area (Table 3). This artery shortened by 31% of its length during this latter process, but did not change in length between the first two.

PLATE I

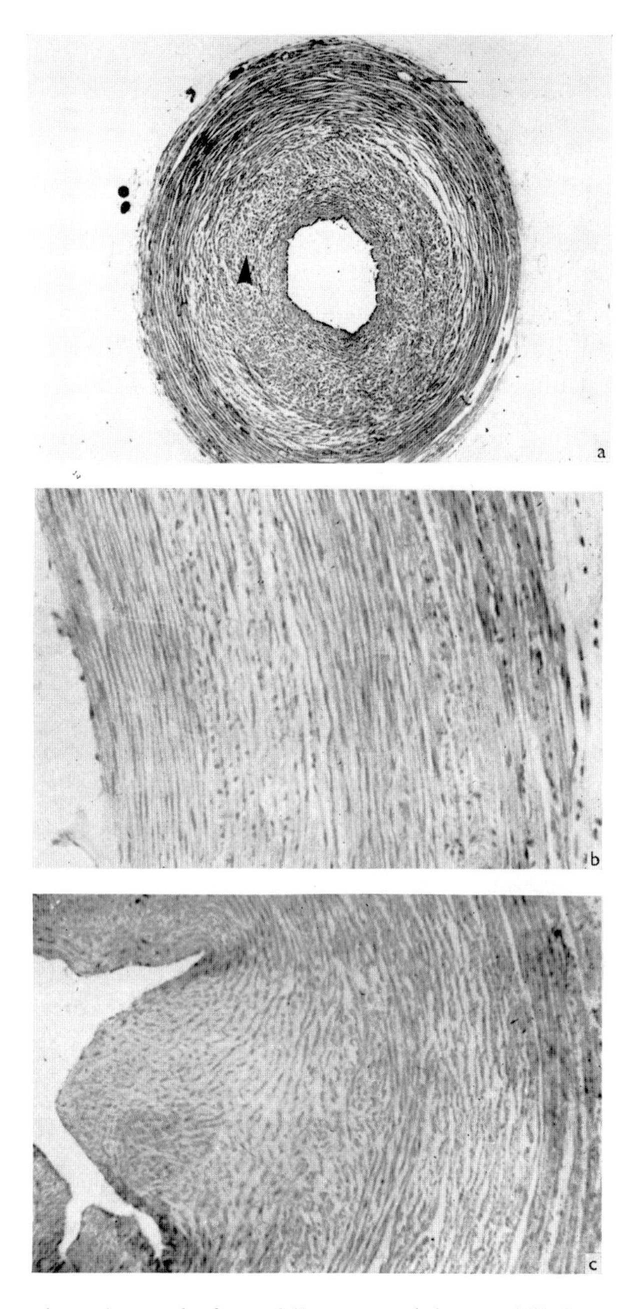

(*a*) Low power photomicrograph of a partially contracted sheep umbilical artery. Note the circular lumen. The arrow shows the level of the vasa vasorum, and the large arrow head the longitudinal muscle.
(*b*) A high power photomicrograph of part of a cross-section of a dilated sheep umbilical artery. Note the five layers of muscle, and the uniform thickness of the inner longitudinal layer.
(*c*) A high power photomicrograph of a cross-section from a contracted sheep umbilical artery. Note the longitudinal muscle protruding into the stellate lumen.

(*Facing p.* 144)

PLATE 2

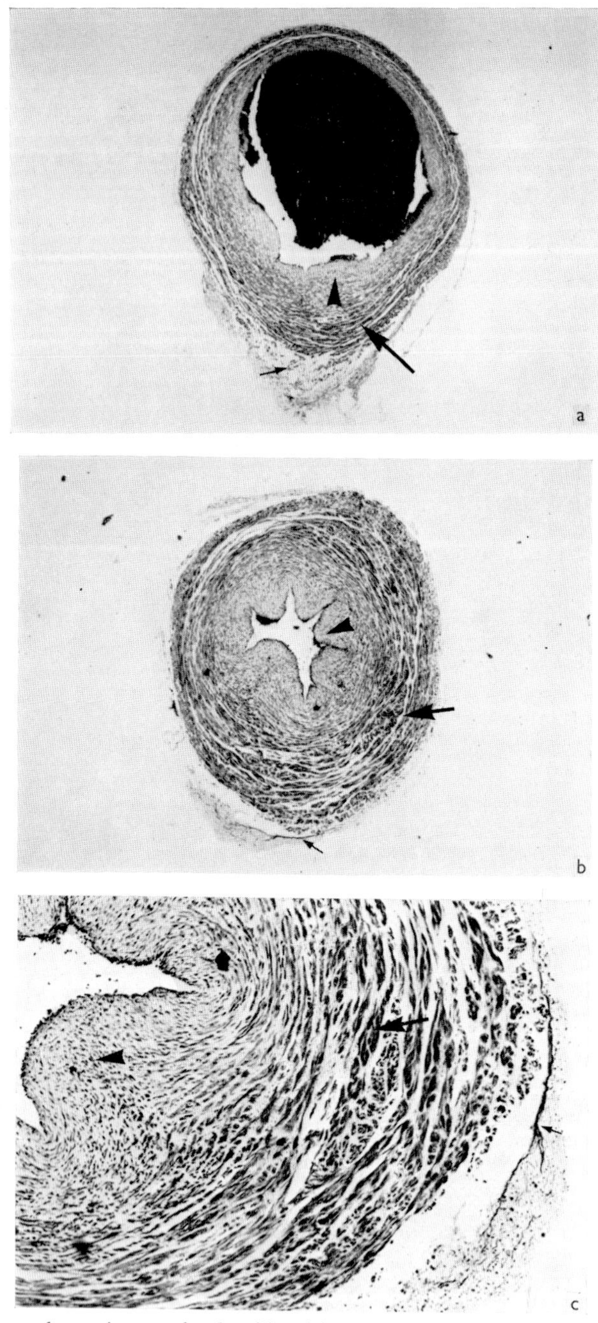

(*a*) Low power photomicrograph of a dilated human umbilical artery with blood in the lumen. The inner longitudinal layer is thin, with a maximum value at the large arrow head. The wall is asymmetrical due to the presence of two coiling muscles – one (large arrow) to coil the cord, and another (small arrow) to coil the artery.

(*b*) A low power photomicrograph of contracted human umbilical artery. The symbols are as in (*a*). Note the stellate lumen, and marked increase in area of the the longitudinal muscle which is pushing into the lumen.

(*c*) A high power photomicrograph of part of (*b*) to show the orientation of the muscle bundles. Symbols as in (*a*).

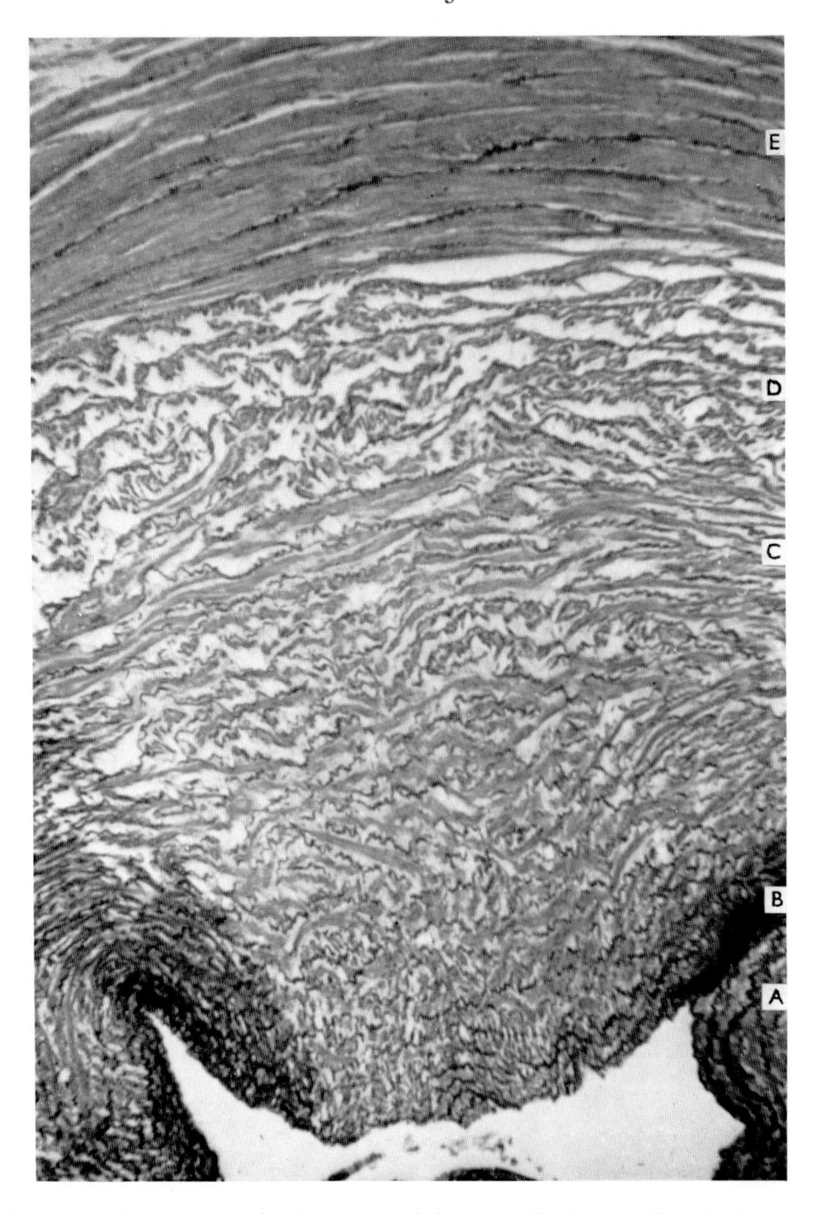

High power photomicrograph of a contracted sheep umbilical artery. Note the five layers of muscle. A – an inner circular layer, pulled out around the stellate protrusions and partially eroded here because the formaldehyde did not reach it. B – an inner longitudinal layer that produces the stellate protrusions. C – a middle circular layer. D – an outer longitudinal layer. E – an outer circular layer which appears to stain differently from the other layers. The black lines are elastin. It appears stretched in the outer layers, and wrinkled in the inner ones. In dilated arteries, it is stretched in all layers.

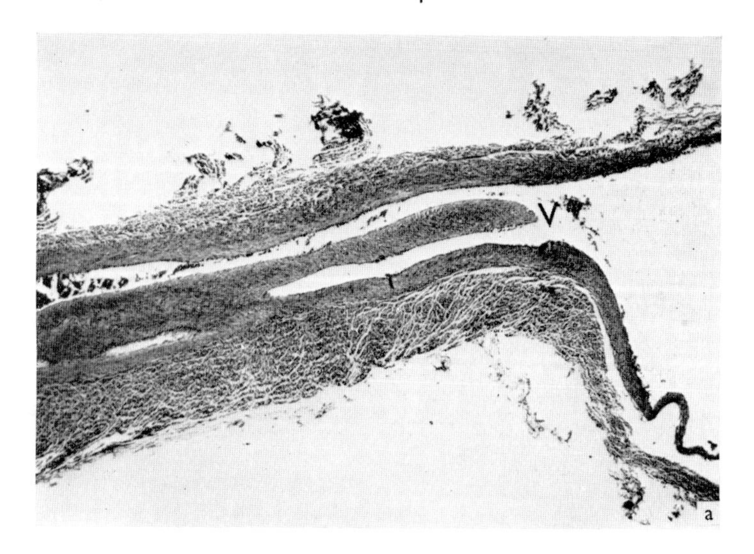

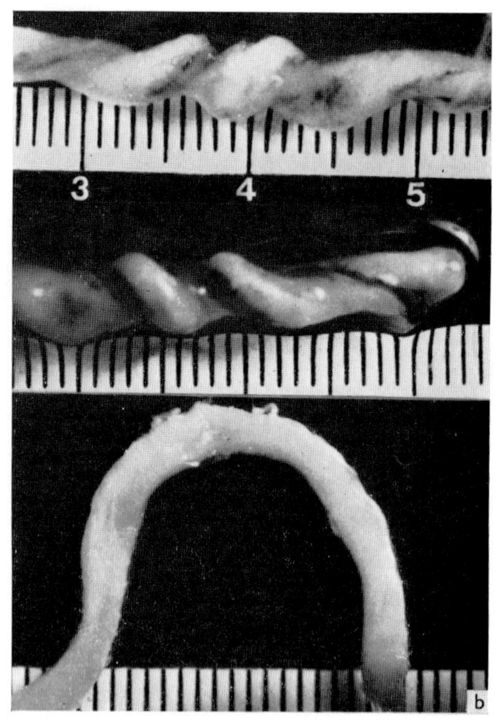

(*a*) A longitudinal section through a human umbilical artery at the junction of a constricted (left) and a dilated portion (right). The 'V' indicates what has been called a valve of Hoboken. For discussion, see text.
(*b*) Photograph of a human umbilical artery to show the role of the small coiling muscle. The top panel shows a coiled artery. The second panel shows the coiling muscle marked with black ink. The third panel shows the dramatic effect of removing this muscle.

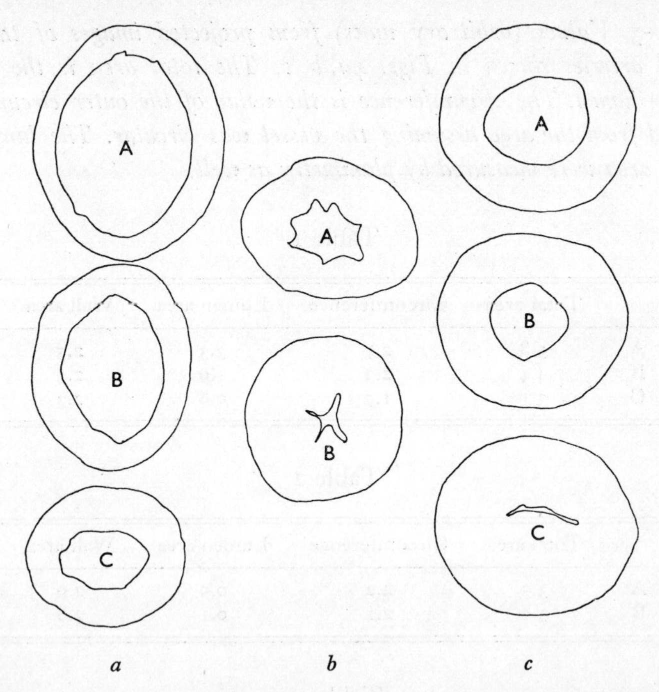

Fig. 2*a*. Line drawings from enlargements of a series of segments of a partially contracted sheep umbilical artery. Only circular fibres appeared to have contracted. Measurements are in Table 1.
Fig. 2*b*. Line drawings from two partially contracted segments of the same artery. Both have stellate lumens with prominent longitudinal muscle bellies. The measurements are in Table 2.
Fig. 2*c*. Line drawings from three segments of another artery. A is widely open; B partially closed but with a circular lumen (circular fibres); and C completely closed (crescentic lumen) due to a large belly of longitudinal muscle. See Table 3.

In sheep umbilical arteries the lumen shape varies as closure occurs. The number of protrusions into the lumen varies from one (crescentic) to five, with most having three or four. In all cases the protrusion was produced by a belly of longitudinal muscle pushing into the lumen. Thus the number of muscle bundles, and their arrangement, determined the shape of the lumen.

Plate 3 shows that the sheep umbilical artery contains five layers of muscle, three circular ones, separated by two longitudinal ones. It is the inner longitudinal layer that appears to contract and produce closure, probably stretching the inner circular layer in the process. The outermost circular layer appears histologically different (darker staining with haematoxylin and eosin) as shown in Plate 3, and may have peculiar contractile or elastic properties which makes it easier for the contracting

Tables 1–3. *Values (arbitrary units) from projected images of the sheep umbilical arteries shown in Figs. 4a, b, c. The total area is the area of the wall + lumen. The circumference is the value of the outer circumference calculated from the area assuming the vessel was circular. The lumen area and wall area were measured by planimetry as well*

Table 1

	Total area	Circumference	Lumen area	Wall area
A	5.8	2.7	3.3	2.5
B	4.4	2.3	1.9	2.5
C	3.1	1.7	0.8	2.3

Table 2

	Total area	Circumference	Lumen area	Wall area
A	3.5	2.2	0.5	2.9
B	3.8	2.1	0.1	3.8

Table 3

	Total area	Circumference	Lumen area	Wall area
A	5.1	2.5	1.3	3.8
B	4.7	2.4	0.8	3.9
C	5.1	2.5	0.04	5.1

longitudinal muscle to push into the lumen where the resistance is less. In this case (Fig. 3) less shortening is required than if the muscle belly bulged in both directions as it shortened.

Comparable measurements have not been made on the human umbilical artery since it contracts segmentally. However, Plate 2 suggests that it also is closed by protrusion of the inner longitudinal layer of muscle into the lumen. We are not sure if there is an absence of this layer in the non-contracted segments, or if there are merely different attachments of the longitudinal muscles in the two species. Carter and I (unpublished) have measured the thickness of the different layers of muscle in serial sections of human umbilical arteries. As shown in Fig. 4, the longitudinal layer is much thicker in the closed zones than in the dilated ones. Theoretically (and we have no proof) this difference could be explained by a fibre arrangement of the type shown in Fig. 5. The overlapping ends in the

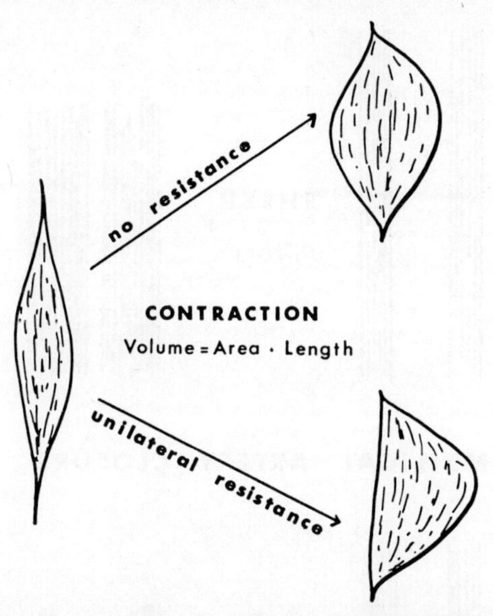

Fig. 3. Schematic diagram of the effect of an external resistance (e.g. the cord substance) on the contraction of a muscle. A 'free' muscle would widen in both directions on shortening, a 'restricted' muscle only in the direction of least resistance (e.g. toward the lumen).

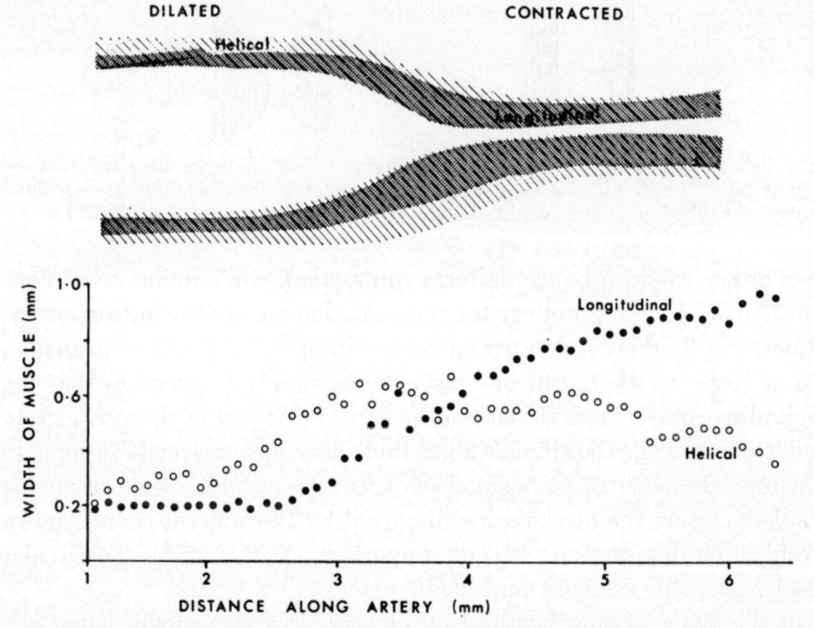

Fig. 4. Measurements of the thickness (width of muscle layer) of different orientations of muscle in a longitudinal section of a human umbilical artery (shown schematically at top). Note that the longitudinal layer (dark circles and cross-hatching) is thin where the artery is dilated and thick where it is constricted. The thickness of the other layers varies randomly along the length of the artery.

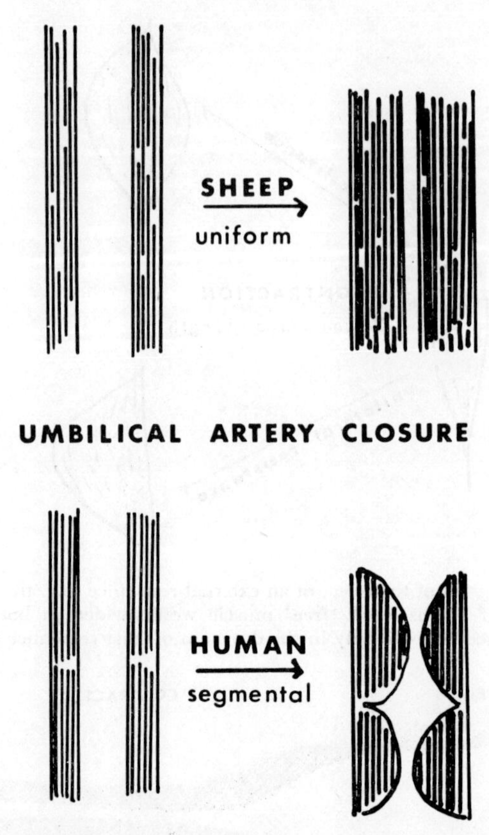

Fig. 5. Schematic diagram of a possible explanation for non-segmental contraction of the sheep umbilical artery with the ends of the fibres randomly arranged, and the segmental contraction of the human artery with the non-contractile ends of the fibres lined up.

sheep artery would produce uniform contraction, while lining up of non-contractile ends could produce the segmentation seen in the human artery.

Energetically there is another advantage in using longitudinal muscle to close a large vessel. It can be shown fairly simply (Appendix) that the longitudinal fibres must develop a tension (T_L) equal to the pressure to begin closure, while the circular fibres must develop a tension (T_C) equal to four times the pressure to begin closure. In the unstable region of almost complete closure the mechanisms discussed by Burton (1965) and known as critical closing pressure become important. At this stage, the circular muscles probably are most important.

On the basis of this biophysical analysis, it seems obvious that the circular muscle probably narrows the vessel (e.g. *in utero*), but the longitudinal muscle closes it. The obvious physiological conclusion is that the stimulus for closure must be one that makes the longitudinal muscle contract. Thus studies using helical preparations, or pressure–flow studies with the segment at fixed length, cannot be accepted as studies which indicate the mechanism for closure.

The valves of Hoboken

For over three hundred years since Hoboken first commented on the 'valves', there has been a debate in the literature about whether true valves exist in the umbilical vessels. Spivack (1936) reviewed the early literature, and concluded that thickenings occurred which were neither true valves nor artefacts. Chacko & Reynolds (1954) also studied them, and reached a similar conclusion. The function of these structures is unknown.

I believe that the valves of Hoboken are produced by the longitudinal muscle bellies described above. If the contraction is uniform (e.g. in the sheep) they appear grossly as folds while if it is segmental (e.g. in the human) they appear as valves. In longitudinal sections, one occasionally sees valve-like structures or tongues projecting into the lumen in the narrowed areas (Plate 4*a*). It is easy to see how this develops. A section through one of the stellate protrusions into the lumen (Plate 2*b*) leaves clear areas on either side. A longitudinal section through this plane produces an appearance similar to that in Plate 4*a*.

Since most valves are short and allow unidirectional flow (e.g. the heart valves), the valves of Hoboken do not appear to be true valves. What have been called 'valves' by others, I accept as an important mechanism for closure of the umbilical arteries. However, I believe that they are produced by contraction of the longitudinal muscle which we have demonstrated to be essential for closure of large vessels.

Stimulus for closure of the human umbilical artery

Since our evidence suggests that the longitudinal muscle is what closes the umbilical artery, experiments must be designed to test this muscle selectively. In the human there are four orientations of muscle: (i) a small circular muscle layer of varying thickness next to the lumen, (ii) a longitudinal layer next to this which we must test, (iii) a large helical bundle which winds around the outside of the artery with the same pitch as the coils of the cord, and with some connection to the cord substance, (iv) a small coiling muscle which coils the artery and appears to have the same pitch as the segments of the artery. To test the longitudinal muscle, it is necessary to inactivate, or minimize the effect, of the other three.

The apparatus used to study contraction of the longitudinal muscles of the human umbilical artery consists of a bath, with a fixed clamp which grips one end of the artery. The other end is fastened in a plastic clamp (two thin plates held together by screws) hooked to a Grass strain gauge (Model FT.03C). This is mounted on a moveable base which allows adjustment of the length. The artery is suspended in Ringer's solution, and can be aerated with various gas mixtures. The bath is kept at 37 ± 0.5 °C. The arteries were obtained within half an hour of delivery, had been cross-clamped to prevent closure and were stored in normal saline. The artery was gently dissected out, and mounted. The small coiling muscle was removed and the large one split. The segments studied were 2–3 cm long.

With this preparation we tested most of the standard drugs used by others (adrenaline, noradrenaline, acetylcholine, bradykinin, histamine, pitressin, and oxytocin) and found no change in tension at fixed length with a resting tension of 1 gm. Temperature (10 to 37 °C) also had no effect

Various gas mixtures (95% N_2+5% CO_2; 10% O_2+85% N_2+5% CO_2; 20% O_2+75% N_2+5% CO_2; and 50% O_2+45% N_2+5% CO_2 and 95% O_2+5% CO_2) were also studied. These produced the pO_2, pCO_2, and pH values shown in Table 4, within 30 seconds.

Table 4. *pO_2, pCO_2 and pH values produced in the bath after* 30–60 *seconds bubbling with the gas mixtures shown*

Gas mixture					
O_2	N_2	CO_2	pO_2	pCO_2	pH
—	95%	5%	10	45	7.25
10%	85%	5%	83	32	7.29
20%	65%	5%	150	35	7.28
95%	—	5%	680	38	7.27

The 95% O_2+5% CO_2 usually produced some increase in tone compared to 95% N_2+5% CO_2, as illustrated by a change in the tension–length curves (Fig. 6). However, it seldom produced a marked contraction unless the artery had been anoxic for one or two hours. In this situation oxygen produced a marked contraction (Fig. 7). The change from 10% to 20% O_2 (roughly equivalent to the change from the *in utero* to the postnatal environment) caused no significant change in either the tension–length curve, or in the degree of contraction (Fig. 7).

The most dramatic response was obtained with stretch. The strain gauge (with attached artery) was moved up by 1–3 mm (in 2–3 cm) for periods of 5–60 seconds. A typical response is seen in Fig. 8, which also illustrates that the process required oxygen. The tension created seemed to be the most important variable. If the artery was stretched until the

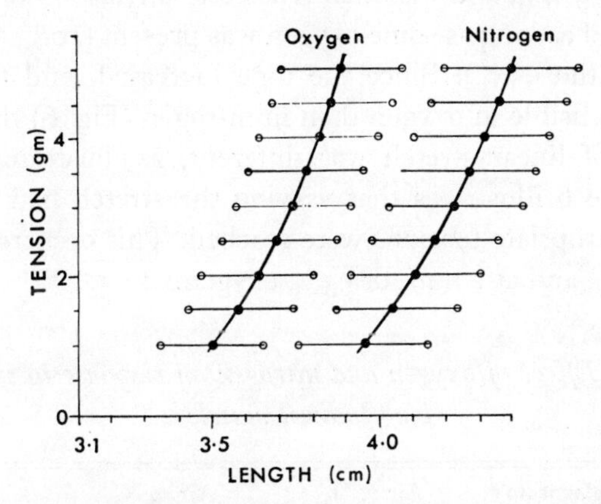

Fig. 6. Average longitudinal tension-length curves with standard error of the mean from twenty human umbilical arteries in 95 % O_2 and 95 % N_2, both with 5 % CO_2. Note that oxygen makes the artery stiffer (due to increased tone).

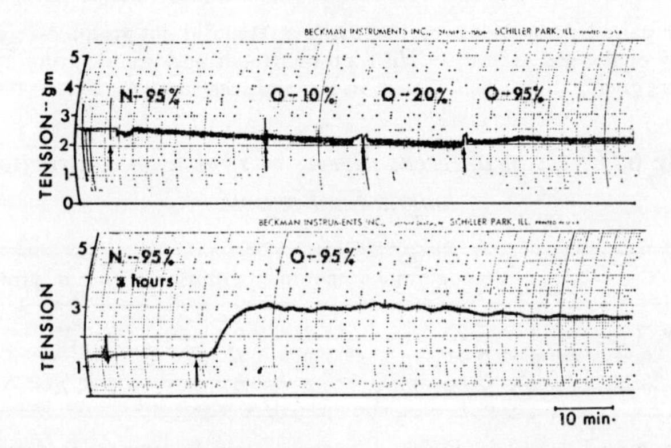

Fig. 7. Response of the longitudinal muscle of a human umbilical artery to varying gas tension. In the top panel, gases were changed every 20–30 min. The tone did not vary. By contrast, in the lower panel, if the artery was completely anoxic (here 95 % N_2+5 % CO_2 for three hours), then oxygen (here 95 % O_2+5 % CO_2) caused a contraction.

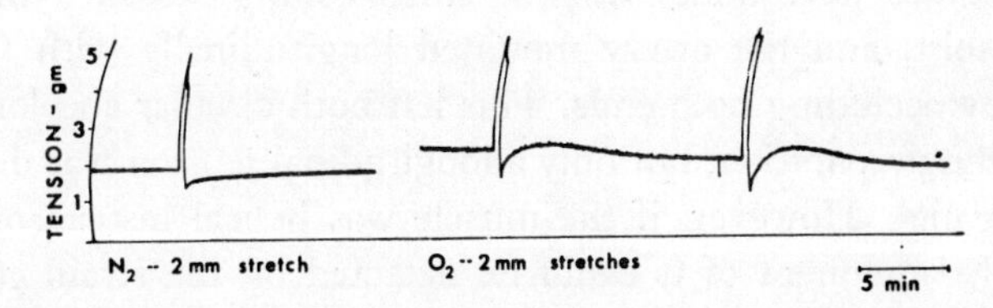

Fig. 8. Effect of a transient stretch on the longitudinal muscle of a human umbilical artery. Note the lack of response in nitrogen compared to oxygen, and the reproducibility of the response.

tension reached 5–6 g, and was then relaxed to a tension not less than 1.5–2 g, it contracted as long as some oxygen was present (10% seemed as useful for this as the 95%). Since the tone increased, and the artery was slightly less extensible in oxygen than in nitrogen (Fig. 6) the effect of the same amount of linear stretch was different, as illustrated in Table 5. However, Table 6 illustrates that varying the stretch had little effect as long as the appropriate tensions were reached. This occurred at 3 mm for air and nitrogen, and at 2 mm for 95% oxygen.

Table 5. *Effect of oxygen and nitrogen on response to stretch of longitudinal muscle*

Stretch	Relaxation	Air *v.* O_2	Air *v.* N_2	O_2 *v.* N_2
2 mm	2 mm	$p < 0.01$	$p < 0.01$	$p < 0.01$
2 mm	3 mm	$p < 0.02$	$p < 0.01$	$p < 0.01$
1 mm	1 mm	$0.5 < p < 0.6$	$0.3 < p < 0.4$	$0.05 < p < 0.1$

p values for paired data as shown. The 1 mm stretch did not produce a significant contraction in any of the gas mixtures. With 2 mm stretch and 2 and 3 mm relaxation the contraction was greatest in 95% O_2, next in air, and least in 95% N_2.

Table 6. *Effect of different degrees of stretch on contraction of longitudinal muscle*

	1 mm *v.* 2 mm	2 mm *v.* 3 mm	3 mm *v.* 4 mm
Air	$p > 0.9$	$p > 0.9$	—
95% O_2	$p < 0.01$	$p > 0.9$	—
95% N_2	$0.5 < p < 0.6$	$p < 0.01$	$0.7 < p < 0.8$

In air, contraction did not occur with < 3 mm stretch. In oxygen, contraction occurred with stretches of 2 and 3 mm but these were not significantly different. In nitrogen, contraction occurred at 3 and 4 mm stretch, but these were not significantly different.

These studies were done with the small coiling muscle removed, the large one split, and the artery mounted longitudinally with the lumen sealed off by occluding both ends. This left both circular and longitudinal muscle in the preparation, but only a longitudinal tension was detected by the strain gauge. However, if the muscle was helical instead of circular, the vertical component of it could be detected by the strain gauge (this would be equal to the helical tension multiplied by the cosine of the angle between this and the edge of the artery). The other problem with this preparation is that we know that the longitudinal muscles tend to protrude into the lumen when they contract, and the resistance to this is increased

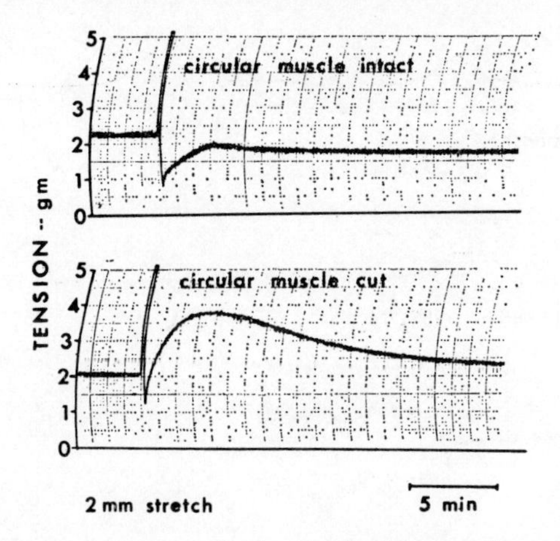

Fig. 9. Comparison of the response of a longitudinal preparation to a stretch of 2 mm in 20 % oxygen with an intact preparation (top), and with the circular muscles cut (bottom). The latter contracted more. See text.

if the lumen is closed at both ends. In an attempt to estimate the magnitude of these errors, we attempted, in sixteen cases, to split one side of the artery longitudinally while it was still mounted, so that we both opened the lumen, and also cut all of the circular fibres in the specimen without damaging more than a few of the longitudinal fibres. The results of the four successful attempts all resembled Fig. 9. Obviously the split preparations developed more tension than the unsplit ones, further confirming that the stretch was affecting the longitudinal muscles, rather than some that were circular or helical.

Oxygen toxicity and spontaneous contractions

A number of authors have commented on the fact that the umbilical arteries sometimes develop spontaneous contractions with a variable periodicity. We, too, have observed this in both our circular and longitudinal preparations, but particularly in the latter. This always occurred while we were using the 95 % O_2 mixture. An example is shown in Fig. 10. This occurred particularly in vessels with high tone (never with a tension less than 1.5 g), and in those exposed to high oxygen tensions for long periods (over one hour). They disappeared promptly if the pO_2 was dropped as shown in Fig. 10. This suggests that the spontaneous contractions are a manifestation of oxygen toxicity, since they occur at pO_2 values which this artery does not normally encounter.

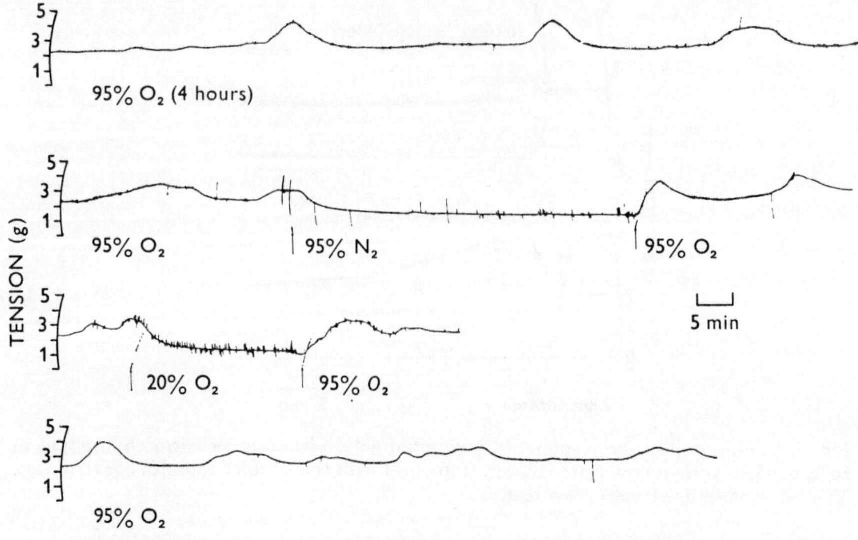

Fig. 10. Oxygen toxicity in the longitudinal muscle of a human umbilical artery. The top panel shows spontaneous contractions. The second one is a continuation and shows that the contractions stop if the 95 % O_2 is switched to 95 % N_2, and reappear if the 95 % O_2 is restarted without a prolonged period of N_2. The third panel shows that 20 % O_2 also stops the spontaneous contractions, suggesting that they are due to oxygen toxicity. The bottom panel shows a recurrence when the 95 % O_2 is restarted.

Practical implications of stretch as a stimulus for closure

Teleologically, stretch is a practical stimulus for closure. In almost all species the cord is longer than the uterus. Thus it is practically impossible to stretch it *in utero* unless it is abnormally short, or wound around the neck. However, after birth this constraint no longer applies. In fact, in animals, my veterinary colleagues tell me that the cord is usually just a little shorter than the distance from the maternal abdominal wall to the ground. Thus if the infant is delivered with the mother standing (as occurs in some sheep, according to Barclay *et al.* 1944), the weight of the infant will pull on the cord because of gravity. If the infant is delivered with the mother lying down (as occurs in others), then the cord will be stretched when the mother stands. This assumes that stretch may be the stimulus for closure in species other than the human. This assumption should be tested, and can be done readily with our technique.

EVIDENCE FOR OTHER TYPES OF VASCULAR SMOOTH MUSCLE IN THE HUMAN UMBILICAL ARTERY

The human umbilical artery, unlike that of the sheep, contains four different orientations of muscle. The longitudinal muscle has already been discussed. There is an *inner circular layer* which seems quite small. The work of Lewis (1968) suggests it may respond to oxygen and drugs. We studied it with pressure-flow studies as described by Lewis, but with the coiling muscle removed, and holding it at fixed length. We found that oxygen (20%–95%) produced contraction, while low oxygen (0–10%) caused relaxation. CO_2 had little effect (0–12%). Adrenaline (0.01–100 μg), noradrenaline (0.01–100 μg), and pitressin (0.05–5 units) produced some contraction. Acetylcholine (0.5–100 μg) and oxytocin (0.05–5 units) caused relaxation. The maximum change in resistance with all of these was about 25%, but in most cases was closer to 10%. The results were less consistent than for the longitudinal muscle, probably because there was more contraction of the circular muscle at the beginning of our experiments. Temperature (10–37 °C) had no effect. Stretch (due to a pressure increase from 60 mm Hg to 300 mm Hg) produced no contraction. The response to all agents was greatest in 95% O_2 + 5% CO_2 and least with 95% N_2 + 5% CO_2. There was no difference in the responses with 10% and 20% O_2.

The human artery also contains a large asymmetric muscle bundle in cross-section (Plate 2, Fig. 2 – large arrow). Histological study reveals that it sends fibres (collagen?) out into the Wharton's jelly. This muscle seems to have the same pitch as the coils of the cord, and probably produces *the coiling of the cord.* Preliminary studies suggest it contracts in response to some mixture of stretch and oxygen. We have had trouble isolating it without damaging it, so that our results may be wrong.

The fourth muscle is a small one, about eight fibres in size, which has not been described previously. It appears to have the same pitch as the segments of the artery (Plate 4*b*). *Coiling of the isolated artery* disappears if this muscle is removed (Plate 4*b*, lowest panel). It was tested with the same apparatus as the longitudinal muscle, and preliminary studies suggest it responds only to temperature changes. In eight preparations it contracted (reversibly) at 27 ± 0.2 °C. The elastic properties of this muscle do not change with gas tension, but do with temperature. It is interesting that it contracts at temperatures between those *in utero* and those at room temperature. Again one is tempted to think teleologically and suggest it may provide a twist around the narrow area to complete the closure.

The importance of this muscle *in vivo* has not been assessed, but I suspect the torque produced may be insufficient to overcome the resistance of the Wharton's jelly and cord substance. These experiments, while intriguing, would be extremely difficult technically, and so have not been attempted.

These observations have a number of important implications.

(1) They demonstrate that not all vascular smooth muscle is the same, and suggest that some previous studies with other arteries which are innervated may have responded to the different stimuli because of a direct effect on the muscle, rather than a neural effect.

(2) They demonstrate the importance of designing experiments to test directions of force, not just contraction in general.

(3) They may explain why other authors have obtained such variable results. The preparations used obviously could have contained one or more of these types of muscle, and the response to a given stimulus would depend on the percent of each type of muscle present.

(4) They suggest that longitudinal muscle may be much more important in vessel closure normally (umbilical artery and ductus) and abnormally (spasm) than previously suspected (see also Roach, 1970).

OXYGEN REQUIREMENTS OF THE UMBILICAL ARTERY

Most types of vascular smooth muscle function more efficiently aerobically than anaerobically. The oxygen reaches the muscle by diffusion from the lumen (intimal side) or via vasa vasorum (adventitial side). In both cases the oxygen actually reaches the muscle by diffusion, and so presumably obeys Fick's law, i.e.

$$\frac{dm}{dt} = -D \cdot A \cdot \frac{dc}{dx}$$

where dm/dt = the rate of transfer of oxygen per unit time, D = diffusion constant, A = area, and dc/dx is the concentration gradient per unit distance.

Our experiments, as well as those of others (e.g. Eltherington *et al.* 1968; Somlyo *et al.* 1965, etc.) show that oxygen (particularly in large concentrations) potentiates the effect of many other stimuli. This has two interesting implications. First, the pO_2 of the blood in the umbilical artery *in utero* is low; probably about 20–25 mm Hg. However, the wall is thin as the vessel is stretched; probably about 3 mm in the sheep and 2 mm in the human. One might question if this allows enough oxygen to reach the wall to permit contraction.

After delivery, the infant begins breathing and the pO_2 in the umbilical artery may rise to arterial levels of 80–100 mm Hg after a few breaths. If the umbilical artery has not closed, this increased oxygen may potentiate the response to whatever the stimulus is for closure. It is conceivable, but unlikely, that oxygen stimulates closure of the umbilical artery as it does for the ductus arteriosus.

The real problem arises, theoretically at least, in the days following delivery, and before the artery fibroses and seals completely. During this period there is no flow in the artery (its end is sealed), there are few if any functional vasa vasorum, and the thick covering of the cord, and the remains of Wharton's jelly, presumably prevent diffusion from the atmosphere.

Most muscle relaxes when it becomes anoxic, as oxygen is required to maintain tension. Colthart & Roach (1970) found that the oxygen consumption of isolated human umbilical arteries at 37 °C was 70–80 μl/g/h. Making a number of assumptions, we predicted that all available stored oxygen would be used up after two to three days in the human artery. This raises the interesting question of why the artery does not open at that time in response to the force generated by the arterial pressure. This rarely happens. However, my paediatric colleagues tell me that if they cannulate the artery several days after birth it rarely closes spontaneously or develops spasm, although both of these events occur frequently immediately after delivery.

We measured the oxygen consumption of human umbilical arteries for up to 24 hours postpartum and found no appreciable change with time. There was also no difference in different parts of the cord. Temperature had a marked effect (Fig. 11) and calculations (Colthart & Roach, 1970) indicated that the Q_{10} varied dramatically between 10 and 40 °C. This may indicate that the type of metabolic process is different at different temperatures. Fig. 11 suggests that contracted muscle may use less oxygen than relaxed muscle at 30 °C. However, we could not measure the energy required during the process of contraction, nor that of relaxation.

Colthart and I also found (unpublished) that sodium amytal, which decreases the Q_{O_2} of human umbilical arteries to 3.5 μl/g/h will cause relaxation of partially contracted arteries, but has no effect on closed arteries. This raises the question of whether the umbilical artery may be a 'catch' muscle similar to those of molluscs.

Ideally, we should measure the Q_{O_2} of each layer of the artery, and also do a full series studying the effects of tension and degree of contraction. So far neither of these studies have been technically feasible.

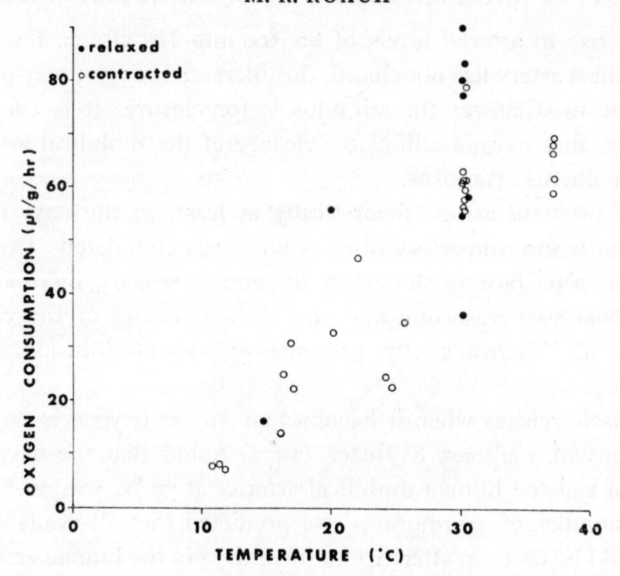

Fig. 11. Oxygen consumption of isolated human umbilical arteries. Note the marked increase with temperature. At 30 °C, the contracted arteries (open circles) seem to require less oxygen than the relaxed arteries (solid circles). The values were not significantly different ($p < 0.1$).

Vasa vasorum

Barclay, Franklin & Prichard (1944) note that most authors feel that there are few vasa vasorum in the umbilical vessels. Our own studies show they are common in the cow which has a very thick artery, and also present, though sparse, in arteries from sheep (Plate 1), pig and man. Presumably they reach the artery from Wharton's jelly, since vessels are commonly seen in it. We do not know where flow in these vessels originates, nor whether it arises in the placenta, or from the baby. Thus it is hard to predict what might happen to the flow in them after delivery. Since we have never seen vasa vasorum reaching the inner longitudinal layer of muscle (i.e. the one that closes the artery), the diffusion distance for oxygen to this muscle, even if the vasa vasorum remain patent, still seems great enough to create potential problems.

IS THE MUSCLE IN THE UMBILICAL ARTERIES A 'CATCH' MUSCLE?

Many molluscan muscles are different from other types of smooth muscle in that they appear to develop contractures which are different both physically and chemically (Johnson & Twarog, 1960) from the contractions which develop in other types of muscle. Twarog (1967) has discussed some of the unique features of these muscles, including their characteristic electron misroscopic features which include an absence of the *Z*-band which is replaced with separate dense bodies and vesicles. The muscle fibres are large, usually about 5 μm in diameter, compared with other smooth muscles which are only about 2 μm in diameter. There are two sizes of filaments instead of the single size range of about 50 Å diameter seen in vertebrate smooth muscles. There do not seem to have been comparable studies on the fine structure of the various muscles of the umbilical arteries. These obviously should be done. Spiteri, Nguyen, Anh & Panigel (1966) have done a study on human cord arteries, but do not state which part of the wall, and hence which muscle, was studied.

Twarog (1967) also notes that the maximum tension developed with tetanus in catch muscles is 10–12 kg/cm^2, which is three or four times as great as in any known non-catch muscle. We found in tightly contracted segments of sheep umbilical arteries, 3 cm long and with an estimated luminal perimeter of 0.3 cm, that 600 mm Hg would not make them dilate. This is equivalent to a force of about 200 kg/cm^2. Since the artery did not open with this force we must conclude that the tension in its wall is adequate to resist this magnitude of force. This supports my idea that these muscles may be similar to the catch muscles of molluscs.

We also found a very dramatic difference between the static elastic properties of contracted and relaxed segments of sheep umbilical arteries. Typical curves obtained by the method of Roach & Burton (1957) are shown in Fig. 12. Obviously the two curves are quite different. Crude calculations (making many assumptions which could be incorrect) suggest the bulk modulus or coefficient of elasticity in the physiological pressure range is about 10^4 dynes/cm^2 in the relaxed preparation and between 10^8 and 10^9 dynes/cm^2 in the contracted preparation. The latter values are comparable to those of collagen, and much higher than any previously reported for muscle. The vessels often appeared completely closed, and so it is difficult to say how strong the surface tension forces were at low pressures. The results in human arteries were even harder to interpret because of the segmental closure. Both human and sheep umbilical arteries showed marked hysteresis in the pressure–volume curves of relaxed (dilated) arteries, but none in the contracted arteries. The latter observation again supports the idea of a contracture, since most authors believe contracted muscle still displays viscous properties and hence hysteresis. A dynamic study of the type described by Bergel (1961) should be done to assess this viscous effect which is characteristic of muscle. To my knowledge, no studies of this type have been done on molluscan muscle, and this too should be done.

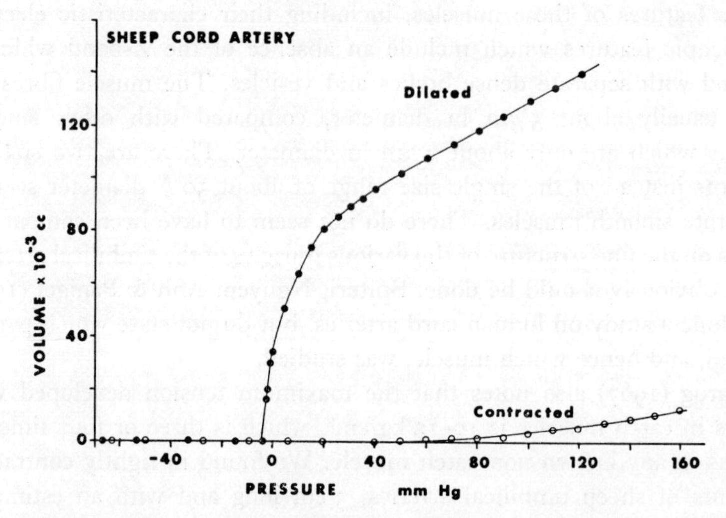

Fig. 12. Volume–pressure curves for a sheep extra-abdominal umbilical artery. The dilated artery was very distensible compared to the contracted one which was extremely resistant to opening. The length–pressure curves were similarly different.

Twarog (1967) and others have also pointed out that the catch muscles of *Mytilus* and other molluscs contain large amounts of a specific protein, paramyosin, as well as actin and myosin. They feel this may be involved in producing the contracture. In view of the apparent differences in response of the different muscle layers of the human umbilical artery, I feel it would be useful to assess the protein content of each of these layers to determine if any, or all, of them contain paramyosin.

While there is a substantial amount of presumptive evidence to suggest that the muscle in the umbilical arteries may be like a catch muscle, specific studies must be done to confirm this.

OTHER STRUCTURAL COMPONENTS OF THE ARTERY

The umbilical arteries are probably the most muscular arteries that occur in mammals. Almost all, if not all other arteries, contain an internal elastic membrane, and most of the extracranial ones also contain numerous circular layers of *elastin* in the media. The umbilical arteries in most species contain very little elastin. The pig arteries seem to contain several heavy elastin layers near the lumen, while the elastin in cow and sheep (Fig. 5) cord arteries is much finer. In the human most of the elastin seems to be oriented longitudinally. The role the elastin plays is unknown. The ductus arteriosus has much heavier elastin which wrinkles when it closes, and may play a role in delaying closure (in comparison to the time required for closure of the umbilical artery).

Collagen does not appear to exist in the umbilical vessels, at least at the light microscope level. There is considerable collagen in the cord substance, and scattered through the jelly however.

Ground substance seems present in much larger amounts in these arteries than in others I have studied. In most cases it seems present in larger amounts on the luminal side. It may act as a lubricant to allow the longitudinal muscle to slide up between the other muscle layers when it shortens to cause closure. This is pure speculation on my part, and must also be tested experimentally.

SUMMARY

The umbilical arteries are intriguing structures for many reasons. They are long, with few branches. They contain almost pure muscle with small amounts of elastin and little if any collagen. They appear to have no nerve supply. They are also the only large vessels to close normally and quickly.

This investigation of umbilical arteries has demonstrated that these arteries are much more complex than previously supposed. I have proved that:

(1) We must not ignore the orientation of fibres in designing physiological experiments. Theory and experiment have demonstrated that only longitudinal muscle can close vessels of this size.

(2) In the human umbilical artery there are four different orientations of muscle. We have evidence that three of these are functionally different as they respond selectively to a number of stimuli. Thus the type of preparation used will determine the most effective stimulus. Interpretation of data from helical strips of unknown pitch seems extremely complicated under these conditions.

(3) The oxygen consumption of human umbilical arteries is comparable to that of other types of vascular smooth muscle. Since oxygen is required for contraction, this poses an unanswered question as to how these arteries remain contracted when all of the stored oxygen is used up.

(4) It is important to ask specific questions before doing experiments. This supports Sir Joseph Barcroft's advice that: 'The worker in the field of foetal physiology must constantly battle against . . . the tendency to obtain more data from an experiment than he is justified in doing' (Barcroft, 1946; p. 188). Since all of the structures in the cord are interrelated, each one must be studied separately to test its individual properties, and then assessed *in situ* to see if the surroundings modify these

properties. While I realize that physiologists feel that only *in vivo* experiments are trustworthy, I believe as a biophysicist that one must understand the parts (e.g. the electrical components) before assessing how their attachments (e.g. the wiring) create the end result (e.g. the black box).

This work was begun while I was a Postdoctoral Fellow at the Nuffield Institute for Medical Research in Oxford. Dr G. S. Dawes and Miss M. M. L. Prichard spent many hours listening to my arguments about 'eliminating holes' and thus speeded up my discovery of the importance of the longitudinal muscle. In Canada, my colleagues in Biophysics, with their arguments, have also helped to clarify my ideas as outlined here. Dr Peter Canham helped with the analysis in the Appendix. Misses Carol Colthart and Mary Humeniuk, and Mrs Phyllis Carter, helped with many of the experiments. The Delivery Room staff at Victoria and St Joseph's Hospitals in London collected the cords. Mrs Vera Jordan typed the manuscript, and Miss Mary Jane Pender did the photography. This work was supported by the Medical Research Council of Canada.

APPENDIX: ENERGETICS OF CLOSURE OF THICK-WALLED ARTERIES

(*a*) *Longitudinal fibres*

Assume shortening occurs inside a constant diameter sheath and that muscle volume is constant.

Work done on the lumen equals $P \cdot \Delta V$ where P = pressure, and ΔV = change in volume.

The work to shorten fibres with a tension $T_L = T \cdot A \cdot \Delta L$

$\Delta V = A \cdot \Delta L$

$\therefore P \cdot \Delta V = T_L \, A \cdot \Delta L$

$\therefore P = T_L$

$\therefore$ tension = pressure to initiate contraction.

(*b*) *Circumferential fibres*

Work = $F \cdot D$ where F = force, and D = diameter

$= 2T_c(R_o - R_i) = P \cdot \pi \cdot D_i$ where R_o and R_i are the outside and inside radii, and T_c is the circumferential tension

$\therefore 2T_c(R_o - R_i) = P \cdot D_i$

$$\therefore T_c = \frac{PR_i}{R_o - R_i}$$

If $R_i = 0.8R_o$, then $T = 4P$.

Note that the tension is constant in the longitudinal direction but a function of geometry in the circular fibres. This is fundamental to understanding the critical tension which creates a critical closing pressure (see Burton, 1965).

REFERENCES

ALTURA, B. M., MALAVIYA, D., REICH, C. F. & ORKIN, L. R. (1972). *Am. J. Physiol.* **222**, 345–55.

BARCLAY, A. E., FRANKLIN, K. J. & PRICHARD, M. M. L. (1944). *The Foetal Circulation.* Oxford: Blackwell Scientific Pub. Pp. 71–5: 119–25; 154–61.

BARCROFT, J. (1946). *Researches on Pre-Natal Life.* Oxford: Blackwell Scientific Pub. P. 188.

BERGEL, D. H. (1961). *J. Physiol.* **156**, 458–69.

BURTON, A. C. (1965). *Physiology and Biophysics of the Circulation.* Chicago: Year Book Med. Pub. Pp. 78–83.

CHACKO, A. W. & REYNOLDS, S. R. M. (1954). *Contrib. to Embryology* **35**, 135–50.

COLTHART, C. & ROACH, M. R. (1970). *Can. J. Physiol. Pharmacol.* **48**, 377–81.

DAVIGNON, J., LORENZ, R. R. & SHEPHERD, J. T. (1965). *Am. J. Physiol.* **209**, 51–9.

DAWES, G. S. (1968). *Foetal and Neonatal Physiology.* Chicago: Year Book Med. Pub. Pp. 66–78.

EHINGER, B., GENNSER, G., OWMAN, C., PERSSON, H. & SJÖBERG, N. O. (1968). *Acta Physiol. Scand.* **72**, 15–24.

ELTHERINGTON, L. G., STOFF, J., HUGHES, T. & MELMON, K. L. (1968). *Circ. Res.* **22**, 747–52.

FOX, H. & JACOBSON, H. N. (1969). *Am. J. Obstet. Gynec.* **103**, 384–9.

GOKHALE, S. D., GULATI, O. D., KELKAR, L. V. & KELKAR, V. V. (1966). *Br. J. Pharmacol.* **27**, 332–46.

HÜLSEMANN, M. (1971). *Z. Zellforsch.* **120**, 137–50.

JOHNSON, W. H. & TWAROG, B. M. (1960). *J. gen. Physiol.* **43**, 941–60.

LEWIS, B. V. (1968). *J. Obstet. Gynaecol. Br. Commonw.* **75**, 87–91.

NADKARNI, B. B. (1970). *Am. J. Obstet. Gynec.* **107**, 303–12.

ROACH, M. R. (1970). *Thromb. Diath. Haemorrh. Suppl.* **40**, 59–77.

ROACH, M. R. & BURTON, A. C. (1957). *Can. J. Biochem. Physiol.* **35**, 681–90.

SOMLYO, A. V., WOO, C. & SOMLYO, A. P. (1965). *Am. J. Physiol.* **208**, 748–53.

SPITERI, M., NGUYEN, J., ANH, H. & PANIGEL, M. (1966). *Pathol. Biol.* **14**, 348–57.

SPIVACK, M. (1936). *Anat. Rec.* **66**, 127–48.

SPIVACK, M. (1943). *Anat. Rec.* **85**, 85–109.

TWAROG, B. M. (1967). *J. gen. Physiol. Suppl.* **50**, 157–69.

PLACENTAL TRANSFER AND FOETAL EFFECTS OF DRUGS WHICH MODIFY ADRENERGIC FUNCTION

By G. R. VAN PETTEN

Division of Pharmacology and Therapeutics,
University of Calgary, Alberta, Canada

Current evidence indicates that most drugs, when given to the pregnant female, cross the placenta and enter the foetal circulation. This view is largely based upon biochemical measurement of drugs in cord or foetal blood following their maternal administration. In contrast, few data are available on the pharmacodynamic effects of drugs in the foetus following maternal administration. Such information is important for two main reasons. First, the foetal sensitivity or responsiveness to drugs may differ from the adult. Second, the blood concentration of drugs is not always directly related to the pharmacodynamic effect. In either case knowledge of the concentration of drug in foetal blood represents only an indirect way of assessing the pharmacological or toxic effects on the foetus of maternal drug administration.

In an attempt to circumvent this problem, experiments have been carried out using the chronically cannulated ovine foetus. The placental transfer of β-adrenoceptor blocking agents and tricyclic antidepressives has been assessed by measuring foetal cardiovascular effects following maternal drug administration. These drugs were chosen because of their possible use during pregnancy, the close chemical similarity of the members of each group and because the drugs in both groups modify adrenergic function.

Chronic arterial and venous cannulas and electrocardiograph leads were placed in the ewe and foetus at 110 days of gestation. This permitted the simultaneous measurement of maternal and foetal blood pressures and electrocardiograms in the unanaesthetized animal. The maternal and foetal effects of the β-adrenoceptor blocking agents were assessed by measuring their ability to inhibit the chronotropic response to isoprenaline. For example, the control response to intravenous injection of isoprenaline into the ewe and foetus was determined. The β-adrenoceptor blocking agent was then infused into the ewe and thereafter additional doses of isoprenaline were injected into the ewe and foetus at various times. Similarly, the maternal and foetal effects of the tricyclic antidepressives on the cardiovascular response to noradrenaline were determined. Noradrenaline was administered to the ewe and foetus and control pressor responses recorded. The antidepressives were then infused into the ewe and further doses of noradrenaline administered to the ewe and foetus at various times.

Previous observations (Van Petten & Willes, 1970) revealed that the β-adrenoceptor blocking agent, propranolol, rapidly crossed the ovine placenta and induced a β-adrenoceptor blockade lasting 9 hours in the foetus compared with 3–4 hours in the ewe; this prolonged β-blockade in the foetus did not correlate with foetal blood levels of propranolol. Seven other β-adrenoceptor blocking agents have been studied. The β-adrenoceptor blockade following i.v. infusion of USVP 65-24, bunolol, oxprenolol and butidrine into the pregnant ewe were 5, 8, 3 and 3 hours respectively in the ewe and 5, 8, 8, and 3 hours respectively in the foetus. The β-blockade produced by AY21011 was of 8 hours duration in the ewe but less than 1 hour duration in the foetus; AH3474 and sotalol were effective in producing a β-blockade in the ewe but did not inhibit the effects of isoprenaline in the foetus when they were administered to the ewe (Truelove, Van Petten & Willes, 1972. *Br. J. Pharmac.* In press).

All of the tricyclic antidepressives used (imipramine, desipramine, amitriptyline and nortriptyline) were found to potentiate the pressor response to noradrenaline in the ewe. Following infusion of imipramine, desipramine and nortriptyline into the ewe, the foetal pressor response to noradrenaline was also potentiated; the duration of the potentiation was 5, 11 and 5 hours respectively in the ewe and 9, 5 and 1.5 hours respectively in the foetus. Maternal administration of amitriptyline did not alter the foetal response to noradrenaline.

These pharmacodynamic measurements of the foetal effects of maternally administered β-adrenoceptor blocking agents and tricyclic antidepressives clearly indicate gross differences in the amounts of individual drug reaching the sites of action in the foetus. Thus, within each of these two groups of drugs, members were found which did not have an effect on the foetus, produced a similar effect in the foetus or had a shorter or longer duration of effect in the foetus compared with the ewe. From the standpoint of predicting the safety of administering drugs during pregnancy, it is suggested that such direct measurement of foetal effects of drugs may be much more useful than measurement of levels of drug in foetal or cord blood.

REFERENCE

Van Petten, G. R. & Willes, R. F. (1970). *Br. J. Pharmac.* **38**, 572–82.

THE RENIN–ANGIOTENSIN SYSTEM IN FOETAL AND NEWBORN MAMMALS

By JOAN C. MOTT
Nuffield Institute for Medical Research,
University of Oxford

INTRODUCTION

A systematic investigation of the effect of haemorrhage on arterial pressure showed that infant rabbits and kittens sustained bleeding better than adult rabbits and cats respectively (Mott, 1965, 1968). This superior ability of the young rabbit was unaffected by interference with the autonomic nervous system but was impaired by acute nephrectomy (which removes the supply of the enzyme renin) to a far greater extent than in the adult (Mott, 1969). The possibility thus arose that the renin–angiotensin system plays a larger part in cardiovascular control in immature than in adult animals.

The purpose of this review is to consider this idea, which requires a preliminary summary of the renin–angiotensin system in adult mammals.

In 1940 Kohlstaedt, Page & Helmer demonstrated the indispensability of plasma for the development of the pressor action of the substance in mammalian kidney extracts named renin by Tigerstedt & Bergmann in 1898. Renin acts on an α_2 globulin in plasma to produce a decapeptide, now named angiotensin I (Page, Helmer, Plentl, Kohlstaedt & Corcoran, 1943). Angiotensin I is degraded by converting enzyme, originally described in plasma by Skeggs, Marsh, Kahn & Shumway (1954) to an octapeptide, angiotensin II. Ng & Vane (1967) showed that the pulmonary circulation, where they found 80% of angiotensin I was converted to angiotensin II, far exceeds the plasma in importance as a site of production of angiotensin II. Conversion does, however, also occur in other vascular beds (Aiken & Vane, 1972). Angiotensin II is the physiologically effective endpoint of the renin–angiotensin system (Page & Bumpus, 1961) and is 10–20-fold more active than angiotensin I.

The rise of arterial pressure induced by injection of renin lasts quite a long time, with a half-life in the circulation of about 15 min. In contrast, the action of angiotensin II lasts only a minute or two (Lee, 1969).

Comprehensive elucidation of the kinetics *in vivo* of the complex chain of reactions occurring between the liberation of renin by the kidney and the production of the effector substance, angiotensin II, in the arterial blood, has not yet been achieved. It seems likely, however, that substrate concentration can be an important factor (Skinner, Lumbers & Symonds, 1968). Angiotensin II is largely destroyed (50–70%) in one passage through systemic vascular beds in adult dogs (Hodge, Ng & Vane, 1967).

Consonant with the presence of renin in the kidneys of all vertebrates investigated, other than Agnatha and Chondrichthyes (Sokabe, Ogawa, Oguri & Nishimura, 1969), early workers found evidence of the presence of renal renin (Kaplan & Friedman, 1942) and of sensitivity to injected renin in foetal mammals (Burlinghame, Long & Ogden, 1942).

Among the questions which pose themselves concerning the renin–angiotensin system in immature mammals are: firstly, can renin (MW 40,000–43,000; Lee, 1969) or even the angiotensins (MW $\sim$ 1000) cross the placenta? And secondly, is the foetal lung with its relatively low blood flow fully capable of conversion of angiotensin I to angiotensin II? Arterial pressures and renal blood flows are low in some immature mammals compared with the corresponding adults (rabbit, Jarai, 1969; piglet, Gruskin, Edelman & Yuan, 1970; puppy, Kleinman & Lubbe, 1972); both these factors may well have a bearing on renin secretion.

Although angiotensin was first revealed by its pressor action (it is the most potent natural vasoconstrictor substance known), angiotensin II has several other physiological and pharmacological actions, none of which has yet been investigated in immature mammals. These include effects on:

(*a*) the adrenal cortex (Boyd & Peart, 1971) and medulla (Feldberg & Lewis, 1964);

(*b*) the area postrema and other central sites (Ferrario, Gildenberg & McCubbin, 1972);

(*c*) thirst (Fitzsimons, 1972);

(*d*) membrane transport (Munday, Parsons & Poat, 1972);

(*e*) heart (Dempsey, McCallum, Kent & Cooper, 1971).

Renin-like substances have also been described from tissues other than kidneys (notably some pregnant uteri) but none has yet been shown to be of physiological significance.

Measurement of components of the renin–angiotensin system

(1) Renal renin is measured in terms of the ability of kidney extracts to liberate angiotensin from excess of renin substrate. Renin preparations may exhibit species specificity (Braun-Menendez, Fasciolo, Leloir, Munoz & Taquini, 1946). For example, extracts of foetal lamb kidney which were pressor in foetal lambs were inactive in rats (Mott, unpublished).

(2) Renin is measured in plasma samples in terms of its ability to liberate angiotensin I from substrate. If excess substrate is added to allow maximal production of angiotensin I, the quantity measured is *plasma renin concentration* (PRC). If, however, renin is only allowed to act on endogenous substrate this may limit production of angiotensin I and the renin is thus measured as *plasma renin activity* (PRA) (Skinner, 1967).

The angiotensin I produced by either method is assayed as in (3) below or by its pressor action in a suitable preparation (Peart, 1955) using, preferably, synthetic angiotensin I as standard though synthetic angiotensin II is often employed. The great variety of systems in use renders direct comparisons of renin measurements from different sources difficult or impossible.

(3) Angiotensin I and angiotensin II can be determined by radioimmunoassay after suitable extraction from plasma samples (Catt, Cain & Coghlan, 1967; Waite, 1972) but current procedures require large samples of plasma.

(4) Angiotensin II can also be estimated by the contraction it produces in the isolated rat colon preparation irrigated intraluminally with pronethalol and superfused by an extracorporeal blood flow which is subsequently returned to the animal under investigation (Vane, 1969). Changes of angiotensin II-like activity so observed are calibrated by infusions of synthetic angiotensin II. No nonspecificity of this method has yet been demonstrated to be of practical importance, and it can detect changes of concentration down to about 100 pg/ml or, exceptionally, even less. At the moment it is the only method available for small animals and the only method yielding continuous records.

The kidneys and renal renin

The kidneys contribute about 2% of the body weight of the mouse but only 0.2% of that of an elephant (Kunkel, 1930; Spector, 1956) and in general kidney weight is inversely related to body weight in adult animals. It is well known that the concentration of renal renin is greater in smaller and younger animals (rats and calves, Grossman & Williams, 1938; rabbits, Pickering, Prinzmetal & Kelsall, 1942).

Goormatigh suggested that the granulated cells (Goormatigh, 1945) surrounding the afferent glomerular arteriole were the source of renin. Cook (1960) showed that renin is associated with the vascular pole of the glomerulus and later (1971) that it is contained in preparations of isolated granules. The degree of granulation of the juxtaglomerular apparatus can be expressed as an index (JGI) devised by Hartroft & Hartroft (1953, review Hartroft, 1966) who found that it was raised in salt deficiency and other conditions.

The morphology of the juxtaglomerular apparatus (including the polkissen and macula densa which are peculiar to mammals (Sokabe *et al.* 1969) has been discussed by Bing (1964). It should be noted that Hartroft & Hartroft (1953) specifically excluded macula densa cells from their procedure for the assessment of granulation. The relation of granulation to secretion of renin is not clear. Staining by the use of a fluorescent antibody technique correlated well with the JGI and biological activity of extracts of rabbit kidney (Hartroft & Edelman, 1961).

Circulating renin and angiotensin

The average arterial level of angiotensin II, determined by radioimmunoassay, in normal adult humans is 25 pg/ml (Catt, Cain, Coghlan, Zimmet, Cran & Best, 1970). The difference between angiotensin II-like activity in blood from intact and from nephrectomized young sheep when assayed on the rat colon preparation gave resting levels of $\leqslant 140$ pg/ml (Broughton Pipkin, Lumbers & Mott, 1972). The corresponding figure for adult rabbits was 225 ± 74 pg/ml (Broughton Pipkin, Mott & Roberton, 1971).

The kidney releases renin in response to a fall of intrarenal perfusion pressure (which may itself be mediated through several mechanisms) and to a reduction of Na^{+} levels (Vander, 1967). While there is dispute about the details of the cellular mechanisms concerned, there is little doubt that the renin–angiotensin system contributes to circulatory regulation in adult mammals (Cowley, Miller & Guyton, 1971; Hall & Hodge, 1971). Increased concentration of plasma renin is correlated with an increased rate of generation of angiotensin and an increased concentration of angiotensin in the blood (Brown, Lever, Robertson, Hodge, Lowe & Vane, 1967). It is interesting that renin production by slices of adult rat kidney *in vitro* has been reported to be independent of renin content (de Vito, Gordon, Cabrera & Fasciolo, 1970).

A standard haemorrhage of 25% blood volume (which reduces renal blood flow in conscious adult rabbits by 42% (Korner, Stokes, White & Chalmers, 1967) raised the circulating levels of angiotensin II-like activity in adult sheep by 240 ± 80 pg/ml and in adult rabbits by 340 ± 80 pg/ml respectively (Broughton Pipkin, Kirkpatrick & Mott, 1971; Broughton Pipkin, Mott & Roberton, 1971).

Mother and foetus

Maternal renin levels rise during pregnancy in women (Geelhoed & Vander, 1968) and in ewes (Lumbers & Trimper, 1972). The relation between maternal and foetal renin and angiotensin levels has not been systematically studied. Woodbury, Robinow & Hamilton (1938) found higher arterial pressures in the infants of toxaemic mothers. Discussion of the relation of the renin–angiotensin system to toxaemia is beyond the scope of this review.

The levels of renin found in human umbilical vein plasma were usually higher than those in maternal venous plasma (Brown, Davies, Doak, Lever, Robertson & Tree, 1964). Skinner, Lumbers & Symonds (1968) and Geelhoed & Vander (1968) have found plasma renin levels usually similar in umbilical artery and vein though in one of the latter's cases in which the infant exhibited respiratory distress the arterial level was considerably raised. Comparison of the uterine arterial and venous levels of renin suggests that renin activity is reduced as the blood traverses the maternal side of the placenta (Smith, Selinger & Stevenson, 1969).

The experiments of Hodgkinson, Hodari & Bumpus (1967) purporting to demonstrate transplacental passage of renin in both directions in dogs are open to other interpretations.

FOETAL AND NEWBORN MAMMALS

The degree of development reached by the kidney at birth varies very greatly. In species born in a mature condition like the lamb the histological appearance of the kidney of the newborn is not easily distinguishable from that of the adult. On the other hand in species such as the rabbit, born in a less mature condition, continuing nephrogenesis is obvious at birth and for some days thereafter. Except during the earliest stages of prenatal development, kidney weight forms a decreasing proportion of body weight as growth proceeds, to an extent similar to that found when the kidney weights of species of differing adult size are compared (see above). However, interesting exceptions to this general pattern are found in some species which are born in an immature condition. In the infant rabbit, there is a dramatic but transient increase in relative kidney weight in the first few weeks after birth (Broughton Pipkin, Mott & Roberton, 1971) and this phenomenon also occurs in puppies, (Kleinman & Lubbe, 1972) and kittens (Hall & MacGregor, 1937; F. Broughton Pipkin, unpublished).

Granulated juxtaglomerular cells have been found in foetal pigs of 15 cm crown–rump length upwards (Sutherland & Hartroft, 1964). Ljungqvist & Wagermark (1966) examined kidneys from human foetus and infants and found granulated juxtaglomerular cells at all ages from 15 weeks' gestation upwards. In general, the juxtaglomerular indices were low but in the foetal kidneys granular cells were encountered outside typical localisations; some such cells were clearly associated with developing glomeruli. Granger, Rojo-Ortega, Casado Pérez, Boucher & Genest (1971) found the JGI in newborn puppies to be significantly lower than in normal adult dogs; it was not consistently altered by 30 min peritoneal dialysis.

In newborn rats, careful histological examination (Endes, Dauda, Devenyi & Szucs, 1965) failed to reveal any granules in the kidneys, despite the fact that extracts of kidney cortex of the newborn contained pressor activity comparable with that of the adult in this species. Granules are present in the newborn rat kidney following administration of KCl 3% in the diet of the pregnant rat (Albrecht, 1968).

Bain & Scott (1960) described a series of human infants born alive and anephric (Potter's syndrome); thus the kidney is not necessarily essential to foetal survival but the infants in this series over 34 weeks' gestation were under-weight. The postnatal clinical picture was reported as respiratory distress and death was attributed to asphyxia rather than anuria. Interpretation of these observations is of course complicated by the associated pulmonary hypoplasia. Bilateral nephrectomy in the foetal monkey, sheep or miniature pig (for references see Berton, 1970) is compatible with survival to term. Berton, however, found in rabbits that bilateral nephrectomy at 25–29 days' gestation is fatal and has postulated that the diminution in amniotic fluid volume led to compression of the umbilical cord.

Acute bilateral nephrectomy in newborn rabbits (Broughton Pipkin, 1971) caused arterial pressure to fall 54% on the first two days of life and by lesser amounts in slightly older rabbits. No comparable phenomenon occurred in newborn lambs (F. Broughton Pipkin, S. M. L. Kirkpatrick & J. C. Mott, unpublished).

Renal renin

Kaplan & Friedman (1942) described the long-lasting pressor action of extracts of foetal hog kidney tested in trained dogs. The response was not inhibited by intravenous cocaine but was abolished in dogs showing tachyphylaxis to known hog renin. These properties were consistent with the presence of renin in the extracts; the renin concentration of the whole mesonephros declined with foetal growth from 17 to 99 mm CR length while that of the whole metanephros rose with foetal growth from 25 to 300 mm CR length to a level approaching that of adult kidney cortex.

Saline extracts of foetal kidneys of three lambs (gestation age of the youngest 75 days) caused a prolonged rise of arterial pressure when injected into the jugular vein of nephrectomised foetal lambs which was accompanied by contraction of superfused colon preparations (F. Broughton Pipkin & J. C. Mott, unpublished).

Grossman & Williams (1938) reported that the kidneys of calves contain more renin than those of adult cattle. Also, saline extracts of kidneys of rabbits up to about 1 kg bodyweight have several-fold the pressor activity (in rats) of extracts of adult kidney (J. C. Mott, unpublished).

In the kidney of foetal puppies not only is renin present but its concentration is increased in pups of bitches subjected to uterine ischaemia (Hodgkinson *et al.* 1967) and in pups of nephrectomised bitches (Hodari & Hodgkinson, 1968). Thus foetal renin concentration can be altered by experimental manipulation of the foetal environment, though the exact mechanism and consequences of these changes are unclear. In newborn puppies Granger *et al.* (1971) measured renal renin content and found values on average somewhat higher than in adult dogs.

In summary, it has been known for a long time that extracts of foetal kidney contain renin and the earliest observations (Friedman & Kaplan, 1942) put its genesis far back in gestation. Furthermore the concentration of foetal renal renin (Hodgkinson *et al.* 1967, Hodari & Hodgkinson, 1968) and the degree of juxtaglomerular granulation (Albrecht, 1968) can be altered by experimental means. In newborn and young animals there seems to be a greater concentration of renin in the kidneys, which in some species are also relatively large.

Circulating renin and angiotensin

While the presence of renal renin in foetal and newborn animals must be necessary for the functioning of the renin–angiotensin system it does not necessarily imply that function. However, since measureable levels of angiotensin II-like activity are demonstrable in foetal lambs (see below), it must be presumed that in this species at least, renin substrate and converting enzyme are also available in the foetus. The physiological effects of the angiotensin II produced will depend on the sensitivity of the systems concerned towards it and this is also related to its rate of destruction.

In the newborn rabbit (F. Broughton Pipkin, personal communication) angiotensin I appears to be converted to angiotensin II during passage through the lungs to a rather greater extent than in the adult. The issue is, however, complicated by a recent claim that some of the decapeptide is directly degraded to smaller peptides without intermediate production of the octapeptide (Ryan, Smith & Niemeyer, 1972).

Angiotensin II is largely destroyed in one passage through systemic vascular beds but, in the young rabbit, the liver is relatively inefficient in this respect (Table 1, Broughton Pipkin, Mott & Roberton, 1971; Broughton Pipkin, 1972).

Table 1. % *Inactivation* (± *SE of mean*) *of Hypertensin* (*Ciba*) *by hindquarters and liver in rabbits of different ages* (*from Broughton Pipkin, Mott & Roberton, 1971, and Broughton Pipkin, 1972*)

Age (days)	No. of rabbits	Hindquarters	Liver	
3–9	4	56.4 ± 3.5	—	
5–29	12	—	51.3 ± 3.1	$P < 0.001$
16–29	7	58.6 ± 2.1	—	
Adult	5	60.6 ± 2.2	—	
Adult	5	—	76.3 ± 2.4	

The half-life of the pressor action of injected angiotensin II is only a few minutes, and thus sustained high levels of circulating angiotensin II must imply sustained production.

Arterial pressor responses to angiotensin and renin

Foetus

Behrman & Kittinger (1968) infused 0.5–25 μg angiotensin into the femoral vein of foetal Macaca mulatta (body weight ~ 0.4 kg) in the course of a few minutes and observed large arterial pressor responses. However only the lowest doses employed were likely to have been in the physiological range. Assali, Holm & Sehgal (1962) injected angiotensin 0.1 μg/kg acutely into the femoral vein of foetal lambs and described the cardiovascular effects as less than those in the ewe. In other experiments, Adams, Assali, Cushman & Westersten (1961) had failed to observe any change of foetal blood pressure or flow following injection of a large dose of angiotensin (1.2 μg) into the umbilical vein. It is possible that angiotensin introduced by this route is largely inactivated during its passage through the foetal liver, but this point requires further study.

Burlinghame *et al.* (1942) described pressor responses to a renin preparation in foetal rats and we have obtained prolonged pressor responses in nephrectomised foetal lambs to injections of saline extracts of foetal lamb kidney.

Newborn

The rise of arterial pressure produced by intravenous injection of Hypertensin (Ciba) in lambs is about 15% at 0.03 μg/kg and 35% at 0.48 μg/kg and is of similar magnitude to that seen both in adult sheep and in two foetal lambs (one of which was nephrectomised) in which the jugular route was employed.

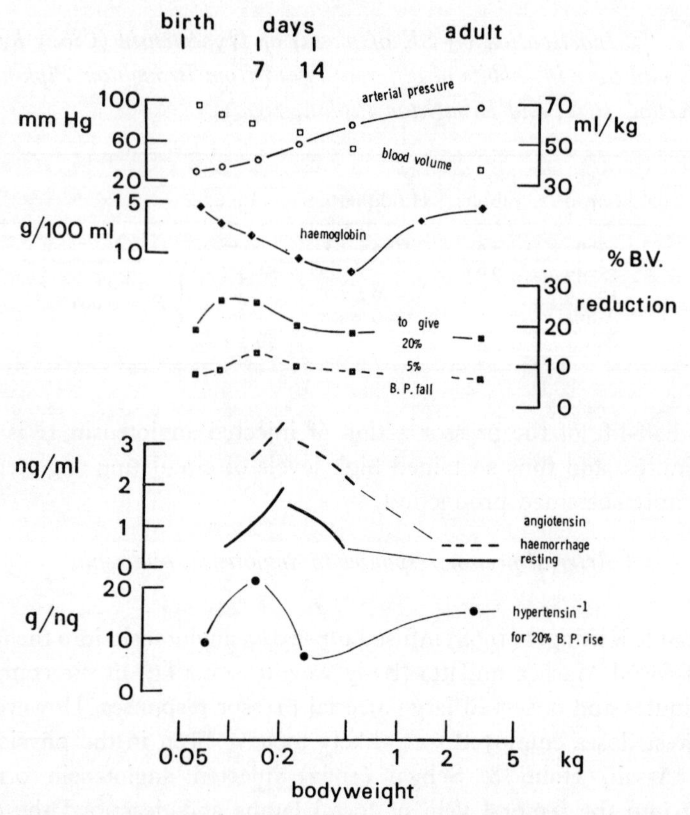

Fig. 1. The figure shows (lowest panel) the doses of hypertensin required to cause a 20% rise in blood pressure and (next panel upwards) the angiotensin II-like concentration in resting rabbits and on haemorrhage by 25% of their blood volume, all variable with age and body weight. Ordinates from above downwards: relative blood volume ml/kg □, arterial pressure, mm Hg, ○, haemoglobin, g/100 ml, ◆, % reduction of blood volume to give 5%, ■, and 20%, ■, fall of arterial pressure, angiotensin II-like activity, ng/ml, after 25% reduction of blood volume (---), and resting level (—), sensitivity, ● (hypertensin g/ng i.v. to give 20% rise of arterial pressure). Abscissae: top, approximate age scale, bottom, body weight (kg). Compiled from Dreyer & Ray (1911), Mott (1965), Broughton Pipkin, Mott & Roberton (1971), and Broughton Pipkin (1971).

Broughton Pipkin (1971 and personal communication) found in the rabbit that the cardiovascular response to injected Hypertensin and also to asp^{1}-val^{5}-angiotensin II free acid (the naturally occurring compound) was significantly increased at 7–10 days of age but decreased at 16–23 days of age compared with both newborn and adult rabbits (Fig. 1, lowest panel). There is no evidence of a general decrease in sensitivity towards angiotensin with postnatal age as occurs in rabbits with adrenaline (Broughton Pipkin, 1971). Grossman & Williams (1938) found that hog renin elicited larger pressor responses in older rats than in those aged 10 weeks.

In summary, there is no evidence of any qualitative difference from the adult in the mechanisms of the renin–angiotensin system in the foetus and newborn in so far as they have been examined. It remains to describe some quantitative differences found in infant rabbits and foetal and newborn lambs.

Foetus

Lumbers & Trimper (1972) have demonstrated that the renin levels (PRC) in arterial blood of foetal lambs (110 days' gestation age upwards) rose substantially in response to administration of the natriuretic, frusemide. Lambs in the last quarter of gestation (full term ~ 147 days) showed an increase in angiotensin II-like activity of 0.2–1.3 ng/ml as assessed by rat colon contraction in carotid arterial blood following a 25% reduction of foeto-placental blood volume (Broughton Pipkin, Lumbers & Mott, 1972). These increments are somewhat lower than in newborn lambs (see below) but direct comparison is inappropriate since the relative contributions of foetal tissues to the inactivation of circulating angiotensin may differ before and after birth.

Newborn

Angiotensin II-like activity in arterial blood of newborn lambs increased by 0.24–1.07 ng/ml when 25% of the blood volume was removed. This is greater than in corresponding experiments in adult sheep (0.17–0.47 ng/ml, Broughton Pipkin *et al.* 1971).

Infant rabbits deprived of 25% of their blood volume showed a bigger increase of arterial angiotensin II-like activity (1.68 ± 0.26 ng/ml) than adult rabbits (0.34 ± 0.08 ng/ml, Broughton Pipkin *et al.* 1971, Fig. 1, middle panels).

Angiotensin II-like activity also increased when kittens were bled (F. Broughton Pipkin, unpublished).

It is therefore clear that the renin–angiotensin system responds to conventional stimuli in the last quarter of gestation in the lamb and that in the newborn lamb and rabbit the increase of arterial angiotensin II-like activity in response to equivalent reduction of blood volume is greater than in the corresponding adults.

Resting levels of Angiotensin II-like activity

Foetus

The carotid arterial blood of an intact lamb foetus in the last quarter of gestation always contained more angiotensin II-like activity as determined by the rat colon assay (see above) than that of an acutely nephrectomised

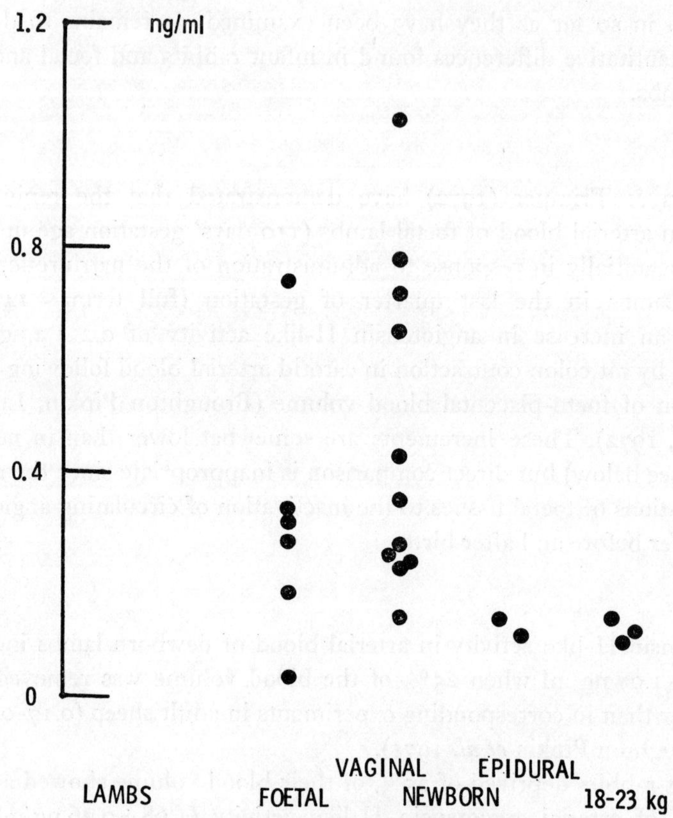

Fig. 2. Ordinate: angiotensin II-like activity in carotid blood in foetal, newborn and older lambs.

twin. Thus it seems unlikely that any maternal contribution to angiotensin II-like activity in the foetus could operate rapidly.

Renin levels were measured in acutely nephrectomised foetal and newborn lambs and were both very low and very much lower than in intact lambs (E. R. Lumbers, personal communication).

The difference between the values of angiotensin II-like activity recorded for the intact and nephrectomised twin has been accepted as the resting level and was in the range <0.04–0.74 ng/ml (Broughton Pipkin *et al.* 1972; Fig. 2).

Newborn lambs

The resting levels of angiotensin II-like activity similarly determined in lambs delivered naturally ranged from 0.14 to 1.03 ng/ml. This was in striking contrast to two lambs delivered under epidural anaesthesia of the ewe in which they were much less than angiotensin 0.14 ng/ml and similar to the levels found in 3 lambs 6–8 weeks old. (Broughton Pipkin *et al.* 1972; Fig. 2).

Rabbits

Infant rabbits had very much more angiotensin II-like activity in their blood (angiotensin 1.16 ± 0.16 ng/ml) than adult rabbits (0.23 ± 0.07). The resting levels increased significantly up to about 1.8 ng/ml at a bodyweight of 220 g (which is reached at about a fortnight old) and thereafter declined to the adult level. The maximum activity followed very closely the transient increase in relative kidney weight seen in this species (see above; Broughton Pipkin *et al.* 1971).

Kittens

In five kittens, resting levels of angiotensin II-like activity showed an increase with bodyweight and age over the first ten days of life concurrently with an increase in relative kidney weight as in rabbits (F. Broughton Pipkin, unpublished).

Puppies

Granger *et al.* (1971) found that plasma renin activity in puppies far exceeded that in adult dogs. It can be calculated from their data that in 14 puppies no more than 48 hours old PRA increased with increasing bodyweight ($P < 0.02$). This increase also appears to be concurrent with a postnatal increase in relative kidney weight (Kleinman & Lubbe, 1972). Kotchen, Strickland, Rice & Walters (1972) have found renin activity in plasma of normal newborn infants greatly to exceed that in normal adult plasma. It was also somewhat higher at 3–6 days of age than on the day of birth.

CONCLUSIONS

Interest in renin revived in the 1930s following the production of hypertension in dogs by renal artery constriction (Goldblatt, Lynch, Hanzal & Summerville, 1934). From quite an early date isolated observations relating to various aspects of the mechanisms concerned in the renin – angiotensin system were made in foetal and immature mammals. The contribution of this system to circulatory regulation in adult mammals is not unimportant (Cowley *et al.* 1971; Hall & Hodge, 1971) but in the newborn rabbit it appears in combination with increased sensitivity to angiotensin II to contribute substantially to their superior resistance to haemorrhage (Mott, 1969; Broughton Pipkin, 1971).

It is interesting that this superior resistance occurs during a phase of falling blood haemoglobin concentration (Fig. 1). In adult rats exposed to reduced air pressure plasma renin substrate levels rose during the first twenty-four hours, but fell as the haematocrit began to rise following an increase in erythropoietin (Gould & Goodman, 1970). It may be that tissue hypoxia arising from the physiological neonatal anaemia of the infant rabbit is responsible for the postnatal rise of resting angiotensin II-like activity. In this species circulating haemoglobin/kg bodyweight falls continuously in the first month of life (Mott, 1965) and the large falls of arterial pressure caused by acute nephrectomy are significantly greater in anaemic animals (Mott, 1969).

Although increased resting levels of angiotensin II-like activity are found in newborn lambs, it seems likely that this is a phenomenon different from that found in infant rabbits, because the resting levels were low in lambs delivered by Caesarian section under epidural anaesthesia of the ewe. The highest values in naturally delivered lambs were found in the three youngest and in one exhibiting respiratory difficulties (Broughton Pipkin *et al.* 1972). Various forms of trauma may accompany natural birth and since the renin–angiotensin system can respond to stimulation *in utero* it would not be surprising if in at least some cases parturition increased circulating angiotensin II-like activity. Whether the increased levels of angiotensin II-like activity promote or hinder survival of the newborn lamb remains to be determined.

Page & Bumpus (1961) in their review of angiotensin remarked that 'These humoral agents seem to us to belong to a humoral control system more primitive than the neural'. Their pronounced activity in immature mammals should not therefore occasion surprise.

I should like to thank my colleagues who have cheerfully endured the tedium involved in carrying out the experiments mentioned in this review, which has also benefited from the informed criticism of Dr G. S. Dawes, Dr E. R. Lumbers and Miss Fiona Broughton Pipkin. We acknowledge gratefully the help of a grant from the Spastics Society.

REFERENCES

ADAMS, F. H., ASSALI, N. S., CUSHMAN, M. & WESTERSTEN, A. (1961). *Pediatrics* **27**, 627–35.

AIKEN, J. W. & VANE, J. R. (1972). *Circulation Res.* **30**, 263–73.

ALBRECHT, I. (1968). *Biol. Neonat.* **12**, 233–8.

ASSALI, N. S., HOLM, L. W. & SEHGAL, N. (1962). *Am. J. Obstet. Gynec.* **83**, 809–17.

BAIN, A. D. & SCOTT, J. S. (1960). *Brit. Med. J.*, **1**, 841–6.

BEHRMAN, R. E. & KITTINGER, G. W. (1968). *Proc. Soc. exp. Biol. Med.* **129**, 305–8.
BERTON, J.-P. (1970). *C. R. Acad. Sci. Paris* **271**, 219–22.
BING, J. (1964). *Dan. med. Bull.* 24–33.
BOYD, G. & PEART, W. S. (1971). *Adv. Metabol. Disorders* **5**, 77–117.
BRAUN-MENENDEZ, E., FASCIOLO, J. C., LELOIR, L. F., MUNOZ, J. M. & TAQUINI, A. C. (1946). *Renal Hypertension*, Transl. Lewis Dexter. Springfield, Illinois: Charles C. Thomas.
BROUGHTON PIPKIN, F. (1971). *Q. Jl exp. Physiol.* **56**, 210–20.
BROUGHTON PIPKIN, F. (1972). *J. Physiol.* **225**, 35–6 *P*.
BROUGHTON PIPKIN, F., KIRKPATRICK, S. M. L. & MOTT, J. C. (1971). *J. Physiol.* **218**, 61P.
BROUGHTON PIPKIN, F., LUMBERS, E. R. & MOTT, J. C. (1972). *J. Physiol.* **226**, 109–10 *P*.
BROUGHTON PIPKIN, F., MOTT, J. C. & ROBERTON, N. R. C. (1971). *J. Physiol.* **218**, 385–403.
BROWN, J. J., DAVIES, D. L., DOAK, P. B., LEVER, A. F., ROBERTSON, J. I. S. & TREE, M. (1964). *Lancet* **2**, 64–6.
BROWN, J. J., LEVER, A. F., ROBERTSON, J. I. S., HODGE, R. L., LOWE, R. D. & VANE, J. R. (1967). *Nature, Lond.* **215**, 853–5.
BURLINGHAME, P., LONG, J. A. & OGDEN, E. (1942). *Am. J. Physiol.* **137**, 473–84.
CATT, K. J., CAIN, M. D. & COGHLAN, J. P. (1967). *Lancet* **2**, 1005–7.
CATT, K. J., CAIN, M. D., COGHLAN, J. P., ZIMMER, P. Z., CRAN, E. & BEST, J. B. (1970). *Circ. Res. Suppl.* 2, **27**, 177–93.
COOK, W. F. (1960). *J. Physiol.* **152**, 27–8*P*.
COOK, W. F. (1971). In *Kidney Hormones*. Ed. Fisher, pp. 117–28. New York and London: Academic Press.
COWLEY, A. W., MILLER, J. P. & GUYTON, A. C. (1971). *Circ. Res.* **28**, 568–81.
DEMPSEY, P. J., MCCALLUM, Z. T., KENT, K. M. & COOPER, T. (1971). *Am. J. Physiol.* **220**, 477–81.
DE VITO, E., GORDON, S. B., CABRERE, R. R. & FASCIOLO, J. C. (1970). *Am. J. Physiol.* **219**, 1036–41.
DREYER, G. & RAY, W. (1911). *Phil. Trans. R. Soc.* **101**, 138–60.
ENDES, P., DAUDA, G., DEVENYI, I. & SZUCS, L. (1965). *Acta. Physiol. Acad. Sci. Hungaricae.* **27**, 349–51.
FELDBERG, W. & LEWIS, G. (1964). *J. Physiol.* **171**, 98–108.
FERRARIO, C. M., GILDENBERG, P. L. & MCCUBBIN, J. W. (1972). *Circ. Res.* **30**, 257–62.
FITZSIMONS, J. T. (1972). *Physiol. Rev.* **52**, 468–561.
GEELHOED, G. W. & VANDER, A. J. (1968). *J. clin. Endocr. Metab.* **28**, 412–15.
GOLDBLATT, H., LYNCH, J., HANZAL, R. F. & SUMMERVILLE, W. W. (1934). *J. exp. Med.* **59**, 347–79.
GOORMATIGH, N. (1945). *Rev. belge Sci. med.* **16**, 65–155.
GOULD, A. B. & GOODMAN, S. A. (1970). *Lab. Inv.* **22**, 443–7.
GRANGER, P., ROJO-ORTEGA, J. M., CASADO PEREZ, S., BOUCHER, R. & GENEST, J. (1971). *Canad. J. Biochem. Physiol.* **49**, 134–8.
GROSSMAN, E. B. & WILLIAMS, J. R. (1938). *Arch. Int. Med.* **62**, 799–804.
GRUSKIN, A. B., EDELMANN, C. M. & YUAN, S. (1970). *Ped. Res.* **4**, 7–13.
HALL, R. C. & HODGE, R. L. (1971). *Am. J. Physiol.* **221**, 1305–9.
HALL, V. E. & MACGREGOR, W. W. (1937). *Anat. Rec.* **69**, 319–31.
HARTROFT, P. M. (1966). *Ann. Rev. Med.* **17**, 113–19.
HARTROFT, P. M. & EDELMAN, R. (1961). *Circ. Res.* **9**, 1069–77.
HARTROFT, P. M. & HARTROFT, W. S. (1953). *J. exp. Med.* **97**, 415–28.

HODARI, A. A. & HODGKINSON, C. P. (1968). *Am. J. Obstet. Gynec.* **102**, 691–701.
HODGE, R. L., NG, K. K. F. & VANE, J. R. (1967). *Nature*, **215**, 138–41.
HODGKINSON, C. P., HODARI, H. A. & BUMPUS, F. M. (1967). *Obstet. Gynec.* **30**, 371–80.
JARAI, I. (1969). *J. Physiol.* **202**, 559–67.
KAPLAN, A. & FRIEDMAN, M. (1942). *J. exp. Med.* **76**, 307–16.
KLEINMAN, L. I. & LUBBE, R. J. (1972). *J. Physiol.* **223**, 411–18.
KOHLSTAEDT, K. G., PAGE, I. H. & HELMER, O. M. (1940). *Am. Heart J.* **19**, 92–9.
KORNER, P. I., STOKES, G. S., WHITE, S. W. & CHALMERS, J. P. (1967). *Circ. Res.* **20**, 676–85.
KOTCHEN, T. A., STRICKLAND, A. L., RICE, T. W. & WALTERS, D. R. (1972). *J. Ped.* **80**, 938–46.
KUNKEL, P. A. (1930). *Johns Hopkins Bull.* **47**, 285.
LEE, M. R. (1969). *Renin and Hypertension.* Fig. 1 & p. 9. London: Lloyd-Luke.
LJUNGQVIST, S. & WAGERMARK, J. (1966). *Acta path. microbiol. Scand.* **67**, 257–66.
LUMBERS, E. R. & TRIMPER, C. E. (1972). *J. Physiol.* **225**, 34–5 *P*.
MOTT, J. C. (1965). *J. Physiol.* **181**, 728–52.
MOTT, J. C. (1968). *J. Physiol.* **194**, 659–67.
MOTT, J. C. (1969). *J. Physiol.* **202**, 25–55.
MUNDAY, K. A., PARSONS, B. J. & POAT, J. A. (1972). *J. Physiol.*, **224**, 195–206.
NG, K. K. F. & VANE, J. R. (1967). *Nature* **216**, 762–6.
PAGE, I. H. & BUMPUS, F. M. (1961). *Physiol. Rev.* **41**, 331–90.
PAGE, I., HELMER, O. M., PLENTL, A. A., KOHLSTAEDT, K. G. & CORCORAN, A. C. (1943). *Science* **98**, 153–4.
PEART, W. S. (1955). *Biochem. J.* **59**, 300–2.
PICKERING, G. W., PRINZMETAL, M. & KELSALL, A. R. (1942). *Clin. Sci.* **4**, 401–20.
RYAN, J. W., SMITH, U. & NIEMEYER, R. S. (1972). *Science* **176**, 64–6.
SKEGGS, L. T., Jr, MARSH, W. H., KAHN, J. R. & SHUMWAY, N. P. (1954). *J. Exptl Med.* **99**, 275–82.
SKINNER, S. L. (1967). *Circulation Res.* **20**, 391–402.
SKINNER, S. L., LUMBERS, E. R. & SYMONDS, M. (1968). *Am. J. Obstet. Gynec.* **101**, 529–33.
SMITH, R. W., SELINGER, H. E. & STEVENSON, S. F. (1969). *Am. J. Obstet. Gynec.* **105**, 1129–31.
SOKABE, H., OGAWA, M., OGURI, M. & NISHIMURA, H. (1969). *Texas Rep. Biol. Med.* **27**, 868–85.
SPECTOR, W. S. (1956). *Handbook of Biological Data.* Philadelphia.
SUTHERLAND, L. & HARTROFT, P. M. (1964). *Anat. Rec.* **114**, 342.
TIGERSTEDT, R. & BERGMANN, P. G. (1898). *Skand. Arch. Physiol.* **8**, 223–71.
VANDER, H. A. (1967). *Physiol. Rev.* **47**, 359–82.
VANE, J. R. (1969). *Br. J. Pharmac.* **35**, 209–42.
WAITE, M. (1972). *J. Physiol.* **222**, 88–9*P*.
WOODBURY, R. A., ROBINOW, M. & HAMILTON, W. F. (1938). *Am. J. Physiol.* **122**, 472–9.

FOETAL MATURATION OF CARDIAC METABOLISM

By KERN WILDENTHAL
The Pauline and Adolph Weinberger Laboratory for Cardiopulmonary Research, Departments of Physiology and Internal Medicine, University of Texas Southwestern Medical School at Dallas, 5323 Harry Hines Boulevard, Dallas, Texas 75235, U.S.A.

By the end of the second trimester of pregnancy the foetal heart has differentiated into a functioning but incompletely developed organ. Its metabolism, like its other properties, is only partially mature at that time. During the last trimester, major changes develop in the intrinsic metabolic capabilities of the heart.

Evaluation of cardiac metabolism in an intact foetus *in utero* is difficult, both for technical reasons and because neural and humoral factors may vary considerably and obscure the intrinsic changes that occur in the developing organ. Accordingly, to avoid the complexity of experimentation *in vivo*, a system for maintaining spontaneously beating hearts in organ culture has been used for studying changes in metabolic capability during late-foetal development. The system, which has been described in detail previously (Wildenthal, 1971), enables intact hearts from 14–22-day foetal mice to be maintained in a stable functional state under precisely controlled conditions. Fibroblastic outgrowth and 'dedifferentiation' of the explants do not occur, nor do the cultured hearts alter their beating rates or metabolic properties significantly over a period of several days. Conditions of culture are obviously different from conditions *in utero*, so that results are best regarded as reflecting general metabolic capability, rather than the actual metabolism of the intact foetus under any given condition *in vivo*.

Glucose and fatty acids are the major substrates used by the heart for providing energy (Opie, 1968–9). Studies of cardiac homogenates, cell fractions, and nonbeating fragments have revealed that activities of the various enzymes and metabolic pathways that are required for utilization of fatty acids and glucose vary considerably in the foetus as compared to the adult (Clark, 1971; Jolley, Cheldelin & Newburgh, 1958; Sippel, 1954; Warshaw, 1969, 1971; Wittels & Bressler, 1965). Mitochondrial number and structure are also different in the foetus (Mackler, Grace &

Duncan, 1971; Sordahl, Crow, Kraft & Schwartz, 1972; Warshaw, 1969). It might be supposed, therefore, that the pattern of glucose and fatty acid utilization of intact, normally-functioning organs would vary with the stage of foetal development.

Table 1. *Influence of foetal age on fatty acid utilization*

	< 16 days	17–19 days	> 20 days
(A) Uptake of medium-chain fatty acids (nmole/mg/hour)	2.6 ± 0.27	1.8 ± 0.26	1.0 ± 0.07
(B) Uptake of long-chain fatty acids (nmoles/mg/hour)	0.1 ± 0.05	0.2 ± 0.06	0.2 ± 0.05

Data were obtained by measurement of substrate concentrations in the medium before and after cultivation of intact, spontaneously-beating hearts of foetal mice for 2 days. Medium = 'medium 199' (Grand Island Biol.) + 35 % calf serum + 50 μg/ml insulin + 500 μmole/l of sodium oleate (B) or sodium octanoate (A). Atmosphere = 95 % O_2 + 5 % CO_2. Temperature = 37.5 °C. Data are reported as means ± 1 standard error from a minimum of 8 hearts.

As shown in Table 1, studies of beating mouse hearts in culture revealed that significant changes in metabolic capability did occur as the foetus matured. At all stages of late-foetal development the hearts were able to metabolize large amounts of octanoate, a medium-chain fatty acid, but could metabolize only relatively small quantities of long-chain acids. To ensure that differences in binding to albumin and cellular availability of the individual acids did not explain the differences observed, large amounts of sodium oleate (> 1500 μmole/l) were added to the medium; uptake of oleate rose significantly but still remained significantly less than that of octanoate. Interestingly, octanoate uptake was much greater in younger (14–15-day) hearts than in those from foetuses near term (20–21-day), whereas the uptake of long-chain acids, although remaining low, actually increased slightly with age. Although medium-chain fatty acids freely enter mitochondria where they are broken down by beta-oxidation, long-chain fatty acids cannot do so and must be bound to carnitine before they can penetrate the mitochondrial membrane and be oxidized (Opie, 1968–9). The results suggest, therefore, that the capacity to metabolize intra-mitochondrial fatty acids is already developed by the end of the second trimester and falls slightly with further development. In contrast, there seems to be a deficiency of the acyl-carnitine transferase system in the mouse heart throughout gestation. The deficiency may become relatively less marked by the end of the last trimester, but it remains present even then.

Utilization of glucose by cultured mouse hearts, like that of octanoic

Table 2. *Influence of foetal age on glucose and lactic acid metabolism*

	< 16 days	17–19 days	> 20 days
(A) Uptake of glucose (nmole/mg/hr)	8.4 ± 1.08	5.4 ± 0.58	4.0 ± 0.43
(B) Release of lactic acid (nmole/mg/hr)	7.1 ± 1.03	3.8 ± 0.22	2.3 ± 0.17
(C) Ratio of lactate release/glucose uptake	0.84 ± 0.047	0.73 ± 0.054	0.55 ± 0.047

Data were obtained as described in Table 1. Medium: '199' (glucose = 5.6 millimole/l, lactic acid = 0) + 50 μg/ml insulin. Atmosphere = 95 % O_2 + 5 % CO_2. Temperature = 37.5 °C.

acid, fell with foetal development (Table 2). Simultaneously, release of lactate from the hearts also fell. The ratio of lactate released to glucose consumed was not constant. As the heart matured, a smaller proportion of the glucose taken up could be accounted for by the production of lactate, suggesting that anaerobic glycolysis was becoming less important and alternative metabolic pathways, such as mitochondrial oxidation, were assuming relatively more importance as the foetus approached term.

The susceptibility of foetal cardiac metabolism to control by insulin was tested by maintaining hearts in the presence or absence of high concentrations of the hormone. High doses of insulin were anabolic at all ages tested, so that hearts cultured in the presence of insulin had improved amino acid balance and increased protein mass compared to controls (Wildenthal, 1972). Utilization of glucose and fatty acids was increased commensurately, but there was no preferential switch to glucose metabolism at any age tested (Table 3).

Table 3. *Influence of insulin on cardiac metabolism at various stages of foetal development*

	< 16 days		> 20 days	
	Insulin absent	Insulin present	Insulin absent	Insulin present
(A) Total cardiac protein content after 2 days in culture (μg)	125 ± 13.2	148 ± 18.0	493 ± 35.2	558 ± 13.1
(B) Glucose uptake (nmole/mg/hr)	6.8 ± 0.21	7.2 ± 0.61	2.5 ± 0.29	2.8 ± 0.16
(C) Octanoic acid uptake (nmole/mg/hr)	2.0 ± 0.13	2.4 ± 0.08	0.8 ± 0.03	0.9 ± 0.08

Data were obtained as described in Table 1. Uptakes were calculated on the basis of cardiac weight before explantation. Medium: '199' + 500 μmole/mg sodium octanoate. Atmosphere: 95 % O_2 + 5 % CO_2. Temperature: 37.5 °C. Matched litter mates were compared at each age.

In summary, studies of beating hearts from 14–22-day foetal mice maintained in organ culture have demonstrated that late foetal maturation is accompanied by progressive changes in cardiac metabolic capabilities. These changes suggest that cardiac glycolytic capacity becomes significantly reduced and other pathways gradually assume relatively greater importance as term approaches. Utilization of long-chain free fatty acids may increase slightly, whereas uptake of medium-chain acids falls, suggesting the possibility of partial maturation of the acyl-carnitine transferase system. Nevertheless, the ability to metabolize long-chain fatty acids remains much lower than in adults, indicating that full maturation of this system in the mouse heart probably occurs after birth as in several other species (Warshaw, 1971; Wittels & Bressler, 1965). Finally, even as late as the last day of gestation insulin causes no specific alteration in the pattern of glucose and fatty acid uptake in the present system, although it can affect protein metabolism as early as the beginning of the third trimester.

It should be emphasized that the exact timing of the precise pattern of foetal metabolic development in a specific organ may vary considerably from species to species. For example, insulin specifically alters glucose uptake in monkey skeletal muscle by early in the last trimester (Bocek & Beatty, 1969) but does not affect muscles from rats until just at or after birth (Blade & Britton, 1970).

Despite such discrepancies of timing, the general pattern of maturation of myocardial metabolism is probably similar in most mammalian species. Organ culture provides a useful system for quantifying maturation of metabolic capability in beating hearts under precisely controlled conditions.

The excellent technical assistance of Mrs Jacqueline Wakeland in performing these experiments is gratefully acknowledged. The work was supported by grants from the US National Heart and Lung Institute and the American Heart Association. Dr Wildenthal holds a USPHS Research Career Development Award.

REFERENCES

Blade, M. J. & Britton, H. G. (1970). *J. Physiol.* **209**, 38–9*P*.

Bocek, R. M. & Beatty, L. H. (1969). *Endocrinology* **85**, 615–18.

Clark, C. M., Jr. (1971). *Am. J. Physiol.* **220**, 583–8.

Jolley, R. L., Cheldelin, V. H. & Newburgh, R. W. (1958). *J. biol. Chem.* **233**, 1289–94.

Mackler, B., Grace, R. & Duncan, H. M. (1971). *Arch. Bioch. Biophys.* **144**, 603–10.

Opie, L. H. (1968–9). *Am. Heart J.* **76**, 685–98; **77**, 100–22, 383–410.

Sippel, T. O. (1954). *J. exp. Zool.* **126**, 205.

SORDAHL, L. A., CROW, C. A., KRAFT, G. H. & SCHWARTZ, A. (1972). *J. molec. cell. Cardiol.* **4**, 1–10.
WARSHAW, J. B. (1969). *J. cell. Biol.* **41**, 651.
WARSHAW, J. B. (1971). *Bioch. biophys. Acta* **223**, 409–15.
WILDENTHAL, K. (1971). *J. appl. Physiol.* **30**, 153–7.
WILDENTHAL, K. (1972). *Nature* **239**, 101–2.
WITTELS, B. & BRESSLER, R. (1965). *J. clin. Invest.* **44**, 1639–46.

FOETAL LUNG LIQUID

By R. E. OLVER, E. O. R. REYNOLDS AND
L. B. STRANG

Department of Paediatrics,
University College Hospital Medical School, London, W.C.1

INTRODUCTION

The lungs of the foetus secrete a specialised liquid that fills the future air spaces. In several ways its properties reflect characteristics of the pulmonary epithelium. The presence of the liquid was known to Preyer (1885) and to Addison & How (1913), but for a long time it was thought to be inhaled amniotic liquid and the true site of origin was not suspected until Jost & Policard (1948) showed that the lungs of the foetal rabbit became abnormally distended following ligation of the trachea. Adams *et al.* (1963) showed that the pattern of electrolyte concentrations in lung liquid differed significantly from that in both plasma and amniotic liquid in ways that excluded the possibility of its containing any significant admixture of amniotic liquid. Foetal lung liquid has been shown to contain surface-active substances (Adams & Fujiwara, 1963) and to contribute a flow to the amniotic sac (Adams *et al.* 1967) where the surface-active lecithins can be detected in samples of amniotic liquid (Gluck *et al.* 1971).

Our own interest in the subject was concerned in the first place with investigating the means by which the liquid is absorbed at the start of breathing (Humphreys *et al.* 1967). Then, having found the composition of the lung liquid to be unlike that of a plasma ultrafiltrate (Adamson *et al.* 1969) we were prompted to use certain experimental opportunities, available because of the existence of the liquid, in order to investigate the permeabilities of the alveolar epithelium to a variety of substances including proteins, and to compare them with capillary permeabilities. These measurements provided information about factors that maintain the integrity of the fine air-spaces and prevent the passive accumulation of liquid from the circulation. More recently we have become interested in the mechanism by which foetal lung liquid is secreted. This turns out to be a process involving the active transport of Cl^- and HCO_3^- ions by alveolar epithelium. Our paper will be mainly concerned with questions related to the secretion of the liquid and the permeability of the alveolar walls. Since these properties of the foetal lung had not previously been investigated, special methods had to be developed which have to be considered first in some detail as a necessary preliminary to understanding the experimental results that bear on these topics.

OUTLINE OF EXPERIMENTAL METHODS

Three extracellular spaces can be identified in electron micrographs of lung – intravascular, interstitial and alveolar – and two complete cell barriers, capillary endothelium and alveolar epithelium. In the exteriorised foetal lamb it is possible to take samples from each of the three spaces.

Through the trachea about half the volume of liquid actually present can be withdrawn into a large syringe, usually between 30 and 100 ml in a mature foetal lamb. Test substances can be added and thoroughly mixed with the liquid by repeated injections and withdrawals. Samples are then taken during the following 2–6 hours at 5–20 min intervals according to a standard procedure in which as much liquid as possible is withdrawn, reinjected and withdrawn a second time; 0.5–1.5 ml is then removed through a two-way tap from the syringe, which now contains a mixed sample of all the liquid, so that the space sampled is the entire internal volume of the lung and bronchial tree. We refer to the sample as 'alveolar liquid', which is justifiable because the alveolar parts of the lung contribute over 90% of the volume and it has also been shown that liquids withdrawn from the trachea, bronchi or peripheral parts of the lung do not differ in composition (Adamson *et al.* 1969).

Samples of liquid from the interstitial space can be obtained as lung lymph by thoracic duct cannulation and ligation of non-pulmonary lymph channels (Humphreys *et al.* 1967). In the mature foetal lamb, lymph flows freely and without artificial encouragements of any kind at a very steady mean rate of 1.29 ml/hr·kg (±0.20 SEM). Water-soluble substances injected into blood appear in lymph with the concentration profiles to be expected from the kinetics of passive transfer across capillary walls. Some idea of the rapidity and consistency with which lung lymph concentrations follow plasma levels may be obtained from the following mean half-times of equilibration between plasma and lung lymph in mature foetal lambs: sucrose = 4.26 (±0.47) min; inulin = 11.6 (±2.2) min; albumin = 69 (±7.1) min. We are not able to assess how well lymph represents the different areas of the lung, but we have no prior reason to expect regional inhomogeneities due to gravity in a liquid-filled lung, nor is the lung normally touched during the dissection so that localised capillary damage is unlikely.

Samples from the vascular space were usually withdrawn from a carotid artery. In tracer experiments carotid concentration sufficiently represents lung capillary concentration only when the test substance is not leaving the capillaries so rapidly as to cause large arterio-venous differences. While this leads to considerable limitations in interpreting certain osmotic transient and single-pass indicator dilution experiments, it has little effect on the particular experiments to be considered in this paper, as the arterio-venous difference after the first few minutes from intravenous injection can be shown by calculation not to exceed 1 % of the plasma level and the analysis depends on following concentration patterns for a matter of hours.

The accessibility of the three spaces allows for several types of experiment. In some, samples of alveolar liquid, lymph and plasma are collected for measurement of electrolytes and proteins including their fractionation by gel-filtration. In others, test substances (usually labelled radioactively) are injected either into blood or into alveolar liquid and followed in the other spaces. These techniques can be used both to measure the permeabilities of alveolar and capillary walls to proteins and non-electrolytes, and also to determine unidirectional fluxes and flux ratios across alveolar walls for individual ions.

COMPARTMENTAL ANALYSIS

In tracer experiments permeabilities of capillary and alveolar walls can be obtained in terms of rate constants using a three-compartment model representing plasma (1), interstitial fluid (2) and alveolar liquid (3). There is a volume flow corresponding to newly formed alveolar liquid, from (2) to (3) which is irreversibly lost from (3) through the trachea. When the contents of the spaces are well mixed so that the only barriers exist at capillary and alveolar walls, we can write the following equations to represent the equilibration of a tracer injected into plasma or alveolar liquid, whether it is a foreign substance such as inulin or a labelled ionic species such as ^{24}Na equilibrating with its unlabelled isotope:

$$\frac{dC_2}{dt} + \frac{V dC_3}{dt} + V\gamma C_3 = K_{ci} C_1 - K_{co} C_2 \tag{1}$$

$$\frac{dC_3}{dt} = K_i C_2 - K_o C_3 - \gamma C_3 \tag{2}$$

(where subscripts 1, 2 and 3 refer to the three spaces; C = tracer concentration; V_2 and V_3 are volumes, $V = V_3/V_2$; $\gamma = J_V/V_3$ and J_V = volume flow from spaces 2 to 3; K's are rate constants (min^{-1}): K_{ci} for transfer from 1 to 2; K_{co} for loss from 2, both to 1 and by drainage to lymph; K_i for transfer from 2 to 3 and K_o for transfer from 3 to 2).

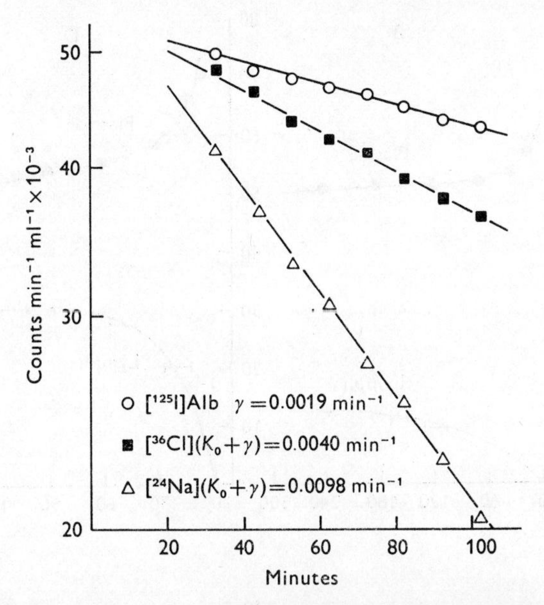

Fig. 1. *Foetal lamb expt 812A.* 39 ml of previously withdrawn alveolar liquid containing [³⁶Cl]sodium chloride (2 μC), [²⁴Na]sodium chloride (6 μC) and [¹²⁵I]albumin (6 μC; 2 mg) were injected via a tracheal cannula at zero time. A value of $\gamma(=J_V/V_3)$ was obtained from the slope of the [¹²⁵I]albumin; values of $(K_o+\gamma)$ from the slopes of [³⁶Cl] and [²⁴Na] and their K_o's by difference. (²⁴Na concentrations multiplied by 3/2 for convenience of diagram.) Points in first 30 min not used in order to avoid mixing effects.

Tracer placed in alveolar liquid

In the simplest and most usual case when a test substance is mixed into alveolar liquid and crosses the alveolar walls slowly enough for the concentrations in lymph and plasma to remain very low, $K_i C_2$ can be neglected and integration of equation (2) yields

$$\ln C_3 = \ln C_{30} - t(K_o + \gamma) \qquad (3)$$

(where $C_{30} = C_3$ at zero time). By introducing into alveolar liquid a substance which does not cross alveolar walls ($K_o = 0$), as has been shown to be the case for inulin and for albumin (Normand *et al.* 1971), we can find γ from the slope of its C_3 values on time and V_3 (alveolar liquid volume) from its intercept at zero time (whence we also get J_V, the secretion rate of newly formed liquid). When permeable tracers are also put in alveolar liquid we can find their values of K_o (alveolar wall rate constant) from the differences between their slopes of C_3 on time and that of the impermeant tracer (see Fig. 1).

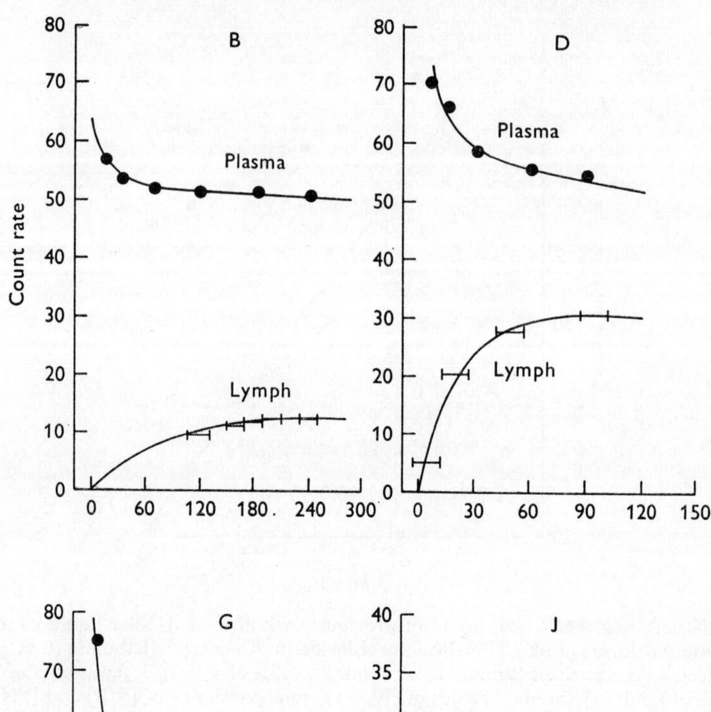

Fig. 2. Count rates per ml and min × 10^{-3} in plasma and lung lymph of 4 gel-filtration fractions B, D, G and J following injection of [^{125}I]PVP at zero time (4 foetal lamb experiments). Steady state lymph/plasma ratios were found by curve fitting. They correspond with ratio at time of peak in lymph when $dC_2/dt = 0$ and agree with values for proteins of similar K_{av}.

Fraction	K_{av}	Mol. rad. (a)
B	0.062	89 Å
D	0.195	58 Å
G	0.412	34 Å
J	0.639	21 Å

(From Boyd *et al.* 1969.)

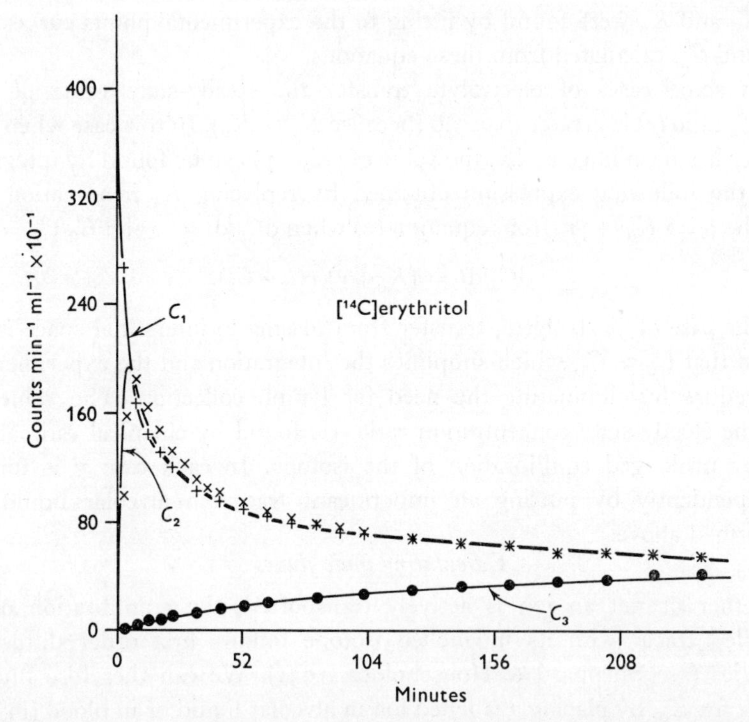

Fig. 3. *Foetal lamb expt 617A.* Count rates (per ml and min) in plasma (C_1), lymph (C_2) and alveolar liquid (C_3) following i.v. injection of [^{14}C]erythritol (17.5 μC; 10 mg) at zero time. Curves of C_2 and C_3 fitted for $K_c = 0.310$ and $K_o = 0.00323$ (min^{-1}) from solutions to equations (1) and (2). (From Normand *et al.* 1971.)

Tracer placed in blood

When a permeant tracer is injected i.v. its plasma concentration can then be described as the sum of exponential terms such that $C_1 = \sum_{i=1}^{j} C_i e^{-n_i t}$ which allows us to obtain integrated solutions for C_2 and C_3. Boyd *et al.* (1969; equation 3A) give a solution for C_2 appropriate for macromolecule transfer across capillary walls, where $K_{ci} \neq K_{co}$ and where no penetration of compartment 3 occurs so that K_i, K_o and C_3 are zero. By fitting curves from their equation to experimental points for plasma and lymph concentration we can obtain values for steady state C_2/C_1 (equivalent to lymph/plasma ratio) and for K_{co} (capillary wall rate constant) (Fig. 2).

Normand *et al.* (1971; equations 3A and 4A) give solutions for C_2 and C_3 appropriate for smaller inert molecules which equilibrate very rapidly between compartments 1 and 2 and also penetrate compartment 3. In their case $K_{ci} = K_{co} = K_c$ and $K_i = K_o$. An example is given in Fig. 3 of an experiment in which [^{14}C]erythritol was injected i.v. and where the values of K_c and K_o were found by fitting to the experimental points curves for C_2 and C_3, calculated from these equations.

In some cases of electrolyte transfer the steady-state value of the C_3/C_2 ratio (r) is greater than 1.0 (because $K_i > K_o$). In this case when the tracer has been injected i.v. the value of $(K_o+\gamma)$ can be found by integrating the following expression obtained by replacing K_i in equation (2) by the term $K_o r+\gamma r$ (from equation (2) when $dC_3/dt = 0$ and $C_3/C_2 = r$):

$$dC_3/dt = (K_o+\gamma)(rC_2-C_3). \tag{4}$$

In the case of electrolytes, transfer from plasma to interstitial space is so rapid that $C_2 \approx C_1$, which simplifies the integration and the experimental procedure by eliminating the need for lymph collection. The value of r – the steady state concentration ratio – is found by chemical estimation or by prolonged equilibration of the isotope. In each case γ is found independently by putting an impermeant tracer in alveolar liquid as described above.

Calculating ionic fluxes

Whether or not an ion is actively transported, the equilibration of a labelled tracer with its unlabelled isotope follows first order diffusion kinetics (see Sheppard & Householder, 1951). We can therefore find a value for K_o, by placing a labelled ion in alveolar liquid or in blood (in the latter case only when the steady state ratio (r) can be determined). Fig. 4 illustrates an example in which K_o values for I^- were obtained simultaneously by the two techniques using two isotopes of iodine. From the K_o value we can obtain the unidirectional transmembrane flux of the ion from alveolar liquid to plasma. $J_{31} = K_o V_3 C_{3T}$ (C_{3T} = total concentration of ion, labelled and unlabelled). Net flux is given by $J_{net} = J_V C_{3T}$ and unidirectional flux from plasma to alveolar liquid by $J_{13} = J_V C_{3T}+K_o V_3 C_{3T}$; whence we obtain

$$\frac{J_{13}}{J_{31}} = \frac{J_V C_{3T}+K_o V_3 C_{3T}}{K_o V_3 C_{3T}} = \frac{\gamma+K_o}{K_o}. \tag{5}$$

PERMEABILITIES OF LUNG CAPILLARIES AND ALVEOLI TO NON-ELECTROLYTES

Capillaries

In experiments on foetal and newborn lambs and sheep, Boyd *et al.* (1969) measured steady-state lymph/plasma ratios for three different fractions of the animal's own proteins, as obtained by gel-filtration on columns of Sephadex G 200. They also calculated steady-state ratios and values of K_{co} from experiments of the type illustrated in Fig. 2, in which

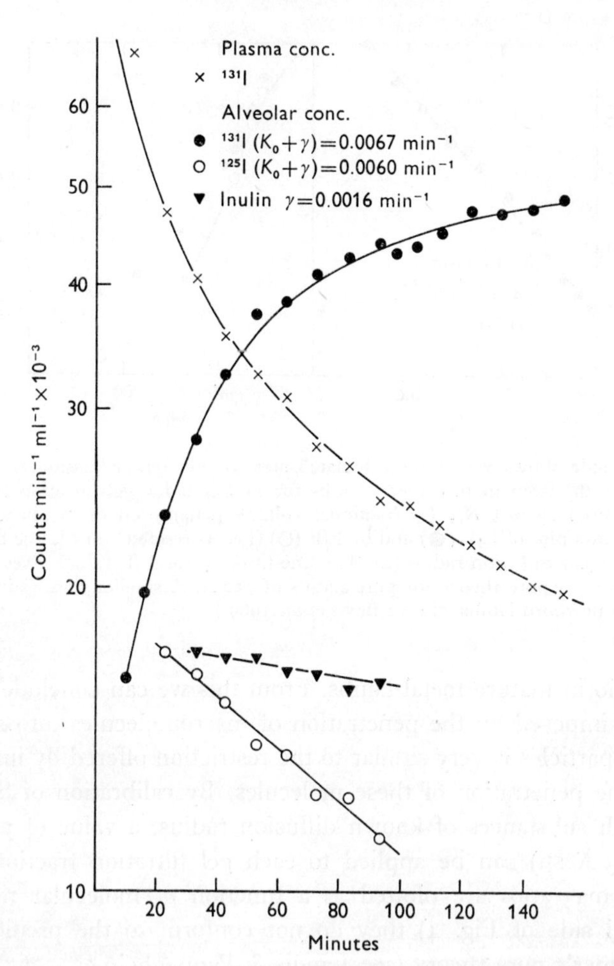

Fig. 4. *Foetal lamb expt 725A*. At zero time [¹³¹I]NaI (25 μC; 5 mg) was injected i.v. and simultaneously [¹²⁵I]NaI (0.5 μC; 2 mg) with 50 mg unlabelled inulin (both previously mixed with 36 ml of alveolar liquid) were injected via trachea. $(K_o+\gamma)$ was found for ¹³¹I results by curve fitting of solution for equation (4); and for ¹²⁵I results by fitting equation (3). γ found from inulin results and equation (3). The values of K_o for the two isotopes of I^- agree within 15 %.

a polydisperse solution of [¹²⁵I]PVP was injected i.v. then separated by gel filtration of lymph and plasma samples into 11 fractions of between 110 Å and 17 Å in estimated diffusion radius (Stokes–Einstein radius). Gel filtration achieves a separation of molecules differing in size because the penetration of gel particles becomes more and more limited as molecular radius increases. For a given molecule the fractional volume of gel particles which it can enter, is expressed as its K_{av}. Fig. 5 gives data from Boyd *et al.* (1969) which show a linear agreement between K_{av} and lymph/

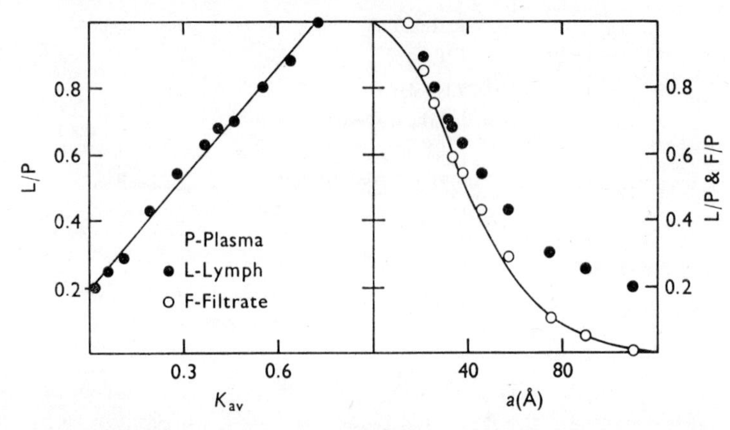

Fig. 5. Left side shows values for calculated steady-state lymph/plasma concentration ratios (L/P = ●) from mature foetal lambs for 11 Sephadex gel-filtration fractions of [^{125}I]PVP plotted against K_{av} (= fractional volume penetration of Sephadex G 200). Right side shows plot of L/P (●) and of F/P (○) (i.e. 'corrected' for plasma contamination) on molecular diffusion radius (a). The line fitted to the F/P values gives prediction of Pappenheimer's pore theory for pore radius of 150 Å. A smaller pore radius of 90 Å was found in newborn lambs. (From Boyd *et al.* 1969.)

plasma ratio in mature foetal lambs. From this we can conclude that the restriction imposed on the penetration of macromolecules into sephadex G 200 gel particles is very similar to the restriction offered by lung capillaries to the penetration of these molecules. By calibration of Sephadex G 200 with substances of known diffusion radius, a value of molecular radius (±5 Å SD) can be applied to each gel filtration fraction. When lymph/plasma ratios are plotted as a function of molecular radius (a) (right-hand side of Fig. 4) they do not conform to the predictions of Pappenheimer's pore theory (see Landis & Pappenheimer, 1963) for any single pore radius, and there is probably no unique solution to the data for any two or more pore radii. On the other hand if we allow, as Grötte (1956) did, that the lymph has been contaminated by the addition of a small amount of unmodified plasma escaping from leaks in some part of the microcirculation other than capillaries (e.g. venules), and that the degree of contamination is defined by the lymph/plasma ratio of the largest PVP fraction (a $\geqslant$ 110 Å), then we can allow for it and show that the modified lymph/plasma ratios (filtrate/plasma, F/P ratios) are appropriate for a pore radius of 150 Å in the mature foetal lamb (Fig. 4). A smaller value of 90 Å was calculated for newborn lambs, the difference from the foetal animals possibly being due to a higher capillary pressure in the mature foetus than in the newborn causing pore-stretching in the way suggested by Shirley *et al.* (1957). (See Boyd *et al.* 1969, for details of calculations used in modifying lymph/plasma ratios.) We would emphasise that the allowance for leaks and therefore the estimates of a uniform pore radius are not unique solutions for the data but represent the simplest of a variety of possibilities. For example, there may be graded porosity in capillaries with larger pores occurring towards the venules as suggested by Rous *et al.* (1930).

When considering the contributions of lung capillary walls to maintaining the structural integrity of lung, the important and inescapable conclusion from these experiments is that even large plasma solutes can pass from plasma to the interstitial space. This transfer may be useful in allowing the supply of complex substrates and antibodies to alveolar cells. It will be seen, however, that protection of the alveolar space against penetration by these large molecules has to depend on properties of the alveolar wall.

Alveolar walls

Sneeberger-Keeley & Karnovsky (1968), using an electron microscope technique for following intravenously injected horseradish peroxidase, found that whereas this relatively large molecule (MW = 40,000) crossed the capillary walls of the adult mouse lung it did not penetrate their alveoli. This agreed with the findings of Normand *et al.* (1970) in the foetal lamb that proteins and [^{125}I]PVP fractions down to 17 Å in molecular radius did not cross alveolar walls. It was then shown by Normand *et al.* (1971), in experiments of the type shown in Fig. 3 using small water-soluble non-electrolytes, that the alveolar epithelium was completely impermeable to inulin (a = 14 Å) and that in the case of water-soluble substances only molecules of 5.1 Å and less in molecular radius passed into the alveoli. Lipid-soluble substances such as ethyl-thiourea crossed the alveolar walls much more rapidly than water-soluble molecules of similar size.

Fig. 6 gives the results of K_0 determinations for water-soluble non-electrolytes and water itself. The striking dependence of penetration rate on molecular size is appropriate for a pore radius of 5.5 Å. This radius, which can be compared with values of 4.3–6.2 Å for erythrocyte walls (Goldstein & Solomon, 1960; Rich *et al.* 1967) reflects the properties of an intact epithelium and their structural counterpart may well be hydrophilic areas in cell walls rather than any kind of permanent structure. Whether or not the very small pores exist, it is concise and also a convenience to express the permeability findings in terms of pores or 'equivalent pores', to use Solomon's (1968) terminology.

These experiments certainly explain the low concentration of protein in alveolar liquid. They also imply that some kind of break-down in the intactness of the alveolar epithelium is a factor in the formation of hyaline

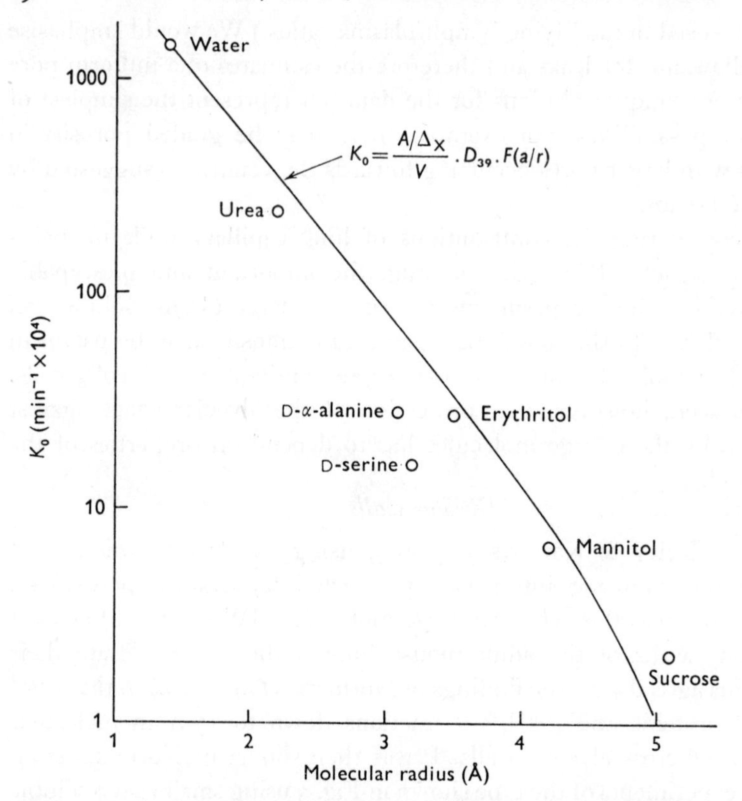

Fig. 6. Values of rate constant for alveolar walls (K_o) for non-electrolytes, plotted against diffusion radius of test substances. The line gives predictions of pore theory for pore radius of 5.5 Å. (From Normand *et al.* 1971.)

membranes in newborn infants as these have been shown by histological techniques to contain plasma proteins (Gitlin & Craig, 1956; Gajl-Peczalska, 1964). In support of this idea Normand *et al.* (1970) found evidence that the epithelium of the immature lung is more easily broken down by the stresses of ventilation than that of the mature lung.

There is some evidence that the relative impermeability of alveoli as compared with capillaries to water-soluble substances is also a feature of the post-natal lung (Courtice & Phipps, 1946; Chinard, 1966; Wangensteen *et al.* 1969; Taylor & Gaar, 1970). Certainly the very low permeabilities of alveoli to this class of substance would explain why extracellular fluids are normally excluded from alveoli and why the pulmonary oedema of heart failure is at first interstitial, invading alveoli only in extreme circumstances (Staub *et al.* 1967). The low permeabilities to solutes, coupled with the high permeability to water itself (see Fig. 6), explains the effects of instilling fresh-water and sea-water into the lung, when the alveolar epithelium behaves like a semipermeable membrane (Colin, 1873; Swan & Spafford, 1951). The much greater permeabilities of alveolar walls to lipid-soluble substances found by Normand *et al.* (1971) explain the characteristically rapid exchange of respiratory and anaesthetic gases and alcohol.

SECRETION OF ALVEOLAR LIQUID

The volume of liquid in the foetal lungs cannot be estimated reliably by aspiration because the amount which can be withdrawn is limited by bronchial closure. Nor can the secretion rate be measured by collecting the tracheal outflow over a short period as has been attempted (Setnikar *et al.* 1959; Adams *et al.* 1963), because it is not possible by this technique to distinguish between the secretion of new liquid and passive emptying of the lung. But both the volume of liquid and its secretion rate can be established with accuracy from the dilution of an impermeant tracer placed in alveolar liquid as described in the Methods section.

From measurements of inulin dilution, Normand *et al.* (1971) found the volume of alveolar liquid/kg body weight to be 30 ml ($\pm$1.4 SE) and its rate of secretion/kg 0.036 ml/min ($\pm$0.003). The secretion rate was not affected by overloading the circulation over a period of one hour by repeated blood or saline transfusions, a procedure which caused pulmonary lymph-flow to increase by a factor of two to three. Potassium cyanide (10–25 mg) mixed into alveolar liquid stopped its secretion and initiated absorption (Fig. 7). These findings certainly suggest that the flow of alveolar liquid is generated not by the hydrostatic pressure in blood capillaries, but by the active transport of ions – a conclusion which is also supported by measurements of composition and ionic flux. When the secretion is stopped by cyanide, the absorption of liquid is presumably due mainly to the now unopposed gradient of protein osmotic pressure between alveolar liquid on the one hand and interstitial fluid and plasma on the other.

Alveolar liquid composition

The liquid withdrawn through the trachea of the foetal lamb is a watery material, but contains occasional fragments of mucus and it is usually a little cloudy due to the presence of epithelial debris; both can be removed by centrifugation, leaving a clear colourless liquid. Concentrations of proteins and electrolytes are unaffected by centrifuging and the total concentration of solids $\approx$ 1% – similar to that of CSF. Despite the presence of a little obvious mucus, the sialic acid concentration* (a marker

* Measured as *N*-acetyl neuraminic acid: we are grateful to Mr Arthur Leigh (Dept of Bacteriology, UCH Medical School) for these measurements.

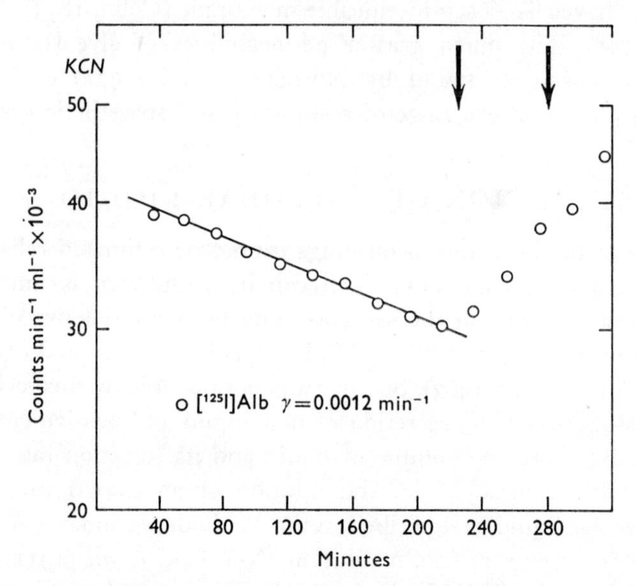

Fig. 7. *Expt 816B.* [^{125}I]albumin (2 μC; 1 mg) was mixed into alveolar liquid at zero time. O = counts in alveolar liquid. The slope of the regression fitted to the points from 32.5 to 212.5 min gives value for = 0.00120 min^{-1} (intercept at zero time gives value of alveolar volume (V_3) = 178 ml and rate of secretion, J_V = 0.22 ml min^{-1}). At 225 min, 23 mg KCN and at 280 min 10 mg KCN were added to alveolar liquid, following which [^{125}I]albumin increased indicating the reversed flow (absorption) of alveolar liquid at about 0.85 ml min^{-1}.

for glycoproteins) is between 30 and 40 μg/ml which is about 7% of the level in the lamb's buccal secretions. This result, the low concentration of protein (0.027 g/100 ml) together with the lack of an effect of centrifugation on protein or ionic concentrations, make us believe it to be unlikely that any significant proportion of the liquid is produced by bronchial glands or that its composition is modified by mucus secretions or desquamated epithelial cells; but there is as yet no definite or direct evidence as to which cells in the pulmonary epithelium are responsible for secretion of the liquid. Since substantial differences in ionic composition (Table 1) have to be defended against equilibration with plasma across the whole alveolar epithelium of the lung, we can safely conclude that the secretory process is likely to depend on the activity of the entire epithelium or at least on a large proportion of its cells. If the alternative explanation were correct, that liquid with a specialised composition were secreted by bronchial glands and then equilibrated with plasma in more distal parts of the lung, we should expect its composition to differ according to the part of the lung (proximal or distal) from which it is removed. This is not the case, the composition remaining constant within very narrow limits irrespective of

Table 1. *Values of distribution ratios* $R_{P/al}$ *and* $R_{P/uf}$

Substance	$R_{P/al}$*	$R_{P/uf}$†
Na^+	1.03	1.05
K^+	0.72	1.04
Ca^{2+}	4.13	1.53
Cl^-	0.68	0.96
HCO_3^-	8.61	0.96
H_2CO_3	1.08	1.00

$R_{P/al}$ = [plasma]/[alveolar liquid].
$R_{P/uf}$ = [plasma]/[ultrafiltrate].
* From Adamson *et al.* (1967).
† From Davson (1967).

the amount first withdrawn through the trachea before taking the sample (Adamson *et al.* 1970).

Table 1 gives values for plasma/alveolar liquid concentration ratios ($R_{P/al}$) from the data of Adamson *et al.* (1969) and for plasma/ultrafiltrate of plasma ratios ($R_{P/uf}$) from Davson (1967) (in both cases concentrations were expressed per kg H_2O). The values of $R_{P/al}$ for [K^+], [Ca^{2+}], [Cl^-] and [HCO_3^-] differ very significantly from the values of $R_{P/uf}$ – particularly in the low ratio for [Cl^-] and the high ratio for [HCO_3^-]. The low concentration of organic substances in the liquid virtually excludes ion binding as an influence in determining ionic concentrations and there is no way in which the Donnan law could account for these values of $R_{P/al}$, even if the epithelium were impermeable to one of the ions measured (which it is not) – so it is almost certain from these measurements alone that one or more ionic species are actively transported across the pulmonary epithelium. Adamson *et al.* (1969) found virtually no titratable fixed-acid in alveolar liquid; its low pH (6.27) was due entirely to the equilibration of an unbuffered solution, ([HCO_3^-] = 2.8 mEq/kg H_2O) at P_{CO_2} = 40 mm Hg. Nor did careful measurement reveal any significant difference in P_{CO_2} between plasma and alveolar liquid.

Ion fluxes

Flux ratios

The flux of an ionic species across a biological membrane is determined by processes that can be identified as either 'active' or passive. Since a passive flux depends in a definite way on forces that can be measured, it is possible to distinguish a flux as active rather than passive by detecting deviations from that dependence. The forces determining passive flux are chemical activity, a (i.e. concentration × activity coefficient) electrical potential, ψ (usually generated by active transport of another ion), and solvent drag (a force exerted in the direction of any transmembrane volume flow, J_V). The ratio of the two unidirectional fluxes, J_{13} and J_{31}, which determines the passive distribution of an ion on the two sides of a membrane, can be related to these forces by Ussing's well-known flux-ratio equation (Koefoed-Johnson & Ussing, 1953):

$$\ln\left[\frac{J_{13}}{J_{31}}\right] = \ln\left[\frac{a_1}{a_3}\right] + \frac{ZF}{RT}(\psi_1 - \psi_3) + \frac{J_V}{D_s}\int_0^x \frac{1}{A}\,dx \tag{5}$$

(where D_s = diffusion coefficient; A = area of membrane available for transport; x = thickness of membrane; Z = valency; F = Faraday's constant; R = gas constant; T = absolute temperature).

In our experiments we have attempted to decide whether Na^+, Cl^- and two other halides, I^- and Br^-, are passively distributed or actively transported, by comparing our experimentally determined values of the flux ratio J_{13}/J_{31} (obtained as described in the Methods section) with the terms on the right-hand side of the equation which represent the forces determining passive flux. These we have attempted to evaluate in the following ways:

In the case of Na^+ and Cl^- the ratios of chemical activities were determined by chemical estimation using a value of 1.0 for the ratios of their activity coefficients. In the case of Br^- and I^- the ratio was obtained from the count rates in plasma and alveolar liquid following i.v. injection of a labelled isotope at the point in time when the concentration in alveolar liquid is at its peak (i.e. when $dC_3/dt = 0$) and the concentration ratio corresponds with that in the steady state.

The electrical potential term, $(\psi_1 - \psi_2)\ ZF/RT$, was evaluated by measuring the potential difference between plasma and alveolar liquid using KCL–agar bridges inserted into the circulation and alveolar liquid and connected to calomel half-cells and a high-impedance voltmeter. Asymmetry potentials, usually 0.1–0.2 mV and always <0.5 mV, were allowed for. A small potential difference between 1 and 10 mV (mean = -4.3 mV ± 0.45 SEM) was recorded in each case, alveolar liquid always being negative with respect to plasma. Evidence as to the significance of the potential was found in the fact that it consistently declined to a slightly reversed value of $+0.2$ to $+0.5$ mV (alveolar liquid positive) over 5–10 min after the foetus was killed.* Nor was a potential difference ever found with both electrodes in different parts of the circulation. In the bronchial tree the potential tended to be more strongly negative as the electrode was

* A potential of this sign and size was found by measurement across an alveolar liquid/plasma interface *in vitro*, as would be predicted by Tasaki & Singer (1968).

advanced distally. For all these reasons and because the potential always had the same sign, we do not think it is an artifact, although it is difficult to be quite sure with such a small potential. The sign of the potential agrees with our finding that Cl^- is actively transported from plasma to alveolar liquid, but that deduction does not depend at all critically on the accuracy of the potential measurements. We found no obvious relationship between the size of the potential and the rate of secretion, probably because the flux of Cl^- inwards is linked in some way to the flux of HCO_3^- outwards.

Evaluation of the drag term $\frac{J_V}{D_s}\int_0^x \frac{1}{A}\,dx$ gives rise to more difficulty, partly because it is unclear on theoretical grounds whether the value D_s should refer to free diffusion in solution or to restricted diffusion in the pores of a membrane (Anderson & Ussing, 1957). Analysis using irreversible thermodynamics by Hoshiko & Lindley (1964) yields a solution (equation 22 in their paper) which is approximately equivalent to the relevant term being the free diffusion coefficient rather than the restricted coefficient. The drag term introduces a much smaller effect if we use the free value of D_s rather than its restricted form. We have evaluated the drag term in our own experiments by generating osmotic flows with sucrose placed in alveolar liquid. Preliminary results confirm that the effect on ion transport is indeed a small one and of the right order for the use of the free diffusion coefficient, which we have therefore used in our calculations.* It makes only a small contribution to the forces determining transport – in electrical units less than 1.0 mV for Na^+. In Table 2 our measured values of flux ratio can be compared with our estimate of forces determining passive transfer as expressed by the sum of the terms on the right-hand side of the flux ratio equation. For each ion the comparison can be made from the difference between the results in columns 4 and 5, or from column 6 which gives the difference in mV (F′). In the case of Na^+ the flux ratio corresponds well with expectations for a passive distribution. In the case of each halide ion there is a large difference between the measured flux ratio and that expected for a passive distribution, which demonstrates that these ions are actively transported against a gradient of electrochemical potential. In the case of Cl^- this active

* An estimate for $\frac{1}{D}\int_0^x \frac{1}{A}\,dx$ can be found from the K_0 of the test substance and the theory of restricted diffusion. According to this theory $K_0V_3 = F(a/r)D_s\int_0^x A\frac{1}{dx}$ $F(a/r)$ is a function of molecular radius (a) and pore radius (r) (see equation 4 in Normand *et al.* (1971) from Renkin (1954)) and could be calculated from the hydrated ionic radius and pore radius = 5.5 Å.

Table 2. *Mean values* ($\pm$ *SE of mean*) *for flux ratios* (J_{13}/J_{31}), *activity ratios* (a_1/a_3), *log$_e$ flux ratios, sum of measured forces* [$ln\ (a_1/a_3) + EZF/RT + F(J_V)$]. (*Where* $E = \psi_1 - \psi_3$ *and* $F(J_V) = (J_V/D_s)\int_0^x \cdot (1/A\ dx)$ *and difference between log$_e$ flux ratios and sum of measured forces in mV* = ($F'mV$)

Ion	$[J_{13}/J_{31}]$	$[a_1/a_3]$	$\ln [J_{13}/J_{31}]$	$\ln [a_1/a_3] + \frac{EZF}{RT} + F(J_V)$	F' mV
Na^+	1.18	1.03	0.163	+0.193	+0.8
(n = 25)	($\pm$0.02)	($\pm$0.005)	($\pm$0.013)	($\pm$0.004)	($\pm$0.4)
Cl^-	1.65	0.68	0.491	−0.472	+25.5
(n = 22)	($\pm$0.05)	($\pm$0.01)	($\pm$0.029)	($\pm$0.009)	($\pm$0.8)
Br^-	1.37	0.50	0.314	−0.790	+29.4
(n = 5)	($\pm$0.06)	($\pm$0.03)	($\pm$0.041)	($\pm$0.007)	($\pm$1.0)
I^-	1.34	0.40	0.286	−1.022	+34.8
(n = 9)	($\pm$0.04)	($\pm$0.02)	($\pm$0.030)	($\pm$0.006)	($\pm$0.7)
1	2	3	4	5	6

transport system generates a force equivalent to a potential difference of 26 mV and the force is still larger for Br^- (29 mV) and I^- (35 mV). The order of F' mV $I^- > Br^- > Cl^-$ was confirmed in two experiments in which they were simultaneously measured in the same animal. It is a fair assumption that the same transport system is available for all three halides as it is in gastric mucosa (Hogben & Green, 1958).

Bicarbonate flux

The negative potential in alveolar liquid, generated by Cl^- transport, is not nearly large enough to account for the [HCO^-_3] distribution ratio ($R_{P/al} = 9.91$); a passively determined distribution ratio of this size would require a potential difference of about −130 mV (alveolar liquid negative). Hence an active transport of HCO_3^- in the reverse direction to Cl^- seems likely. Investigation of bicarbonate transfer by the measurement of unidirectional tracer fluxes was avoided because it is difficult to distinguish ^{14}C transfer as [^{14}C]CO_2 from that as [^{14}C]HCO_3^-. We were, however, able to study the net flux of HCO_3^- from alveolar liquid to plasma. By adding measured amounts of an isomolar solution of $NaHCO_3$ to alveolar liquid we could raise [HCO_3^-] from the steady-state value of 2.8 mEq/kg H_2O to levels as high as 75 mEq/kg H_2O.* We then found that the concentration decreased with time in the manner shown in the

* This raised the pH of alveolar liquid up to 7.6, but caused no change in the permeability characteristics of the membrane as judged by ^{24}Na and ^{36}Cl rate constants.

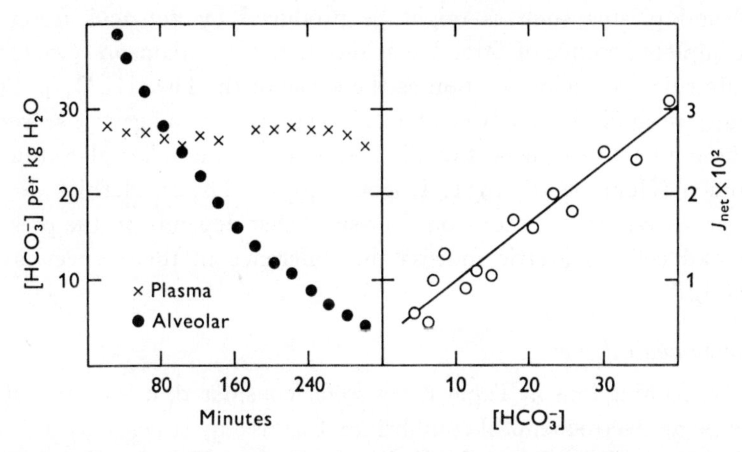

Fig. 8. *Expt 824.* At zero time 31 ml of isotonic $NaHCO_3$ were added to alveolar liquid. ● = $[HCO_3^-]$/kg H_2O in alveolar liquid. × = $[HCO_3^-]$/kg H_2O in plasma (standardised to P_{CO_2} = 50 mm Hg). Right-hand graph shows net HCO_3^- flux (in mEq $min^{-1} \times 10^2$) from alveolar liquid to plasma (J_{net}) as a function of $[HCO_3^-]$ in alveolar liquid.

left-hand panel of Fig. 8 until a resting value of 2–4 mEq/kg H_2O was attained. This net transport was achieved without significant alterations in plasma $[HCO_3^-]$ and P_{CO_2}. It is evident from the data in Fig. 8 and in similar experiments, that when $[HCO_3^-]$ in alveolar liquid drops below the plasma level, net transport is taking place against an increasing gradient of electrochemical activity.

The right-hand panel in Fig. 8 shows that the net flux out of the lung (J_{net}) increased linearly with $[HCO_3^-]$ in alveolar liquid, suggesting that the carrier mechanism responsible for the transport is unsaturated at all levels of alveolar $[HCO_3^-]$. A similar result was obtained in other experiments in which alveolar $[HCO_3^-]$ was raised as high as 75 mEq/kg H_2O. It was also found that increasing plasma $[HCO_3^-]$ by infusion from about 24 mEq/kg H_2O to over 40 mEq/kg H_2O for a period of one hour had no effect on alveolar $[HCO_3^-]$. This lack of effect may reflect a very low passive permeability to HCO_3^- or it could be an effect of unsaturation of the transport system, so that inward diffusion is fully met by active transport outwards.

We also had to consider the alternative explanation for these observations, that they were the result of net HCl secretion; but to explain the result in that way we would have to suppose that the acid was secreted at a constant rate which was exactly balanced by the inward diffusion of HCO_3^- from plasma, no titratable acid ever accumulating as it does in urine and gastric juice, two body fluids into which net acid secretion does occur. (Careful measurement detected no gradient for P_{CO_2} between alveolar liquid and plasma such as might be produced by net acid secretion.) The complete absence of titratable acidity noted by Adamson *et al.* (1969) virtually rules out acid secretion as the cause of the low $[HCO_3^-]$. There certainly are similarities between foetal lung secretion and gastric secretion, as in both an active transport of Cl^- and other halides takes place towards the lumen (Heinz *et al.* 1954; Hogben, 1955). The similarities do not, however, extend to H^+ secretion; but since that depends on the presence of parietal cells in gastric mucosa the difference in this respect is not surprising.*

Potassium and calcium

Of the remaining ions in Table 1, not so far considered, it is clear that K^+ cannot be at electrochemical equilibrium. Ca^{2+} is almost certainly passively distributed; we have found that it has a very low K_0 in keeping with its large ionic radius (hydrated radius = 4.1 Å). Flux ratio calculations agree with predictions for passive transfer within 3 mV if we assume that ionised plasma calcium is 0.67 of the total.

CONCLUSIONS

We have reviewed experimental evidence which goes some way to explaining the secretion of lung liquid and its unusual electrolyte composition. The ion transport systems identified must depend on properties of the alveolar epithelium, which are in some respects similar to those of gastric mucosa, perhaps reflecting the common origin of the two epithelia from the embryonic foregut. The similarity also extends to certain structural characteristics (Buckingham *et al.* 1968) and to the effects of corticosteroids on its maturation (Kikkawa *et al.* 1971). We have also seen that the alveolar epithelium forms a barrier to the penetration of all but the smallest water-soluble substances and protects the alveoli from accumulations of fluid and protein.

In the future investigators will probably continue to find alveolar liquid useful because of the access it gives to alveolar cells, particularly as evidence accumulates that it is the maturation of these cells that determines the capacity of the lung to accomplish satisfactory gas exchange. In particular it may prove possible to study the biosynthesis of surface-active substances by detecting exchanges between alveolar cells and alveolar liquid.

Although the functions served by alveolar liquid in the foetal lung are

* In the rabbit foetus Wright (1962) found that H^+ secretion was first detectable at 23 days (term = 30 days) and coincided with first appearance of parietal cells.

not really known, it would certainly be difficult to imagine how the complicated network of air spaces could be formed if they were all collapsed and not filled-out with liquid during development. It is probably appropriate that the degree of distension should be controlled by secretory characteristics of the cells that are undergoing development in the process of morphogenesis. Towers (1968) has presented histological evidence that different parts of the foetal lung accumulate liquid in a cyclical manner that may reflect variations in regional secretory activity.

At the start of positive pressure ventilation in the lamb, Humphreys *et al.* (1967) showed that there was an uptake of liquid from alveoli to the interstitial space, from where a substantial amount was drained away in lymphatics. They speculated that the liquid could be displaced from alveoli to the interstitial space by pressures generated by ventilatory movements, but no direct evidence has been produced that this is what actually happens, and we have shown that alveolar liquid can be absorbed in the absence of respiratory movements when the secretory process is stopped by cyanide. We also have to explain what becomes of the secretion of alveolar liquid in the air-breathing lung. We can probably conclude that the secretion is stopped in some way after birth because the air-filled lung which collapses behind an obstruction, does not then fill up with liquid; whereas the foetal lung distal to a congenital obstruction becomes grossly distended with liquid (Potter & Bohlender, 1941; Griscom *et al.* 1969). The factors responsible for 'switching on and off' the secretion of alveolar liquid have not yet been investigated; they are likely to be relevant both to the means by which alveolar liquid is initially absorbed and to the subsequent function of the lung.

Much of the work described in this paper and many of the ideas were worked out with collaborators to whom we owe a special debt of gratitude. These include Michael Adamson, Robert Boyd, Valerie Cole, June Hill, Peter Humphreys, Colin Normand and Keesley Welch. We are also grateful to Dr Hugh Davson for helpful comments. Mr Michael Bright gave valuable technical assistance. The work received financial support from the Medical Research Council, the Wellcome Trust, the Sir Halley Stewart Trust and the Children's Research Fund.

REFERENCES

Adams, F. H. & Fujiwara, T. (1963). *J. Pediat.* **63**, 537–42.

Adams, F. H., Moss, A. J. & Fagan, L. (1963). *Biologia Neonat.* **5**, 151–8.

Adams, F. H., Desilets, D. T. & Towers, B. (1967). *Resp. Physiol.* **2**, 302–9.

Adamson, T. M., Boyd, R. D. H., Platt, H. S. & Strang, L. B. (1969). *J. Physiol.* **204**, 159–68.

Addison, W. H. F. & How, H. W. (1913). *Am. J. Anat.* **15**, 199–214.

Anderson, B. & Ussing, H. H. (1957). *Acta Physiol. Scand.* **39**, 228–39.

Boyd, R. D. H., Hill, J. R., Humphreys, P. W., Normand, I. C. S., Reynolds, E. O. R. & Strang, L. B. (1969). *J. Physiol.* **201**, 567–88.

BUCKINGHAM, S., MCNARY, W. F., SOMNERS, S. C., ROTHSCHILD, J. (1968). *Fed. Proc.* **27**, 328.

CHINARD, F. P. (1966). In *Advances in Respiratory Physiology*, pp. 106–47, ed. Caro, C. G. Baltimore: Williams & Wilkins.

COLIN, G. (1873). In *Traité de Physiologie Comparée des Animaux*, 2nd edn, vol. 2, pp. 109–10. Paris: Ballière et Fils.

COURTICE, F. C. & PHIPPS, P. J. (1946). *J. Physiol.* **105**, 186–90.

DAVSON, H. (1967). *Physiology of the Cerebrospinal Fluid*, p. 40. London: J. & A. Churchill.

GAJL-PECZALSKA, K. (1964). *Arch. Dis. Child.* **39**, 226–31.

GITLIN, D. & CRAIG, J. M. (1956). *Pediatrics, Springfield* **17**, 64–71.

GLUCK, L., KULOVICH, M. V., BORER, R. C., Jr, BRENNER, P. H., ANDERSON, G. G. & SPELLACY, W. N. (1971). *Am. J. Obst. Gynecol.* **109**, 440–5.

GOLDSTEIN, D. A. & SOLOMON, A. K. (1960). *J. gen. Physiol.* **44**, 1–17.

GRISCOM, N. T., HARRIS, G. B. S., WOHL, M. E. B., VAWTER, G. F. & EKRALIS, A. J. (1969). *Pediatrics, Springfield* **43**, 383–9.

GRÖTTE, G. (1956). *Acta Chir. Scand. Suppl.* **211**, 1–84.

HEINZ, E., OBRUNK, K. J. & ULFENDAHL, H. (1954). *Gastroenterology* **27**, 98–112.

HOGBEN, C. A. M. (1955). *Am. J. Physiol.* **180**, 641–9.

HOGBEN, C. A. M. & GREEN, N. D. (1958). *Fed. Proc.* **17**, 72.

HOSHIKO, T. & LINDLEY, B. H. (1964). *Biochem. Biophys. Acta.* **79**, 301–17.

HUMPHREYS, P. W., NORMAND, I. C. S., REYNOLDS, E. O. R. & STRANG, L. B. (1967). *J. Physiol.* **193**, 1–29.

JOST, A. & POLICARD, A. (1948). *Archs. Anat. Microsc.* **37**, 323–32.

KIKKAWA, Y., KAIBARA, M., MOTOYAMA, E. K., ORZALESI, M. M. & COOK, C. D. (1971). *Am. J. Path.* **64**, 423–32.

KOEFOED-JOHNSON & USSING, H. H. (1953). *Acta Physiol. Scand.* **28**, 60–76.

LANDIS, E. M. & PAPPENHEIMER, J. R. (1963). In *Handbook of Physiology*, section 2, vol. II, ch. 29, Circulation. Ed. Hamilton, W. F. & Dow, P. Washington: American Physiological Society.

NORMAND, I. C. S., REYNOLDS, E. O. R. & STRANG, L. B. (1970). *J. Physiol.* **210**, 151–64.

NORMAND, I. C. S., OLVER, R. E., REYNOLDS, E. O. R., STRANG, L. B. & WELCH, K. (1971). *J. Physiol.* **219**, 303–20.

POTTER, E. L. & BOHLENDER, G. P. (1941). *Am. J. Obstet. Gynec.* **42**, 14–22.

PREYER, W. (1885). *Specielle Physiologie des Embryo*, p. 148. Leipzig: Th. Grieben's Verlag (L. Fernan).

RENKIN, E. M. (1954). *J. gen. Physiol.* **38**, 225–43.

RICH, G. T., SHA'AFI, R. I., BARTON, T. C. & SOLOMON, A. K. (1967). *J. gen. Physiol.* **50**, 2391–405.

ROUS, P., GILDING, H. P. & SMITH, F. (1930). *J. exp. Med.* **51**, 807–30.

SCHNEEBERGER-KEELEY, E. E. & KARNOVSKY, M. J. (1968). *J. cell. Biol.* **37**, 781–93.

SETNIKAR, I., AGOSTONI, E. & TAGLIETTI, A. (1959). *Proc. Soc. Exp. Biol., N.Y.* **101**, 842–5.

SHEPPARD, C. W. & HOUSEHOLDER, A. S. (1951). *J. appl. Physics* **22**, 510–20.

SHIRLEY, H. H., Jr, WOLFRAM, C. G., WASSERMANN, K. & MAYERSON, H. S. (1957). *Am. J. Physiol.* **190**, 189–93.

SOLOMON, A. K. (1968). *J. gen. Physiol.* **51**, 335–64.

STAUB, N. C., NAGANO, H. & PEARCE, M. L. (1967). *J. appl. Physiol.* **22**, 227–40.

SWANN, H. G. & SPAFFORD, N. R. (1951). *Tex. Rep. Biol. Med.* **9**, 356–62.

TASAKI, I. & SINGER, I. (1968). *Ann. N.Y. Acad. Sci.* **148**, 36–53.

TAYLOR, A. E. & GAAR, A. K. (1970). *Am. J. Physiol.* **218**, 1133–40.
TOWERS, B. (1968). In *Biology of Gestation*, Vol. 2, pp. 189–223. Ed. Assali, N. S. New York and London: Academic Press.
WANGENSTEEN, O. D., WITTMERS, L. E. & JOHNSON, J. A. (1969). *Am. J. Physiol.* **216**, 719–27.
WRIGHT, G. M. (1962). *J. Physiol.* **163**, 281–93.

THE PRODUCTION AND COMPOSITION OF LUNG LIQUID IN THE IN-UTERO FOETAL LAMB

By T. M. ADAMSON, V. BRODECKY,
T. F. LAMBERT, J. E. MALONEY,
B. C. RITCHIE AND A. WALKER

Department of Paediatrics & Department of Medicine,
Monash University;
Department of Physiology, Melbourne University,
Melbourne, Australia

The foetal lung in mammals develops as a liquid-filled organ. The solute composition of this liquid suggests it is a secretion, or an ultrafiltrate of plasma with selective ionic absorption (Adamson *et al.* 1969), the liquid being produced in the foetal lung. Most studies reported have been in acute animal preparations (Adams, 1966), and reliable rates of flow and production of lung liquid over long periods are not available. Goodlin & Rudolph (1970) have suggested that the rate of tracheal fluid flow is low, while Dawes *et al.* (1972) showed in chronic foetal lamb preparations only a small volume of fluid flowed out of the lungs during foetal respiratory movements. Recently Normand *et al.* (1971) in acute preparations of 5 hours' duration gave rates of formation of 0.036 ml/min·kg.

To study lung liquid production more critically we developed an in-utero foetal lamb preparation in which lung liquid production could be followed for days. Five ewes, 105–115 days pregnant (term 145 days) were anaesthetized with halothane and oxygen. The throats of the lambs were delivered aseptically and their tracheas were ligated proximal to the larynx. Distally a catheter was tied into the trachea draining the lung. This catheter was exteriorized through the side of the ewe and attached to a collecting bag forming a foetal tracheal fistula. As lung liquid was formed it drained into the bag, allowing its daily volume to be measured and solute composition analysed. Prior to being returned to the uterus, the lambs had a foetal electrocardiograph electrode implanted, and a second catheter attached, to enable amniotic liquid monitoring of pressure and electrolyte composition.

The preparations lasted 13–27 days (mean 20.3 days), three having an elective Caesarean section at 133 days. In all preparations the cumulative daily volume was remarkably high, exceeding 2 litres in the first 14-day

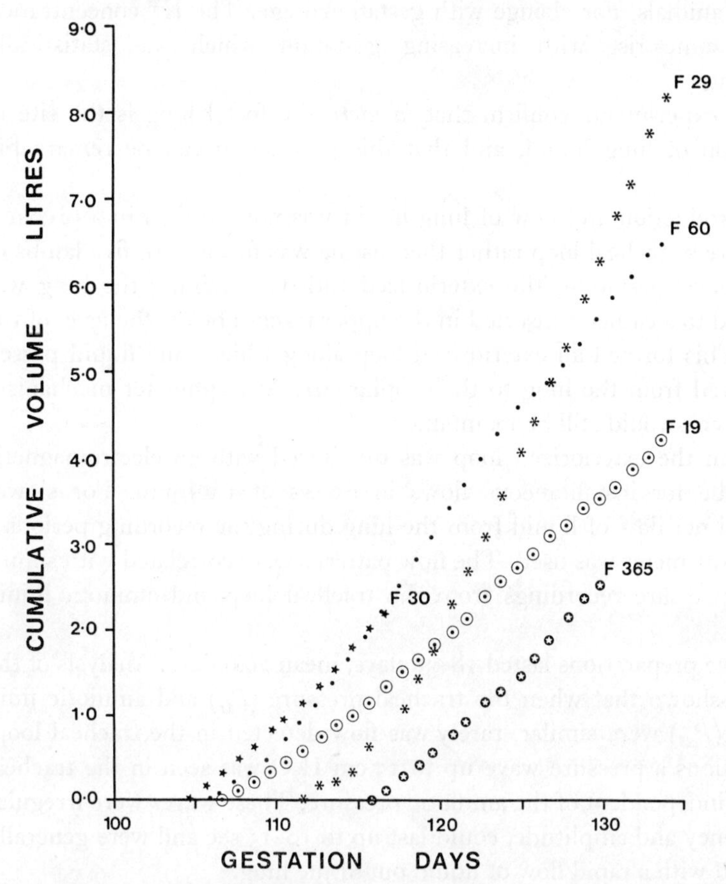

Fig. 1. The volumes obtained from daily collections of lung liquid in five lambs *in utero* in whom a free-draining tracheal fistula had been formed. The results are expressed as cumulative volumes plotted against gestational age for each animal.

period (Fig. 1). In one animal 8.2 litres were produced over 21 days. In five animals the total volume of lung liquid produced was equivalent to a mean flow of 9.4 ml/h (range 6.6–15.4) over the total collecting period.

In most preparations after an initial rise in flow over the first 5–7 days, there was little increase in rate of flow thereafter despite increasing gestational age and foetal growth.

Analysis of the solute composition of lung liquid showed mean concentrations $\pm$ SD for Na^+ 148 $\pm$ 7.3 mEq/l; K^+ 5.7 $\pm$ 1.0 mEq/l; Ca^{2+} 2.45 $\pm$ 1.55 mg %; Mg^{2+} 0.39 $\pm$ 0.34 mg %; Cl^- 146 $\pm$ 7.1 mEq/l; and urea 34 $\pm$ 9.0 mg %. These are similar to those reported for acute preparations (Adamson *et al.* 1969) except that the Cl^- and K^+ are lower. The concentrations of Na^+, Ca^{2+}, Mg^{2+}, Cl^- and urea showed no significant difference between animals, nor change with gestational age. The K^+ concentration showed some rise with increasing gestation which was statistically significant.

These experiments confirm that *in utero* the foetal lung is the site of production of lung liquid, and that this production can be remarkably high.

The production and flow of lung liquid was now studied in a preparation where a tracheal loop rather than fistula was formed. In five lambs of 94–105 days' gestation, the exteriorized catheter draining the lung was connected to a catheter inserted in the upper trachea below the level of the larynx. This formed an exteriorized loop along which lung liquid passed as it flowed from the lung to the oropharynx. Any sphincter mechanism at the larynx would still be maintained.

Flow in the exteriorized loop was monitored with an electromagnetic flow probe for instantaneous flows in excess of 1 ml/min. For slower flows and net flow of liquid from the lung during the recording periods a bubble flow meter was used. The flow patterns were correlated with simultaneous pressure recordings from the tracheal loop and amniotic liquid cavity.

The five preparations lasted 18–38 days, mean 26.6 days. Analysis of the data has shown that when the tracheal pressure (P_{tr}) and amniotic fluid pressure (P_{af}) were similar, rarely was flow detected in the tracheal loop. On occasions a pressure wave up to 15 cm H_2O was seen in the tracheal pressure independent of the amniotic pressure. These waves were irregular in frequency and amplitude, could last up to 10–15 sec and were generally associated with a rapid flow of liquid out of the lung.

On occasions the tracheal pressure (P_{tr}) was markedly different from the amniotic pressure (P_{af}) with a sharp rise or fall of pressure in the tracheal line up to 12–15 cm H_2O. These could occur singly or in combination with a periodicity of 1–5/sec (Fig. 2). Although these patterns were always associated with rapid movements of liquid in the tracheal line as seen in the electromagnetic flow probe recording $\dot{Q}_{tr}$ (Fig. 2), when monitored with a bubble flow meter there was a variable net flow of lung liquid from the lung. On occasions a large net flow was observed from the lungs, while other times no net flow or even a small retrograde flow. These tracheal pressure recordings are similar to those of Dawes *et al.* (1972) and represent foetal respiratory movements.

Although peak flows up to 600 ml/h have been recorded in the exteriorized loop these are of short duration (0.5 sec). When lung liquid flow was monitored over a 6-h period with a bubble flow meter, recordings similar to Fig. 3 were obtained. In each record lung liquid flow was intermittent

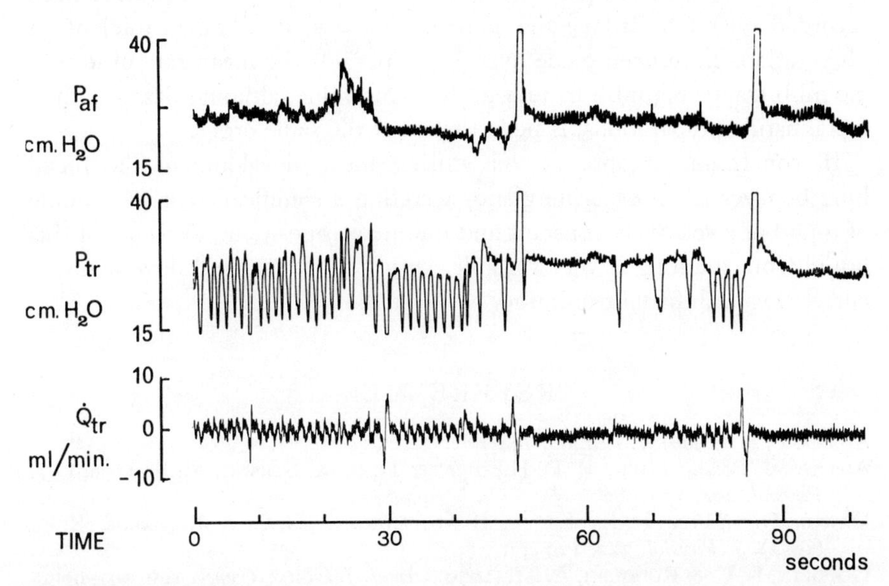

Fig. 2. Lamb of 126 days' gestation *in utero*. Records from above downwards of amniotic fluid (P_{af}) and tracheal (P_{tr}) pressures and of tracheal fluid flow ($\dot{Q}_{tr}$) from an electromagnetic flow-meter probe around the exteriorized tracheal loop.

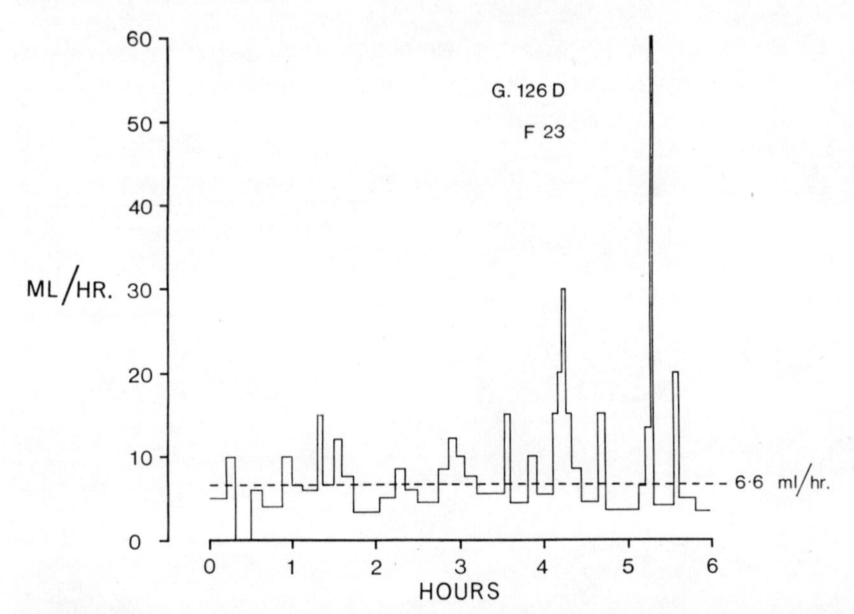

Fig. 3. A 6-h record of lung liquid flow from a lamb *in utero* at 126 days. Flow measured with a bubble flow meter inserted into the exteriorized tracheal loop. Mean flow over the 6-h period was 6.6 ml/h.

and irregular. In this record at 126 days' gestation the mean flow was 6.6 ml/h, while in some preparations mean flows up to 9.0 ml/h have been recorded over 6 h. In the five animals analysed in whom a total of 38 observations have been made over 3–9-h periods the mean rate of flow is 5.0 ml/h (range 0.3 ml/h to 11.1 ml/h). This value, although less than the foetal fistula preparations, is nevertheless of the same order.

In conclusion, it appears that during foetal development the foetal lung behaves as an exocrine gland, secreting a significantly large volume of liquid of a relatively constant and unique composition. The flow of this liquid from the lung is very variable, with periods of rapid flow at times, correlating with foetal respiratory movements.

REFERENCES

ADAMS, F. H. (1966). *J. Ped.* **68**, 794–801.

ADAMSON, T. M., BOYD, R. D. H., PLATT, H. S. & STRANG, L. B. (1969). *J. Physiol.* **204**, 159–68.

DAWES, G. S., FOX, H. E., LEDUC, B. M., LIGGINS, G. C. & RICHARDS, R. T. (1972). *J. Physiol.* **222**, 119–43.

GOODLIN, R. C. & RUDOLPH, A. M. (1970). *Amer. J. Obstet. Gynec.* **106**, 597–606.

NORMAND, I. C. S., OLVER, R. E., REYNOLDS, E. D. R., STRANG, L. B. & WELCH, K. (1971). *J. Physiol.* **219**, 303–30.

INITIATION OF RESPIRATORY MOVEMENTS IN FOETAL SHEEP BY HYPOXIC STIMULATION OF FOETAL CENTRAL CHEMORECEPTOR

By V. CHERNICK and A. H. JANSEN
Children's Hospital, Winnipeg, Manitoba, Canada

An interaction between hypoxia and hypercapnia at the time of the first breath of the anaesthetized foetal sheep at term has been demonstrated previously using a technique of cross-circulating a foetal lamb with a newborn lamb (Pagtakhan, Faridy & Chernick, 1971). During cross-circulation the foetus could be maintained in the apnoeic state with normal blood gas tensions despite absent placental circulation for up to 30 minutes. In order to initiate respiratory movements foetal blood gases were altered by adjusting the alveolar ventilation and composition of the inspired gas of the newborn lamb. Liquid ventilation was monitored by a plethysmograph and foetal femoral arterial blood was sampled at the time of the first breath. At the time of the first breath there was a linear relationship between the arterial P_{O_2} and P_{CO_2} ($P_{aO_2} = 0.14, Pa_{CO_2} - 0.31; r = 0.83, p < 0.001$).

Further studies have been undertaken with the same experimental technique following bilateral section of sinus and vagus nerves. Fig. 1 illustrates the mean ($\pm$ SE) P_{aO_2} measurements at three levels of P_{aCO_2} (<40, 40–100 and >100 mm Hg) obtained at the time of the first breath in intact foetuses and in those following bilateral carotid sinus nerve transection (Pagtakhan *et al.* 1969). In the intact foetus, as the arterial P_{CO_2} increased, the P_{aO_2} at which breathing was initiated also increased, indicating an interaction between hypoxaemia and hypercapnia. However, following denervation of the carotid body the P_{aO_2} required to initiate the first breath was independent of P_{aCO_2}. Thus, the effect of P_{CO_2} was mediated by the carotid body, but this effect had only a minor modulating influence on the level of hypoxaemia required to initiate breathing. At a P_{aCO_2} of less than 40 mm Hg the P_{aO_2} required to initiate respiratory movement was less than 5 mm Hg in intact foetuses, but about 10 mm Hg following carotid body denervation. Thus an arterial P_{CO_2} of less than 40 mm Hg had an inhibitory effect on the initiation of respiratory movement in intact foetuses, whereas a P_{aCO_2} above 100 mm Hg was stimulatory. Additional studies in foetuses following total peripheral chemoreceptor denervation indicated that a mean P_{aO_2} of about 8 mm Hg initiated respiratory move-

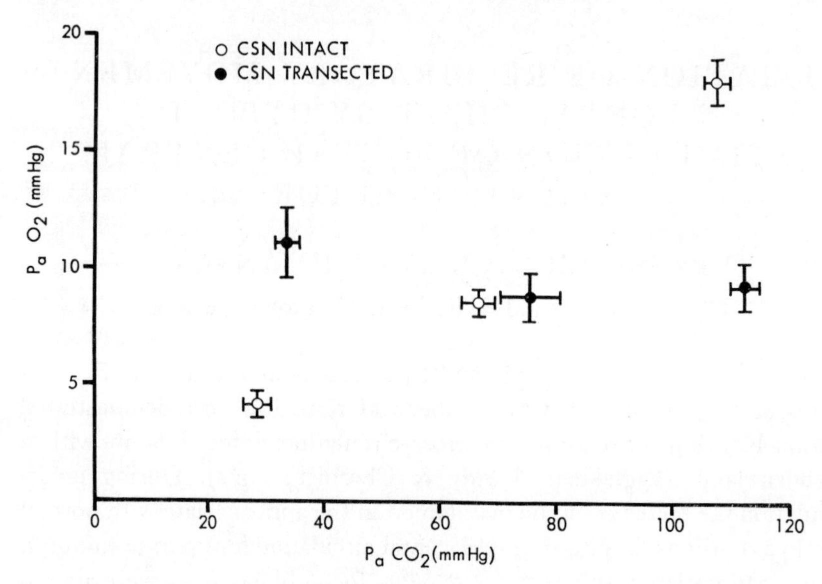

Fig. 1. Relationship between arterial P_{O_2} and P_{CO_2} at the time of the first breath in foetal sheep with intact carotid sinus nerve (CSN) and following bilateral transection of CSN.

ment, and that this level was independent of P_{aCO_2} ranging from 20 to 130 mm Hg.

Additional studies were undertaken in both anaesthetized and non-anaesthetized exteriorized term foetal sheep. Increasing concentrations of sodium cyanide (2 ml volume) were injected into a femoral vein at 15-minute intervals in intact foetuses, and following bilateral vagotomy and total peripheral chemodenervation. The dose required to initiate breathing movement was independent of the presence of peripheral chemoreceptors (Fig. 2). Pentobarbital anaesthesia was associated with about a five-fold increase in the dose of NaCN/kg body weight required to initiate respiratory movements.

Sodium cyanide (0.1 ml) in increasing concentrations was also injected into the area of the ventral brain stem of the anaesthetized term foetus with intact peripheral chemoreceptors and following vagotomy and total peripheral chemodenervation (Table 1). This technique involved removal of the dorsal portion of the first cervical vertebra and placement of two catheters into the subdural space of the ventral medulla, so that injection did not increase intracranial pressure. Approximately 1 mg NaCN initiated respiratory movement, even in the absence of peripheral chemoreceptors. Mock CSF had no effect. Application of cyanide to the dorsal medulla did not initiate respiratory movement when diffusion along the subdural space to the ventral medulla was prevented.

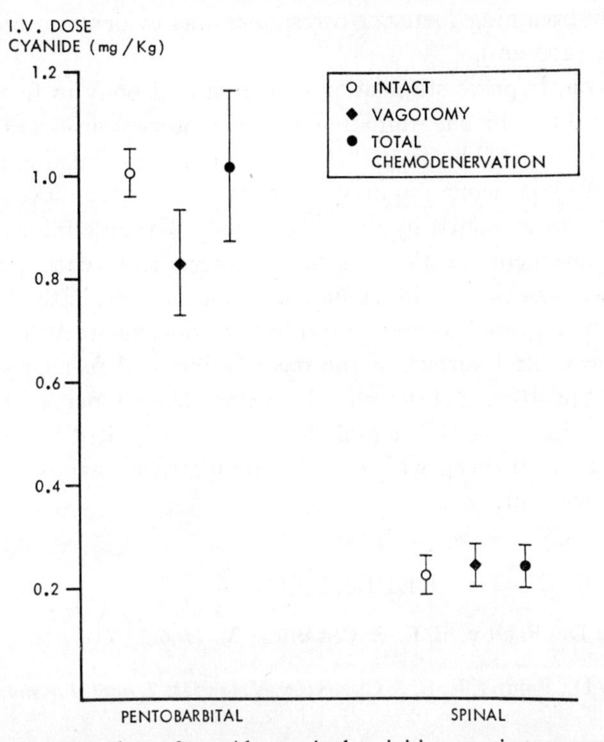

Fig. 2. Intravenous dose of cyanide required to initiate respiratory movement in intact foetal sheep following vagotomy and total peripheral chemoreceptor denervation. Maternal administration of pentobarbital was used in one group of animals and the other group was delivered following spinal anaesthesia of the ewe.

Table 1. *Mean (± SE) dose of cyanide required to initiate respiratory movement in anaesthetized foetal sheep*

Cyanide was injected into the subdural space of the ventral brain stem.

Cyanide dose (mg)	
Intact (5 foetuses, 11 injections)	1.03 ± 0.16
Vagotomy (5 foetuses, 11 injections)	0.92 ± 0.13
Total chemodenervation (8 foetuses, 15 injections)	1.10 ± 0.18

Ventral brain stem injections of dinitrophenol (0.4–0.5 mg) were associated with variable results ranging from no respiratory response to a rapid onset of respiratory movement. EGTA (4 μm) injections were associated with rapid respiratory responses but increasing doses were required to obtain a response. $CaCl_2$ (2 μm) injections did not prevent the response to cyanide or EGTA. Xylocaine (2%) did not prevent the cyanide response or cause cessation of respiratory movement in the spontaneously breathing foetus. No response was evident to injection of noradrenaline (200 μm).

In conclusion, hypoxic stimulation of the carotid body in foetal sheep did not contribute to the initiation of respiratory movement despite profound hypoxaemia. The arterial P_{CO_2} had a minor modulating influence on the level of hypoxaemia required to initiate respiratory movement and this influence was mediated by the carotid body. Cyanide initiated foetal respiratory movement in the absence of peripheral chemoreceptors. Anaesthesia was associated with about a five-fold increase in the threshold dose of cyanide required to initiate respiratory movement. Application of cyanide to the ventral surface of the medulla initiated foetal respiratory movement in the absence of peripheral chemoreceptor innervation. These results suggest that there is a central chemoreceptor located in the ventral medulla of the foetal sheep which, when stimulated by hypoxia, initiates respiratory movement.

REFERENCES

Pagtakhan, R. D., Faridy, E. E. & Chernick, V. (1969). *The Physiologist* **12**, 321.

Pagtakhan, R. D., Faridy, E. E. & Chernick, V. (1971). *J. appl. Physiol.* **30**, 382.

THE ONSET AND CONTROL OF BREATHING AFTER BIRTH

By P. JOHNSON, J. S. ROBINSON and D. SALISBURY

The Nuffield Institute for Medical Research, University of Oxford

It has long been thought that the foetus is inhibited from breathing *in utero*. Harned, Herrington & Ferreiro (1970) and Tchobroutsky, Merlet & Rey (1969) suggested that immersion in a liquid environment inhibits breathing. They found that foetal lambs, placed in a water bath and with a tracheotomy open to air, did not establish effective ventilation when their umbilical cords were tied; they died. They also showed that the newborn lamb, breathing spontaneously through a tracheotomy, always open to air, ceased breathing when immersed in water. Further experiments suggested that this reflex inhibition was mediated via the upper airway. However, Dawes *et al.* (1970) and Merlet *et al.* (1970) have shown that episodic breathing movements in foetal lambs are normally present in intrauterine life. It was therefore decided to examine the nature of the reflex inhibition of breathing on immersion of the foetal or newborn lamb. A preliminary account was given elsewhere (Dawes, Johnson & Robinson, 1971).

METHODS

Seven foetal lambs of 142 days' gestation were delivered by Caesarean section into a warm saline bath alongside the ewe. An arterial catheter and P_{aO_2} probe were inserted into the carotid artery. A tracheotomy was performed, tracheal fluid was removed and a Fleisch pneumotachograph was attached to the outlet; 50% O_2 in N_2 or air was made available. The umbilical cord was then ligated and continuous observations were made on the circulation, breathing and the blood gas values.

In 60 newborn lambs from 1 hour to 21 days the trachea was divided. A pneumotachograph was attached to the lower end. Liquids were introduced through the upper end into the upper airway. The oesophagus was divided and an electromagnetic flowmeter was inserted into the upper end to record swallowing; the outlet was attached to a reservoir. A P_{aO_2} probe and catheter were inserted into the carotid artery. In 12 lambs biparietal cortical electrodes were implanted.

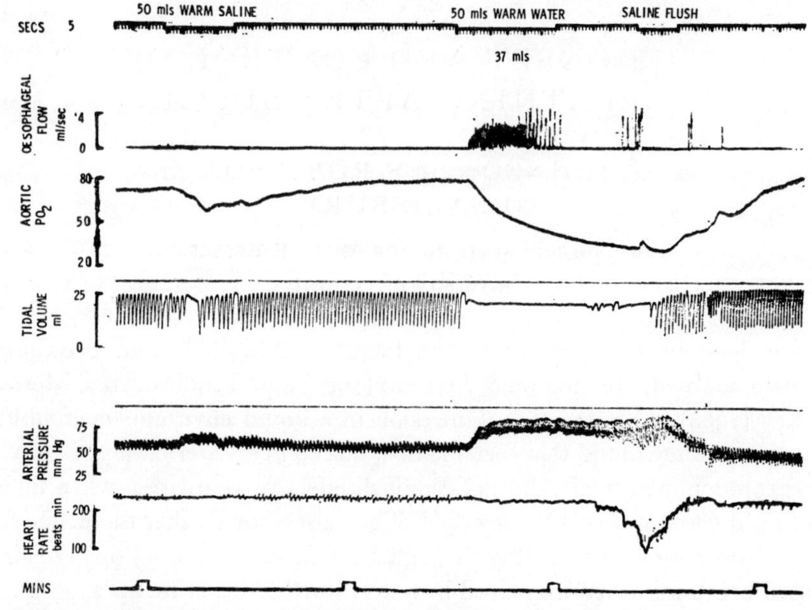

Fig. 1. Newborn lamb. Introduction of warm saline into the upper airways caused only a transient disturbance; introduction of warm water caused a brief episode of swallowing (above), apnoea and a large fall in aortic P_{O_2} (next two records) which lasted until the upper airways were flushed with saline. The consequent rise of arterial pressure and, in this instance, delayed bradycardia are shown in the lower two records.

Thirty lambs were subjected to a decreased concentration of inspired O_2 (22%, 12% or 8% O_2 in N_2) or to 3% CO_2 in air, in order to examine the ventilatory response to hypoxia and hypercapnia with age.

RESULTS

Seven mature foetal lambs were delivered by Caesarean section under maternal epidural anaesthesia and were placed in a warm saline bath (39 °C) with their placental circulation intact. They all established effective rhythmic ventilation through a tracheotomy when their umbilical cords were ligated, while immersed in the liquid except for their tracheal cannula.

THE REFLEX INHIBITION OF BREATHING AND AGE FROM BIRTH

All the lambs (and five ewes) tested ceased breathing and began swallowing when water was introduced into the upper airway, but not when 0.15 M saline was introduced. The apnoea was not always associated with swallowing.

In a group of 30 lambs there was an increased ventilatory response to hypoxia with age. There appeared to be a breed variation since newborn Border-Leicester lambs had an inadequate response to hypoxia, when compared with Dorsets. All the lambs examined increased their ventilation with hypercapnia.

Eighteen lambs remained apnoeic so long as water was present in the upper airway. Of these eight were newborn Border-Leicester lambs, which had a poor ventilatory response to hypoxia; these remained almost totally apnoeic for 30–40 minutes, yet recovered spontaneously upon removal of the water. In three the water was not removed from the upper airway; they died.

Ten lambs (Dorset, or Border-Leicester more than two days old) remained apnoeic until their arterial P_{aO_2} fell below 10 mm Hg; the electrocorticogram became isoelectric. About 10–15 seconds later there was a convulsive episode associated with one or two deep gasps and a short burst of swallowing; arterial P_{aO_2} rose and the electrocortical activity returned. If the water was removed or the superior laryngeal nerves were sectioned at any time except during an isoelectric phase there was immediate restoration of breathing with a vigorous hyperpnoea. If the arterial P_{aO_2} was below 10 mm Hg and the electrocorticogram isoelectric, the lamb did not respond to removal of the water or to section of the superior laryngeal nerves; spontaneous recovery then occurred only after a convulsive movement associated with a gasp. If the water was not removed and the superior laryngeal nerves were intact the events just described were repeated episodically for several minutes until the water was removed. Prolonged apnoea in older lambs was tolerated for only 3–4 min before serious deterioration occurred (i.e. central respiratory failure and/or cardiovascular collapse).

Twelve newborn lambs (Dorset, or Border-Leicester more than two days old) exhibited only transient apnoea when water was introduced into the upper airway. Six of these lambs had a good ventilatory response to hypoxia. After bilateral crushing of the carotid nerves the response to hypoxia was abolished; the apnoea on introduction of water to the upper airway was still transient. Thus rapid recovery in these lambs was not a consequence of effective respiratory stimuli from the carotid chemoreceptors.

LOCALIZATION AND NATURE OF THE RECEPTORS

Experiments on five lambs in which the different parts of the upper airway (nasopharynx, oropharynx, laryngopharynx, larynx and trachea) were isolated showed that the receptors were located at the entrance to the larynx. Bilateral section of the superior laryngeal nerves or application of local anaesthetic solution to the ary-epiglottic folds abolished the apnoea caused by introduction of water into the larynx.

A variety of physiological fluids were tested. Normal amniotic fluid, tracheal fluid and sheep's milk did not cause apnoea, whereas allantoic fluid, cow's milk, human and artificial milks all caused apnoea and swallowing similar to that induced by water. Hydrochloric acid (0.01 N), quinine HCl (0.02 M) and sucrose (1 M) also induced apnoea. Breathing was not arrested by NaCl solutions of 0.077 M to 0.616 M, but was arrested by glucose solutions ranging from 0.111 to 2.22 M. Altering the pH of the sheep or cow's milk between 2 and 8 did not change the responses. Altering the temperature of the solutions from 20 to 42 °C did not appreciably modify the responses.

Table 1

Perfusing fluid	Apnoea	Swallowing
0.154 M Saline	−	−
Water	+	+
0.01 N HCl	+	+
0.02 M Quinine HCl	+	+
1 M Sucrose	+	+
0.111 M Glucose	+	+
2.22 M Glucose	+	+
0.077 M Saline	−	−
0.616 M Saline	−	−
Amniotic fluid	−	−
Tracheal fluid	−	−
Sheep's milk	−	−
Cow's milk	+	+
Allantoic fluid	+	+

Examination of serial histological sections of the region revealed approximately 2000 taste buds spread uniformly and almost exclusively over the pharyngeal surface of the ary-epiglottic folds and the laryngeal surface of the epiglottis.

DISCUSSION

The experiments show that certain liquids inhibit breathing in mature foetal, newborn and adult sheep by stimulating a reflex whose receptors are located at the entrance to the larynx. Yet normal amniotic fluid and tracheal fluid do not cause inhibition of respiration. Foetal breathing is unlikely to be inhibited by this means during normal intrauterine life.

The afferent nervous pathway is through the superior laryngeal nerves. The nature and range of the stimuli used suggest taste receptors and not osmo-, pH, thermal, tactile or pressure receptors. The abundance of taste buds concentrated in this region supports this theory (Lalonde & Eglitis, 1961; Wilson, 1905). The length of the apnoea that is induced shows that the receptors do not adapt rapidly.

It is concluded that there is a reflex pathway from laryngeal taste receptors functional throughout life which, when stimulated by specific fluids, causes apnoea. Recovery on removal of the stimulus was always associated with an hyperpnoea proportionate to the degree of asphyxia. This shows that the CO_2 and O_2 drive to respiration, even when well developed, was overridden by the reflex. Analysis of the length of apnoea induced in relation to age from birth suggests that there are stages in the development of respiratory control which take time to mature. Apnoea is most readily induced in the newborn lamb. These have an inadequate response to hypoxia, but a vigorous response to hypercapnia. However, these lambs can withstand the resultant asphyxia longer than the older animal.

The type of reflex described has also been observed in calves, newborn and adult monkeys. The role of such a reflex in apnoea of the newborn, in sudden death in infancy, and even in drowning, is worthy of further investigation.

The authors are indebted to Dr G. S. Dawes for his advice and to Mr A. Stevens and Mr N. Green for their technical assistance. This work was carried out with the aid of a grant from the Medical Research Council.

REFERENCES

DAWES, G. S., FOX, H. E., LEDUC, B. M., LIGGINS, G. C. & RICHARDS, R. T. (1970). *J. Physiol.* **210**, 47–8*P*.

DAWES, G. S., JOHNSON, P. & ROBINSON, J. S. (1971). The onset of respiration in the newborn. Abstract: European Society of Pediatric Research/Neonatal Society. June 1971.

HARNED, H. S., HERRINGTON, R. T. & FERREIRO, J. I. (1970). *Pediatrics* **45**, No. 4.

LALONDE, E. R. & EGLITIS, J. A. (1961). *Anat. Rec.* **140**, 91–3.

MERLET, C., HOERTER, J., DEVILLENEUVE, C. & TCHOBROUTSKY, C. (1970). *C. r. hebd. Seanc. Acad. Sci., Paris* **270**, 2462–4.

TCHOBROUTSKY, C., MERLET, C. & REY, P. (1969). *Respiration Phys.* **8**, 108–17.

WILSON, J. G. (1905). *Brain* **28**, 339–51.

SESSION 3

PLACENTAL FUNCTION

SOME PRINCIPLES GOVERNING MATERNAL-FOETAL TRANSFER IN THE PLACENTA

By R. E. FORSTER, II

Department of Physiology, School of Medicine,
University of Pennsylvania, Philadelphia, Pennsylvania 19104

I was asked to discuss the mechanisms of placental exchange, particularly of gases. However, there are excellent and recent, comprehensive and scholarly reviews of this subject, several of the authors of which are actually at this Symposium (Assali, Dilts, Plentl, Kirschbaum & Gross, 1968; Bartels, 1970; Dawes, 1968; Longo, 1972; Metcalfe, 1967; Metcalfe, Bartels & Moll, 1967). Discretion dictates therefore that I limit my paper very strictly. I will discuss three discrete aspects of placental gas exchange which are of current interest, and restrict any review of the field to the framework necessary to support them.

A major function of the placenta is to transfer gas between the circulations of the mother and of the foetus, most critically to transfer oxygen to the foetus at the rate required by the foetal metabolism and at the P_{O_2} necessary to provide the requisite diffusion gradient from the foetal peripheral capillary blood to the tissues. Researches into placental oxygen exchange were pioneered by Barcroft (1947) and have been continued by his students many of whom are present. I shall not even attempt to summarize this work but refer you for further information to the several reviews and monographs above.

Fig. 1 is an extremely simplified diagram of the placenta. Gas exchange takes place by diffusion under a gradient of partial pressure between the capillary blood in the foetal and maternal placental microcirculations. The rate of gas movement will be directly proportional to this pressure difference, to the area of contiguity of capillaries, that is the area of placental membrane separating the microcirculations, and the permeability of this membrane, which is in turn equal to the product of the solubility and the diffusion coefficient of the particular gas in the material making up the

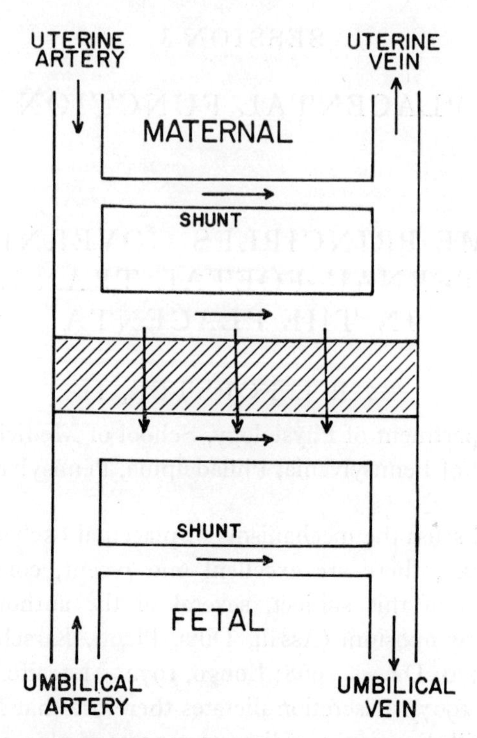

Fig. 1. A diagram, very simplified, of the placenta as an exchanger. Concurrent flow of the maternal and foetal blood has been assumed; however, it is likely that the exchange volumes of capillary blood resemble, at least on one side, mixed pools, so that the flow is strictly neither concurrent nor counter-current. The shaded region represents all of the layers of the placental membrane.

membrane. In the case of those physiological gases which react chemically with blood elements, the rate of the pertinent chemical reactions in the blood must also be considered as well as the volume of blood in the capillaries (V_c).

The physical exchange characteristics of the placenta, as for any other type of exchange mechanism, can be expressed as a transfer coefficient equal to the rate of gas flow across the membrane mean pressure difference between the capillaries. This was called the *diffusion constant* by Barcroft (1947) and the *diffusion coefficient* by Barron & Alexander (1952). We have called it the *placental diffusing capacity*, D_p, admittedly a poor term, in analogy to the diffusing capacity of the lung, which has the virtues only of being well established. It is generally possible to measure the flux, that is the transport in ml of gas at STP/min, across the placental membrane from the product of the placental blood flow and the arterio-venous content, or from a change in blood concentration times the foetal, or maternal, blood volume. It is much more difficult to estimate the mean difference in gas partial pressure between the maternal and foetal circulations in part because of the complexity of placental circulation and our lack of knowledge. This difficulty becomes greater the more nearly the gas achieves complete equilibrium, because under these conditions the partial pressure difference between foetal and maternal blood is small and the relative importance of errors rises.

In order to interpret measured values of gas partial pressure in maternal arterial, uterine venous, and umbilical arterial and venous bloods, it is necessary to know the geometrical relationship of the capillaries and the directions of flow in them. Microscopic studies of the anatomy of the placenta suggested to many investigators (Barcroft, 1947) that the gas exchanges took place primarily in regions where the maternal and foetal capillary flows were parallel but in opposite directions; so-called counter-current flow. This is to be contrasted with a concurrent flow exchanger in which the maternal and foetal capillaries are also parallel but their flows are in the same direction. The counter-current exchanger has the theoretical advantage that it maintains the partial pressure difference between the two bloods more uniform along the capillaries, and the outflowing foetal blood P_{O_2} can be greater than the outflowing maternal blood. The concurrent exchanger permits the two bloods to approach gaseous equilibrium so that very little gas movement may occur towards the end of the capillary. To be effective functionally a counter-current exchanger requires extremely exact geometrical relations; exchange of gases in regions where the flow is not counter-current leads to a decreased efficiency. At least for mammals, sheep, dog and man, the evidence for a functional counter-current system is not convincing at this time and I have assumed concurrent flow. The irregularity of the spatial relations of the two capillary beds, presence of shunts and small lacunae of blood, all make it more likely that the placenta functions as a mixed exchanger (Metcalf *et al.* 1967), approaching a model which has diffusion exchange between fairly well mixed foetal and maternal blood pools.

Functional shunts of inflowing blood across the exchange surfaces are known to exist (Longo, Power & Forster, 1969) on both or either sides of the membrane, increasing the difficulty of studying the performance of the placenta. This acts to reduce the efficiency of gas exchange.

If the maternal and foetal placental capillary blood flows are not matched at each point of diffusion exchange, the transfer of gases from one circulation to the other is less complete than if the maternal blood flow/foetal capillary blood flow is the same throughout (Longo, 1972). The presence of non-uniform capillary blood flow ratios can produce a difference in gas tensions of the outflowing mixed maternal and mixed foetal placental capillary blood in spite of the fact that the gas partial pressures are the same in blood leaving each associated maternal and foetal capillary. There is some experimental evidence for the existence of such non-uniformity; in fact it would be very unlikely if it did not exist. Because the placenta is not indefinitely distensible, a change in volume of one capillary system will tend to change the volume of the other system in the opposite direction; a type of Starling resistance regulation. Non-uniformity of capillary blood flow will produce the same effect on uterine venous and umbilical venous blood gases as a shunt, and can be represented by a shunt, as in the diagram of Fig. 1. It may be technically difficult in fact to discriminate between anatomical shunt and non-uniform placental blood flow in a given placenta. The shunt on the maternal side of the placenta is made up of possible microscopic shunts in the actual placenta as well as some blood flow from uterine muscle, because the venous drainage on the maternal side is not anatomically distinct.

It is convenient to consider the placental exchange characteristics of four classes of gases: inert gases, CO, O_2 and CO_2.

INERT GAS EXCHANGE

An *inert gas* is defined as one which does not react chemically with blood or tissues, is neither consumed nor produced in the placenta and is therefore carried in the blood only as dissolved gas. Examples are N_2, H_2, He and anaesthetic gases. Inert gases appear to reach complete diffusion equilibrium between the maternal and foetal placental capillary circulation in the region where exchange takes place. This might be expected intuitively because the amount of gas that must be transported across the placental membrane to produce equilibrium is only the amount that dissolves in the blood. Experimental evidence supporting this conclusion has been obtained in perfused placenta (Meschia *et al.* 1967; Metcalfe *et al.* 1965). In addition when two gases of considerably different diffusion coefficients, SF_6 and H_2, are introduced into the umbilical artery, they appear in the same ratio in umbilical and uterine venous blood (Longo, Delivoria-Papadopoulos, Power, Hill & Forster, 1970). This could only occur if both gases were at least nearly completely equilibrated between foetal and maternal placental venous blood.

CARBON MONOXIDE EXCHANGE

CO has about 250 times the affinity for Hb that O_2 does and is a 'physiological' gas, in that it is produced and catabolized normally in the body. These characteristics qualify it for the measurement of the diffusing capacity of the placenta because if the experimental conditions are properly chosen the estimation of the mean P_{CO} difference between maternal and foetal circulations is more easily and precisely measured than is the case for O_2, and the result is much less influenced by the presence of shunts or non-uniform capillary blood flow ratios. In the measurement CO is added to the maternal circulation, time allowed for the HbCO to mix with the mother's blood and equilibrate with myoglobin and other sites of CO binding, and then the transfer of CO to the foetus obtained by following the change of maternal [HbCO] times the mother's effective blood volume. The mean P_{CO} of the maternal placental capillaries is calculated from measured values of P_{O_2} and HbO_2 saturation in maternal arterial and uterine venous blood, using the Haldane relationship

$$P_{CO} = \frac{P_{O_2} \times [HbCO]}{M[HbO_2]}. \qquad (1)$$

M is a constant equal to about 210. HbCO concentration will be practically constant along the capillary; since the affinity is so high, only a small proportion of the CO present crosses the membrane. While P_{O_2} and HbO_2 saturation change along the capillary, their ratio is more nearly constant. Therefore one can obtain a relatively reliable value of average maternal placental capillary P_{CO}. Mean foetal placental P_{CO} is much less than maternal, particularly at the start of exchange, so that errors in it are of less significance.

The placental diffusing capacity $D_{p,CO}$, has been measured in acutely anaesthetized sheep and dogs, and values of 0.54 ml/(mm Hg × min × kg of foetus) and 0.57 ml/(min × mm Hg × kg of foetus) respectively obtained (Longo, Power & Forster, 1969). If these are compared with estimates of D_{p,O_2} calculated from experimental measurements of $[HbO_2]$ and P_{O_2} in uterine and umbilical arterial and venous blood, the values obtained with CO are about five times greater. As yet no measurements of D_p have been obtained in chronic unanaesthetized animals.

An additional complication is the fact that CO exchange is not simply a matter of diffusion across the placental membrane, but also involves chemical reaction and diffusion within the maternal and foetal blood, particularly with the red cells. Thus the total diffusion resistance to the movement of CO from maternal to foetal haemoglobin equals the sum, the resistance in the maternal blood + the resistance in the membrane + the resistance in the foetal blood. In terms of diffusing capacity, which is the reciprocal of resistance,

$$1/D_{p,CO} = 1/\theta_M V_{c,M} + 1/D_{p,mem} + 1/\theta_F V_{c,F}. \qquad (2)$$

$D_{p,CO}$ is the overall diffusing capacity, and decreases with increasing P_{O_2}, and $D_{p,mem}$ is the diffusing capacity of the membrane alone, both in ml/(min × mm Hg × kg of foetus). $V_{c,M}$ and $V_{c,F}$ are the volumes of blood in the placental capillaries of the mother and foetus respectively, in ml per kg of foetus. θ_M and θ_F are the diffusing capacities of a ml of maternal and foetal blood, respectively, in ml/(min × mm Hg × ml blood). They are obtained *in vitro* with a rapid-reaction apparatus and decrease in a predictable way with increasing P_{O_2} because of the competition between O_2 and CO for the available sites on the haemoglobin molecule. Breathing air, about 60 % of the total diffusion resistance is in the placental membrane and 40 % in the foetal and maternal capillary blood. These calculations were made assuming the maternal and foetal capillary blood volume, $V_{c,F}$ and $V_{c,M}$, were equal. Their value was calculated to be 4 ml.

Dr Delivoria-Papadopoulos and Dr Coburn have applied this method to the measurement of $D_{p,CO}$ in pregnant women, albeit in a more approximate form. Fig. 2 shows one mother's blood HbCO saturation as a function of time while she was rebreathing on the closed circuit. Blood carboxyhaemoglobin concentration rose during the first 120 minutes owing to the combined CO production and catabolism of the mother, foetus and placenta. Fifty ml of CO was added to the rebreathing circuit at the time indicated by the arrow. After 30 minutes had been allowed for this CO to distribute itself in the mother's circulation and equilibrate with maternal tissues, blood [HbCO] rose to about four times its previous level, that is to 5.9 % saturation, and then fell slowly over the next two hours. The step change in [HbCO] provided a measure of the effective dilution volume for CO in the mother as well as producing a P_{CO} gradient from mother to foetus. The rate of CO loss from the mother, which is considered the same as the rate of transfer to the foetus, equalled the total rate of CO production before the addition of CO, less the rate afterwards. That is 1.70 ml/h + 0.64 ml/h = 2.34 ml/h.

The mean placental capillary P_{CO} has to be calculated from the Haldane relationship (Equation 1) and the important additional datum required is $P_{O_2}/[HbO_2]$. No experimental measurements of arterial or uterine placental venous blood were obtained, although it can be safely assumed that the arterial value of the ratio was 100/97 %. $P_{O_2}/[HbO_2]$ can only vary by a factor of about 2, as seen from inspection of the O_2–Hb equilibrium curves,

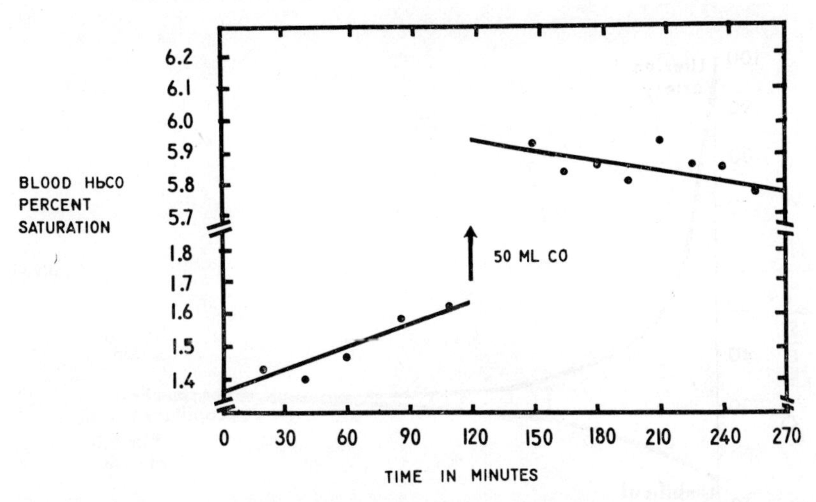

Fig. 2. A graph of blood HbCO saturation in a normal near-term pregnant woman rebreathing on a closed circuit with O_2 being added to maintain the P_{O_2} near 300 mm Hg and CO_2 being absorbed. At the arrow 50 ml of 100 % CO were added to the rebreathing circuit for the measurement of placental diffusing capacity.

and the mean value of the ratio probably lies close to its value at the outflow end of the capillary (Longo, Power & Forster, 1967). Thus we estimate average $P_{O_2}/[HbO_2]$ to equal 41/65 %. Mean [HbCO] during the period of observation was 5.85 %. Maternal P_{CO} is therefore $41/65 \times 5.85/210 = 0.018$ mm Hg.

Our knowledge of average foetal placental P_{CO} is, if anything, less than that of the maternal value. However, we would expect that foetal mean placental P_{CO} would be nearly equal to the maternal value at the start of the period of observation, except for a small P_{CO} gradient needed to transfer the normal CO production of the foetus into the maternal circulation. Foetal mean placental capillary P_{CO} should thus be only one quarter of the maternal value after the addition of the 50 ml of CO. If we simply neglect foetal P_{CO}, the resulting calculated value of $D_{p,CO}$ will be about 25 % lower than the true value. A minimal value of $D_{p,CO}$, $2.34/0.018 \times 60) = 2.2$ ml/(min × mm Hg), is obtained for the example in Fig. 2. The average value for four near-term pregnant women was 2.7 ml/(min × mm Hg). With an average foetal weight (assumed) of 3 kg, this gives a $D_{p,CO}$ of 0.9 ml/(min × mm Hg × kg of foetal weight) for comparison with the value of 0.54 ml/(min × mm Hg × kg of foetus) for sheep. Using Equation 2, the placental diffusing capacity for O_2, D_{p,O_2}, should be 0.68 ml/(min × mm Hg × kg of foetus) (Longo, Power & Forster, 1969); with a foetal O_2 consumption of 7.4 ml/(min × kg of foetus) the average P_{O_2} difference between maternal and foetal placental capillaries is 10.9 mm Hg.

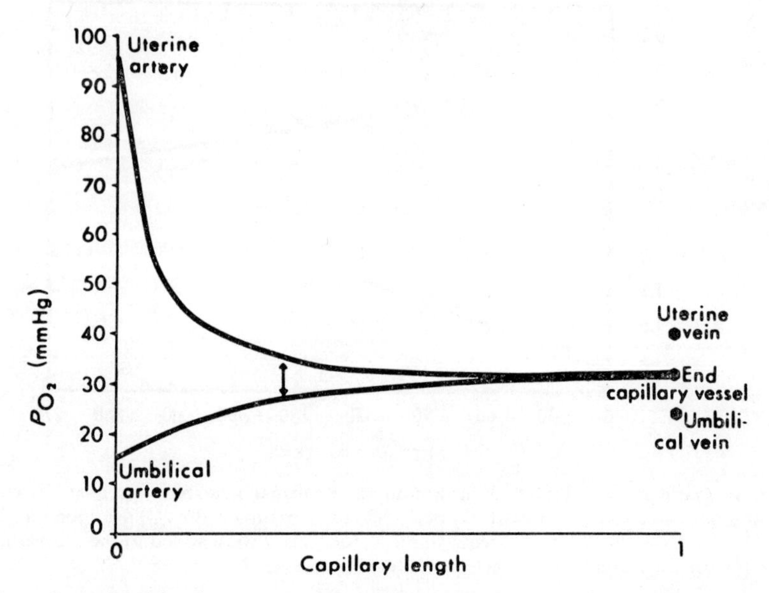

Fig. 3. A graph of the time course of P_{O_2} in the maternal and foetal placental vessels. The P_{O_2} and HbO_2 of uterine and umbilical arteries and veins represent values from experimental measurements. The actual curves and the end capillary values were calculated by a numerical integration of the diffusion equations assuming a value of 0.67 ml/(min × mm Hg × kg of foetus) for D_{p, O_2} derived from experimental measurement of $D_{p, CO}$. (Taken from Longo, 1972.)

There are clearly many assumptions in this method and its usefulness will depend on the as yet undetermined variations in $D_{p, CO}$ with physiological and pathological parameters of interest.

The level of maternal [HbCO] reached during the procedure was less than that seen in smoking mothers. However it would be better to alter the procedure and use smaller amounts of CO labelled with a stable isotope.

OXYGEN EXCHANGE

Oxygen movement across the placenta and the variation in maternal and foetal capillary P_{O_2}, assuming concurrent flow, are shown in Fig. 3 for the human. Assuming a D_{p, O_2} of 0.67 ml/(min · mm Hg · kg of foetus), which derives from measurements of $D_{p, CO}$ in the sheep and in the dog (Longo, Power & Forster, 1967), the P_{O_2} would be the same in the end capillary maternal and foetal placental capillary blood. The P_{O_2} difference between end capillary and uterine venous umbilical venous blood is assumed produced by shunts. In addition to the presence of these shunts and non-uniform maternal capillary blood flow/foetal capillary blood flow, the placenta itself consumes O_2, which further complicates the interpretation of P_{O_2} measurements in uterine and umbilical blood.

The time course has been computed by numerical integration of the diffusion equations and represents an application of a Bohr integration to the placenta (Metcalfe, Bartels & Moll, 1967.) If the experimentally determined uterine venous and umbilical venous P_{O_2} are considered to represent the true end capillary values and D_{p, O_2} calculated from them, its value is only 0.17 ml/(min × mm Hg × kg of foetus), one quarter of the value calculated from the $D_{p, CO}$.

The O_2 equilibrium curve for foetal blood is generally to the left (higher affinity) of that for maternal blood (Barcroft, 1947) and it appears to be a general belief that this helps placental O_2 exchange and is an advantage for the foetus. I should like to question this assumption.

The ultimate purpose of the placenta and circulatory systems of the mother and foetus is to provide O_2 to the foetal peripheral tissue at the required partial pressure. The critical P_{O_2} of mitochondria is less than 1 mm Hg; that is, their oxygen consumption falls if the P_{O_2} is lower than this, ATP/ADP decreases and anaerobic metabolism increases (Chance, 1964). Oxygen moves from the blood in the peripheral capillaries to the mitochondria only by diffusion, which requires a P_{O_2} gradient determined by the number of open capillaries, the oxygen consumption of the tissue and its diffusion characteristics for O_2. Thus the maintenance of the mean P_{O_2} of the foetal peripheral capillaries is the primary purpose of the maternal respiration and circulation, the placenta and the foetal circulation. The HbO_2 saturation or content in these capillaries is of relatively little importance, since the protein cannot leave the red cells, let alone the capillaries. We have no good method of estimating foetal peripheral capillary P_{O_2}. The most reasonable available index of it, or at least changes in it, is umbilical arterial P_{O_2}. I conclude that in spite of its disadvantages we should judge the efficacy of the placenta by the level of umbilical arterial P_{O_2}.

Recent work by Coburn & Mayers *et al.* (1971) and Whalen (1971) suggests that the P_{O_2} of tissue is normally lower than many have assumed; from 1 to 6 mm Hg. Therefore a decrease of only a few mm Hg in mean capillary P_{O_2}, which would produce an equivalent drop in P_{O_2} at the mitochondria if no regulatory adjustments in the microcirculation were made, could lead to tissue anoxia. A further remarkable experimental finding was that the tissue P_{O_2} did not rise as arterial P_{O_2} was increased from about 50 to over 600 mm Hg, implying that the body regulates peripheral capillary flow to keep tissue P_{O_2} from rising.

In Fig. 4 I have plotted the O_2–Hb equilibrium curves for maternal

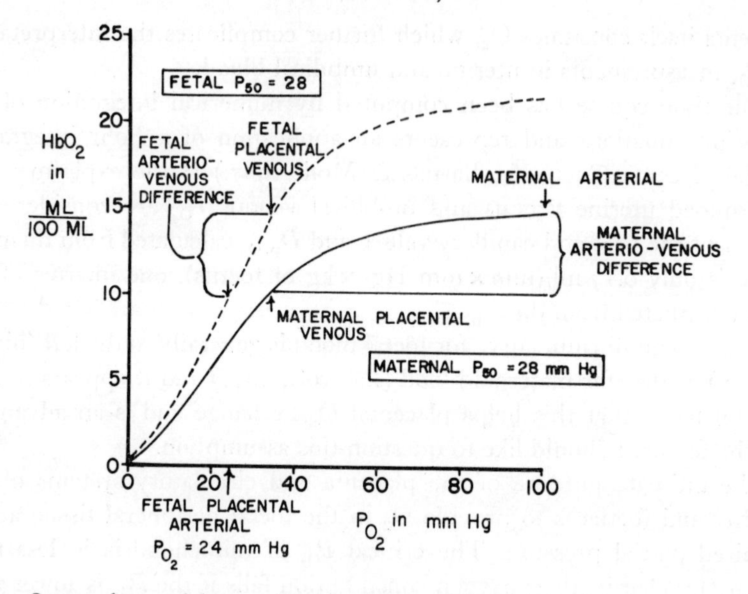

Fig. 4. Oxygen–haemoglobin equilibrium curves of foetal (----) and maternal (——) whole blood showing the effect of a position, or affinity, of the two curves on umbilical arterial P_{O_2}. Complete diffusion equilibrium for O_2 is assumed at the end of the placental capillaries, and shunts and non-uniform maternal capillary blood flow/foetal capillary blood flow are neglected. The maternal and foetal placental blood flows are considered equal. The arterial–venous O_2 content difference is assumed equal to 30% saturation. Oxygen capacity of foetal blood is assumed 22 ml/100 ml; that of maternal blood, 15 ml/100 ml. Maternal arterial P_{O_2} = 100 mm Hg. P_{50} of maternal O_2–Hb equilibrium curve is 28 mm Hg; foetal P_{50} also = 28 mm Hg.

blood with a capacity of 15 ml/100 ml and a P_{50} of 28 mm Hg and for foetal blood with a capacity of 22 ml/100 ml and the same P_{50}. For simplicity the foetal placental blood flow is assumed equal to maternal placental blood flow, so the arterio-venous O_2 content difference is the same on both sides of the placenta. The P_{O_2} of uterine placental venous blood is assumed equal to placental foetal venous blood. The foetal placental arterial blood P_{O_2} in this case is 24 mm Hg.

In Fig. 5 the conditions are assumed to be the same, but the P_{50} of foetal blood is chosen as 34 mm Hg; that is the foetal curve would be shifted to the *right* of the maternal curve if the Hb content of the two bloods were the same. The foetal placental arterial P_{O_2} is now 23 mm Hg, slightly, but not significantly lower than when the foetal curve had the same affinity as the maternal curve.

In the third example, Fig. 6, the P_{50} of the foetal blood is 12 mm Hg. This value was chosen because it is the lowest seen by Dr M. Delivoria-Papadopoulos in human newborns in respiratory distress. Far from helping O_2 transfer to the foetus, the foetal arterial P_{O_2} fell 10 mm Hg to

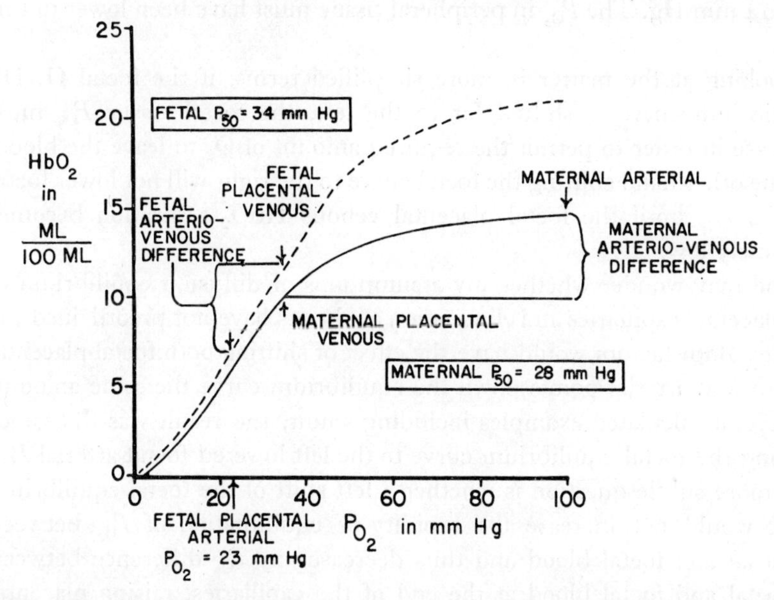

Fig. 5. A graph of maternal and foetal oxygen–haemoglobin equilibrium curves. Maternal P_{50} = 28 mm Hg. Foetal P_{50} = 34 mm Hg, a right shift. The other conditions and assumptions are the same as in Fig. 4.

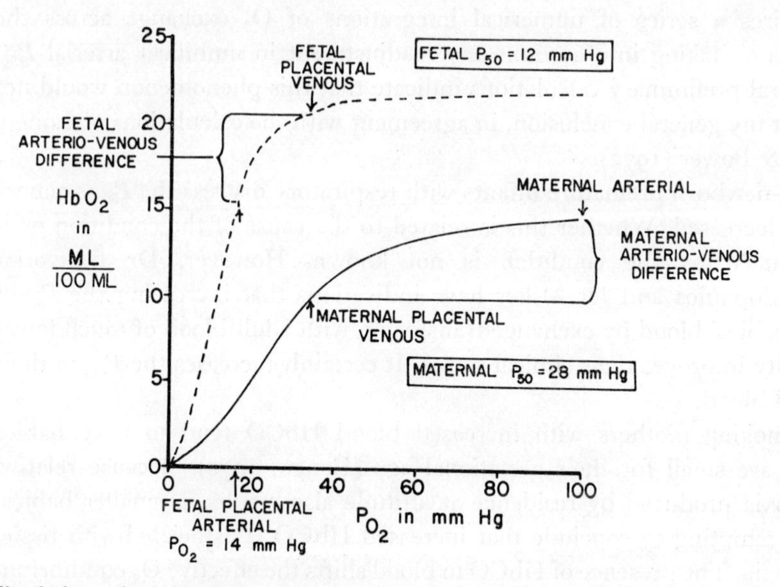

Fig. 6. A graph of maternal and foetal oxygen–haemoglobin equilibrium curves. Maternal P_{50} = 28 mm Hg. Foetal P_{50} = 12 mm Hg, a left shifted curve. The other conditions and assumptions are the same as in Fig. 4.

only 14 mm Hg. The P_{O_2} in peripheral tissue must have been lower in this case.

Looking at the matter in more simplified terms, if the foetal O_2–Hb equilibrium curve is shifted far to the left, the foetal tissue P_{O_2} must decrease in order to permit the required amount of O_2 to leave the blood. On the other hand shifting the foetal curve to the right will not lower foetal arterial P_{O_2} until the foetal placental venous HbO_2 saturation becomes extremely low.

One may wonder whether my assumptions of diffusion equilibrium in the placental capillaries and elimination of shunts have not preordained the results. Both factors would have the effect of shifting both foetal placental venous and arterial points down the equilibrium curve the same amount. In several calculated examples including shunt, the result was the same; shifting the foetal equilibrium curve to the left lowered foetal arterial P_{O_2}.

A more subtle question is whether a left shift of the foetal equilibrium curve would not increase the rapidity of equilibration of P_{O_2} between maternal and foetal blood and thus decrease any P_{O_2} difference between maternal and foetal blood at the end of the capillaries, raising placental venous P_{O_2} and, in the end, foetal umbilical arterial P_{O_2} as well. My assumption of complete diffusion equilibrium between the two capillary flows eliminates this possibility in the examples. To answer this question requires a series of numerical integrations of O_2 exchange across the placenta, taking into account any readjustment in umbilical arterial P_{O_2}. Several preliminary calculations indicate that this phenomenon would not affect my general conclusion, in agreement with the calculations of Longo, Hill & Power (1972).

In newborn premature infants with respiratory distress the P_{50} is generally decreased. Whether this is related to the cause of the condition or is produced by the condition is not known. However, Dr Delivoria-Papadopoulos and Dr Miller have indications that increasing the P_{50} of the babies' blood by exchange transfusion with adult blood of much lower affinity improves their clinical status. It certainly increases the P_{O_2} in their caval blood.

Smoking mothers with increased blood HbCO tend to have babies who are small for their gestational age (Longo, 1970). Because relative hypoxia produced by residence at altitude also results in smaller babies, it is tempting to conclude that increased HbCO is associated with tissue hypoxia. The presence of HbCO in blood shifts the effective O_2 equilibrium curve to the left; the O_2–Hb curve becomes more hyperbolic. The foetal HbCO concentration is generally greater than that in the maternal blood, exaggerating the effect of any exposure of the mother to CO.

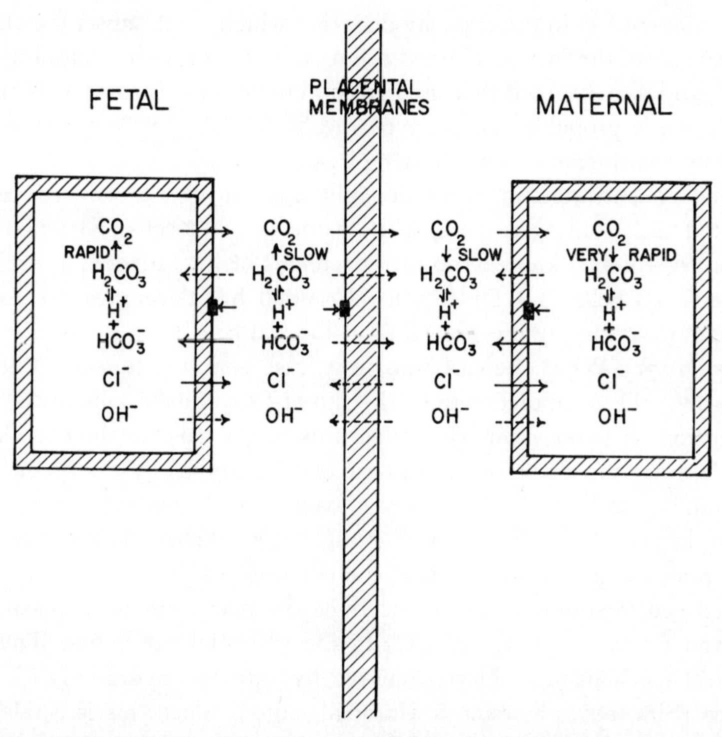

Fig. 7. A simplified diagram of the exchange of CO_2 between foetal and maternal blood in the placenta. The reactions and exchanges of O_2 are neglected, as is the formation of carbamino-haemoglobin. All membranes were considered impermeable to H^+. The placental membrane is considered relatively impermeable to HCO_3^-, but highly permeable to CO_2. The hydration–dehydration of CO_2 in foetal red cells is assumed to be much slower than in the maternal red cells, but still catalysed.

CARBON DIOXIDE EXCHANGE

I would like to conclude by making a brief mention of several aspects of the placental exchange of CO_2 which may be of practical importance. Fig. 7 is a diagram of the exchanges of CO_2 between foetal and maternal blood, neglecting other exchanges, in particular that of O_2. The major processes were outlined in a review by Roughton in 1935 and include the rapid catalysed hydration–dehydration of CO_2 inside the cells and the exchange of HCO_3^- for Cl^- across the cell wall. There are several new points of emphasis in the figure. While it is assumed that the membrane is relatively highly permeable to CO_2, it is also assumed that the membrane is relatively impermeable to HCO_3^-, at least for a net transfer of the anion (Meschia, Battaglia & Bruns, 1967; Longo, 1972). It is intuitively reasonable that the mechanism of CO_2 transfer be the diffusion of the gas itself, since the movement of significant amounts of HCO_3^- would require equal movements of Cl^- in the opposite direction which would upset the electrolyte balance of the foetus. The red cell membrane is relatively impermeable to H^+, and it is assumed that the placental membrane is as well. While the membrane is probably relatively permeable to OH^-, its concentration is very low and the net amount moving is small.

The concentration of carbonic anhydrase in the foetal red cells is reported to be only 5 to 10% of its value in adult cells (Kirschbaum & DeHaven, 1968). I know of no measurements of the rate of CO_2 exchange of the foetal cells, but Dr Edward Crandall has developed a computer program describing these processes in the red cell as have Hill, Power & Longo (1972). When the carbonic anhydrase concentration is reduced to 5 to 10%, the computed rate of the initial CO_2 uptake falls in the same proportion. It is not clear what effect this might have on the overall process, because the slower second phase of CO_2 uptake by the blood is not affected.

One important modification of Roughton's schema, which may be of great practical importance, is that the exchange of HCO_3^- for Cl^- across the red cell membrane will not produce the readjustment of plasma pH required by the exchange of CO_2 in the placental capillaries. The only practical mechanism is the uncatalysed hydration–dehydration of CO_2 in plasma (Sirs, 1970; Forster & Crandall, 1972). Since this is a relatively slow process with a half-time of the order of 5 seconds, it appears certain that at the end of the placental capillaries the pH of maternal and foetal plasma is not in chemical equilibrium with the interior of the respective red cells, nor are the plasma pH's in equilibrium with each other. After the blood leaves the placental capillaries, these readjustments will take place. Thus the chemical composition of placental blood samples which have had time to reach chemical equilibrium may be different from when they were in the capillaries *in vivo*. A delayed reaction of this type could produce experimentally measured differences in maternal and foetal placental P_{CO_2} which did not exist *in vivo*.

I have attempted a brief sketch of the general mechanism of gas exchange in the placenta and touched in more detail on three topics which may be of interest at this time to those working in the field of perinatal physiology. It appears possible to obtain an estimate of the diffusing capacity of the placenta in the pregnant human. I tried to show that a shift of the O_2–Hb equilibrium curve of the foetus to the left is not advantageous but disadvantageous. Finally, the chemical and exchange readjustments that must take place in the capillary blood after passing through the placenta, may require many seconds to achieve a new equilibrium and can give an erroneous impression of the state of the blood *in vivo*.

REFERENCES

ASSALI, N. S., DILTS, P. V. Jr, PLENTL, A. A., KIRSCHBAUM, T. H. & GROSS, S. J. (1968). In *Biology of Gestation. Vol. 1. The Maternal Organism.* Ed. N. S. Assali. New York and London: Academic Press. Pp. 186–289.

BARCROFT, J. (1947). *Researches on Pre-Natal Life.* Springfield, Ill.: Charles C. Thomas. Vol. 1.

BARRON, D. H. & ALEXANDER, G. (1952). *Yale J. Biol. Med.* **25**, 61–6.

BARTELS, H. (1967). In *Development of the Lung. CIBA Foundation Symposium.* Ed. A. V. S. de Reuck. Boston: Little, Brown and Co. Pp. 276–91.

BARTELS, H. (1970). *Prenatal Respiration.* Amsterdam: North-Holland Publishing Co.

CHANCE, B., SCHOENER, B. & SCHINDLER, F. (1964). In *Oxygen in the Animal Organism.* Ed. F. Dickens and E. Neil. Pergamon Press, London. Pp. 367–92.

COBURN, R. F. & MAYERS, L. B. (1971). *Am. J. Physiol.* **220**, 66–74.

DAWES, G. S. (1968). *Foetal and Neonatal Physiology.* Chicago: Year Book Medical Publishers.

FORSTER, R. E. & CRANDALL, E. D. (1972). *Fed. Proc.* **31**, 347 abs.

HILL, E. P., POWER, G. G. & LONGO, L. D. (1972). *Fed. Proc.* **31**, 237 abs.

KIRSCHBAUM, T. H. & DEHAVEN, J. C. (1968). *Biology of Gestation. Vol. II. The Fetus and Neonate.* Ed. N. S. Assali. London: Academic Press. Pp. 143–87.

LONGO, L. D. (1970). *Ann. N.Y. Acad. Sci.* **174** (article 1), 1–430.

LONGO, L. D. (1972). In *Pathophysiology of Gestation. Vol. II. Fetal–Placental Disorders.* Ed. N. S. Assali. New York and London: Academic Press. Pp. 1–65.

LONGO, L. D., DELIVORIA-PAPADOPOULOS, M., POWER, G. G., HILL, E. P. & FORSTER, R. E. II. (1970). *Am. J. Physiol.* **219**, 561–9.

LONGO, L. D., HILL, E. P. & POWER, G. G. (1972). *Am. J. Physiol.* **222**, 730–9.

LONGO, L. D., POWER, G. G. & FORSTER, R. E. II. (1967). *J. Clin. Invest.* **46**, 812–28.

LONGO, L. D., POWER, G. G. & FORSTER, R. E. II. (1969). *J. Appl. Physiol.* **26**, 360–70.

MESCHIA, G., BATTAGLIA, F. G. & BRUNS, P. D. (1967). *J. Appl. Physiol.* **22**, 1171–8.

METCALFE, J. (1967). In *Development of the Lung. CIBA Foundation Symposium.* Ed. A. V. S. de Reuck. Little, Brown and Co., Boston, pp. 271–5.

METCALFE, J., BARTELS, H. & MOLL, W. (1967). *Physiol. Rev.* **47**, 782–838.

METCALFE, J., MOLL, W., BARTELS, H., HILPERT, P. & PARER, J. T. (1965). *Circulation Res.* **16**, 95–101.

ROUGHTON, F. J. W. (1935). *Physiol. Rev.* **15**, 241–96.

SIRS, J. A. (1970). In *Blood Oxygenation.* Ed. D. Hershey. New York: Plenum Press. Pp. 116–36.

WHALEN, W. J. (1971). *Physiologist* **14**, 69–82.

THE EFFECT OF PERMEABILITY ON PLACENTAL OXYGEN TRANSFER

By JOHN H. G. RANKIN
University of Wisconsin Medical School,
Madison, Wisconsin, U.S.A.

INTRODUCTION

The relationships between the structure and the function of heat exchangers have often been usefully applied to the study of placental exchange. The application requires that the variables relating to heat transfer be changed into equivalent variables relating to mass transfer. The relationships between heat and mass transfer are well defined. When the appropriate substitutions are made, it can be shown that the effectiveness of mass transfer (E) is a function of the flows, the architecture of the exchanging areas and the ability of the substance in question to move from one side of the exchanger to another, i.e.

$$E = f(\text{flows, architecture, permeability}).$$

The functional relationships between effectiveness and flows and architecture in the sheep placenta have been discussed elsewhere (Rankin, 1972).

There is considerable doubt in the literature as to whether placental oxygen transfer is limited by diffusion or not limited by diffusion. If the effectiveness of placental oxygen transfer is equal to that of very diffusible substances, then oxygen transfer is not diffusion-limited, but if the effectiveness of placental oxygen transfer is less than that of very diffusible substances then placental oxygen transfer will, in part, be diffusion-limited. The relationships between placental permeability and the effectiveness of placental oxygen transfer in the sheep are described in this publication.

METHODS

The exchanging characteristics of a placenta can be described with three dimensionless variables which are as follows (Bartels & Moll, 1964):

$$E \text{ (effectiveness)} = (\text{MA} - \text{MV})/(\text{MA} - \text{FA}) \text{ or } (\text{FV} - \text{MA})/(\text{MA} - \text{FA})$$

where MA and MV are the maternal arterial and venous analogues of temperature and FA and FV are the respective analogues of temperature in the blood in the umbilical artery and the umbilical vein.

R (capacity rate ratio) = (Foetal fluid capacity rate)/(Maternal fluid capacity rate), where the fluid capacity rates are the mass flow rates if the exchanging substance is only in physical solution. In this variable, the words maternal and foetal refer to the uterine and umbilical circulations respectively.

d (the number of transfer units) = permeability/(one of the fluid capacity rates).

The question of oxygen is of particular interest because the partial pressure of oxygen in blood is not linearly related to the oxygen content of blood. In the case of oxygen, *E* can be defined in terms of partial pressure. It would be quite acceptable to use concentration in physical solution as the analogue of temperature but the variables become somewhat cumbersome. The variable *d* is defined as the diffusing capacity for oxygen (D_{PO_2}) divided by $Q_m S_m$ where Q_m is the uterine placental blood flow and S_m the effective solubility of oxygen in maternal blood, *R* is defined as $Q_F S_F / Q_m S_m$, where *F* refers to foetal blood. The effective solubility is the change in content that accompanies a unit change in partial pressure and is expressed as ml/(ml × mm Hg), i.e. $10^{-2} \times V\%$/mm Hg. With experimental values for these variables, it is possible to determine the effectiveness of placental oxygen transfer.

The effective solubility of oxygen in blood with a given haemoglobin concentration and type and at a predetermined saturation was obtained by calculating the change in the oxygen content of the blood that would occur if the partial pressure of oxygen was increased by 1 mm Hg. Consideration was given to both bound and physically dissolved oxygen. Calculations were performed at several values of pH using the mathematical description of the appropriate oxygen dissociation curves in the form of modified Hill equations provided by Meschia *et al.* (1965). The oxygen-carrying capacity was set at 13 $V\%$ and the solubility of oxygen in sheep blood was taken to be 0.023 ml/(ml × atmosphere).

The value of *d* corresponding to each value of effective solubility was calculated using a D_{PO_2} of 2.1 ml/(min × mm Hg) and a Q_m of 800 ml/min.

The relationship between *d* and the effectiveness of exchange was determined by calculating the percentage decrease in effectiveness that would be observed in a sheep placenta with concurrent exchange and a capacity rate ratio of 1 if *d* were changed from infinity to the calculated value. The quality of the conclusions is not dependent on the type of exchanger. For this reason, a concurrent type was selected as the simplest example. All calculations were performed on a UNIVAC 1108 computer via a remote teletype terminal in our laboratories.

RESULTS

The capacity rate ratio (*R*)

The uterine blood flow in the near-term sheep has been reported to be about 1000 ml/min and the umbilical blood flow is approximately 500 ml/min (Metcalfe *et al.* 1967). Approximately 20% of each flow does not participate in exchange (Rankin, 1972). The ratio of the umbilical to uterine blood flow is, therefore, 0.5. The capacity rate ratio is not 0.5 because the effective solubility of oxygen is very different in the two bloods. The effective solubility of oxygen in foetal blood or maternal blood is not constant. It will change in a manner that depends on the position on the oxygen dissociation curve. The effective solubility of oxygen in sheep blood containing haemoglobin type A, type B and foetal haemoglobin (S_A, S_B and S_F) was calculated at various percent saturations. These results are shown in Table 1.

Table 1. *Effective solubility* ($10^{-2} \times V\%/mm\ Hg$) *of oxygen in sheep blood containing haemoglobin F, A and B* (S_F, S_A, S_B) *and the ratio of* S_F *to* S_A *and* S_B *at various levels of oxygen saturation at a pH of 7.4 and an oxygen-carrying capacity of 13 V%*

% Sat.	S_A	S_B	S_F	S_F/S_A	S_F/S_B
20	0.0035	0.0026	0.0060	1.7	2.3
30	0.0037	0.0027	0.0062	1.7	2.3
40	0.0036	0.0026	0.0060	1.7	2.3
50	0.0032	0.0023	0.0053	1.7	2.3
60	0.0026	0.0019	0.0043	1.7	2.3
70	0.0019	0.0014	0.0032	1.7	2.3
80	0.0012	0.0009	0.0020	1.7	2.2

It can be seen that S_F is approximately twice S_A and S_B. For this reason, the capacity rate ratio of the sheep placenta approximates a value of 1. All further calculations are performed with a capacity rate ratio of 1.

The number of transfer units (*d*) *for oxygen*

The number of transfer units for oxygen exchange is equal to the placental diffusing capacity for oxygen divided by a fluid capacity rate. As the fluid capacity rate of the uterine blood flow approximates that of the umbilical blood flow, then either of these two parameters can be used. In the following calculations, the capacity rate for the uterine flow has been used. The diffusing capacity for oxygen of the sheep placenta is approximately 2.1 ml/(min × mm Hg) (Longo *et al.* 1967). The placental

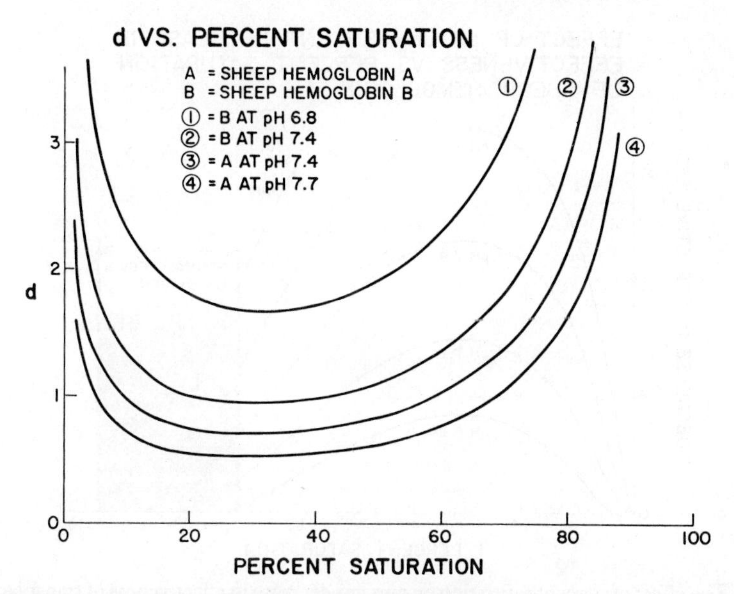

Fig. 1. The effect of haemoglobin type and pH on the relationship between d and percent saturation of sheep blood. Constants used in the calculation are, uterine placental blood flow = 800 ml/min, D_{PO_2} = 2.1 ml/(min × mm Hg), oxygen capacity = 13 V%, solubility of oxygen in sheep blood = 0.023 ml/(ml × atmosphere).

diffusing capacity is constant but the fluid capacity rate will change with the effective solubility. For this reason, d for oxygen will change and will be dependent upon the saturation of the blood. The value of d for oxygen has been calculated at various percent saturations and pH's of maternal blood containing haemoglobin types A and B. These results are shown in Fig. 1. It can be seen that at low and high oxygen saturations, the value of d is relatively large but in the intermediate range of oxygen saturations, relatively low values of d are obtained. At the same pH, type A haemoglobin produces lower values of d than type B haemoglobin. Alkalosis lowers the value of d and acidosis results in high values of d.

The percent decrease in effectiveness (Dec)

The relationship between E and d for concurrent exchange when R is unity is:

$$E = 0.5(1 - e^{-2d})$$

E can be seen to have a maximum value of 0.5 when d is very large, and a minimum of zero when d tends to zero. Dec has a value of 1% when d is equal to 2.30 and a value of 10% when d has a value of 1.14. The effect of percent saturation of sheep haemoglobin A on Dec is illustrated in Fig. 2. The normal region of oxygen transfer is shown and the effect of pH

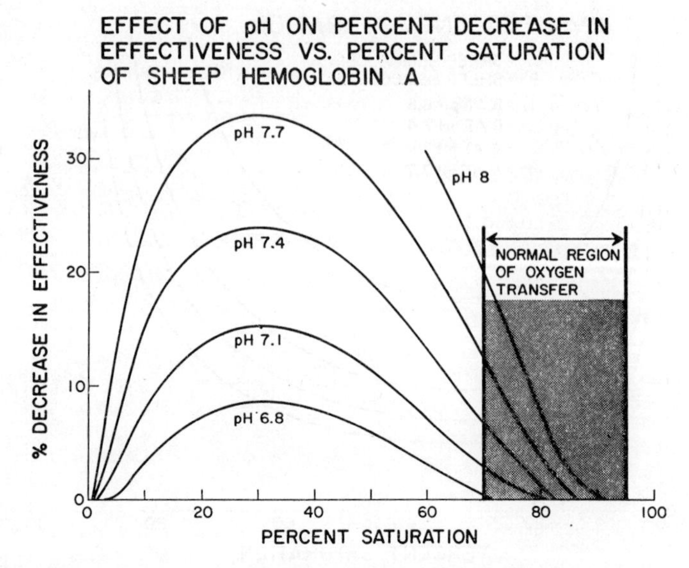

Fig. 2. The effect of percent saturation on percent decrease in effectiveness of transplacental oxygen exchange (Dec) in sheep carrying haemoglobin A at several values of pH. The shaded area shows the normal region of oxygen transfer.

changes on the relationship is also apparent. It can be seen that the normal placental exchange of oxygen is on the borderline between significant diffusional limitation and no diffusional limitation. Sheep with haemoglobin B would be less likely to suffer a diffusional limitation to oxygen transfer than sheep carrying haemoglobin A, but both types could suffer a diffusional limitation to placental oxygen transfer if subjected to left shifts of the oxygen dissociation curve or abnormal unsaturations during the placental transit.

With maternal hypoxia, low oxygen saturations are obtained in the uterine blood during the placental transit. It can be seen from Fig. 2 that with both haemoglobin types, the exchange at low saturation is much less effective than exchange during normoxia. Maternal hypoxia would therefore tend to decrease the effectiveness of placental transfer and increase the probability of there being a diffusional limitation to the transfer of oxygen.

A most interesting situation is shown in Figs. 1 and 2 with respect to the changes in pH. It can be seen that maternal acidosis decreases the slope of the dissociation curve (Bohr effect), decreases the effective solubility of oxygen in blood and increases the effectiveness of oxygen transfer under all conditions. On the other hand, maternal alkalosis changes the shape of the oxygen dissociation such that the effective solubility of oxygen is increased. The effectiveness of transfer during maternal alkalosis is therefore abnormally low. A diffusional limitation to oxygen transfer could, therefore, be expected during maternal hypoxia and alkalosis.

DISCUSSION

The results presented in this paper correlate very well with the existing experimental literature. Rankin *et al.* (1971) examined the relationship between the uterine and umbilical venous P_{O_2} in sheep and concluded that the exchange of oxygen during maternal hypoxia appears to be limited, to some extent, by the placental diffusing capacity for oxygen. This result is predicted from the theoretical study presented here. From Figs. 1 and 2, it can be seen that the effective solubility of oxygen in maternal blood is increased at low saturation. The value of d for placental oxygen exchange during maternal hypoxia may be less than that required for flow-limited oxygen exchange. It can, therefore, be concluded that the component of placental oxygen transfer observed by Rankin *et al.* during maternal hypoxia that appears to be limited by diffusion, is limited by diffusion because of the increased effective solubility of oxygen in maternal blood during hypoxia.

There is a great deal of literature pertaining to the effect of maternal alkalosis on placental oxygen exchange. This is summarized by Dawes (1968). In this review, Dawes points out that several investigators have found that maternal hypocapnia is frequently associated with foetal hypoxia and acidosis. These results are clearly supported by the predictions in this paper. It can be seen from the data displayed in Fig. 2 that maternal alkalosis increases the effective solubility of oxygen in the maternal blood, and decreases d below the level necessary for flow-limited oxygen transfer. During maternal alkalosis, one would therefore predict that the effectiveness of placental oxygen transfer could drop to levels such that the foetus was unable to extract sufficient oxygen for its needs. In this situation, foetal hypoxia and secondary acidosis would be observed. In a more recent publication, (Bailie *et al.* 1971), Dawes' group re-examined the problem of maternal hyperventilation in the pregnant sheep and concluded that, although maternal hyperventilation caused a small fall in mean foetal carotid P_{O_2}, this was readily reversible and there was no evidence of progressive acidaemia. Dawes' group concludes that the effects of maternal alkalosis have been exaggerated in the past. It should be noted that the results from Dawes' laboratory and those of previous workers are not necessarily conflicting. The results presented here predict that sheep with haemoglobin type A will show a greater decrease in the effectiveness of placental oxygen transfer during maternal hypocapnia than sheep with haemoglobin type B. The small discrepancy in results between Dawes' group and earlier workers could therefore be explained on the grounds that the two groups were working with sheep of different haemoglobin types. It is of some interest to note at this point that data pertaining to oxygen transfer in the sheep should specify the maternal haemoglobin type before comparisons are made with the results of other laboratories.

The effect of maternal alkalosis on placental oxygen transfer is of particular importance during delivery because a hypocapnic alkalosis may be induced during anaesthesia for Caesarean section. Furthermore, a popular method of natural childbirth (LaMaze) employs panting during labour. It appears that many patients, in the emotional stress of the moment, hyperventilate to some extent during LaMaze deliveries. This would induce a maternal alkalosis and would be expected to decrease the effectiveness of placental oxygen transfer by increasing the effective solubility of oxygen in the maternal blood. This being the case, it would certainly seem to be in order to advise such patients not to hyperventilate. Maternal acidosis, on the other hand, should cause an increase in the effectiveness of placental oxygen transfer.

CONCLUSIONS

The overall effectiveness of placental oxygen transfer is not constant and can be decreased during maternal hypoxia and maternal alkalosis. These situations should, therefore, be avoided during near-term pregnancies. These conclusions were obtained from theoretical considerations and are supported by the experimental literature. The prediction can be made that maternal acidosis may, in some circumstances, increase the effectiveness of placental oxygen transfer.

REFERENCES

BAILIE, P., DAWES, G. S., MERLET, C. L. & RICHARDS, R. (1971). *J. Physiol.* **218**, 635–50.

BARTELS, H. & MOLL, W. (1964). *Pflugers Archiv.* **280**, 165–77.

DAWES, G. S. (1968). *Foetal and Neonatal Physiology*. Chicago: Year Book Medical Publishers.

LONGO, L. D., POWER, G. G. & FORSTER, R. E., II (1967). *J. Clin. Invest.* **46**, 812–28.

MESCHIA, G., COTTER, J. R., BREATHNACH, C. S. & BARRON, D. H. (1965). *Quart. J. Exp. Physiol.* **50**, 466–80.

METCALFE, J., BARTELS, H. & MOLL, W. (1967). *Physiol. Rev.* **47**, 782–838.

RANKIN, J. H. G., MESCHIA, G., MAKOWSKI, E. L. & BATTAGLIA, F. C. (1971). *Am. J. Physiol.* **220**, 1688–92.

RANKIN, J. H. G. (1972). In *Respiratory Gas Exchange in the Placenta*. Ed. L. D. Longo & H. Bartels. (In press.)

PLACENTAL EXCHANGE AND MORPHOLOGY IN RUMINANTS AND THE MARE

By MARIAN SILVER, D. H. STEVEN AND R. S. COMLINE

Physiological Laboratory and Department of Anatomy, University of Cambridge, England

INTRODUCTION

In 1674 the young Oxford physician John Mayow drew attention to the respiratory function of the placenta in the following words: '. . . the blood of the embryo, conveyed by the umbilical arteries to the placenta . . ., brings not only nutritious juices, but along with this a portion of nitro-aerial particles [oxygen] to the foetus for its support . . . And therefore I think that the placenta should no longer be called a uterine liver but rather a uterine lung' (Mayow, 1674).

In 1876 the situation was much as Mayow had left it two hundred years before. Turner (1876) then remarked, 'Undoubtedly there are many facts on record which show that the foetus *in utero* needs to respire, and that the placenta is the organ in which respiration is conducted. But there is no evidence that the respiratory changes during intra-uterine life are actively carried on.' It was not until 1927, when Huggett made the first comprehensive study of blood gases on both foetal and maternal sides of the placenta, that the hypothesis expressed by Mayow received direct experimental support.

Huggett's experimental technique involved the immersion of anaesthetised goats in a bath of warm saline. The foetus was delivered by Caesarian section into the bath, where its head was kept constantly immersed. Blood samples were taken when required from umbilical artery and vein, uterine vein, and foetal and maternal carotid arteries (Huggett, 1927). This technique was adopted and refined by Barcroft and his colleagues (Barcroft, 1946), whose many contributions to our knowledge of the physiology of the foetus formed the starting point of much of the subsequent research in this field.

Perhaps the most outstanding contribution to foetal physiology since this pioneer work has been that of sampling the conscious foetus *in utero* by means of implanted catheters (Meschia, Cotter, Breathnach & Barron, 1965). This technique enabled animals to be followed sequentially under conditions which most nearly approached normality. Barcroft himself was acutely aware of the problems associated with anaesthesia and exteriorisation of the foetus and would probably have been the first to adopt this method. There are, of course, still many aspects of foetal physiology which cannot be studied in this way and for which some form of anaesthesia, surgery and partial exteriorisation are necessary. However, problems of placental exchange are ideally suited to the indwelling catheter technique and much new information has been gained in recent years by this means.

Since the subject of placental gas transfer and particularly the factors which affect the passage of blood gases across the placenta have been extensively reviewed in the last few years (Metcalfe, Bartels & Moll, 1967; Dawes, 1968; Bartels, 1970), it seemed more useful to limit the present review to those species which have been studied chronically *in utero*, to look at placental exchange in a wider sense, and where possible to compare these findings with the earlier work in which the more traditional experimental methods were used. This will necessarily limit the number of species to be considered in any detail to the various ruminants and the mare. However, before examining any findings on blood gas or metabolite exchange in these species it is logical to look first at the general morphology of the ruminant and equine placentae, and in particular at the various vascular arrangements which may be found in both maternal and foetal tissue in these species.

PLACENTAL MORPHOLOGY

The placenta of the mare is usually classified as diffuse, while the placentae of the cow and sheep are both cotyledonary in arrangement (Fig. 1). In the mare, however, the placenta is considerably more specialised than is at first apparent, for small tufts of chorionic villi project into corresponding invaginations of the endometrium to form small globular structures known as microcotyledons. The definitive placenta consists of many thousands of such microcotyledons which are closely packed together in the superficial endometrial layers (Fig. 1).

Chorionic villi in the sheep and cow are developed under normal circumstances only in those areas of the chorion which overlie the maternal caruncles; the placentomes in both species are formed by growth and interdigitation of foetal villi and caruncular crypts (Fig. 1). The placentomes of the cow are convex and pedunculated: in the sheep they are usually concave and sessile, though the degree of concavity may vary from breed to breed and from place to place in the same placenta (Fig. 1).

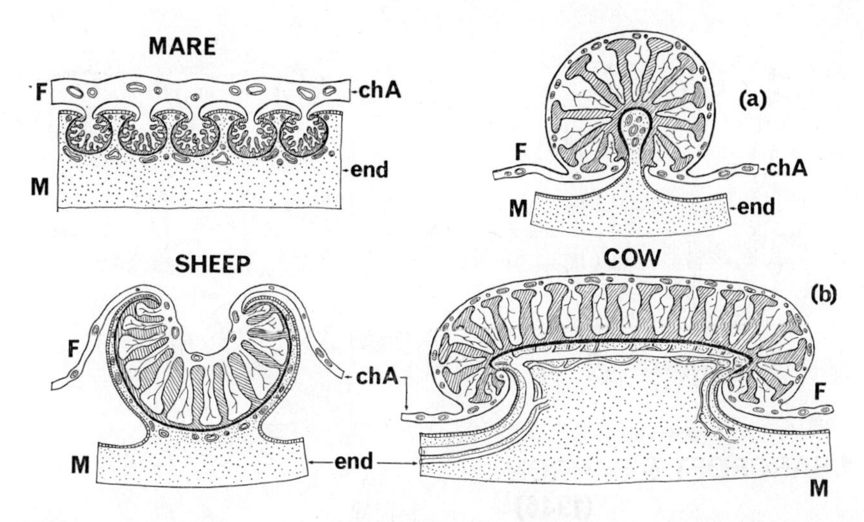

Fig. 1. Macroscopic structure of the definitive placenta of the mare, the Welsh Mountain sheep and the Jersey cow, based on formalin fixed material. The placenta of the mare is diffuse and microcotyledonary; those of the sheep and cow are cotyledonary in structure. chA, chorio-allantois; end, endometrium; F, foetal; M, maternal. (*a*) transverse section; (*b*) longitudinal section.

Vascular arrangement

The placentomes of the cow and sheep, like the microcotyledons of the mare, are very highly vascularised on both foetal and maternal sides. The arrangement of the vessels within the ovine placentome was first described by Barcroft & Barron (1942, 1946) and from it they derived their theory of countercurrent flow in maternal and foetal capillary nets (Fig. 2).

It can be seen from the diagram that the disposition of the veins, both foetal and maternal, is important in the assessment of directions of flow. While Barcroft & Barron were aware that veins appear within the villus as pregnancy proceeds (Barcroft, 1946, p. 7) they were more concerned with the reduction of the pressure head required to perfuse the expanding capillary net than with the effect on the direction of capillary flow. A few years later Wimsatt (1950) demonstrated the presence of veins within the foetal villus of the sheep placenta at term, but made no comment on their possible functional significance. When Metcalfe, Moll, Bartels, Hilpert & Parer (1965) showed that in the last third of pregnancy it was possible to reverse the direction of flow in the umbilical circulation without greatly altering the efficiency of transport of certain gases across the placental membrane, they deduced from their experiments that the system of capillary flow could not be countercurrent, for reversal of umbilical flow would convert it into a concurrent system of lower efficiency. They there-

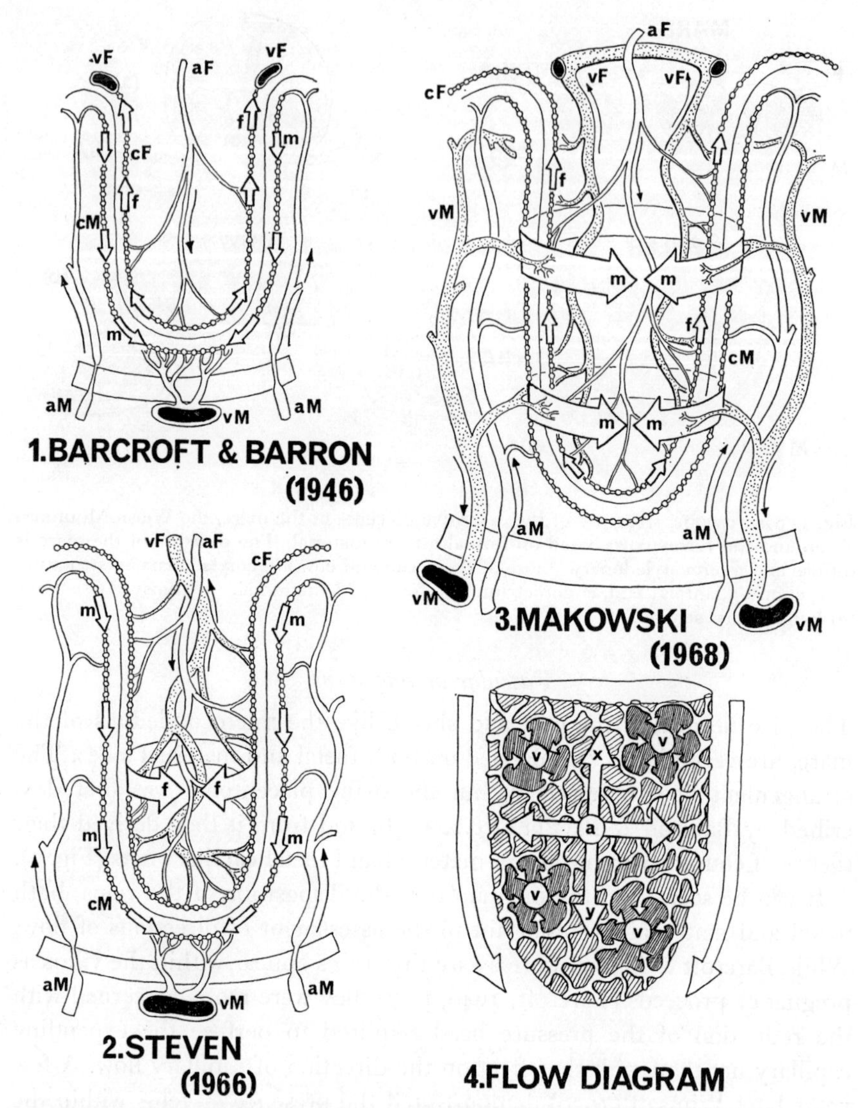

Fig. 2. Diagram to show possible directions of capillary flow in the placenta of the sheep. aM, maternal artery; vM, maternal vein; aF, foetal artery; vF, foetal vein; cF, foetal capillaries; cM, maternal capillaries; m, maternal capillary flow; f, foetal capillary flow. (1) Countercurrent flow in maternal and foetal capillary nets, after Barcroft & Barron (1946). (2) Crosscurrent flow, after Steven (1966). (3) Crosscurrent flow, after Makowski (1968). (4) Flow diagram based on (2): for explanation see p. 249. a, foetal arterial input; v, foetal venous outflow; x, y, flows towards the base and the tip of the villus respectively. Small open arrows represent flow in the foetal capillary net. Large curved arrows represent maternal capillary flow.

fore concluded that the placenta probably consisted of equal numbers of concurrent and countercurrent flow units.

The presence of veins within the foetal villus in the last $\frac{1}{3}$ of pregnancy was confirmed by Steven (1966), who suggested that while foetal capillary blood might flow locally from artery to vein in any direction (small open arrows, Fig. 2.4) the mean direction of flow would nevertheless be at right angles to the long axis of the villus, and therefore crosscurrent with respect to maternal blood flow (Fig. 2.2). The argument, which was not developed in full, depends on unidirectional maternal flow. If the flows on the foetal side are distributed equally towards the tip and the base of the villus the system will effectively be crosscurrent with respect to maternal blood (x, y, Fig. 2.4). If on the other hand these flows are unequal in distribution, there will be a bias either towards concurrent or countercurrent flow. If such a bias is present, the complexities of the vascular arrangements preclude its prediction on purely morphological grounds.

The findings of Makowski (1968) differed from those of previous authors in that he described veins as well as arteries extending far into the maternal crypts (Fig. 2.3). Makowski also suggested a system of crosscurrent flow, but with foetal blood flowing parallel to the long axis of the villus and maternal blood at right angles to it. He drew attention to localised constrictions of the maternal arteries at the periphery of the placentome, and suggested that these structures might be under hormonal control.

There are several possible explanations for these divergencies of view. Barcroft & Barron's countercurrent flow diagram was based on vascular arrangements of the placenta at 45 days of gestation. By the 60th day, however, veins have appeared in the stem of the villus and are no longer restricted entirely to its base (Steven, unpublished observations). It is quite possible therefore that the pattern of flow on the foetal side may change with the changing pattern of axial vessels. It is, however, more difficult to account for the conflict of evidence on the arrangement of the maternal vessels within the placentome: the explanation may well lie in morphological variations between different breeds of sheep.

In the cow the vascular arrangements of the placentome have been studied by Tsutsumi (1962) and Tsutsumi & Hafez (1964). Both papers describe arteries and veins within the walls of the maternal crypts and suggest that foetal and maternal capillary blood may flow in opposite directions. These observations are summarised in Fig. 3. It should perhaps be pointed out that the findings of Tsutsumi & Hafez (1964) for the cow bear some resemblance to those of Makowski (1968) for the sheep, yet the interpretations of the directions of capillary flow are quite different

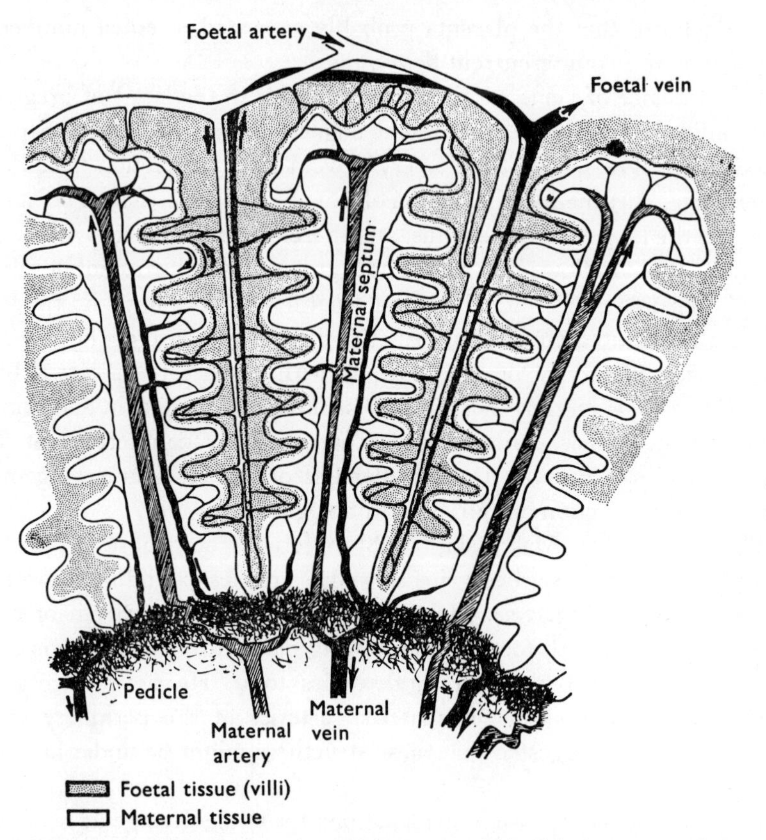

Fig. 3. Diagram to show the vascular arrangements within the placentome of the cow, from Tsutsumi & Hafez (1964). Arrows show the postulated direction of flow in foetal and maternal vessels.

in each case. This effectively demonstrates the difficulty of deducing haemodynamic properties from the examination of structure alone, especially when dealing with problems of microcirculation.

The vascular supply of the microcotyledons of the mare has been described in detail by Tsutsumi (1962) and more briefly by Steven (1968). On the maternal side long straight arteries from the sub-endometrial vascular plexus pass between the uterine glands to the deep surface of the endometrial epithelium, where they divide into a number of branches. Each branch passes over the rim of the nearest microcotyledon where it gives rise to a dense capillary network in the walls of the maternal crypts. This network is drained by branches of a single vein, which courses directly from the base of the microcotyledon to the veins of the sub-endometrial plexus (Fig. 4).

On the foetal side the chorionic villi are supplied by branches of the

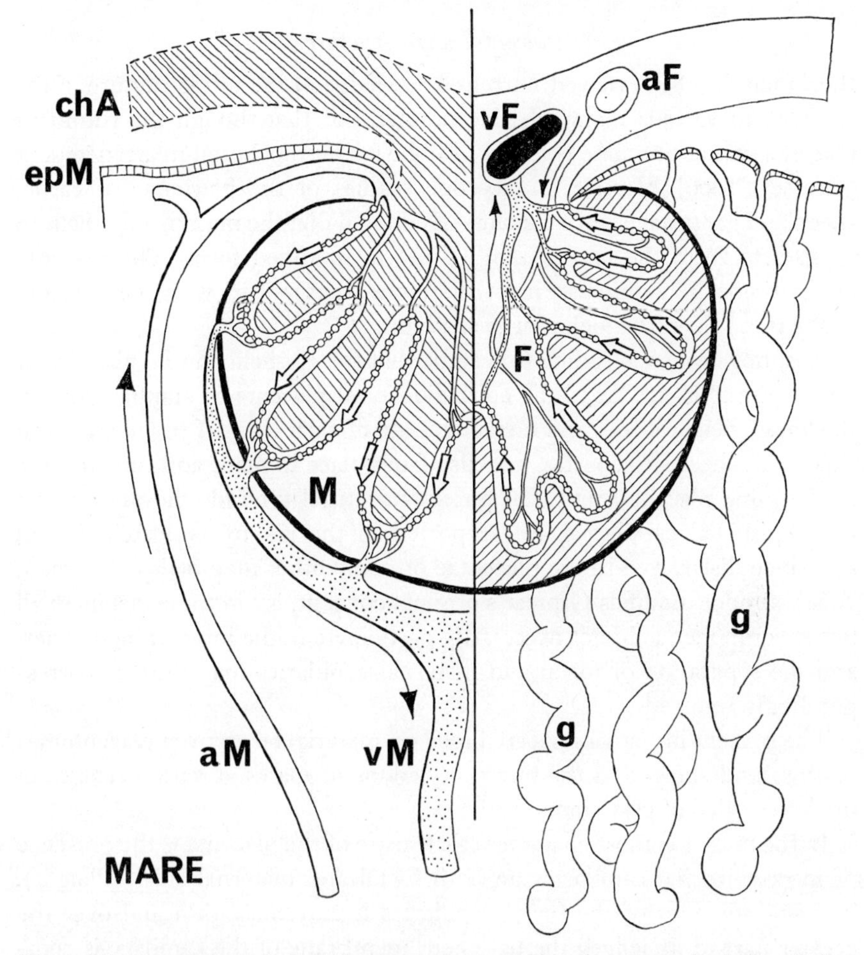

Fig. 4. Diagram to show the vascular arrangements within the placental microcotyledons of the mare, after Tsutsumi (1962). Small open arrows show the postulated directions of flow in foetal and maternal capillaries. aM, maternal artery; vM, maternal vein; chA, chorio-allantois; epM, uterine epithelium; g, g, endometrial glands; aF, foetal artery; vF, foetal vein; F, foetal; M, maternal.

umbilical arteries and veins. If, as Tsutsumi (1962) depicts, the veins do not extend as far as the arteries into the foetal villi, then the direction of capillary flow will be from the tip to the base of the villus and thus in the opposite direction to that of maternal capillary blood (Fig. 4). This interpretation is supported by evidence obtained from electron micrographs of the terminal branches of foetal villi which show a single axial vessel surrounded by surface capillaries (Steven, unpublished observations).

Placental ultrastructure

Björkman (1968) reviewed current knowledge of the ultrastructure of the placenta in various mammals. He pointed out that though the ruminant placenta was at one time considered to be syndesmochorial in arrangement (Grosser, 1909), electron microscope studies of the placentae of cattle, sheep and goats have since conclusively shown that the maternal epithelium persists to form the crypt lining, albeit in modified form. The placenta of the domestic ruminants must therefore be classified with the placenta of the mare as epitheliochorial in structure.

It is, however, misleading to assume that the epitheliochorial placenta of one mammal is similar to the epitheliochorial placenta of another, for the thickness, permeability and metabolic rate of each layer of tissue may well vary from species to species, and also from place to place and time to time in the same placental unit. On purely structural grounds the presence of intra-epithelial capillaries, first reported in the pig by Goldstein (1926) and since discovered in the placentae of many other mammals (Björkman, 1968), at once modifies Grosser's original concept, for in these instances all tissue components do not necessarily participate in the intervening barrier, and the separation of foetal and maternal capillaries may thus be correspondingly reduced.

The present investigations are based on material taken from placentomes of cows and sheep and the microcotyledons of mares at various stages in the latter third of gestation.

In the mare the most characteristic feature of the placenta is the presence of intra-epithelial capillaries on both foetal and maternal sides (Plate 1). The endothelium of the foetal capillaries is extremely thin and over the greater part of its length the basement membrane of the capillary is combined with the basement membrane of the chorionic epithelium to form a single lamina (Plate 2). The maternal capillary endothelium is thicker than that of the foetal side, and collagen fibres separate the basement membrane of the maternal epithelium from that of the underlying capillary.

In the cow the foetal capillaries also indent the chorionic epithelium, but the majority do so to a lesser extent than is usual in the mare (Plate 3). On the maternal side the capillaries are separated from the microvillous junctional zone not only by the full thickness of the uterine epithelium, but also by fibrocytic processes and varying amounts of collagen. In most places the maternal epithelium is thick and cellular, but tenuous areas of incomplete syncytium are occasionally found (Plate 4).

In the sheep the foetal capillaries lie close to the microvillous junctional zone: they indent the chorionic epithelium in a similar manner to the foetal capillaries of the mare. On the maternal side, however, the capillaries lie at some distance from the incomplete syncytium which forms the maternal epithelial boundary (Plate 5). The intervening space is filled with a scattering of cell processes which are said to be derived from the maternal syncytium (Lawn, Chiquoine & Amoroso, 1969): between the processes are incomplete laminae of basement membrane-like material (Plate 6).

It would be rash in a survey of this kind to quantify the separation of foetal and maternal capillaries, except in specific instances which may not be typical of the placenta as a whole. In the mare, however, it is relatively easy to obtain micrographs which show foetal and maternal capillaries in relation to the microvillous junctional zone. Because of the greater separation of the respective vessels, it is less easy in the cow and extremely difficult in the sheep to obtain comparable photographic records. Measurements have therefore been omitted from Fig. 5 which summarises the present findings in diagrammatic form.

PLACENTAL EXCHANGE

It now seems generally agreed that while the permeability of placental tissue may affect the transfer of many substances, this is unlikely to be a limiting factor for blood gases under normal circumstances, whereas a number of other factors are of prime importance for all types of placental transfer. These include: concentration gradients across the exchange areas; the rate and distribution of blood flow on either side; the vascular architecture, diffusion distances and total surface area for diffusion; usage by placental tissue itself; and finally, any special features which appear to facilitate the transfer of a given substance. Theoretical analyses of transplacental diffusion and the relative importance of factors which affect placental exchange have been discussed by Meschia, Battaglia & Bruns (1967) and by Longo, Hill & Power (1972). In this review attention will be focused on measurements which have been made in the chronic preparation on (*a*) concentration differences between maternal and foetal blood, (*b*) changes during the course of gestation, and (*c*) dynamic aspects of transfer where such information is available.

Validity of sampling procedures

Ideally, any studies on placental exchange should be made on blood which enters and leaves the actual exchange areas on either side of the placenta. This is still not feasible and samples withdrawn from the umbilical artery or lower aorta, the umbilical vein, a large uterine vein and a maternal artery

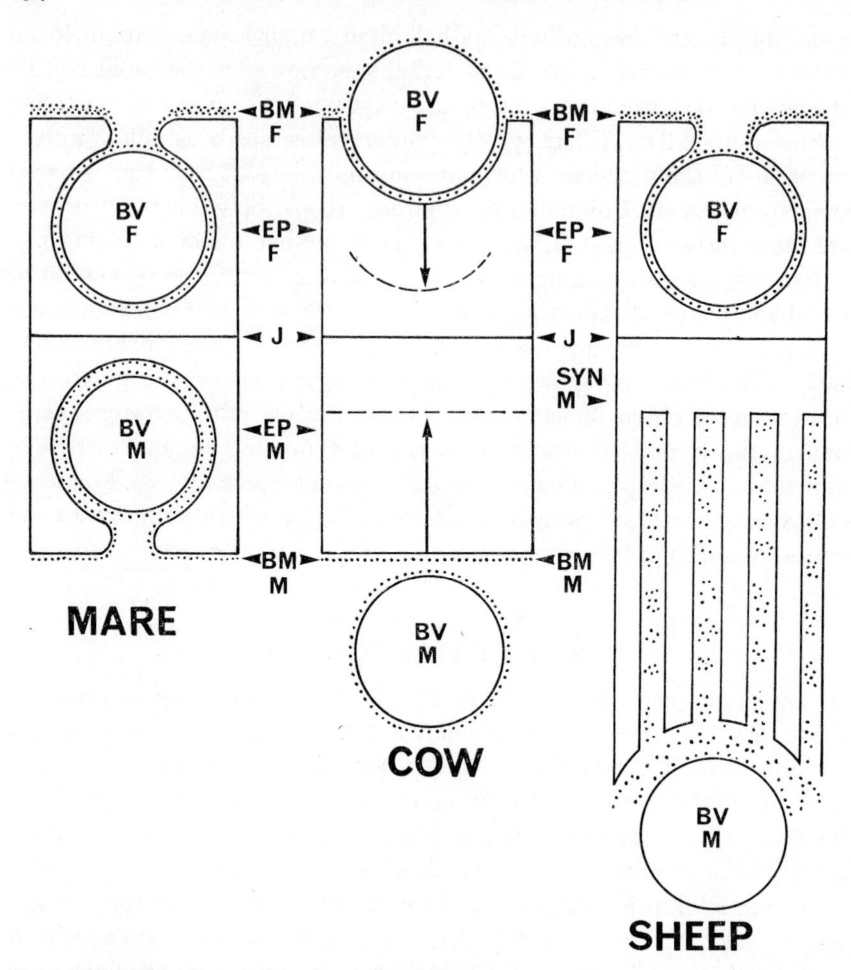

Fig. 5. Diagrammatic representation of the ultrastructural characteristics of the placental barrier in the mare, cow and sheep. The dashed lines (Cow) represent variations in arrangement which are occasionally found (cf. Plate 5). BV, capillary blood vessel; BM, basement membrane; EP, epithelium; J, junctional zone; SYN, syncytium. (F, foetal; M, maternal.)

are generally used. Since over 95% of the umbilical circulation goes through the placenta in the sheep (Makowski, Meschia, Droegemueller & Battaglia, 1968*a*) and probably an even greater percentage does likewise in the mare (Comline & Silver, unpublished observation) the use of umbilical blood to provide representative samples of foetal placental blood seems reasonably justifiable. However, this is not necessarily true for uterine blood, since the venous drainage of the uterus comes from the placenta, endometrium and myometrium, and while the placental fraction is certainly high, it can vary with species and with gestational age. In the sheep at term 85% of the uterine blood flow is placental in origin (Makowski *et al.* 1968*b*) and in the mare the fraction may be higher.

Thus the first approximation which must be accepted in the absence of more precise sampling methods is that of the possible admixture of placental venous blood with that from non-placental sites. The second point which requires consideration is whether the presence of catheters in the umbilical vessels and in a uterine vein causes any restriction of flow, for a major factor which affects placental transfer of any given substance is its rate of passage through the placental vascular bed. Both umbilical and uterine blood flows have been measured extensively in the conscious sheep and goat with indwelling catheters and compared with earlier observations in acute preparations. A critical assessment of the various values obtained under different experimental conditions is given by Crenshaw, Huckabee, Curet, Mann & Barron (1968), and there seems little doubt that the highest values for both umbilical and uterine blood flow have been obtained in the conscious animal. So that it seems unlikely that the presence of catheters in the placental circulation impedes blood flow significantly.

One last point should be considered before an assessment of the results obtained by the indwelling catheter technique can be made, particularly in relation to previous work: namely, the criterion for normality of the preparation. The birth of live, fully developed young at term is obviously the best evidence for this but there are many instances in the literature where catheters appear to have remained patent for only a short period and the foetus has apparently died some time later *in utero*. All too often, insufficient details are given and no criteria by which the authors judge the foetus to be normal are stated. It is probable that very short-term chronic experiments are no more valuable than good acute preparations in which the indwelling catheter technique is used, the foetus remains undisturbed *in utero* and the mother is maintained under optimal conditions (see Comline & Silver, 1970*b*).

Blood gas levels in foetus and mother during normal gestation

The work of Huggett (1927) and the more extensive investigations of Barcroft (1946) indicated the existence of large P_{O_2} gradients across the placenta of sheep and goats. More recently the problem of placental gas exchange in the ruminants and the mare has been examined in the conscious animal in a number of ways: by direct measurement of blood gas tensions and pH in the uterine and umbilical vessels; by following the changes in these vessels as gestation proceeds; and also by experimental manipulation of the levels or the rate of supply.

Table 1. *Mean blood gas and pH levels in uterine (MU) and umbilical (FU) venous blood of ruminants and mare*

Species	P_{O_2} (mm Hg) MU	FU	diff.	P_{CO_2} (mm Hg) MU	FU	diff.	pH MU	FU	diff.
Sheep*	53	35	18	38	42	4	7.44	7.40	0.04
Sheep†	49	32	17	38	42	4	7.42	7.37	0.05
Goat‡	45	27	18	—	—	—	7.38	7.34	0.04
Cow§	60	39	21	39	43	4	7.44	7.39	0.05
Mare§	51	48	3	38	39	1	7.408	7.404	0.004

* Comline & Silver (1970*a*, *b*) (chronic).
† Meschia, Makowski & Battaglia (1970) (chronic).
‡ Prystowsky, Meschia & Barron (1960) (acute).
§ Present investigation (chronic).

Table 1 summarises the data available for umbilical and uterine blood gas tensions and pH levels in the ruminants and mare near the end of gestation. When vein-to-vein gradients are compared, major differences clearly exist between the ruminants and the horse. In the former the minimum P_{O_2} gradient between foetus and mother is 17–21 mm Hg whereas in the mare this difference is only 3 mm Hg as measured between uterine and umbilical venous blood. P_{CO_2} gradients in all species are much smaller, as might be expected from the higher diffusibility of this gas, but differences between the ruminant and equine placental P_{CO_2} gradient still exist. Thus vein-to-vein differences of 4 mm Hg are found in the ruminant whereas in the foal there is virtually no P_{CO_2} gradient at all. A similar situation is found with respect to pH. Thus umbilical and uterine venous pH does not differ significantly in the mare but in the ruminant the difference is of the order of 0.045 pH units.

The mean values for different species (Table 1) give no indication of the overall changes which may occur during the latter part of gestation. Such information can only be obtained from sequential sampling in conscious animals. Meschia *et al.* (1965) were the first to show by means of the indwelling catheter technique that foetal and maternal blood gas levels in the pregnant sheep and goat remained stable during the last third of pregnancy. In a more detailed study of daily changes in the same species (Comline & Silver, 1970*a*) it was shown that small fluctuations in umbilical blood gas levels were associated with similar variations in uterine venous levels (Fig. 6), but there was no evidence for any long-term changes in either mother or foetus. Indeed, blood gas levels in the umbilical blood were generally very stable and contrasted sharply with the frequent changes in maternal and foetal hexose levels. Other prolonged studies on the con-

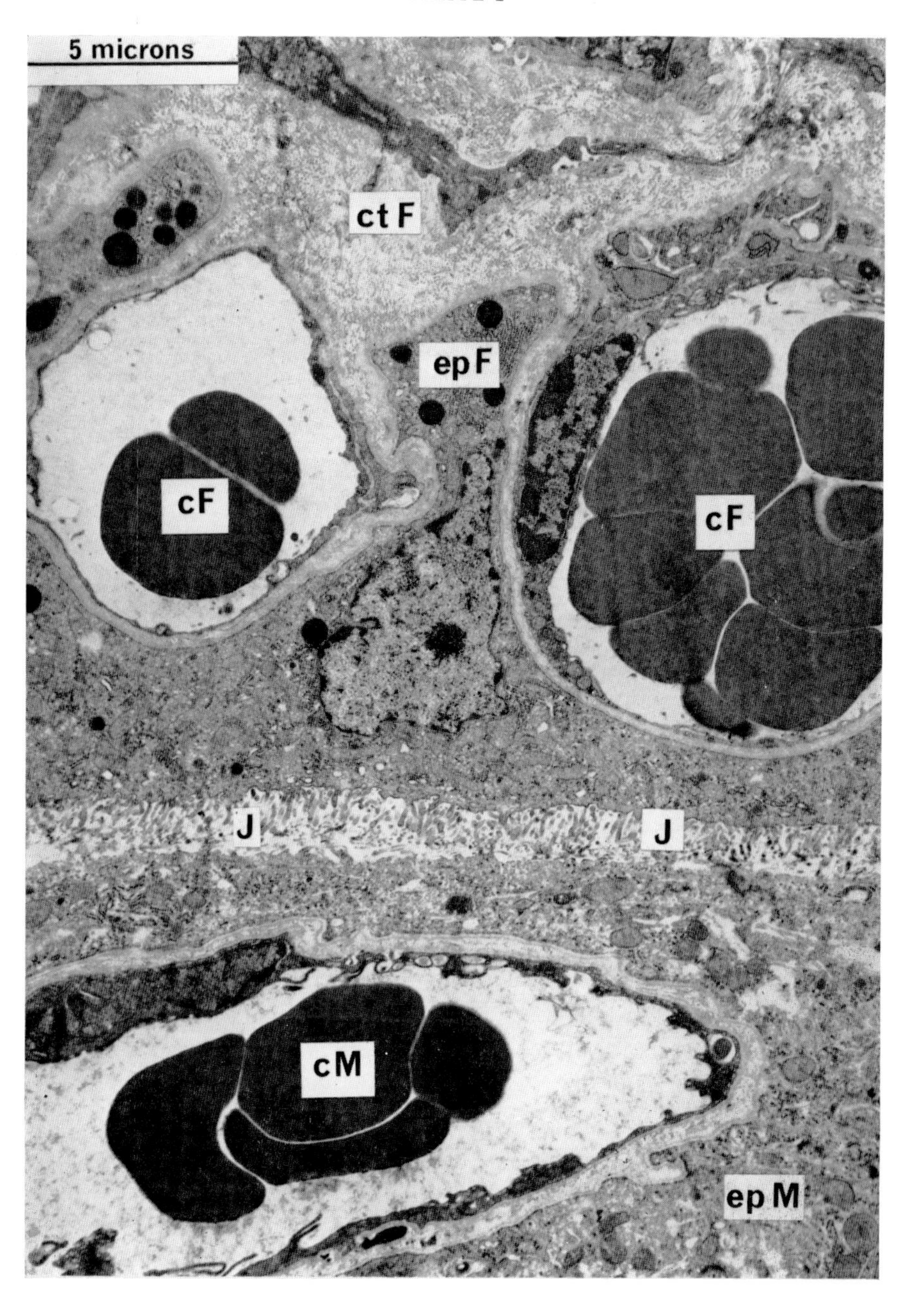

Intraepithelial capillaries on both foetal and maternal sides of the placenta of the mare. The endothelium of the foetal capillaries (cF) is extremely thin, and over the greater part of its length the basement membrane of the capillary is combined with the basement membrane of the chorionic epithelium (epF) to form a single lamina. The maternal capillary endothelium is thicker than that of the foetal side. Collagen fibres separate the basement membrane of the maternal epithelium (epM) from that of the underlying capillary (cM). J, microvillous junctional zone; ctF, foetal connective tissue. Mare: placenta in the last week of pregnancy.

(*Facing p.* 256)

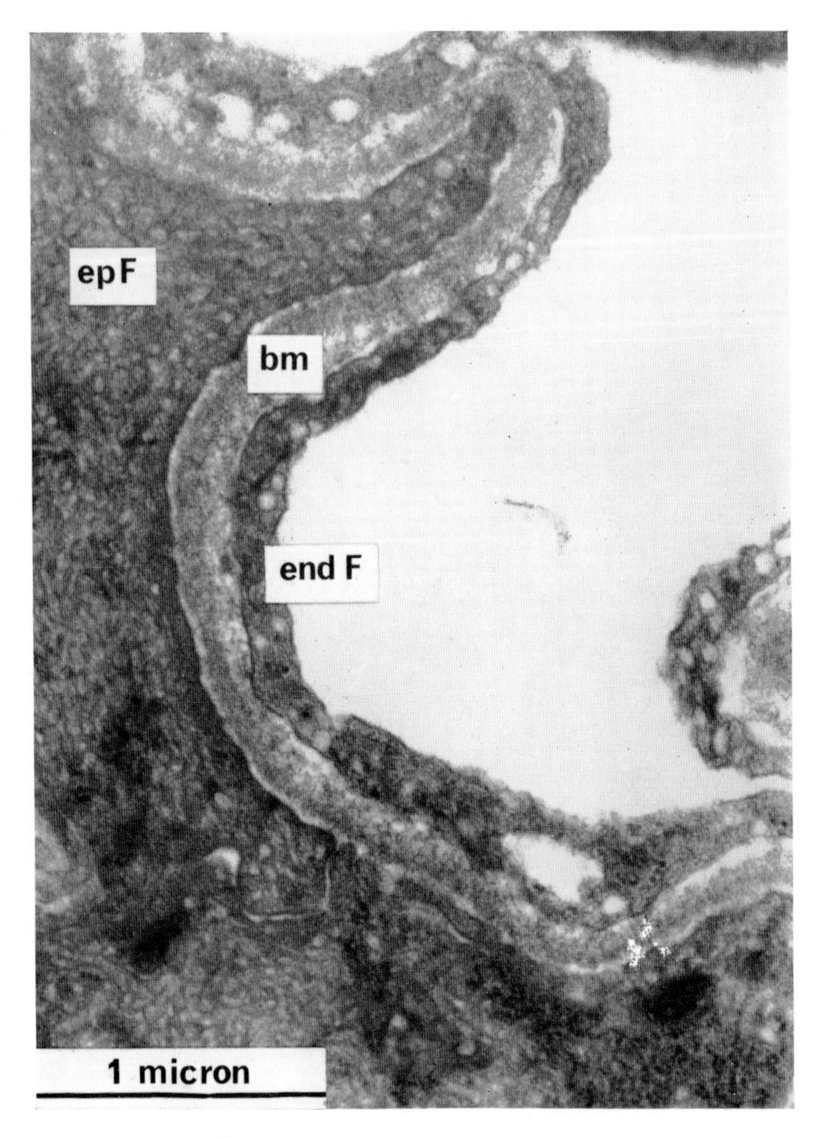

Endothelium (endF) of foetal capillary separated from chorionic epithelium (epF) by a single lamina of basement membrane (bm). The combined thickness of endothelium and basement membrane is in the region of 0.25–0.5 micron. Mare: placenta at 7 months.

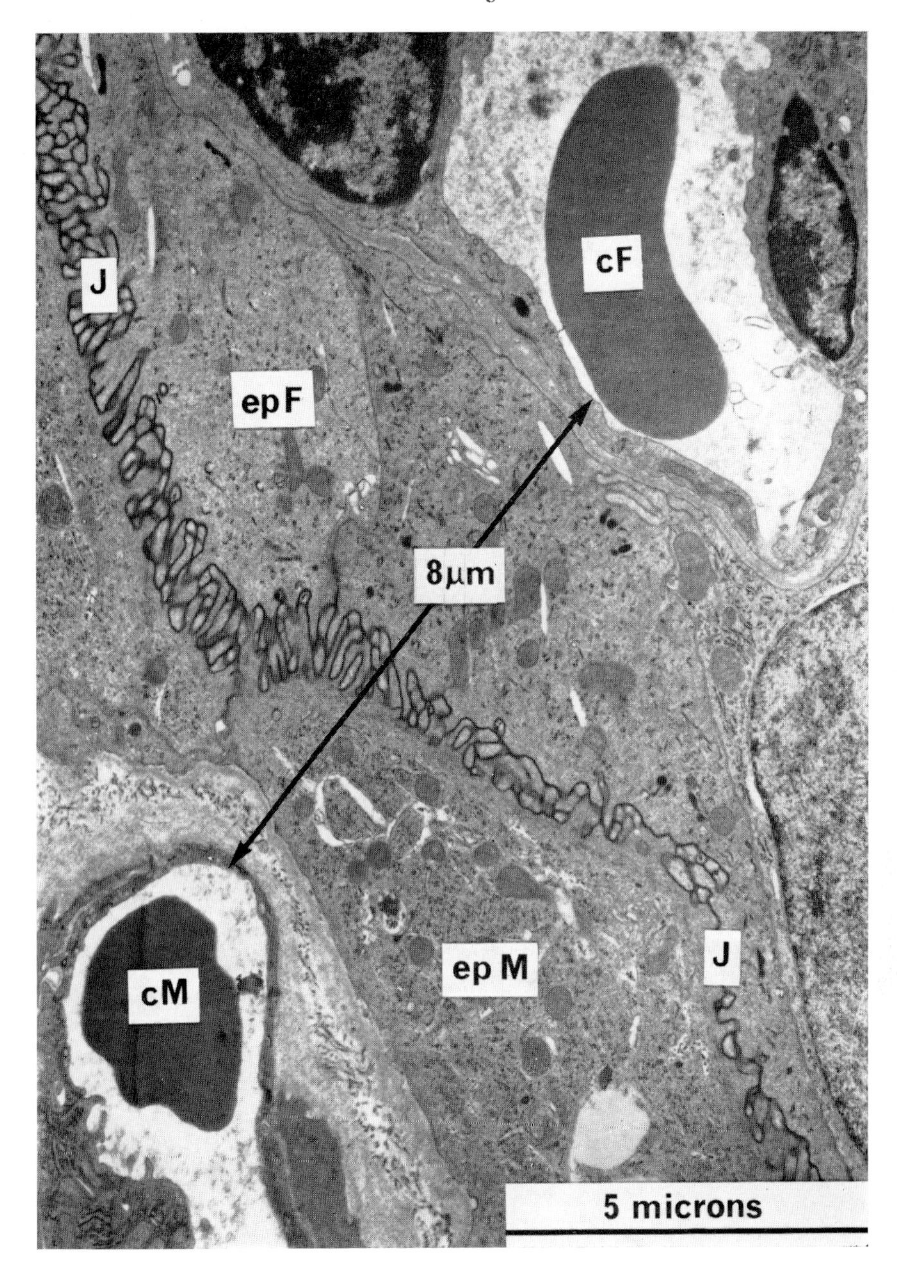

Vascular relations in the placenta of the cow. The foetal capillaries (cF) indent the chorionic epithelium (epF) to a lesser extent than is usual in the mare. On the maternal side the capillaries (cM) are separated from the microvillous junctional zone (J) not only by the full thickness of the maternal epithelium (epM), but also by a much greater thickness of connective tissue. Cow: placenta at 265 days.

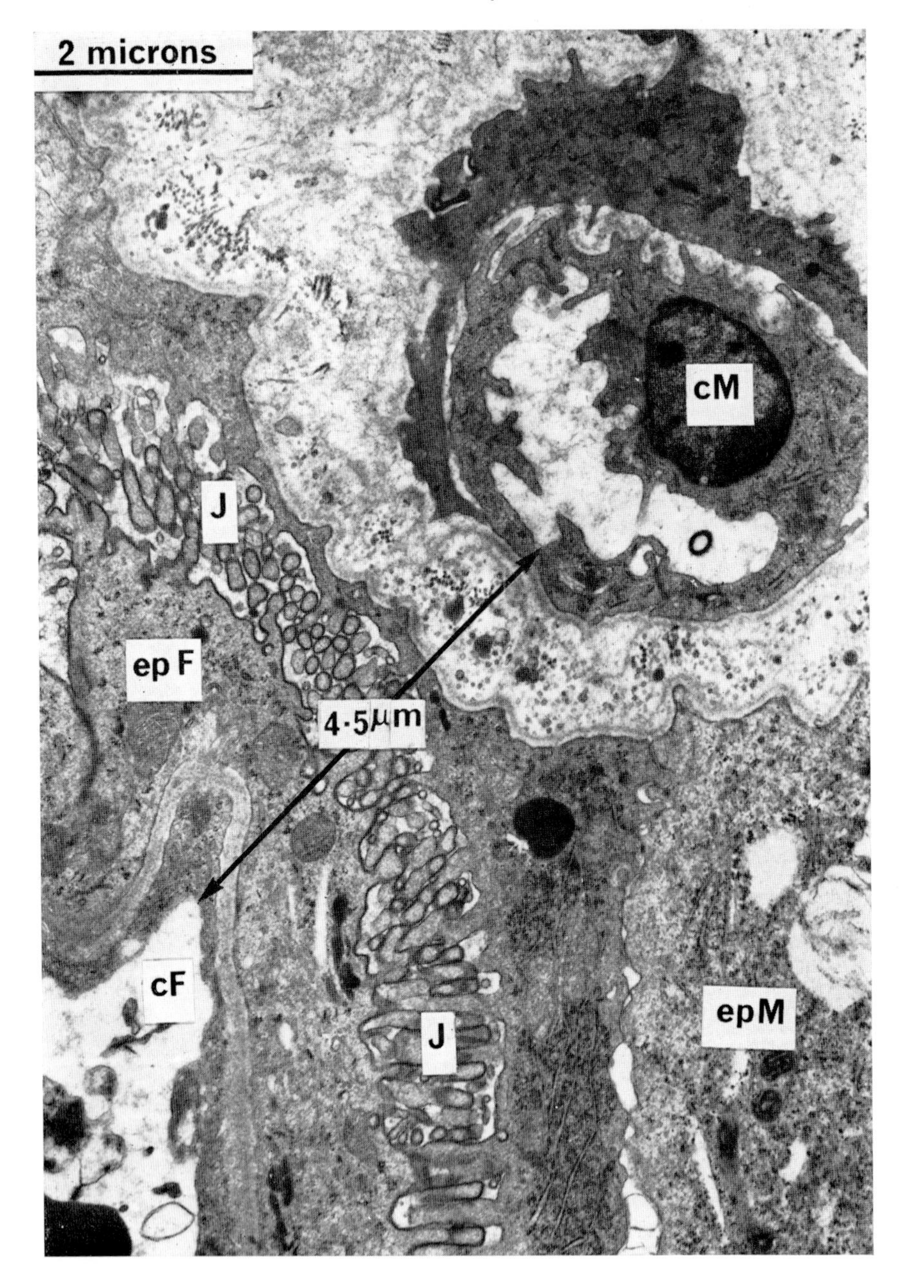

The maternal epithelium in the placenta of the cow is usually thick and cellular, but in some places tenuous areas of incomplete syncytium are occasionally found. This figure shows an unusually close approximation of foetal and maternal capillaries. cF, foetal capillary; cM, maternal capillary; epF, foetal epithelium; epM, maternal (synctial) epithelium; J, junctional zone. Cow: placenta at 265 days.

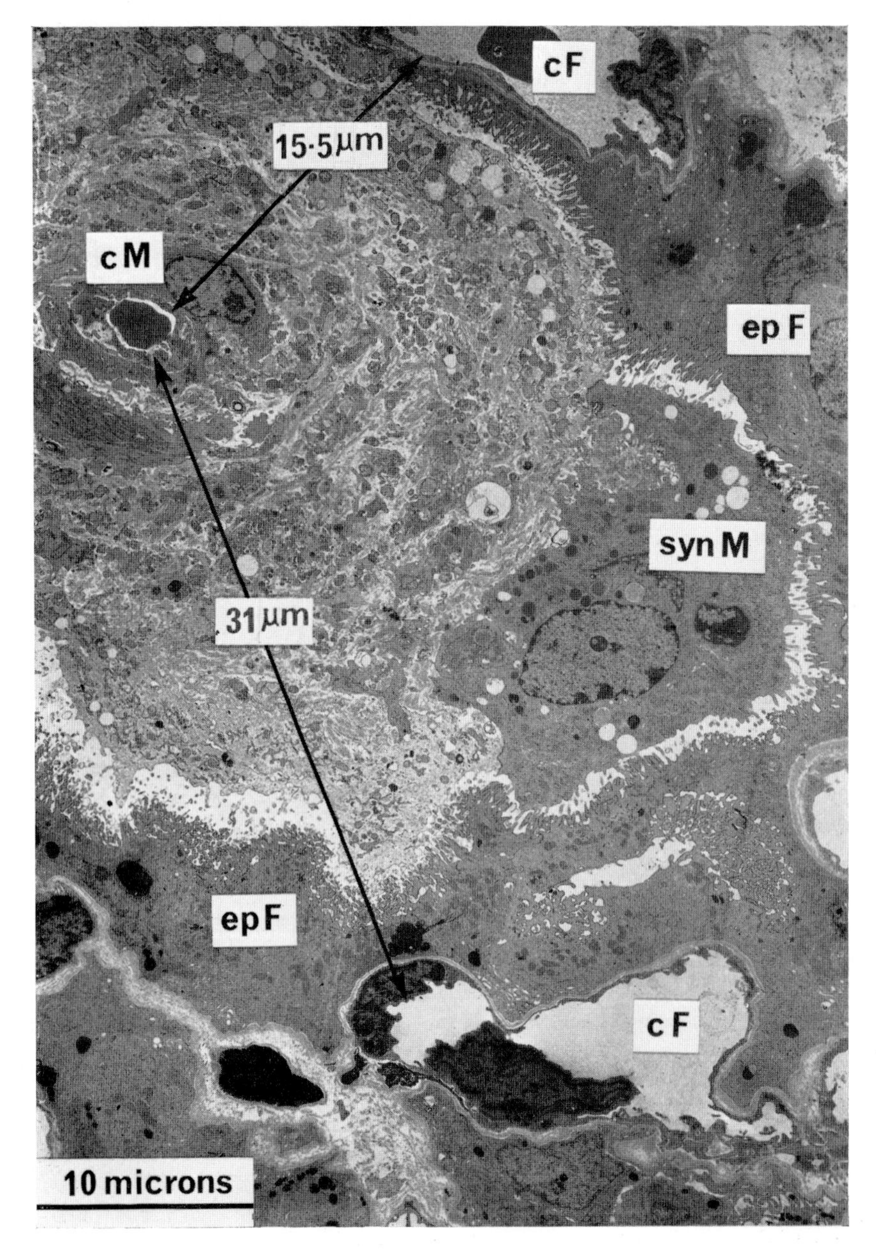

Vascular relations in the placenta of the sheep. The foetal capillaries (cF) lie close to the microvillous junctional zone: they indent the chorionic epithelium (epF) in a similar manner to those of the mare. On the maternal side, however, the maternal capillaries (cM) lie at some distance from the incomplete syncytium (synM) which forms the maternal epithelial boundary. Sheep: placenta at 130 days.

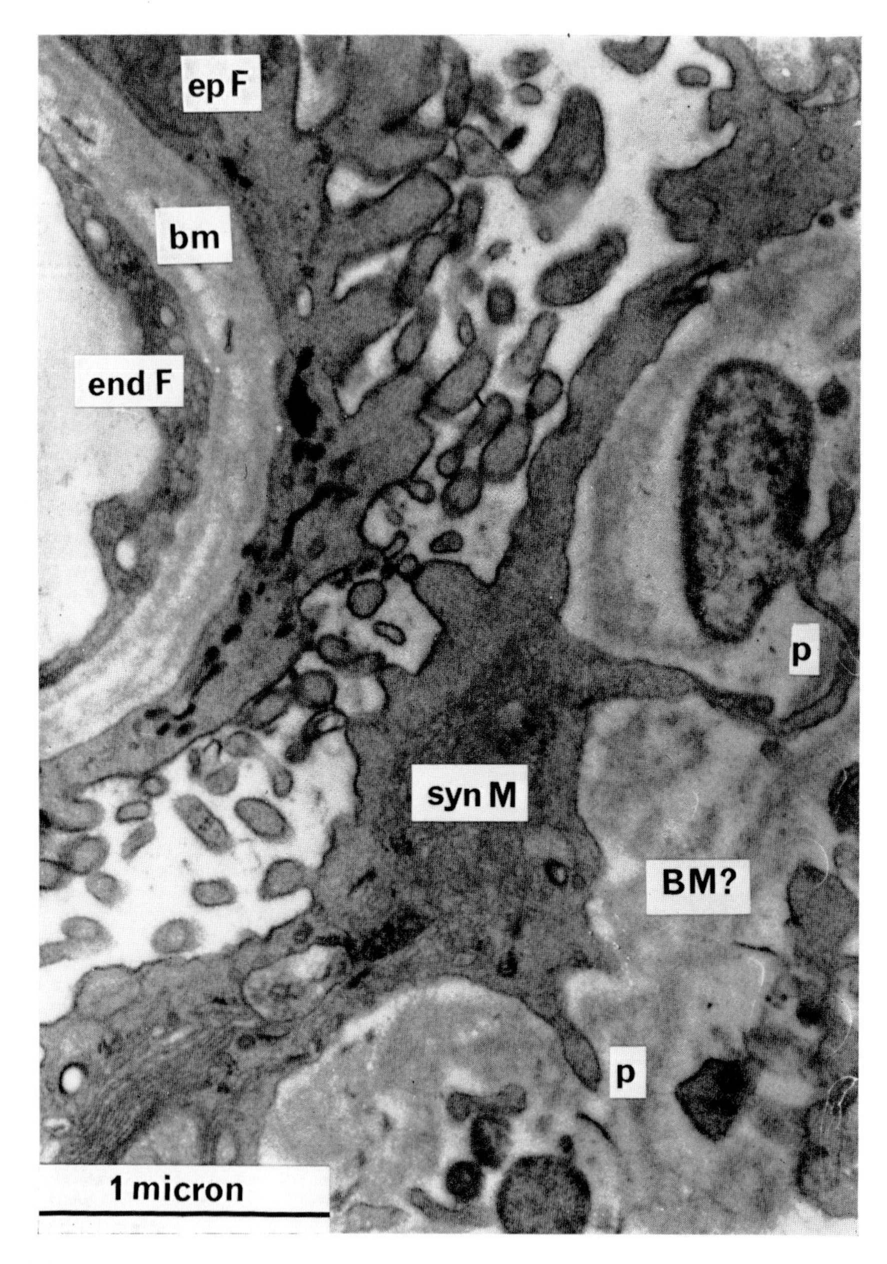

The space between the maternal syncytium (synM) and the maternal capillary is filled with a scattering of cell processes (p,p) separated by large quantities of basement membrane-like material (BM?). Neither the syncytium nor the maternal capillaries appear to possess definite basement membranes of their own. endF, foetal endothelium; bm, basement membrane; epF, chorionic epithelium. Sheep: placenta at 130 days.

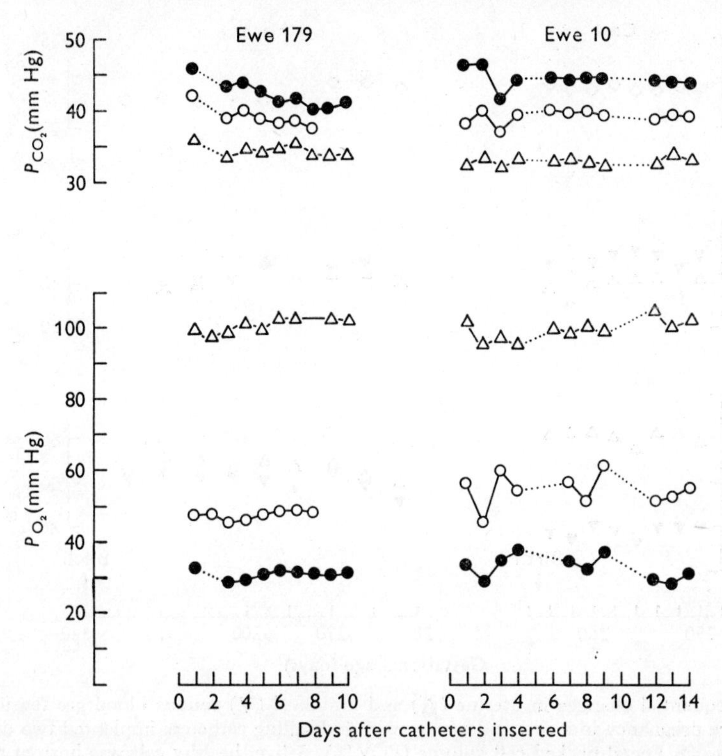

Fig. 6. Daily changes in maternal and foetal blood P_{CO_2} and P_{O_2} in two ewes. △, maternal artery; ○, uterine vein; ●, umbilical vein. Ewe 179, 123 days' gestation; Ewe 10, 99 days' gestation, at time of operation (Comline & Silver, 1970*a*). Reproduced by permission of the *Journal of Physiology*.

scious sheep have since been carried out, in which vessels in the foetal limb rather than in the umbilical circulation have been used for sampling; results from these experiments also indicate no change in foetal blood gases, pH, PCV, O_2 saturation or O_2 capacity during the last third of pregnancy (Joelsson, Barton, Daniel, James & Adamsons, 1970). In the two other species (cow and horse) examined in detail for long periods during gestation, there is also no indication of any obvious changes in foetal and maternal blood gas levels or PCV as term approaches. This is illustrated in Fig. 7 in which values from a cow and a mare and their foetuses sampled sequentially during the last few weeks of gestation are plotted.

The picture which emerges from this type of investigation suggests that during a normal uneventful pregnancy, few changes occur particularly with respect to blood gas levels in mother and foetus and that the vein-to-vein gradients across the placenta do not usually alter significantly. This stability is maintained throughout the first stage of labour and only at or

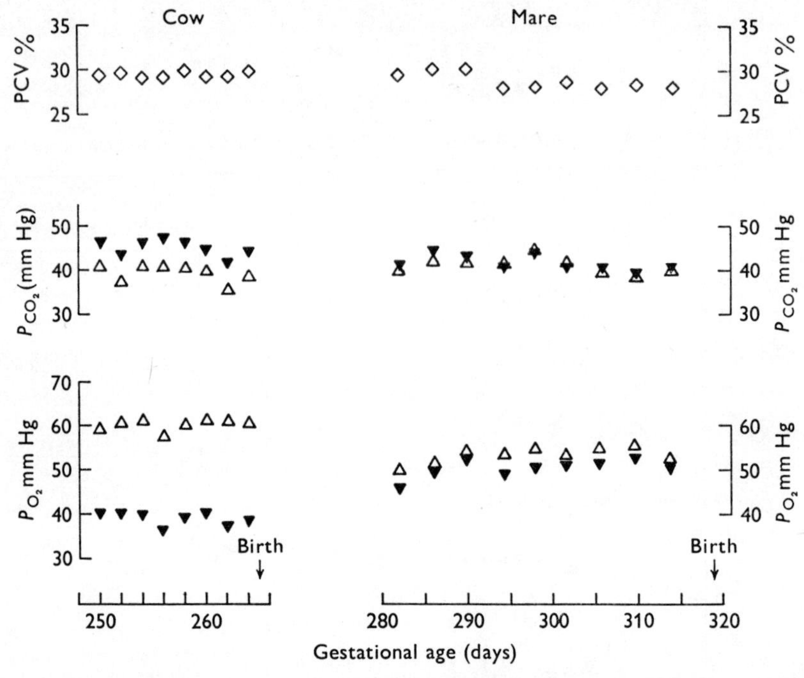

Fig. 7. Sequential changes in uterine (△) and umbilical (▼) venous blood gas tensions during late pregnancy in a cow and a mare with indwelling catheters implanted two days previously. ◇, Foetal packed cell volume (PCV %). A live, healthy calf was born at 264 days (term ~ 281 days); a still-born foal was delivered at 318 days (term ~ 340 days).

mmediately before delivery in both the sheep and cow is there any fall in foetal P_{O_2} and pH or rise in P_{CO_2} and PCV (Comline & Silver, 1972*a*). On the other hand, it could be argued that experimental animals with indwelling catheters are usually kept under the best possible conditions and samples are taken when the animal is resting quietly. It may well be that under more realistic field conditions foetal stability is not always maintained and that under such circumstances at least some of the many foetal defence mechanisms may be called into play: many of these are centred around circulatory re-adjustments which maintain umbilical blood flow and placental exchange (Dawes, 1968).

It is difficult to test such a hypothesis directly but there are situations both natural and experimental in which either maternal or foetal blood gas levels and other parameters may be altered. It is clearly of interest to examine such data to see whether the normal vein-to-vein gradients are maintained under adverse or other circumstances in the various species studied so far. The results may also throw some light on the differences between ruminants and the mare outlined above.

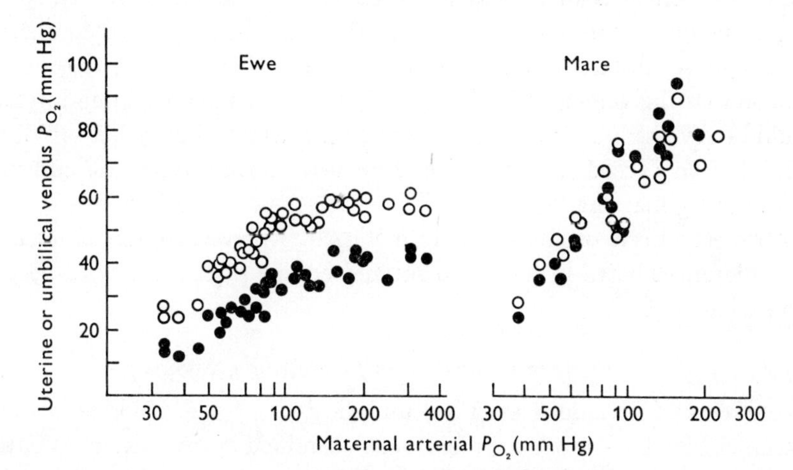

Fig. 8. The relation between the P_{O_2} in maternal arterial blood (log scale) and that in the uterine vein (○) and umbilical vein (●) in seven ewes and seven mares. (Data from Comline & Silver, 1970*b*).

(i) *Experimental alterations in blood gas levels*

In the ruminant it has been known for some time that during very wide fluctuations in maternal arterial P_{O_2} there is comparatively little change in foetal P_{O_2} levels. This phenomenon has now been investigated further in the sheep in both acute (Comline & Silver, 1970*b*) and chronic (Rankin, Meschia, Makowski & Battaglia, 1971) preparations by varying the composition of the maternal inspired air. The results from both types of experiment confirm and extend the original observations, and in each case the large uterine vein–umbilical vein gradient of 15–20 mm Hg P_{O_2} was maintained irrespective of the maternal arterial levels (Fig. 8).

The effects of maternal hypoxia have also been investigated in another way in sheep which were transported from 5000 to 14,000 feet after operation (Makowski, Battaglia, Meschia, Behrman, Schruefer, Seeds & Bruns, 1968). It was found that after an initial drop in P_{O_2} in both foetus and mother, the umbilical and uterine vein P_{O_2} levels gradually rose, so that eventually uterine venous P_{O_2} was only 5–10 mm Hg below that in the maternal artery. Nevertheless the 20 mm Hg P_{O_2} vein-to-vein gradient between umbilical and uterine circulations was unaltered at all stages of acclimatisation and the authors concluded that an increase in uterine blood flow had occurred during the process.

These results for the ruminant are in marked contrast to our findings on the mare during maternal hypoxia and hyperoxia (Comline & Silver, 1970*b*). The experiments were carried out on acute preparations and it was found that changes in maternal arterial P_{O_2} were accompanied by much larger changes in umbilical venous P_{O_2} than had been seen in ruminants. Furthermore, these foetal changes were greater than those in the uterine vein so that the vein-to-vein gradient did not remain constant and indeed could be reversed. This is illustrated in Fig. 8 which shows that umbilical vein P_{O_2} can exceed that in the uterine vein during hyperoxia and vice versa during maternal hypoxia.

All these effects of alterations in maternal P_{O_2} levels further emphasise the differences between the ruminant and equine placenta with respect to gas exchange.

(ii) *Effect of alterations in blood flow on placental gas exchange*

The effects of variations in either umbilical or uterine blood flow upon placental blood gas levels have not been examined systematically in either ruminant or mare. Experimentally reduced changes in umbilical flow are difficult to produce when the foetus is *in utero*, yet during normal pregnancy some occlusion of the umbilical cord may well occur. This condition has recently been simulated in the chronic preparation by means of an inflatable cuff placed round the cord and left *in situ* (Towell & Salvador, 1971). During mild occlusion pronounced falls in O_2 saturation, P_{O_2} and pH of the foetal arterial blood occurred. However, no measurements of umbilical or uterine venous blood gas levels were made in this study and so only the overall effect on blood gas transfer can be gauged.

The deleterious effects of removal of the foetus from the uterus are well known, and the resultant changes in umbilical blood flow and blood gas levels have been investigated by Heymann & Rudolph (1967). They showed very clearly that a drastic fall in umbilical blood flow occurred under these circumstances together with an increased umbilical A–V difference in P_{O_2}. A similar situation has been observed in the foetal foal on exteriorisation (Silver & Comline, 1972). Thus, initially umbilical venous P_{O_2} rises after exteriorisation and the normal vein-to-vein gradient is decreased as a result of slowed umbilical flow, but eventually the condition of the foetus deteriorates since insufficient oxygen is being supplied to the foetal tissue.

Large variations in uterine blood flow are probably more widespread than alterations in umbilical flow. Measurements of uterine flow in conscious animals under optimal experimental conditions have yielded data which cover a wide range of normal resting values in any one species (Huckabee, Crenshaw, Curet & Barron, 1972). It therefore seems likely that under stressful conditions even greater variations in flow may occur.

Certainly during spinal anaesthesia with hypotension uterine blood flow falls; it also decreases when the circulatory catecholamine levels are high (Assali, Dilts, Plentl, Kirschbaum & Gross, 1968). However, the effects of experimentally induced uterine flow changes on placental gas exchange do not appear to have yet received much attention, although the effects of changes in uterine blood flow at the time of parturition, particularly during increased uterine motility, have been investigated to some extent. As might be expected total uterine blood flow falls during uterine contractions in the sheep (Greiss, 1965) and in the monkey measurements of the relative distribution of uterine flow show that placental flow decreases more than myometrial flow during contractions (Lees, Hill, Ochsner, Thomas & Novy, 1971). Changes in foetal oxygen tension with uterine contractions have been reported for both human and sheep (see Comline & Silver, 1972, for references) but the data are largely qualitative and no vein-to-vein measurements have been made.

Metabolite levels in foetus and mother

The studies on blood gas tensions in umbilical and uterine blood of the ruminant and the mare indicate that fundamental differences in the mechanism of placental blood gas transfer may exist between these two groups of animals. A number of explanations may be advanced to account for these differences but before these are discussed it seems relevant to examine the data available on metabolite exchange across ruminant and equine placentae to see whether similar differences in transfer are found with substances less freely diffusible than oxygen and carbon dioxide.

(i) *Glucose*

A very large maternal-to-foetal gradient in blood glucose in the sheep was noticed in acute experiments some years ago; later the relation between maternal and foetal plasma hexoses was investigated in the chronic preparation by Comline & Silver (1970*a*). Large day-by-day variations in maternal glucose levels were often seen and these were reflected in similar, though smaller, fluctuations in foetal plasma glucose (Fig. 9). In a recent series of experiments on the sheep, cow and mare, in which both acute and chronic preparations were used, a wide variety of maternal glucose levels were encountered; the relationship in each species between the corresponding maternal and foetal values is shown in Fig. 10. Thus in the sheep, the foetal plasma glucose is only 26% of that in the ewe; in the cow the percentage is 32, while in the mare foetal levels are as high as 60% of those in maternal plasma. These various gradients appear to be maintained over a wide range of resting maternal levels.

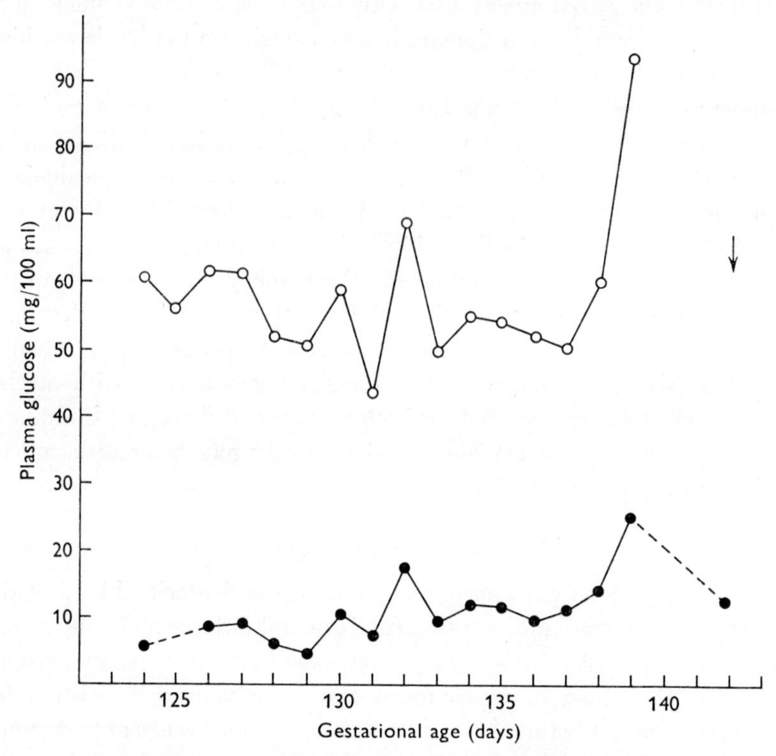

Fig. 9. Daily change in maternal (○) and foetal (●) plasma glucose in a sheep between 124 days' gestation and its delivery under sodium pentobarbitone anaesthesia at 144 days. (Data from Comline & Silver, 1970*a*.)

In some preliminary experiments designed to investigate more dynamic aspects of glucose transfer during maternal hyperglycaemia, further differences between the ruminant and mare have been observed. Infusions of glucose (10–15 min duration) into the maternal lower aorta, in amounts sufficient to cause an immediate two-fold increase in uterine venous plasma levels, were given to a conscious sheep and an anaesthetised mare. Samples were taken simultaneously from the uterine and umbilical veins during and after the infusion period and the results are plotted in Fig. 11. In the foetal foal there was an immediate, rapid rise in umbilical plasma glucose at the same time as the major uterine venous change occurred. In the sheep, on the other hand, the rate of change in the foetus was much slower than that in the maternal plasma; the foetal values only reached a maximum at the end of the maternal glucose infusion.

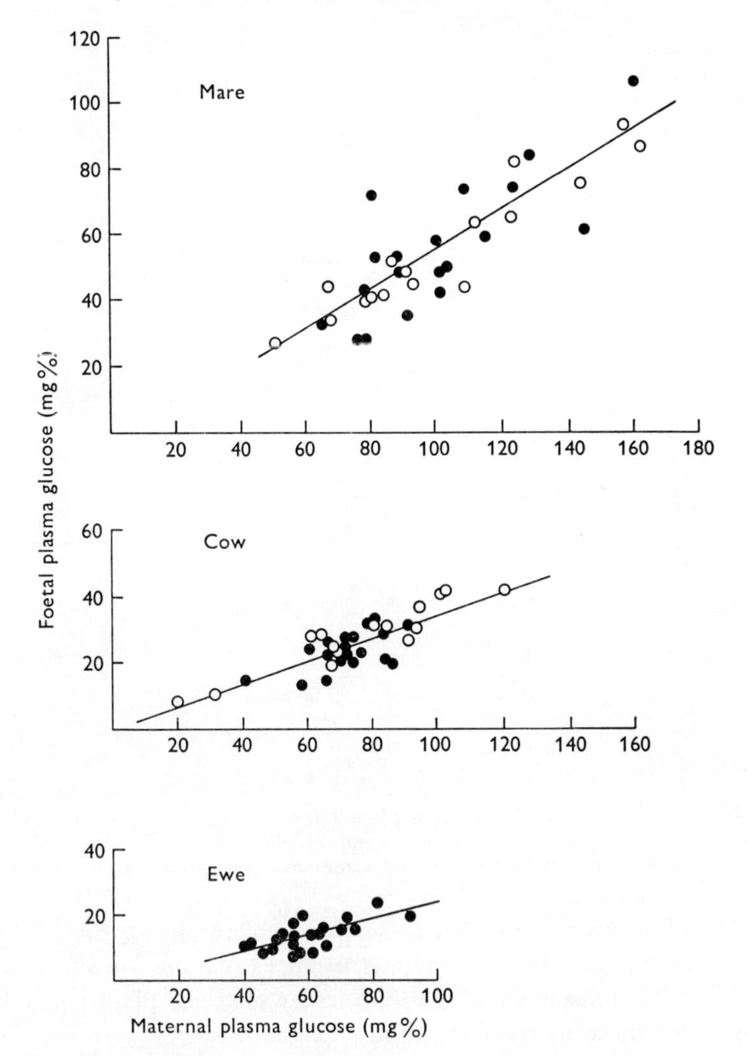

Fig. 10. The relation between maternal and foetal plasma glucose levels in mare, cow and ewe. ●, Values from chronic preparations; ○, values from acute preparations (chloralose anaesthesia) with foetus undisturbed *in utero*. Two to three values are included per experiment. Regression lines were fitted by the method of least squares.
Ewe, $\hat{Y} = 0.26X - 2.0$; Cow, $\hat{Y} = 0.32X - 0.19$; Mare, $\hat{Y} = 0.60X - 5.6$.

(ii) *Free fatty acids (FFA)*

Relatively little work has been done on the uptake of FFA by the foetus. In the foetal lamb extremely low plasma FFA levels have been reported by a number of workers in both acute and chronic preparations (see Comline and Silver, 1972, for references). Low plasma FFAs have now been found in the foetal calf under similar chronic and acute condi-

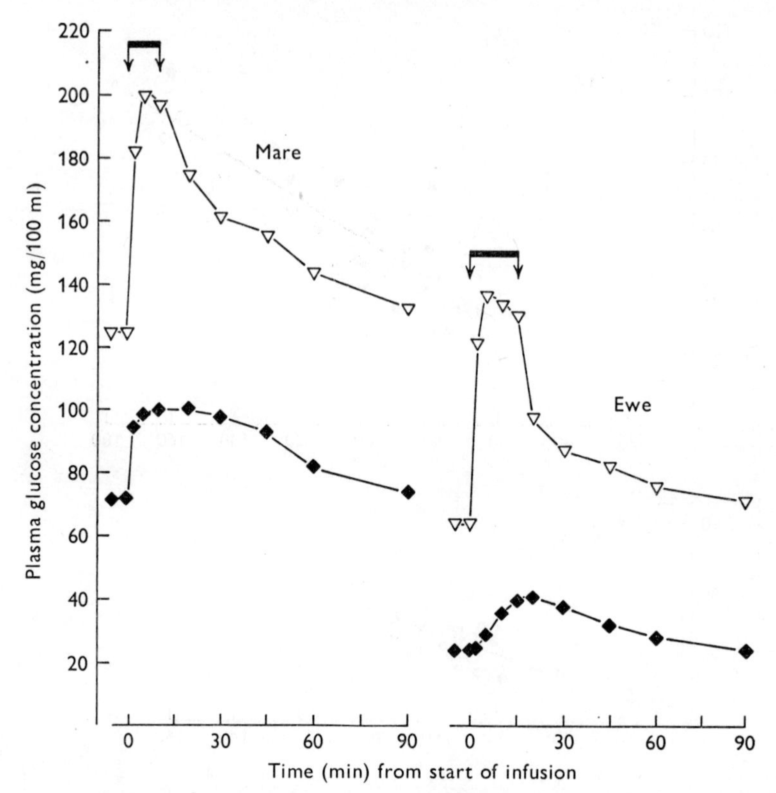

Fig. 11. The effect of 10–15 min infusions (double arrows) of glucose into the uterine circulation (via the lower aorta) on uterine (▽) and umbilical (◆) venous plasma glucose levels in an anaesthetised mare (chloralose) and a conscious ewe with indwelling catheters.

tions. Fig. 12 shows that in both these species of ruminant, there is no apparent relation between maternal and foetal FFA levels. In the sheep under certain circumstances FFA values in the maternal plasma may be very high, but those in the foetus rarely exceed 0.1 mEq/l. Attempts to raise the foetal levels in the ruminant have largely failed; injections of catecholamines into the foetus result in very small foetal plasma FFA changes (James, Meschia & Battaglia, 1971; Comline & Silver, 1972).

The mare again presents a complete contrast to the ruminant in that foetal plasma FFA appears to be directly related to the maternal levels. Considerable individual variation was encountered, particularly in acute experiments in which the maternal FFA values were often very high at the beginning of an experiment but later decreased; occasionally in such experiments the foetal FFA levels exceeded those in the maternal plasma. Nevertheless there was a highly significant correlation between maternal and foetal FFA ($r = 0.88$, $p < 0.01$). These results suggest that FFAs can

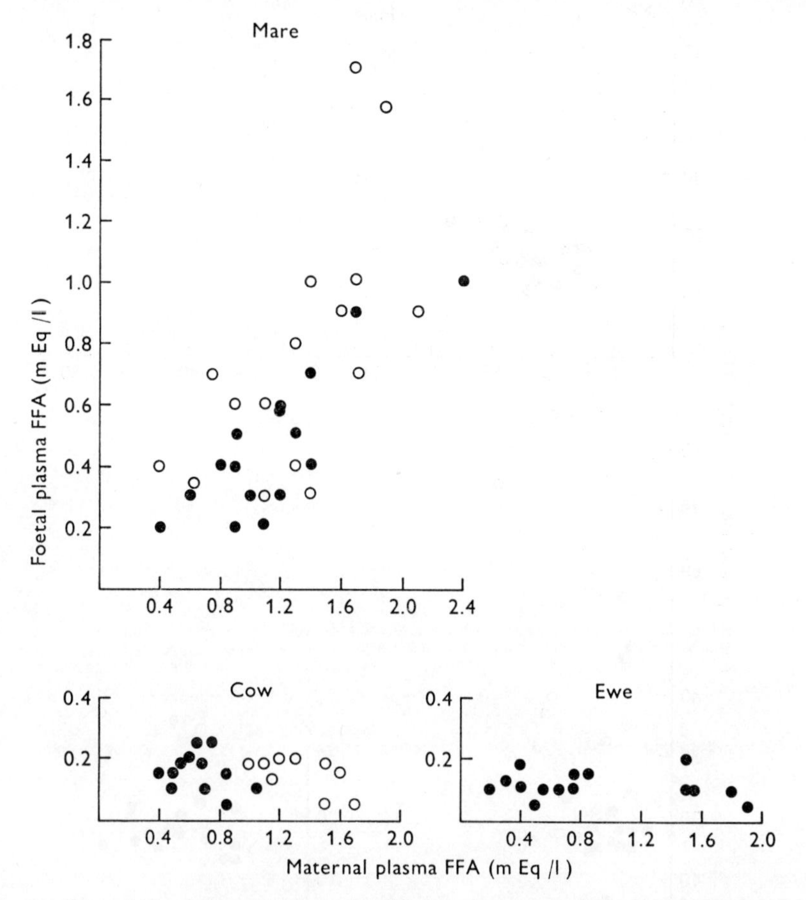

Fig. 12. The relation between maternal and foetal plasma free fatty acids (FFA) in mare, cow and ewe. Symbols and other details as in Fig. 10.

pass across the equine placenta with relative ease and it may well be that they play a more important role in foetal metabolism than in the ruminant.

(iii) *Lactic acid*

The problem of lactic acid transfer across the placenta has long been the subject of debate but little positive evidence has been presented. The production of lactic acid by the foetus in times of distress and the persistance of a low foetal pH long after the asphyxial episode has passed are well known, yet the way that the foetus eventually overcomes its metabolic acidosis is still not clear. Blechner, Meschia & Barron (1960) first showed that pH changes in both mother and foetus could occur independently of

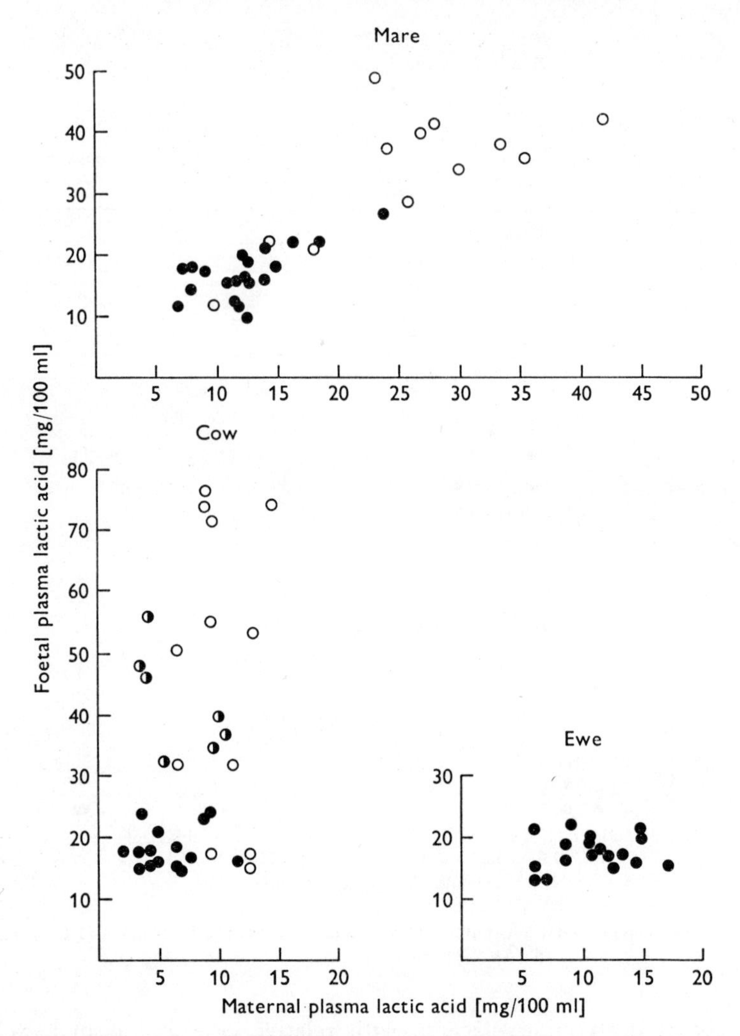

Fig. 13. The relation between maternal and foetal plasma lactic acid in mare, cow and sheep. ◑, Values for three conscious animals 1–2 days before parturition in which foetal arterial pH was low (7.29–7.32). Other symbols and details as in Fig. 10.

one another under chronic conditions in the sheep. These studies were extended and confirmed by Comline & Silver (1970*b*) who showed that in the sheep, at least, maternal pH changes of respiratory origin were readily reflected by similar foetal pH changes, whereas those of metabolic origin were not. Later they showed that infusion of lactic acid into the foetus led to very profound pH changes which were relatively slow to recover (Nathanielsz, Comline, Silver & Paisey, 1972); these changes were not accompanied by any maternal lactacidaemia even in the uterine venous blood. The lack of any relation between foetal and maternal plasma lactate in both sheep and cow is shown in Fig. 13. Lactic acid levels were particularly variable in the foetal calf under anaesthesia but there were no corresponding maternal changes. In the mare, on the other hand, foetal lactate levels followed those in the maternal blood more closely, particularly in acute preparations (Fig. 13); the correlation between maternal and foetal values was statistically significant ($r = 0.66$, $p < 0.01$). Yet even in this species, in which a foetal-to-maternal transfer of lactate is implicated, artificially induced lactacidaemia in the foal led to negligible changes in lactate levels in the maternal uterine venous blood. A similar, and seemingly close, relation between foetal and maternal lactate levels has also been shown for the primate (Derom, 1964). Furthermore, in the pregnant monkey, labelled lactate injected into the foetus has been detected in the maternal blood soon afterwards (Friedman, Gray, Grynfogel, Hutchinson, Kelly & Plentl, 1960). It remains to be shown in both ruminant and mare whether labelled lactic acid can be detected in the uterine venous blood after administration to the foetus.

DISCUSSION

In 1970 we suggested, on the basis of studies on blood gas tensions and pH levels in umbilical and uterine blood of the mare and sheep, that the equine placenta might be more efficient than that of the ruminant. The fact that umbilical venous P_{O_2} exceeded that in the uterine vein at high maternal arterial levels whilst, conversely, uterine venous P_{CO_2} could sometimes exceed umbilical vein P_{CO_2}, indicated that there might be a countercurrent mechanism in operation in the microcotyledons of the mare. The anatomical studies on the vasculature of the equine placenta were consistent with this idea (Tsutsumi, 1962). In the sheep, on the other hand, the presence of a maternal capillary cascade within the placentome (Steven, 1966), with the possibility of considerable venous admixture within the system, appeared to provide a simple anatomical explanation of the wide umbilical vein–uterine vein differences in blood gas tensions in this species (see Comline & Silver, 1970*a*).

The large uterine-to-umbilical vein P_{O_2} gradient found in the cow was not unexpected and in fact the vein-to-vein differences in P_{O_2}, P_{CO_2} and pH were similar to those in sheep and goats. However, the placentome of the cow is very different in size and shape from that seen in most breeds of sheep and its vascular arrangements do not appear to include a maternal capillary cascade (see Fig. 3).

The present observations on glucose, FFA and lactate levels in foetal and maternal plasma in the ruminants and mare show that there are obvious parallels between metabolite and blood gas exchange mechanisms in these three species. Each of the three metabolites examined appeared to pass much more freely across the equine placenta than across that of the ruminant. The question inevitably arises as to whether the length of the diffusion pathway between maternal and foetal vessels in the three species can account for the various differences in placental transfer. Since the inter-capillary distance between foetus and mother is usually shortest in the equine placenta (see Fig. 5) it is tempting to explain the relatively high foetal plasma glucose and FFA levels, the rapid placental transfer of glucose and the similarity of umbilical and uterine vein blood gases, pH and lactate in the mare on this basis. Conversely, the larger foetal-to-maternal plasma concentration differences of the various metabolites in the ruminants may be related to the much greater diffusion distances found between the placental vessels of these species. This is almost certainly an over-simplification of the problem since there are clearly many other factors which can affect placental transfer: the rate of supply and removal on both sides (placental perfusion), the possibility of facilitated transfer on the one hand or barriers to diffusion on the other, and also the extent of placental metabolism, will all affect the final concentration differences observed in the uterine and umbilical blood.

Some of these difficulties can be overcome by examining the equilibration of inert substances across the placenta since here there should be no consumption or special transfer mechanisms involved. In some preliminary studies on the diffusion of antipyrine across the equine placenta during constant infusion into the foetus using the method of Meschia, Cotter, Makowski & Barron (1967), the concentrations in uterine and umbilical venous blood soon became virtually identical and remained so for as long as the infusion was continued. In comparable experiments on the ruminant (Meschia *et al.* 1967; Meschia, Cotter *et al.* 1967) a vein-to-vein difference was consistently found during equilibrium conditions (Fig. 14). These observations lend further support to the view that placental transfer in the mare is more efficient than in the ruminant.

There are obviously many questions to be answered before a clear picture can be given of the way in which the equine placenta functions as compared with that of the sheep. While the direction of flow in the microcotyledons appears from both anatomical and physiological studies to be countercurrent, direct verification of this is needed. In addition, information on the rates of uterine and umbilical blood flows and their distribution under chronic conditions is required; as yet only a few observations on acute preparations have been made in the mare (Comline & Silver, 1972).

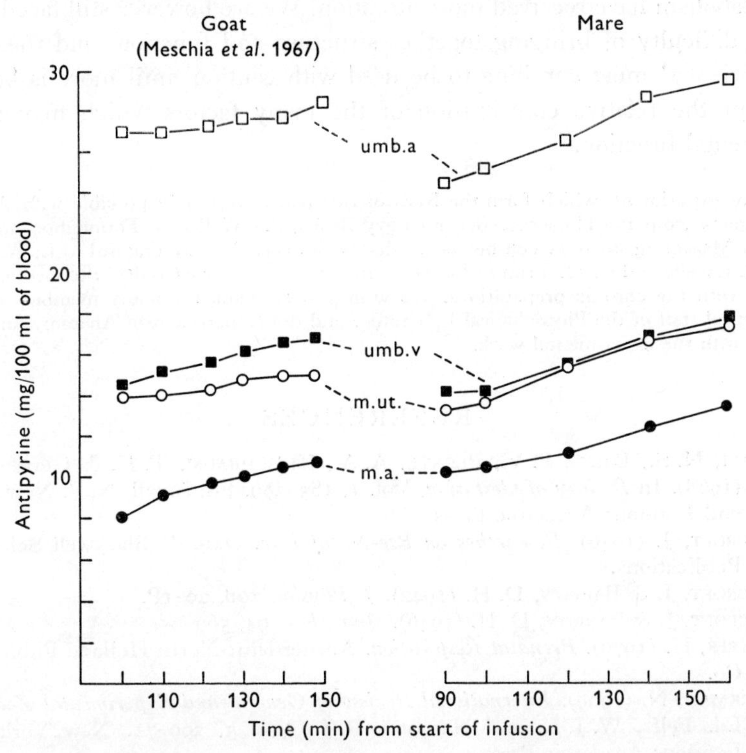

Fig. 14. A comparison between the antipyrine levels in the umbilical and uterine circulations in the mare and goat during constant infusion of antipyrine into the foetal circulation for simultaneous umbilical and uterine flow measurement (Meschia, Cotter, Makowski & Barron, 1967). Umb. a., umbilical artery; umb. v., umbilical vein; m.ut., uterine vein; m.a., maternal artery.

A comparison between the metabolism of the equine placenta and its ruminant counterpart is also needed in order to assess the relative contribution made by the placental tissue in the total metabolism of the uterus, the placenta and the foetus. The overall effect of the type of placental exchange seen in the mare on the nutrition and growth of the foetus is a more general problem. In this context, the inability of the mare to continue twin pregnancies until term, in contrast to the ruminant, may be another example of the more delicate balance between foetus and mother in this species.

The interpretation of the observations given here and their fusion into a general scheme to define the role of the placenta in the regulation of the foetal blood composition poses the same problems as those considered by Barcroft in 1946 when he discussed the relation between permeability and efficiency; at that time the permeability of the placental barrier dominated thought in this field whereas recently problems of placental perfusion and metabolism have received more attention. We are however still faced with the difficulty of bringing together structure and function, and the term 'efficiency' must continue to be used with caution until more is known about the relative contribution of the many factors which may affect placental function.

The experiments which form the basis of this paper were only possible with the aid of grants from the Horserace Betting Levy Board, the Wellcome Foundation and the Milk Marketing Board which helped to defray the cost. We are grateful to L. W. Hall, R. B. Lavelle and C. M. Trim of the Department of Veterinary Clinical Studies for their help with the chronic preparations. We wish also to thank the many members of the technical staff of the Physiological Laboratory and the Department of Anatomy for their help with the experimental work.

REFERENCES

ASSALI, N. S., DILTS, P. V., PLENTL, A. A., KIRSCHBAUM, T. H. & GROSS, S. J. (1968). In *Biology of Gestation*, Vol. 1, 185–289. Ed. Assali, N. S. New York and London: Academic Press.

BARCROFT, J. (1946). *Researches on Pre-Natal Life*. Oxford: Blackwell Scientific Publications.

BARCROFT, J. & BARRON, D. H. (1942). *J. Physiol.* **100**, 20–1P.

BARCROFT, J. & BARRON, D. H. (1946). *Anat. Rec.* **94**, 569–95.

BARTELS, H. (1970). *Prenatal Respiration*. Amsterdam: North Holland Publishing Co.

BJÖRKMAN, N. (1968). *International Review of General and Experimental Zoology*. Ed. Felts, W. J. L. and Harrison, R. J., Vol. 3, 309–71. New York and London: Academic Press.

BLECHNER, J. N., MESCHIA, G. & BARRON, D. H. (1960). *Q. Jl exp. Physiol.* **45**, 60–71.

COMLINE, R. S. & SILVER, M. (1970*a*). *J. Physiol.* **209**, 567–86.

COMLINE, R. S. & SILVER, M. (1970*b*). *J. Physiol.* **209**, 587–608.

COMLINE, R. S. & SILVER, M. (1972). *J. Physiol.* **22**, 233–56.

CRENSHAW, C., HUCKABEE, W. E., CURET, L. B., MANN, L. & BARRON, D. H. (1968). *Q. Jl exp. Physiol.* **53**, 65–75.

DAWES, G. S. (1968). *Foetal and Neo-Natal Physiology*. Chicago: Year Book Medical Publishers.

DEROM, R. (1964). *Am. J. Obstet. Gynec.* **89**, 241–51.

FRIEDMAN, E. A., GRAY, M. J., GRYNFOGEL, M., HUTCHINSON, D. L., KELLY, W. T. & PLENTL, A. A. (1960). *J. clin. Invest.* **39**, 227–35.

GOLDSTEIN, S. R. (1926). *Anat. Rec.* **34**, 25–35.

GREISS, F. C. (1965). *Am. J. Obstet. Gynec.* **93**, 917–23.

GROSSER, O. (1909). *Vergleichende Anatomie und Entwicklungsgeschichte der Eihäute und der Placenta*. Vienna and Leipzig: W. Braumüller.

HEYMANN, M. A. & RUDOLPH, A. M. (1967). *Circulation Res.* **21**, 741–5.

HUCKABEE, W. E., CRENSHAW, C., CURET, L. B. & BARRON, D. H. (1972). *Q. Jl exp. Physiol.* **57**, 12–23.

HUGGETT, A. St G. (1927). *J. Physiol.* **62**, 373–84.

JAMES, E., MESCHIA, G. & BATTAGLIA, F. C. (1971). *Proc. Soc. exp. Biol. Med.* **138**, 823–6.

JOELSSON, I., BARTON, M. D., DANIEL, S., JAMES, L. S. & ADAMSONS, K. (1970). *Am. J. Obstet. Gynec.* **107**, 445–52.

LAWN, A. M., CHIQUOINE, A. D. & AMOROSO, E. C. (1969). *J. Anat.* **105**, 557–78.
LEES, M. H., HILL, J. D., OCHSNER, A. J., THOMAS, C. J. & NOVY, M. J. (1971). *Am. J. Obstet. Gynec.* **110**, 68–81.
LONGO, L. D., HILL, E. P. & POWER, G. G. (1972). *Am. J. Physiol.* **222**, 730–9.
MAKOWSKI, E. L. (1968). *Am. J. Obst. Gynec.* **100**, 283–8.
MAKOWSKI, E. L., BATTAGLIA, F. C., MESCHIA, G., BEHRMAN, R. E., SCHRUEFER, J., SEEDS, E. A. & BRUNS, P. D. (1968). *Am. J. Obstet. Gynec.* **100**, 852–61.
MAKOWSKI, E. L., MESCHIA, G., DROEGEMUELLER, W. & BATTAGLIA, F. C. (1968*a*). *Circulation Res.* **23**, 623–31.
MAKOWSKI, E. L., MESCHIA, G., DROEGUMUELLER, W. & BATTAGLIA, F. C. (1968*b*). *Am. J. Obstet. Gynec.* **101**, 409–12.
MAYOW, J. (1674). *Medico-Physical Works.* Edinburgh: The Alembic Club, 1907.
MESCHIA, G., BATTAGLIA, F. C. & BRUNS, P. D. (1967). *J. Appl. Physiol.* **22**, 1171–8.
MESCHIA, G., COTTER, J. R., BREATHNACH, C. S. & BARRON, D. H. (1965). *Q. Jl exp. Physiol.* **50**, 185–95.
MESCHIA, G., COTTER, J. R., MAKOWSKI, E. L. & BARRON, D. H. (1967). *Q. Jl exp. Physiol.* **52**, 1–18.
MESCHIA, G., MAKOWSKI, E. L. & BATTAGLIA, F. C. (1970). *Yale J. Biol. Med.* **42**, 154–65.
METCALFE, J., BARTELS, H. & MOLL, W. (1967). *Physiol. Rev.* **47**, 782–838.
METCALFE, J., MOLL, W., BARTELS, H., HILPERT, P. & PARER, J. T. (1965). *Circulation Res.* **16**, 95–101.
NATHANIELSZ, P. W., COMLINE, R. S., SILVER, M. & PAISEY, R. B. (1972). *J. Reprod. Fertil.* Suppl. 16, 39–59.
PRYSTOWSKY, H., MESCHIA, G. & BARRON, D. H. (1960). *Yale J. Biol. Med.* **32**, 441–8.
RANKIN, J. H. G., MESCHIA, G., MAKOWSKI, E. L. & BATTAGLIA, F. C. (1971). *Am. J. Physiol.* **220**, 1688–92.
SILVER, M. & COMLINE, R. S. (1972). In *Placental Gas Exchange.* Ed. Longo, L. D. US Public Health Publication (in press).
STEVEN, D. H. (1966). *J. Physiol.* **183**, 13–15P.
STEVEN, D. H. (1968). *J. Physiol.* **196**, 24–6P.
TOWELL, M. E. & SALVADOR, H. S. (1971). *Fetal Evaluation During Pregnancy and Labor*, pp. 143–56. Ed. Grosignani, P. G. and Pardi, G. New York and London: Academic Press.
TSUTSUMI, Y. (1962). *J. Fac. Agric. Hokkaido (imp.) Univ.* **52**, 372–482.
TSUTSUMI, Y. & HAFEZ, E. S. E. (1964). *Fifth International Congress for Animal Reproduction and Artificial Insemination*, Trento, Vol. **2**, 195–200.
TURNER, W. M. (1876). *Lectures on the Comparative Anatomy of the Placenta.* Edinburgh: A. & C. Black.
WIMSATT, W. A. (1950). *Amer. J. Anat.* **87**, 391–458.

ACUTE CHANGES OF OXYGEN PRESSURE AND THE REGULATION OF UTERINE BLOOD FLOW

By G. MESCHIA and F. C. BATTAGLIA
University of Colorado Medical Center,
4200 East 9th Avenue, Denver, Colorado

Several investigators have assumed that the respiratory gases are important regulators of placental blood flow. This is a reasonable assumption in view of the fact that the placenta functions as the respiratory organ of the foetus. The simplest hypothesis is that O_2 and CO_2 pressures control the magnitudes of maternal and foetal placental blood flows, but a different, subtler type of regulation that controls the efficiency of transplacental exchange by altering the degree of uneven perfusion of the placenta has been proposed also (Power *et al.* 1967).

In one of our studies, we had defined experimentally the concept and the technique for the measurement of the flow-limited clearance of inert molecules across the sheep placenta (Meschia, Battaglia & Bruns, 1967). This study showed that the placental clearances of antipyrine and tritiated water are flow-limited by demonstrating that they are equal and maximal (Table 1) despite large physico-chemical differences of the two substances.

On the basis of this information, we have used antipyrine clearance as a tool for measuring the overall efficiency of placental perfusion by maternal

Table 1. *A comparison of the placental clearances of antipyrine, tritiated water and urea in sheep (data of Meschia* et al. 1967)

Foetal weight (g)	Clearances (ml/min) of: Antipyrine	Tritiated water	Urea
642	60	65	8
694	42	39	22
950	92	84	19
1458	137	145	40
1824	143	130	50
2912	209	201	64
3134	235	220	46
3552	261	264	81
3670	268	264	60
3760	285	283	73

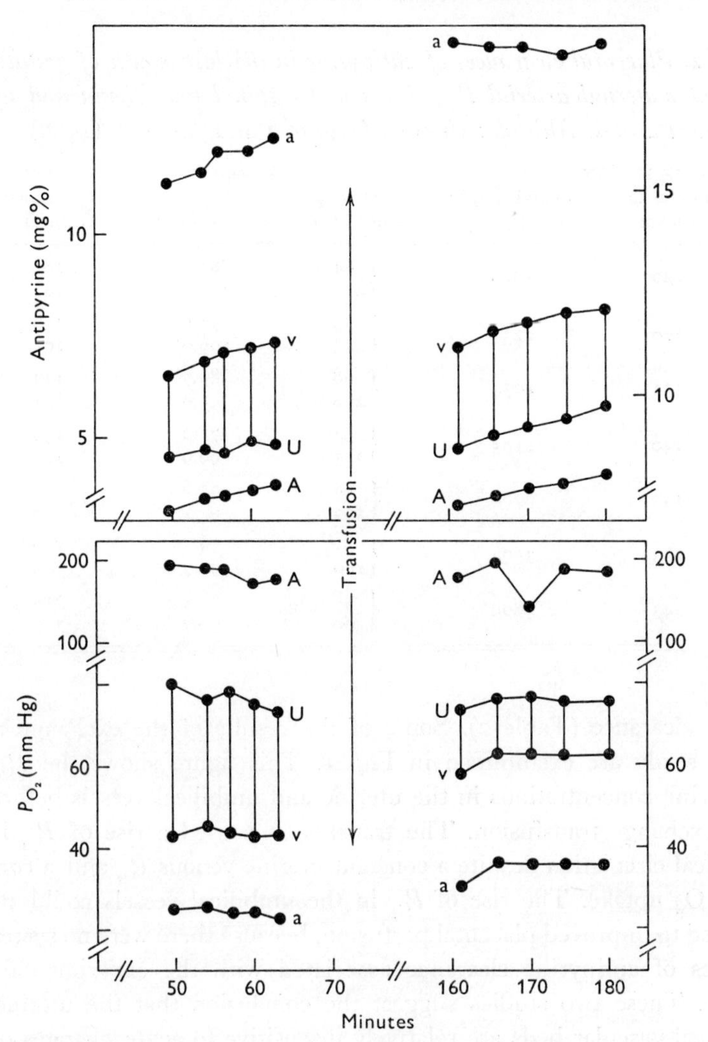

Fig. 1. Antipyrine concentration and P_{O_2} in the maternal artery (A), uterine vein (U), umbilical vein (v) and umbilical artery (a), before and after exchange transfusion of adult blood. Mother inhaling 100 % O_2 from the start of the experiments. Antipyrine infused at constant rate in the foetal circulation from time zero. Note that the narrower U-v P_{O_2} difference after transfusion is not associated with narrower a-A or v-U concentration difference of antipyrine. Reproduced from Meschia *et al.* (1969).

and foetal blood in two experimental situations: (1) maternal inhalation of 100% O_2 (Battaglia, Meschia, Makowski & Bowes, 1968) and (2) exchange transfusion of adult blood with low O_2 affinity in foetal lambs (Meschia, Battaglia, Makowski & Droegemueller, 1969). In the maternal hyperoxia study, we observed marked elevations of oxygen tension and saturation in the uterine and umbilical vessels with no significant changes of anti-

Table 2. *Placental clearances of antipyrine in the last month of gestation at different maternal arterial* P_{O_2}. *Ewes under spinal anaesthesia and sedated with barbiturates. Altitude:* 1600 *m. Data of Battaglia* et al. (1968)

Foetal age (days)	Foetal weight (g)	Art. P_{O2} (mm Hg)	Art. % O_2 sat.	Clearance (ml/min)
120	1828	64	83.6	181
		215	100.0	181
120	2233	55	78.6	130
		243	100.0	101
131	2034	68	85.7	124
		270	99.9	121
140	2165	52	71.2	139
		315	99.9	129
140	2700	60	85.6	175
		303	99.1	190
143	3504	57	79.4	300
		190	99.7	309
143	4809	48	68.5	312
		80	90.7	392

pyrine clearance (Table 2). Some of the results of the exchange transfusion study are exemplified in Fig. 1. This figure shows the P_{O_2} and antipyrine concentrations in the uterine and umbilical vessels before and after exchange transfusion. The transfusion caused a rise of P_{O_2} in the umbilical circulation despite a constant uterine venous P_{O_2} and a constant foetal O_2 uptake. The rise of P_{O_2} in the umbilical vessels could not be ascribed to improved placental perfusion, because there were no systematic changes of antipyrine clearance associated with the experimental procedure. These two studies suggest the conclusion that the uterine and umbilical vascular beds are relatively insensitive to acute changes of P_{O_2}. It seemed desirable to verify part of this conclusion by a different methodology suitable for detecting small and rapid changes of uterine blood flow in the nonanaesthetized, unstressed animal (Makowski, Hertz & Meschia, 1972). Pregnant sheep, homozygous for haemoglobin B, were equipped with electromagnetic flow probes chronically implanted around the uterine arteries and a permanent femoral arterial catheter. The non-anaesthetized animals were exposed to alternate periods of (1) air breathing, (2) inhalation of 100% O_2 and (3) inhalation of 15% O_2 in N_2. A separate study (Rankin, Meschia, Makowski & Battaglia, 1971) had demonstrated that in Denver (altitude 1600 m) the inhalation of 15% O_2 in N_2 mixture by sheep with B haemoglobin produced a decrease in maternal arterial saturation from 89% to 60% and in umbilical vein saturation from 80% to 45%. The flow data were analysed as paired observations and expressed as percent change of the control value. Hypoxia caused a very small drop of uterine flow significant at the 0.05 P level ($-2.4\% \pm 1.0$) and hyperoxia did not produce any significant change ($-1.0\% \pm 1.2$). There were significant changes ($P < 0.01$) of haemoglobin concentration in the arterial blood ($+6.4\%$ and -7.0% in hypoxia and hyperoxia, respectively) and a 5% drop of blood pressure in hyperoxia. The end results of these changes were a rise of haemoglobin flow to the pregnant uterus in hypoxia ($+3.7\% \pm 1.2$) and a fall of haemoglobin flow in hyperoxia ($-8.0\% \pm 1.3$). There is no inconsistency between these results and the observation that severe hypoxia may induce a large fall of uterine blood flow in anaesthetized animals (Dilts, Brinkman, Kirschbaum & Assali, 1969; Duncan, 1969). It seems likely that the latter phenomenon is secondary to the action of catecholamines or other vasoactive substances released by the hypoxic stimulus rather than a direct effect of hypoxia on the uterine vasculature.

Because changes of maternal P_{O_2} may affect uterine blood flow indirectly, experiments of maternal hypoxia and hyperoxia do not provide conclusive evidence about the local effect of respiratory gases on the uterine circulation. In order to explore this question of local control, we have performed some experiments on the effect of foetal death on uterine blood flow (Raye, Killam, Battaglia, Makowski & Meschia, 1971). The experiments were on nonanaesthetized sheep, two to seven days after surgery and the blood flow through the uterine arteries was measured before and after foetal death caused by the injection of a bolus of air into a vein of the lamb. In late pregnancy, abrupt foetal death caused a 75% reduction of total uterine O_2 consumption. Despite this large and sudden drop in metabolic rate and a rise of uterine venous P_{O_2} from 46 to 59 mm Hg there were no appreciable changes of uterine blood flow in the first hour after foetal demise (see Fig. 2 and Table 3). Additionally, in these experiments we were able to verify the observation (Greiss, 1966) that the release of maternal aortic occlusion is not followed by reactive hyperaemia of the pregnant uterus.

In the evaluation of experiments on the regulation of uterine blood flow in pregnancy, we must consider the fact that this flow is a mixture of placental, endometrial and myometrial flows and that these flows may react differently to the same stimulus. Fortunately, it is possible to measure these flows as separate entities by the radioactive microsphere technique (Makowski, Meschia, Droegemueller & Battaglia, 1968). In Fig. 3, placental, myometrial and endometrial flows are plotted against foetal

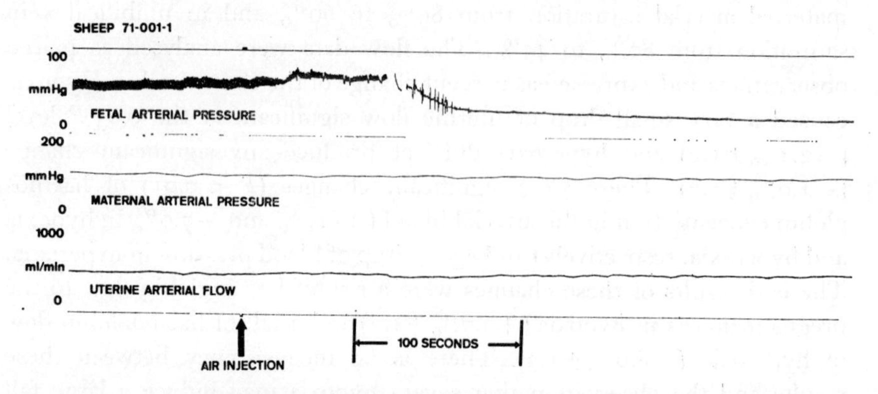

Fig. 2. Recording of foetal arterial pressure, maternal arterial pressure and blood flow through one uterine artery before and after foetal death caused by the injection of a bolus of air into the foetal inferior vena cava. Note the absence of an acute response of uterine blood flow to abrupt foetal demise. Reproduced from Raye *et al.* (1971).

Table 3. *Comparison of mean blood flow through one uterine artery in the control period (the hour preceding foetal death) and in the experimental period (the hour following foetal death). Data from Raye* et al. (1971)

Foetal age (days)	Foetal weight (kg)	Blood flow F_{con} (ml/min)	Blood flow F_{exp} (ml/min)	F_{exp}/F_{con}
141	3.30	816	825	1.01
134*	2.50	787	752	0.96
129	3.10	426	372	0.87
143	2.50	408	426	1.04
134	4.10	648	690	1.06
112	1.45	143	162	1.13
112	1.71	154	164	1.06
119	2.37	304	328	1.08
138	4.34	608	560	0.92
				Mean 1.01

* One of twins.

weight. It is evident that in sheep there is a marked growth of placental flow during pregnancy and that in late pregnancy placental flow is the primary constituent (~84%) of total uterine flow. This knowledge is useful in drawing inferences about placental blood flow from measurements of total uterine flow. For example, late in pregnancy it is impossible for the placental flow to rise more than 20% without a detectable rise in uterine blood flow. Hence we may conclude from the relative constancy of uterine blood flow in hypoxia and hyperoxia that neither of these two

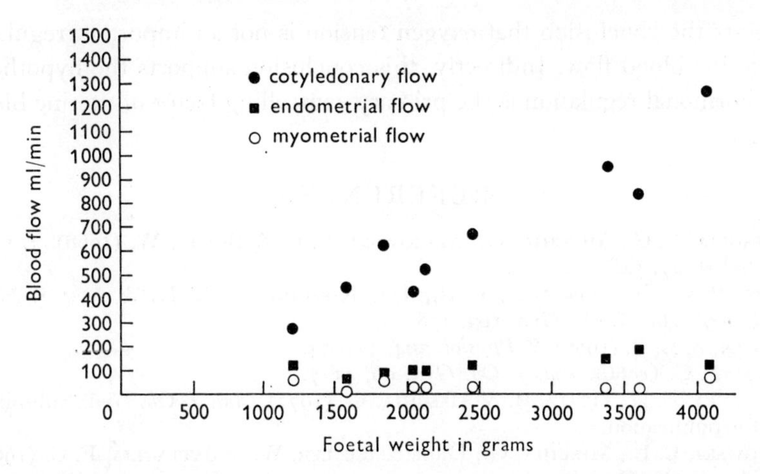

Fig. 3. Concomitant growth of foetal weight and blood flow to the cotyledons of the sheep placenta during the last 40 days of gestation. Reproduced from Makowski *et al.* (1968).

experimental conditions did cause a large increase of placental blood flow in the normal animal. On the other hand, it is theoretically possible to have a large decrease of placental blood flow that would go undetected by measurements of uterine flow because of a concomitant rise of myometrial and endometrial flows.

Measurements of placental flow are macroscopic measurements that cannot reveal subtle changes in the pattern of flow within the placenta. A resolution of placental flow into flows to each placental cotyledon is possible by means of the microsphere technique (Rankin, Meschia, Makowski & Battaglia, 1970). Contrary to the results of an earlier report (Power *et al.* 1967) it would seem that the placental cotyledons in sheep have a remarkably small degree of uneven perfusion, down to the level of one gram placental slices. Beyond this level of resolution measurements of flow become unreliable and it is necessary to use measurements of flow-limited clearance in order to evaluate the overall efficiency of placental perfusion.

In summary, we have described several experiments which were centred around the problem of uterine and placental blood flow regulation by the respiratory gases. The uterine blood flow appears to be insensitive to acute changes of respiratory gases caused by maternal hyperoxia and hypoxia, occlusion of the maternal abdominal aorta and foetal death. Furthermore, the overall efficiency of placental perfusion is not affected by a rise of P_{O_2} in the uterine and umbilical vessels. More experiments need to be done acutely and chronically especially in the area of maternal hypoxia, hyper- and hypocapnia. Nevertheless, the evidence at hand suggests the conclusion that oxygen tension is not an important regulator of uterine blood flow. Indirectly, this conclusion supports the hypothesis that hormonal regulation is the primary controlling factor of uterine blood flow.

REFERENCES

BATTAGLIA, F. C., MESCHIA, G., MAKOWSKI, E. L. & BOWES, W. (1968). *J. Clin. Invest.* **47**, 548–55.

DILTS, P. V., Jr, BRINKMAN, C. R., III, KIRSCHBAUM, T. H. & ASSALI, N. S. (1969). *Am. J. Ob. Gyn.* **103**, 138–57.

DUNCAN, S. L. B. (1969). *J Physiol.* **204**, 421–34.

GREISS, F. C. (1966). *Am. J. Ob. Gyn.* **96**, 41–7.

MAKOWSKI, E. L., HERTZ, R. & MESCHIA, G. (1972). *Am. J. Ob. Gyn.* Submitted for publication.

MAKOWSKI, E. L., MESCHIA, G., DROEGEMUELLER, W. & BATTAGLIA, F. C. (1968). *Am. J. Ob. Gyn.* **101**, 409–12.

MESCHIA, G., BATTAGLIA, F. D. & BRUNS, P. D. (1967). *J. Appl. Physiol.* **22**, 1171–8.

MESCHIA, G., BATTAGLIA, F. C., MAKOWSKI, E. L. & DROEGEMUELLER, W. (1969). *J. Appl. Physiol.* **26**, 410–16.

POWER, G. G., LONGO, L. D., WAGNER, H. W., KUHL, D. E. & FORSTER, R. E., II. (1967). *J. Clin. Invest.* **46**, 2053–63.

RANKIN, J., MESCHIA, G., MAKOWSKI, E. L. & BATTAGLIA, F. C. (1970). *Am. J. Physiol.* **219**, 9–16.

RANKIN, J. H. G., MESCHIA, G., MAKOWSKI, E. L. & BATTAGLIA, F. C. (1971). *Am. J. Physiol.*, **220**, 1688–92.

RAYE, J. R., KILLAM, A. P., BATTAGLIA, F. C., MAKOWSKI, E. L. & MESCHIA, G. (1971). *Am. J. Ob. Gyn.* **111**, 917–24.

DIURNAL PATTERNS IN UTERINE DYNAMICS

By GUY M. HARBERT, Jr
Department of Obstetrics and Gynecology, School of Medicine, University of Virginia, Charlottesville, Virginia, USA

A characteristic of spontaneous activity of uterine muscle is variation in the intrinsic contractile pattern. Previous studies have demonstrated that in the nonpregnant subhuman primate this phenomenon is related, in part, to diurnal variation in average intrauterine pressure and in average contraction frequency (Harbert, Cornell & Thornton, 1970). The present study was undertaken in an attempt to delineate a diurnal pattern in the spontaneous activity of the nonpregnant human and the pregnant subhuman primate uterus.

Intrauterine pressure was recorded through an open-end, fluid-filled catheter connected by way of a pressure transducer to a recording polygraph. For studies in the human subject, the patient was kept at bedrest and light cycles were controlled to give 14 hours of light and 10 hours of darkness. For the subhuman primate studies, unanaesthetized monkeys at known stages of gestation were confined to restraining chairs and maintained in a 12-hour cycle of light and dark. In each instance, the pressure transducer was positioned to remain at the approximate level of the uterus.

Total uterine activity was derived from on-line electronic integration of the pressure signal. Average pressure was calculated by dividing total uterine activity by a selected time internal (Harbert, Cornell, Wax & Thornton, 1971). Average contraction frequency is expressed as the number of complete contraction–relaxation complexes occurring per minute. In selected animals, uterine blood flow was recorded by an electromagnetic flow probe implanted on the uterine artery. Periodic determination of zero flow was made possible by inflation of an occluder placed around the hypogastric artery. Average blood flow was calculated from data obtained by electronic integration of the flow signal. All experimental variables were monitored continuously until completion of the observations.

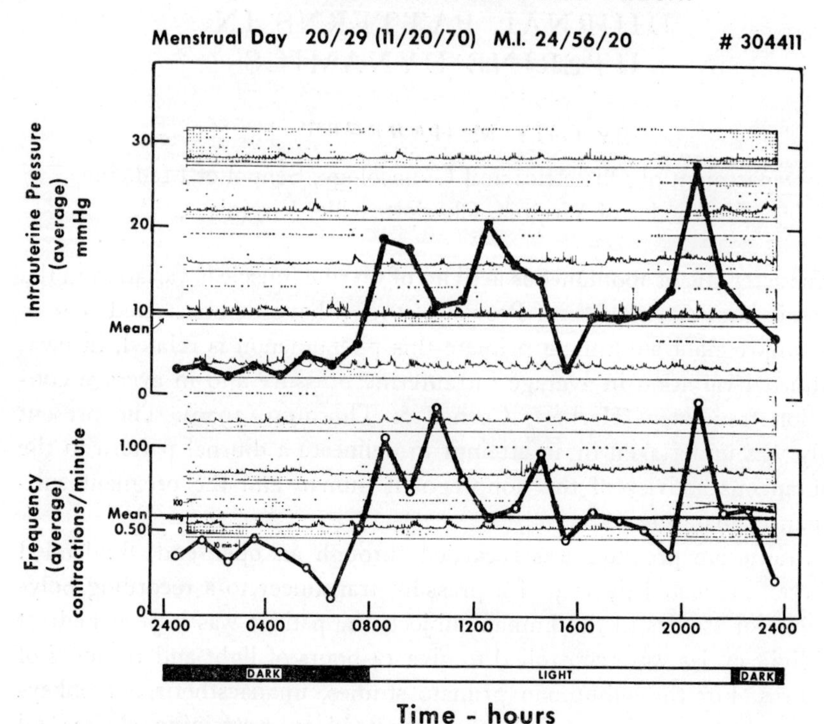

Fig. 1. Hourly averages of intrauterine pressure and contraction frequency recorded on day 20 of a 29-day menstrual cycle are plotted against a background of the eight three-hour segments of polygraph tracings from which they were calculated. The arithmetical means for each parameter for the entire 24-hour period are indicated by the heavy horizontal lines. The maturation index (MI) is shown.

NONPREGNANT HUMAN

Intrauterine pressure was measured for five days on three separate occasions in a healthy 41-year-old female. The pattern of uterine activity recorded during the secretory phase of the menstrual cycle is represented in Fig. 1. The hour-to-hour variation in average values about the mean for the day was rather marked. Average intrauterine pressure was lowest during the early morning hours. It then progressed to higher levels during the late morning, afternoon and early evening hours. The variation in average pressure values resulted from alterations in amplitude, frequency, and duration of contraction and shifts in base-line pressure.

To evaluate the differences in hourly values, the data obtained on days 20 through 24 of the menstrual cycle were subjected to the analysis of variance. The results show that the greater variability is due to differences in pressure ($F_{4,92} = 18.40$, $p < 0.001$) and frequency ($F_{4,92} = 7.13$, $p < 0.001$) on various days. A smaller but significant variability is due to differences in pressure ($F_{23,92} = 7.56$, $p < 0.001$) and frequency ($F_{23,92} = 3.11$, $p < 0.05$) between hours of the day.

Spontaneous uterine activity was measured also during the midproliferative phase of the menstrual cycle. Despite distinct differences in configuration of the individual contractile complexes and the absolute values of pressure and frequency between the two phases of the cycle there was still diurnal variation. Hourly averages were lowest during early morning hours in the period of darkness and highest during the late morning or early afternoon hours in the period of light. The analyses of variance for Fourier terms indicate that in both phases of the menstrual cycle the daily variation of intrauterine pressure and of contraction frequency takes the form of a highly significant sine curve.

Uterine motility is controlled by a variety of processes. However, the occurrence of this spontaneous variation in both proliferative and secretory phases of the menstrual cycle would suggest that it is not due to the alterations in the amounts of oestrogen and progesterone during the cycle.

Among other factors which may account for these diurnal changes are adrenal hormones and neural mechanisms. To test these factors, forty-two urine fractions collected over ten days of monitoring were analysed for content of catecholamines (Robinson & Watts, 1965). The urinary excretion rates of norepinephrine ranged from 0.84 to 2.98 μg/h while the rates for excretion of epinephrine varied from 0.002 to 0.280 μg/h. Statistical analyses indicate that the diurnal variation in urinary excretion rate of norepinephrine is significantly correlated with the average values of intrauterine pressure ($r = 0.456$) and contraction frequency ($r = 0.363$) measured during corresponding time periods. Comparisons of average pressures and frequencies with urinary excretion rate of epinephrine were not statistically significant. Although a statistical correlation between two variables known to exhibit diurnal patterns does not necessarily indicate a cause-and-effect relationship, the role of the adrenal cycle in the modulation of certain peripheral rhythms and the effect of norepinephrine and epinephrine on uterine contractility have been well established.

Another synchronizing factor noted frequently in the study of physiological rhythms has been the relationship of light and dark phases. During one of the 5-day periods of monitoring, exposure to a 24-hour period of light at the end of the first day shifted the 14 hours of light and 10 hours of dark 180 degrees. On the first day of recording, prior to the light–dark cycle change, the acrophase of the sine curves calculated for intrauterine pressure and contraction frequency occurred at 16.52 and 18.34 hours, respectively. On the fifth day, the phase angle for pressure had shifted by 127 degrees and for frequency by 97 degrees to conform more closely with the hours of light and dark. The maximum amplitude occurred at 08.22 hours for pressure and at 12.45 hours for frequency.

Corresponding relationships of urinary excretion of catecholamines and of the effect of alterations of exposure to light and dark on the diurnal pattern of spontaneous uterine activity have been demonstrated in non-pregnant subhuman primates. These observations substantiate further the similarities of reproductive physiology between the rhesus monkey and man.

PREGNANT SUBHUMAN PRIMATE

Evaluation of the presence of diurnal patterns in the dynamics of the sub-human primate uterus during pregnancy was based on measurement of uterine activity in nine monkeys for a total of 59 days. Successful measurement of uterine artery blood flow was accomplished in four of the animals for a total period of 40 days. Fig. 2 is representative of measurements obtained during late gestation. A diurnal pattern of uterine activity was present. The lowest values of intra-amniotic pressure were observed during the period of darkness and the highest values during the period of light. The variation in average pressure was due primarily to spontaneous alterations in the amplitude and duration of the contractile complexes.

During the day, average values of uterine artery blood flow varied by as much as 65%. Highest flow rates were recorded during the hours of darkness when average intra-amniotic pressure was lowest and decreased as average pressure values approached their apogee. The slope of the regression line calculated by comparison of blood flow and intra-amniotic pressure is negative and the correlation coefficient is statistically significant, indicating that the diurnal variation in uterine blood flow is inversely related as a linear function to the diurnal variation in intra-amniotic pressure. Since uterine blood flow is governed, in part, by extrinsic vascular resistance which in turn is a function of myometrial activity, this reciprocal relationship of the diurnal variation of uterine blood flow and intra-amniotic pressure was expected.

In midgestation the infrequent and low amplitude contractions resulted in less variation of uterine activity about the daily mean than observed during late gestation or in the nonpregnant uterus. However, decreases in myometrial activity during the morning hours and increases in average values during the late afternoon and evening hours still occurred. The analyses of variance indicate that the variation between the hours of each

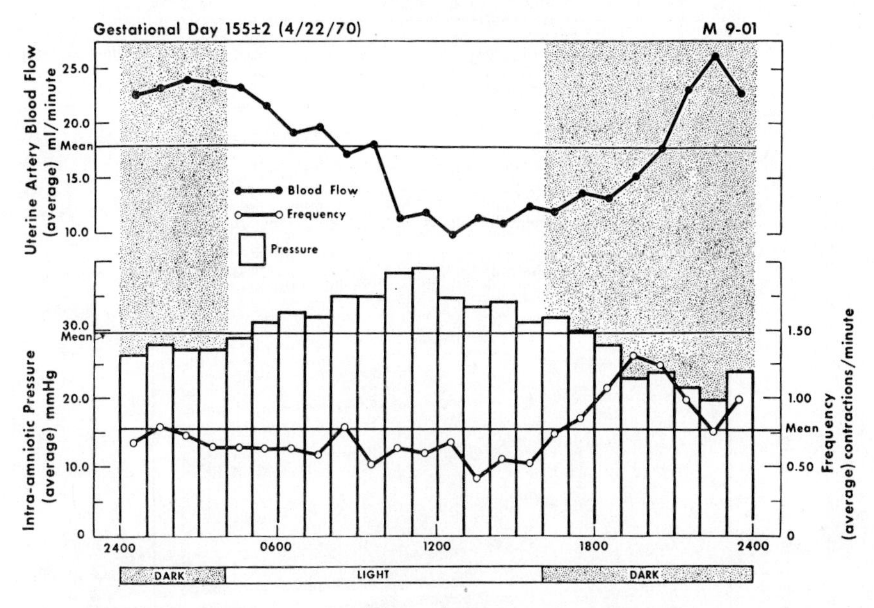

Fig. 2. Hourly average values of intra-amniotic pressure, contraction frequency, and uterine artery blood flow measured during late gestation in a rhesus monkey. The 12-hour periods of light and dark are represented by the clear and stippled background areas, respectively. Calculation of the regression line by the method of least squares ($y = 1.87 - 0.63x$) indicates that the reciprocal relationship between intra-amniotic pressure and uterine artery blood flow is statistically significant ($r = 0.665$, $p < 0.001$).

day for pressure ($F_{23,\,322} = 2.25$, $p < 0.05$) and for frequency ($F_{23,\,322} = 5.05$, $p < 0.05$) is statistically significant. Expression of these data in Fourier terms indicates that the variation of hourly values about the mean for the day conforms to a sine curve.

Measurement of uterine artery blood flow in midgestation indicated that while the mean flow rate for each day ranged from 9.4 to 21.6 ml/min significant variability due to differences in flow between hours of the day occurs ($F_{23,\,253} = 4.57$, $p < 0.05$). Even during the uterine inactivity characteristic of this period of gestation, average flow rates varied as much as 34% and were related inversely to intra-amniotic pressure values. As in the nonpregnant uterus, when the periods of exposure to light and dark were altered, the diurnal variation of uterine activity and of uterine artery blood flow tended to shift in phase.

Parturition modified the diurnal pattern of uterine activity to a two-term Fourier curve and could override the synchronizing effect of light

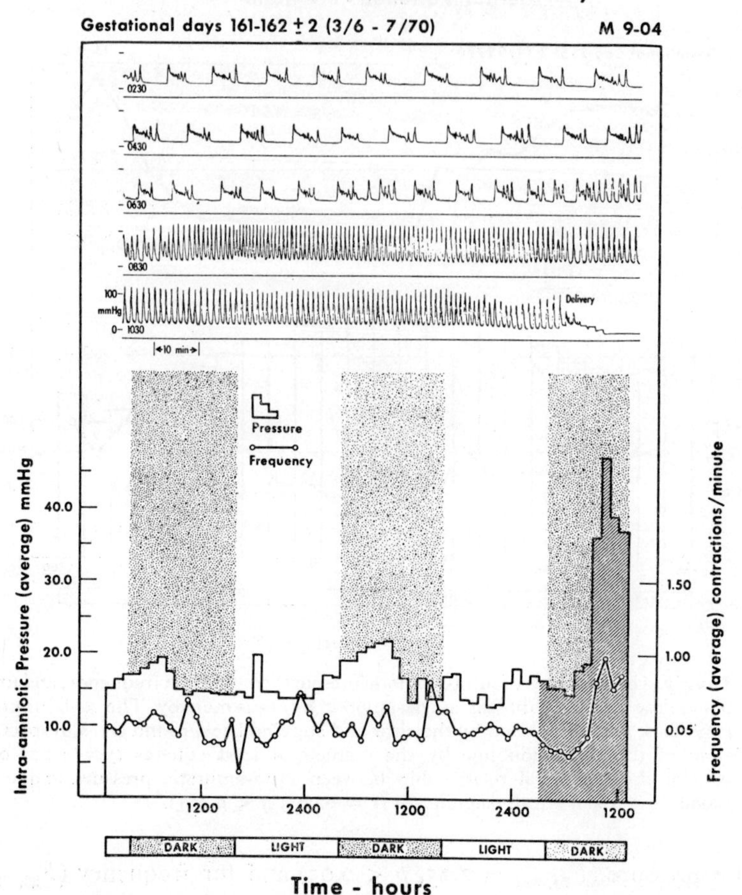

Fig. 3. Pattern of uterine activity recorded during late pregnancy, labour, and delivery. The polygraph tracings illustrate the evolution from the irregular contractions of pre-labour into the high pressure, smooth contraction pattern of labour culminating in delivery of the foetus. The bars and lines present the hour-to-hour measurements of pressure and frequency for a 50-hour period preceding and including labour. The shaded bars were calculated from the 10 hours of polygraph tracing shown.

and dark on the period of maximum uterine activity. The lowest average values of pressure and frequency often occurred during the period of light while the highest values would be observed during the period of dark. However, the hour-to-hour variation in average values of intra-amniotic pressure and contraction frequency followed a distinct and predictable sinusoidal pattern that merged into the even greater average values generating labour (Fig. 3).

Regardless of the light-to-dark relationships, a decrease in average intra-amniotic pressure was observed consistently in the hours immediately prior to the onset of labour and was accompanied by a 30 to 50% increase in uterine artery blood flow. These changes occurred during the period of evolution from the irregular contractions of prelabour into the high pressure, smooth contraction pattern of labour. As labour progressed, hourly average values of uterine artery blood flow successively decreased to levels below the mean flow rates recorded during the preceding days.

The inherent hour-to-hour variation observed in the dynamics of the primate uterus may modify placental and foetal homeostasis. This probability should be weighed in studies of placental function and parturition.

Supported in part by Research Grants HD-02798, National Institute of Child Health and Human Development and RR-104, General Clinical Research Center Programs of the Division of Research Resources, United States Public Health Service.

REFERENCES

HARBERT, G. M., Jr, CORNELL, G. W. & THORNTON, W. N., Jr (1970). *Science*, **170**, 82.

HARBERT, G. M., Jr, CORNELL, G. W., WAX, S. H. & THORNTON, W. N., Jr (1971). *Obstet. Gynec.* **37**, 487.

ROBINSON, R. L. & WATTS, D. T. (1965). *Clin. Chem.* **11**, 968.

EFFECTS OF PROGESTERONE ADMINISTRATION ON THE RATE OF UTERINE BLOOD FLOW OF EARLY PREGNANT SHEEP

BY D. CATON, R. M. ABRAMS, L. K. LACKORE, G. B. JAMES AND D. H. BARRON

Department of Obstetrics and Gynecology and Anesthesiology, University of Florida College of Medicine, Gainesville, Florida 32601

The rate of uterine blood flow (ml/kg/min) in sheep is high in early pregnancy (full term, 145–147 days). Between the 60th and 90th days of gestation, however, the rate falls from a mean of 544 to 247 ml/kg/min (Huckabee *et al.* 1972). At this same time in gestation there is a three-fold rise in levels of plasma progesterone (Bassett *et al.* 1969). The association in time of these two aspects of gestation, together with the observation that the administration of progesterone to castrate non-pregnant ewes which are receiving oestrogen is followed by a comparable decrease in the rate of uterine blood flow (D. Caton, R. Abrams, J. Clapp & D. H. Barron, unpublished data), pointed to the possibility that the fall in rate in pregnancy is the result of the increase in plasma levels of progesterone.

To test the possibility that the rate of uterine blood flow falls as a consequence of the rising levels of progesterone, this hormone was given to 26 mixed breed ewes, which were between the 30th and 55th days of gestation. Rates of uterine blood flow and of oxygen consumption were estimated by the diffusion equilibrium technique using antipyrine as the test material and by measurements of arterial and uterine vein oxygen content. Twelve ewes were studied as controls. The remaining fourteen animals received intramuscular injections of progesterone, 9–24 mg/kg body weight, on each of the two days prior to the experiment. Measurements of the rates of uterine blood flow and of oxygen consumption were made on another seven ewes which were prepared surgically for chronic studies. Measurements on these animals were made before, during and after administration of progesterone starting on the day of surgery and continuing for as long as 14 days thereafter.

In acute experiments the mean rate of uterine blood flow was 864 ml/kg/min in the control series; in ewes treated with progesterone it was 354 ml/kg/min (the difference is statistically significant at the 0.001 level of confidence). In chronic studies the mean rate of blood flow in control periods, that is before and after treatment, was 629 ml/kg/min; it was 398 ml/kg/min in the same animals while they were receiving progesterone (the difference is statistically significant at the 0.005 level of confidence).

The arteriovenous oxygen difference of the uterus appeared to be unchanged by the administration of progesterone. The mean difference of control animals in acute experiments was 0.47 mM/l; in the treated ewes it was 0.61 mM/l. In the chronic studies the mean differences were 0.60 mM/l in control periods, and 0.66 mM/l while the ewes were receiving progesterone. That is to say that the rate of oxygen consumption (ml/kg/min) of the pregnant uterus and its contents in early gestation appeared to be reduced by progesterone administration and the decrease in oxygen consumption was proportional to the change in the rate of uterine blood flow.

The data suggest that administration of progesterone to ewes early in pregnancy results in a decrease in the uterine blood flow. They are also consistent with the view that the fall in rate, which normally occurs in these animals after the 60th day of gestation, is related to, and may be the direct result of, the increase in the plasma progesterone level that takes place at or about that time.

This paper was supported in part by grants from NIH (No. 1-F3-GM-39,82401) and the Association for the Aid of Crippled Children.

REFERENCES

Bassett, J. M., Oxborrow, T. J., Smith, I. D. & Thorburn, G. D. (1969). *J. Endocr.* **45**, 449–57.

Huckabee, W. E., Crenshaw, C., Curet, L. B. & Barron, D. H. (1972). *Q. Jl Exp. Physiol.* **57**, 12–23.

EFFECT OF OESTROGENS ON PLACENTAL AND MYOMETRIAL BLOOD FLOWS IN RABBITS

By N. W. BRUCE

The Nuffield Institute for Medical Research,
University of Oxford

In non-pregnant (Huckabee, Crenshaw, Curet, Mann & Barron, 1970) and pregnant ewes (Greiss & Marston, 1965) oestrogens have been shown to increase total uterine blood flow. However the distribution of uterine blood flow between the placentas and the remainder of the uterus can vary under different physiological conditions (Duncan, 1969; LeDuc, 1972). In the rabbit, the proportion of total flow to the uterus and adnexae which is distributed to the placentas is 22% at Day 16, increasing to 69% at Day 28; term is around Day 31 (Bruce & Abdul-Karim, 1973). Thus total uterine flow can be a misleading guide to maternal placental flow. This report deals with (1) the acute effects of oestrogens on uterine and maternal placental blood flow in ovariectomized mid-term rabbits; (2) the acute effects of oestrogens in normal, late-term rabbits; and (3) the chronic effects of a lack of oestrogens in ovariectomized late-term rabbits.

New Zealand White and Chinchilla rabbits were used: the day of mating was called Day 0 of pregnancy. Blood flow measurements were made with radioactive microspheres (Duncan, 1969; Bruce & Abdul-Karim, 1973). Rabbits were anaesthetized with sodium pentobarbitone, 30 to 40 mg/kg intravenously, supplemented with ether when necessary. The left ventricle was catheterized, for microsphere injection, and the left femoral artery was catheterized to record arterial pressure. The right femoral artery was catheterized for blood withdrawal at a constant rate of 5 ml/min during microsphere injection. A known quantity of 15 μm or 25 μm diameter microspheres was injected into the left ventricle; the rabbit was then killed, dissected and its organs were measured for radioactivity. The organ radioactivities were compared with the radioactivity of the total injection dose of microspheres, to determine the proportionate distribution of cardiac output, and with the 5 ml/min arterial blood sample to determine absolute organ blood flows in ml/min.

The first experiment (Adbul-Karim & Bruce, 1972) was carried out to examine the acute effect of oestrogens in oestrogen-depleted rabbits around mid-term. Twenty-six rabbits were bilaterally ovariectomized on

Day 10 and given a progesterone supplement, 5 mg medroxy-progesterone acetate (Depoprovera, Upjohn) intramuscularly to maintain pregnancy. On Days 16 or 18 they were anaesthetized and injected intravenously with either 150 μg/kg conjugated equine oestrogens (Premarin, Ayerst) (treated groups) or its vehicle (control groups). Forty-five minutes later microspheres were injected.

Table 1. *Effects of oestrogen on the rate and distribution of uterine blood flow in ovariectomized, mid-term pregnant rabbits*

	16 Days' gestation		18 Days' gestation	
	Control	Oestrogen	Control	Oestrogen
No. in group	6	7	5	8
Total uterine flow (ml/min)	16±2	14±1	16±2	17±2
Myometrial flow* (ml/min per 100 g)	28±4	37±4	22±2	42±6
Vaginal flow† (ml/min per 100 g)	16±4	54±13	13±3	51±11
Placental flow† (ml/min per 100 g)	22±3	13±1	21±1	15±2

Mean values ± SE. Effects of treatment: * $P < 0.01$; † $P < 0.005$.

The results (Table 1) were similar in both 16- and 18-Day pregnant rabbits and were analysed to determine the specific effects of oestrogen treatment. Total uterine blood flow (ml/min) was similar in treated and control groups but the distribution of flow within the uterus differed greatly. Maternal placental blood flow (ml/min per 100 g tissue) in oestrogen treated groups was only about 65% of that in control groups ($P < 0.005$). Conversely, in oestrogen treated groups, myometrial flow was 160% ($P < 0.01$) and vaginal flow 360% ($P < 0.005$) of the comparable flows in control groups.

The second experiment was carried out to determine whether oestrogens would have a similar effect on the distribution of uterine blood flow in normal intact rabbits near term. A naturally occurring oestrogen, oestradiol 17-β, was administered intravenously in doses from 2 to 40 μg (0.5 to 10 μg/kg). Blood flow measurements were made 30 minutes later.

Total uterine blood flow in oestradiol treated rabbits was greater, but not significantly so, than in ten control rabbits from this and similar previous experiments (Table 2). However oestradiol affected the distribution of blood flow within the uterus. Myometrial and vaginal blood flows (ml/min per 100 g tissue) were significantly greater in oestradiol treated

Table 2. *Foetal weights, placental weights and the rate and distribution of uterine blood flow in normal, oestradiol-treated and ovariectomized, oestrogen-depleted 28-day pregnant rabbits*

	Normal oestradiol-treated	Normal control	Ovariectomized progesterone-supplemented
No. in group	10	10	14
Mean foetal weight (g)	34.0 ± 1.0	34.2 ± 1.6	40.6 ± 2.0*
Mean placental weight (g)	5.03 ± 0.38	5.61 ± 0.33	9.01 ± 0.43†
Total uterine flow (ml/min)	38.2 ± 3.6	32.6 ± 4.4	36.4 ± 4.6
Myometrial flow			
(ml/min per 100 g)	25.7 ± 3.0†	11.2 ± 1.2	13.5 ± 1.9
Vaginal flow			
(ml/min per 100 g)	41.5 ± 7.4*	21.3 ± 3.9	10.9 ± 2.1*
Maternal placental blood flow			
(ml/min)	2.54 ± 0.25	2.96 ± 0.37	4.39 ± 0.47*
(ml/min per 100 g)	52.1 ± 5.3	52.7 ± 6.5	49.7 ± 5.6

Mean values ± SE. Significantly different to controls: * $P < 0.05$; † $P < 0.001$.

rabbits although maternal placental blood flow was unchanged. The increase in vaginal and myometrial flows appeared to be dose dependent. The correlation coefficient for dose of oestradiol and vaginal flow was $r = 0.852$ ($P < 0.001$) and for myometrial flow, $r = 0.602$ ($P < 0.05$). Oestradiol had no apparent effect on placental blood flow at any of the dose rates examined ($r = -0.040$).

In the third experiment, the effect of a chronic lack of oestrogen was examined in fourteen 28-Day pregnant rabbits. These rabbits were bilaterally ovariectomized on Day 7 and given a progesterone supplement, 5 mg medroxy-progesterone acetate, on Day 6 to maintain pregnancy.

In the ovariectomized rabbits both placental and foetal weights were significantly increased above controls (Table 2). Total uterine blood flow was slightly but not significantly higher in ovariectomized rabbits, vaginal blood flow (ml/min per 100 g tissue) was significantly less and myometrial blood flow was not changed. The mean maternal placental blood flow (ml/min) was significantly greater in ovariectomized rabbits although, expressed in ml/min per 100 g tissue, placental flow was unchanged. The increase in placental flow in ovariectomized rabbits thus reflected the increase in placental mass. These effects may have been due to a lack of oestrogens, excessive progesterone supplementation or to an imbalance between oestrogen and progesterone. Similar increases in both foetal and placental weights following ovariectomy and progesterone supplementation were reported previously by Abdul-Karim, Nesbitt, Drucker &

Rizk (1971). The present work shows that maternal placental blood flow also is increased, so the increase in foetal weight may well be due to a greater availability of nutrients.

The three experiments have shown that both the acute administration of oestrogens and the chronic lack of oestrogens can greatly alter the distribution of blood flow within the uterus. Oestrogens significantly increased both vaginal and myometrial blood flows above controls in two experiments while oestrogen lack caused a decrease in vaginal blood flow below controls. Oestrogens reduced placental blood flow in ovariectomized rabbits at Days 16 and 18, but had little effect in normal rabbits near term. Chronic lack of oestrogens was associated with an increased maternal placental blood flow in ovariectomized rabbits near term.

The hormonal control of uterine and particularly maternal placental blood flow is far from being understood. The hormone preparation, the dose rate, route of administration, stage of gestation, preconditioning and balance of other hormones present, might all affect the outcome of experiments. However, by manipulating the hormonal status of the rabbit, placental weights and maternal placental blood flows can be altered. The effect of these changes on foetal growth and development may well prove interesting.

I acknowledge with gratitude the advice of Dr G. S. Dawes and the technical assistance of Mrs J. McCairns. This work was supported by a grant from the Medical Research Council.

REFERENCES

Abdul-Karim, R. W. & Bruce, N. W. (1972). The regulatory effect of oestrogens on fetal growth. II. Uterine and placental blood flow in rabbits. *J. Reprod. Fert.* **30**, 477.

Abdul-Karim, R. W., Nesbitt, R. E. L., Jr, Drucker, M. H. & Rizk, P. T. (1971). The regulatory effect of oestrogens on fetal growth. I. Placental and fetal body weights. *Am. J. Obstet. Gynec.* **109**, 656.

Bruce, N. W. & Abdul-Karim, R. W. (1973). Relationships between fetal weight, placental weight and maternal placental circulation in the rabbit at different stages of gestation. *J. Reprod. Fert.* **32**, 15.

Duncan, S. L. B. (1969). The partition of uterine blood flow in the pregnant rabbit. *J. Physiol., Lond.* **204**, 421.

Greiss, F. C., Jr & Marston, E. L. (1965). The uterine vascular bed: effect of estrogens during ovine pregnancy. *Am. J. Obstet. Gynec.* **93**, 720.

Huckabee, W. E., Crenshaw, C., Curet, L. B., Mann, L. & Barron, D. H. (1970). The effect of exogenous oestrogen on the blood flow and oxygen consumption of the uterus of the non-pregnant ewe. *Q. Jl Exp. Physiol.* **55**, 16.

Leduc, B. (1972). The effect of hyperventilation on maternal placental blood flow in pregnant rabbits. *J. Physiol. Lond.* **225**, 339.

MYOGLOBIN IN FOETAL MUSCLE AND ITS ROLE IN OXYGENATION

By LAWRENCE D. LONGO AND BRIAN J. KOOS

Departments of Physiology and Obstetrics and Gynecology,
School of Medicine, Loma Linda University,
Loma Linda, California 92354, USA

A number of factors may play a vital role in delivery of an adequate amount of oxygen to the developing foetus. These include: maternal and foetal arterial O_2 tensions; placental blood flows; maternal and foetal blood O_2 affinity and capacity; the diffusing capacity of the placenta; blood flows to the various foetal tissues; the relative capillarity of foetal tissues; and the amount of foetal myoglobin.

Myoglobin, a respiratory pigment in muscles, is important physiologically in man, but its role in foetal oxygenation has been relatively unexplored. In the adult, it facilitates the diffusion of O_2 in muscle cells and serves as an O_2 reservoir in these tissues. Myoglobin concentrations, [Mb], are increased as an acclimatization response to high altitudes, and in diving mammals. The question arises as to what extent myoglobin is present in the foetus, and whether it serves as an important adaptive mechanism to maintain oxygenation of foetal muscles. In an effort to understand the importance of myoglobin for foetal oxygenation, we measured [Mb] in foetal lamb cardiac, diaphragmatic and skeletal muscle and compared the values with those of the adult. We have calculated the extent to which myoglobin may facilitate the diffusion and store of oxygen. We also calculated the mean foetal myocardial intracellular oxygen tension using the competitive binding of O_2 and carbon monoxide with myoglobin.

METHODS

Muscle was obtained from the left and right ventricles, diaphragm and posterior thigh of 14 near-term foetal lambs and their mothers. Muscle samples were also obtained from several human foetuses and infants. Approximately 1.5 g of pulverized muscle was homogenized in phosphate buffer, and centrifuged. The supernatant liquid containing myoglobin was transferred to a tonometer where it was gently rotated with carbon monoxide and a small amount of dithionite converting myoglobin to carboxymyoglobin. The carboxymyoglobin solution was transferred anaerobically to a covered cuvette, and the absorbance read at 538 and 568 nanometres (millimicrons) in a Cary model 16 spectrophotometer. [Mb] was determined from the absorbance based on the molar absorptivities of carboxymyoglobin and carboxyhaemoglobin at these wavelengths. These constants were determined from standard solutions of chromatographically and electrophoretically pure sheep myoglobin and haemoglobin.

RESULTS

In sheep [Mb] averaged 0.079 (±0.018 SD) mM/kg wet weight (1.41 mg/g) in the near-term lamb heart, a value to be compared with 0.17 mM/kg in the ewe heart (Table 1). [Mb] averaged 0.046 mM/kg in lamb diaphragm compared with 0.24 mM/kg in the adult. More strikingly, skeletal muscle [Mb] was less than 0.001 mM/kg in the foetus, a scarcely detectable quantity, and far less than the adult level of 0.26 mM/kg.

Table 1. *Sheep myoglobin concentrations*

	Foetus (mM/kg)	Adult (mM/kg)
Heart	0.079 (±0.018)	0.17
Diaphragm	0.046	0.24
Skeletal muscle	< 0.001	0.26

In three human foetuses at about 24 weeks' gestation cardiac muscle [Mb] averaged 0.036 mM/kg and skeletal muscle [Mb] was less than 0.001 mM/kg. By four months after birth the cardiac [Mb] had increased to 0.11 mM/kg in two infants. Adult values of myoglobin in cardiac and skeletal muscle are 0.19 and 0.50 mM/kg, respectively (Biörck, 1949).

DISCUSSION

Comparison of present results with previous studies

While the presence of a respiratory pigment in muscle has been postulated for many years (Kölliker, 1850), Mörner first established that the pigment in muscle was a compound distinct from haemoglobin (Mörner, 1897). Myoglobin was first isolated and crystallized by Theorell in 1932, from horse myocardium (Theorell, 1932). A number of workers have reported the presence of myoglobin in various foetal muscles, but relatively little quantitative data are available and the results are contradictory. Jonxis was apparently the first to detect myoglobin in the foetus (see Millikan,

1939). He found 'considerable' amounts of myoglobin in diaphragm, but failed to find it in foetal heart muscle. Biörck (1949), to the contrary, found relatively high levels of newborn cardiac [Mb]. He also noted foetal skeletal muscles contain low [Mb], but cautioned that his values were uncertain due to the small size of his samples. Lawrie (1950) reported cardiac, diaphragmatic and skeletal muscle [Mb] was low in single measurements from the foetal horse, cow and sheep. Exact values were not given, but estimates can be made from his plots. Kagan & Christian (1966) found that skeletal muscle contained from 0.001 to 0.031 mM myoglobin per kg wet muscle weight in human foetuses from 20 to 40 weeks' gestational age, values at most only 12% of adult levels. Myoglobin has also been reported as being present, but not quantitated, by other workers in foetal calves (Jonxis & Wadman, 1952; Rossi-Fanelli *et al.* 1959) and human foetuses (Benoit *et al.* 1964; Kossman *et al.* 1964; Perkoff, 1964, 1966, 1968; Rossi-Fanelli *et al.* 1959; Schneiderman, 1962; Singer *et al.* 1955; Sudaka *et al.* 1970; Worton *et al.* 1961, 1963; Wolfson *et al.* 1967).

Although several workers reported that the myoglobin in foetal muscles was a distinct compound from that in the adult (Jonxis & Wadman, 1952; Singer *et al.* 1955; Rossi-Fanelli *et al.* 1954; Perkoff, 1966), it is now generally recognized that foetal and adult myoglobin are chemically the same (Timmer *et al.* 1957; Rossi-Fanelli, 1959; Schneiderman, 1962; Kossman *et al.* 1964; Wolfson *et al.* 1967; Perkoff, 1968).

Physiological role in foetal oxygenation

Since the pioneering studies of Hurtado (Hurtado *et al.* 1937) it has been known that [Mb] increases, and may double, as an acclimatization response to high altitude. The concentration is 20–30 times higher in some diving mammals than in terrestrial mammals (Wittenberg, 1970). Skeletal muscle [Mb] may also increase several-fold in response to exercise (Whipple, 1926; Lawrie, 1953). These observations suggest that myoglobin may serve as an O_2 store in muscle and facilitate O_2 diffusion through muscle tissues at altitude and during exercise. A direct causal effect of hypoxia or exercise on increased [Mb] is very difficult to prove, however.

The myoglobin concentration of about 0.079 mM/kg in the foetal lamb heart muscle is about 50% of the adult value. Using this value, one can make a rough calculation of the role of myoglobin in supplying O_2 to the contracting myocardial cells. This [Mb] would have an O_2 capacity of 1.83×10^{-3} ml/g. Assuming a tissue O_2 consumption of 12 ml/(kg × min) (twice resting value), the entire stored O_2 in myoglobin would be utilized 6 times per minute $[(1.83 \times 10^{-3}\ \text{ml/g})/(12 \times 10^{-3}\ \text{ml/(g} \times \text{min))}]$. Of course this assumes complete saturation of the myoglobin, which is unlikely. Mean cardiac intracellular O_2 tension in foetal lambs is about 3 mm Hg (unpublished observations). Since the partial pressure at which myoglobin is 50% saturated at 37 °C, pH 7.4 is 2.75 mm Hg (Rossi-Fanelli & Antonini, 1958), the mean saturation of cardiac myoglobin is probably closer to 50%, and the O_2 store of the term foetal lamb heart would last for 5 sec, rather than 10 sec, as calculated above. At any rate, the amount of myoglobin in foetal myocardium, while less than that in the adult, is probably an adequate O_2 store for short periods of relative hypoxia. This could be of advantage, for example, during cardiac contractions when coronary blood flow is reduced sharply during systole.

Wittenberg (1970) has shown that the facilitated flux of O_2 is proportional to the concentration of myoglobin. Since the foetal cardiac [Mb] is about 50% that of the adult, this suggests a similar reduction in the facilitation of O_2 in the foetal heart. As Wittenberg has pointed out, it might be that, since the diffusivity of molecular O_2 is much greater than the diffusivity of oxymyoglobin, consequently the flux of myoglobin bound O_2 would be negligible compared with the flux of free O_2. This is not necessarily so, however, because of the abundance of myoglobin relative to the amount of O_2 in solution. The concentration of O_2 in foetal tissue water at 37 °C and an assumed mean intracellular P_{O_2} of 3 mm Hg (see above) is 4.22×10^{-6} M. The foetal cardiac [Mb] of 8×10^{-5} M is about 19-fold greater than the O_2 concentration. The diffusivity of molecular O_2 in water at 25 °C is about 20 times that of the diffusivity of myoglobin. The abundance of myoglobin, relative to dissolved O_2, thus may compensate for its smaller diffusion coefficient and the fluxes of free and molecular bound O_2 should be of a similar order of magnitude.

[Mb] in foetal diaphragmatic muscle was only 20% of adult values. This value is only an approximation, because the amount of fibrous connective tissue in diaphragm was quite variable. The diaphragm probably contracts periodically *in utero* in association with respiratory-like movements by the foetus (Dawes *et al.* 1972).

The virtual absence of myoglobin in foetal lamb skeletal muscle is also of interest and probably reflects the small amount of work done by these muscles suspended in a gravity-free environment. This low level may also be additional evidence that foetal cells are not hypoxic *in utero*, despite the relatively low arterial O_2 tension of foetal blood. The differing levels of myoglobin in the heart, diaphragm and skeletal muscles of the foetus thus appear, as a first approximation, to parallel the activities of these muscles *in utero*.

Finally, we are measuring intracellular O_2 tensions of foetal cardiac muscle using myoglobin as an indicator. Since O_2 and CO compete for binding sites on Mb, the relative concentrations of oxymyoglobin and carboxymyoglobin indicate the P_{O_2} in that region (Coburn & Mayers, 1971). Preliminary studies in five near-term foetal lambs indicate myocardial intracellular P_{O_2} averages about 3 mm Hg, a value essentially the same as in the myocardial cells of the adult ewe.

CONCLUSIONS

We conclude that: (1) myoglobin concentrations of about 50% of adult values in foetal heart, 20% in diaphragm, but less than 1% in skeletal muscles probably reflect the relative activities of these muscles *in utero*; (2) myoglobin probably acts as a short-term O_2 store in the foetal heart. It also probably facilitates O_2 diffusion in foetal cardiac muscle and accounts for 50% of the O_2 supplied to that organ; (3) myoglobin is probably not a significant factor in oxygenation of foetal skeletal muscle despite a relatively low foetal arterial oxygen tension; (4) the mean foetal myocardial intracellular P_{O_2} is about 3 mm Hg, a value similar to that of the adult.

This study was supported by USPHS Grant HD 03807. L. D. Longo is recipient of USPHS Career Development Award 2-K4 HD 23,676.

REFERENCES

Benoit, F. L., Theil, G. G. & Watten, R. H. (1964). *Ann. Int. Med.* **61**, 1133.
Biörck, G. (1949). *Acta Medica Scandanavica* **133** (Supp. 226), 1–216.
Coburn, R. F. & Mayers, L. B. (1971). *Amer. J. Physiol.* **220**, 66–74.
Dawes, G. S., Fox, H. E., Leduc, B. M., Liggins, G. C. & Richards, R. T. (1972). *J. Physiol.* **220**, 119–43.
Hurtado, A., Rotta, A., Merino, C. & Pons, J. (1937). *Amer. J. Med. Sci.* **194**, 708–13.
Jonxis, J. H. P. & Wadman, S. K. (1952). *Nature* **169**, 884–6.
Kagen, L. J. & Christian, C. L. (1966). *Amer. J. Physiol.* **221**, 656–60.
Kölliker, A. (1850). *Microskopische Anat.* (*Leipzig*) **2**, 1.
Kossman, R. J., Fainer, D. C. & Boyer, S. H. (1964). *Cold Spring Harbor Symp. Quant. Biol.* **29**, 375–85.
Lawrie, R. A. (1953). *Nature* **171**, 1069–70.
Lawrie, R. A. (1950). *J. Agri. Sci.* **40**, 356–66.
Millikan, G. A. (1939). *Physiol. Rev.* **19**, 503–23.
Mörner, K. A. H. (1897). *Nord. Med. Archiv.* **30**, 1–8.
Perkoff, G. T. (1966). *J. Lab. Clin. Med.* **67**, 585–600.
Perkoff, G. T. (1968). *J. Lab. Clin. Med.* **71**, 610–17.
Perkoff, G. T. (1964). *New Eng. J. Med.* **270**, 263–9.
Rossi-Fanelli, A. & Antonini, E. (1958). *Arch. Biochem. Biophys.* **77**, 478–92.
Rossi-Fanelli, A., Antonini, E., Demarco, D. & Benerecetti, S. (1959). In *Biochemistry of Human Genetics*, pp. 144–50, ed. by G. E. W. Wolstenholme and C. M. O'Connor. Boston: Little, Brown Co.

SCHNEIDERMAN, L. J. (1962). *Nature* **194**, 191–2.
SINGER, K., ANGELOPOULOS, B. & RAMOT, B. (1955). *Blood* **10**, 987–98.
SUDAKA, P. & CIAUDO, J. (1970). *Rev. Europ. Etudes. Clin. Et. Biol.* **15**, 220–5.
THEORELL, A. H. T. (1932). *Biochem. Zeit.* **252**, 1–7.
TIMMER, R., VAN DER HELM, H. J. & HUISMAN, T. H. J. (1957). *Nature* **180**, 239–40.
WHIPPLE, G. H. (1926). *Amer. J. Physiol.* **76**, 693–707.
WHORTON, C. M., HUDGINS, P. C. & CONNERS, J. J. (1961). *New England J. Med.* **265**, 1242–5.
WHORTON, C. M., HUDGINS, P. C., CONNERS, J. J. & NADAS, A. S. (1963). *Southern Med.* **56**, 583–7.
WITTENBERG, J. B. (1970). *Physiol. Rev.* **50**, 559–636.
WOLFSON, R., YAKULIS, V., COLEMAN, R. D. & HELLER, P. (1967). *J. Lab. Clin. Med.* **69**, 728–36.

ESTIMATIONS OF THE UMBILICAL UPTAKE OF GLUCOSE BY THE FOETAL LAMB

By CARLYLE CRENSHAW, Jr, ROBERT CEFALO,* DAVID W. SCHOMBERG, L. B. CURET† and DONALD H. BARRON‡

The Departments of Obstetrics and Gynecology and Physiology, Duke University Medical Center, and the Department of Physiology, Yale University

In a previous report we (Crenshaw *et al.* 1968) demonstrated that the diffusion-equilibrium method, using antipyrine as the test substance, could be used to estimate the umbilical blood flow in unstressed sheep and goats. In that small series of animals we found that the umbilical flow serially estimated over a period of days could vary quite widely in a single animal, i.e. 298–491 ml/kg/min, and from animal to animal, i.e. 204–491 ml/kg/min. However, the metabolic rate of the foetus as estimated by its oxygen consumption varied very little, 7.8–8.4 ml/kg/min in a single foetus and 7.2–10.8 ml/kg/min between foetuses. Because of the stability of this foetal sheep preparation and the demonstrated ability to estimate repeatedly umbilical blood flow in a single animal without impairing the growth of the foetus, we felt that this preparation provided an excellent model for the study of foetal glucose metabolism.

The studies to be reported were begun at Yale University with Drs Barron and Curet and subsequently continued at Duke University with Drs Cefalo and Schomberg. The purpose of these studies was to determine the rate at which the sheep foetus received glucose from the mother under unstressed conditions.

MATERIALS AND METHODS

The methods of surgical preparation and estimation of umbilical flow were quite similar to those previously reported (Crenshaw *et al.* 1968). Dorset

* Present address: Department of Obstetrics and Gynecology, United States Naval Medical Center, Bethesda, Maryland.

† Present address: Department of Gynecology and Obstetrics, University of Wisconsin, Madison, Wisconsin.

‡ Present address: Department of Obstetrics and Gynecology, University of Florida, Gainesville, Florida.

and Western ewes with known breeding dates were selected for study. Although in our previous report the animals were given preoperative and postoperative progesterone in oil, intramuscularly, it was not used in these experiments because of its unknown effect upon the glucose metabolism of the sheep foetus. The animals were starved for 12–18 hours prior to surgery but allowed food and water *ad libitum* immediately postoperatively and thereafter. No preoperative sedative was used. Surgical anaesthesia was provided with either a spinal anaesthetic (10 mg Pontocaine, Winthrop), or halothane (Fluothane, Ayerst) using endotracheal tube, depending upon the capabilities of the laboratory at the time. The surgical procedures for insertion of polyvinyl catheters (0.584 mm i.d. and 0.965 mm o.d.) into the umbilical artery and vein and into the maternal iliac artery and vein were identical to those previously reported.

The animals were allowed a 48 to 72 hour postoperative recovery period prior to starting an experiment; experiments were not repeated more often than every other day. The experiments were repeated at intervals until the umbilical catheters stopped flowing, the foetus was delivered, or the foetus was sacrificed for another purpose. Only those foetuses delivered alive or those alive at sacrifice are included in this report.

Prior to each experiment the ewes were placed in a small portable stall and allowed to continue eating. The experiments were performed in the animal room in the presence of other sheep in an attempt to disturb the experimental ewe as little as possible.

Antipyrine (2 g) was infused into the maternal iliac vein over a 5 to 7 minute interval. At the start of the antipyrine infusion, blood was withdrawn from the umbilical artery and vein at a rate of 0.2 ml/min into oiled, heparinized syringes using a Harvard infusion–withdrawal pump. This withdrawal was continued for 25 minutes at which time the pump was stopped and 2 ml of blood was immediately withdrawn from the umbilical artery into a separate syringe. About halfway through the 25 minute period of withdrawal of umbilical blood, 2 ml of maternal arterial blood were withdrawn.

Immediately at the end of the experiment the concentration of glucose was estimated in triplicate in the 'integrated' umbilical arterial and the 'integrated' umbilical venous and the maternal arterial bloods by the glucose oxidase method (Ultra Micro Glucostat, Worthington Laboratories) after precipitation with zinc sulphate. The concentration of antipyrine was estimated in duplicate in the 'integrated' umbilical arterial, the 'integrated' umbilical venous, and the 'spot' umbilical arterial (T_{26}) bloods by the Autoanalyzer method (Meschia, 1964).

The umbilical blood flows were then estimated from the antipyrine concentrations of the bloods using the equation:

$$\text{Flow (ml/kg/min)} = \frac{T_{26}}{\int_{T_0}^{T_{25}} (V-A) \times 25}.$$

The umbilical uptake of glucose was estimated from the product of the umbilical veno-arterial difference of glucose and the umbilical blood flow. At the end of each experiment, the foetus was transfused with a volume of maternal arterial blood equal to the amount of umbilical blood withdrawn during the experiment. Also, the foetus was given 100,000 units of aqueous penicillin through the umbilical vein.

RESULTS

We have been successful in estimating the umbilical uptake of glucose 27 times in nine foetuses (121 to 146 days' gestational ages). The data are presented in Table 1. Only one experiment was carried out in three foetuses prior to termination while the largest number of experiments carried out in a single foetus (No. 146) over a three-week period was seven.

The mean concentration of glucose in maternal arterial whole blood was 36.5 (SE $\pm$ 0.7) mg/100 ml and 10.1 (SE $\pm$ 0.7) mg/100 ml in umbilical arterial whole blood. These concentrations are not significantly different from the means of a larger series studied for other reasons in our laboratory (unpublished data).

As in the studies of 'chronically' catheterized animals (Crenshaw *et al.* 1968) the umbilical blood flow varied widely (Fig. 1). In this study, flow varied from 111 ml/kg/min in one foetus to 624 ml/kg/min in another. However, the mean rate of umbilical blood flow in this series of foetuses, 295 (SE $\pm$ 24) ml/kg/min, was not significantly different from that of our first series, 302 ml/kg/min. The mean rate of umbilical uptake of glucose was 4.5 (SE $\pm$ 0.4) mg/kg/min. In contradistinction to the estimations of umbilical oxygen uptake in our earlier studies, the rate of glucose uptake varied considerably in a single foetus from day to day, i.e. 1.1 to 8.8 mg/kg/min, and from foetus to foetus, i.e. 0.7 to 8.8 mg/kg/min (Fig. 2). This variability of glucose uptake, when compared to oxygen uptake, is not surprising when one considers that glucose can be stored (as glycogen) and oxygen cannot, and that the foetus appears to obtain significant sources of energy from substances other than glucose (Tsoulos *et al.* 1971). There was no statistically significant correlation between the rate of umbilical

Table 1. *Chronic umbilical glucose uptake*

Animal	Gestational age (days)	Glucose concentrations MA (mg %)	UV (mg %)	UA (mg %)	V to A diff. (mg %)	Blood flow (ml/kg/min)	Glucose uptake (mg/kg/min)
6140	128	53.6	8.8	7.5	1.3	216	2.8
50	129	36.8	16.4	13.9	2.5	182	4.6
24	145	—	8.9	7.2	1.7	430	7.3
6551	134	58.6	6.4	5.8	0.6	111	0.7
	142	—	9.2	8.3	0.9	278	2.5
	144	—	19.2	16.4	2.8	146	4.1
14	121	21.1	7.7	5.9	1.8	246	4.4
	125	24.8	11.1	8.6	2.5	307	7.7
	128	36.3	17.4	15.0	2.4	265	6.4
	135	18.1	7.0	5.7	1.3	481	6.3
	140	14.4	4.8	4.3	0.5	394	2.0
96	127	28.2	11.8	10.6	1.2	624	7.5
	136	50.0	15.3	14.2	1.1	328	3.6
714	122	42.5	13.6	10.3	3.3	167	5.5
	131	43.9	12.1	10.0	2.1	124	2.6
	133	32.9	9.4	7.7	1.7	231	3.9
13	139	—	10.4	9.6	0.8	452	3.6
	141	—	15.1	14.5	0.6	298	1.8
	143	—	10.1	9.0	1.1	379	4.2
	146	—	6.8	5.8	1.0	491	4.9
146	121	38.5	20.3	18.2	2.1	261	5.5
	125	25.2	7.8	7.5	0.3	360	1.1
	129	31.9	8.6	7.5	1.1	267	2.9
	132	38.0	13.4	10.6	2.8	316	8.8
	136	43.3	16.1	13.3	2.8	289	8.1
	139	42.9	14.9	11.5	3.4	194	6.6
	142	49.8	15.0	13.2	1.8	141	2.5
Mean and standard error		36.5 ±2.7	11.8 ±0.8	10.1 ±0.7	1.7 ±0.2	295 ±24	4.5 ±0.4

blood flow and the rate of umbilical uptake of glucose. Nor was there any consistent change in uptake as gestation progressed from 121 days to term.

DISCUSSION

To our knowledge no previous estimations of the umbilical uptake of glucose in chronically catheterized sheep foetuses have been published.

Alexander and co-workers (1969) utilizing the exteriorized foetal sheep preparation found that 6 mg of glucose/kg/min was taken up by the umbilical circulation of their animals. Mann and co-workers (1970) used the antipyrine 'steady state' method for estimating flow, and recently reported that, in the acute foetal sheep preparation with the foetus *in utero*, the

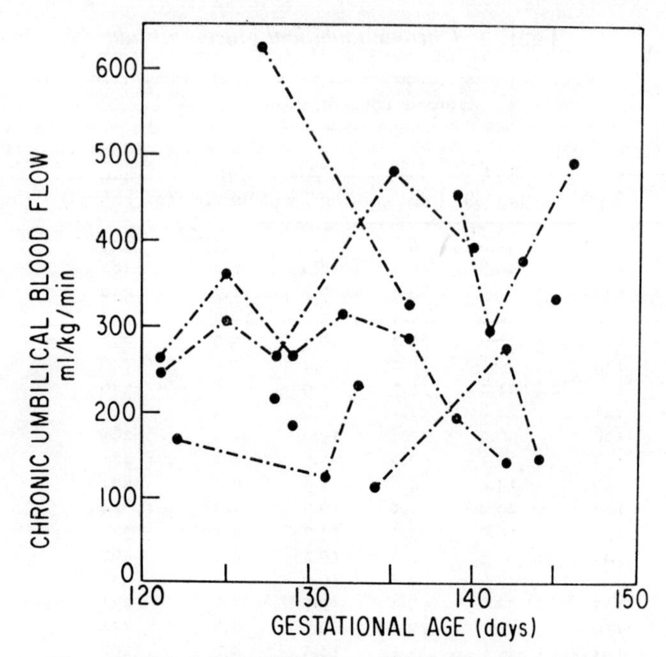

Fig. 1. Serial umbilical blood flows (ml/kg/min) in 9 foetuses at various gestational ages (121–146 days).

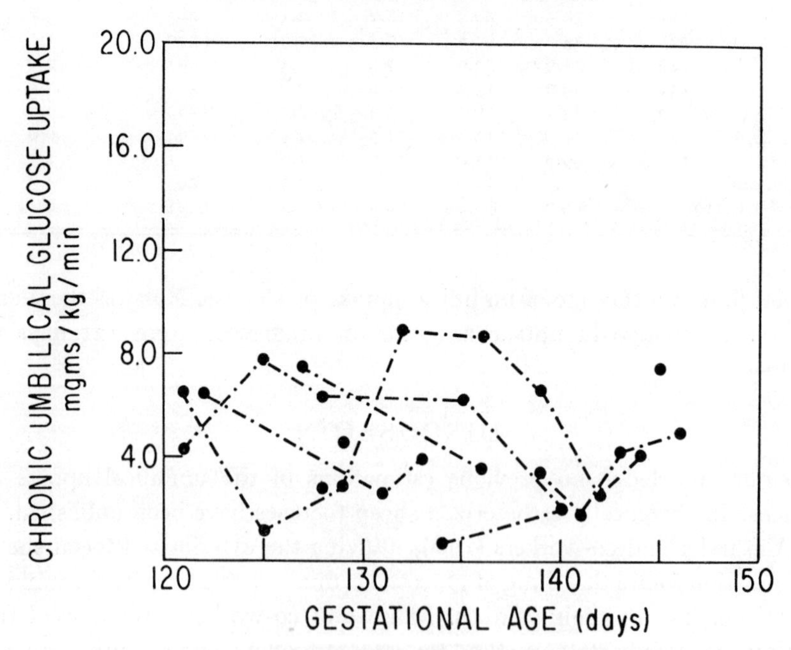

Fig. 2. Serial umbilical glucose uptakes (mg/kg/min) in 9 foetuses at various gestational ages in the unstressed, fed state.

foetal glucose consumption was 10.7 mg/kg/min. These experiments were conducted with the ewe and her foetus under the stresses of anaesthesia and surgery. Tsoulos and colleagues (1971) utilizing the chronic foetal sheep preparation compared the umbilical veno-arterial differences of glucose and oxygen and found that 48% of the oxygen was needed to metabolize the glucose to CO_2 and H_2O. They did not estimate the rate of oxygen or glucose uptake. However, if we assume that the oxygen uptake of their foetuses was similar to that of our previously reported foetuses (8.5 ml or 0.383 mM/kg/min) the glucose uptake of their foetuses would have been 5.6 mg/kg/min $\left(\frac{0.48 \times 0.383}{6} \times 180 = 5.6\right)$. In an attempt to conserve umbilical blood for other purposes we did not estimate the veno-arterial oxygen differences in our foetuses, and, consequently, cannot calculate a glucose/oxygen quotient – $\frac{6 \times \Delta \text{ glucose}}{\Delta \text{ oxygen}}$.

Jarrett and colleagues (1964) found that newborn lambs utilized glucose at a rate of 5 mg/kg/min.

The exact regulation of foetal glucose metabolism is unknown. It is assumed that it is the result of complex interactions between glucose concentration on both sides of the placenta and the foetal endocrine system. It is known that artificial elevations of glucose concentrations on the maternal side of the placenta lead to increased transfer of glucose to the foetus (Alexander *et al.* 1955). It is also known that hyperglycaemia resulting from infusions of glucose into the foetal circulation increases the foetal concentration of insulin (Davis *et al.* 1971, and personal observations), and that insulin administered to the unstressed sheep foetus increases the umbilical uptake of glucose (Colwill *et al.* 1970, and personal observations). It has been appreciated for quite some time that foetal pituitary hormones and corticosteroids are necessary for glycogen accumulation (Shelley, 1961).

Under the circumstances of our investigations, i.e. unstressed, fed ewes, there was no correlation between the foetal arterial concentrations of glucose and the umbilical uptake of glucose (Fig. 3). Also, there was no correlation between the maternal concentrations of glucose and the umbilical uptake of glucose (Fig. 4).

SUMMARY

Using the chronic umbilical catheter preparation and the diffusion–equilibrium method for estimating umbilical blood flow, serial estimations (27) of the umbilical uptake of glucose have been made in nine foetal

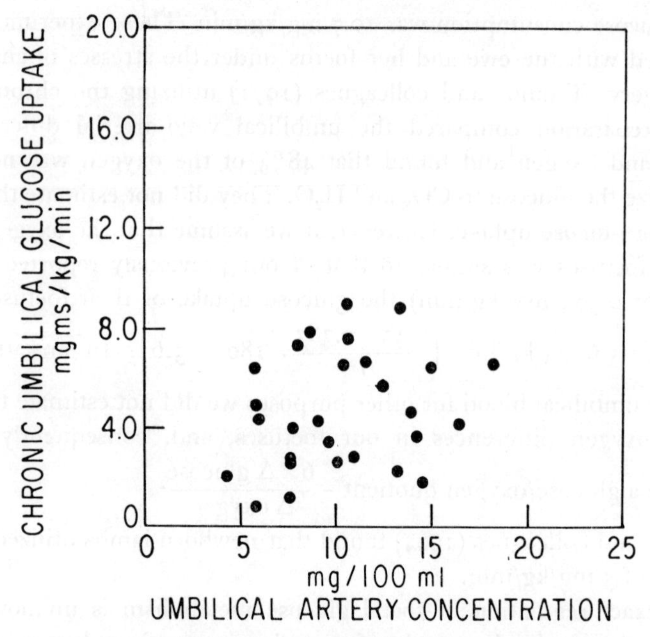

Fig. 3. Comparison of the umbilical glucose uptake and the umbilical arterial whole blood glucose concentration.

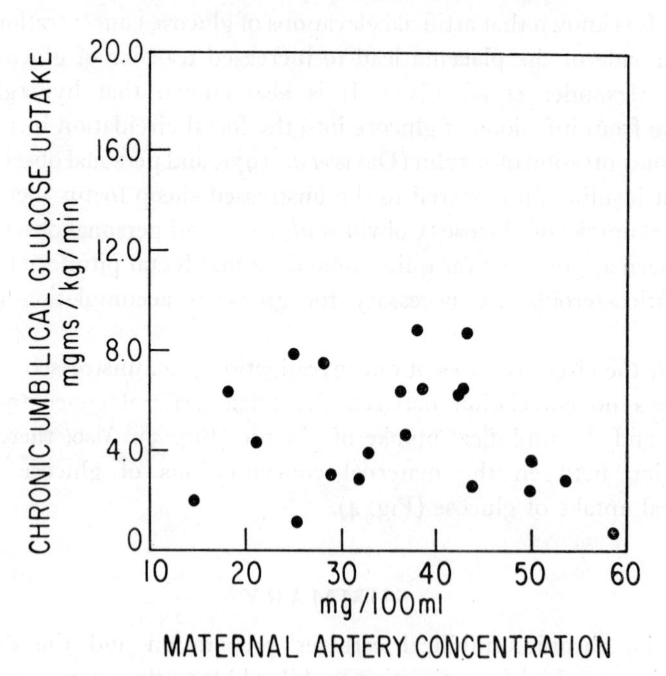

Fig. 4. Comparison of the umbilical glucose uptake and maternal arterial whole blood glucose concentration.

lambs, gestational ages 121 to 146 days. The mean umbilical blood flow was 295 ml/kg/min. The mean umbilical uptake of glucose was 4.5 mg/kg/min.

There was no statistically significant correlation between the uptake of glucose and the umbilical blood flow, the gestational age, the maternal arterial whole blood glucose concentrations or the umbilical arterial whole blood glucose concentrations.

The authors wish to thank Mr Lawrence Kodack, Research Associate, Duke University Medical Center, for his excellent technical help.

This work was supported by grants from the National Institutes of Health (HD 02300, HD 0234, HD 0628) and the Duke University Research Council.

REFERENCES

Alexander, D. P., Andrews, R. D., Huggett, A. St G., Nixon, D. A. & Widdas, W. F. (1955). The placental transfer of sugars in the sheep: studies with radioactive sugar. *J. Physiol.* **129**, 352.

Alexander, D. P., Britton, H. G., Cohen, N. M. & Nixon, D. A. (1969). Foetal metabolism. In *Foetal Autonomy*, G. E. W. Wolstenholme and M. O'Connor (Ed.), J. & A. Churchill Ltd, London, pp. 95–113.

Colwill, J. R., Davis, J. R., Meschia, G., Makowski, E. L., Beck, P. & Battaglia, F. C. (1970). Insulin-induced hypoglycemia in the ovine fetus in utero. *Endocrinology* **87**, 710.

Crenshaw, C., Huckabee, W. E., Curet, L. B., Mann, L. & Barron, D. H. (1968). A method for the estimation of the umbilical blood flow in unstressed sheep and goats with some results of its application. *Quart. J. Exp. Physiol.* **53**, 65.

Davis, J. R., Beck, P., Colwill, J. R., Makowski, E. L., Meschia, G. & Battaglia, F. C. (1971). Insulin response to fructose and glucose infusions into the sheep fetus. *Proc. Soc. Exp. Biol. Med.* **136**, 972.

Jarrett, I. G., Jones, G. B. & Potter, B. J. (1964). Changes in glucose utilization during development of the lamb. *Biochem. J.* **90**, 189.

Mann, L. I., Prichard, J. W. & Symmes, D. (1970). The effect of glucose loading on the fetal response to hypoxia. *Am. J. Obst. Gynec.* **107**, 610.

Meschia, G. (1964). Proceedings of the Technicon International Symposium on automation in analytical chemistry. Chauncey, New York: Technicon Instruments Corporation.

Shelley, H. J. (1961). Glycogen reserves and their changes at birth and in anoxia. *Brit. Med. Bull.* **17**, 137.

Tsoulos, N. G., Colwill, J. R., Battaglia, F. C., Makowski, E. L. & Meschia, G. (1971). Comparison of glucose, fructose, and O_2 uptakes by fetuses of fed and starved ewes. *Am. J. Physiol.* **221**, 254.

DIFFUSIONAL EXCHANGE BETWEEN FOETUS AND MOTHER AS A FUNCTION OF THE PHYSICAL PROPERTIES OF THE DIFFUSING MATERIALS

By J. JOB FABER
Department of Physiology,
University of Oregon Medical School,
Portland, Oregon 97201

DIFFUSION THROUGH TISSUES

Diffusion is a flux of dissolved molecules through an imaginary surface in the solvent, solely under the influence of random motion of the molecules. If diffusion is to pursue its course undisturbed, no other form of transport of particles may be superimposed on it. Such situations are rare except when the fluid in which diffusion is taking place is constrained into a motionless state by a rigid surrounding structure of small dimensions. Interstitial fluid and intracellular fluid meet this requirement reasonably well. The equation describing the diffusional flux of dissolved material through a motionless layer whose surface area is A and whose thickness is l is Fick's first law of diffusion:

$$\dot{N} = A \cdot D \cdot (C_1 - C_2)/l \text{ millimoles/second.} \quad (1)$$

This law states that the transfer rate $\dot{N}$ (millimoles per second) is proportional to the concentration difference across the layer $(C_1 - C_2)$, and to the surface area of the layer (A) and inversely proportional to its thickness (l). It is valid only if C_1 and C_2 do not change (steady state).

D, the coefficient of free diffusion, is a constant whose value depends on the properties of the fluid (in this case water), and the properties of the diffusing molecules. The coefficient of free diffusion shows how the rate of diffusional transfer of one substance compares to the rate of transport of another substance. Einstein derived a formula to predict its value, assuming that the conversion of chemical energy into heat could be calculated from Stokes' formula for viscous drag on a spherical particle (Kotyk & Janáček, 1970):

$$D = R \cdot T/(6\pi N \cdot \eta \cdot r) \text{ cm}^2\text{/sec} \quad (2)$$

(R is the gas constant, T the absolute temperature, N Avogadro's number, η the viscosity of water, and r the radius of the diffusing particles). This formula shows that the coefficient of free diffusion depends on temperature

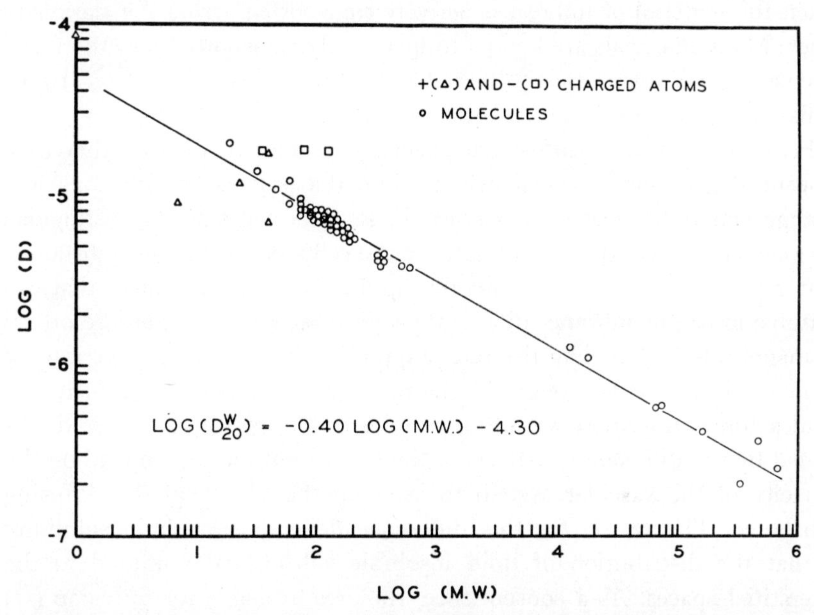

Fig. 1. Plot of 65 experimentally determined coefficients of free diffusion in water (where necessary corrected to 20 °C by use of equation 2) and molecular weights on a double logarithmic scale. Numerous points cannot be shown because of overlap. The coefficient of correlation is −0.972. The 99% confidence limits of the slope (−0.40) are −0.37 to −0.43. (Sources: Longsworth, 1954, 1957; Andrews, 1965; Atlas & Farber, 1956; Edelhoch, 1960; Keckwick, 1938; Largier, 1958; Wagner & Scheraga, 1956; Rothen, 1944; Shulman, 1953).

in two ways, directly through the absolute temperature, and indirectly through the influence of temperature on the viscosity of water. It also shows that D depends on the molecular radius (r), and hence depends in some way on molecular weight.

Because of the usefulness of knowing D, and the scarcity of published determinations of its value, we made a graph relating D to molecular weight for some 65 substances for which published data could be found (Fig. 1). The graph is on logarithmic axes in order to linearize the relationship. The least squares fit predicting D from molecular weight is:

$$D = 5.1 \cdot 10^{-5}/(\text{molecular weight}^{0.40})\ \text{cm}^2/\text{sec} \qquad (3)$$

at a temperature of 20 °C. The standard error of estimate is a factor of 1.29. To convert to 37 °C use $D_{20}^{w} = 0.649\ {}^{3}D_{37}^{w}$.

When there is an electrical potential difference across the stationary layer, the diffusion of electrically charged particles is grossly affected. Although a number of equations exist to deal with this problem (Kotyk & Janáček, 1970), equation 1 no longer applies. Another complication exists when the material of interest is actively transported across the stationary layer. Many materials are known to be actively transported into the foetal circulation (Widdas, 1961; Reynolds & Young, 1971; Young, 1971). This article will not deal with these processes.

Previous studies of diffusional exchange in animal tissues (other than placentas) give some crucial insights into the process of diffusional exchange across the placental barrier. First, it is necessary to distinguish between substances that can dissolve in the cell membranes and substances that cannot. The latter are generally lipid insoluble. Substances that can dissolve in cell membranes diffuse through tissues at rates that are orders of magnitude higher than the rates found for lipid insoluble materials of comparable molecular weight (Renkin, 1952). The rates of diffusion of lipid soluble substances are often so high that they are no longer determined by the diffusion resistance of the tissues but instead depend on the capacity of the vascular system to keep up the supply of the diffusing substance. The reason for this difference depending on lipid solubility is that the distribution of lipid insoluble substances is limited to the interstitial spaces. As a consequence, the area available for diffusion (A) in equation 1 is much less than the surface area of the tissue. The interstitial space may be very narrow indeed. In between endothelial cells, it may be narrow enough to all but preclude diffusional exchange of lipid insoluble materials. In addition, the length l is very much greater than the thickness of the tissue layers, due to the tortuosity of the interstitial spaces. There are great differences between tissues, and in tissues between their constituent layers with respect to the dimensions of the interstitial spaces, and corresponding differences in the observed permeabilities of lipid insoluble materials (Landis & Pappenheimer, 1963; Renkin, 1964; Schafer & Johnson, 1964).

Second, for lipid *in*soluble materials, there is a well established proportionality between tissue permeability and coefficient of free diffusion (Landis & Pappenheimer, 1963; Renkin, 1964). The relationship breaks down, however, at very large molecular dimensions. Obviously, when the diameter of a diffusing particle is equal to, or greater than, the diameter of the narrowest tissue space through which it has to pass, the particle can no longer penetrate. This effect becomes noticeable when the diameter of the diffusing particles is greater than 10% of the diameter of the bottlenecks in the interstitial spaces and is known as *restricted diffusion* or *steric hindrance*. In practice, the effect is comparable to that caused by a reduction in the available area for diffusion (A in equation 1); this matter is reviewed by Landis & Pappenheimer (1963) and by Renkin (1964).

Third, diffusional transfers proceed at rates that are determined by the random motions of the diffusing particles. The rate is therefore different for a free particle and for a particle that is bound to some other, usually larger, molecule. The concentrations in equation 1 are the concentrations of free, *physically dissolved*, material. Diffusion of lipid insoluble materials (and also of lipid soluble materials) can be grossly affected by their attachment to other molecules such as plasma proteins.* The diffusion of oxygen is greatly influenced by its binding to haemoglobin. Computation of the oxygen permeability is a complicated process, that must take into account that the binding between oxygen and haemoglobin cannot be described by a single equilibrium constant. It will be dealt with by Forster in this Symposium.

DIFFUSIONAL PERMEABILITY OF PLACENTAS OF DIFFERENT SPECIES

I do not think there is an organ whose diffusional characteristics are more easily measured than those of the placenta. When the concentrations in the maternal and foetal plasmas flowing in and out of the placenta are measured, the average concentration difference between maternal and foetal plasma ($\overline{\Delta C}$) can be calculated with reasonable accuracy (Fig. 2) for all but the most permeable substances. The rate of transfer can be determined by application of Fick's principle (flow times arteriovenous concentration difference) to the maternal or to the foetal circulation. The compound variable (AD/l) (ml/sec) is the ratio of the rate of transfer $\dot{N}$ and the average concentration difference $\overline{\Delta C}$. Since it is difficult and as yet uninformative to try to estimate the values of the surface area A and the path length l, we are satisfied knowing the value of (AD/l) which is called the placental permeability,† and denoted by the letter P. To compare different placentas,

* Plasma protein binding is a frequent occurrence with drugs. It is not necessary, however, for plasma binding of drugs in the maternal circulation to reduce foetal exposure to the drug, the opposite may be true. It depends on whether the effectiveness of a drug for the mother is determined by its concentration in free solution, or by its content, in the maternal plasma. And also, of course, on whether its effects on the foetus are determined by concentration in free solution or by content. If a drug exists in two forms, A and B, of which only B is protein bound in foetal and maternal plasma, then similar *contents* of A and B in the maternal plasma give a lesser concentration in free solution of B than of A. Even in that case, the advantage may be small, if placental transfer of the drug is relatively fast in comparison to foetal degradation of the drug and the duration of therapy. On the other hand, if plasma binding necessitates correspondingly higher maternal plasma contents of the drug for effective therapy, plasma binding actually increases the exposure of the foetus. The reason for this is, that according to the physics of placental transfer, plasma binding is equivalent to an increase in placental blood flow (for the drug in question) with a corresponding increase in placental transfer (Faber, 1969).

† This is a deviation from standard terminology. Usually, *permeability* is defined per unit surface area of the capillaries; in that case, the units of P are cm min^{-1}. This practice

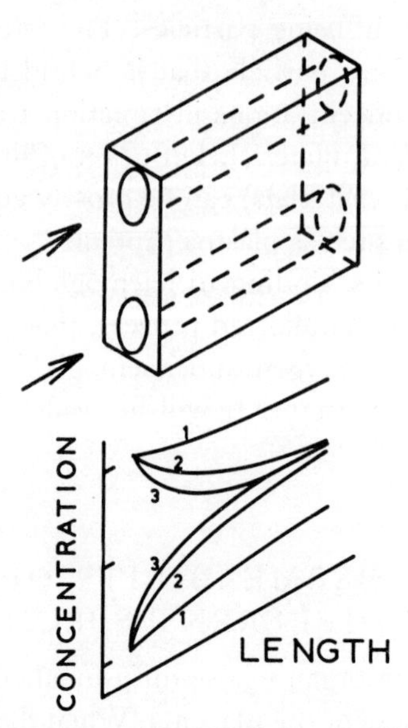

Fig. 2. A capillary pair (maternal and foetal) is shown, together with plots of the plasma concentrations, along the capillaries for fairly impermeable (1), fairly permeable (2), and very permeable (3) materials. In the case of relatively impermeable materials (1), the mean concentration difference along the capillaries ($\overline{\Delta C}$) is about the same as the concentration difference between the arterial (inflowing) bloods. Although the mean concentration difference along the capillaries is almost twice as high for a fairly permeable substance (2), as for a very permeable substance (3), the venous (outflow) concentrations are almost the same. It is difficult, therefore, to estimate the mean concentration difference from the venous concentrations, except for relatively impermeable substances, and the calculated permeabilities are correspondingly uncertain.

P is usually divided by placental weight. It is often expressed per minute rather than per second: ml/(min/g). The diffusional resistance is the inverse of the permeability.

Although the measurements outlined here are simple enough, an even simpler method can often be used. If the substance of interest is sufficiently impermeable for its concentration in the foetal blood to remain close to zero for several hours after its introduction into the maternal blood stream, the average concentration difference across the placental barrier is accurately approximated by the concentration in the maternal plasma. The rate of transfer is obtained by analysis of the foetus; the amount found in the foetus at the end is divided by the duration of the experiment.

Experiments of this kind were first done on a profitable scale by Flexner's group in 1941 (Flexner & Pohl, 1941 *a*, *b*, *c*; Pohl, Flexner & Gell-

is questionable since the area S of the capillaries is not easy to measure. Many investigators, therefore, are satisfied, knowing the ***permeability surface area product***, (PS), which is the same as our permeability P, and has the same units (ml min^{-1}).

horn, 1941; Flexner & Gellhorn, 1942; Flexner, Cowie, Hellman, Wilde & Vosburgh, 1948). They studied the permeability of radioactive sodium ions, a substance which is sufficiently impermeable to satisfy all of the theoretic requirements, which is not noticeably bound by plasma protein and which is unmistakably recognizable even if it takes part in a metabolic process after its transfer to the foetus. Its only disadvantage is that it *may* be actively transported in some placentas since there are a number of placentas that are known to have an electrical potential difference across their barrier: guinea pig and rat (Mellor, 1969), sheep and goats (Meschia, Wolkoff & Barron, 1958; Mellor, 1970). There is no positive evidence for active transport of sodium in any of these. The electrical potential will modify the diffusional rate of transport across the barrier, however, and this makes the interpretation of the results uncertain. No other substance of interest in the context of this paper has been studied in so many species before or since the experiments of Flexner.

Table 1. *Diffusional resistance of placentas of different animal species (All data collected in last trimester of pregnancy)*

		Diffusional resistance of one gram of placenta in last trimester, in min/ml for		
Species	Type of placenta	Cl^-	Na^+	Urea
Man	Haemo*mono*chorial	—	30 (2)	—
Rhesus	Haemo*mono*chorial	26 (3)	40 (3)	15 (3)
Guinea pig	Haemo*mono*chorial	—	33 (1)	—
Rabbit	Haemo*di*chorial	—	29 (1)	19 (6)
		23 (5)	46 (5)	18 (5)
Rat	Haemo*tri*chorial	—	24 (1)	—
Cat	Endotheliochorial	—	290 (1)	—
Goat	Syndesmochorial	—	500 (1)	—
Sheep	Syndesmochorial	≫ 65 (3)	≫ 65 (3)	7 (4)
Pig	Epitheliochorial	—	7700 (1)	—

≫ Much greater than. — Insufficient information.
(1) Flexner *et al.* (1942). (2) Flexner *et al.* (1948). (3) Battaglia *et al.* (1968). (4) Meschia *et al.* (1967). (5) Our own work. (6) Štulc *et al.* (1969).

Table 1 summarizes the results. All values are expressed per gram placenta, and to make comparison with the numbers of layers of these placentas easier, the results are given as resistances ($1/P$) instead of permeabilities (P). Other measurements are shown in addition to Flexner's, some made with chloride ion, and some with urea.

It is at once apparent that the diffusional resistances to sodium ion increase with the number of layers in the placental barrier although the

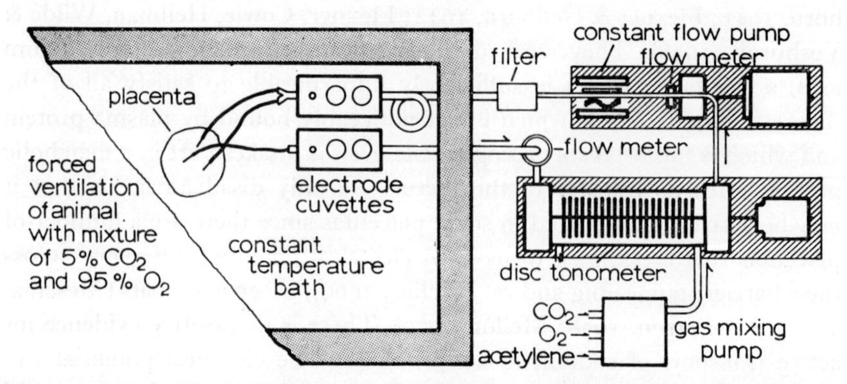

Fig. 3. Schematic diagram of the artificially perfused rabbit placenta. The anaesthetized doe is submersed with open abdomen in 39 °C Ringer's solution. The foetal side of the in-situ placenta is perfused with rabbit blood of controlled gas concentration. (From Faber & Hart, 1966. By permission of the American Heart Association, Inc.

haemochorial group is fairly homogeneous regardless of the number of chorionic layers. It is also apparent that the diffusion resistance of urea in the sheep is an anomaly in this table; nevertheless we feel that its value is firmly established.

In view of the values obtained in Table 1, it appeared worthwhile to study an animal with a placenta consisting of a limited number of layers and without a transplacental potential (Wright, 1966) and to use a large range of molecular weights. The rabbit was in all respects the ideal candidate. A source of variability was removed by limiting the investigation to gestational ages of 27 to 29 days (term is 30 to 31 days).

LOCALIZATION OF THE DIFFUSION RESISTANCE IN THE THREE HISTOLOGICAL LAYERS OF THE RABBIT PLACENTA

In a rabbit, it is relatively easy to open the uterus, to cannulate the two umbilical arteries and the one umbilical vein and to perfuse the foetal side of the placenta from a miniature 'heart lung machine' after removal of the foetus (Fig. 3). It is necessary to keep the uterus stretched underneath the placenta to prevent placental separation. But if bad preparations are discarded, the ones that remain show normal oxygen uptakes and pressure-flow curves when perfused with rabbit blood; they were therefore considered to be representative of intact placentas (Faber & Hart, 1966).

To study placental transfer, a radioactive substance is added to the reservoir or injected into the maternal blood stream. With very impermeable substances, best accuracy is achieved with the latter alternative and by not *re*circulating the foetal perfusate. The mean concentration difference across the placental barrier is practically equal to the difference in concentrations between the maternal and 'foetal' arterial bloods since the arterio-venous concentration differences are only a few percent of the arterial concentration difference, at least for lipid insoluble substances (Fig. 2). It is not necessary, therefore, to take into account that the rabbit placenta is a counter-current exchanger (Mossman, 1926) in the computation of the mean concentration difference across the barrier ($\overline{\Delta C}$) except when very permeable substances are concerned (Faber & Hart, 1966; Faber, 1969). The rate of transfer $\dot{N}$ is calculated from the gain, or loss, from the 'foetal' circulation.

Table 2. *Permeability of the rabbit placenta. Values given are millimoles per minute transferred per gram placenta per millimole per millilitre mean concentration difference between foetal and maternal plasma*

Substance	D_{20}^{W} (cm²/sec)	Permeability ± 1 SEM (ml/(min·g))	
Acetylene	$1.2 \cdot 10^{-5}$†	100*	
Oxygen	$1.2 \cdot 10^{-5}$†	100*	
Water	$2.0 \cdot 10^{-5}$	(THO) 0.8*	±0.16
Urea	$1.2 \cdot 10^{-5}$	0.054	±0.0065
Chloride ion	$1.8 \cdot 10^{-5}$	0.044	±0.0079
Sodium ion	$1.2 \cdot 10^{-5}$	0.022	±0.0055
Inulin	$1.4 \cdot 10^{-6}$†	0.0016	±0.00024
Albumin	$5.8 \cdot 10^{-7}$	0.00065	±0.00010

* Order of magnitude estimate.
† Estimated from molecular weight and Fig. 1.
D_{20}^{W} is the coefficient of free diffusion in water at 20 °C.
From Faber & Hart (1966, 1967) and some further experiments (Faber, 1970).

The results are shown in Table 2. It is clear that, in accord with our expectations, lipid soluble substances are very much more permeable than lipid insoluble substances. In fact it is not possible to assess accurately the permeabilities of the former since minute errors in the measurements of venous concentrations would cause order of magnitude errors in the computed permeabilities (Faber, 1969). Lipid insoluble substances show permeabilities that decrease with increasing molecular weights. The one unexpected finding is that plasma albumin is permeable in proportion to its coefficient of free diffusion in water, whereas in other tissues plasma albumin is already too large to fit loosely in the bottlenecks of the interstitial spaces (Pappenheimer, Renkin & Borrero, 1951; Grotte,

1956; Renkin, 1964; Schafer & Johnson, 1964; Garlick & Renkin, 1970).

The rabbit placenta consists of three histological layers, all of foetal origin (Larsen, 1962; Enders, 1965). The foetal blood is contained in a foetal endothelium (Plate 1). Around these vessels, there are two basement membranes of undetermined material. The outer basement membrane is loosely apposed to a thin monocellular layer of cytotrophoblast, and the cytotrophoblast is covered by a syncytiotrophoblast. The syncytiotrophoblast forms folds of capillary dimensions that contain the maternal blood. Where it is close to a foetal capillary, the syncytiotrophoblast is very thin and punctuated with clear cut holes of some 500 Å in diameter (Plate 1). Knowledge of the individual contributions made to the total resistance to diffusion by each of these layers could be a first step in an interspecies analysis of placental transfer characteristics.

Diffusion resistance can be proven to obey the law of electrical resistance; the resistance of a series is equal to the sum of the individual resistances. Martín de Julián & Yudilevich (1964) made a straightforward analysis with a reasonable minimum of assumptions that showed that the diffusion resistance of the first layer confronting a blood stream can be calculated from results obtained with a multiple indicator dilution experiment. This opened the door to the determination of the diffusion resistances of the endothelium of the foetal circulation and the syncytiotrophoblast of the maternal circulation in the rabbit placenta. The resistance of the middle layer could then be obtained by subtraction of the resistances of the outer layers from the total resistance of the barrier.

The principle of the double indicator dilution method is shown in Fig. 4. The method consists of a sudden injection of a mixture of two tracers into the arterial circulation. One of the tracers must be a reference tracer which is essentially impermeable in the wall of the capillaries. A labelled form of plasma albumin is often used for this purpose. The reference tracer appears in the outflow in concentrations that reflect the distribution of transit times in the vascular bed of the tested organ (Meier & Zierler, 1954). These tracer particles are depicted as heavy black dots in Fig. 4. In contrast, some of the permeable tracer particles diffuse out of the capillary into the tissue spaces behind the capillary endothelium (fine dots in Fig. 4). After the main body of tracer material has passed through the capillary, the concentration of permeable tracer in the interstitial spaces exceeds that in the capillary plasma and the direction of the diffusional process reverses inwards (lower schematic of Fig. 4). The result is that the outflow of permeable tracer is delayed in comparison with the outflow of the reference tracer. The concentration time-curves recorded in such experiments allow the computation of the permeability of the endothelial

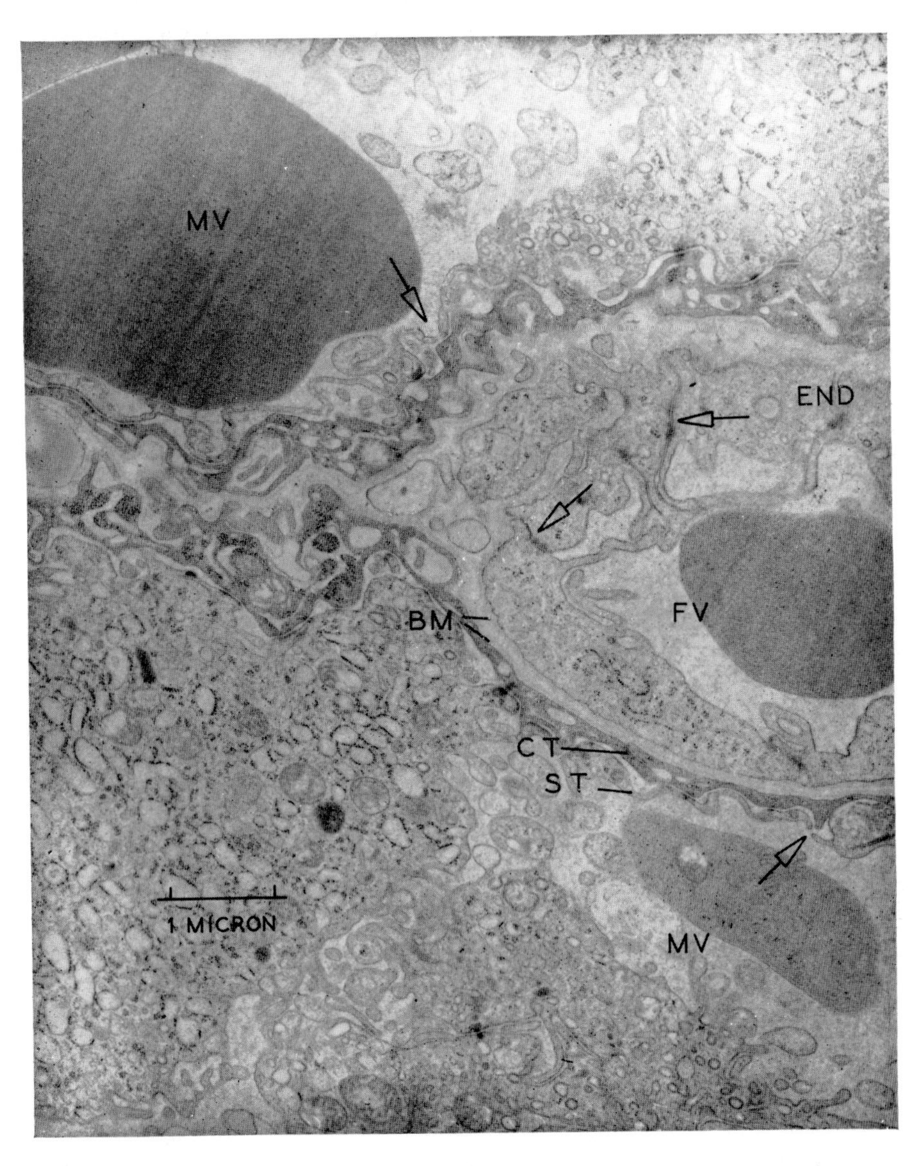

Electronmicrograph of a 28 day rabbit placenta. BM are the two basement membranes, FV and MV are foetal and maternal vessels respectively, all show parts of red blood cells. END is the endothelium of the foetal capillary, CT and ST are the cyto- and syncytio-trophoblasts respectively. Arrows indicate openings between endothelial cells and 'holes' in the syncytiotrophoblast. The fine stipples, many in the foetal capillary and some in between the basement membranes, are ferritin molecules injected into the foetal circulation 30 minutes before fixation. (Brooks, Faber, Green & Thornburg, unpublished experiments.)

(*Facing p.* 314)

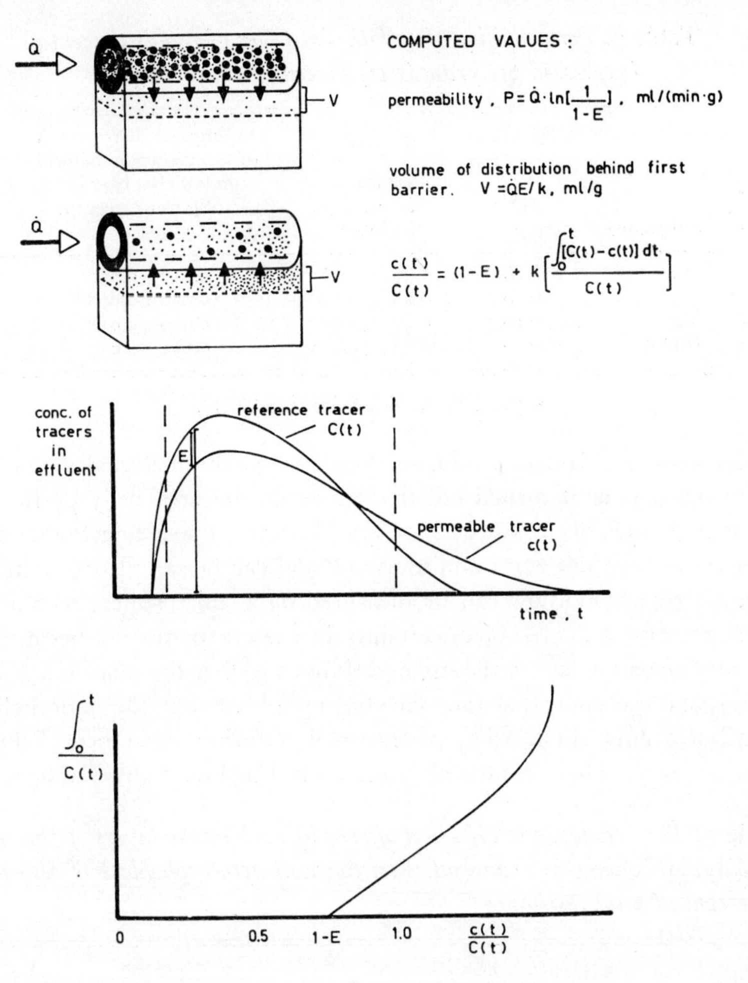

Fig. 4. Principle of the double tracer dilution method for estimating permeability of endothelial layer. See text. Lowest graph illustrates method for determining volume of distribution of tracer behind first barrier (Martin de Julián & Yudilevich, 1964).

wall, provided that certain fairly liberal requirements are met (Levitt, 1970).

The results of double tracer dilution experiments on the perfused foetal side of the rabbit placenta show that the endothelium contributes only about 10% of the total resistance to diffusion of the placental barrier (Faber, Hart & Poutala, 1968). The permeability of the endothelial layer is proportional to the coefficient of free diffusion of the permeating molecule as is expected, for it would be inconsistent to find steric hindrance in one layer of the placenta and not in the whole barrier (Table 3).

Table 3. *Permeability of endothelial layer of rabbit placenta (expressed per gram (wet) placental tissue weight)*

Substance	D_{20}^{W} cm²/sec	Permeability (ml/(min ·g)) ± SEM	Volume of distribution of test molecule behind endothelial barrier* (ml/g wet placenta) ± SEM
Urea	$1.2 \cdot 10^{-5}$	0.36 ± 0.06	0.117 ± 0.057
Cl^-	$1.8 \cdot 10^{-5}$	0.28 ± 0.02	0.030 ± 0.008
Na^+	$1.2 \cdot 10^{-5}$	0.42 ± 0.08	0.050 ± 0.018
Inulin	$1.4 \cdot 10^{-6}$	0.053 ± 0.007	0.094 ± 0.045

* See Faber, Hart & Poutala (1968).

The same experiment could not be done on the maternal side of the rabbit placenta as it turned out that we could not artificially perfuse the rabbit uterus (Faber & Stearns, 1969). However, tracer injections can be made into the rabbit aorta and uterine blood can be sampled in a uterine vein. Uterine blood flow can be measured by a 'microsphere technique'. There remains a degree of uncertainty in these experiments because the uterine circulation serves the myometrium as well as the placenta. Nevertheless, one can show that the syncytiotrophoblast, like the endothelium, contributes only about 10% of the total diffusion resistance (Faber & Stearns, 1969). These results are summarized in Fig. 5 and Table 4.

Table 4. *The resistance to diffusion offered by each of the layers of the rabbit placenta. All values in g·min/ml. Resistance of cytotrophoblast is also given as percent of total resistance*

Substance	Whole barrier (Mean ± SEM)	Endothelium (Mean ± SEM)	Syncytiotrophoblast (Lower and upper limit)	Cytotrophoblast (By subtraction)
Water	1.36 ± 0.29	< 0.6 ± 0.014	—	—
Urea	18.5 ± 2.2	2.8 ± 0.45	3.3–10.0	10 (54%)
Chloride ion	22.7 ± 4.1	3.6 ± 0.26	2.0– 3.3	17 (75%)
Sodium ion	45.4 ± 11.3	2.4 ± 0.46	4.0– 5.0	39 (86%)
Inulin	625.0 ± 94.0	19.0 ± 2.50	20.0–25.0	584 (93%)
Albumin	1540.0 ± 230.0	—	—	—

From Faber (1970).

We are led to the conclusion that the diffusion resistance of the rabbit placenta is localized mainly in the middle layers. It is believed (Karnovsky, 1968) that basement membranes are coarse filters that offer little resistance

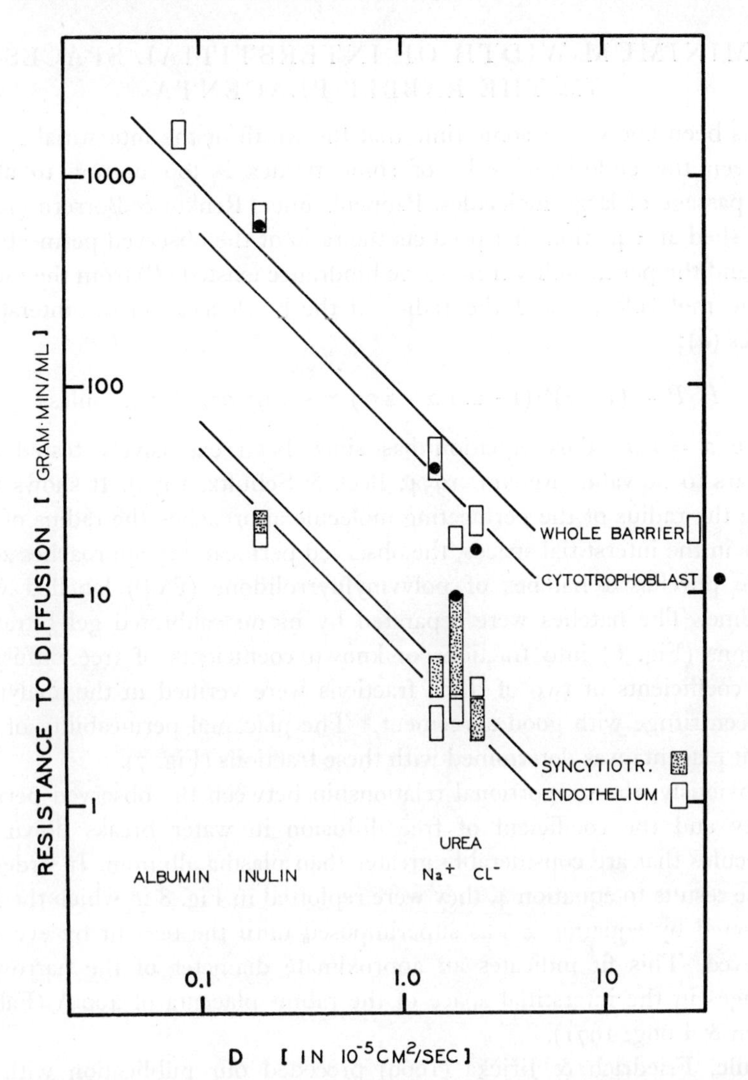

Fig. 5. Resistance to diffusion (the inverse of the permeability) plotted on double logar i th mic scales as a function of the coefficient of free diffusion in water (20 °C) for each of the three layers of the hemo*di*chorial rabbit placenta. Note the approximate 45° slopes. Courtesy *Pflügers Archiv für die gesamte Physiologie.*)

to particles of molecular dimensions. It may be that the resistance of the rabbit placenta is in the junctions between the cells of the cytotrophoblast. These junctions are hard to identify in electronmicrographs (Plate 1). It is also possible that the resistance resides in the long paths in between cell layers since the 'pores' in these layers do not appear to be lined up (Plate 1). I do not know the answer to this important puestion.

MINIMUM WIDTH OF INTERSTITIAL SPACES IN THE RABBIT PLACENTA

It has been known for some time that the width of the interstitial spaces between the endothelial cells of some tissues is too narrow to allow free passage of large molecules. Pappenheimer, Renkin & Borrero (1951) published an equation that predicts the ratio of the observed permeability (P') and the permeability if no steric hindrance existed (P) from the radius of the molecule (r) and the radius of the bottlenecks in the interstitial spaces (a):

$$P'/P = (1-\alpha)^2 \cdot (1-2.1\,\alpha+2.09\,\alpha^3-0.95\,\alpha^5) \text{ dimensionless} \quad (4)$$

where $\alpha = r/a$. This equation has since been extensively tested and appears to be valid (Renkin, 1954; Beck & Schultz, 1970). It shows that when the radius of the permeating molecule approaches the radius of the pores in the interstitial spaces, the observed permeability approaches zero.

We purchased batches of polyvinylpyrrolidone (PVP) labelled with ^{125}iodine. The batches were separated by us on calibrated gel filtration columns (Fig. 6) into fractions of known coefficients of free diffusion. The coefficients of two of these fractions were verified in the analytical ultracentrifuge with good agreement.* The placental permeability of the rabbit placenta was determined with these fractions (Fig. 7).

Obviously, the proportional relationship between the observed permeability and the coefficient of free diffusion in water breaks down for molecules that are considerably greater than plasma albumin. In order to fit the results to equation 4, they were replotted in Fig. 8 in which the line predicted by equation 4 was superimposed until the best fit by eye was achieved. This fit indicates an approximate diameter of the narrowest passages in the interstitial space of the rabbit placenta of 400 Å (Faber, Green & Long, 1971).

Štulc, Friedrich & Jiřička (1969) preceded our publication with an estimate of pore radii in rabbit placentas, based on perfusions of the foetal circulation with plasma. There are differences between the results obtained by them and the results shown in Table 2 that cannot be explained by the statistical uncertainty of random variation. For inulin these investigators found a permeability three times larger, and for plasma albumin a permeability three times less, than the ones found by us. Štulc and his co-workers fitted their data to a system containing two populations of pores, one of 6 Å and one of 200 Å diameter. The diameter of the first

* Dr D. A. Rigas of the Department of Biochemistry kindly made these determinations for us.

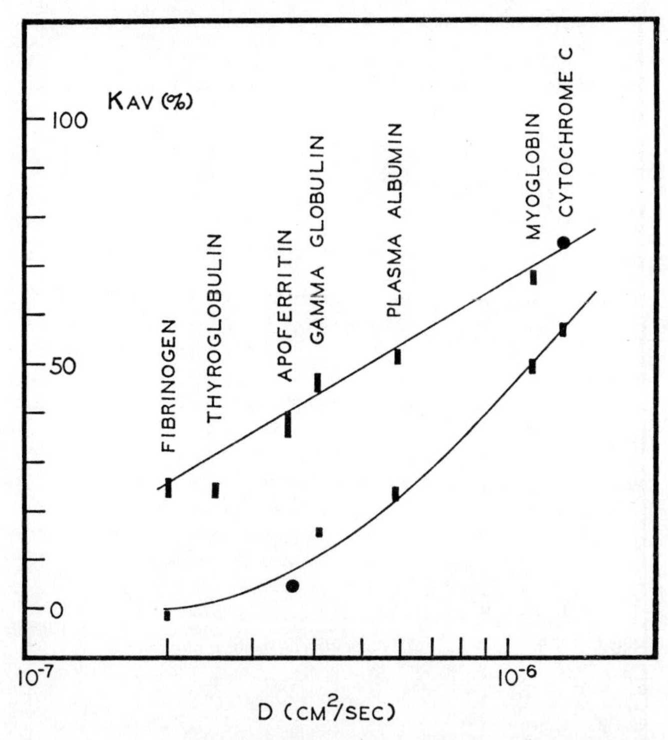

Fig. 6. Calibration graph of gel filtration columns. (From Faber, Green & Long, 1971. Courtesy *American Journal of Physiology*.)

population was calculated from the placental permeability to water. Whether water traverses a pore system or dissolves into the cell membrane (like oxygen or acetylene) is not clear, however (Table 2), and until the existence of pores of about 6 Å can be proven by use of some other small molecules, I feel that their existence is uncertain. Solomon (1968) discusses some candidates (ethylene glycol, glycerol, and erythritol) for such jobs. The results with polyvinylpyrrolidone are inconsistent with a pore diameter of only 200 Å. These differences should not be allowed to obscure the agreement of Štulc's and our studies in pointing to a pore diameter that is substantially greater than the 70–80 Å found by others for the pores, or slits, in capillary endothelium of muscle (Landis & Pappenheimer, 1963; Renkin, 1964; Perl, 1971), and is more like the 250–300 Å found for foetal pulmonary capillaries (Boyd, Hill, Humphreys, Normand, Reynolds & Strang, 1969).

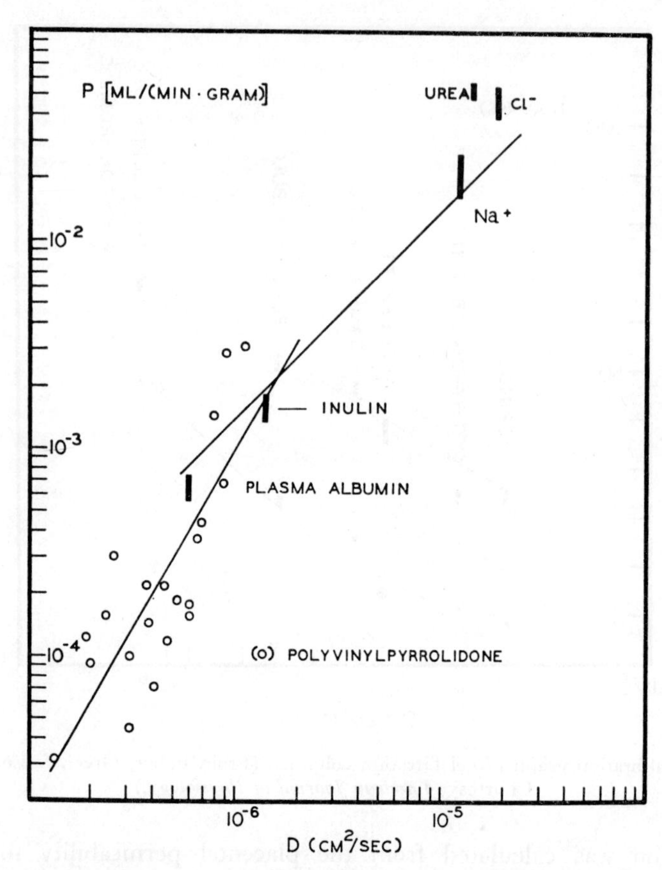

Fig. 7. Permeability of the rabbit placenta (all layers) as a function of the coefficients of free diffusion (D_{20}^{W}). Mean values ±S.E.M. for all substances except polyvinylpyrrolidone for which individual data are shown. Note logarithmic scales. Line fitted to PVP data is a least squares fit. (From Faber, Green & Long, 1971. Courtesy *American Journal of Physiology*.)

DIRECT OBSERVATION OF TRANSPLACENTAL TRANSMISSION

Karnovsky and Palade (Majno, 1965) have shown that transmembrane travel of particles can be visualized in electron micrographs by the use of particles large enough to be recognizable at high magnification, or by the use of particles whose reaction products with some stain are electron opaque. Tillack (1966) publishes electron micrographs of the rat placenta showing the transfer of ferritin from the maternal to the foetal circulation and also in the opposite direction. Ferritin is a large iron-containing molecule of 240,000 to 480,000 molecular weight, depending on iron content, and a molecular diameter of about 120 Å (Haggis, 1965). Tillack (1966)

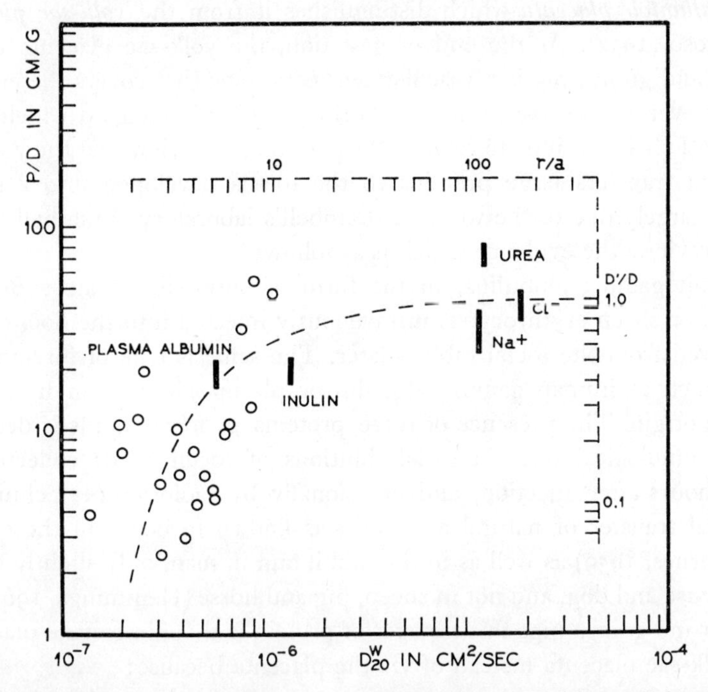

Fig. 8. Same data as shown in Fig. 7, but permeabilities divided by coefficients of free diffusion. If no steric hindrance exists, the ratio P/D should be the same for all substances. It clearly diminishes for large substances (small values of D^{w}_{20}). Dotted line is obtained by use of equation 4 and fitted by eye to the data. (From Faber, Green & Long, 1971. Courtesy *American Journal of Physiology*.)

concludes that transfer from foetus to mother is easier than transfer in the reverse direction, a conclusion I do not find convincing after considering the evidence. The study has nevertheless the great merit of showing that, at least in this haemo*tri*chorial placenta, the passages must be more than 120 Å wide. Plate 1 shows a rabbit placenta fixed about half an hour after the injection of ferritin into the intact foetal circulation. We believe that we see ferritin in the basement membranes outside the foetal endothelium, but this will need a great deal of better work.

THE YOLK-SAC PLACENTA

I tacitly assumed that whenever a particle injected into the maternal blood stream is recovered in the foetus it must have crossed 'the' placenta. However, in rabbits, guinea pigs, rats, and mice, but not in man, 'the' placenta is not the only connection between foetal and maternal circulations. For this reason, anatomists refer to it more explicitly as the *chorioallantoic placenta* which distinguishes it from the *yolk-sac placenta* (Amoroso, 1952). At the end of gestation, the yolk-sac placenta of the rabbit and guinea pig is a vascularized membrane that covers the embryo almost everywhere except at the chorioallantoic placenta, to which from now on I shall continue to refer as the placenta. The view that the yolk-sac placenta transfers large proteins to the foetus developed into a sound theory largely due to the work in Brambell's laboratory. Brambell (1958, 1970) reviews the evidence which is as follows.

Rabbit gamma globulins, in the form of antibodies against *Brucella abortus* or sheep erythrocytes, intravenously injected into the doe, can be recovered from the foetal rabbits later. The same is true of foreign proteins such as human gamma globulin or plasma albumin of human or bovine origin. The presence of these proteins in foetal fluids is detected by immunological tests in serial dilutions of foetal fluids collected 24 or 48 hours after injection, and occasionally by radioisotope techniques. Prenatal transfer of natural antibodies is known to occur in the guinea pig (Barnes, 1959) as well as in the rabbit and in man, only slightly in the rat, mouse and dog, and not in sheep, pig and horse (Hemmings, 1961). In rabbits and guinea pigs this transfer of protein is thought to take place via the yolk-sac placenta instead of via the placenta because:

1. Protein in the maternal circulation can also be detected in the intrauterine fluid of pregnant rabbits (Kulangara & Schechtman, 1962);
2. Immunologically recognizable protein injected into the lumen of the pregnant rabbit uterus is partially recovered from foetuses in the injected horn, but not from the maternal circulation and to a lesser degree only from foetuses in the non-injected horn (Hemmings, 1958; Barnes, 1959);
3. After intrauterine injection, injected protein can be detected in intact foetuses, but only to a lesser degree in foetuses whose yolk-sac circulations were tied off by previous careful, and properly controlled, surgical intervention (Hemmings, 1956, 1961; Barnes, 1959);
4. Foreign protein injected intravenously into the doe can be detected 24 hours later in the foetuses, but not if the yolk-sac circulation had been tied off (Kulangara & Schechtman, 1962; Barnes, 1959);
5. Transfer of intrauterine protein to the rabbit foetus is selective on a basis other than molecular weight (Hartley, 1951; Brambell, Hemmings, Henderson & Rowlands, 1950; Brambell, Hemmings, Oakley & Porter, 1960);
6. The uptake of injected intrauterine gamma globulin by the yolk-sac can be visualized by immunofluorescence (Slade & Wild, 1971);
7. And finally, protein uptake from the uterine lumen by the yolk-sac can be demonstrated also in electronmicrographs (King & Enders, 1970; Slade, 1970).

Those who hold the view that the yolk-sac transfers protein to the foetal rabbit and guinea pig stand with confidence on a position of strength. If this position is interpreted to mean that large molecules, or at least large proteins, are mainly transported from the maternal blood stream to the foetal blood stream via the yolk-sac placenta, it would be in direct conflict with results obtained by us on several counts.

Fig. 7 shows the (apparent?) permeability of the rabbit placenta to human plasma albumin. If this molecule is transported by an active (transcellular) process in the yolk-sac placenta, instead of by diffusion through the placenta, one is faced with the coincidence that the activity of the process in the yolk-sac placenta is almost exactly of the right magnitude to place the apparent permeability of albumin where it would be expected to be in Fig. 7, on the basis of the coefficient of free diffusion of this molecule. Moreover, the same coincidence would exist for polyvinylpyrrolidone (Fig. 7), unless one assumes that albumin is transferred by the yolk-sac placenta but polyvinylpyrrolidone is not. In that case, however, one cannot understand why passages in the placenta that are wide enough to permit diffusion of polyvinylpyrrolidone do not permit diffusion of the much smaller albumin molecule. If serum albumin diffuses through the placenta, the contribution made to its transfer by the yolk-sac placenta cannot but be small (Fig. 7).

We will deal here only with transport between the maternal blood stream and the foetal blood stream since the maternal blood stream is the source of all materials taken up by the foetus no matter what route they follow. The experiments that consist of intrauterine injections cannot prove that transport by the placenta does not exist, or even that it is not the main avenue of supply since the placenta does not have an opportunity to demonstrate its capabilities in these experiments. Such experiments do demonstrate, of course, the ability of the yolk-sac placenta to transport materials into the foetal circulation.

There are at present enough reasons to accept that diffusion of gamma globulin through the placenta, if it occurs, is supplemented by a flow of this material by way of the yolk-sac placenta. For the materials tested in our experiments (human plasma albumin and polyvinylpyrrolidone) the main route appears to be diffusion through the placenta. Evidence in conflict with this view is published by Kulangara & Schechtman (1962) who show that after tying off the yolk-sac circulation, transfer of human albumin and some other foreign proteins no longer occurs in the rabbit foetus. These investigators do not specify whether their foetuses survived the operation, which they often do not (Brambell, Hemmings & Henderson, 1951). Fig. 9 shows a plot of placental permeability (P) as a function of

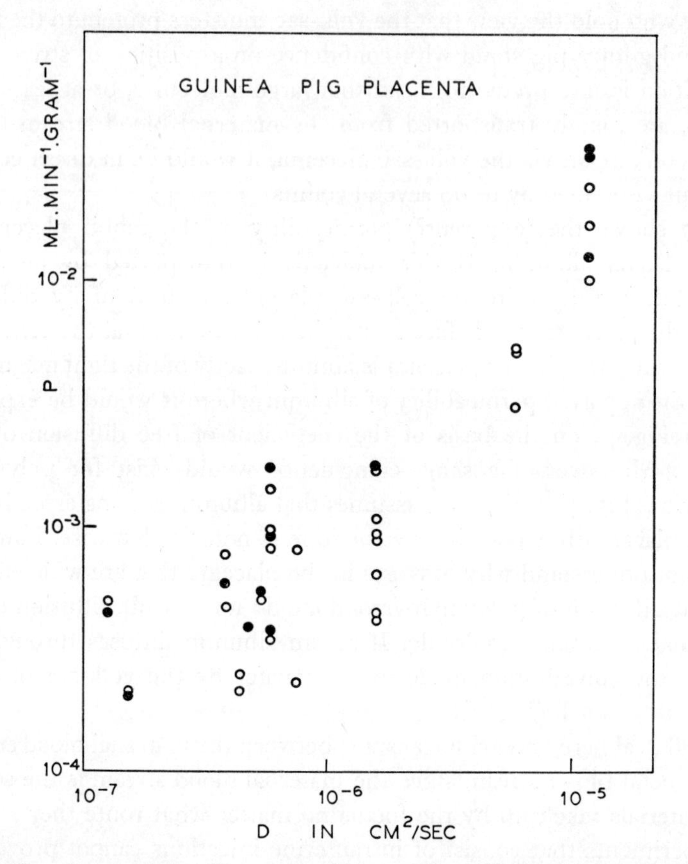

Fig. 9. Plot of permeability P versus coefficient of free diffusion in water (D^{w}_{20}) for data obtained on guinea pig placentas. Black dots indicate data from experiments where the yolk-sac circulation had been tied off on the day preceding the experiment. (Faber, Green, Gault & Thornburg, unpublished data.)

coefficient of free diffusion (D) in the guinea pig. The plot is analogous to Fig. 7 for the rabbit. The black dots are from foetuses whose yolk-sac circulation was tied off a day before the experiment. We injected Na-ion, mannitol, inulin, human plasma albumin, or polyvinylpyrrolidone intravenously into the sows and analysed only foetuses who were alive at the end of the experiment (respiring or kicking).

There is nothing in Fig. 9 to suggest that tying off the yolk-sac circulation interferes with transfer of these materials from mother to foetus.

CONCLUSIONS

Lipid soluble materials diffuse rapidly across the placenta (e.g. Kayden, Dancis & Money, 1969; Hershfield & Nemeth, 1968). Their rate of transfer is not so much determined by their permeability in the placental barrier as by the rate of their supply and removal by the maternal and foetal blood flows in the placenta.

The transfer of lipid insoluble materials depends on their molecular size and on the histological characteristics of the placenta. Transfer of lipid insoluble materials is proportional to the coefficient of free diffusion in water. However, very large molecules may not be able to diffuse across at all, due to 'bottlenecks' in the interstitial spaces of the placenta. In the rabbit and the guinea pig near term, molecules larger than plasma albumin can still diffuse across the (chorioallantoic) placenta, the 'bottlenecks' of the rabbit placenta are estimated to be 400 Å wide. The sheep placenta, on the other hand, is very tight.

Most of the diffusion resistance of the rabbit placenta appears to be localized in between, and not in, the outer layers, the endothelial layer and the syncytiotrophoblast. Whether it is localized in the cytotrophoblast junctions or in the interlaminar fluid spaces between layers is not known.

The contribution made to the transfer of plasma albumin of human origin by the yolk-sac placenta of the rabbit and the guinea pig appears to be small, and the same is true for polyvinylpyrrolidone. The yolk-sac placenta, however, may transfer significant amounts of plasma gamma globulin.

In all diffusional transfer, the rate of transport is proportional to the mean concentration difference between the maternal and foetal plasmas confronting the barrier. The relevant concentrations are the concentrations of physically dissolved material, not the contents. *At a given content*, plasma protein binding of the diffusing material decreases the concentrations of physically dissolved material and decreases the rate of transport. For *a given concentration in free solution*, plasma protein binding increases the rate of placental transfer.

I write for a laboratory, the names of its members appear in the list of references, and the work would not have been possible without them. We thank an embarrassingly large number of colleagues who helped us, without naming them individually. Almost all of our own work was done with a grant entitled '*Oxygen transfer through the placenta*'. We express our appreciation to the officials of the National Institute of Child Health and Human Development who administered this grant with patience and discretion.

The author is an Established Investigator of the American Heart Association. This work was supported by grants from the National Institute of Child Health and Human Development, HD 2313 and HD 6689.

REFERENCES

AMOROSO, E. C. (1952). In *Marshall's Physiology of Reproduction*, ed. Parkes, A. S. Longman, London.

ANDREWS, P. (1965). *Biochem. J.* **96**, 595–606.

ATLAS, S. M. & FARBER, E. (1956). *J. biol. Chem.* **219**, 31–7.

BARNES, J. M. (1959). *J. Path. Bact.* **77**, 371–80.

BATTAGLIA, F. C., BEHRMAN, R. E., MESCHIA, G., SEEDS, A. E. & BRUNS, P. (1968). *Am. J. Obst. Gynec.* **102**, 1135–43.

BECK, R. E. & SCHULTZ, J. S. (1970). *Science, N.Y.* **170**, 1302–5.

BOYD, R. D. H., HILL, J. R., HUMPHREYS, P. W., NORMAND, I. C. S., REYNOLDS, E. O. R. & STRANG, L. B. (1969). *J. Physiol.* **201**, 567–88.

BRAMBELL, F. W. R. (1958). *Biol. Rev.* **33**, 488–531.

BRAMBELL, F. W. R. (1970). In *Frontiers of Biology*, Vol. 18, North Holland, Amsterdam.

BRAMBELL, F. W. R., HEMMINGS, W. A. & HENDERSON, M. (1951). *Antibodies and Embryos*. Athlone Press, London.

BRAMBELL, F. W. R., HEMMINGS, W. A., HENDERSON, M. & ROWLANDS, W. T. (1950). *Proc. R. Soc. Ser. B* **137**, 239–52.

BRAMBELL, F. W. R., HEMMINGS, W. A., OAKLEY, C. L. & PORTER, R. R. (1960). *Proc. R. Soc. Ser. B* **151**, 478–82.

EDELHOCH, H. (1960). *J. biol. Chem.* **235**, 1326–34.

ENDERS, A. C. (1965). *Am. J. Anat.* **116**, 29–68.

FABER, J. J. (1969). *Circulation Res.* **24**, 221–34.

FABER, J. J. (1970). In *Capillary Permeability*. Eds. Crone, C. & Lassen, N. A. Munksgaard, Copenhagen.

FABER, J. J. & HART, F. M. (1966). *Circulation Res.* **19**, 816–33.

FABER, J. J., GREEN, T. J. & LONG, L. R. (1971). *Am. J. Physiol.* **220**, 688–93.

FABER, J. J. & HART, F. M. (1967). *Am. J. Physiol.* **213**, 890–4.

FABER, J. J., HART, F. M. & POUTALA, A. C. (1968). *J. Physiol.* **197**, 381–93.

FABER, J. J. & STEARNS, R. S. (1969). *Pflügers Arch. ges. Physiol.* **310**, 337–53.

FLEXNER, L. B., COWIE, D. B., HELLMAN, M. L. M., WILDE, W. S. & VOSBURGH, G. J. (1948). *Am. J. Obstet. Gynec.* **55**, 469–80.

FLEXNER, L. B. & GELLHORN, A. (1942). *Am. J. Obst. Gynec.* **43**, 965–74.

FLEXNER, L. B. & POHL, H. A. (1941*a*). *Am. J. Physiol.* **134**, 344–9.

FLEXNER, L. B. & POHL, H. A. (1941*b*). *J. Cell. Comp. Physiol.* **18**, 49–59.

FLEXNER, L. B. & POHL, H. A. (1941*c*). *Am. J. Physiol.* **132**, 594–606.

GARLICK, D. G. & RENKIN, E. M. (1970). *Am. J. Physiol.* **219**, 1595–1605.

GROTTE, G. (1956). *Acta chir. Scand.* **111**, 419–20.

HAGGIS, G. H. (1965). *J. mol. Biol.* **14**, 598–602.

HARTLEY, P. (1951). *Proc. R. Soc., Ser. B* **138**, 499–513.

HEMMINGS, W. A. (1956). *Proc. R. Soc., Ser. B* **145**, 186–95.

HEMMINGS, W. A. (1958). *Proc. R. Soc., Ser. B* **148**, 76–83.

HEMMINGS, W. A. (1961). *Brit. med. Bull.* **17**, 96–101.

HERSHFIELD, M. S. & NEMETH, A. M. (1968). *J. Lipid Res.* **4**, 460–8.

KARNOVSKY, M. J. (1968). *J. gen. Physiol.* **52**, 64–95.

KAYDEN, H. J., DANCIS, J. & MONEY, W. L. (1969). *Am. J. Obst. Gynec.* **104**, 564–72.

KECKWICK, R. A. (1938). *Biochem. J.* **32**, 552–60.

KING, B. F. & ENDERS, A. C. (1970). *Am. J. Anat.* **129**, 261–88.

KOTYK, A. & JANÁČEK, K. (1970). *Cell Membrane Transport*. Plenum Press, New York & London.

KULANGARA, A. C. & SCHECHTMAN, A. M. (1962). *Am. J. Physiol.* **203**, 1071–80.

Landis, E. M. & Pappenheimer, J. R. (1963). In *Handbook of Physiology*, Vol. 2, *Circulation II*. Ed. Hamilton, W. F. American Physiological Society, Washington, D.C.
Largier, J. F. A. (1958). *Arch. Biochem. Biophys.* **77**, 350–67.
Larsen, F. J. (1962). *J. Ultrastruct. Res.* **7**, 535–49.
Levitt, D. G. (1970). *Circulation Res.* **27**, 81–95.
Longsworth, L. G. (1954). *J. Phys. Chem.* **58**, 770–2.
Longsworth, L. G. (1957). American Institute of Physics Handbook. Ed. Gray, D. E. McGraw-Hill, New York.
Majno, G. (1965). In *Handbook of Physiology*, Vol. 2, *Circulation III*. Ed. Hamilton, W. F. American Physiological Society, Washington, D.C.
Martín de Julián, P. & Yudilevich, D. (1964). *Am. J. Physiol.* **207**, 162–8.
Meier, P. & Zierler, K. L. (1954). *J. appl. Physiol.* **6**, 731–44.
Mellor, D. J. (1969). *J. Physiol.* **204**, 395–405.
Mellor, D. J. (1970). *J. Physiol.* **207**, 133–50.
Meschia, G., Breathnach, C., Cotter, J. R., Hellegers, A. & Barron, D. H. (1965). *Quart. J. exp. Physiol.* **50**, 23–41.
Meschia, G., Wolkoff, A. S. & Barron, D. H. (1958). *Proc. Nat. Acad. Sci.* **44**, 483–9.
Mossman, H. W. (1926). *Am. J. Anat.* **37**, 433–97.
Pappenheimer, J. R., Renkin, E. M. & Borrero, L. M. (1951). *Am. J. Physiol.* **167**, 13–46.
Perl, W. (1971). *Microvascular res.* **3**, 233–51.
Pohl, H. A., Flexner, L. B. & Gellhorn, A. (1941). *Am. J. Physiol.* **134**, 338–43.
Renkin, E. M. (1952). *Am. J. Physiol.* **168**, 538–45.
Renkin, E. M. (1954). *J. gen. Physiol.* **38**, 225–43.
Renkin, E. M. (1964). *The Physiologist* **7**, 13–27.
Reynolds, M. L. & Young, M. (1971). *J. Physiol.* **214**, 583–97.
Rothen, E. (1944). *J. biol. Chem.* **152**, 679–93.
Schafer, D. E. & Johnson, J. A. (1964). *Am. J. Physiol.* **206**, 985–91.
Shulman, S. (1953). *J. Am. chem. Soc.* **75**, 5846–52.
Slade, B. S. (1970). *J. Anat.* **107**, 531–45.
Slade, B. S. & Wilde, A. E. (1971). *Immunology* **20**, 217–23.
Solomon, A. K. (1968). *J. gen. Physiol.* **51**, 335–64.
Štulc, J., Friedrich, R. & Jiřička, Z. (1969). *Life Sci.* **8**, 167–80.
Tillack, T. W. (1966). *Lab. Invest.* **15**, 896–909.
Wagner, M. L. & Scheraga, H. A. (1956). *J. Phys. Chem.* **60**, 1066–76.
Widdas, W. F. (1961). *Brit. med. Bull.* **17**, 107–11.
Wright, G. H. (1966). *Biol. Neonat.* **10**, 193–9.
Young, M. (1971). In: *Metabolic Processes in the Foetus and Newborn Infant*. Eds. Jonxis, J. H. P., Visser, H. K. A. & Troelstra, J. A. Kroese, Leiden.

SESSION 4

METABOLISM

FREE AMINO ACID TRANSFER ACROSS THE PLACENTAL MEMBRANE

By MAUREEN YOUNG AND PENNY M. M. HILL
Department of Gynaecology,
St Thomas's Hospital Medical School,
London, S.E.1

In Joseph Barcroft's chapter 'Crossing the Placental Barrier' in *Researches on Pre-Natal Life* his first concern was for the daily net increment of some of the elements by the foetus. His calculations for the 115-day foetal lamb showed, to his surprise, that water took pride of place over oxygen for the quantity transferred from mother to foetus at this age. About 3.3 Eq. (60 g) of water are laid down by the foetus each day, while only 0.93 Eq. (15 l) of oxygen is used. The rate of accumulation of many other elements can now be added to his short list and characterise the net transfer capacity of different placentas (Widdowson, 1968). Blaxter (1964) has reviewed the evidence for the rate of accumulation of nitrogen and the recent studies of Sykes & Field (1972) also show that the amount of nitrogen in the foetal lamb is directly related to the ewe's protein intake. The 0.11 Eq. nitrogen (6 g protein) laid down in 24 hours, in the foetal lamb, represents the placental transfer of 50 μmole N min^{-1} kg, in comparison with 14 mmole O_2 min^{-1} kg.

It is of interest that the relative amounts of the individual amino acids are very similar in the carcasses of the foetal, neonatal and adult rat (Williams *et al.* 1954 and Southgate, 1971) and similar to that in the pig and in the chick. Amino acid requirements are, therefore, the same at different ages and amongst the species. The mechanisms for transferring amino acids across the foetal membrane and for protein synthesis are also likely to be similar at all ages and in different species; but in the foetus and young animal the overall rate of protein synthesis must be faster and correspond with the growth rate. There is no evidence to suggest that nitrogen reaches the foetus other than by the transfer of free amino acid (Dancis & Shafran, 1958). Similarity between the transport mechanisms of amino acids in the foetal placenta and adult tissues has been shown by

Christensen & Streicher (1948), using the characteristics of selective transfer of the L-isomers, competitive inhibition between members of the same transport groups and saturation of the transport mechanisms. Some recent observations on the mechanisms which ensure the accumulation of nitrogen by the foetus will be described, and the particular characteristics of the free amino acid pattern in the foetal and maternal plasmas discussed in the light of recent evidence obtained concerning their metabolism in the non-pregnant adult.

EXPERIMENTAL PREPARATIONS

The placenta of the foetal guinea pig perfused *in situ* has proved particularly useful for studying the control of transfer on the foetal side of the placental membrane (Money & Dancis, 1960); the perfusion fluid contains a low molecular weight dextran, physiological salts and glucose together with amino acids, in concentrations similar to those occurring in foetal plasma. Changes in the maternal blood flow through the placenta are followed by measuring the transfer of antipyrine from the maternal circulation into the perfusate; with a good blood flow the concentration in the perfusate should be only slightly below that of the maternal plasma. The placental membrane is capable of active transport in this preparation, for following the maternal injection of a mixture of L- and D-leucine, the ratio of the L/D isomers in the perfusate rises four or five times above the ratio injected (Reynolds & Young, 1971). Studies have also been made in ewes, 125–140 days pregnant, under spinal anaesthesia with catheters in the maternal carotid artery, a large branch of one uterine vein and a main umbilical artery and vein. Some observations were also made in chronic preparations in which the foetal jugular vein had been catheterised for one to two weeks.

The individual amino acids were separated by Moore & Stein's method, using a single-column gradient elution technique for ion-exchange chromatography, with estimation by the ninhydrin colour reaction in a Technicon Amino Acid Analyser.

INFLUENCE OF MATERNAL BLOOD FLOW AND PLASMA AMINO CONCENTRATION

The free amino acid differences between the umbilical vein and artery are normally small (Fig. 1) as they are across other tissues, but the supply to the foetus must be relatively large because of the high blood flow per unit weight of placenta. The effects of a fall in the maternal blood flow through the placenta have been studied in the guinea pig by bleeding the mother.

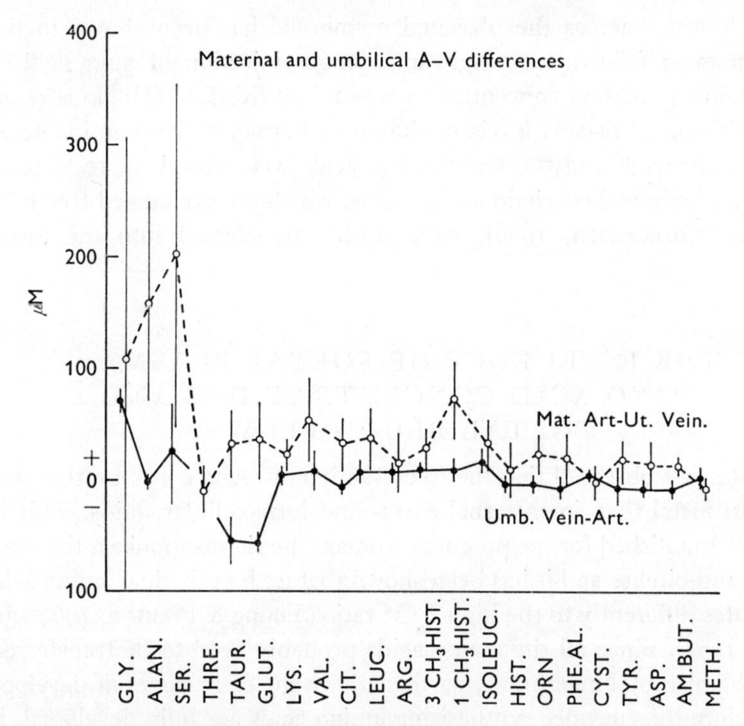

Fig. 1. Maternal arterio–venous and foetal umbilical vein–artery differences (±SE) across the placenta in the sheep.

Each is small, except for glycine, alanine and serine on the maternal side, but demonstrates placental uptake from the mother and transfer to the foetus. The umbilical V–A differences for taurine and glutamate suggest transfer from the foetus to the mother. (Drawn from Hopkins, McFadyen & Young, 1971.)

The transfer of α-amino nitrogen decreased as the placental flow fell, but a reduction below 50% of that of the control level was not observed, even when maternal arterial pressure was only 30 mm Hg. This indicates a considerable placental reserve and may explain why infants with placental blood flows of only 40% of normal may be born alive and survive. This is also indicated in a different type of experiment, related both to placental flows and transport capacity. When a single injection of an unmetabolisable amino acid, α-aminoisobutyric acid (AIB) was given intravenously to an unanaesthetised pregnant guinea pig, the animal sacrificed after 30 min and the AIB extracted with water from the foetuses and maternal organs, the analysis showed that if the litter contained foetuses of even weight about 8% of the injected AIB was present in each. In litters with conceptuses of uneven weight, however, the runt, half the size of its litter mate, sometimes contained only 2% of the injected AIB (Young, 1969).

The relative unimportance of the maternal plasma level of free amino acid on transfer across the placental membrane has been shown in the pregnant ewe; following a single intravenous injection of 5 mg kg^{-1} of amino acid the plasma concentration was raised five-fold (Hopkins *et al.* 1971). The foetal plasma levels of the neutral straight-chain amino acids were not changed, and the basic amino acids were raised by 10% temporarily. The branched-chain amino acids, which are exchanged freely by diffusion (Christensen, 1969), were readily transferred into the foetal plasma.

THE INFLUENCE OF FOETAL PLASMA AMINO ACID CONCENTRATIONS AND FOETAL BLOOD FLOW

Van Slyke first observed that the free α-amino nitrogen concentration was higher in foetal than in maternal extra- and intra-cellular fluids. This is now well established for the pregnant woman, the rhesus monkey, the ewe, rodents and bitches and it has been shown that each individual amino acid contributes differently to the high F:M ratio (Young & Prenton, 1969 and Young, 1971). Since all the amino acids probably need to be transferred across the placental membrane, particularly in the early stages of development before the enzymes synthesising amino acids are fully developed, it was important to prove that the free amino acids in the maternal plasma could be transferred against the high F:M concentration gradients which exist across the placental membrane; evidence derived from the small umbilical vein–artery differences was inadequate. Fig. 2 shows that in the foetal placenta of the guinea pig perfused 'closed circuit', the placental membrane was capable of transferring amino acids against large concentration gradients; the aminogram of the initial concentration in the perfusion fluid, equivalent to foetal plasma, is compared with the concentration of each amino acid found after 2 hours perfusion, when a steady state was reached. The concentration of 15 of the 20 amino acids had risen by 1.3 and 4.7 times the initial perfusate concentration. There was little difference between the ratios attained by the neutral straight-chain and branched-chain amino acids or the basic amino acids. When the final concentrations were compared with the levels found in the maternal plasma the mean F:M ratio was 8.4 with a range of 4.0–31.5, tyrosine and citrulline reaching the highest ratios (Hill & Young, 1972). These results provide evidence for the accumulation, on the foetal side of the placenta, of six of the eight amino acids essential for growth in young animals: the concentration of threonine did not rise in the perfusate and no data were obtained for tryptophan. The majority of the non-essential amino acids were

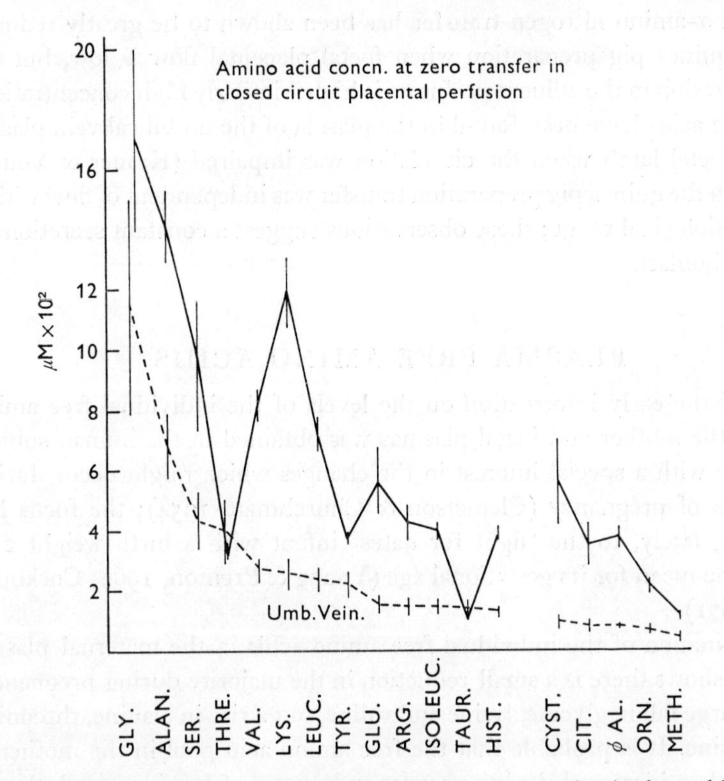

Fig. 2. A comparison of the plasma free amino acid concentration in the umbilical vein of the foetal guinea pig, with the final amino acid concentration achieved when the foetal placenta is perfused 'closed circuit' for two hours.

The placenta is able to increase the concentration of some of the amino acids in the perfusate by a factor of four; the average foetal:maternal concentration ratio achieved is 11.0.

accumulated in the perfusate, and evidence for the transport of cystine and for tyrosine is important because it shows that their supply is ensured before the foetal liver enzymes are developed for their conversion from methionine and phenylalanine respectively (Sturman *et al.* 1970 and Jakubovic, 1971). Indirect evidence for enzyme development has also been obtained in chronic foetal lamb preparations near term (McFadyen *et al.* 1972): the normal placental supply of amino acids to the foetus was supplemented by a constant intravenous infusion of a mixture of all the essential amino acids, together with glycine and sorbitol. Plasma changes which were not directly related to the amino acids infused occurred; immediate increases in plasma free serine, alanine and cystine were seen, with a return to the control level within 24 hours. Plasma tyrosine also rose, but more slowly, and remained elevated 24 hours after the infusion.

Total α-amino nitrogen transfer has been shown to be greatly reduced in the guinea pig preparation when foetal placental flow is low, but the concentration in the effluent perfusate is high; similarly high concentrations of amino acids have been found in the plasma of the umbilical vein plasma in the foetal lamb when the circulation was impaired (Krauer & Young, 1972). In the guinea pig preparation transfer was independent of flow within the physiological range; these observations suggest a constant secretion by the trophoblast.

PLASMA FREE AMINO ACIDS

Most of the early information on the levels of the individual free amino acid in the mother and foetal plasmas was obtained in the human subject by those with a special interest in the changes which might occur during toxaemia of pregnancy (Clemetson & Churchman, 1954); the focus has changed, lately, to the 'light for dates' infant with a birth weight 2 SD below the mean for its gestational age (Young & Prenton, 1969; Cockburn *et al.* 1971).

Examination of the individual free amino acids in the maternal plasma (Fig. 3) shows there is a small reduction in the majority during pregnancy, with a large fall in glycine, lysine and valine, but a rise in alanine, threonine and cystine. It is probable that the free amino acid pool in the mother is nevertheless increased during pregnancy because of her increased plasma volume. These changes in the maternal free amino acids occur early in gestation, before the conceptus's demand is large, and can be induced in the male and non-pregnant ewe by giving oestrogen and progesterone together (Zinneman *et al.* 1967; Curet *et al.* 1970); small changes in plasma amino acids are also seen during the luteal phase of the menstrual cycle. The relatively high plasma alanine levels in pregnancy suggest that gluconeogenesis is depressed, for alanine has been shown to be the main precursor for gluconeogenesis in the non-pregnant subject (Felig *et al.* 1970). In contrast, however, Freinkel *et al.* (1972) provide evidence for a limitation in the supply of this substrate as the cause for the hypoglycaemia occurring during starvation in pregnancy.

In the term foetus, the pattern of the individual free amino acid in the plasma is similar to that found in the non-pregnant woman, although most of the levels are slightly higher, alanine, lysine and threonine are particularly elevated, together with the waste product taurine. The high alanine levels again suggest a low gluconeogenic activity and it is known that hepatic uptake of a model amino acid, related to alanine, is not active during foetal life (Christensen & Clifford, 1963). It is probable that gluconeogenesis is

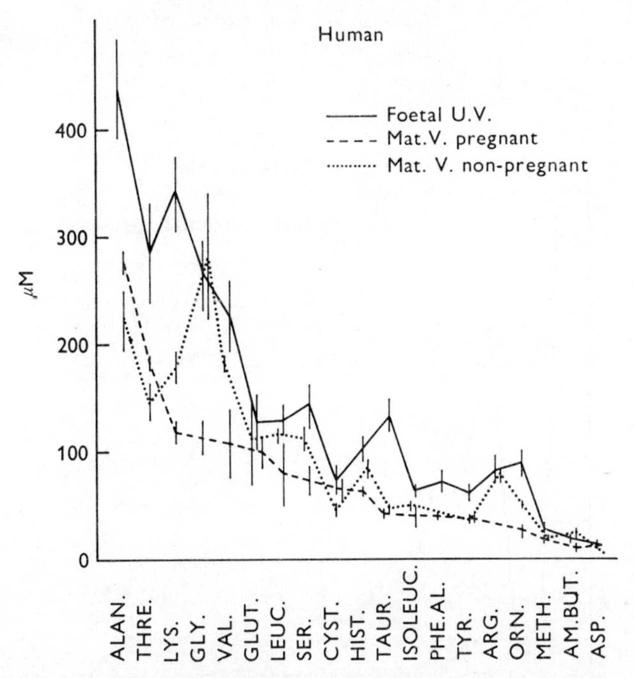

Fig. 3. Venous plasma free amino acid concentrations in the term-pregnant woman and umbilical vein at delivery, and in the non-pregnant woman. The aminograms are arranged in order of magnitude of the concentrations occurring in the pregnant woman.

The foetal levels are higher than the maternal with different concentration ratios for each amino acid. The foetal and non-pregnant levels and pattern are similar except for the high foetal, alanine, lysine and threonine.

not necessary in the foetus with a good glucose supply from the mother, across the placenta.

It is not clear why the free amino acid concentrations should be high in the foetal plasma and intracellular fluid, but high intracellular concentrations are also found to accompany nitrogen retention in regenerating liver and tumours; they would therefore appear to be characteristic of the increased protein turnover rate in growing animals (Waterlow & Stephen, 1968). An indication of the part played by foetal metabolism in causing the differences in concentration between the free amino acids in the maternal and foetal plasma was demonstrated in the guinea pig placenta perfused after removal of the foetus. Fig. 4 shows the aminograms of the maternal and foetal plasmas, and foetal placental venous effluent, when no amino acids were added to the perfusion fluid. The differences between the maternal and foetal plasma were large as in other species and glycine, alanine, serine and threonine were the greatest. The pattern of amino acids transferred into the perfusate is similar to that in the maternal plasma,

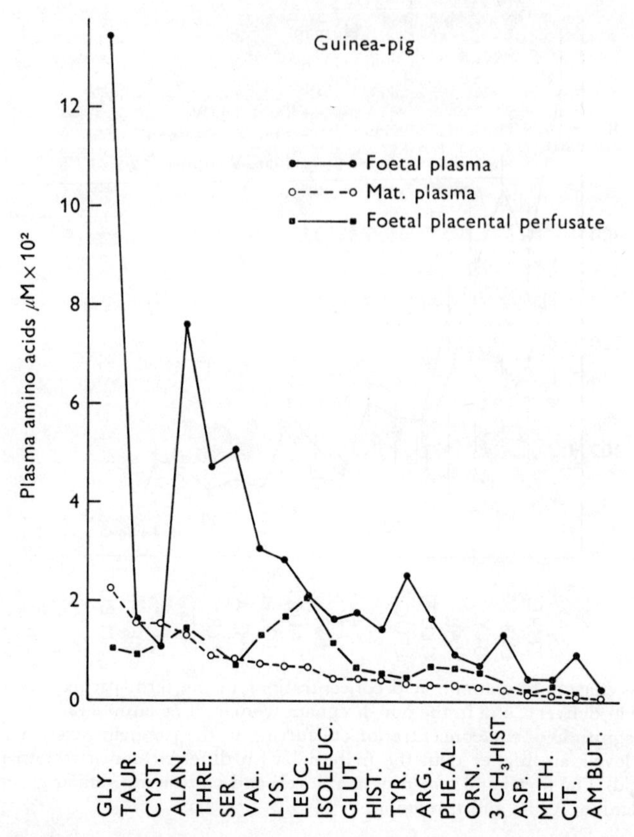

Fig. 4. Aminograms of the maternal and foetal plasmas and foetal placental perfusate in the guinea pig.

The foetal plasma levels are much higher than the maternal. In the perfusate the levels are similar to the maternal plasma, with the exception of the branched-chain amino acids and lysine. (Young, 1971.)

except that lysine and the branched-chain amino acids have been more freely exchanged from mother to foetus. This would suggest that the free amino acid pattern of the foetal plasma is dominated by its own metabolism. The reverse situation which occurs at birth, when the constant infusion of amino acids ceases across the placental membrane, is of little help in sorting out the causes of the differences between maternal and foetal concentration of free amino acid in the plasma: the withdrawal of nutrients, however temporary, induces its own trail of metabolic changes and alterations in free amino acid pattern (Mestyn *et al.* 1969; Lindblad, 1970).

Table 1. *A comparison of* $t_{\frac{1}{2}}$ *and metabolic clearance rate for amino acids in the ewe and foetal lamb, using the single injection technique*

		Neutral straight-chain	Neutral branched-chain	Basic
MCR	Mat.	(5) 14.4 ± 4.3	(6) 11.5 ± 1.6	(7) 10.1 ± 1.0
(ml/min kg)	Foet.	—	(2) 26	—
$t_{\frac{1}{2}}$ min	Mat.	(14) 6.2 ± 1.7	(19) 5.3 ± 0.7	(18) 3.8 ± 0.2
	Foet.	(3) 10.1 ± 5.0	(4) 7.4 ± 2.8	(4) 7.2 ± 1.6

The $t_{\frac{1}{2}}$ for the basic amino acids in the mother were significantly different from the branched chain, $P < 0.05$, and for the straight chain, $P < 0.02$.

METABOLIC CLEARANCE RATE

In order to obtain more information about the differences in metabolism between mother and foetus, utilisation rates of the amino acids were compared in the two. Plasma half-lives ($t_{\frac{1}{2}}$) and metabolic clearance rates (MCR) were studied using the decay curves, following single intravenous injections of amino acids, in the pregnant ewe (Gurpide, 1972). The results are shown in the Table 1. The maternal $t_{\frac{1}{2}}$ of 7–4 min is comparable with values obtained in other non-pregnant animals using radioactively labelled amino acids (Henriques *et al.* 1955); the value is significantly shorter for the basic amino acids than for the neutral straight-chain and branched-chain amino acids, but foetal $t_{\frac{1}{2}}$s are not significantly different from the maternal values for any of the amino acids. The maternal MCR, 10–14 ml min^{-1} kg, is about half the value for the hormone insulin, and no significant differences were observed between the groups. The two foetal values, of 26 ml min^{-1} kg, are high but agree with some preliminary figures for alanine and leucine in the newborn lamb (Soltesz, 1972). Both plasma half-life and metabolic clearance measurements are crude, and depend upon the rate of uptake by a variety of tissues with different relative mass in the mother and foetus and, probably, different uptake capacities and blood flows. It may, therefore, not be possible to demonstrate metabolic differences between mother and foetus without making tissue uptake studies, using labelled amino acid.

Many good colleagues have helped: Lynda Clarke, I. R. McFadyen, Jenny Joyce and Gyula Soltesz.

REFERENCES

BLAXTER, K. L. (1964). In *Mammalian Protein Metabolism*, Vol. 2, p. 173, ed. H. N. Munro and J. B. Allison. New York and London: Academic Press.

CHRISTENSEN, H. N. (1968). In *Protein Nutrition and Free Amino Acid Patterns*, p. 40, ed. J. H. Leathem. Rutgers University Press.

CHRISTENSEN, H. N. & CLIFFORD, J. B. (1963). *J. biol. Chem.* **238**, 1743.

CHRISTENSEN, H. N. & STREICHER, J. A. (1948). *J. biol. Chem.* **175**, 95.

CLEMETSON, C. A. B. & CHURCHMAN, J. (1954). *J. Obstet. Gynec. Brit. Cwlth* **61**, 364.

COCKBURN, F., BLAGDEN, A., MICHIE, E. A. & FORFAR, J. O. (1971). *J. Obstet. Gynec. Brit. Cwlth* **78**, 215.

CURET, L. B., MANN, L., ABRAMS, R., CRENSHAW, M. C. & BARRON, D. H. (1970). *J. appl. Physiol.* **28**, 1.

DANCIS, J. & SHAFRAN, M. (1958). *J. clin. Invest.* **37**, 1093.

FELIG, P., POZEFSKY, T., MARLISS, E. & CAHILL, G. F. (1970). *Science* **167**, 1003.

FREINKEL, N., METZGER, B. E., NITZAN, M., HARE, J. W., SHAMBAUGH, G. E., MARSHALL, R. T., SURMACZYNSKA, B. Z. & NAGEL, T. C. (1972). *Israel J. Med. Sci.* **8**, 426.

GURPIDE, E. (1972). In *Perfusion Techniques*, p. 26, ed. E. Diczfalusy. Stockholm: Karolinska Institutet.

HENRIQUES, O. B., HENRIQUES, S. B. & NEUBERGER, A. (1955). *Biochem. J.* **60**, 409.

HILL, P. M. M. & YOUNG, M. (1972). *J. Physiol.* (In press.)

HOPKINS, L., MCFADYEN, I. R. & YOUNG, M. (1971). *J. Physiol.* **215**, 11P.

JAKUBOVIC, A. (1971). *Biochem. biophys. Acta* **237**, 469.

KRAUER, F. & YOUNG, M. (1972). *J. Physiol.* (In press.)

LINDBLAD, B. S. (1971). In *Metabolic Processes in the Foetus and Newborn Infant*, p. 111, ed. J. H. P. Jonxis, H. K. A. Visser and J. A. Troelstra. Leiden: Stenfert Kroese.

MCFADYEN, I. R., NOAKES, D., SOLTESZ, G. & YOUNG, M. (1972). *J. Physiol.* (In press.)

MESTYAN, J., FEKETE, M., JARAI, I., SULYOK, E., IMHOF, S. & SOLTESZ, GY. (1969). *Biol. Neonate* **14**, 153.

MONEY, W. L. & DANCIS, J. (1960). *Am. J. Obstet. Gynec.* **80**, 209.

PRENTON, M. A. & YOUNG, M. (1969). *J. Obstet. Gynec. Brit. Cwlth* **76**, 404.

REYNOLDS, M. L. & YOUNG, M. (1971). *J. Physiol.* **214**, 583.

SOUTHGATE, D. A. T. (1971). *Biol. Neonate* **19**, 272.

STURMAN, J. A., GAULL, G. & RAIHA, M. C. A. (1970). *Science* **169**, 74.

SYKES, A. R. & FIELD, A. C. (1972). *J. agric. Sci., Camb.* **78**, 119.

WATERLOW, J. C. & STEPHEN, J. M. L. (1968). *Clin. Sci.* **35**, 287.

WIDDOWSON, E. M. (1968). *Proc. Nutr. Soc.* **28**, 17.

WILLIAMS, H. H., CURTIN, L. V., ABRAHAM, J., LOOSLI, J. K. & MAYNARD, L. A. (1954). *J. biol. Chem.* **208**, 277.

YOUNG, M. (1969). In *Foetus and Placenta*, p. 139, ed. A. Klopper and E. Diczfalusy. Oxford: Blackwell.

YOUNG, M. (1971). In *Metabolic Processes in the Foetus and Newborn Infant*, p. 97, ed. J. H. P. Jonxis, H. K. A. Visser and J. A. Troelstra. Leiden: Stenfert Kroese.

YOUNG, M. & PRENTON, M. A. (1969). *J. Obstet. Gynec. Brit. Cwlth* **76**, 333.

ZINNEMAN, H. H., SEAL, U. S. & DOE, R. (1967). *J. clin. Endocr.* **27**, 397.

SULPHUR AMINO ACIDS, FOLATE AND DNA: METABOLIC INTERRELATIONSHIPS DURING FOETAL DEVELOPMENT

By GERALD E. GAULL

Department of Pediatric Research, New York State Institute for Basic Research in Mental Retardation, Staten Island, New York, and the Department of Pediatrics and Clinical Genetics Center, Mount Sinai School of Medicine of the City University of New York

Cystathionase, the last enzyme in the pathway of transsulphuration of methionine to cysteine (Fig. 1), is not measurable in the liver and brain of the human foetus (Sturman, Gaull & Raiha, 1970; Gaull, Raiha & Sturman, 1972). Human foetal liver does not contain enzymatically-inactive precursor protein, immunologically related to mature human hepatic cystathionase, suggesting that human foetal liver is not competent to synthesize more than trace amounts of the intact enzyme (Pascal, Gillam & Gaull, 1972). We suggested that cyst(e)ine may be an essential amino acid for the liver and brain in the developing human, until sometime after birth, when mature activities of cystathionase are attained.

Therefore, we examined the concentrations of methionine and cysteine in human foetal and maternal plasma, as well as their transfer into foetal plasma after intravenous infusion into the mother (Gaull, Raiha, Saarikoski & Sturman, 1972). Methionine was transferred into foetal plasma against a threefold concentration gradient; leucine, which is transported by the same carrier system in most tissues, behaved similarly. To our surprise, concentrations of cystine and cysteine in foetal plasma were equal to or lower than those in maternal plasma. Cystine was unique amongst plasma amino acids in not showing higher concentrations in foetal plasma than in maternal plasma. In addition, in the human female at term (Young & Prenton, 1969) and in the monkey in late gestation (Sturman & Gaull, 1971), cystine is the only plasma amino acid which is higher in the pregnant female than in the non-pregnant female. Following a load of cystine or cysteine, the concentration of these amino acids in foetal plasma exhibited a slower increase and remained lower than in maternal plasma throughout the experimental period. Ornithine, which is transported by the same carrier system as cystine in most tissues, was similar to methionine and leucine, further suggesting that the transfer of cyst(e)ine from mother to foetus is under a separate and special control mechanism. The transfer of the D-cystine was

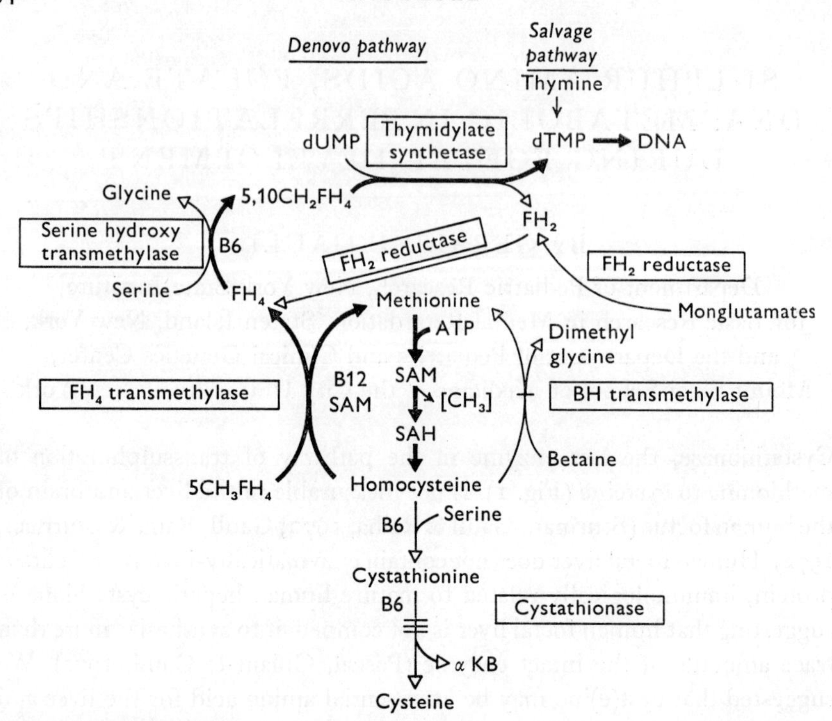

Fig. 1. Interrelationships of metabolism of sulphur-containing amino acids, folate and DNA. Abbreviations: FH_2, dihydrofolate; FH_4, tetrahydrofolate; $5CH_3FH_4$, ^5N-methyltetrahydrofolate; $5,10CH_2FH_4$, $^{5,10}N$-methylenetetrahydrofolate; dUMP, deoxyuridylate; dTMP, deoxythymidylate; SAM, *S*-adenosylmethionine; SAH, *S*-adenosylhomocysteine; αKB, alpha-ketobutyrate.

far slower than that of L-cystine, giving evidence that the transfer was not by simple diffusion, but rather by a facilitated or carrier-mediated transfer.

Experiments with ^{35}S-methionine and ^{35}S-cystine in pregnant Rhesus monkeys (whose transfer of methionine and cystine across the placenta resembles that of the human) gives evidence that the inability of the placenta to transfer cyst(e)ine in the foetus against a concentration gradient is not because of a rapid uptake or metabolism of cyst(e)ine by the foetal organs or by the placenta.

Since the transsulphuration pathway was shut off with the accumulation of cystathionine, but without accumulation of homocysteine, the methyltransferases involved in alternative pathways of homocysteine metabolism (Fig. 1) have been studied (Gaull, von Berg, Raiha & Sturman, 1972). The activity of ^5N-methyl-tetrahydrofolate-homocysteine methyltransferase is 4-fold *higher* in human foetal liver (2nd trimester) than in mature liver. In foetal brain, this activity has been high and *decreases* 3-fold to reach the nadir of mature activity during the same period of gestation. Serine hydroxymethyltransferase, the enzyme which converts tetrahydrofolate to [5]*N*-methylenetetrahydrofolate (a precursor for the *de novo* synthesis of thymidylate for incorporation into DNA), also shows a simultaneous 3-fold *decrease* in foetal brain, reaching a mature nadir of activity at the end of this period. In contrast, the other major pathway open to homocysteine, the betaine–homocysteine methyltransferase, is 4-fold *lower* in the liver and 2-fold *lower* in the brain during the same period.

The relative activities of these methyltransferases in the foetal liver and brain, considered with the low activity of cystathionase, suggest that the transsulphuration pathway is turned off in the foetus for the further metabolism of homocysteine in favour of the folate-B_{12} remethylation pathway. This pathway converts [5]*N*-methyltetrahydrofolate, the major monoglutamic folate in liver and serum, to tetrahydrofolate. The latter form of folate reacts with serine on serine hydroxymethyltransferase to form [5,10]*N*-methylenetetrahydrofolate, a precursor uniquely required for synthesis of DNA, but not RNA. We suggest that the β-carbon of serine is shunted into DNA synthesis at periods of rapid cellular multiplication, rather than having the entire carbon skeleton accept the sulphur from homocysteine to form cysteine. The latter thus becomes an essential amino acid in foetal liver and brain.

Cyst(e)ine is the end product of the transsulphuration pathway of methionine, and is a known inhibitor of the first enzyme on this pathway, the methionine-activating enzyme (Fig. 1). In partially purified extracts of rat liver, we have demonstrated that this enzyme is allosteric (Tallan & Gaull, unpublished), a kinetic property often found in regulatory enzymes. We postulate that the slower transfer of cystine across the placenta during foetal growth may be an adaptation to more rapid DNA synthesis during this period.

The expenses of these studies were shared by the New York State Department of Mental Hygiene, the Lalor Foundation, and the Association for the Aid to Crippled Children.

REFERENCES

GAULL, G. E., RAIHA, N. C. R., SAARIKOSKI, S. V. & STURMAN, J. A. (1972). *Pediat. Res.* **6**, 76 (Abstract).

GAULL, G. E., RAIHA, N. C. R. & STURMAN, J. A. (1972). *Pediat. Res.* **6**, 538–47.

GAULL, G. E., VON BERG, W., RAIHA, N. C. R. & STURMAN, J. A. (1972). *Pediat. Res.* (In the Press.)

PASCAL, T. A., GILLAM, B. & GAULL, G. E. (1972). *Pediat. Res.* (In preparation.)

STURMAN, J. A., GAULL, G. E. & RAIHA, N. C. R. (1970). *Science* **169**, 74–5.

STURMAN, J. A., NIEMANN, W. K., & GAULL, G. E. (1970). *Biochem. J.* **125**, 78P (Abstract).

YOUNG, M. & PRENTON, M. A. (1969). *J. Obstet. Gynaec. Brit. Cwlth* **76**, 333–44.

THE EFFECT OF PROTEIN DEPRIVATION ON FOETAL SIZE AND SEX RATIO

By L. B. CURET
Department of Gynecology–Obstetrics,
University of Wisconsin School of Medicine,
Madison, Wisconsin 53706

In 1945 L. R. Wallace reported on a series of experiments carried out to determine the effect of maternal protein deprivation on the size of the offspring. Wallace used sheep as the experimental model and as a part of the study he fed a group of pregnant ewes a diet low in protein beginning shortly after breeding. He performed Caesarean sections at different stages of the pregnancies and compared the weight of the lambs born of the diet group of ewes to that of control lambs. His findings that the 'diet' lambs delivered at 144 days were smaller than the controls is not surprising, as other investigators had demonstrated that in rats, sheep and other animals protein deprivation resulted in smaller offspring. What was interesting, however, was the fact that before 100 days the weight of the lambs in the experimental groups did not differ from the weight of control lambs.

We designed a series of experiments to find out if a difference in foetal weight before 100 days could be produced by depriving the ewe of protein before breeding.

METHODS

We selected ten purebred Dorset ewes and fed them a diet which contained only 8–10% of their daily protein requirements for two months, at the end of which time the ten ewes were bred. Ten other similar ewes which were fed *ad libitum* were also bred and served as controls. The same ram serviced all twenty ewes within a short period of time.

Caesarean sections were performed under sterile conditions at several stages of pregnancy up to 90 days and the number, weight and sex of the lambs were recorded.

The next year the groups were switched around so that the controls served as the diet group and vice versa. As some of the ewes failed to resume ovulation and some of them actually died new ewes were comparable to the ones already used. The same ram was used in the second year.

The experimental protocol was repeated with the interchanged groups after which the data for both years were analysed. We were able to carry out twelve paired experiments. In seven of them the same ewe served both as experiment and control.

RESULTS

Foetal weight

A. There were four experiments in which the litter size was the same in both groups so that comparisons were obvious. As seen in Table 1, in all these four cases the diet lambs were smaller than the controls.

Table 1

Days gestation	Sheep number	Group	Weight of lambs (g)
64	390	Control	83.0
	304	Diet	79.0
70	393	Control	149.0
		Diet	132.0
77	385	Control	330.0
		Diet	220.0
90	383	Control	575.0–564.0
	380	Diet	512.0–467.0

In eight experiments the litter sizes were different. In seven instances the diet ewes had singletons and the control ewes twins, while in one instance the control ewe had triplets and the diet ewe had twins.

In three of these experiments the diet ewes weighed less than the controls:

1.	Control	41 g	46 g	50 g
	Diet	29 g	33 g	
2.	Control	284 g	270 g	
	Diet	249 g		
3.	Control	445 g	380 g	
	Diet	354 g		

In three of the remaining five experiments the diet singleton weighed more than the lighter but less than the heavier of the control twins. In the remaining two experiments the diet singleton weighed as much as the heavier of the control twins.

Foetal sex

B. These data are summarised in Table 2. Of the fourteen lambs delivered of the diet ewes only three were female. Thus, a male to female ratio of 3.66 to 1 was obtained. In the control group there were eleven male and eleven female lambs. The difference between these two ratios is significant ($p < 0.05$).

Table 2

Days Gestation	Sheep number	Group	Sex of lambs
58	398	Control	M–M–M
	320	Diet	F–M
60	386	Control	F–F
	320	Diet	M
64	390	Control	F
	304	Diet	M
6	395	Control	M–F
	381	Diet	M
68	388	Control	F–F
		Diet	M
	391	Control	F–M
		Diet	M
70	384	Control	M–M
		Diet	F
	393	Control	M
		Diet	M
77	394	Control	F–M
		Diet	M
	385	Control	F
		Diet	M
84	387	Control	F–M
		Diet	M
90	383	Control	F–M
	380	Diet	F–M

DISCUSSION

It appears from these observations that maternal protein deprivation, begun before breeding, will result in smaller offspring before 90 days' gestation. The explanation for such observations, however, is beyond the time limits of this presentation.

Much has been written on the effect of diet on gonadotrophin secretion, ovarian steroidogenesis, ovulation rates, implantation, and embryonal loss. However, little is found in relation to sex of the foetus.

In their book *The Evolution of Sex* Geddes & Thomson (1901) quote several interesting experiments:

1. It was found by Yung that by adding extra protein to tadpoles he could increase the percentage of females born.

2. Another investigator, Von Siebold, demonstrated that in wasps the production of females was increased by increasing the food supply.

3. Treat showed that if caterpillars were starved the resultant butterflies or moths were males while others of the same brood but highly nourished came out females.

4. Rolph showed the same effect in crustaceans.

Regarding mammals, the influence of nutrition upon sex has been more difficult to prove. Yet some important observations have been made by two investigators. Girou (in Geddes & Thompson, 1901) divided a flock of 300 ewes into two equal groups, one of which was well fed while the other was kept poorly fed. The proportion of female lambs was 60% in the well fed and 40% in the poorly fed group. He advanced the theory that the better nourished parent tends to determine the same sex in the offspring.

In 1943, Hadzel *et al.* tested this hypothesis in rats and found that mating of normal and high protein males with low protein females resulted in a male to female sex ration of 145/100. The groups were then switched and the male to female ratio was 92/100.

Lastly, in the human species the influence of nutrition, though hard to estimate, has been hinted at. Recently Dobson & Williamson (1970) have found a 2/1 male to female ratio in infants with PKU where the inability to metabolize phenylalanine is tantamount to deprivation of that amino acid.

In conclusion, then, we have shown that maternal protein deprivation may have altered the sex ratio of the offspring in favour of the male. The mechanism for such alteration remains to be elucidated.

REFERENCES

Dobson, J. & Williamson, M. (1970). *New Eng. J. Med.* **282**, 1104.

Geddes, P. & Thomson, J. A. (1901). *The Evolution of Sex.* Rev. Ed. Walter Scott, London.

Hadzel, F., Da Costa, E. & Carlson, A. J. (1939). *Proc. Soc. Exp. Biol. Med.* **40**, 334–5.

Wallace, L. R. (1948). *J. Agr. Sci.* **38**, 93–153, 243–302, 367–401.

FOETAL DECAPITATION AND THE DEVELOPMENT OF INSULIN SECRETION IN THE RABBIT

BY PATRICIA M. B. JACK AND R. D. G. MILNER
The Department of Child Health, University of Manchester,
St Mary's Hospital, Manchester 13 OJH

INTRODUCTION

In 1970 Kervran, Jost and Rosselin indicated that the head might affect the foetal development of insulin secretion in the rabbit. Foetuses decapitated on day 24 or 26 and harvested on day 29 had higher plasma insulin levels and a lower pancreatic insulin content than littermate controls. Indirect evidence of adenohypophyseal influence on the development of the human endocrine pancreas came from the observation that failure of the endocrine pancreas to hypertrophy in anencephalic foetuses born to women with impaired glucose tolerance occurred only if there was disruption of the foetal hypothalamo-hypophyseal axis (Van Assche, Gepts & De Gasparo, 1970). In the present experiments the effect of decapitation on the development of the foetal β cell has been explored further. Some of the work has been reported briefly (Jack & Milner, 1972).

METHODS

One foetus in a litter was decapitated on day 24 by the technique described by Bearn (1968). The litter was harvested by laparotomy on day 29 and blood was collected from the neck veins of the control foetuses after decapitation and from the posterior vena cava of the experimental foetus. Plasma was stored at -20 °C until assayed for insulin content as described previously (Milner, 1969*a*). After decapitation the control foetuses were reweighed and the one with the headless body weight nearest to that of the experimental foetus was used for the study of insulin secretion *in vitro*.

Insulin secretion *in vitro* was studied as described previously (Milner, Barson & Ashworth, 1971). In some experiments the pancreas of the control and experimental animals was divided into pieces which were incubated separately and in other experiments the pancreas was incubated whole. Insulin secretion was studied for 30-min periods in media containing the following substances: 0.6 mg glucose/ml, 3.0 mg glucose/ml, 0.6 mg glucose + 60 μmoles KCl/ml, 3.0 mg glucose + 5 μg glucagon/ml (Eli Lilly

& Co., Basingstoke). In one experiment the experimental and control pancreata were incubated in medium containing 0.6 mg glucose/ml for a 5-min and a 25-min period followed by three 30-min periods and then in medium containing 3.0 mg glucose/ml for a 5- and a 25-min period followed by two 30-min periods. This experiment was designed to investigate whether a difference occurred between the control and experimental pancreas in the 'early' and 'late' phases of insulin release. In some experiments the pancreatic insulin content was determined as described previously (Milner, 1969*a*).

RESULTS

Plasma and pancreatic insulin concentrations

In 21 experiments the plasma insulin concentration of the decapitated foetus was the highest or equal highest in the litter on 15 occasions. The mean (± SE of mean) plasma insulin level of the decapitated foetuses was 4.0 ± 0.6 ng/ml ($n = 21$) and that of the control foetuses was 1.9 ± 0.1 ng/ml ($n = 113$). There was wide variation between foetuses within a litter and between litters, but the differences between the plasma insulin concentrations of the control and decapitated foetuses when tested statistically by the Mann–Whitney U test was highly significant ($P < 0.001$).

The concentration of insulin in the experimental and control pancreas was calculated in nine experiments. The mean (± SE of mean) pancreatic insulin content of the experimental pancreas was 252 ± 72 ng/mg and that of the control pancreas, 246 ± 49 ng/mg. In six experiments the pancreas had been used for secretion studies. The fraction of pancreatic insulin content secreted varied between 1.6 and 20.2 per cent.

Insulin secretion in vitro

Insulin release from the pancreas of decapitated foetuses was persistently greater than that from control foetuses in medium containing 0.6 or 3.0 mg glucose/ml in the incubation periods 31–120 min. In medium containing 0.6 mg glucose/ml there was a significant and similar fractional fall in insulin release in period 61–90 min compared with period 31–60 min. No other significant changes in insulin release were observed within pieces in the experiments summarized in Table 1. In the period 31–120 min the mean (± SE of mean) insulin released from the pancreas of the decapitated foetus was 45 ± 15 per cent greater than that from the control in 0.6 mg glucose/ml and 70 ± 23 per cent greater in 3.0 mg glucose/ml. The difference in insulin secretion between experimental and control animals in 3.0 mg glucose/ml persisted for 4.5 h. Incubation of pancreas for 5 and 25 min in medium containing 0.6 or 3.0 mg glucose/ml showed that

Table 1. *Insulin secretion (means ± SE of mean) from the pancreas of 29-day-old decapitated rabbit foetuses and their littermate controls in incubation medium containing 0.6 or 3.0 mg/ml glucose*

	Insulin secretion (pg/mg pancreas/30 min) Incubation period (min)		
	31–60	61–90	91–120
+0·6 mg/ml glucose (n = 8)			
Control	872 ± 76	676 ± 74	590 ± 44
Decapitate	1194 ± 112	852 ± 90	918 ± 108
Decapitate ÷ Control	1.43 ± 0.17*	1.43 ± 0.25	1.55 ± 0.11‡
+3.0 mg/ml glucose (n = 7)			
Control	2768 ± 650	2806 ± 740	2200 ± 426
Decapitate	4968 ± 1048	3554 ± 566	3746 ± 672
Decapitate ÷ Control	2.05 ± 0.38*	1.53 ± 0.23*	1.89 ± 0.27†

Significance of the increment in insulin release from the pancreas of the decapitated foetus compared with the control: * $P < 0.05$, † $P < 0.01$, ‡ $P < 0.001$.

increased release of insulin from the experimental pancreas in 3.0 mg/ml glucose occurred in both the 'early' and 'late' phases of insulin secretion (Fig. 1). The fractional increase in insulin released in 3.0 mg glucose/ml from 121 to 210 min compared with that released in 0.6 mg glucose/ml from 31 to 120 min revealed no difference between decapitate and control pancreas in four experiments.

Pancreas was incubated for four 30-min periods (0–120 min) in 3.0 mg glucose/ml and then for three 30-min periods in 3 mg glucose plus 5 μg glucagon/ml (121–210 min) (n = 12). When the insulin released from 121–180 min was divided by that released from 61–120 min, glucagon stimulated insulin secretion 98 ± 26 per cent (mean ± SE of mean) from control pancreas and 46 ± 8 per cent from experimental pancreas. The absolute increase in insulin release from experimental or control pancreas was similar, the difference in the fractional increase being due to higher insulin release from the experimental pancreas in 3.0 mg glucose/ml.

In experiments of similar design pancreas was incubated for four 30-min periods (0–120 min) in 0.6 mg glucose/ml and then for three 30-min periods (121–210 min) in 0.6 mg glucose + 60 μmoles KCl/ml (n = 23). The mean (± SE of mean) stimulation of insulin secretion by potassium was 552 ± 110 per cent in the experimental pancreas and 640 ± 82 per cent in the control pancreas. Again the absolute stimulation was similar, the difference in the fractional rise being due to a higher basal insulin release from the pancreas of the decapitated foetus.

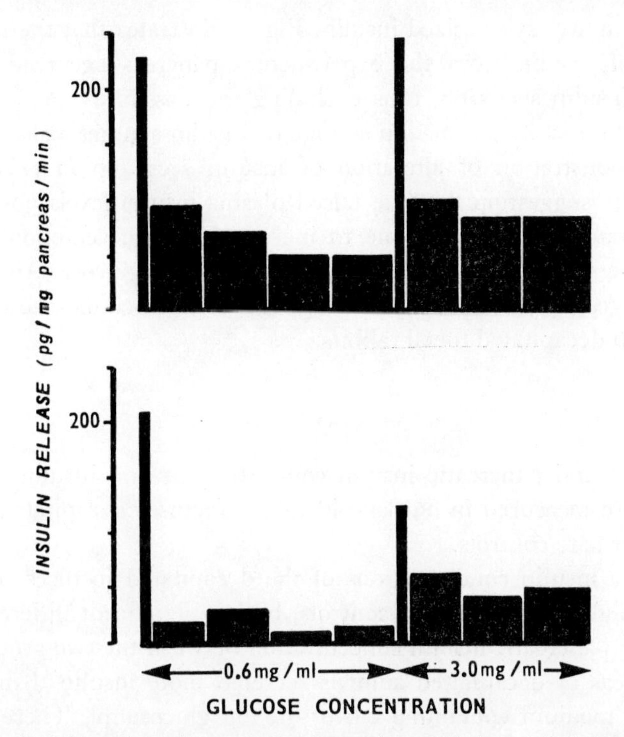

Fig. 1. Insulin release from the pancreas of a 29-day rabbit foetus decapitated on day 24 (above) and from that of a littermate control (below) in medium containing 0.6 mg glucose/ml or 3.0 mg glucose/ml. Apart from the initial 5-min period of incubation, insulin release from the experimental pancreas was consistently greater than that from the control.

DISCUSSION

A raised plasma insulin concentration in 29-day-old rabbit foetuses decapitated on day 24 could be due to increased insulin secretion or a decreased rate of destruction. The present experiments have shown that there is increased insulin secretion *in vitro* from the experimental pancreas under basal conditions (0.6 mg glucose/ml) and that the effect is demonstrable also in media containing a high glucose concentration (3.0 mg/ml). In order to test if the alteration in insulin release was linked specifically to glucose stimulation insulin secretion was evoked also by potassium and glucagon. Both stimuli caused similar increases in insulin secretion, indicating that the effect of decapitation was not a generalized one. The results are compatible with the hypothesis that decapitation alters the way in which glucose stimulates insulin release. Insulin secretion induced by glucose occurs biphasically. There is an initial spike, thought to be the secretion of stored insulin, followed by a wave thought to be the release of, in part, newly synthesized insulin. Fig. 1 illustrates that the increased secretion of insulin from the experimental pancreas occurred in both phases of insulin secretion, thus excluding the possibility that the effect of decapitation is one on insulin storage or insulin synthesis alone.

The demonstration of alteration of insulin secretion *in vitro* makes plausible the suggestion that the raised plasma insulin levels observed in the decapitated foetuses are due to increased insulin secretion *in utero* and that the stimulus to increased secretion may be glucose. An effect of these changes may be to stimulate lipogenesis which is known to be increased in decapitated foetal rabbits.

SUMMARY

1. Plasma and pancreatic insulin concentrations and insulin secretion *in vitro* were measured in 29-day-old rabbit foetuses decapitated on day 24 and littermate controls.

2. Plasma insulin concentrations of the decapitated foetuses were significantly higher than those of controls, but no significant difference was observed in pancreatic insulin concentration between the two groups.

3. Pancreas of decapitated animals secreted more insulin than that of controls in medium containing 0.6 or 3.0 mg glucose/ml. There was no significant difference between the experimental and control animals in the stimulation of insulin secretion *in vitro* by glucagon (5 μg/ml) or potassium (60 μmoles/ml).

4. The results of these experiments suggest that cephalic factors influence the development of β cell function in the foetal rabbit.

We are grateful to Professor J. A. Davis for his encouragement. This work was supported by the British Diabetic Association and the Medical Research Council.

REFERENCES

BEARN, J. G. (1968). *Brit. J. Exp. Path.* **49**, 136.
JACK, P. M. B. & MILNER, R. D. G. (1972). *J. Endocr.* **55**, P XXVI.
KERVRAN, A., JOST, A. & ROSSELIN, G. (1970). *Diabetologia* **6**, 51.
MILNER, R. D. G. (1969*a*). *J. Endocr.* **43**, 119–24.
MILNER, R. D. G. (1969*b*). *J. Endocr.* **44**, 267–72.
MILNER, R. D. G., BARSON, A. J. & ASHWORTH, M. A. (1971). *J. Endocr.* **51**, 323–32.
VAN ASSCHE, F. A., GEPTS, W. & DE GASPARO, M. (1970). *Biol. Neonat.* **14**, 374–88.

FURTHER STUDIES ON THE REGULATION OF INSULIN RELEASE IN FOETAL AND POST-NATAL LAMBS: THE ROLE OF GLUCOSE AS A PHYSIOLOGICAL REGULATOR OF INSULIN RELEASE *IN UTERO*

By J. M. BASSETT, DENISE MADILL, DIANNE H. NICOL and G. D. THORBURN

The Ian Clunies Ross Animal Research Laboratory,
CSIRO, Division of Animal Physiology,
Prospect, P.O. Box 239, Blacktown, NSW 2148, Australia

Insulin is present in the pancreas and plasma of the foetal lamb during most of gestation (Alexander, Britton, Cohen, Nixon & Parker, 1968; Willes, Boda & Stokes, 1969) and there is evidence that the rate of glucose utilization in foetal lambs is sensitive to insulin (Colwill, Davis, Meschia, Makowski, Beck & Battaglia, 1970). Administered glucose may stimulate insulin release (Alexander, Britton, Cohen & Nixon, 1969; Alexander, Britton, Mashiter, Nixon & Smith, 1970; Bassett & Thorburn, 1971; Davis, Beck, Colwill, Makowski, Meshia & Battaglia, 1971), yet it is still uncertain whether glucose is involved in the physiological regulation of insulin release in the lamb *in utero*.

During foetal life the plasma glucose concentration of the lamb is only 10–30 mg/100 ml. However, in-vitro studies on the insulin secretory response to glucose by pieces of pancreas or islets from adult rats and mice show that the response curve is sigmoidal, with negligible insulin being released below a threshold glucose concentration of approximately 70–90 mg/100 ml (Malaisse, Malaisse-Lagae & Wright, 1967; Ashcroft, Bassett & Randle, 1972). In-vitro studies on pieces of pancreas from adult sheep indicate a similar threshold glucose concentration of 60–80 mg/100 ml for stimulation of insulin release (Bassett & Madill, unpublished observations) and the relationship between plasma glucose and insulin concentrations in lambs (Bassett & Alexander, 1971) is consistent with this. Therefore, if glucose is to be a physiologically important regulator of insulin release in the lamb *in utero* the threshold for glucose stimulation of insulin release must be far lower than it is during post-natal life.

Significant positive correlations between glucose and insulin levels in plasma samples obtained from acutely exposed foetal lambs (Alexander *et al.* 1968) and from cannulated foetal lambs at surgery (Bassett & Thorburn, 1971) suggest that fluctuations in glucose concentration within the physiological range for foetuses may stimulate insulin release, but the validity of these relationships cannot be judged, because of the stressful situations in which the foetuses were placed. However, i.v. injections or infusions of glucose which stimulated insulin secretion in foetal lambs (Alexander *et al.* 1969, 1970; Bassett & Thorburn, 1971; Davis *et al.* 1971) have usually involved increases in the glucose concentration to values that are outside the normal range encountered in foetuses and are more like those of post-natal lambs.

In this communication we present further evidence bearing on the question of whether glucose can act as a physiological regulator of insulin release in the foetal lamb and whether the stimulatory mechanism involved differs from that involved during post-natal life.

IN-VIVO OBSERVATIONS OF THE ROLE OF GLUCOSE AS A PHYSIOLOGICAL REGULATOR OF INSULIN RELEASE IN THE FOETAL LAMB

Examination of the relationship between glucose and insulin concentrations in lamb plasma suggests that this relation changes shortly after birth (Fig. 1). Plasma samples were obtained from 48 naturally born and reared crossbred lambs bled once or more during the first 24 h after birth and then once daily. Fig. 1 shows that plasma insulin levels in many samples obtained during the first 24 h of post-natal life were substantially above 20 μU/ml at glucose concentrations between 30 and 80 mg/100 ml, yet in lambs more than 2 days old, insulin concentrations were rarely above 20 μU/ml unless the glucose concentration was more than 95 mg/100 ml. This difference appears to indicate a marked increase during the first few days after birth in the threshold for glucose stimulation of insulin release, implying that it may also be low before birth.

Evidence that small increases in the plasma glucose concentration can stimulate insulin release in foetal lambs has been obtained more directly by studies on cannulated lambs using procedures similar to those previously described (Bassett & Thorburn, 1971). In 15 glucose infusions on 8 foetal lambs (125 days' gestation or more) plasma insulin increased from a basal value of 14.9 ± 1.8 μU/ml to 31.7 ± 3.5 and 42.0 ± 4.8 μU/ml after 15 and 60 min of infusion respectively, while plasma glucose increased from 15.7 ± 1.1 to 42.6 ± 1.9 and 62.5 ± 2.7 mg/100 ml. Longer-term infusions of glucose which increased plasma glucose to values between 40 and 80 mg/100 ml also clearly increased plasma insulin in 4 cannulated

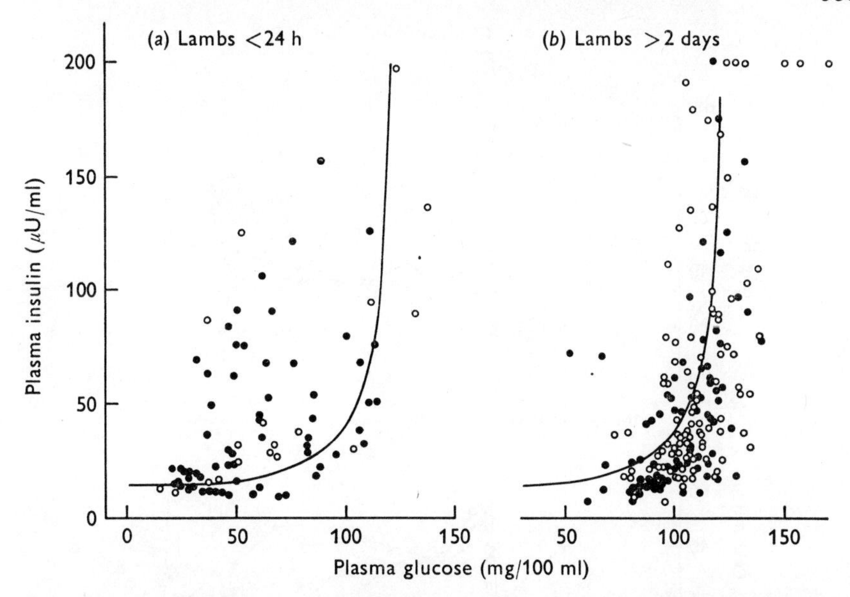

Fig. 1. The relationship between glucose and insulin concentrations in plasma samples obtained from 48 crossbred lambs, (*a*) within 24 h of birth, and (*b*) when they were more than 2 days old. The continuous line (same in both *a* and *b*) was drawn arbitrarily to indicate the glucose threshold for insulin release in lambs older than 2 days.

foetal lambs (Fig. 2). However, during these longer infusions plasma fructose increased from values of 100 to 120 mg/100 ml to values in the range 200–350 mg/100 ml. Partial correlation analysis indicated that glucose was the principal regulator of the insulin level and fructose had no significant role. Despite this, some uncertainty remains about the interpretation of in-vivo studies because of the close interrelation of glucose and fructose concentrations, and because fructose, though less effective than glucose, can stimulate insulin release from the pancreas of the foetal lamb (Bassett & Thorburn, 1971; Davis *et al.* 1971).

We have therefore turned to in-vitro systems in an effort to assess the sensitivity of the foetal pancreas to glucose alone and to ascertain whether cooperation between glucose and fructose may be of importance in determining the rate of insulin release by the pancreas of the foetal lamb.

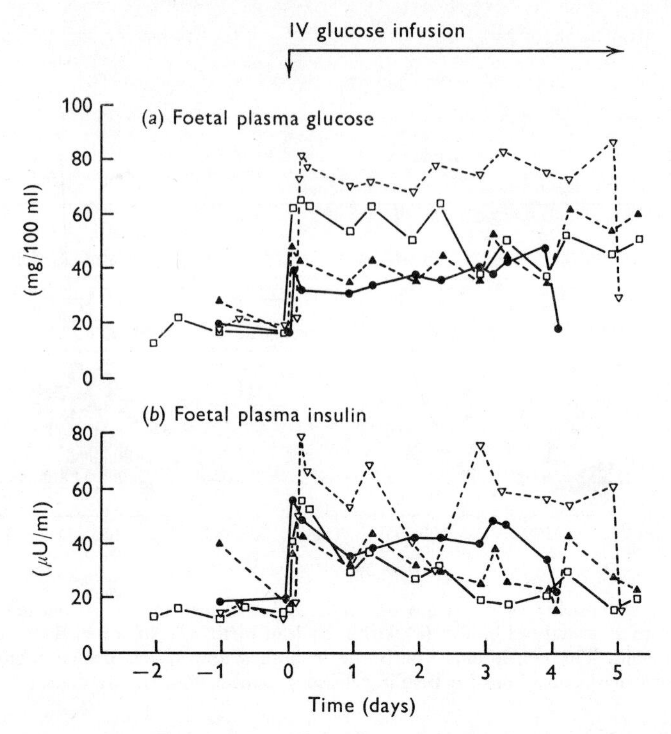

Fig. 2. Changes in (*a*) plasma glucose and (*b*) plasma insulin in 4 chronically cannulated foetal lambs (gestational age > 130 days) during prolonged i.v. infusion of glucose *in utero* at a rate of 70 mg/min.

IN-VITRO STUDIES ON THE ROLE OF GLUCOSE AS A REGULATOR OF INSULIN RELEASE FROM THE PANCREAS OF FOETAL AND POST-NATAL LAMBS

To study insulin release *in vitro*, small pieces of pancreas from foetal lambs of 120–145 days' gestation were incubated either in a conventional incubation system or in a continuous flow incubation (perifusion) system using a buffered bicarbonate medium containing 0.5% bovine albumin, 5 mM pyruvate, 5 mM fumarate, 5 mM glutamate and 5 mM caffeine. The insulin secretory responses of foetal pancreas pieces to glucose stimulation differed markedly from those of post-natal lamb pancreas in both systems (Figs. 3 and 4). In the conventional incubation system, glucose-stimulated insulin release from post-natal lamb pancreas was directly comparable to that seen with pancreas tissues from other species (Malaisse *et al.* 1967; Ashcroft *et al.* 1972). The dose response curve for glucose-stimulated insulin release was sigmoidal with a threshold between 50 and 100 mg/100 ml

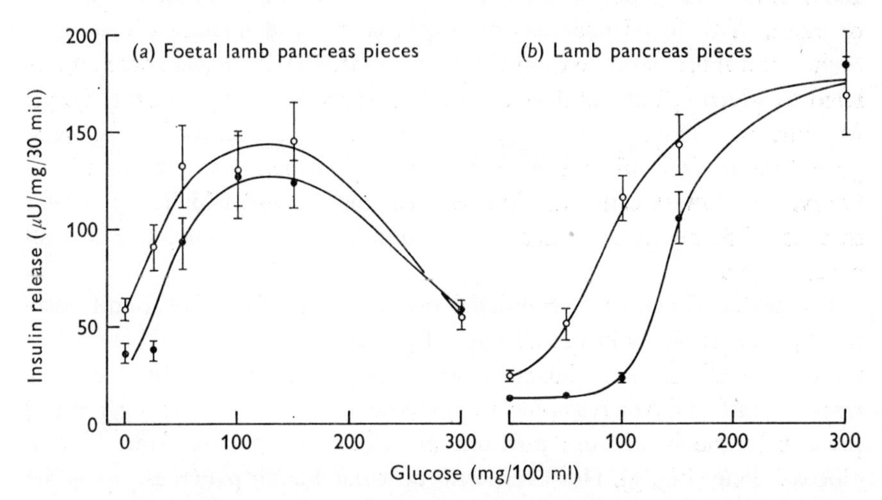

Fig. 3. In-vitro insulin release by pancreas pieces from (*a*) 3 foetal (> 140 days' gestation) and (*b*) 3 post-natal lambs (2–7 days of age) during incubation in medium containing glucose alone (●) or glucose + 5.5 mM fructose (○). Incubation was carried out for 30 min in a bicarbonate medium containing 5 mM caffeine following preincubation for 60 min in the same medium without glucose or fructose. Each point is the mean ± SE of 10–15 separate incubations.

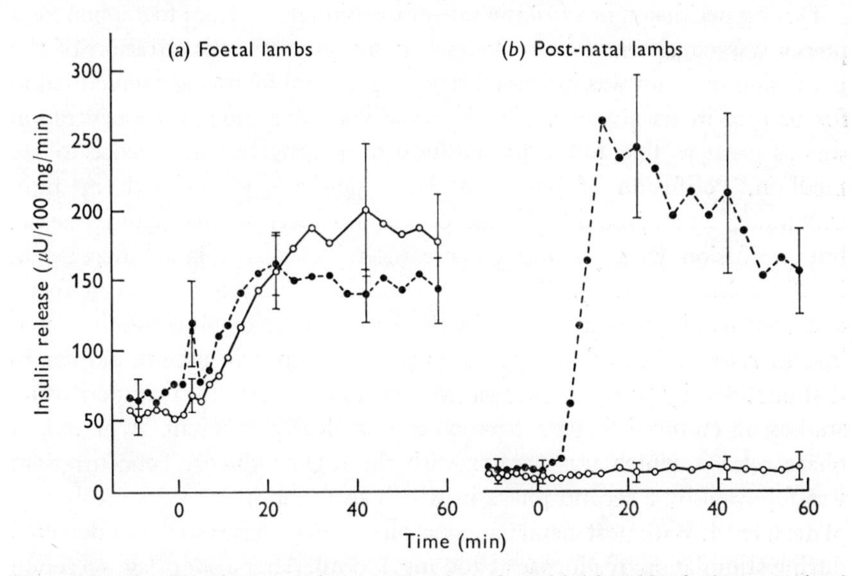

Fig. 4. Mean insulin release rates by pancreas pieces from (*a*) 6 foetal lambs (120–147 days' gestation) and (*b*) 6 post-natal lambs (2–7 days of age) during perifusion *in vitro* with a bicarbonate medium containing 25 mg/100 ml (○) or 300 mg/100 ml (●) glucose. Perifusion with stimulatory medium was begun at time 0 following perifusion for 60 min in glucose-free medium. Vertical bars show standard errors.

and maximal rates of release at glucose concentrations of 300 mg/100 ml or more. With foetal pancreas pieces, basal rates of insulin release were higher and more variable than those of lambs. Insulin release was stimulated by glucose, but the dose response curve was not sigmoidal (Fig. 3). Maximal rates of insulin release were obtained with concentrations between 50 and 150 mg/100 ml, and concentrations higher than this were less stimulatory. The lowest concentration used (25 mg/100 ml) which was similar to that of foetal plasma increased insulin release in some, but not all experiments.

Fructose and mannose stimulated insulin release from foetal and post-natal pancreas pieces in the absence of glucose, but their effects were less than those of glucose. Galactose was ineffective. The insulin secretory response of foetal pancreas pieces to glucose with fructose (100 mg/100 ml) present in the incubation medium did not differ greatly from that to glucose alone (Fig. 3). However, with post-natal lamb pancreas pieces the threshold for glucose-stimulated insulin release was greatly reduced when fructose (100 mg/100 ml) was present in the medium (Fig. 3).

In the absence of caffeine, insulin release was not significantly increased by glucose from either foetal or post-natal lamb pancreas pieces, even though caffeine had no effect on insulin release in the absence of glucose.

During perifusion *in vitro* the rate of insulin release from foetal pancreas pieces was consistently increased when the glucose concentration of the perifusing medium was increased to 25 mg/100 ml following preincubation for 60 min in its absence. The increase with this glucose concentration was as great as that following addition of 300 mg/100 ml glucose to the medium. Perifusion of post-natal lamb pancreas pieces with medium containing 25 mg/100 ml glucose caused no increase in insulin release, but perifusion with 300 mg glucose/100 ml caused a large increase in insulin release (Fig. 4). There was a marked difference between foetal and post-natal pancreas pieces in the time course of glucose-stimulated insulin release. With foetal pancreas pieces a biphasic pattern similar to that obtained by Burr, Kanazawa, Marliss & Lambert (1971) in perifusion studies on cultured foetal rat pancreas was clearly apparent. A brief first phase release, which was greater with the higher glucose concentration, was followed by a second phase in which secretion increased steadily to a plateau level. With post-natal pancreas pieces, two phases were not detected during stimulation by glucose at 300 mg/100 ml. After a short lag, secretion increased rapidly to a maximum and then began falling steadily (Fig. 4). The presence of fructose (100 mg/100 ml) in the medium permitted post-natal pancreas to respond significantly to 25 mg glucose/100 ml, but the magnitude of the response was small. It did not alter the response of foetal pancreas to glucose at 25 or 300 mg/100 ml, although it increased the basal rate of insulin release during the preincubation period when glucose was absent.

Observations on the relation of insulin and glucose concentrations in the plasma of newborn lambs (Fig. 1) suggest that the change in responsiveness of the pancreas occurs after birth; incubation and perifusion studies on pancreas pieces from two lambs less than 12 h old indicate that this is so, since the responses of tissues from these animals were identical with those of foctal lambs and unlike those of older post-natal lambs.

DISCUSSION

The in-vivo and in-vitro observations reported provide strong evidence that insulin release from the pancreas of the foetal lamb can be stimulated by glucose at the concentrations present in plasma and imply that glucose is a physiological regulator of insulin release in the lamb *in utero*. Further studies on cannulated foetal lambs will be necessary to demonstrate the functional significance of this regulatory mechanism in relation to foetal nutrition and growth.

In-vitro studies on post-natal lamb pancreas pieces as well as studies on other species (Ashcroft *et al.* 1972) indicated that cooperation between glucose and fructose could account for the sensitivity of the foetal pancreas to low concentrations of glucose. However, while such cooperative effects and other similar ones probably contribute to the overall control of insulin release in the foetal lamb, our in-vitro observations on foetal pancreas show that glucose is capable of achieving this regulation independently of fructose.

To obtain insulin secretory responses to glucose *in vitro* it was necessary to include caffeine in the incubation media as it has been with rat tissue (Lambert, Junod, Stauffacher, Jeanrenaud & Renold, 1969). However, the in-vivo responses of insulin release to glucose seen here and in earlier studies (Bassett & Thorburn, 1971), indicate clearly that the presence of inhibitors of phosphodiesterase activity such as caffeine is not essential *in vivo* for effects of glucose on insulin release to be demonstrated although this has been necessary in some other species (Grasso, Messina, Saporito & Reitano, 1970; Chez, Mintz & Hutchinson, 1971). The close similarity between time courses of glucose-stimulated insulin release in perifusion studies with 5 mM caffeine present (Fig. 3) and those observed *in vivo* in its absence (Bassett & Thorburn, 1971) suggests also that differences in phosphodiesterase activity do not explain the difference between foetal and post-natal lambs in the pattern or magnitude of these responses.

The slight stimulatory effect of high glucose concentrations on insulin release from foetal pancreas *in vitro* suggests a possible explanation for the failure of glucose to stimulate insulin release in in-vivo studies such as those of Willes, Manns & Boda (1969). It probably reflects feedback inhibition of glucose metabolism within the β cells, but studies on glucose metabolism of isolated islets will be necessary to ascertain whether this is so.

The in-vitro observations reported here demonstrate a striking change in the kinetics of glucose-stimulated insulin release during the first 24–48 h after birth. This change also occurs shortly after birth in lambs born prematurely following corticotrophin infusion *in utero* suggesting that it is consequent on parturition (Bassett, Thorburn & Nicol, 1973). Birth induces a multiplicity of changes that might influence pancreatic β cell function. In many other tissues the changes at birth involve changes in enzyme patterns. There are, for instance, striking changes in the enzymes regulating carbohydrate metabolism within the liver and other tissues (Walker, 1968). It seems probable that there are also changes in the enzymes regulating glucose metabolism within the β cell at this time. However, these changes have yet to be detected and the mechanisms involved in the changed sensitivity of insulin secretion to glucose stimulation remain unknown.

REFERENCES

Alexander, D. P., Britton, H. G., Cohen, N. M., Nixon, D. A. & Parker, R. A. (1968). *J. Endocr.* **40**, 389–90.

Alexander, D. P., Britton, H. G., Cohen, N. M. & Nixon, D. A. (1969). *Biol. Neonat.* **14**, 178–93.

Alexander, D. P., Britton, H. G., Mashiter, K., Nixon, D. A. & Smith, F. C. (1970). *Biol. Neonat.* **15**, 361–7.

Ashcroft, S. J. H., Bassett, J. M. & Randle, P. J. (1972). *Diabetes* **21** (Suppl. 2), 538–45.

Bassett, J. M. & Alexander, G. (1971). *Biol. Neonat.* **17**, 112–25.

Bassett, J. M. & Thorburn, G. D. (1971). *J. Endocr.* **50**, 59–74.

Bassett, J. M., Thorburn, G. D. & Nicol, D. H. (1973). *J. Endocr.* (In press.)

Burr, I. M., Kanazawa, Y., Marliss, E. B. & Lambert, A. E. (1971). *Diabetes*, **20**, 592–7.

Chez, R. A., Mintz, D. H. & Hutchinson, D. L. (1971). *Metabolism* **20**, 805–15.

Colwill, J. R., Davis, J. R., Meschia, G., Makowski, E. L., Beck, P. & Battaglia, F. C. (1970). *Endocrinology*, **87**, 710–15.

Davis, J. R., Beck, P., Colwill, J. R., Makowski, E. L., Meschia, G. & Battaglia, F. C. (1971). *Proc. Soc. Exp. Biol. Med.* **136**, 972–5.

Grasso, S., Messina, A., Saporito, N. & Reitano, G. (1970). *Diabetes* **19**, 837–41.

Lambert, A. E., Junod, A., Stauffacher, W., Jeanrenaud, B. & Renold, A. E. (1969). *Biochim. Biophys. Acta* **184**, 529–39.

Malaisse, W. J., Malaisse-Lagae, F. & Wright, P. H. (1967). *Endocrinology* **80**, 99–108.
Walker, D. G. (1968). In *Carbohydrate Metabolism and its Disorders*, Vol. 1, pp. 465–96. Eds. F. Dickens, W. J. Whelan & P. J. Randle. New York and London: Academic Press.
Willes, R. F., Boda, J. M. & Stokes, H. (1969). *Endocrinology* **84**, 671–5.
Willes, R. F., Manns, J. G. & Boda, J. M. (1969). *Endocrinology* **84**, 520–7.

THE USE OF CHRONICALLY CATHETERIZED FOETAL LAMBS FOR THE STUDY OF FOETAL METABOLISM

BY HEATHER J. SHELLEY*
Nuffield Institute for Medical Research,
University of Oxford

Place me in the sheep's position,
Spare me from the goat's condition,
Bring me to complete fruition!
(*Dies Irae*)

By far the most exciting recent development in foetal physiology has been the elaboration of techniques for studying the foetal lamb *in utero* for extended periods during the second half of gestation by the implantation of indwelling catheters in various foetal compartments. As described elsewhere in this volume, this has made possible the continuous recording of foetal blood pressure and heart rate, foetal breathing and swallowing movements and their relationship to the foetal electrocorticogram, in conscious animals, not only under near-normal conditions but also under various experimental conditions and during parturition. Samples of foetal body fluids for biochemical analysis can be obtained from the appropriate catheters and, in so far as changes in their composition reflect changes in foetal metabolism, these preparations can, within limits, be used for the study of foetal metabolism. During the last two years we have collected more than 2000 carotid blood samples for multiple assay from 102 chronically catheterized foetal lambs and their mothers at ages varying from 90 to 146 days' gestation (term is at 147 days). The purpose of this article is to describe our findings, to compare them with those of other groups studying chronically-catheterized foetal lambs, to reappraise the data obtained from acute preparations and to make suggestions for future work.

THE NORMAL FOETAL ENVIRONMENT

One of the pre-requisites for any study of foetal physiology is the establishment of criteria for assessing the state of the preparation at the time of

* The results to be presented in this paper were obtained with the help of numerous colleagues, notably Dr G. C. Liggins, who introduced us to the technique, and Drs K. Boddy and J. S. Robinson who carried on where he left off. My thanks are also due to Dr R. D. G. Milner of the Department of Child Health, University of Manchester, who was responsible for the insulin assays.

the experiment. The study of foetal lambs with indwelling vascular catheters is comparatively new, it is only seven years since Meschia, Cotter, Breathnach & Barron (1965) published their first paper on the subject, and although there is now fairly good agreement about acceptable levels for the P_{O_2}, P_{CO_2} and pH of foetal blood from different sites, there is less information about other parameters. One might postulate that the ideal preparation should go to term and deliver a healthy lamb of normal birth weight, but there are practical difficulties in achieving this. Not only is the incidence of premature labour high in such preparations, but it may be desirable to terminate the experiment before term, and in metabolic experiments it may be necessary to take such large volumes of foetal blood for analysis that subsequent survival is jeopardized. Moreover foetal survival is not necessarily a good index of foetal normality for, in common with other species, the foetal lamb is able to tolerate a surprising degree of hypoxia and hypoglycaemia. A case is already on record of a foetal lamb with a carotid P_{O_2} consistently below 14 mm Hg and a carotid plasma glucose of 6–12 mg/100 ml which was still alive and in reasonably good condition when the experiment was terminated at 132 days' gestation, nine days after the implantation of catheters (Dawes, Fox, LeDuc, Liggins & Richards, 1972 and unpublished data).

Post-operative changes

Most authors would agree that the foetus is unlikely to be in a normal state during the immediate post-operative period and many stress that no observations were made for at least 48 hours after operation. But recent more extensive studies suggest that the foetus may require 5–7 days to recover from the effects of surgery, even in the best preparations. Bassett & Thorburn (1969) found the plasma corticosteroid levels to be grossly elevated ($\times 6$ initially) in the mother during the first few days post-operatively, and Mellor & Slater (1971) showed that the plasma urea concentrations in mother and foetus fell to less than half the post-operative level during the first 3–5 days. Gresham, Rankin, Makowski, Meschia & Battaglia (1970) reported that the rate of foetal urine secretion was reduced post-operatively and did not stabilize for 3–6 days, and Mellor & Slater (1971) demonstrated changes in amniotic and allantoic fluid composition during this period.

In our own study the P_{O_2}, P_{CO_2} and pH, packed cell volume (PCV), plasma glucose and lactate were determined on all blood samples and in foetal samples the plasma fructose, which is synthesized from glucose in the placenta, was also determined. Figs. 1 and 2 summarize results obtained from the 13 foetuses of 10 ewes where blood samples were taken each morning for the first 6 days post-operatively. The ewes had been fasted overnight before operation and carotid and jugular catheters had been implanted in all 10 ewes and in two sets of twins under maternal epidural anaesthesia. In the other nine foetuses catheters were inserted not only into a carotid artery but also into the trachea and amniotic sac as described by Dawes *et al.* (1972), electromagnetic flowmeters were inserted in the trachea of two and the oesophagus of one and, in two, scalp electrodes were implanted biparietally to record the foetal electrocorticogram. Thus the type of preparation varied considerably, as did their gestational ages, 90–129 days on the day of operation. After operation all ewes were kept in individual cages and allowed water and food (concentrates,* hay and fresh grass) *ad libitum*, and the most important thing that these ewes had in common was that all ten made excellent post-operative recoveries. They looked healthy and alert, stood and ate within two hours of operation, there was no evidence of infection, and, as shown in Fig. 1, the maternal and foetal blood gases and pH had stabilized at acceptable levels within 24 hours of operation. Nevertheless it was 3–5 days before the other parameters had stabilized, similar changes being observed in all ten ewes (Fig. 2). In all but one ewe, where the level on day 1 was low, the maternal plasma glucose (□) was lowest on the 2nd or 3rd day post-operatively, after which it rose steadily and usually exceeded the initial level. These changes in maternal glucose were paralleled by smaller changes in the foetal plasma glucose (○) and there was a large post-operative fall in foetal plasma fructose (◑). There was a tendency for the plasma lactate level to be slightly higher than subsequently on the first post-operative day in the mother (■) and on the first two post-operative days in the foetus (●).

The changes in maternal and foetal plasma glucose were most probably related to changes in maternal food intake. Adequate food records were available for only 5 of the 10 ewes but, as shown in Fig. 2, their consumption of concentrates rose steadily during the first four days and the quantity of hay (200–800 g) consumed by these sheep (body weight 65–85 kg) also tended to rise. In sheep, the plasma glucose is almost wholly endogenous in origin, for the dietary carbohydrate is converted to short-chain fatty acids in the rumen (Elsden & Phillipson, 1948), and it is well established that when the food intake of a pregnant ewe is severely restricted, her blood glucose drops to half the normal level within 24–48 h (Parry & Shelley, 1958). This effect can be demonstrated from 80 days' gestation onwards and voluntary food restriction is common among ewes placed in

* A mixture of barley, beet pulp, flaked maize, beans and molassine meal in the proportions by weight of 10:1.5:1.5:1:1, fortified with added minerals.

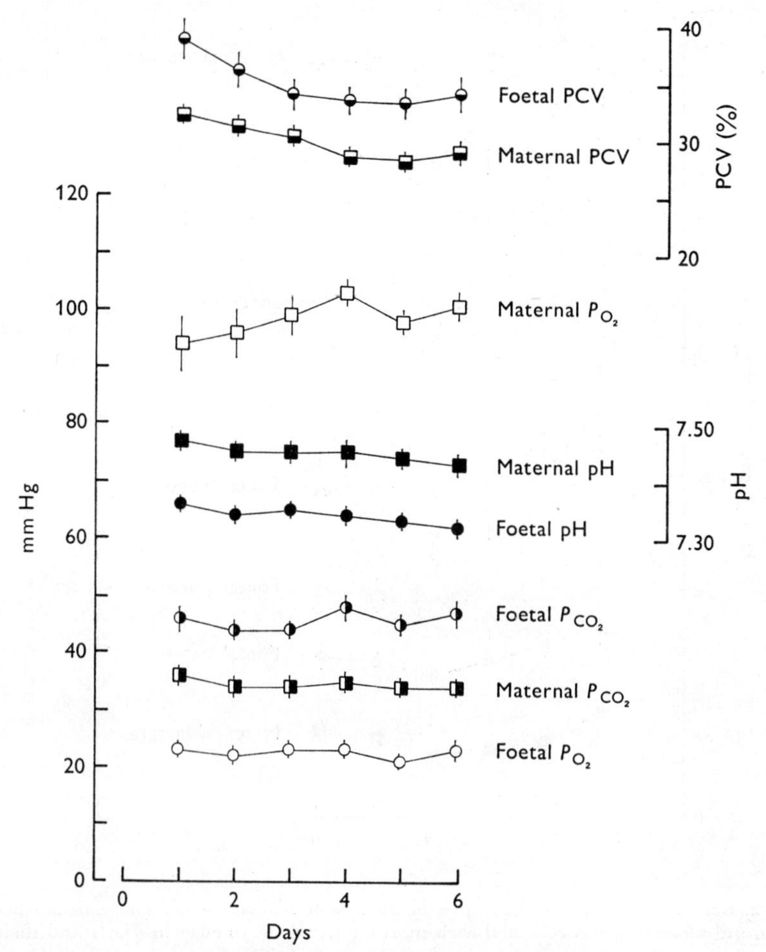

Fig. 1. Resting maternal and foetal PCV and pH in carotid blood samples collected each morning from 10 ewes and their 13 foetuses during the post-operative period; means ± SE. The operations were performed on day 0 at 90–129 days' gestation; for details see text.

strange surroundings (H. B. Parry and H. J. Shelley, unpublished). When sheep are placed in metabolism cages it is usually 4–5 days before their food intake stabilizes, even when there has been no major surgery, and if they are disturbed unduly their food intake may drop again. Thus fluctuations in daily food consumption are likely to explain some of the fluctuations in maternal and foetal plasma glucose.

Materno–foetal relationships

The post-operative changes in Figs. 1 and 2 are likely to be the minimum ever encountered, for not all sheep make such good post-operative recoveries.

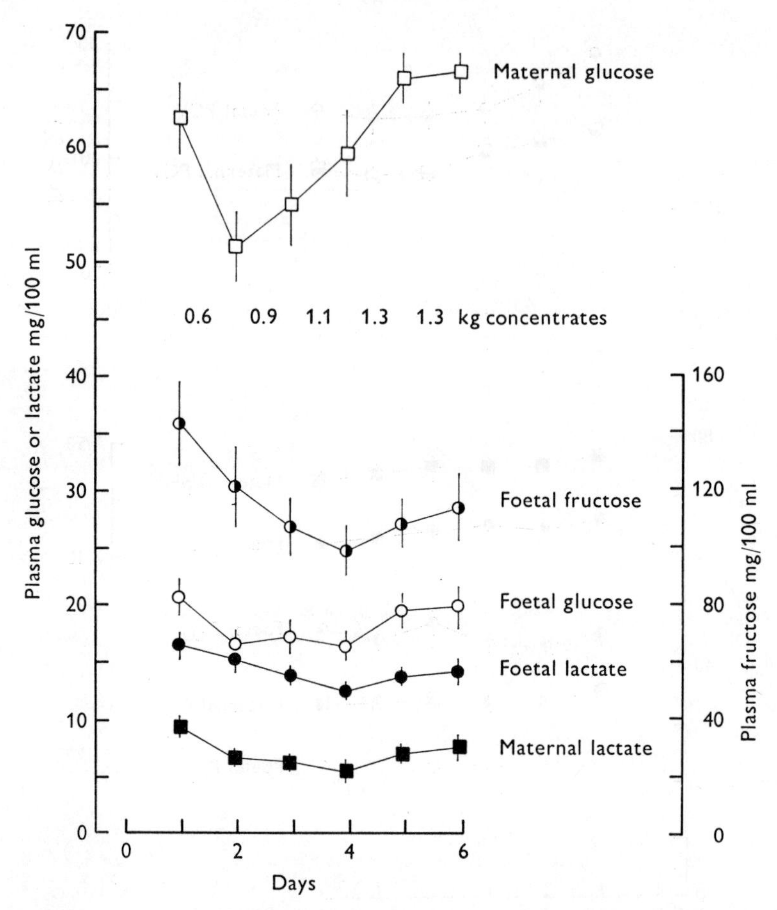

Fig. 2. Resting maternal and foetal plasma glucose and lactate, and foetal plasma fructose in carotid blood samples collected each morning from the 10 ewes in Fig. 1 and their 13 foetuses during the post-operative period; means ± SE. The figure also shows the mean daily food intake of 5 of these ewes.

Daily food consumptions of 200 g concentrates or less are not uncommon post-operatively and some such ewes may never eat normally again, their plasma glucose falls below 40 mg/100 ml and there is a large fall in foetal plasma glucose and fructose. The almost linear relationship between maternal and foetal plasma glucose is illustrated in Fig. 3, which summarizes results obtained from 76 well-oxygenated foetuses (plasma lactate < 20 mg/100 ml) of 93–146 days' gestation. Results obtained on the first two post-operative days have been omitted because, at a given maternal glucose concentration, the foetal level tended to be higher then than subsequently. As in earlier work on acute preparations (Shelley, 1960) and in the chronic preparations of Comline & Silver (1970), the maternal

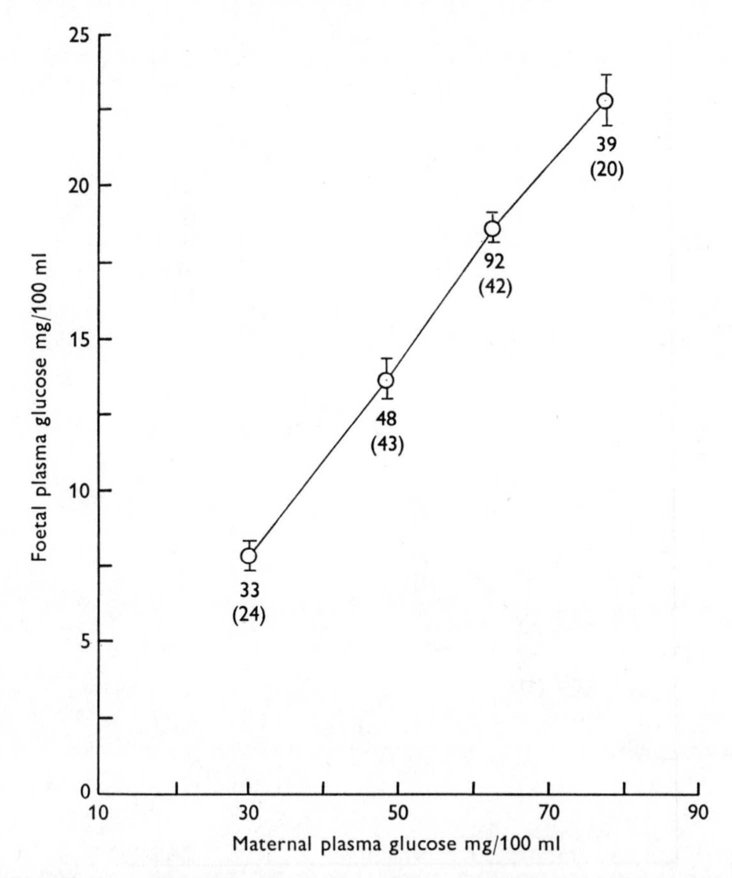

Fig. 3. The relationship between the resting foetal and maternal carotid plasma glucose levels in 76 well-oxygenated foetal lambs of 93–146 days' gestation; data from the first two days after operation have been omitted. Means ± SE with the number of samples (and foetuses) in parentheses.

and foetal blood and plasma glucose concentrations were independent of gestational age.

Fig. 4 shows the relationship between foetal plasma fructose and foetal plasma glucose in the same well-oxygenated foetuses. Again results from the first two days have been omitted, because of the large post-operative fall (◑, Fig. 2) which was also noted by Comline & Silver (1970) and Bassett & Thorburn (1971), and results from the two days immediately preceding the onset of labour have also been omitted. Comline & Silver (1970) described an abrupt fall in plasma fructose in the few days before delivery which they later ascribed to deterioration of their preparation (Comline & Silver, 1972), but we have noticed a slower fall in foetuses where there was no evidence of deterioration and no change in plasma

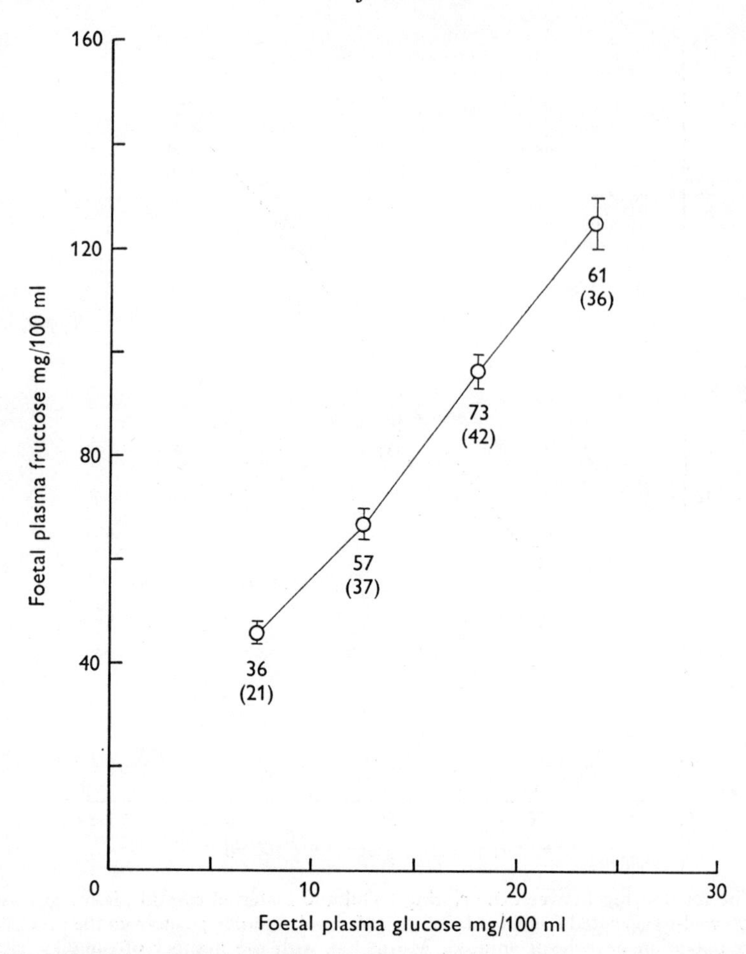

Fig. 4. The relationship between the resting foetal carotid plasma glucose and fructose in 77 well-oxygenated foetal lambs of 93–141 days' gestation; data from the first two days after operation and the two days preceding the onset of labour have been omitted. Means ± SE with the number of samples (and foetuses) in parentheses.

glucose or lactate. Somewhat surprisingly the relationship in Fig. 4 also appeared to be independent of gestational age and it seems likely that the fall in plasma fructose during the last third of gestation which has been noted in acute and chronic preparations (Huggett, Warren & Warren, 1951; Shelley, 1960; Comline & Silver, 1970) may have been due to (*a*) the much higher plasma fructose concentrations which *are* seen in the younger lambs (< 110 days' gestation) in the immediate post-operative period and (*b*) a higher incidence of maternal hypoglycaemia in ewes near term when they are housed under laboratory conditions. The relationship between glucose and fructose can be obscured in short-term experiments, for the changes in plasma fructose in response to changes in glucose are comparatively slow. When large amounts of glucose are administered to the ewe, as in the acute experiments of Huggett *et al.* (1951) and in our own chronic preparations, peak fructose concentrations are achieved four hours after the peak glucose concentration, and when the glucose change is within the physiological range, as in fasting and re-feeding experiments, it may be several days before the plasma fructose stabilizes.

A few samples were also assayed for plasma ketones, free fatty acids and total esterified fatty acids, total plasma protein and (through the kindness of Dr. Maureen Young) plasma amino acids. The results were compatible with data in the literature (see elsewhere in this volume) and foetal levels were unrelated to maternal levels. Briefly, the foetal plasma ketones (< 0.5 mM) and free fatty acids (< 0.3 m-equiv/l) were always low, regardless of the nutritional status of the ewe, the foetal plasma esterified fatty acids (usually 1–2 m-equiv/l) and total protein (3–4 g/100 ml) were about half the maternal levels, and the foetal plasma amino acid levels were higher than in the mother, though there were indications of a large post-operative fall.

Effects of hypoxia

Throughout this study, the foetal plasma lactate has been used as an index of foetal oxygenation. This method has its limitations for, as emphasized by Huckabee (1958), any factor which causes a rise in tissue pyruvate will automatically cause a rise in tissue lactate, and, as shown in Fig. 5, the administration of glucose to a pregnant ewe produces a transitory rise in foetal plasma lactate which is independent of any change in foetal P_{O_2}. But this effect is only seen when the foetal plasma glucose is raised above the normal physiological range and, in practice, the prognosis was usually poor if the foetal plasma lactate rose above 20 mg/100 ml after the immediate post-operative period. Such a rise was usually associated with a foetal carotid P_{O_2} of < 20 mm Hg, excluding foetuses where the lactate rise was sufficient to cause a fall in arterial pH and the Bohr effect was operative, but the converse was not the case. In some foetuses, a carotid P_{O_2} of 18–20 mm Hg was accompanied by plasma lactates below 15 mg/100 ml and one must assume that in spite of the low P_{O_2}, the rate of blood flow must have been sufficient to supply the oxygen needs of the tissues. Whereas over short time intervals the foetal arterial P_{O_2} can vary by several mm Hg, the foetal plasma lactate level is usually more stable and if elevated deliberately, e.g. by giving the ewe 9% oxygen to breathe, it usually takes several hours to return to the pre-hypoxic level. Thus the occasional raised plasma lactate which is accompanied by a normal arterial P_{O_2} may be

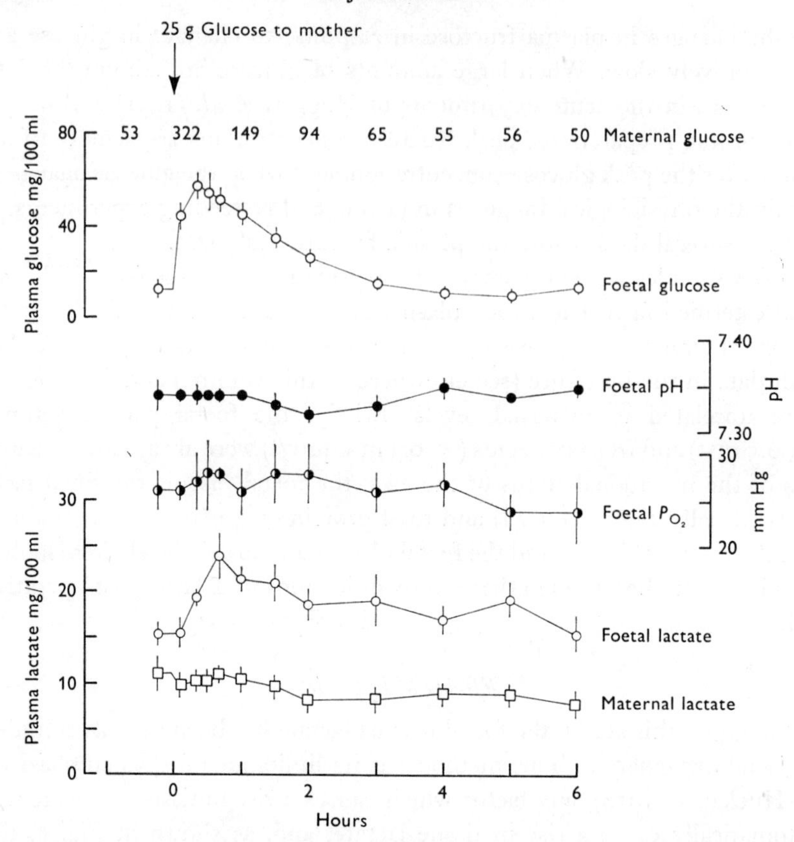

Fig. 5. The maternal and foetal carotid plasma glucose and lactate, foetal pH and P_{O_2} in 4 ewes of 132–138 days' gestation injected with 50 ml 50 % glucose via a maternal jugular vein. Means ± SE. Two ewes were breathing 50 % oxygen during the first two hours, hence the higher foetal arterial P_{O_2} at this time.

indicative of reduced blood flow to the tissues, but is more likely to be evidence of a previous disturbance of foetal metabolism.

An indication of whether the plasma lactate is likely to be elevated can be obtained quickly and with very little trouble by measuring the foetal PCV. As shown in Fig. 1 (⬓, ◒), both the maternal and foetal PCV are slightly elevated during the first two post-operative days, at a time when the plasma lactate is also slightly raised (■, ●, Fig. 2), and Fig. 6 shows the relationship between the foetal carotid PCV and plasma lactate in the later post-operative period. As indicated by the accompanying data for foetal P_{O_2}, P_{CO_2} and pH, grossly hypoxaemic foetuses (where the PCV was sometimes 50 % or more) were excluded from Fig. 6 and all plasma lactates were below 40 mg/100 ml. The relationship was independent of gestational age and maternal PCV. A comparable rise in foetal PCV occurs

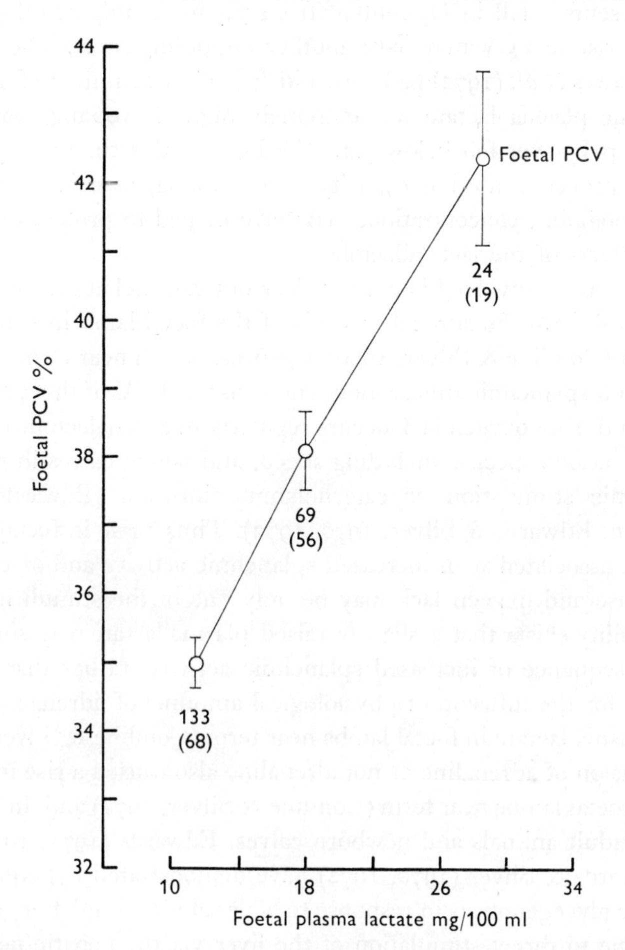

Fig. 6. The relationship between the resting foetal carotid PCV and plasma lactate in 89 foetal lambs of 93–143 days' gestation; data from the first two days after operation have been omitted. Means ± SE with the number of samples (and foetuses) in parentheses. The figure also shows the corresponding mean value ± SE for maternal carotid PCV and foetal carotid P_{O_2}, P_{CO_2} and pH.

when lambs in mid-gestation are asphyxiated by tying the umbilical cord (Dawes, Mott, Shelley & Stafford, 1963) and in the present work a prompt rise of several per cent usually followed the administration of 9% O_2 to the ewe. This rise in PCV partially compensates for the fall in P_{O_2} for if the oxygen content of the blood is calculated from the data in Fig. 6, using the formula of Cassin, Dawes, Mott, Ross & Strang (1964)* and a value of 0.28 for the ratio of haemoglobin content to PCV (McFadyen, Boonyaprakob & Hutchinson, 1968), the fall in P_{O_2} from 21 to 17 mm Hg (a 20% drop) represents a fall in O_2 content from 7.4 to 6.7 ml/100 ml (a 9.5% drop). The rise in PCV may have another important effect. The hypoxic foetus of Dawes *et al.* (1972) had a carotid P_{O_2} below 10 mm Hg for several days and the plasma lactate was extremely high, > 100 mg/100 ml, yet the carotid pH never fell below 7.2. The PCV at this time was 54–59% and the improved buffering capacity of the blood, associated with the raised haemoglobin concentration, may have helped to protect the foetus from the effects of the lactacidaemia.

The mechanism involved is not yet clear but catecholamines are known to be released from the adrenal medulla of the foetal lamb in response to oxygen lack (Comline & Silver, 1966), a process which near term is largely dependent on splanchnic innervation, and a rise in PCV of the same order as that seen during oxygen lack occurs regularly in adrenalectomized adult animals of various species, including sheep, and young calves in response to splanchnic stimulation or catecholamine infusion (Edwards, 1971, 1972*a* and *b*; Edwards & Silver, 1970, 1972). Thus a rise in foetal PCV is likely to be associated with increased splanchnic activity and/or catecholamine release and oxygen lack may be only one of the stimuli involved. The possibility exists that a slightly raised plasma lactate may sometimes be the consequence of increased splanchnic activity, rather than oxygen lack *per se*, for the infusion of physiological amounts of adrenaline causes a rise of plasma lactate in foetal lambs near term (Comline & Silver, 1972).

The infusion of adrenaline or noradrenaline also caused a rise in plasma glucose in foetal lambs near term (Comline & Silver, 1972) and, in adrenalectomized adult animals and newborn calves, Edwards (1971, 1972*a* and *b*) and Edwards & Silver (1970, 1972) have demonstrated hyperglycaemia and hepatic glycogenolysis in response to splanchnic stimulation, an effect which is due to direct stimulation of the liver via the hepatic nerves. In our experiments, the administration of 9% oxygen to the ewe caused a rise in foetal plasma glucose of about 5 mg/100 ml at 120–130 days' gestation but, as shown in Fig. 7, at 140 days there was a much larger rise. This rise was independent of any change in maternal plasma glucose and was almost certainly due to foetal hepatic glycogenolysis for, twelve hours after the experiment in Fig. 7, the foetal liver glycogen was only 6 mg/g compared with normal levels of 100 mg/g or more.

In these experiments the rise in foetal plasma glucose was followed by

* Log $P_{O_2} = 3.63 - 0.32\,\text{pH} + 0.42 \log S/100S$ where S is the O_2 saturation of the blood.

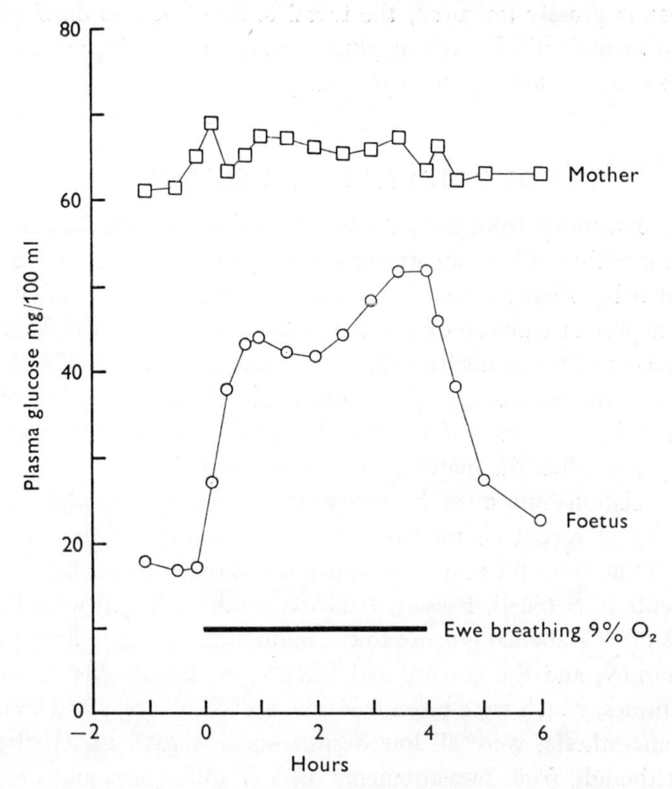

Fig. 7. Changes in maternal and foetal carotid plasma glucose during acute hypoxia in a ewe at 140 days' gestation. The ewe was breathing a mixture of 9 % O_2 + 3 % CO_2 + 88 % N_2 during the period indicated (black bar), the CO_2 being added to maintain the P_{CO_2} during maternal hyperventilation.

a rise in foetal plasma fructose, similar to that seen after administering glucose to the foetus, but in cases of spontaneous hypoxaemia, where there was presumably some failure of placental function, the foetal plasma fructose fell, even when there was a rise in foetal plasma glucose. Thus in moderate spontaneous hypoxaemia (foetal plasma lactate 20–40 mg/100 ml), the foetal plasma fructose at a given foetal plasma glucose concentration tended to be 10–20 mg/100 ml below the values shown in Fig. 4. The foetal plasma glucose was often well maintained, but if the foetus was grossly hypoxaemic, the plasma glucose was usually lower than would be predicted from the relationship in Fig. 3. Thus the hypoxaemic foetus cited by Dawes *et al.* (1972) had plasma glucose levels of 6–12 mg/100 ml, yet its mother was never hypoglycaemic and the plasma glucose in its better-oxygenated twin was 14–17 mg/100 ml for most of the period studied. It seems clear that if placental function has failed to such an extent that the placental transfer of glucose is grossly impaired, the foetal lamb of < 130 days' gestation is unable to maintain its plasma glucose at normal levels, even though it has some capacity for hepatic glycogenolysis.

FOETAL ENERGY SOURCES

It can be concluded from the data in the previous section that the normal pregnant ewe should have an arterial plasma glucose level of 60–70 mg/100 ml and that her foetus should have a carotid plasma glucose of 18–22 mg/100 ml, a plasma fructose of about 100 mg/100 ml and a plasma lactate concentration of 10–15 mg/100 ml. If the maternal glucose falls, the foetal glucose and fructose should fall as indicated in Figs. 3 and 4, but placental failure can be suspected if the foetal plasma glucose and fructose are unusually low when the maternal glucose is normal.

These relationships must be taken into consideration when assessing the validity of recent claims that the glucose uptake of the foetal lamb *in utero* can account for only a small part of its oxidative metabolism. The experiments of Setchell, Bassett, Hinks & Graham (1972) were based on infusions of ^{14}C-labelled glucose to the mother and foetus within 24 hours of laparotomy, and the glucose and fructose concentrations in the foetal blood samples, which were taken from the umbilical artery and vein under general anaesthesia, were all low despite some normal maternal glucose levels. Although, from measurements of A–V differences and blood flow in ewes with cannulae in a femoral artery and uterine vein, they demonstrated an uptake of glucose by the pregnant uterus which was more than sufficient to account for its oxygen consumption, they calculated from the specific activities of blood glucose and CO_2 that glucose oxidation could account for rather less than half its CO_2 production. In a further experiment they infused ^{14}C-glucose into the foetus via a foetal cotyledonary vein and calculated that less than a quarter of the foetal CO_2 production was derived from blood glucose. But in this experiment the glucose concentration in the umbilical artery was higher than in the vein and, in this and in four similar preparations, the foetal liver glycogen was lower than in animals where only the maternal vessels were cannulated. It seems likely that these foetuses were hypoxic, a situation which would upset not only their glucose homeostasis but also their CO_2 production, and the situation was complicated further by the presence of the placenta and the finding that ^{14}C-fructose was utilized to a much greater extent than in the isolated foetuses of Alexander, Britton & Nixon (1970).

But Tsoulos, Colwill, Battaglia, Makowski & Meschia (1971) came to a similar conclusion using a somewhat different approach. Their experiments and the substrates available for oxidative metabolism in the foetus will be discussed in detail elsewhere in this volume (paper by Battaglia *et al.*) but some comment is relevant here. Their technique is to compare the umbilical A–V difference for oxygen with that of the substrate under consideration in ewes with chronically implanted catheters in a branch of the umbilical vein and a foetal femoral artery. In the work already published (Tsoulos *et al.* 1971) the foetal survival rate was good and all but one foetus, which was subsequently stillborn, were well-oxygenated as judged by the oxygen content of the blood, though, as discussed above, this is not a sensitive index of slight degrees of oxygen lack. In the hypoglycaemic foetuses of fasted ewes the glucose difference across the cord was equivalent to less than one-fifth of the oxygen uptake, and even in ewes fed *ad libitum*, the uptake of glucose appeared to account for not more than half that of oxygen. But the 'fed' ewes had been fasted for 48 hours before operation, some had been fasted post-operatively, and all were moved to restraining cages the day before samples were taken. Although the blood glucose was sometimes normal, the fructose was always low, < 45 mg/100 ml, and some caution should be exercised in extrapolating these findings to the truly normal foetus. Not only is it likely that the nutritional status of the ewes was not entirely normal, but the very high blood glucose levels in some experiments, equivalent to plasma glucose levels of > 100 mg/100 ml in the mother and up to 40 mg/100 ml in the foetus, suggest that the ewes were disturbed by the sampling procedure. Near term large amounts of glycogen are present not only in the foetal liver but also in the foetal skeletal muscles (Shelley, 1960; Setchell *et al.* 1972) and though it is not yet known if the foetus can utilize these in response to hypoglycaemia, the response to catecholamines has already been described.

STUDIES WITH INSULIN

The close relationship between maternal and foetal plasma glucose in foetal lambs in good condition (Fig. 3) suggests a lack of control by the foetus, yet insulin is present in the plasma as early as 42 days' gestation in amounts which are independent of the maternal concentration (Alexander, Britton, Cohen, Nixon & Parker, 1968). Although Bassett & Thorburn (1971) found no correlation between plasma glucose and insulin in chronically catheterized foetal lambs, they were more successful in later work (see paper by Bassett in this volume) and Fig. 8 summarizes our own findings where insulin was determined by the double antibody method of Hales & Randle (1963). Although the plasma glucose in the foetus was so much lower than in the mother, the plasma insulin was in the adult range and

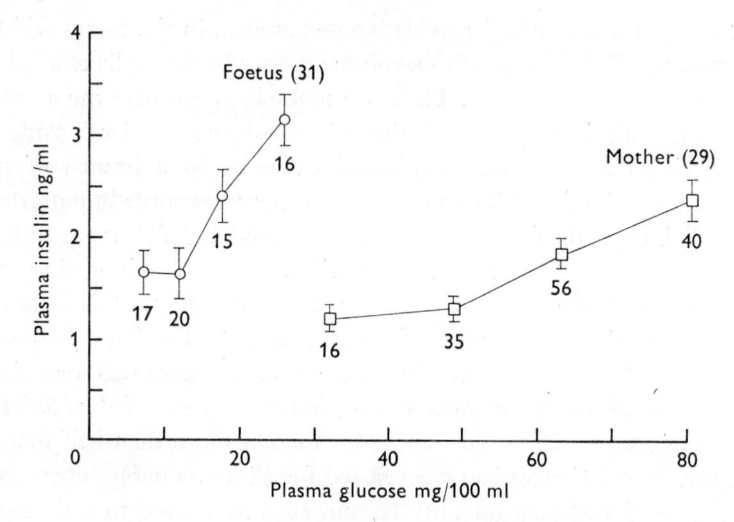

Fig. 8. The relationship between the resting carotid plasma insulin and glucose in 31 well-oxygenated foetuses and their mothers at 102–146 days' gestation. Means ± SE with the number of samples in parentheses.

appeared to vary over a much smaller range of glucose concentrations than in the adult.

This was surprising, for there is a large body of evidence suggesting that insulin release in response to glucose administration is sluggish or absent in the foetus of several species including sheep (Willes, Boda & Manns, 1969; Alexander, Britton, Cohen & Nixon, 1969; Alexander, Britton, Mashiter, Nixon & Smith, 1970; Davis, Beck, Colwill, Makowski, Meschia & Battaglia, 1971; Bassett & Thorburn, 1971; Mintz, Chez & Horger, 1969; Chez, Mintz, Hutchinson & Horger, 1972; Obenshain, King, Adam, Raivio, Teramo, Räihä & Schwartz, 1969; Oakley, Beard & Turner, 1972), and the response, when seen, was variable even in the same animal on different occasions. It is likely that catecholamine release or hypoxia can account for some of the variation, for in-vivo and in-vitro experiments have demonstrated conclusively that these are potent inhibitors of insulin release in all the species studied, including sheep and newborn lambs (Porte, Graber, Kuzuya & Williams, 1966; Baum & Porte, 1969; Milner & Hales, 1969; Bassett, 1970; Bassett & Alexander, 1971). In our own experiments a rise in plasma insulin following the injection of glucose to the ewe was seen in mother and foetus only when both were in good condition and undisturbed, and there was no change in foetal plasma insulin in the hypoxia experiment in Fig. 7, despite the rise in foetal plasma glucose. Moreover the relationship between resting plasma glucose and insulin (Fig. 8) was only seen in foetal lambs with plasma

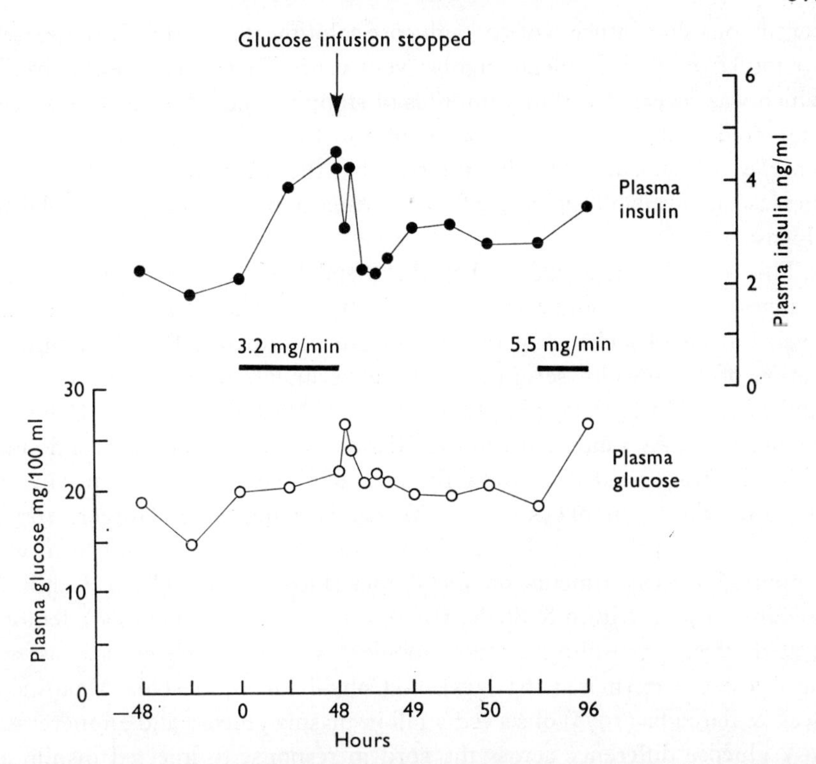

Fig. 9. The effect of infusing glucose via a foetal jugular vein on the foetal carotid plasma insulin and glucose in a well-oxygenated foetal lamb. The experiment covered a period of 7 days, from 107 to 113 days' gestation; note the scale expansion at 48–50 hours. Glucose was infused at the rates indicated (black bar) and the lamb weighed 2.3 kg at 114 days' gestation. The small rise in plasma glucose immediately after stopping the glucose infusion was associated with flushing the infusion catheter.

lactates below 25 mg/100 ml. But, as discussed by Bassett (paper in this volume), another factor responsible for the poor response in foetal lambs, if not in other species, may have been the large amounts of glucose given. In most investigations the foetal plasma glucose was raised to 100–200 mg/100 ml or more, concentrations which Bassett found to inhibit insulin release from foetal ovine pancreas *in vitro* and which we have observed to cause a rise in foetal plasma lactate of 100 mg/100 ml or more, sufficient to cause a fall in arterial pH and similar to that observed when the ewe is breathing 9% oxygen.

Such experiments are useless as a means of validating the relationship in Fig. 8, but in a few preliminary experiments we have investigated the response to smaller amounts of glucose and have observed changes in foetal plasma insulin in response to changes in plasma glucose within the physiological range. In one particularly successful experiment (Fig. 9) the continuous slow infusion of 50% glucose solution at a rate of approximately 1.5 mg/kg min via a foetal jugular vein caused a rise in plasma insulin which was reversed within 5 minutes of stopping the infusion. It has been suggested that the glucose uptake of the foetal lamb is approximately 6 mg/kg min and it is of some interest that the infusion rate in Fig. 9, an increase in available glucose of 25%, caused no rise in carotid plasma glucose.

This experiment suggests that the foetal lamb can respond to small changes in glucose entry rate not only by changing the plasma insulin level, but also by changing the rate of glucose uptake. But the administration of anti-insulin serum in acute experiments on exteriorized foetal lambs caused little change in plasma glucose (Alexander, Britton, Cohen & Nixon, 1970; Alexander, Britton & Nixon, 1971), and very large amounts of insulin were needed to lower the plasma glucose in acute experiments on foetal rhesus monkeys (Chez, Mintz, Hutchinson & Horger, 1972). The suggestion that 'insulin sensitivity' does not develop until term was supported by experiments on foetal rats (Picon, 1967; Clark, Cahill & Soeldner, 1968; Britton & Blade, 1970), yet Bocek & Beatty (1969) demonstrated effects of insulin on rhesus monkey skeletal muscle as early as 100 days' gestation (term is at 165 days) and Colwill, Davis, Meschia, Makowski, Beck & Battaglia (1970) observed a fall in plasma glucose and an increased A–V glucose difference across the cord in response to injected insulin as early as 122 days' gestation in chronically catheterized foetal lambs.

In our own experiments we have tried to determine the 'sensitivity' to insulin of foetal lambs by giving a continuous slow infusion of insulin to one of twins, using the untreated twin as a control to compensate for any change in maternal plasma glucose. As shown in Fig. 10, the carotid plasma glucose level could be maintained at about half that in the control twin for the duration of the infusion, 7–8 days in three experiments, and there was a large progressive fall in plasma fructose. We have not yet 'caught' the point at which the glucose starts to fall, but as shown in Fig. 11, a maximal depression of the foetal plasma glucose had been achieved by the time the plasma insulin had risen to 15 ng/ml, and on stopping the infusion there was no change in plasma glucose until the plasma insulin had fallen below 5 ng/ml. In similar experiments on adult sheep, maximal depression of the plasma glucose was achieved at a plasma insulin concentration of 5–8 ng/ml and it is possible that the insulin sensitivity of the foetal lamb near term is similar to that of the adult.

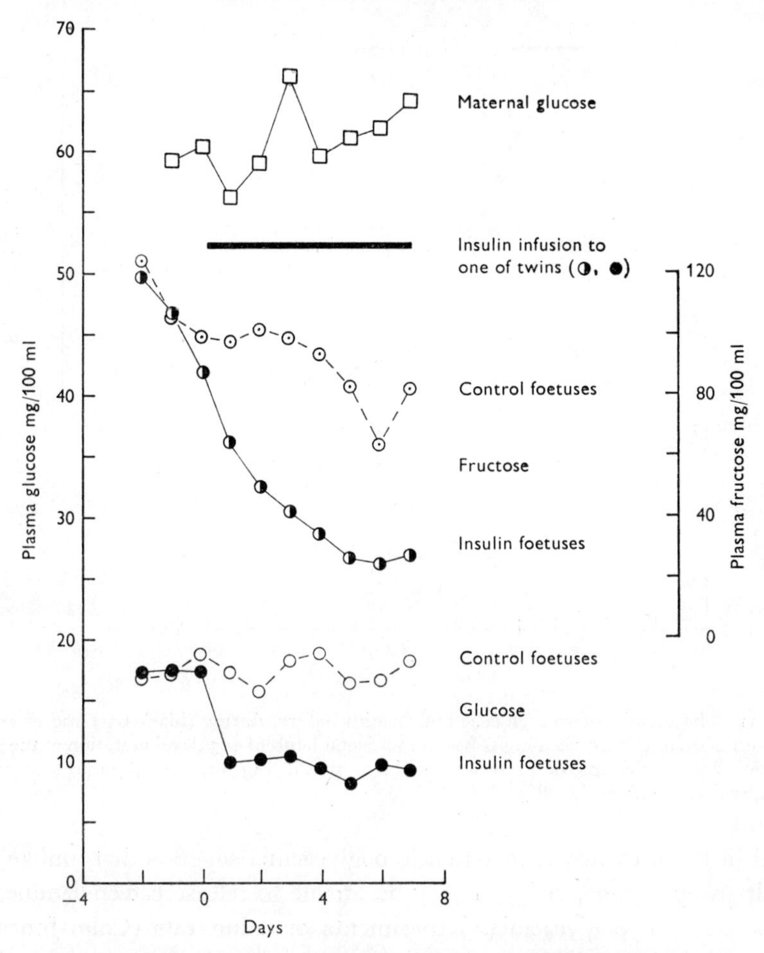

Fig. 10. The effect of a continuous slow infusion of insulin (black bar) via a foetal carotid artery on the carotid plasma glucose and fructose in 6 foetal lambs of 104–122 days' gestation at the start of the infusion. Each lamb was one of twins and the figure also shows the carotid plasma glucose and fructose in the control twin, which received saline, and the maternal plasma glucose. The insulin infusion rate was approximately 16 U/day. The SE of the means have been omitted because they indicate the variation between the different pregnancies, not that between the control and insulin-treated foetuses.

CONCLUSIONS

The experiments described above suggest that the foetal lamb in the last third of gestation can regulate its plasma glucose level to some extent by raising or lowering the plasma insulin level in response to changes in glucose entry rate. It can respond to oxygen lack by increasing the output of catecholamines from the adrenal medulla and mobilizing its hepatic glycogen reserves, but its singular inability to maintain its plasma glucose

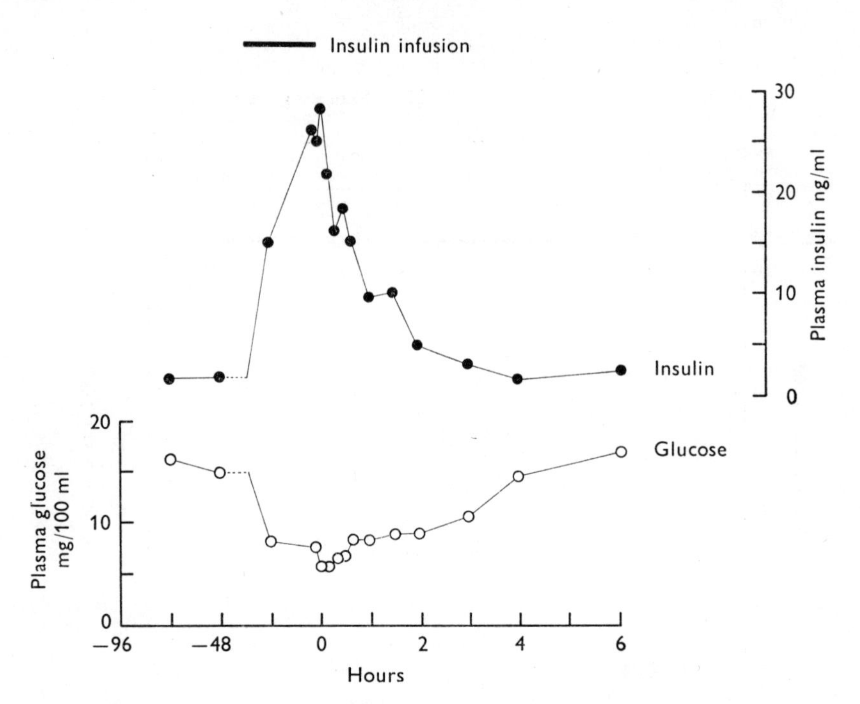

Fig. 11. The carotid plasma glucose and insulin before, during (black bar) and after the slow infusion of insulin via a jugular vein in a foetal lamb of 123 days' gestation at the start of the infusion. The infusion rate was similar to that in Fig. 10, 4 mU/kg min (160 ng/kg min), and the lamb weighed 3.2 kg.

level in the presence of maternal hypoglycaemia suggests that, unlike the adult sheep (Crone, 1965), it may be unable to release catecholamines in response to hypoglycaemia. Experiments on adult rats (Colin-Jones & Himsworth, 1970; Himsworth, 1970) suggest that the centres responsible for stimulating the secretion of catecholamines during hypoglycaemia may lie bilaterally in the lateral regions of the hypothalamus, and though the existence of such centres in the sheep has been challenged by Crone (1965), on the basis of studies with decerebrate and spinal preparations, his experiments cannot be regarded as conclusive. It is possible that the ability of these centres to respond to hypoglycaemia is not acquired until very late in foetal development or after birth, a factor which might contribute to the high incidence of hypoglycaemia in the immediate newborn period in babies and animal species, and some support for such a view was provided by our experiments on the effects of insulin infusion. When insulin was infused for 24–48 h in adult sheep and then stopped, there was a rapid rise in plasma glucose to grossly hyperglycaemic levels which were maintained for several hours, but, as indicated in Fig. 11, this was never seen in similar experiments on the foetus. It is likely that the post-insulin hyperglycaemia was due to adaptive changes in enzymes concerned with gluconeogenesis and hepatic glucose release, and catecholamines are known to be effective agents in this respect in both adult and foetal rats (Yeung & Oliver, 1968; Greengard, 1969; Reshef & Hanson, 1972).

But cortisol and glucagon are also effective agents which are likely to be released in the adult in response to hypoglycaemia, and it is of some interest that the increased ability of the foetal lamb to raise its plasma glucose in response to oxygen lack (Fig. 7) was only seen within a few days before the onset of labour, at a time when the plasma cortisol is rising (Nathanielsz, Comline, Silver & Paisey, 1972). The release of ACTH in response to hypoglycaemia, and possibly that of glucagon, could also be mediated via the hypothalamic centres, and because of the role of ACTH and cortisol of foetal origin in the initiation of parturition (Liggins, Grieves, Kendall & Knox, 1972; Thorburn, Nicol, Bassett, Shutt & Cox, 1972) it is important to see whether this response is present in the foetus. The role of the central nervous system in the control of insulin release should also receive some attention for, in contrast to the inhibitory effects of catecholamines, acetylcholine is an effective stimulant of insulin release and it is possible to establish a conditioned release of insulin in adults (Woods, Hutton & Makous, 1970; Woods, Alexander & Porte, 1972).

One other phenomenon of practical importance which was observed in our experiments was the rise in plasma lactate which was observed in foetal lambs on injecting glucose (Fig. 5), a response which occurred even when the mother was breathing 50% oxygen, and which was never seen in the adult in spite of the much higher maternal plasma glucose levels. This effect was also seen in well-oxygenated newborn lambs and has been observed in newborn premature babies (Soltész, Mestyán, Járai, Fekete & Schultz, 1971). The rise in lactate was roughly proportional to the rise in glucose and some doubt must be cast on the wisdom of the wide-spread practice of administering glucose to women in labour for, in our experience, if the foetal lamb became hypoxic following the administration of glucose, the resultant more rapid rise in plasma lactate and the fall in arterial pH were greater than in normoglycaemic hypoxic controls, and whereas the controls usually recovered when their P_{O_2} was raised again, the glucose-treated foetuses usually died. These investigations are in an early stage, but they suggest that the increased intra-uterine mortality in the baby of the diabetic mother, which has been attributed to the effects of hypoglycaemia (Oakley *et al.* 1972), may in fact be associated with hyperglycaemia in mothers where the diabetes is poorly controlled.

The purpose of this paper was to illustrate the type of metabolic experiment for which the chronically catheterized foetal lamb is especially suitable and to outline some of the hazards involved in work on conscious animals. It is clear that the possibilities are almost endless and that, though the economic status of the sheep is declining, its role in foetal physiology is assured for many years to come.

This work was made possible by a grant from the Medical Research Council. I also acknowledge the help of Mrs Janet Inman.

REFERENCES

ALEXANDER, D. P., BRITTON, H. G., COHEN, N. M., MASHITER, K., NIXON, D. A. & SMITH, F. G., Jr (1971). *Biol. Neonat.* **17**, 381–93.

ALEXANDER, D. P., BRITTON, H. G., COHEN, N. M. & NIXON, D. A. (1969). *Biol. Neonat.* **14**, 178–93.

ALEXANDER, D. P., BRITTON, H. G., COHEN, N. M. & NIXON, D. A. (1970). *Biol. Neonat.* **15**, 142–55.

ALEXANDER, D. P., BRITTON, H. G., COHEN, N. M., NIXON, D. A. & PARKER, R. A. (1968). *J. Endocr.* **40**, 389–90.

ALEXANDER, D. P., BRITTON, H. G., MASHITER, K., NIXON, D. A. & SMITH, F. G., Jr (1970). *Biol. Neonat.* **15**, 361–7.

ALEXANDER, D. P., BRITTON, H. G. & NIXON, D. A. (1970). *Q. Jl exp. Physiol.* **55**, 346–62.

ALEXANDER, D. P., BRITTON, H. G. & NIXON, D. A. (1971). *Biol. Neonat.* **19**, 451–8.

BASSETT, J. M. (1970). *Aust. J. biol. Sci.* **23**, 903–14.

BASSETT, J. M. & ALEXANDER, G. (1971). *Biol. Neonat.* **17**, 112–25.

BASSETT, J. M. & THORBURN, G. D. (1969). *J. Endocr.* **44**, 285–6.

BASSETT, J. M. & THORBURN, G. D. (1971). *J. Endocr.* **50**, 59–74.

BAUM, D. & PORTE, D., Jr (1969). *J. clin. Endocr.* **29**, 991–4.

BOCEK, R. M. & BEATTY, C. H. (1969). *Endocrinology* **85**, 615–18.

BRITTON, H. G. & BLADE, M. (1970). *Biol. Neonat.* **16**, 370–5.

CASSIN, S., DAWES, G. S., MOTT, J. C., ROSS, B. B. & STRANG, L. B. (1964). *J. Physiol.* **171**, 61–79.

CHEZ, R. A., MINTZ, D., HUTCHINSON, D. L. & HORGER, E. O. (1972). In *Physiological Biochemistry of the Fetus*, ed. A. A. Hodari and F. G. Mariona, pp. 117–28. Springfield: Charles C. Thomas.

CLARK, C. M., Jr, CAHILL, G. F., Jr & SOELDNER, J. A. (1968). *Diabetes* **17**, 362–8.

COLIN-JONES, D. G. & HIMSWORTH, R. L. (1970). *J. Physiol.* **206**, 397–409.

COLWILL, J. R., DAVIS, J. R., MESCHIA, G., MAKOWSKI, E. L., BECK, P. & BATTAGLIA, F. C. (1970). *Endocrinology* **87**, 710–15.

COMLINE, R. S. & SILVER, M. (1966). *Br. med. Bull.* **22**, 16–20.

COMLINE, R. S. & SILVER, M. (1970). *J. Physiol.* **209**, 567–86.

COMLINE, R. S. & SILVER, M. (1972). *J. Physiol.* **222**, 233–56.

CRONE, C. (1965). *Acta physiol. scand.* **63**, 213–24.

DAVIS, J. R., BECK, P., COLWILL, J. R., MAKOWSKI, E. L., MESCHIA, G. & BATTAGLIA, F. C. (1971). *Proc. Soc. exp. Biol. Med.* **136**, 972–5.

DAWES, G. S., FOX, H. E., LEDUC, B. M., LIGGINS, G. C. & RICHARDS, R. T. (1972). *J. Physiol.* **220**, 119–43.

Dawes, G. S., Mott, J. C., Shelley, H. J. & Stafford, A. (1963). *J. Physiol.* **168**, 43–64.
Edwards, A. V. (1971). *J. Physiol.* **213**, 741–59.
Edwards, A. V. (1972*a*). *J. Physiol.* **220**, 315–34.
Edwards, A. V. (1972*b*). *J. Physiol.* **220**, 697–710.
Edwards, A. V. & Silver, M. (1970). *J. Physiol.* **211**, 109–124.
Edwards, A. V. & Silver, M. (1972). *J. Physiol.* **223**, 571–93.
Elsden, S. R. & Phillipson, A. T (1948). *A. Rev. Biochem.* **17**, 705–26.
Greengard, O. (1969). *Biochem. J.* **115**, 19–24.
Gresham, E. L., Rankin, J. H. G., Makowski, E. L., Meschia, G. & Battaglia, F. C. (1970). *Pediat. Res.* **4**, 445–6.
Hales, C. N. & Randle, P. J. (1963). *Biochem. J.* **88**, 137–46.
Himsworth, R. L. (1970). *J. Physiol.* **206**, 411–17.
Huckabee, W. E. (1958). *J. clin. Invest.* **37**, 244–54.
Huggett, A. St G., Warren, F. L. & Warren, N. V. (1951). *J. Physiol.* **113**, 258–75.
Liggins, G. C., Grieves, S. A., Kendall, J. Z. & Knox, B. S. (1972). *J. Reprod. Fert.* Suppl. **16**, 85–103.
McFadyen, I. R., Boonyaprakob, U. & Hutchinson, D. L. (1968). *Am. J. Obstet. Gynec.* **100**, 686–95.
Mellor, D. J. & Slater, J. S. (1971). *J. Physiol.* **217**, 573–604.
Meschia, G., Cotter, J. R., Breathnach, C. S. & Barron, D. H. (1965). *Q. Jl exp. Physiol.* **50**, 185–95.
Milner, R. D. G. & Hales, C. N. (1969). *Biochem. J.* **113**, 473–9.
Mintz, D. H., Chez, R. A. & Horger, E. O., iii (1969). *J. clin. Invest.* **48**, 176–86.
Nathanielsz, P. W., Comline, R. S., Silver, M. & Paisey, R. B. (1972). *J. Reprod. Fert.* Suppl. **16**, 39–59.
Oakley, N. W., Beard, R. W. & Turner, R. C. (1972). *Br. med. J.* **1**, 466–9.
Obenshain, S., King, K., Adam, P., Raivio, K., Teramo, K., Räihä, N. & Schwartz, R. (1969). *Pediat. Res.* **3**, 380–1.
Parry, H. B. & Shelley, H. J. (1958). *J. Physiol.* **140**, 48–9P.
Picon, L. (1967). *Endocrinology* **81**, 1419–21.
Porte, D., Graber, A. L., Kuzuya, T. & Williams, R. H. (1966). *J. clin. Invest.* **45**, 228–36.
Reshef, L. & Hanson, R. W. (1972). *Biochem. J.* **127**, 809–18.
Setchell, B. P., Bassett, J. M., Hinks, N. T. & Graham, N. McC. (1972). *Q. Jl exp. Physiol.* **57**, 257–66.
Shelley, H. J. (1960). *J. Physiol.* **153**, 527–52.
Soltész, G., Mestyán, J., Járai, I., Fekete, M. & Schultz, K. (1971). *Biol. Neonat.* **19**, 118–31.
Thorburn, G. D., Nicol, D. H., Bassett, J. M., Shutt, D. A. & Cox, R. I. (1972). *J. Reprod. Fert.* Suppl. **16**, 61–84.
Tsoulos, N. G., Colwill, J. R., Battaglia, F. C., Makowski, E. L. & Meschia, G. (1971). *Am. J. Physiol.* **221**, 234–7.
Willes, R. F., Boda, J. M. & Manns, J. G. (1969). *Endocrinology* **84**, 520–7.
Woods, S. C., Hutton, R. A. & Makous, F. (1970). *Proc. Soc. exp. Biol. Med.* **133**, 964–8.
Woods, S. C., Alexander, K. R. & Porte, D., Jr (1972). *Endocrinology* **90**, 227–31.
Yeung, D. & Oliver, I. T. (1968). *Biochem. J.* **108**, 325–31.

FOETAL METABOLISM AND SUBSTRATE UTILIZATION

By F. C. BATTAGLIA and G. MESCHIA
Division of Perinatal Medicine, University of Colorado Medical Center, Denver, Colorado 80220, USA

This is a review of studies which have sought to measure feotal metabolic rate and substrate utilization. For the most part these studies have been carried out in sheep foetuses.

FOETAL OXYGEN CONSUMPTION

Oxygen consumptions of foetal lambs ranging from 4 to 10 ml/min·kg have been reported in various studies (James, Raye, Gresham, Makowski, Meschia & Battaglia, 1972). Since these studies have involved animals in different physiological states, this variability is perhaps not surprising. The biological preparations have varied all the way from exteriorized foetuses with the mothers under acute operative stress to animals studied in a steady state, four or more days after foetal surgery. For metabolic studies, this time period of four days represents a minimum for full recovery from operative and anaesthetic stress. Despite the stability of the physiological state provided by chronic preparations, we have found in such animals a variation in foetal oxygen consumption from one foetus to the next of approximately 5 to 9 ml/min·kg (James *et al.* 1972; Boyd, Morriss, Meschia, Makowski & Battaglia, 1972). When the same foetus is studied several times, the variation in metabolic rate is less. Although the factors contributing to the differences in metabolic rate from one foetus to another are not yet known, there is a suggestion that twin pregnancies account for the lower range of oxygen consumptions. In a series of 13 twin pregnancies in sheep, in which 21 measurements of foetal oxygen consumption were made, we found a mean ± SEM oxygen consumption of the foetus of 6.53 ± 0.29 ml/min·kg. In contrast, the mean ± SEM foetal oxygen consumption of 47 measurements in 24 singleton pregnancies was 7.85 ± 0.26 ml/min·kg. Twins are known to be smaller in body weight at the same gestational age than singletons. The possibility that a lower oxygen consumption may be a characteristic of foetuses with some degree of intrauterine growth retardation is strengthened by an observation made in a foetus with marked intrauterine growth retardation (James *et al.* 1972). This foetal lamb was one of discordant twins, with a birthweight of 2.0 kg compared to his twin, which had a birthweight of 3.87 kg. His oxygen consumption per kilogram per minute was 4.81 and 4.41 on the two occasions on which it was measured. These have been the lowest values measured by us in chronic preparations.

In sheep (Dawes & Mott, 1959) and in man (Hill, 1968) there is a marked rise in the resting oxygen consumption during the first month of postnatal life. In both man and sheep the oxygen consumption of the newborn increases by approximately 100% during the first month of postnatal life. There is still some disagreement in the literature on the validity of the observation that infants with intrauterine growth retardation have higher metabolic rates when compared to infants of the same age or size. Should this observation be confirmed by further studies, it would point to a striking change in oxygen consumption per kilogram per minute from foetal to postnatal life in such infants.

FOETAL RESPIRATORY QUOTIENT

Until recently the respiratory quotient of the foetus had not been investigated by direct techniques. There had been an attempt to estimate foetal respiratory quotient indirectly by determining the maternal respiratory quotient before and after occlusion of the umbilical cord (Bohr, 1931). This work suggested a foetal respiratory quotient of one and was interpreted as implying that carbohydrate was the sole metabolic fuel of the developing foetus. Alexander, Britton & Nixon (1966) determined the amount of glucose required to maintain a stable glucose concentration during perfusion of isolated sheep foetuses. From these measurements and measurements of umbilical venous arterial glucose differences, it was concluded that glucose is virtually the only energy source of the sheep foetus. However, simultaneous foetal O_2 consumption measurements were not made in those studies. Recently we have determined the CO_2 production rate and the oxygen consumption of sheep foetuses and have shown that the respiratory quotient is significantly less than one. This observation challenges the assumption that carbohydrate is indeed the principal metabolic fuel of the foetus (James *et al.* 1972). However, it should be pointed out that even if the foetal respiratory quotient were equal to one, the interpretation that this finding implies metabolism of carbohydrate *requires* the assumption that the accumulation of carbon for growth is negligible with respect to the amount of carbon excreted by the foetus. As we shall see later in this review, that assumption is incorrect. Clearly, a more direct approach to foetal substrate utilization is needed than merely determining respiratory quotient.

FOETAL GLUCOSE/OXYGEN QUOTIENT

The glucose/oxygen quotient is defined as:

$$\frac{6 \times \Delta \text{glucose}}{\Delta O_2}$$

where Δglucose equals umbilical venous-arterial differences of glucose (mM blood) and ΔO_2 equals umbilical venous–arterial differences of oxygen (mM blood). Thus defined, the quotient represents the fraction of foetal oxygen consumption required to metabolize aerobically the umbilical glucose uptake. In a total of 43 pregnant ewes in the fed state the glucose/oxygen quotient has been measured 74 times, with a mean of 0.46 (95% confidence limits 0.42 to 0.51) (Tsoulos, Colwill, Battaglia, Makowski & Meschia, 1971; James *et al.* 1972; Boyd *et al.* 1972). Thus, approximately half of the foetal oxygen consumption could be accounted for by glucose metabolism. In starvation sufficiently prolonged to produce maternal hypoglycaemia, glucose uptake by the foetus falls, as does the glucose/oxygen quotient. After one week of starvation of a pregnant ewe, the glucose/oxygen quotient can be less than 0.2; that is, glucose accounting for less than 20% of the foetal oxygen consumption. These data on the umbilical glucose/oxygen quotient do not support the widely held view that glucose is the predominant and obligatory metabolic fuel of the sheep foetus.

UMBILICAL FRUCTOSE, FATTY ACID AND GLYCEROL UPTAKE

The umbilical venous–arterial differences of fructose (the principal carbohydrate in sheep foetal blood), free fatty acids of carbon chain length greater than C_6, and glycerol, have been measured in our laboratories and compared with simultaneous measurements of the umbilical venous–arterial differences for oxygen (Tsoulos *et al.* 1971; James, Meschia & Battaglia, 1971; James *et al.* 1972). In this way the umbilical uptake of these solutes could be expressed in oxygen equivalents, just as was done for the glucose/oxygen quotient (James *et al.* 1972; Boyd *et al.* 1972; Tsoulos *et al.* 1971; James, Meschia & Battaglia, 1971). No detectable venous–arterial differences were found across the umbilical circulation for fructose and free fatty acids. Free fatty acid concentrations are very low in the foetal blood of sheep. Thus it would appear that these substrates are not major metabolic fuels for the sheep foetus when the mother is in the fed state. During starvation the concentrations of free fatty acid and ketoacids increase in both the maternal and foetal circulations. Fig. 1 presents the FFA and glycerol

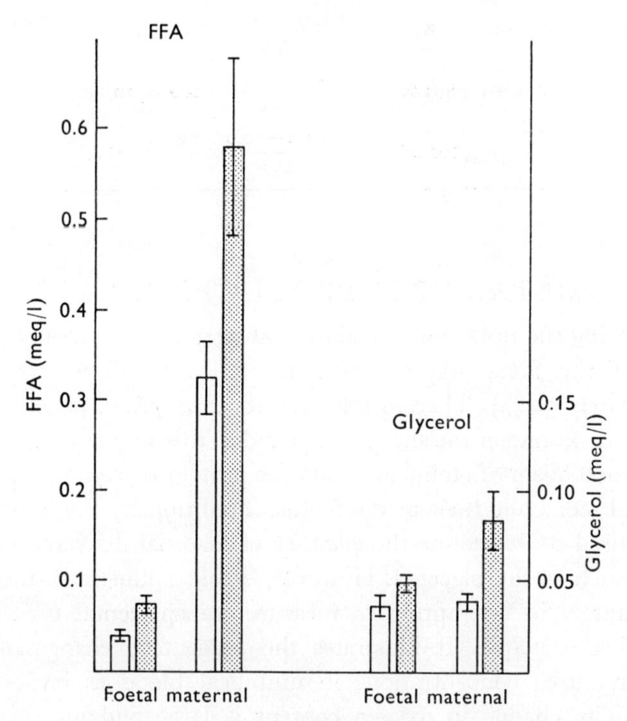

Fig. 1. Blood free fatty acids (FFA) and glycerol in fed (open bars) and fasted (stippled bars) ewes.

concentrations in umbilical arterial and maternal arterial bloods during the fed (open bars) and fasted (stippled bars) states. There is a significant increase in blood levels in both maternal and foetal circulations, but even with the elevated concentrations in foetal blood induced by maternal starvation no significant umbilical venous–arterial differences could be detected.

Glycerol was the only one of the above substrates that showed a small but significant umbilical venous–arterial difference (mean ± SEM = 6 ± 1 μM/l). Since 1 mole of glycerol requires 3.5 moles of oxygen for complete aerobic oxidation, this represents a glycerol/oxygen quotient of 0.012; that is, only a small proportion (1.2 %) of the foetal oxygen consumption could be accounted for by glycerol metabolism.

Table 1. *Comparison of umbilical A–V differences*

O_2 (mM)	$\frac{5.2-3.5}{3.5} \times 100 = 49\%$
Glucose (mg/100 ml)	$\frac{27.5-25.0}{25.0} \times 100 = 10\%$
Urea (mg/100 ml)	$\frac{29.6-30.0}{29.6} \times 100 = -1.3\%$

FOETAL UREA PRODUCTION RATE

After studying the uptake of the above substrates it was clear that a large fraction of the foetal oxygen consumption was still unaccounted for (approximately 50%). Therefore it seemed appropriate to determine the urea production rate of the sheep foetus and in this way evaluate the extent to which catabolism of amino acids and/or protein represents a proportion of the total metabolic fuels of the foetus. Traditionally the Fick principle has been used to determine the *quantity* of material delivered to or from an organ such as the placenta. However, Table 1 illustrates the problem one encounters in attempting to measure transplacental urea excretion by the Fick principle. It illustrates the percentage change of oxygen, glucose and urea concentrations in umbilical blood as it perfuses the placenta. The change in oxygen content is large and measurable with precision. It should be emphasized that the precision with which the umbilical uptake of oxygen, glucose or other solutes can be measured is a function of both the analytical and biological sampling errors. We have found it essential, in order to maintain these errors within tolerable limits, to use the mean of at least five sets of arteriovenous differences as the difference used in calculations involving the Fick principle. When the steady-state diffusion technique for antipyrine is used in the measurement of umbilical blood flow, the same precautions for minimizing biological sampling errors must be applied. Fig. 2 presents the umbilical arterial and venous concentrations of antipyrine in a foetus during the second and third hours of a constant infusion into the foetal circulation. The regression lines are calculated by the method of least squares for the concentrations in each vessel with time, and these are used for the calculation of the mean arteriovenous differences during the period of study. From these six sets of arterial and venous concentrations a single umbilical blood flow is determined which is used in the application of the Fick principle for the calculation of umbilical glucose or oxygen uptakes.

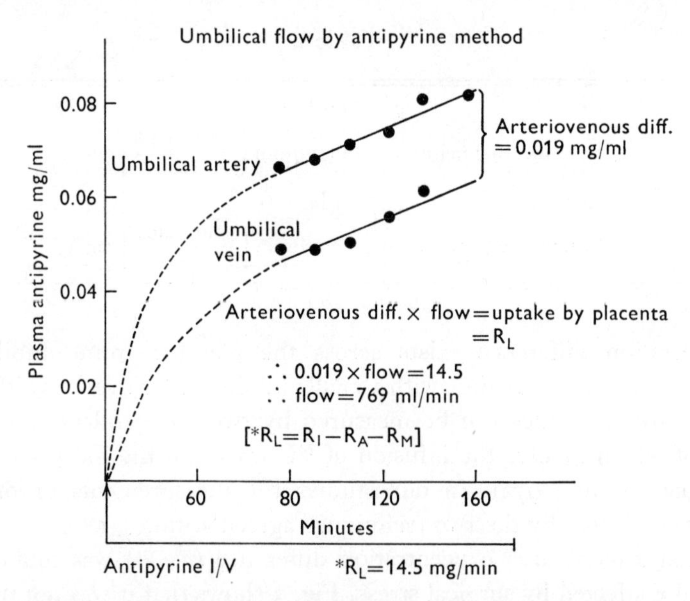

Fig. 2. Umbilical arterial and venous concentrations of antipyrine in foetal blood.

However, even with suitable precautions the percentage change in glucose concentration in the umbilical circulation already begins to approach the limits of precision, since there is only a 10% change that one is attempting to measure precisely. The change in urea concentration (approximately 1%) precludes the measurement of urea arteriovenous differences with any acceptable degree of precision. Fortunately, the transplacental excretion rate of urea ($\dot{q}_p$) can be measured in another way. The placental clearance of urea (C_u) has been defined as ($\dot{q}_p$) divided by (a_a–A_A), where a = umbilical arterial concentration of urea, A_A = maternal arterial concentration of urea. Thus if an independent measurement of C_u is made, the transplacental excretion rate can be calculated from the *arterial* concentration difference across the placenta, which is much larger than the umbilical arterial–venous difference and can be measured with reasonable precision. Table 2 compares the two approaches to the determination of the transplacental excretion rate of urea, using representative concentrations of urea in umbilical arterial (a) and venous (v) blood and the maternal arterial blood (A). The calculation above the dotted line is based upon the Fick principle and that below the line is based upon the use of placental clearance. The essential difference is the larger ($a-A$) concentration difference for urea compared to the umbilical arteriovenous difference. This approach may be useful for other solutes where a large

Table 2

$$\dot{q} = \text{flow}\ [a-v]$$

$$\text{Example: } 0.7\ \text{mg/min} = 180\ \text{ml/min}\left[\frac{30.0-29.6}{100}\right]\text{mg}\ \%$$

$$\dot{q} = C_u\,[a-A]$$

$$\text{Example: } 0.7\ \text{mg/min} = 18.6\ \text{ml/min}\left[\frac{30.0-26.0}{100}\right]\text{mg}\ \%$$

concentration difference exists across the placenta from umbilical to maternal uterine circulation, with a small umbilical arteriovenous difference.

The urea clearance can be measured by two independent techniques, both of which involve the infusion of ^{14}C urea into the foetal circulation (Gresham *et al.* 1972). In our studies the measurements of placental clearance of urea by the two techniques agreed within $\pm 5\%$. The foetal–maternal arterial urea concentration difference $(a-A)$ was found to be markedly affected by surgical stress. Fig. 3 shows that it was not until the fourth postoperative day that a stable $(a-A)$ difference was established across the placenta. This was probably due to a change in the urea production rate of the foetus, since the urea clearance measurements made in animals under operative stress were not significantly different from urea clearance measurements made in animals free of anaesthetic or surgical stress. The $(a-A)$ from the fourth postoperative day on was 0.034 ± 0.003 mg/ml plasma water. The urea excretion rate of the sheep foetus via the placenta was approximately 0.54 mg/min·kg foetal weight (or 0.25 mg urea nitrogen per kg foetal weight). It can be estimated that the production of 0.54 mg of urea requires 1.6 ml of oxygen consumption. Thus, this urea excretion rate would account for approximately 25% of the foetal oxygen consumption through the catabolism of amino acids. In the last third of gestation the lamb accumulates approximately 0.6 grams of nitrogen per kilogram per day, or 4 mg/min. Hence it would appear that approximately 60% of the nitrogen crossing the placenta from the mother is retained by the sheep foetus for growth.

These observations in the foetus are new and contrary to the common belief that nitrogen catabolism proceeds slowly in foetal life. This interpretation was based upon the observations of a low nitrogen excretion in newborn infants (McCance & Widdowson, 1954) and upon the erroneous conclusion that glucose is the major metabolic fuel of the sheep foetus (Alexander, Britton & Nixon, 1966). It is worth noting, perhaps, that the foetal urea excretion rate we have measured in normal foetuses, while

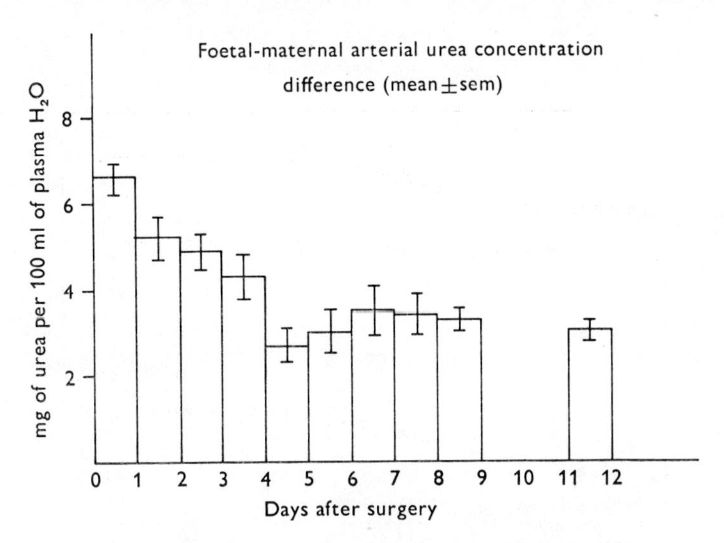

Fig. 3. Umbilical arterial–maternal arterial concentration differences of urea in the first 12 days after surgery.

quite high, is well below the capability of the foetal liver. We have demonstrated that the rate of urea synthesis by the foetus can be increased threefold by infusion of ammonium lactate into the foetal circulation. Fig. 4 shows the increase in $(a-A)$ concentration difference induced in a sheep foetus by the infusion of ammonium lactate. The placental clearance of urea at the end of the infusion was 17.2 ml/min·kg, and the $(a-A)$ difference was 0.09 mg/ml plasma water. Thus the urea production rate of this foetus was 1.5 mg/min·kg, approximately three times the normal value (Gresham *et al.* 1972). It should be emphasized that the placental excretion rate of urea is somewhat lower than the urea production rate of the foetus, since some urea accumulates within the amnion and allantois during gestation via foetal renal excretion, and some urea could diffuse directly across from the foetal liquors into the maternal circulation.

It is likely that glucose represents a larger portion of the total metabolic fuels in primates than in sheep. However, a high urea production rate by the human foetus is suggested by the following data. The placental clearance of urea has been measured across the subhuman primate placenta, and its mean value (15 ml/min·kg foetal weight) is not very different from that of the sheep placenta (17 ml/min·kg foetal weight). The $(a-A)$ concentration difference in man was found to be 2.5 ± 0.3 mg/100 ml plasma water (Gresham, Simons & Battaglia, 1971). These data would lead us to estimate the transplacental excretion rate in man at approximately 0.38 mg/min·kg foetal weight. While this rate is considerably lower than that for

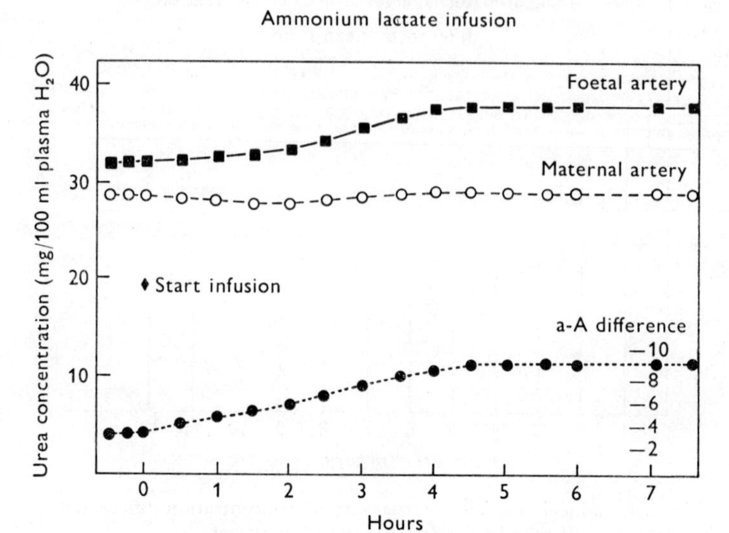

Fig. 4. Concentrations of urea in umbilical arterial and maternal arterial blood during the constant infusion of ammonium lactate into the foetus. Infusion rate = 50 μM/min.

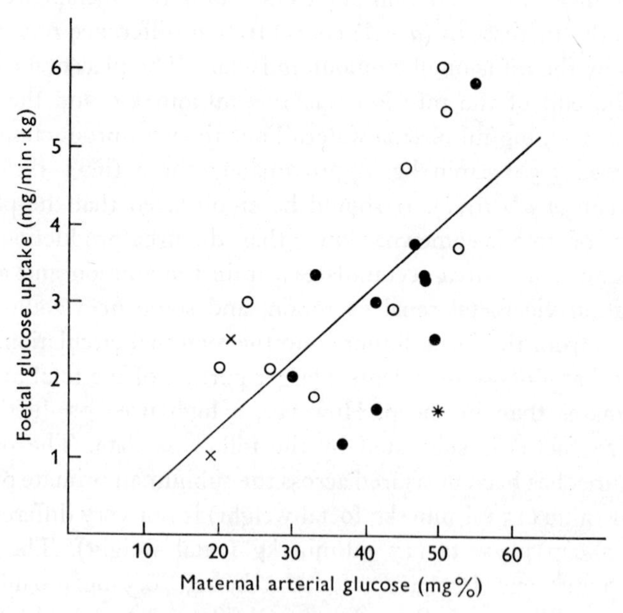

Fig. 5. Relationship of foetal glucose uptake to maternal arterial glucose concentration (mg/100 ml of blood). The asterisk is the result of measurements on a growth-retarded twin foetus. ● Twin; ○ Single; × UA 26; *UA 20

Table 3. *'Metabolic balance sheet' (foetal lamb)*

Substrate	% Total O_2 Uptake
Glucose	46
Fructose	Not measurable
FFA (C_6)	Not measurable
Glycerol	1
'Amino acids'	25
	Total 72

sheep (0.54 mg/min·kg), it is still much higher than the urea excretion rate of newborn infants, which in all studies has been estimated at less than 0.1 mg/min·kg.

In summary, these studies in unstressed, chronic animal preparations on the umbilical uptake of various solutes compared to the umbilical uptake of oxygen permit us to draw up the balance sheet shown in Table 3. It points out that anywhere from 20% to 30% of the foetal oxygen consumption is still unaccounted for.

FOETAL GLUCOSE UPTAKE VERSUS MATERNAL ARTERIAL GLUCOSE CONCENTRATION

In the past, the techniques were not available for measuring umbilical glucose uptake repetitively in animals free of operative or anaesthetic stress. For obvious reasons, it would be desirable to be able to measure umbilical glucose uptake in subhuman primates. However, at this time no one has succeeded in establishing preparations in primates which permit repeated measurements of umbilical arteriovenous differences four or more days after surgery.

In sheep, such studies are now feasible. When foetal umbilical glucose uptake is plotted against maternal arterial glucose concentration, a linear relationship is evident (Fig. 5). Over the concentration range we have studied thus far there has been no suggestion of a plateau or of maximum foetal umbilical glucose uptake; that is, there has been no evidence of saturation of a specific carrier system. However, the concentration range studied has not been extended to particularly high maternal glucose concentrations. Of interest is the relationship between umbilical arterial and maternal arterial glucose concentration. Figs. 6 and 7 compare the relationship between umbilical arterial and maternal arterial concentrations of urea and of glucose. It is clear that the foetal arterial urea concentration is

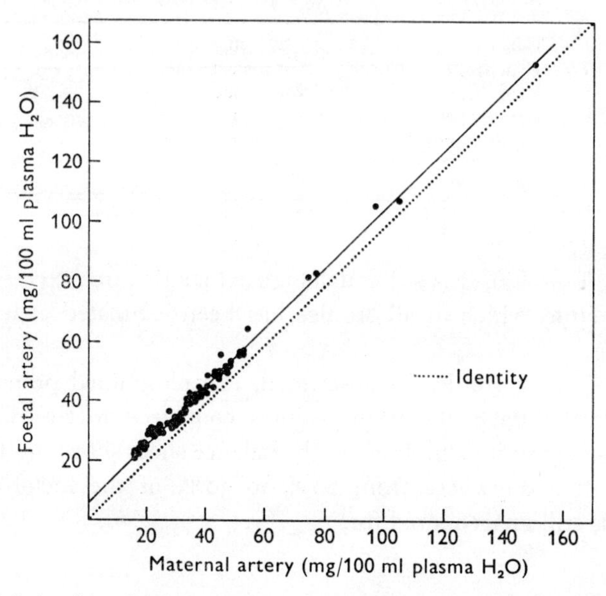

Fig. 6. Relationship of foetal arterial urea concentration to the concentration of urea in maternal blood. Observations made 4 or more days after surgery.

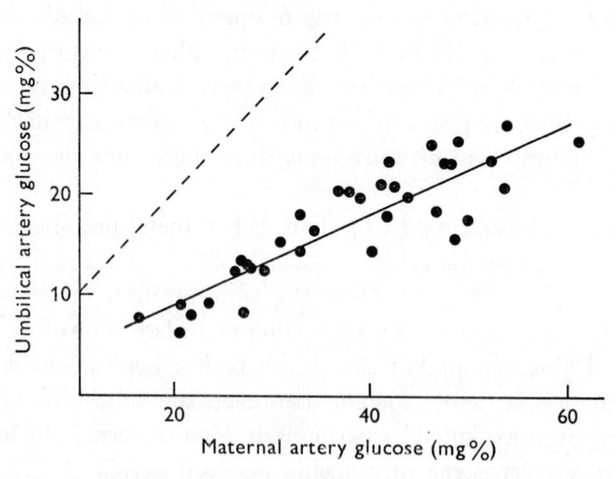

Fig. 7. Relationship of foetal arterial glucose concentration to maternal arterial concentration (mg/100 ml of blood). Note the divergence from the identity line (---).

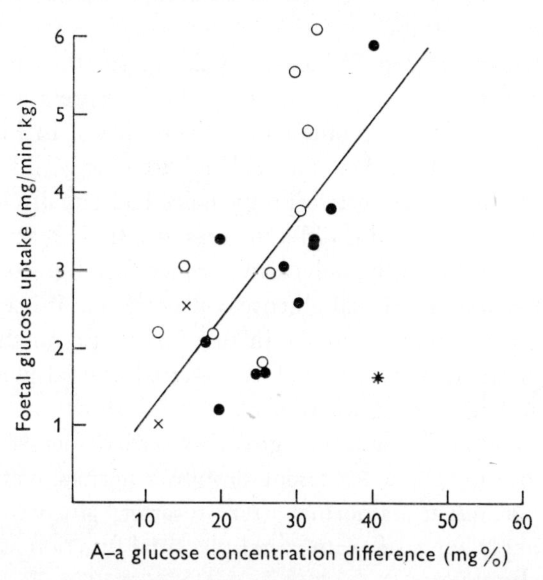

Fig. 8. Relationship of foetal glucose uptake to the concentration difference of glucose between maternal arterial and foetal arterial blood (mg/100 ml of blood). Symbols as in Fig. 5.

always higher than the maternal, and the foetal glucose concentration always lower than the maternal. However, in the case of urea, the artery-to-artery difference is fairly constant over a wide range of maternal urea concentrations. In contrast, the glucose difference increases as the maternal arterial glucose concentration increases. This fact is particularly interesting in the light of the following fact. Foetal umbilical glucose uptake is linearly related to artery-to-artery glucose difference (see Fig. 8). Thus the following important characteristic of placental glucose transfer can be demonstrated; namely, that foetal glucose uptake is dependent upon maternal arterial glucose concentrations. The higher the maternal glucose level the greater the transplacental transfer to the foetus. While this relationship has often been hypothesized in man to account for some of the features of the infants born to mothers with diabetes, this is the first demonstration of such a relationship in any species.

These studies are at variance with observations reported by Crenshaw (1970). He has found no relationship between foetal glucose uptake and either maternal or foetal glucose concentration. There is no apparent reason for this discrepancy. Others have reported a wide range of uptakes, but these studies were carried out in a wide variety of different biological preparations under surgical and anaesthetic stress, and often foetal glucose uptake was not compared with simultaneous determinations of maternal arterial glucose concentration.

The measurement of foetal glucose uptake versus the glucose concentration difference across the placenta puts the interpretation of the very low glucose uptake of some foetuses on a firmer basis. In Fig. 8 the data obtained on two animals (UA 26 and UA 20) are of special interest in this regard. The foetuses in those two pregnancies had equally low umbilical glucose uptakes, but for quite different reasons. In UA 26 the pregnant ewe was ill and developed hypoglycaemia. Since foetal glucose uptake is a function of the maternal arterial glucose concentration, it is not surprising that there was a low glucose uptake in this foetus. Fig. 8 shows that the uptake was consistent with the level of maternal arterial glucose concentration, and there is no suggestion of placental dysfunction. On the other hand, UA 20 was an intrauterine growth-retarded foetus. Here, a low umbilical glucose uptake was present despite a normal maternal arterial glucose concentration and a normal artery-to-artery glucose concentration difference, strongly suggesting impaired placental function as the cause of the low glucose uptake.

METABOLIC BALANCE OF THE SHEEP FOETUS

From the measurements of oxygen consumption, carbon dioxide production, glucose uptake, and urea production of the sheep foetus, together with carcass analyses for carbon and nitrogen, the following tentative balance can be calculated. The sheep foetus is increasing its weight in the latter third of gestation by approximately 35 g/kg·day. This is a more rapid increase than one sees in primates. From carcass analyses we have found that carbon and nitrogen increase within the foetal carcass during this time by the following amounts:

$$\text{carbon} \cong 3.15 \text{ g carbon/kg·day}$$
$$\text{nitrogen} \cong 0.6 \text{ g nitrogen/kg·day}$$

Carbon balance

Carbon accumulating in the carcass =	3.15 g/kg·day
Carbon excreted as CO_2:	
5.65 ml CO_2/min·kg × 0.539 mg carbon/ml CO_2 × 1440 =	4.38 g/kg·day
Carbon excreted as urea:	
0.54 mg urea/min·kg × 0.2 mg carbon/mg urea × 1440 =	0.16 g carbon/kg·day
Total	7.69 g carbon/kg·day

This balance illustrates that the carbon accumulated in the carcass each day is almost equal to the carbon excreted as CO_2. For this reason one cannot interpret a respiratory quotient of one as reflecting the metabolism of carbohydrate alone in any rapidly growing organism.

Nitrogen balance

Urea N excreted	0.36 g/kg·day
N accumulating in the carcass	0.60 g/kg·day
Total	0.96 g/kg·day

This balance illustrates that the sheep foetus utilizes approximately 60% of its nitrogen intake for growth.

Glucose uptake versus total carbon uptake

Umbilical glucose uptake accounts for the following quantity of carbon acquired by the foetus in one day:

$$3.06 \text{ mg glucose/min}\cdot\text{kg} \times 72 \text{ mg carbon/180 mg glucose} \times 1440 = 1.76 \text{ g/kg}\cdot\text{day}$$

This quantity represents 23% of the estimated total carbon crossing the placenta from mother to foetus in one day.

Is the bulk of the remaining carbon (5.9 g/kg·day) crossing the placenta in the form of amino acids? Given a total N balance of 0.96 g/kg·day, the C/N ratio of the hypothetical amino acid mixture crossing the placenta would be:

$$C/N = \frac{5.9}{0.96} = 6.1.$$

This is a relatively high ratio (C/N for alanine = 2.57, C/N for an 'average' protein = 3.3). Thus it would seem that substrates other than glucose, fructose, long-chain fatty acids and amino acids cross the sheep placenta from mother to foetus in relatively large amounts.

There are virtually no studies that have attempted to define the metabolic rate and substrate utilization for individual foetal organs. Recently, however, we have begun to study the blood flow to the brain of the foetus (Schneider, Tsoulos, Colwill, Makowski, Battaglia & Meschia, 1972) and to determine glucose/oxygen quotients across the foetal cerebral circulation (Tsoulos, Schneider, Colwill, Meschia, Makowski & Battaglia, 1972). Using the radioactive microsphere technique we described for the measurement of total arterial flow to an organ (Makowski, Meschia, Droegemueller & Battaglia, 1968) we found in acute animal preparations a mean brain blood flow of 111 ± 7.8 ml/100 g/min. This represented approximately 40% of the total cephalic flow. The observation that brain flow in

Table 4. *Human cerebral glucose/oxygen quotients calculated from data from normal man*

	Δ oxygen mM/l	Δ glucose mM/l	Glucose/ Oxygen
Wortis *et al.*	3.08	0.50	0.97
Gibbs *et al.*	2.99	0.56	1.12
Scheinberg *et al.*	2.69	0.55	1.23
Kety	2.83	0.50	1.06
Sokoloff *et al.*	2.62	0.50	1.15
	(95 % limits 1.23–0.99)		1.10
Our data, sheep foetus	(95 % limits 1.12–1.02)		1.06

Table 5

	Cerebral blood flow ml/100 g/min	Cerebral uptake: O_2 ml/100 g/min	Cerebral uptake: Glucose mg/100 g/min	Cerebral uptake: Glucose/O_2	Arterial concentrations: O_2 mM/l	Arterial concentrations: Glucose mg %
Sheep foetus (present study)	111	4.0	5.3	1.06	3.4	16.6
Human adult	53	3.5	5.1	1.10	8.8	90.0

the sheep foetus represents less than half the total cephalic flow, plus the fact that there are marked differences in the regulation of flow to different tissues, suggests that changes in carotid flow, measured by electromagnetic flow probes, may be an unreliable index of changes in cerebral flow.

The oxygen and glucose uptakes from the cerebral circulation were 4.04 ± 0.17 ml/100 g·min and 5.31 ± 0.42 mg/100 g·min, respectively. The glucose/oxygen quotient was determined 21 times in 10 animals, with a mean of 1.06 and 95 % confidence limits of 1.02 to 1.12. Table 4 compares our data on the sheep foetus with studies in normal man. Clearly in both species the glucose/oxygen quotient is slightly but significantly greater than one, in sharp contrast with the umbilical glucose/oxygen quotient of 0.46.

Table 5 compares our data for the sheep foetus on cerebral blood flow, cerebral glucose and oxygen uptakes, and arterial concentrations versus data in the literature for adult man. The striking fact brought out in the table is that the sheep foetus must meet oxygen and glucose requirements per 100 grams brain tissue that are equal to that of adult man, despite markedly lower arterial inflow concentrations. To some extent this is compensated for by the higher foetal cerebral blood flow, but it does emphasize the narrower margin of safety in foetal life to meet cerebral metabolic requirements.

REFERENCES

Alexander, D. P., Britton, H. G. & Nixon, D. A. (1966). *J. Physiol.* (*London*) **185**, 382.

Bohr, C. (1931). In Needham, J., *Chemical Embryology*, pp. 728–9. New York: Macmillan Co.

Boyd, R. H., Morriss, F. H., Meschia, G., Makowski, E. L. & Battaglia, F. C. (1972). In the Press.

Crenshaw, C., Jr (1970). *Clin. Obstet. Gynec.* **13**, 579.

Dawes, G. S. & Mott, J. C. (1959). *J. Physiol.* **146**, 295.

Gresham, E. L., James, E., Raye, J. R., Battaglia, F. C., Makowski, E. L. & Meschia, G. (1972). *Pediatrics*. In the Press.

Gresham, E. L., Simons, P. S. & Battaglia, F. C. (1971). *J. Pediat.* **79**, 809.

Hill, J. (1968). *J. Physiol.* **199**, 693.

James, E., Meschia, G. & Battaglia, F. C. (1971). *Proc. Soc. Exp. Biol. Med.* **138**, 823.

James, E., Raye, J. R., Gresham, E. L., Makowski, E. L., Meschia, G. & Battaglia, F. C. (1972). *Pediatrics*. In the Press.

McCance, R. A. & Widdowson, E. M. (1954). *Cold Springs Harbor Symposia on Quantitative Biology* **19**, 961.

Makowski, E. L., Meschia, G., Droegemueller, W. & Battaglia, F. C. (1968). *Circ. Res.* **23**, 623.

Schneider, J. A., Tsoulos, N. G., Colwill, J. R., Makowski, E. L., Battaglia, F. C. & Meschia, G. (1972). *Amer. J. Ob.-Gyn.* In the Press.

Tsoulos, N. G., Colwill, J. R., Battaglia, F. C., Makowski, E. L. & Meschia, G. (1971). *Amer. J. Physiol.* **221**, 234.

Tsoulos, N. G., Schneider, J. A., Colwill, J. R., Meschia, G., Makowski, E. L. & Battaglia, F. C. (1972). *Pediat. Res.* **6**, 182.

PATHOPHYSIOLOGICAL CHANGES IN THE FOETAL LAMB WITH GROWTH RETARDATION

By ROBERT K. CREASY, MICHAEL DE SWIET, KARI V. KAHANPÄÄ, WILLIAM P. YOUNG AND ABRAHAM M. RUDOLPH

The Cardiovascular Research Institute, and the Departments of Obstetrics and Gynecology, and Pediatrics, University of California, San Francisco

A review of the aetiological factors of foetal growth retardation both in the human and in animal experiments reveals that a chronic reduction in maternal uteroplacental blood supply is frequently encountered. Other than alterations in total foetal weight and length, organ weight and haematocrit changes, the responses of the foetus to this reduction in uteroplacental blood flow are largely unknown, particularly in large animals. We present herein some of our studies on the physiology of the foetal lamb in which growth retardation was produced by injury of the uteroplacental circulation.

Using singleton time-dated pregnancies, at approximately 110 days, catheters were placed in the foetal femoral artery and vein, and a forelimb vein. A catheter was also placed retrograde into the main uterine artery of the pregnant horn, from a distal branch. Growth retardation of the foetal lamb was produced by means of embolization of the maternal uteroplacental vascular bed of the pregnant horn with 15 μm diameter microspheres (non-radioactive) over a number of days, as previously described (Creasy, Barrett, de Swiet, Kahanpää & Rudolph, 1972). In 11 animals, approximately 2 million microspheres were injected daily into the uterine artery catheter, if foetal arterial blood gases were normal; 8 additional animals served as controls. No significant changes in maternal blood pressures occurred during the whole course of the study, nor were there any acute changes observed in foetal heart rate, blood pressure, or blood gases with the daily embolization of the uteroplacental circulation.

At approximately 139 days of gestation, radioactive microspheres, 15 μm diameter, were used to study the foetal circulation. Foetal cardiac output and the distribution of output were measured by injecting, simultaneously, different radio-nuclides into the forelimb vein and the hindlimb vein for each study (Rudolph & Heymann, 1967; Makowski, Meschia, Droegemueller & Battaglia, 1968). Foetal heart rates and arterial pressures did not change with the radioactive microsphere injection.

At the completion of the distribution studies, hysterotomy was performed, the foetus weighed and dissected, and the tissues prepared for analysis of radioactivity, and DNA and protein content. All flows were calculated in reference to the known femoral arterial sample. In view of the relative inaccuracies in the measurement of local blood flow when the total number of microspheres does not exceed 400 (Buckberg, Luck, Payne, Hoffman, Archie & Fixler, 1971), blood flows to organs containing fewer than 400 microspheres are not reported.

There were no significant differences in foetal heart rate or arterial pressure between the control and experimental groups when the blood flows were studied. Foetal arterial pH was between 7.31 and 7.38 in all animals and did not differ significantly between the two groups. However, mean foetal arterial P_{O_2} at the time of the circulatory studies was 17 mm Hg in the embolization group and 23 mm Hg in the controls, a statistically significant difference. Indeed this difference in P_{O_2} was present during the last two weeks of experimental observation.

Table 1. *Mean foetal organ weights* (g ± 1 SE)

	Control	Embolized	% Control
Brain	55±2	48±1*	87
Liver	149±4	91±6*	61
Heart	33±1	23±2*	69
Lung	117±10	86±5*	73
Thymus	18±1	7±1*	41
Gut	195±11	160±12	82
Kidney	33±2	29±3	88
Adrenal	0.5±0.05	0.5±0.06	95
Placenta	472±56	279±24*	59
Spleen	10±1	6±1*	64
Thyroid	1.3±0.1	1.0±0.1*	73
Upper carcass	1649±94	1178±88*	71
Lower carcass	1872±111	1271±72*	68

* Differences between control and embolized significant ($P < 0.05$).

Foetal weight in the embolization group was 3.20 kg, which was significantly lower than that in the control group, 4.54 kg. Mean foetal crown-rump length was 44.7 cm in the embolization group, a significant difference from 52 cm in the controls. As shown in Table 1, there is a variable reduction in organ weights of the growth-retarded foetuses. The most marked reduction occurred in the thymus, placenta, liver, heart and lung.

Although the brain weight was significantly ($P < 0.05$) reduced, the difference from the controls was not as marked. If organ-to-body-weight ratios are examined, the brain-to-body-weight ratio was actually significantly higher, and the liver- and thymus-to-body-weight ratios significantly lower in the embolization group. As has been suggested (Dawkins, 1964), if one uses the brain-to-liver-weight ratio as an index of foetal growth retardation, there was a significant increase from 0.31 in the controls to 0.52 in the embolization group.

Table 2. *Measurement of foetal cardiac output and distribution*

	Control	Embolized
Cardiac output		
ml/min	1807 ± 168	1171 ± 89*
ml/min/kg foetus	406 ± 30	371 ± 19
% to Brain	3.4 ± 0.4	6.8 ± 1.0*
Heart	2.3 ± 0.3	4.5 ± 0.5*
Gut	6.6 ± 0.8	10.8 ± 0.8*
Liver	0.5 ± 0.1	0.5 ± 0.1
Kidney	3.1 ± 0.3	5.0 ± 0.6*
Lung	5.4 ± 0.6	1.7 ± 0.4*
Placenta	41.9 ± 0.5	29.1 ± 1.6*

* Differences between control and embolized significant ($P < 0.05$).
Values ± SE.

Measurements of foetal cardiac output and the percent of cardiac output distributed to various organs are listed in Table 2. The thymus and adrenals are excluded due to inadequate numbers of spheres. Total cardiac output was significantly lower in the embolization group, 1171 ml/min as compared with 1807 ml/min in the controls, but when examined per kilogram of foetal weight the difference was not significant. In the embolization foetuses the percent of cardiac output distributed to the brain and heart was approximately doubled, 3.4 to 6.8 and 2.3 to 4.5: the percent of the cardiac output distributed to the gut also increased, as did that to the kidney. Both the lungs and the placenta received a significantly decreased proportion of the reduced cardiac output in the growth-retarded group.

When organ blood flow was examined in the embolization group (Table 3), umbilical blood flow was significantly reduced when expressed either as actual flow, or flow per kilogram of foetal weight. Blood flow (ml/min per 100 g) to the brain and heart was significantly increased, with a trend of an increase in flow to the kidney in these foetuses. The flow to the lungs was significantly decreased.

Table 3. *Measurement of foetal organ blood flow*

	Control	Embolized
Umbilical flow		
ml/min	717 ± 78	339 ± 28*
ml/min/kg foetus	158 ± 14	109 ± 7*
Organ flow ml/min/100 g		
Brain	96 ± 18	158 ± 18*
Heart	126 ± 18	238 ± 25*
Lung	82 ± 8	26 ± 6*
Gut	79 ± 16	88 ± 9
Liver	6 ± 1	6 ± 1
Kidney	153 ± 11	188 ± 14

* Differences between control and embolized significant ($P < 0.05$). Values in ml ± SE.

In preliminary studies related to the cellular aspects of the differences between the controls and the growth-retarded foetuses, we determined DNA and protein contents of various organs. The results have shown a moderate degree of variability, which was probably related to the degree of growth retardation. An analysis of the amount of DNA (mg/g of tissue) and of protein (mg/g of tissue) revealed no significant differences between the two groups. The total DNA content was not significantly reduced in the brain, heart and lung, whereas there was a significant reduction in total protein content of these organs in the growth-retarded foetuses. This would indicate that cells in the brain, heart and lung were reduced in size in these foetuses. DNA and protein content were both reduced in the liver and thymus, indicating a decreased number of cells in these organs.

The changes in the circulation of the experimental growth-retarded foetus raise the question as to whether the alteration in blood flow is in response to the decrease in organ weights, or is actually the cause of the alteration in organ weight. There appears to be some preferential flow to certain vital structures of these foetuses. Similar alterations in organ blood flow have been observed with acute hypoxia in the foetal lamb and in the subhuman primate (Cohn, Sacks, Heymann & Rudolph, 1972; Behrman, Lees, Peterson, de Lannoy & Seeds, 1970). We can only speculate that hypoxaemia on a chronic basis may have caused changes in organ blood flow with a resultant change in organ weights. Also it would appear that mild hypoxaemia, as present in these studies, may have a profound influence on organ blood flow, and organ weights. The variability of the DNA and protein contents probably relates not only to the various degrees of growth retardation present in the experimental procedure, but also to the particular phase of organ growth, be it hyperplasia, hypertrophy, or a combination of both at the time of interference with the maternal utero-placental vasculature. The redistribution of cardiac output may thus not only help to preserve the weight of certain vital organs, but more importantly the cellular composition of these organs.

REFERENCES

BEHRMAN, R. E., LEES, M. H., PETERSON, E. N., DE LANNOY, C. W. & SEEDS, A. E. (1970). *Am. J. Obstet. Gynec.* **108**, 956–69.

BUCKBERG, G. C., LUCK, J. C., PAYNE, B., HOFFMAN, J. I. E., ARCHIE, J. P. & FIXLER, D. E. (1971). *J. Appl. Physiol.* **31**, 598–604.

COHN, H. E., SACKS, E. J., HEYMAN, M. A. & RUDOLPH, A. M. (1972). *Pediat. Res.* **6**, 342.

CREASY, R. K., BARRETT, C. T., DE SWIET, M., KAHANPÄÄ, K. V. & RUDOLPH, A. M. (1972). *Am. J. Obstet. Gynec.* **112**, 566–73.

DAWKINS, M. J. R. (1964). *Proc. R. Soc. Med.* 1063.

MAKOWSKI, E. L., MESCHIA, G., DROEGEMUELLER, G. W. & BATTAGLIA, F. C. (1968). *Circ. Res.* **23**, 623–31.

RUDOLPH, A. M. & HEYMAN, M. A. (1967). *Circ. Res.* **21**, 163–84.

THE DEVELOPMENT OF LIPOGENESIS IN THE FOETAL GUINEA PIG

By C. T. JONES

The Nuffield Institute for Medical Research, University of Oxford

Several mammalian species possess a large amount of lipid at birth (Widdowson, 1950). However, it is not known to what extent the lipid is the result of foetal synthesis *de novo* or transfer from the maternal circulation. While there is an increase in the deposition of lipid in the adipose tissue of most species as term approaches there is also a large accumulation of lipid in the foetal liver in the rabbit (Popjak, 1946) and particularly in the guinea pig (Flexner & Flexner, 1950; Hershfield & Nemeth, 1968; Bohmer, Havel & Long, 1972; Jones, unpublished). In the foetal guinea pig (term 68 to 72 days) there is an approximately 6–7-fold rise in liver triglyceride between days 55 to 60; just before term triglyceride represents approximately 30% of the dry weight of the liver. The triglyceride accumulation is associated with high foetal plasma free fatty acid levels (Jones, unpublished). During this same period the perirenal adipose tissue of the foetal guinea pig also is growing rapidly.

Foetal liver is a very active lipogenic tissue (Popjack, 1954; Villee & Loring, 1961; Ballard & Hanson, 1967; Taylor, Bailey & Bartley, 1967) and in the foetal guinea pig the rate of lipogenesis is markedly higher than in the maternal liver, with a maximum rate of lipid synthesis from acetate at 45 to 50 days and a significantly lower rate at 55 to 60 days (Jones & Ashton, unpublished).

The present study is concerned with the site of and possible precursors for foetal lipid synthesis in the guinea pig. The rate of incorporation of acetate into the lipids of the foetal liver and kidney *in vivo* was maximal at about 45 days and declined with progressing foetal maturity (Table 1). Acetate incorporation into the foetal adipose tissue lipid was maximal at 57 days and markedly higher than in the foetal liver. On injection of acetate in the maternal guinea pig the time course of the change in the ^{14}C-acetate concentration in the maternal and foetal plasma suggested that the placenta was permeable to acetate. After maternal injection of radioactive acetate, incorporation into the lipids of the foetal liver was approximately constant from 10 min to 60 min, while incorporation into the lipids of the foetal lung and kidney was significantly less than into the liver. The foetal adipose

Table 1. *The incorporation of ^{14}C-acetate into the tissue lipids of the foetal and neonatal guinea pig. Foetal weight was estimated from a growth curve and a direct foetal injection of ^{14}C-acetate was made at* 10 *μCi/kg. At the end of* 30 *min the lipids were extracted by chloroform:methanol* (2:1), *saponified and radioactivity was determined in a petroleum ether extract*

	Incorporation (dpm/g wet wt)				
	Foetal			Neonatal	
Days...	45–7	58–7	67–8	0.25	3
Number	4	4	4	4	4
Liver	6298 ± 1021	1066 ± 104	1110 ± 296	1302 ± 373	886 ± 115
Lung	2074 ± 414	1471 ± 107	1859 ± 117	792 ± 192	474 ± 46
Kidney	2365 ± 786	1805 ± 298	331 ± 184	464 ± 70	463 ± 117
Adipose	23,141 ± 2470	44,498 ± 4942	6086 ± 1500	403 ± 89	385 ± 24
Placenta	353 ± 182	771 ± 78	279 ± 13	—	—

Means ± SD in this and all subsequent tables.

tissue showed a progressive increase in acetate incorporation into lipid with time (Table 2). With the exception of the lung acetate incorporation into the lipids of the maternal tissues was very low. After direct injection of ^{14}C-acetate into the foetus incorporation into the foetal liver lipid was 2- to 3-fold higher than after maternal injection, and incorporation into the foetal lung, kidney and adipose lipids at least an order of magnitude higher. The incorporation of acetate into the foetal liver lipid was maximal after 10 min and declined with time (Table 2).

These observations support the view that foetal and not maternal tissues are the major site of synthesis of foetal lipids and that the foetal liver synthesises lipids that may be transported to other foetal tissues. In addition the foetal liver of the guinea pig probably removes most of the acetate from the umbilical vein blood after maternal injection, thus that organ and not the adipose tissue may be the major site of synthesis of lipid from acetate.

In maternal tissues the pattern of synthesis of lipid from glucose was similar to that observed with acetate as precursor (Table 3). However after maternal or foetal injection of glucose there is rapid incorporation into lipid not only in the foetal liver but also in the foetal adipose tissue (Table 3). The incorporation of glucose into the lipids of the foetal adipose tissue and liver at 57 to 59 days was higher than that of foetuses 43 to 45 days old (Table 4). Therefore the foetal tissues, particularly adipose tissue, may be the major site of the synthesis of foetal lipids from glucose. Results similar to those with glucose were obtained with amino acids (Table 5).

Table 2. *The incorporation of ^{14}C-acetate into the tissue lipids of the foetal guinea pig* in vivo. *Details as for Table* 1. *Maternal and foetal injections were at* 10 *μCi/kg, so that assuming rapid equilibration of radioactive label, the foetus received the same quantity of ^{14}C-acetate irrespective of the method of administration. Foetal age* 45–47 *days.*

	Incorporation (dpm/g wet wt)		
	10 min	30 min	60 min
Maternal injection			
Number	6	8	4
Foetal			
Liver	3030 ± 341	2364 ± 463	2822 ± 281
Lung	299 ± 87	241 ± 36	458 ± 201
Kidney	381 ± 147	329 ± 72	774 ± 114
Adipose	961 ± 225	1590 ± 236	5063 ± 419
Placenta	802 ± 217	677 ± 169	686 ± 140
Maternal			
Liver	293 ± 44	556 ± 144	380 ± 99
Lung	2209 ± 775	3410 ± 1020	4631 ± 571
Kidney	282 ± 109	378 ± 42	407 ± 110
Adipose	238 ± 43	418 ± 216	214 ± 58
Foetal injection			
Number	4	4	4
Foetal			
Liver	10,218 ± 1047	6298 ± 1021	5805 ± 1139
Lung	3546 ± 175	2074 ± 414	1766 ± 521
Kidney	3421 ± 390	2365 ± 786	1470 ± 451
Adipose	16,302 ± 1520	23,141 ± 2470	12,575 ± 1404
Placenta	463 ± 171	353 ± 182	369 ± 101

Table 3. *The incorporation of ^{14}C-glucose into tissue lipids of the foetal and maternal guinea pig* in vivo. *Details as for Tables* 1 *and* 2. *Foetal age* 43–45 *days*

	Incorporation (dpm/g wet wt)	
	Maternal injection	Foetal injection
Number	4	4
Foetal		
Liver	829 ± 37	903 ± 113
Lung	113 ± 63	96 ± 56
Kidney	99 ± 32	203 ± 45
Adipose	7879 ± 1037	2895 ± 217
Placenta	67 ± 8	77 ± 10
Maternal		
Liver	55 ± 7	—
Lung	186 ± 43	—
Kidney	60 ± 8	—
Adipose	85 ± 13	—

Table 4. *The incorporation of* ^{14}*C-glucose into tissue lipids of the foetal and maternal guinea pig* in vivo. *Details as for Tables* 1 *and* 2. *Foetal age* 57–59 *days*

	Incorporation (dpm/g wet wt)	
	Maternal injection	Foetal injection
Number	4	4
Foetal		
Liver	1472 ± 295	1763 ± 317
Lung	261 ± 27	315 ± 72
Kidney	212 ± 93	333 ± 100
Adipose	20,761 ± 3512	10,899 ± 1271
Placenta	41 ± 10	58 ± 11
Maternal		
Liver	62 ± 10	—
Lung	204 ± 35	—
Kidney	49 ± 9	—
Adipose	92 ± 17	—

Table 5. *The incorporation of* ^{14}*C-alanine into tissue lipids of the foetal and maternal guinea pig* in vivo. *Details as for Tables* 1 *and* 2. *Foetal age* 45–46 *days*

	Incorporation (dpm/g wet wt)	
	Maternal injection	Foetal injection
Number	4	4
Foetal		
Liver	1197 ± 172	1123 ± 177
Lung	51 ± 26	22 ± 6
Kidney	39 ± 13	28 ± 7
Adipose	2230 ± 180	3036 ± 556
Placenta	22 ± 6	125 ± 177
Maternal		
Liver	49 ± 1	—
Lung	254 ± 36	—
Kidney	15 ± 4	—
Adipose	42 ± 6	—

In vitro the incorporation of glucose and a range of amino acids into the lipid of liver slices is much less than that of acetate (Table 6). While the rate of glucose incorporation is higher at 58 days than at 45 days (Fig. 1), it is still comparatively slow, although closer to the rate of acetate incorporation. Reduction of the glucose concentration of the medium to a level closer to the physiological range further reduced the rate of incorporation of glucose into liver lipid (Table 7).

Table 6. *The incorporation of amino acids, glucose and acetate into the lipid of slices of liver from the foetal and maternal guinea pig. Liver slices (50–100 of mg) were incubated for 90 min in Krebs' bicarbonate saline, with 1 μCi of radioactive precursor. Amino acids were present at 1 mM, glucose 10 mM and acetate at 3 mM. Other details as in Table 1. Foetal age 45–47 days*

	Incorporation (nmol/g wet wt)	
	Maternal liver	Foetal liver
Glucose	3.9 ± 2.4	100 ± 10.5
Glycine	0.66 ± 0.23	8.4 ± 1.5
Aspartate	0.5 ± 0.12	10.6 ± 1.5
Alanine	0.64 ± 0.29	47 ± 1.5
Serine	2.5 ± 0.28	14.2 ± 1.6
Glutamate	6.1 ± 0.8	13.4 ± 2.8
Acetate	21.4 ± 2.9	972 ± 87

Number of observations 4–8.

Table 7. *The influence of the glucose concentration on the incorporation of glucose, alanine and acetate into the lipid of liver slices from the foetal guinea pig. Details as for Tables 1 and 6*

	Incorporation (nmol/g wet wt)	
	Glucose 1 mM	Glucose 10 mM
Number	6	6
Alanine	41 ± 4.2	29 ± 5.3
Acetate	625 ± 142	1150 ± 119
Glucose	46 ± 6.1	149 ± 27

Despite the low level of foetal and maternal plasma acetate concentration (approximately 0.05 mM), if it is effectively removed by the foetal liver the incorporation of acetate into hepatic lipids may represent an important synthetic pathway. This is supported by the fact that two enzymes associated with lipid synthesis, glucose 6-phosphate dehydrogenase and ATP-citrate lyase, show changes in foetal liver activity with gestational age very similar to the changes in acetate incorporation into lipid (Jones & Ashton, unpublished). Foetal adipose tissue may be another major site of foetal lipid synthesis from glucose and possibly amino acids, particularly during maximal adipose tissue growth between 57 and 60 days.

The foetal liver (using acetate, glucose and amino acids) is probably the important site of the synthesis of foetal lipids early in gestation, but

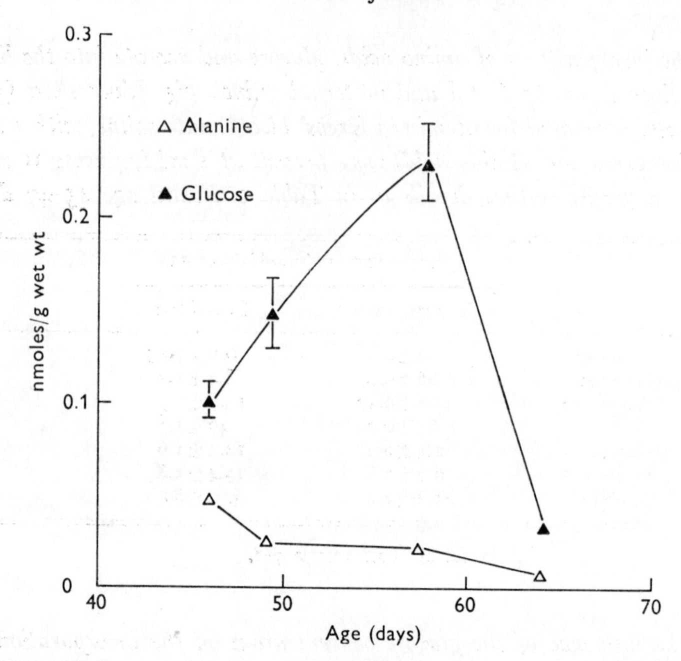

Fig. 1. The effect of gestational age on the incorporation of glucose and alanine into liver slices from the foetal guinea pig. Details as for Tables 1 and 6. Means ± SD of 4 observations for each point.

the rate is inadequate to explain the accumulation of triglycerides in the foetal liver between days 55 to 60. This accumulation occurs at a time when the plasma free fatty acids are high in the foetal and maternal plasma, and at the end of which the plasma triglycerides and very low-density lipoproteins are very high (Hershfield & Nemeth, 1968; Bohmer, Havel & Long, 1972; Jones, unpublished). The only foetal or maternal tissue exhibiting a lipogenic rate capable of supplying precursors for the deposition of triglycerides in the foetal liver is the foetal adipose tissue. However, the level of total triglyceride lipase in the foetal adipose tissue falls towards term, maternal adipose activity (while somewhat lower) shows a small rise and the foetal liver activity remains very high throughout the latter part of gestation (Jones, unpublished). Also, as already stated, the foetal adipose tissue is rapidly increasing in size at this time and is not being depleted.

The lipolytic action of glucocorticoids and pituitary hormones is well known (Rudman, Brown & Malkin, 1963; Rudman & Shank, 1966; Fain, Kovacev & Scow, 1965; Jeanrenaud, 1967). In the foetal guinea pig plasma the level of cortisol only rises after day 60, while there is a large rise in the maternal plasma cortisol concentration starting after day 50, and reaching a level at least six times higher than that before day 50 by day 56. This rise in cortisol and possibly ACTH (Jones & Pucklavec, unpublished), may release fatty acids from the maternal adipose tissue which pass across the placenta (Hershfield & Nemeth, 1968). Rises in liver triglycerides associated with elevated plasma free fatty acid levels have been reported (Van Harken, Dixon & Heimberg, 1969; Kohout, Kohoutova & Heimberg, 1971).

In summary, the foetal liver and adipose tissue may be important sites of foetal lipid synthesis, not necessarily from glucose, in species with a low rate of maternal lipogenesis. However, towards term, a large influx of lipid from the maternal adipose tissue may occur.

This work was supported by a grant from the Medical Research Council.

REFERENCES

Ballard, F. J. & Hanson, R. W. (1967). *Biochem. J.* **102**, 952.
Bohmer, T., Havel, R. J. & Long, J. A. (1972). *J. Lipid Res.* **13**, 371.
Fain, J. N., Kovacev, K. P. & Scow, R. O. (1965). *J. Biol. Chem.* **240**, 3522.
Flexner, J. B. & Flexner, L. B. (1950). *Anat. Rec.* **106**, 413.
Hershfield, M. S. & Nemeth, A. M. (1968). *J. Lipid Res.* **9**, 460.
Jeanrenaud, B. (1967). *Biochem. J.* **103**, 627.
Kohout, M., Kohoutova, B. & Heimberg, M. (1971). *J. Biol. Chem.* **246**, 5067.
Popjak, G. (1946). *J. Physiol.* **105**, 236.
Popjak, G. (1954). *Cold Spring Harbor Symp. Quant. Biol.* **19**, 200.
Rudman, D., Brown, S. J. & Malkin, M. F. (1963). *Endocrinol.* **72**, 527.
Rudman, D. & Shank, P. W. (1966). *Endocrinol.* **79**, 565.
Taylor, C. B., Bailey, E. & Bartley, W. (1967). *Biochem. J.* **105**, 717.
Van Harken, D. R., Dixon, C. W. & Heimberg, M. (1969). *J. Biol. Chem.* **244**, 227.
Villee, C. A. & Loring, J. M. (1961). *Biochem. J.* **81**, 488.
Widdowson, E. M. (1950). *Nature, Lond.* **166**, 626.

THERMOGENESIS IN PREMATURELY DELIVERED LAMBS

By G. ALEXANDER, D. NICOL and G. THORBURN
CSIRO, Division of Animal Physiology,
P.O. Box 239, Blacktown, NSW, Australia

INTRODUCTION

The temperature regulating mechanisms in lambs are remarkably mature at birth (Alexander, 1970). Even within minutes of delivery the lamb can elevate its metabolic rate to four or five times the minimum resting level, and there is little improvement with increasing age, except through increasing insulation of the growing fleece.

In man, 5 to 10 per cent of infants are born prematurely (earlier than the 37th week of a normally 40-week gestation), and most survive, but in sheep less than 1 per cent of lambs can be considered premature (Dawes & Parry, 1965) and most of these die. Further, lambs delivered by Caesarian section earlier than the 140th day of the normally 150-day gestation, are not usually viable. It has, therefore, not been feasible to study the pre-natal development of temperature regulating mechanisms in sheep. However, the recent development of methods of inducing premature delivery of viable lambs by ACTH infusion into the foetus, while still *in utero* (Liggins, 1968, 1969), has now provided a means of studying the maturation of physiological processes at about the time of birth. This paper reports on the development of shivering and non-shivering thermogenesis associated with metabolism of brown adipose tissue in lambs, and also reports on the patency of the foetal circulatory shunts in prematurely delivered lambs. The early phases of the work have been reported in detail elsewhere (Alexander *et al.* 1972).

METHODS

Ten days before the desired date of premature delivery, catheters of polyvinyl chloride were inserted into the recurrent tarsal vein of foetal lambs (Bassett, Thorburn & Wallace, 1970). Infusion of 0.24 mg/day of synthetic ACTH was started 5 to 8 days later, i.e. 2 to 5 days before the desired time of delivery, the shorter periods of infusion being used during the later stages of gestation; infusion continued until the lambs were born.

After birth lambs were kept in a room held at about 31 °C and were fed initially on ewes' colostrum, that had been stored deep-frozen, and subsequently on fresh goat's milk. The lambs drank readily through a rubber teat, and consumption was virtually *ad libitum*.

Metabolic rate of the lambs was measured at various ages, using a closed-circuit respiration chamber with the lambs resting under thermoneutral conditions ('basal' metabolism) or exposed to low temperature and wind sufficient to cause a slow but perceptible fall in rectal temperature (summit metabolism) (Alexander & Williams, 1968).

Work on the first few lambs was restricted to observations on growth, survival and metabolic rate. Later observations were extended to include estimations of the concentration of glucose, lactate, free fatty acids and cortisol in the blood, examination for patency of the ductus arteriosus and foramen ovale and, finally, stimulation of the lamb with noradrenaline while under basal conditions to provide an estimate of the potential for non-shivering thermogenesis (Alexander & Williams, 1968).

RESULTS

Survival and growth

In all, twenty-two lambs were born; conceptual ages ranged from 123 to 144 days. One died soon after birth (123 days) and four died between one and two days of age; the others survived and grew at rates approximating the expected rate of intrauterine growth. This survival rate contrasts with the death, soon after birth, of all of seven lambs delivered by Caesarian section between conceptual ages 123 and 141 days, during the period of the study.

Metabolic rate

The minimum resting metabolic rate, measured within 1 to 4 days of delivery, was unrelated to the conceptual age at delivery, over the range 130 to 143 days, and approximated 1.0 l O_2/(kg h) which is approximately normal for full-term lambs (Fig. 1).

In contrast, the maximum metabolic response to cold (summit metabolism), measured in the various lambs within three days of delivery, clearly increased with the conceptual age at delivery, from about 2.5 l O_2/(kg h) at 130 days to levels approaching the normal full-term value of about 3.8 l O_2/(kg h) by 143 days (Fig. 1). However, when individual lambs were studied at different ages, there was little or no increase towards normal levels with advancing post-natal age, and beyond conceptual age of 150 days summit metabolism tended to fall, but at a much slower rate than in normal full-term lambs (Fig. 2).

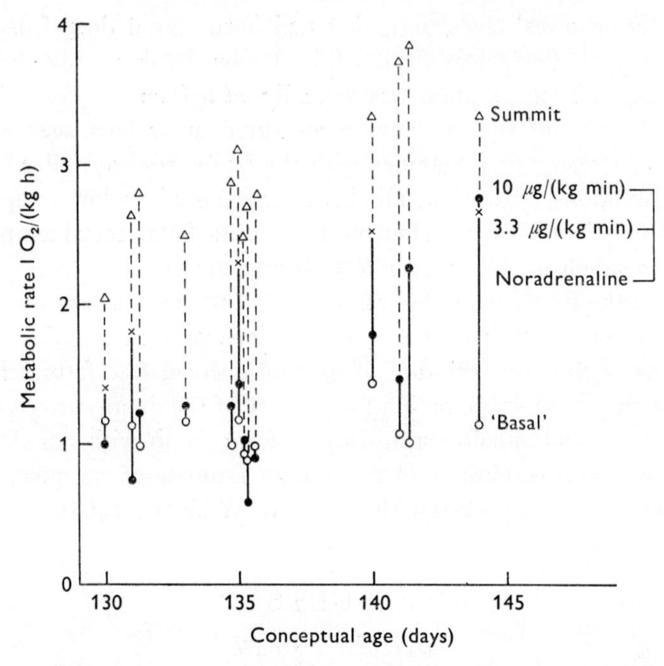

Fig. 1. Effect of infusion of noradrenaline on basal metabolic rate of premature lambs. The lower of the two infusion rates produces the higher responses (× compared with ●) particularly in the younger lambs. The basal metabolic rate (○) is independent of age but the noradrenaline response (×) and summit metabolism (△) increase with age. (Noradrenaline was infused at 10 µg/(kg·min) following a priming dose of 20 µg/kg.) Lambs were examined within 4 days of delivery at the conceptual ages shown.

Effect of noradrenaline on basal metabolic rate

Initially eight premature lambs less than four days old were infused with noradrenaline at the rate of 10 µg/(kg min) following a single priming injection of 20 µg/kg, doses that are effective in increasing metabolic rate in normal full-term lambs by 1 to 2 l O_2/(kg h). There was little response except in one of the oldest lambs delivered at 141 days' conceptual age (Fig. 1). Indeed in at least one lamb noradrenaline appeared to depress metabolic rate, and hence for the remaining few lambs the dose was reduced to 3.3 µg/(kg min) following a single priming injection of 6.7 µg/kg. Each of these lambs showed a clear response (Fig. 1) indicating stimulation of metabolism of brown adipose tissues. However, when the original higher dose was given following a period of infusion at the lower rate, the response was reduced, particularly in the younger lambs (Fig. 1).

Although summit metabolism changed little between delivery and conceptual age about 150 days, the response to noradrenaline given under resting conditions declined to less than 0.6 l O_2/(kg h) during this time,

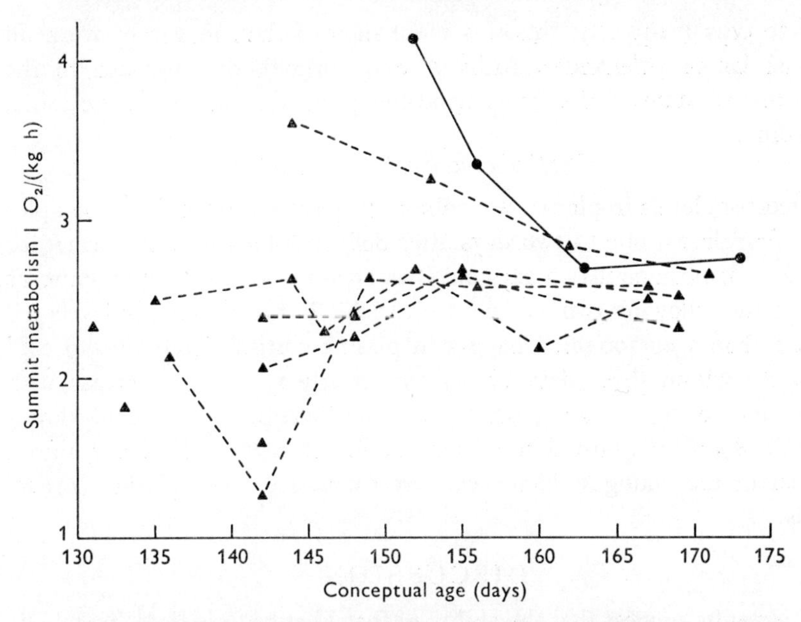

Fig. 2. Changes in summit metabolism of premature lambs (▲) compared with changes in one normal full-term lamb (●).

and was virtually absent (> 0.3 l O_2/(kg h)) by 170 days' conceptual age. The dose dependence apparent at delivery was not apparent during the subsequent measurements at about 150 and 170 days.

Foetal shunts

The magnitude of shunts through the ductus arteriosus and foramen ovale were estimated by dye dilution methods (Alexander & Williams, 1970) three to six days and two to three weeks after delivery. Small shunts (< 10 per cent of cardiac output) through one or both structures were detected in some lambs at the initial examination, but none was detected at the second examination.

Glucose, lactate and free fatty acid concentrations in blood

The concentrations of glucose and lactate in blood drawn prior to the first determination of summit metabolism, and the marked increase shown in samples drawn immediately after the period of cold exposure (approx. 40 min duration), were very similar to those found in normal newborn lambs subjected to the same routine of cold exposure. The mean initial and final concentrations were 150 and 250 mg/100 ml for glucose and 35 and 80 mg/100 ml for lactate. The concentrations of free fatty acids likewise showed a consistent rise due to cold exposure (from a mean of

0.45 to 0.90 mequiv/l); this was more marked than in earlier work on normal lambs (Alexander, Mills & Scott, 1968), due perhaps to the undisturbed state of the tame, hand-fed premature lambs at the initial sampling.

Cortisol concentration in blood

The cortisol levels in plasma were about 5 μg/100 ml, when blood samples were first drawn one to two days after delivery. This level is normal for full-term lambs about two days old (Bassett & Alexander, 1971). Patterns during the following two weeks were variable but subsequently levels fell to less than 1 μg/100 ml. Changes in plasma cortisol levels due to cold exposure within three days of delivery clearly tended to increase with increasing conceptual age (Fig. 3), and the increase was generally lower than the 8 μg/100 ml usual in full-term lambs (Bassett & Alexander, 1971); indeed in the youngest lambs cold exposure depressed plasma cortisol levels.

DISCUSSION

These results suggest that the ability of foetal lambs to increase metabolic rate in response to cold is already developing some 20 days prior to full term. However, the experimental lambs were exposed to adrenal stimulation for several days prior to birth, as indeed are normal newborn lambs, and it could be argued that the normal foetus, at say 130 days, does not possess the ability to increase metabolic rate in response to cold. Evidence on this point could be obtained by examining Caesarian-delivered lambs, but unfortunately their viability is low, so a satisfactory comparison has not been made.

The development of the cold response between conceptual age 130 days and normal term appears to depend on continued intrauterine existence; despite satisfactory growth and despite maintenance in a warm environment as close to intrauterine temperature as seemed possible without the risk of heat stress, the metabolic response tended to remain constant between delivery and 150 days' conceptual age. Presumably the events leading to birth terminate the development phase.

During this period (130 days to full term) the non-shivering response, as assessed by stimulation with noradrenaline, declined, so that the shivering response appears to have increased (Fig. 4), but this is difficult to measure with accuracy. Hence the failure of the thermogenic response, as a whole, to increase after delivery appears largely due to failure of brown fat to continue to develop post-natally. Presumably, the replacement of brown fat with white fat, that normally commences at term (Gemmell, Bell & Alexander, 1972) begins at the time of delivery, regardless of the

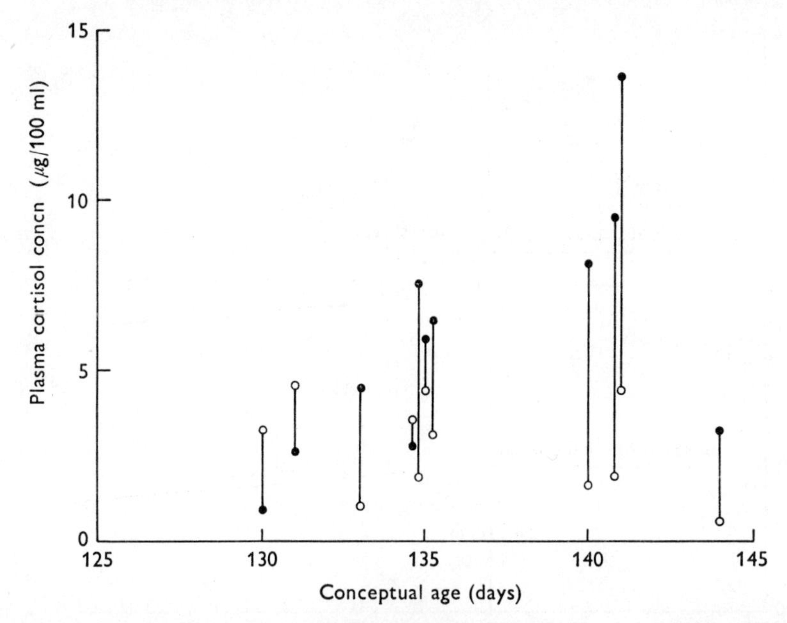

Fig. 3. Change in plasma cortisol concentration induced by approximately 40 minutes cold exposure of the premature lamb (○ – before exposure, ● – immediately after exposure). The response increases with increasing conceptual age at delivery. Lambs were examined once, within 3 days of delivery at the conceptual ages shown.

conceptual age of the lamb. The mechanisms controlling the development of brown fat and its conversion to white fat are not known.

The nature of the immaturity of the thermogenic mechanisms, whether hormonal or neurological, is not clear. Despite this immaturity the premature lambs appeared normal in other respects such as the cold-induced changes in the level of blood metabolites, the resting (basal) level of metabolism, and the low incidence of shunts through foetal circulatory channels. However, the sensitivity of the non-shivering thermogenic mechanisms to noradrenaline was clearly different from that in full-term lambs. Reasons for the observed dose dependence of the noradrenaline response in the premature lambs are not clear, but it is suggested that the infusion rate thought necessary to stimulate brown fat in full-term lambs causes vasoconstriction in the brown adipose tissue of premature lambs; in the mature lambs there is some evidence that this dose causes vasodilation in brown adipose tissue (Alexander *et al.* 1972).

A further difference between premature and full-term lambs lies in the response of plasma cortisol levels to cold exposure; the response was particularly low in the younger lambs. Indeed some of the early deaths occurred within 24 hours of exposure to the very cold summit conditions,

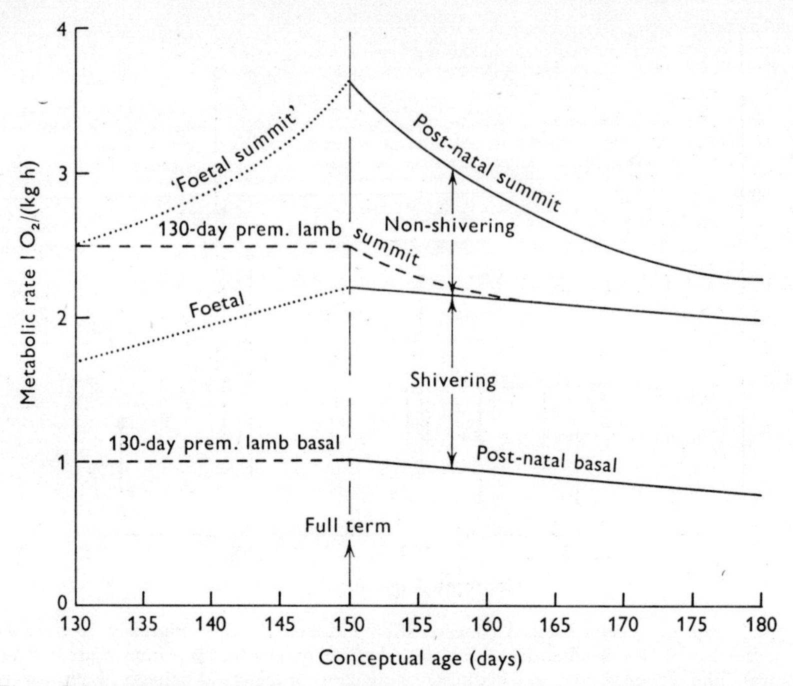

Fig. 4. Schematic diagram of the contribution of shivering and non-shivering thermogenesis to summit metabolism in lambs according to conceptual age. The dotted lines represent the situation for a *series* of premature lambs examined soon after birth at ages indicated on the abscissa. The broken lines represent the *progressive changes* in summit metabolism of a premature lamb born at conceptual age 130 days. The solid lines represent the progressive changes in a lamb born at full term (150 days).

and could have been due to adrenal insufficiency, but there is no evidence to indicate whether this poor response is due to immaturity of the hypothalamus, pituitary or of the adrenal cortex itself.

The lower viability of lambs delivered by Caesarian section, compared with ACTH-induced lambs, appears to be due to immaturity of the respiratory system; the Caesarian-delivered lambs made feeble attempts to breathe, but failed to establish respiration, and the ACTH-treated lambs that survived for only 1–2 days showed evidence of pulmonary failure. These phenomena are presumably related to the effects of corticosteroids in accelerating the appearance of pulmonary surfactant (Delemos *et al.* 1971). Whether the ACTH-treated lambs that died could have been salvaged by the type of intensive treatment given to premature human infants is not clear; and it remains to be determined whether survival would have been improved by continuing ACTH or corticosteroid treatment postnatally.

PLATE I

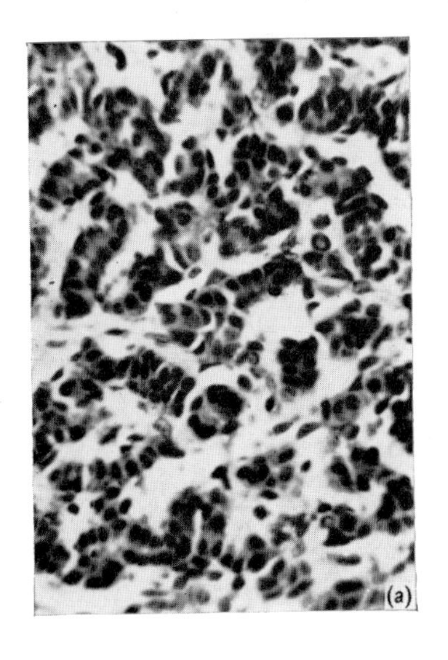

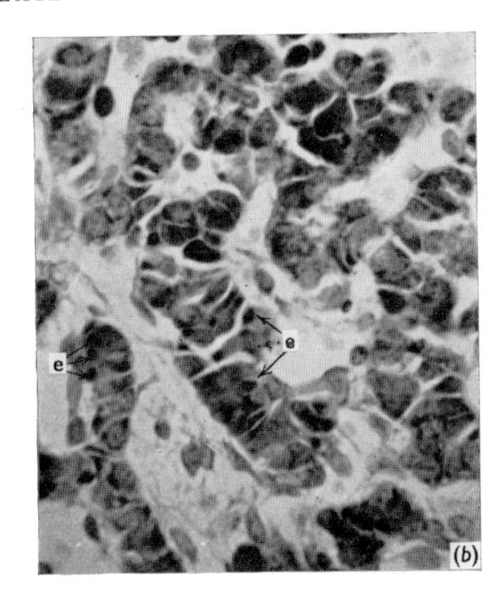

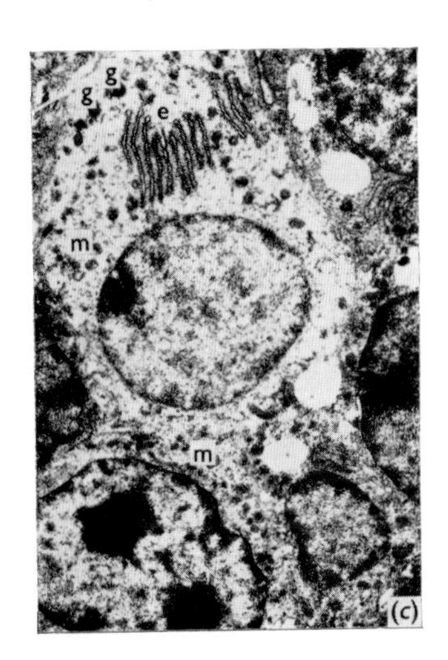

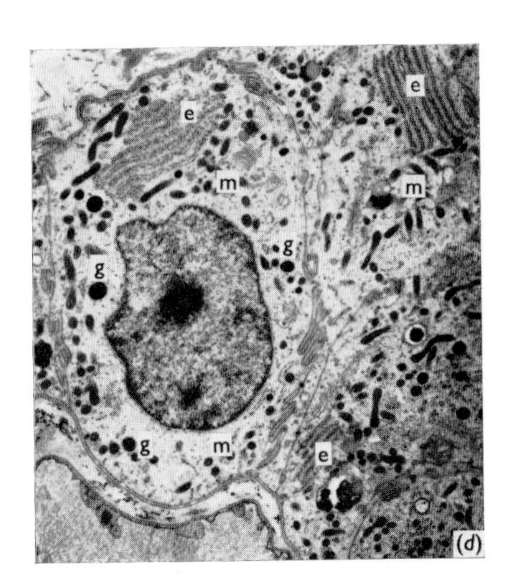

(*a*) Parathyroid of 128-day male foetus. H & E × 400.

(*b*) Similar to (*a*) but stained pyronin/methyl green × 700. e, RNA–protein masses.

(*c*) EM of 128-day foetal parathyroid. × 6800 (This shows some post-mortem damage.) e, granular endoplasmic reticulum; m, mitochondria; g, secretion granule.

(*d*) EM of parathyroid of normal adult. × 5200. Abbreviations as in (*c*).

(*Facing p.* 416)

REFERENCES

ALEXANDER, G. (1970). In *Physiology of Digestion and Metabolism in the Ruminant*, ed. Phillipson, A. T. Newcastle-upon-Tyne: Oriel Press.

ALEXANDER, G. & WILLIAMS, D. (1968). *J. Physiol., Lond.* **198**, 251–76.

ALEXANDER, G., BELL, A. W. & SETCHELL, B. P. (1972). *J. Physiol., Lond.* **220**, 511–28.

ALEXANDER, G., MILLS, S. C. & SCOTT, T. W. (1968). *J. Physiol., Lond.* **198**, 277–89.

ALEXANDER, G., THORBURN, G. D., NICOL, D. H. & BELL, A. W. (1972). *Biol. Neonate* **20**, 1–8.

BASSETT, J. M. & ALEXANDER, G. (1971). *Biol. Neonate* **17**, 112–25.

BASSETT, J. M., THORBURN, G. D. & WALLACE, A. L. C. (1970). *J. Endocr.* **48**, 251–63.

DAWES, G. S. & PARRY, H. B. (1965). *Nature, Lond.* **207**, 330.

DELEMOS, R., SHERMETA, D., KNELSON, S., KOTAS, R. & AVERY, M. (1971). *Am. Rev. resp. Dis.* **102**, 459–61.

GEMMELL, R. T., BELL, A. W. & ALEXANDER, G. (1972). *Am. J. Anat.* **133**, 143–63.

LIGGINS, G. C. (1968). *J. Endocr.* **45**, 323–9.

LIGGINS, G. C. (1969). *J. Endocr.* **45**, 515–23.

ACTIVE FAT

By DAVID HULL AND MARTIN HARDMAN
Hospital for Sick Children, Great Ormond Street, London

In the newborn there are two forms of adipose tissue: unilocular white adipose tissue, which is primarily concerned with the storage and supply of fatty acids, and multilocular brown adipose tissue, which is a major site of thermoregulatory heat production. Sympathetic nervous activity or noradrenaline administration stimulate lipolysis in both tissues. In the former, this leads to the net release of fatty acids into the circulation. In the latter it leads to the intracellular oxidation of fatty acids with the net production of heat.

In the adult it is difficult to distinguish between the two tissues. With increasing age the fat content of brown adipose tissue rises and the cells become unilocular and therefore to the naked eye and under light microscopy the tissue resembles white adipose tissue. Nevertheless in the adult, adipose lobes with the same general anatomical relationships and the same blood and nervous supply as the lobes of brown adipose tissue in the newborn, are present.

With increasing age non-shivering thermogenesis falls and brown adipose tissue appears no longer to produce heat. What then is its function? In function as well as structure does brown adipose tissue change into 'white' adipose tissue?

Experiments on rabbits one week after birth (body weight 110 g, cervical and interscapular fat form about 2.5% body weight) showed that brown adipose tissue although it was multilocular and although it still produces heat at the same rate as it did in the newborn, nevertheless released fatty acids into the circulation (Hardman & Hull, 1970). The rate of release of fatty acids depended on the ambient temperature at which the rabbit had been reared and also on its plane of nutrition (Hardman & Hull, 1971). At this age brown adipose tissue has the dual function of heat production and fatty acid supply.

Experiments on three-month-old adolescent rabbits (body weight about 2000 g) showed that their brown adipose tissue no longer produced heat. In the week-old animal the resting tissue blood flow was 0.9 ml/(g tissue wet weight·min) whereas in the three-month-old animal it was 0.35 ml/(g tissue wet weight·min). Nevertheless noradrenaline infusion caused a large increase in blood flow which was accompanied by a large increase in the net release of fatty acids. Acute starvation also led to a large increase in the net release of fatty acids. There was an increase in tissue blood flow and also in the arterio-venous difference across the tissue.

In six-month-old adult rabbits (body weight 4000 kg, cervical and interscapular fat form about 2.5 % body weight) the tissue blood flow was only 0.15 ml/(g tissue wet weight·min). This is ten times greater than that observed in inguinal white adipose tissue of adult rabbits (Lewis & Matthews, 1968). Again in contrast to white adipose tissue, the cervical adipose tissue responded to noradrenaline infusion with a large increase in tissue blood flow accompanied by a large increase in the net release of fatty acids.

It would appear that although the cervical adipose tissue's histological features in the adult resemble those of white adipose tissue, it still differs from white adipose tissue in that it releases larger amounts of fatty acid per gram tissue at rest and responds to noradrenaline with a large increase in the rate of fatty acid release. These differences could be largely explained by the greater vascularity of persisting, unilocular, non-thermogenic 'brown' adipose tissue.

REFERENCES

Hardman, M. J. & Hull, D. (1970). *J. Physiol.* **206**, 263–73.
Hardman, M. J. & Hull, D. (1971). *J. Physiol.* **214**, 191–9.
Lewis, G. P. & Matthews, J. (1968). *Br. J. Pharmacol.* **34**, 564–72.

SESSION 5

ENDOCRINOLOGY

CALCIUM, PARATHYROID HORMONE AND CALCITONIN IN THE FOETUS

By D. PAULINE ALEXANDER,* H. G. BRITTON,* D. A. NIXON,* E. CAMERON,† C. L. FOSTER,† R. M. BUCKLE‡ AND F. G. SMITH, JR§

* Department of Physiology and † Department of Cellular Biology and Histology, St Mary's Hospital Medical School, London; ‡ Department of Endocrinology, General Hospital, Southampton; § UCLA School of Medicine, Los Angeles, California

CALCIUM AND PHOSPHATE

In the sheep, foetal plasma calcium concentration is considerably greater than that of the mother in the latter half of gestation, and this difference can only be partly attributed to greater calcium binding by the foetal plasma proteins since the ultrafilterable calcium is also higher (Bawden, Wolkoff & Flowers, 1965; Delivoria-Papadopoulos, Battaglia, Bruns & Meschia, 1967; Bawden & Wolkoff, 1967). Ionised calcium levels have not been reported and it is possible that calcium binding molecules of low molecular weight may be present in the foetal plasma. The citrate concentration in the foetus is somewhat greater than in the mother but is insufficient to bind appreciable amounts of calcium (mean foetal plasma citrate 4.1 mg/100 ml at 87–143 days' gestation, mean maternal 2.0 mg/100 ml; Fenton, personal communication). Not only is calcium greater in foetal plasma but phosphate is also present in higher concentration than in the mother. On the other hand, virtually no phosphate is found in foetal urine (Alexander, Nixon, Widdas & Wohlzogen, 1958; Smith, Adams, Borden & Hilburn, 1966). Foetal calcium and phosphate metabolism thus differs substantially from that of the mother.

In man also at term, foetal plasma calcium and phosphate levels are higher than in the mother, and no phosphate can be found in the foetal urine. In this case binding of calcium by the foetal plasma proteins is not increased (Delivoria Papadopoulos *et al.* 1957; Dean & McCance, 1948; McCrory, Forman, McNamara & Barnett, 1952). Again, in monkey, pig and cow, foetal plasma total calcium concentrations are higher than in the mother in the latter part of gestation. In rabbit (Graham & Porter, 1971), guinea pig (Graham & Scothorne, 1970) and rat (Garel, 1969*a*) large changes in foetal plasma calcium have been reported during development. The plasma calcium falls to its lowest level in these species at the time of onset of skeletal calcification (Graham & Porter, 1971; Graham & Scothorne, 1970; Jost, Moreau & Fournier, 1960) and then rises to reach or exceed the maternal level at the end of gestation. The foetal plasma proteins differ from the mother (Lehrer & Toben, 1965; Graham & Scothorne, 1970) but the calcium binding properties of the foetal plasma proteins in these species are not known. In the guinea pig plasma-citrate concentrations are insufficient to contribute appreciably to binding (*c.* 2.0 mg/100 ml foetal plasma; Fenton, personal communication).

PARATHYROID HORMONE

Secretion in the foetal sheep

In a series of five sheep foetuses of 80 to 142 days' gestational age, exposed at Caesarian section, immunoassayable parathyroid hormone (PTH) was demonstrable in the plasma of two (0.53 and 0.75 ng/ml), and in the maternal plasma of three (0.23–0.50 ng/ml). There was no correlation between maternal and foetal values (Smith, Alexander, Buckle, Britton & Nixon, 1972). When ethylene diamine tetra-acetate (EDTA) was infused into the foetus over periods of 90–180 min, PTH appeared in the foetal plasma of all of the animals. In none of the experiments was there any significant change in maternal calcium or PTH levels. The PTH levels are plotted against calcium levels and compared with data from the three adult ewes in Fig. 1. The PTH response to change in plasma calcium appeared to be essentially similar to that of the adult but set at a slightly higher calcium level.

In the sheep, therefore, maternal and foetal PTH levels are independent, presumably reflecting a lack of placental permeability to PTH. Moreover foetal calcium seems to be a factor controlling PTH secretion in the foetus.

Light microscopy of the foetal parathyroid gland shows anastomosing cords of chief cells (Plate 1*a*) which contain discrete masses of RNA-protein corresponding to granular endoplasmic reticulum (Plate 1*b*). With the electron microscope (Plate 1*c*, *d*) a well-developed endoplasmic reticulum and Golgi apparatus together with abundant mitochondria, and small numbers of membrane-bounded bodies resembling secretory granules are seen. Apart from a lack of glycogen the EM appearances are similar to the human foetal and adult gland (Nakagami, Yamazaki & Tsunoda,

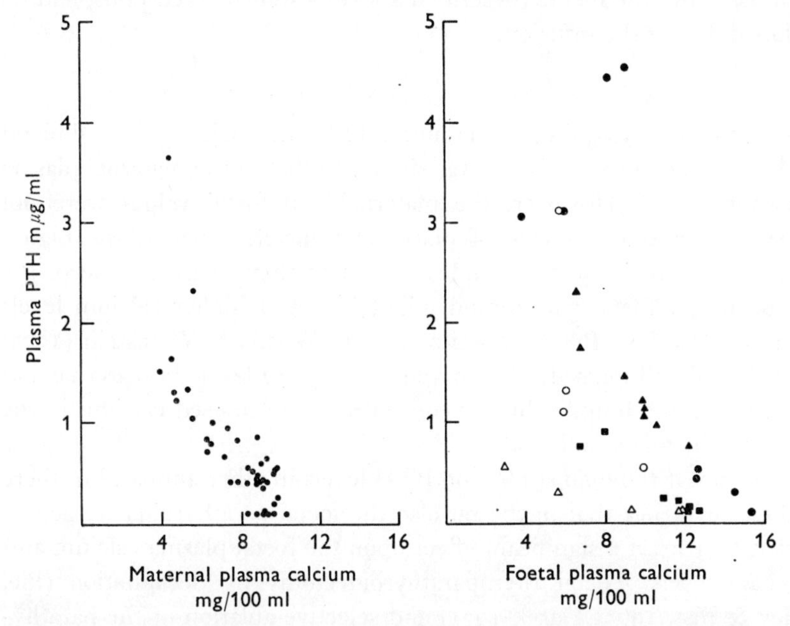

Fig. 1. Comparison of the concentrations of plasma parathyroid hormone (PTH mμg/ml) in relation to plasma calcium (mg/100 ml) in five sheep foetuses and in three adult ewes during EDTA-induced hypocalcaemia. Foetal age in days: ○, 80; ▲, 132; △, 134; ●, 135; ■, 142. The 134-day animal which showed little response to hypocalcaemia was in poor condition.

1968; Munger & Roth, 1963), and suggest that the glands are competent to secrete. Explants of the foetal parathyroid from foetuses of 30 days' gestational age and over, cultivated on chick chorio-allantoic membrane with neonatal rat parietal bone, cause reabsorption of the bone, providing further evidence for PTH secretion by the foetal gland (Scothorne, 1964).

Biological activity of parathyroid hormone in the sheep foetus

Bovine PTH injected into the sheep foetus causes no change in plasma calcium levels or change in calcium excretion by the kidney (Alexander & Nixon, 1969; Smith, Tingloff, Meuli & Borden, 1969). Similarly, bovine PTH has no effect upon the plasma calcium in the adult (Lotz, Talmage & Comar, 1954). In contrast, the infusion of PTH causes a brisk rise in phosphate excretion by the foetal kidney and a probable rise in the foetal plasma phosphate (Alexander & Nixon, 1969; Smith, Tingloff, Meuli & Borden, 1969). The plasma PTH levels were probably high in these experiments and it is not clear whether such a response may be expected at physiological levels. However, in one of the experiments in which EDTA was infused into the foetus (described above) a well-marked phosphaturia developed during the infusion.

Parathyroid hormone in species other than sheep

In a study in man, Lequin (1969) found PTH in the majority of cord blood samples and the range of values was similar to that in the maternal plasma (< 70–330 pg/ml). However, the maternal and foetal values were not correlated, suggesting a lack of placental transfer. Further, the higher calcium levels in the foetal samples suggested that, as in the sheep, the response of the foetal parathyroid might be set at higher calcium levels than in the mother. Recently Reitz, Daane, Woods & Weinstein (1972) reported total and ionised calcium and parathyroid levels in maternal and cord blood. They found a higher concentration of ionised calcium in the foetal than the maternal blood.

There appear to be no studies on PTH levels in other animals but there is indirect evidence that in the rat also the foetal parathyroids are active, and that they exert a significant effect upon the foetal plasma calcium and phosphate levels. Foetal thyroparathyroidectomy by decapitation (Pic, Maniey & Jost, 1965; Garel, 1971) and selective ablation of the parathyroids with some thyroid tissue (Pic, 1970) both lower foetal plasma calcium towards the maternal level whereas unilateral parathyroidectomy is without effect (Pic, 1970). Antibovine PTH serum injected subcutaneously into rat foetuses of $21\frac{1}{2}$ days, appears to lower the plasma calcium and to raise the plasma phosphate after 1–2 h but a large correction had to be applied for spontaneous changes (Garel, 1971). The PTH in the foetal circulation seems to originate within the foetus since ^{125}I PTH injected into the mother does not enter the foetal compartment nor does ^{125}I PTH injected into the foetus pass to the mother (Garel, 1972). The ^{125}I PTH injected into the foetus rapidly disappeared from the foetal circulation with a half-life (31 min) similar to that in the mother (21 min). Substantial quantities of TCA-precipitable radioactivity appeared in the foetal liver with less activity in the placenta, kidney and gastrointestinal tract. In this respect the metabolism differed from the adult where most of the activity appeared in the kidney.

Injection of sufficient PTH into the mother to give a marked increase in maternal plasma calcium in both guinea pigs and rats, gave either no rise or only a small delayed rise in the foetal plasma calcium (Burnett, Simpson, Chandler & Bawden, 1968; Garel, 1969*a*). This may further suggest that the rat and guinea pig placentae are impermeable to PTH. Earlier indirect evidence for placental impermeability to PTH in a number of species is summarised by Krukowski & Lehr (1963).

It has recently been reported that in bovine parathyroid glands a second form of the hormone exists with a higher molecular weight and less biological activity (Arnaud, Tsao & Oldham, 1970; Sherwood, Rodman & Lundberg, 1970; Cohn, Macgregor, Chu & Hamilton, 1972). The chemical nature of the hormone in foetal plasma may therefore require further investigation.

CALCITONIN

There have been no studies of this hormone in the foetal sheep. However, calcitonin has been extracted from pig foetal thyroid tissue (Phillipo, Care & Hinde, 1969), its presence has been shown in the foetal human gland by immunofluorescence (Pearse, private communication) and C cells have been described in the foetal rat (Pearse & Carvalheira, 1967; Welsch, 1971). In the foetal rat of 19½–20½ days' gestation, administration of pork calcitonin to the foetus produces hypocalcaemia and hypophosphataemia but the foetus of 17½ days does not respond (Garel, Milhaud & Jost, 1968). The latter workers also found that administration of calcitonin to the mother causes hypocalcaemia and hypophosphataemia in both mother and foetus, but more recently Wezeman & Reynolds (1971) in similar experiments report that the foetal plasma calcium returns to normal after 2 h despite a continuing maternal hypocalcaemia. Garel, Milhaud & Sizonenko (1969) find that ^{125}I calcitonin does not cross the placenta in either direction and with tissue homogenates these workers found the foetal kidney to be the most active foetal tissue degrading the hormone (Garel, Milhaud & Sizonenko, 1970). A plasma inactivating factor in rat foetal plasma (of somewhat less potency than that found in the adult) has also been described (Wezeman & Reynolds, 1971). Garel (1969*b*) showed that prematurely delivered rats were more sensitive to calcitonin than foetal rats of the same age (21½–22½ days). The change occurred over a period of several hours after delivery and thus did not reflect a stabilising effect of the placenta on the plasma calcium.

ROLE OF THE PLACENTA

The higher concentration of calcium in the umbilical vein compared with the artery in man and cow (see McCance & Widdowson, 1961; Armstrong, Singer & Makowski, 1970) although not in the sheep (Kaiser & Cummings, 1958) may suggest that the placenta is the portal of entry of calcium into the foetus. However, there is insufficient information about the state of the calcium in the foetal and maternal plasma to decide whether placental transfer of calcium is active or passive. Transplacental electrical potentials must also be taken into account if calcium penetrates as the free ion.

Recent reports indicate that such potentials may be substantial (Widdas, 1961; Mellor, 1969, 1970). Whatever transport process may be involved it seems that the calcium concentrations on the two sides are relatively independent of one another and that rapid movements of calcium do not occur in response to changes in concentration of calcium. Maternal and foetal total and ultrafiltrable plasma calcium are independent of one another in the guinea pig and sheep when calcium gluconate is infused into the mother (Greeson, Crawford, Chandler & Bawden, 1968; Bawden *et al.* 1965; Bawden & Wolkoff, 1967), and relatively small changes occur in foetal calcium levels in the rat and guinea pig when PTH or calcitonin (in the rat) is given to the mother (see above). Further, there is a relatively slow rate of isotopic exchange in cow (Plumlee, Hansard, Comar & Beeson, 1952), sheep (Kronfeld, Ramberg & Delivoria-Papadopoulos, 1971), rat and rabbit (Wasserman, Comar, Nold & Lengemann, 1957) and monkey (McDonald, Hutchinson, Hepler & Flynn, 1965; Kronfeld *et al.* 1971). The placenta may therefore have a major regulatory role in controlling foetal calcium metabolism. In this context the experiments shown in Fig. 2 upon perfused sheep placentae of 78–140 days' gestational age may be of interest. The replacement of the foetus with a pump and a reservoir containing 200 ml maternal blood caused a fall in total plasma calcium towards that of the mother but the level returned to the foetal level within 20–30 min, suggesting a homeostatic mechanism within the placenta. The rate of transfer required to return the calcium to the original level was of the order suggested by tracer studies (Kronfeld *et al.* 1971).

DISCUSSION

It seems clear that foetal calcium and phosphate metabolism differ substantially from the adult. In several species including man, sheep and rat, the parathyroid glands appear to be secreting before birth. We have shown that in the sheep, foetal hypocalcaemia is a powerful stimulus to PTH secretion. The response to hypocalcaemia seems to be similar to that in the adult except that it is set at a somewhat higher plasma calcium level. There is also evidence in both the sheep and the rat that PTH is metabolically active. There is no clear evidence for calcitonin secretion in the foetus, but calcitonin is found in the foetal pig thyroid and administration of calcitonin causes metabolic changes in the foetal rat. Neither PTH nor calcitonin appears to cross the placenta in any of the species investigated.

Despite the findings summarised above the overall control of foetal calcium metabolism is still not clearly defined. The possibility must also

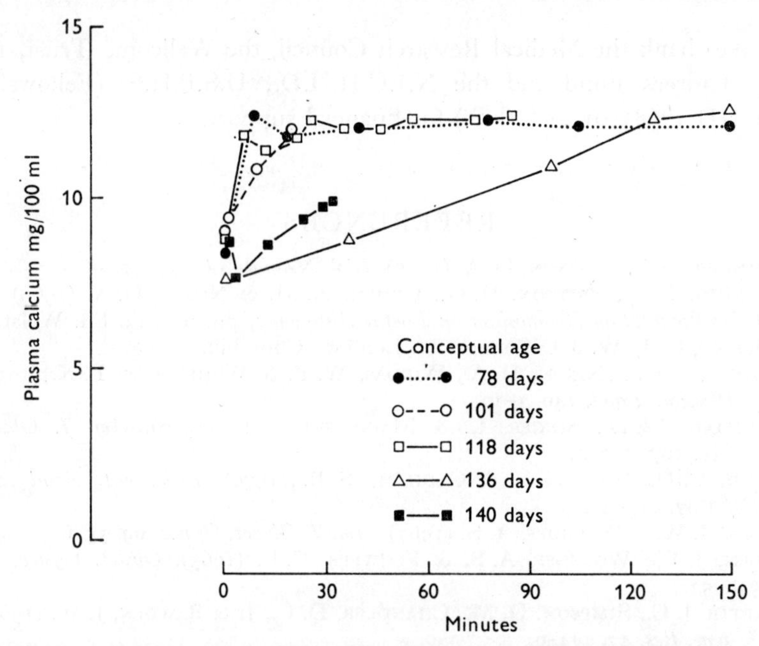

Fig. 2. Change in plasma calcium concentration of perfusate during in-situ recirculating perfusion of sheep placentae.

be considered in this context that control mechanisms appropriate for the adult may be inappropriate in the foetal state (Alexander, Britton, Cohen & Nixon, 1969). For example, mobilisation of foetal skeletal calcium might not be an appropriate response to calcium deficiency. Again a renal response to PTH will be of little or no value in foetal homeostasis. It is possible that the placenta may be the major regulator of foetal calcium metabolism and it would seem especially important to establish how calcium is handled by this organ.

SUMMARY

Certain aspects of calcium metabolism in the foetus are reviewed. Evidence is presented that in the sheep, in the latter half of gestation, the foetal parathyroids are active and respond to hypocalcaemia; these glands also appear to be active in the human and rat. The placenta appears to be impermeable to parathyroid hormone. The metabolic effects of parathyroid hormone in the foetus are considered. Evidence that the placenta may be impermeable to calcitonin and that this hormone may be secreted by the foetus is also discussed. Consideration is given to the role of the placenta and it is concluded that this organ may play a particularly important part in calcium homeostasis.

We thank the Medical Research Council, the Wellcome Trust, the Mary Kinross Fund and the N.I.C.H.H.D., U.S.P.H.S. (Fellowship IF03-HD 37683-01 to F.G.S.) for financial support.

REFERENCES

ALEXANDER, D. P. & NIXON, D. A. (1969). *Biol. Neonate* **14**, 117–30.

ALEXANDER, D. P., BRITTON, H. G., COHEN, N. M. & NIXON, D. A. (1969). In *Ciba Foundation Symposium on Foetal Autonomy*, pp. 95–113. Ed. Wolstenholme, G. E. W. & O'Connor, M. London: Churchill.

ALEXANDER, D. P., NIXON, D. A., WIDDAS, W. F. & WOHLZOGEN, F. X. (1958). *J. Physiol., Lond.* **140**, 1–13.

ARMSTRONG, W. D., SINGER, L. & MAKOWSKI, E. L. (1970). *Am. J. Obstet. Gynec.* **107**, 432–4.

ARNAUD, C. D., TAO, H. S. & OLDHAM, S. B. (1970). *Proc. nat. Acad. Sci. USA* **67**, 415–22.

BAWDEN, J. W. & WOLKOFF, A. S. (1967). *Am. J. Obstet. Gynec.* **99**, 55–60.

BAWDEN, J. W., WOLKOFF, A. S. & FLOWERS, C. E. (1965). *Obstet. Gynec.* **25**, 548–52.

BURNETTE, J. C., SIMPSON, D. M., CHANDLER, D. C., Jr & BAWDEN, J. W. (1968). *J. dent. Res.* **47**, 444–6.

COHN, D. V., MACGREGOR, R. R., CHU, L. L. H. & HAMILTON, J. W. (1972). In *Calcium, Parathyroid Hormone and the Calcitonins*, pp. 173–82. Ed. Talmage & Munson. Amsterdam: Excerpta Medica.

DEAN, R. F. A. & MCCANCE, R. A. (1948). *J. Physiol., Lond.* **107**, 182–6.

DELIVORIA-PAPADOPOULOS, M., BATTAGLIA, F. C., BRUNS, P. D. & MESCHIA, G. (1967). *Am. J. Physiol.* **213**, 363–6.

GAREL, J. M. (1969*a*). Doctorate Thesis, Paris.

GAREL, J. M. (1969*b*). *C. r. Acad. Sci. Paris* **268**, 1525–8.

GAREL, J. M. (1971). *Israel J. med. Sci.* **7**, 349–50.

GAREL, J. M. (1972). *Horm. Metab. Res.* **4**, 131–2.

GAREL, J. M. & DUMONT, C. (1972). *Horm. Metab. Res.* **4**, 217–21.

GAREL, J. M., MILHAUD, G. & JOST, A. (1968). *C. r. Acad. Sci. Paris* **267**, 344–7.

GAREL, J. M., MILHAUD, G. & SIZONENKO, P. C. (1969). *C. r. Acad. Sci. Paris* **269**, 1785–7.

GAREL, J. M., MILHAUD, G. & SIZONENKO, P. C. (1970). *C. r. Acad. Sci. Paris* **270**, 2469–71.

GRAHAM, R. W. & PORTER, G. P. (1971). *Q. Jl exp. Physiol.* **56**, 160–8.

GRAHAM, R. W. & SCOTHORNE, R. J. (1970). *Q. Jl exp. Physiol.* **55**, 44–53.

GREESON, C. D., CRAWFORD, E. G., CHANDLER, D. C. & BAWDEN, J. W. (1968). *J. dent. Res.* **47**, 447–9.

JOST, A., MOREAU, G. & FOURNIER, C. (1960). *Arch. Anat. micr. Morphol. exper.* **49**, 431–58.

KAISER, I. H. & CUMMINGS, J. N. (1958). *Am. J. Physiol.* **193**, 627–33.

KRONFELD, D. S., RAMBERG, C. F., Jr & DELIVORIA-PAPADOPOULOS, M. (1971). In *Cellular Mechanisms for Calcium Transfer and Homeostasis*, pp. 339–49. Eds. Nichols, G. & Wasserman, R. H. New York & London: Academic Press.

KRUKOWSKI, M. & LEHR, D. (1963). *Arch. int. Pharmacodyn.* **146**, 245–65.

LEHRER, S. & TOBEN, H. (1965). *Anat. Rec.* **151**, 378.

LEQUIN, R. M. (1969). Ph.D. Thesis, Rotterdam.

LOTZ, W. E., TALMAGE, R. V. & COMAR, C. L. (1954). *Proc. Soc. exp. Biol. NY* **85**, 292–5.

MCCANCE, R. A. & WIDDOWSON, E. M. (1961). *Brit. med. Bull.* **17**, 132–6.

MCCRORY, W. W., FORMAN, C. W., MCNAMARA, H. & BARNETT, H. (1952). *J. clin. Invest.* **31**, 357–66.

MCDONALD, N. S., HUTCHINSON, D. L., HEPLER, M. & FLYNN, E. (1965). *Proc. Soc. exp. Biol. NY* **119**, 476–81.

MELLOR, D. J. (1969). *J. Physiol., Lond.* **204**, 395–405.

MELLOR, D. J. (1970). *J. Physiol., Lond.* **207**, 133–50.

MUNGER, B. L. & ROTH, S. I. (1963). *J. Cell Biol.* **16**, 379–400.

NAKAGAMI, K., YAMAZAKI, Y. & TSUNODA, Y. (1968). *Z. Zellforsch.* **85**, 89–95.

PEARSE, A. G. E. & CARVALHEIRA, A. F. (1967). *Nature, Lond.* **214**, 929–30.

PHILLIPPO, M., CARE, A. D. & HINDE, F. R. (1969). *J. Endocr.* **43**, 15P.

PIC, P. (1970). D.Sc. Thesis, Paris.

PIC, P., MANIEY, J. & JOST, A. (1965). *C. r. Soc. Biol. Paris*, **159**, 1274–7.

PLUMLEE, M. P., HANSARD, S. L., COMAR, C. L. & BEESON, W. M. (1952). *Am. J. Physiol.* **171**, 678–86.

REITZ, R. E., DAANE, T. A., WOODS, J. D. & WEINSTEIN, R. L. (1972). In *4th International Congress of Endocrinology*, p. 208. Washington, DC: International Congress Series 256, Excerpta Medica.

SCOTHORNE, R. J. (1964). *Ann. NY Acad. Sci.* **120**, 669–76.

SHERWOOD, L. M., RODMAN, J. S. & LUNDBERG, W. B. (1970). *Proc. nat. Acad. Sci.* **67**, 1631–8.

SMITH, F. G., Jr, ADAMS, F. H., BORDEN, M. & HILBURN, J. (1966). *Am. J. Obstet. Gynec.* **96**, 240–6.

SMITH, F. G., Jr, ALEXANDER, D. P., BUCKLE, R. M., BRITTON, H. G. & NIXON, D. A. (1972). *J. Endocr.* **53**, 339–48.

SMITH, F. G., Jr, TINGLOF, B., MEULI, J. & BORDEN, M. (1969). *J. appl. Physiol.* **27**, 276–9.

WASSERMAN, R. H., COMAR, C. L., NOLD, M. M. & LENGEMANN, F. W. (1957). *Am. J. Physiol.* **189**, 91–7.

WELSCH, U. (1971). In *Mem. Soc. Endocrn.* No. 19, pp. 557–66. Ed. Heller, H. & Lederis, K. Cambridge University Press.

WEZEMAN, F. H. & REYNOLDS, W. A. (1971). *Endocrinology* **89**, 445–52.

WIDDAS, W. F. (1961). *Brit. med. Bull.* **17**, 107–11.

STUDIES OF THE NEUROHYPOPHYSIS IN FOETAL MAMMALS

By A. M. PERKS and E. VIZSOLYI
Physiology Group, Department of Zoology,
University of British Columbia,
Vancouver 8, B.C., Canada

The neurohypophysial system of foetal mammals has received relatively extensive histological investigation (see Yakovleva, 1965). Despite species differences, individual variability, and conflicting opinions, some generalisations can be made. In all mammals investigated, except the rat, neurosecretory material has been found during foetal life. Although it is assumed that neurosecretion is produced in the supraoptic and paraventricular nuclei of the hypothalamus, where it is first detected in individual human foetuses (and perhaps in foetal calves), in most prenatal mammals it first reaches stainable levels in the pars nervosa of the pituitary, where it is stored. The quantity of neurosecretion increases steadily with development, but at birth it is still below adult levels in the pars nervosa.

The few physiological and pharmacological studies available show a parallel with the histological work, except where they appear to be more sensitive to small quantities of neurohypophysial secretions. Studies of foetal and newborn cats, dogs and humans, and newborn rats and guinea-pigs, showed that neurohypophysial principles could be detected during prenatal life, but that the quantity present at birth was lower than that in the adult mammal (Heller & Zaimis, 1949; Dicker & Tyler, 1953*a*, *b*; Acher & Fromageot, 1957; Heller & Lederis, 1959). These studies suggested that the neurohypophysial system, although advanced, was not completely mature at time of birth.

In addition, the above investigators made two unexpected observations on the foetal and newborn principles. Firstly, the ratio of vasopressin to oxytocin (V/O ratio) was higher than the value of unity typical of most adult mammals. In the human, V/O ratios were reported as high as 28:1, but they fell to unity at birth. In the cat and dog, the predominance of vasopressor activity declined during late foetal life, but persisted into the first month after birth. In the guinea-pig and rat, studies were confined to the newborn, but a similar fall in V/O ratios was described in the first weeks after delivery. However, the general picture of a decline in V/O ratios during development was based largely on studies of the newborn;

evidence from foetuses themselves was restricted to the excellent, but limited pioneer studies of Dicker & Tyler (1953*a*, *b*). These workers had investigated only late foetuses of cats and dogs, and although their studies of human foetuses had extended through a wider range of gestation, the irregular variations in the V/O ratios obtained might well have reflected post-mortem changes, due to delays in the dissection of human material. Clearly, new work was needed.

The second unusual observation on the development of neurohypophysial peptides was made by Heller & Lederis (1959); they noted that the oxytocic principle of the newborn rat had an unusually high solubility in wet acetone. These workers indicated that the high V/O ratios of foetal and newborn pituitaries might have been artifacts caused by the use of acetone in drying the glands. Their preliminary tests were unable to show any chemical differences between the principles of the newborn and the adult gland, except in the ambiguous question of solubility; however, the methods could not be regarded as conclusive (cf. Perks, 1966).

The work presented below was based on the anomalous characteristics of the foetal and newborn principles, and it was carried out for three main reasons. Firstly, it extended the limited work on foetuses to wider periods of gestation, especially in species obtainable in fresh condition. Secondly, it extended work to lyophilised tissues, which had not been studied by previous workers; such tissues could be assayed in terms of their dry weight, and would show whether there were changes in V/O ratios in pituitaries free of acetone extraction. Thirdly, it seemed possible that the high V/O ratios and the unusual solubility of 'newborn' oxytocin indicated the presence of new neurohypophysial peptides; therefore, the principles of foetal seals were purified from lyophilised glands, and analysed for their chemical structure.

Initial studies used foetal sheep, but later work concentrated on the rich pituitaries of foetal seals (*Callorhinus ursinus*). Preliminary data on foetal pigs and guinea-pigs are unpublished observations of Kontor & Perks. Foetal ages are expressed as a proportion of term; mating (in the seal, implantation) = 0.00, delivery = 1.00. Total gestation times were: seal, 240 days (implantation to delivery); sheep, 147 days; guinea-pigs, 65 days; pigs, 114 days. The neural lobes were dissected rapidly, frozen in liquid nitrogen or dry ice, and lyophilised for 24 hours. In each species, foetal glands of the same age were pooled; the numbers varied with availability (6–30 seals; 5–22 sheep; 40–98 guinea-pigs; 11–60 pigs). The glands were homogenised into 0.25 % acetic acid, and extracted at 100 °C (British Pharmacopoea, 1963). Neutralised extracts were assayed by the rat uterus (oxytocic) method of Holten (1948), and by the rat vasopressor and frog-bladder assays of Sawyer (1960, 1961): the four-point statistical design was used wherever possible (Holten, 1948).

Vasopressor and oxytocic activities were detected in all foetal pituitaries examined; their potencies increased steadily towards term, but never reached adult levels by birth. In the youngest foetuses yet studied, seals at 0.19 of term, the vasopressor activity was 115.7 mU/mg powder, and the oxytocic potency 30.3 mU/mg. Close to birth (0.93 of term) these values had risen to 2824.0 ± 251.0 mU/mg vasopressor activity (52.2 % of the adult value) and 1201.0 ± 58.0 mU/mg oxytocic potency (34.8 % of the adult value). Clearly, oxytocic activity developed slowest. Other species showed a similar pattern. Sheep foetuses at midterm (0.59 of term) gave 738.0 ± 156.0 mU/mg vasopressor, and 52.0 mU/mg oxytocic activity; these values rose to 1035.7 ± 104.0 and 134.7 ± 7.5 mU/mg, respectively, at 0.96 of term, when they represented 72.9 % and 8.3 % of adult values, respectively. Pig foetuses at midterm (0.57 of term) showed 118.0 ± 15.0 mU/mg vasopressor activity (assayed against lysine vasopressin), and 30.0 ± 4.0 mU/mg oxytocic potency. By 0.89 of term these values had increased to 1144.0 ± 122.0 mU/mg vasopressor and 525.0 ± 44.0 mU/mg oxytocic activity – 82.3 % and 30.1 % of adult values, respectively. Late guinea-pig foetuses (0.77 of term) gave 220.0 ± 26.0 mU/mg vasopressor and 14.0 mU/mg oxytocic activity, but by 0.92 of term these values had increased to 641.0 ± 103.0 and 45.0 ± 5.0 mU/mg, respectively – figures which represented 26.0 % and 7.3 % of adult values, in each case. Recently, Burton & Forsling (1972) have obtained results in agreement with these guinea-pig assays.

The V/O ratios are shown in Fig. 1. All foetuses showed higher values than the adults; this was marked in early sheep and guinea-pigs. In the seal, where particularly young foetuses were studied, the ratio remained approximately stable until midterm. Between midterm and birth, all species showed a marked and progressive decline in the ratios, but only the seal and pig approached adult proportions at birth. Since the lyophilised glands had never been treated with acetone, the changes in ratio could not have been artifacts produced by acetone extraction.

An alternative explanation was suggested by histological studies. The hypothalami and pituitaries of foetal seals were sectioned and stained for neurosecretion by the aldehyde fuchsin and Alcian blue–Schiff's orange techniques (Dawson, 1953; van Oordt, personal communication). At 0.3 of term the supraoptic nucleus contained no neurosecretion, but by 0.5 of term, 52 % of the cells showed granules; this proportion increased to 92 % in a 1–3-day-old pup. The quantity of neurosecretion in the nucleus was approximated as a 'neurosecretory index', which was based on the

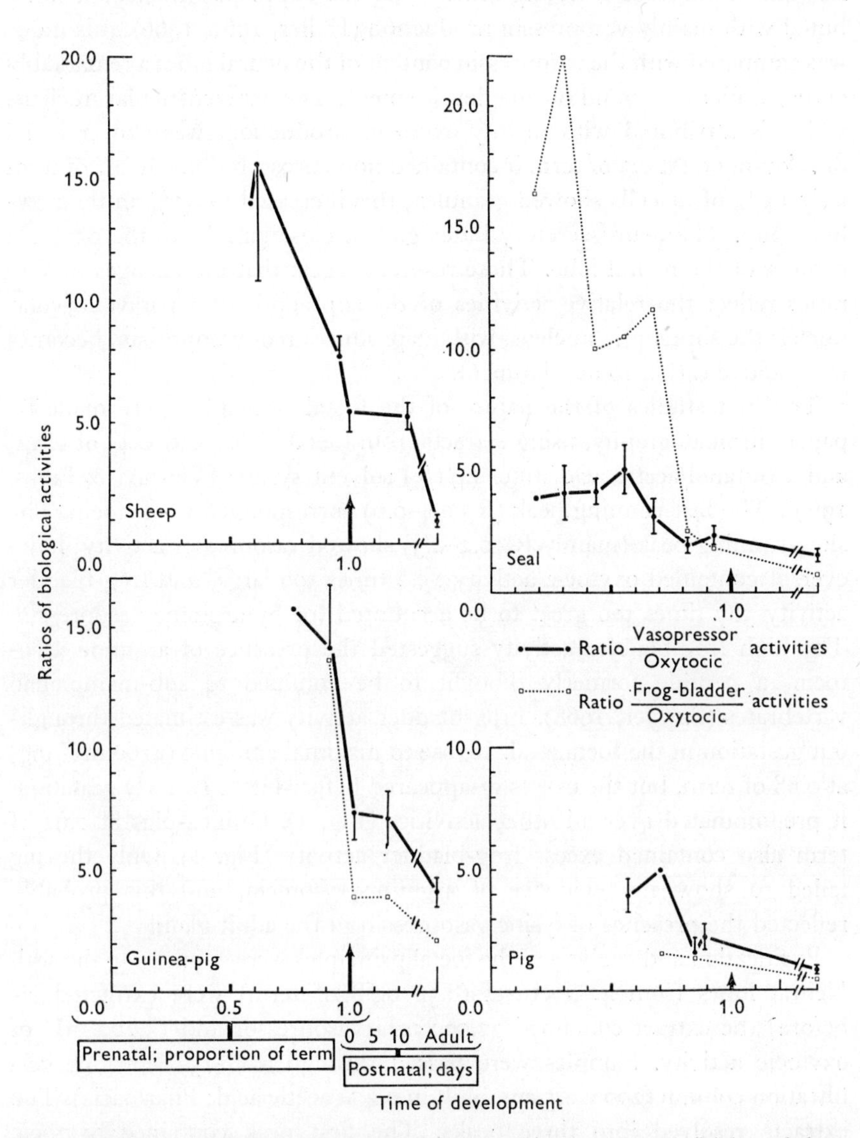

Fig. 1. The ratios of biological activities during development, in neural lobes of sheep, seals, guinea-pigs and pigs. All species plotted to the same scale. Abscissae = the stage of development; prenatal age is expressed as the proportion of term, where 1.00 denotes birth (marked by the arrow); postnatal age is given as days after delivery. Ordinates = the ratios of biological activities; vasopressor/oxytocic activity is shown by solid lines and circles; frog-bladder/oxytocic activity is given by dotted lines and open squares. Vertical bars present fiducial limits at $P = 0.05$.

size and number of secreting cells. Since the supraoptic nucleus is attributed with mainly vasopressin production (Heller, 1961, 1966), this index was compared with the vasopressin content of the neural lobe; a remarkably close parallel was found during development. The paraventricular nucleus, which is attributed with mainly oxytocin production, was slower in its development. At 0.3 of term it contained no neurosecretion; by 0.5 of term only 14% of its cells showed granules; this increased to 85% in the newborn pup. The neurosecretory index gave a close parallel to the oxytocin content of the neural lobe. These results suggest that the changes in V/O ratios reflect the relative activities of the supraoptic and paraventricular nuclei; the supraoptic nucleus, with its production of vasopressin, becomes more active earlier in development.

The first studies of the nature of the foetal principles were made by paper chromatography, using extracts from foetal seals (0.39–0.43 of term) and a butanol/acetic acid/water (4/1/5) solvent system (Vizsolyi & Perks, 1969). The fast-running peak (R_F 0.4–0.6) corresponded to oxytocin. The slow-running peak (mainly R_F 0.2–0.3) showed vasopressor activity; however, it contained oxytocic activity 15.2 times too large, and frog-bladder activity 587 times too great to be accounted for by arginine vasopressin. The high frog-bladder activity suggested the presence of arginine vasotocin, a peptide formerly thought to be confined to sub-mammalian vertebrates (Sawyer, 1968). Frog-bladder activity was estimated throughout gestation in the foetal seal. It rose to maximal amounts (4500 mU/mg) at 0.68 of term, but the excess disappeared before birth. In early gestation, it predominated over all other activities (Fig. 1). Guinea-pigs at 0.92 of term also contained excess frog-bladder activity (Fig. 1). Only the pig failed to show any evidence of arginine vasotocin, and this probably reflected the presence of lysine vasopressin in the adult gland.

Purification and analysis of the foetal principles were confined to the seal. Neural lobes from 42 foetuses (0.56–0.68 of term) were extracted, as before; the extract contained 32,000 mU vasopressor and 15,200 mU of oxytocic activity. Samples were passed through a G-15 Sephadex gel-filtration column (200 × 2.5 cm, built in 0.2 M acetic acid; Pharmacia). The extracts resolved into three peaks. The first peak contained oxytocic activity alone. It was adjusted to contain 0.002 M ammonium acetate, pH 5.0, and adsorbed onto a CM-Sephadex column (15 × 1 cm; Pharmacia). A gradient to 0.1 M ammonium acetate, pH 5.0, eluted a single peak, with 100% recovery. This peak showed the bio-assay properties of oxytocin. Hydrolysis in 6 N HCl and amino acid analysis (Biocal 200) gave the amino acids of oxytocin (Table 1). The third gel-filtration peak showed only vasopressor activity. It was adjusted to contain 0.02 M ammonium

Table 1. *The amino acid analysis of the neurohypophysial peptides of seal foetuses* (0.56–0.68 *of term*)

	Ratio to Glu					
	Oxytocic fraction from CM Sephadex		Vasopressor fraction from Phosphocellulose		Vasotocin fraction from IRC-50	
Amino acid	Seal	Theoretical*	Seal	Theoretical†	Seal	Theoretical‡
Asp	1.08	1.00	0.82	1.00	1.03	1.00
Glu	1.00	1.00	1.00	1.00	1.00	1.00
Pro	0.97	1.00	1.03	1.00	1.00	1.00
Gly	0.96	1.00	0.98	1.00	1.20	1.00
½ Cys	0.86 } 1.20	2.00	0.40 } 1.29	2.00	0.93 } 1.95	2.00
Cys acid	0.34 }		1.89 }		1.02 }	
Ileu	0.75	1.00	—	—	0.69	1.00
Leu	0.96	1.00	—	—	0.13	—
Tyr	1.00	1.00	0.50	1.00	0.43	1.00
Phe	—	—	1.06	1.00	0.15	—
Arg	—	—	0.86	1.00	0.80	1.00
Ratios of additional residues, to Glu	Ser. 0.08, Ala. 0.11, His. 0.15, Lys. 0.41		Ser. 0.30, Ala. 0.04, His 0.30		Ser. 0.10, Ala. 0.20, Val. 0.07, His. 0.06, Lys. 0.16.	

* Oxytocin = Cys-Tyr-*Ileu*-Gln-Asn-Cys-Pro-*Leu*-Gly-NH_2

† Arginine vasopressin = Cys-Tyr-*Phe*-Gln-Asn-Cys-Pro-*Arg*-Gly-NH_2

‡ Arginine vasotocin = Cys-Tyr-*Ile*-Gln-Asn-Cys-Pro-*Arg*-Gly-NH_2

acetate, pH 5.0, and adsorbed onto a phosphocellulose column (16 × 1 cm; Selectacel). A gradient to 0.2 M ammonium acetate, pH 5.0, eluted a single vasopressor peak, with 80% recovery. The eluate was hydrolysed and analysed, as above, and it gave the amino acids of arginine vasopressin (Table 1). The second gel-filtration peak, which was only partly separated from the third peak, showed all the biological activities under assay, but frog-bladder activity predominated. This peak (in 0.2 M acetic acid) was adsorbed directly onto an IRC-50 column (10 × 0.5 cm; Rohm and Haas). The loaded column was equilibrated with 0.1 M ammonium acetate, pH 5.0. A gradient to 0.32 M ammonium acetate, pH 6.0, eluted a peak with a vasopressor/oxytocic activity of approximately 2.0, a value typical of arginine vasotocin. A final wash with 0.75 M ammonium acetate, pH 7.7, displaced residual arginine vasopressin from the column. The supposed arginine vasotocin peak showed pharmacological activities in exact parallel with synthetic arginine vasotocin, when compared directly against it.

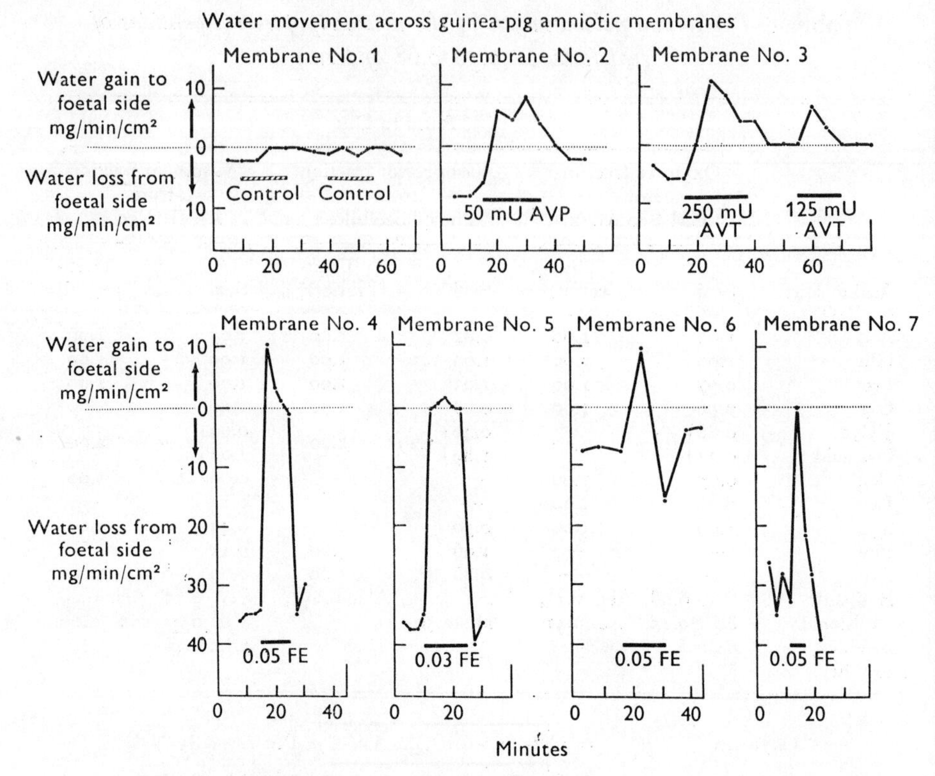

Fig. 2. The effect of neurohypophysial principles on water movement through amniotic membranes. Each graph represents an individual guinea-pig amniotic membrane, supported *in vitro*, at 37 °C; age = 0.38–0.43 of term. Ordinates = water movement, mg/min·cm² of membrane; the dotted line = zero change; values above this line = water gain towards the foetal side (amniotic cavity); values below the line = water loss towards the allantoic side (allantoic cavity). Abscissa = time in minutes.

The hatched bars (control) represent periods of immersion in fresh saline, on both sides of the membrane.

The black bars represent the same method of immersion, but with the following principles added to the 'amniotic' saline on the foetal side of the membrane (saline volume = 2.5 ml):

50 mU AVP = 50 mU vasopressor activity, synthetic arginine vasopressin.
250 mU AVT = 250 mU vasopressor activity (1 μg), synthetic arginine vasotocin.
125 mU AVT = 125 mU vasopressor activity (0.5 μg), synthetic arginine vasotocin.
0.05 Fe = 0.05 ml, 3 mg/ml, extract from neural lobes, foetal seals (0.44 of term) (contained 135 mU frog-bladder, 50 mU vasopressor, and 12.5 mU oxytocic activity).
0.03 Fe = 0.03 ml, same extract, foetal seals.

Hydrolysis and amino acid analysis gave the amino acids of arginine vasotocin (Table 1). It was clear that the foetal pituitaries at around midterm contained three neurohypophysial principles – oxytocin, arginine vasopressin and arginine vasotocin.

The presence of arginine vasotocin in foetal extracts suggested that the function of the foetal principles might lie in the control of water movement through embryonic membranes. Therefore, amniotic membranes from guinea-pigs were set up *in vitro* at 37 °C. The membranes were placed across glass tubes (0.9 cm diameter), foetal side inwards. Each tube was filled with 2.5 ml of a saline of the ionic composition of amniotic fluid, buffered to pH 7.4 with phosphates. The tubes were suspended from a continuous weighing device, in order to record water movement (Beckman). The membrane was immersed in 50 ml of an outer saline of the ionic composition of maternal plasma, but buffered as before, and oxygenated with air. Water passed from the foetal (inner) side to the allantoic (outer) side, under a hydrostatic pressure of 3 cm water, and an osmotic gradient which resulted from the hypotonicity of the amniotic saline. Fig. 2 shows that extracts from foetal seal neural lobes, and either synthetic arginine vasotocin or vasopressin, caused a slowing, or more often a reversal of water flow. Allantoic membranes behaved similarly. These membranes are in contrast to the foetal kidney, which seems to be little affected by neurohypophysial hormones (Alexander & Nixon, 1962). It is speculated that foetal arginine vasotocin and vasopressin pass in the foetal urine to the embryonic membranes, where they cause water uptake into the amniotic cavity; this water might be available to the foetus, by swallowing or other mechanisms.

We are glad to acknowledge the National Research Council of Canada for their generous support (Grant A2584), and Mr Ian MacAskie, Miss Daphne Hards, Dr G. Harry, Jr, Dr Sidney Cassin, and Dr G. S. Dawes for many forms of kindness.

REFERENCES

Acher, R. & Fromageot, C. (1957). In *The Neurohypophysis*, ed. Heller, H., pp. 39–48. London: Butterworths.

Alexander, D. P. & Nixon, D. A. (1962). *Brit. med. Bull.* **17**, 112–17.

Burton, A. M. & Forsling, M. L. (1972). *J. Physiol.* **221**, 6P–7P.

Dawson, A. B. (1953). *Anat. Rec.* **115**, 63–70.

Dicker, S. E. & Tyler, C. (1953*a*). *J. Physiol.* **120**, 141–5.

Dicker, S. E. & Tyler, C. (1953*b*). *J. Physiol.* **121**, 206–14.

Heller, H. (1961). In *Oxytocin*, pp. 3–23. Eds. Caldeyro-Barcia, R. & Heller, H. New York, Oxford, London, Paris: Pergamon Press.

Heller, H. (1966). *Brit. med. Bull.* **22**, 227–31.

Heller, H. & Lederis, K. (1959). *J. Physiol.* **147**, 299–314.

HELLER, H. & ZAIMIS, E. J. (1949). *J. Physiol.* **109**, 162–9.
HOLTEN, P. (1948). *Brit. J. Pharmacol.* **3**, 328–34.
PERKS, A. M. (1966). *Gen. comp. Endocrinol.* **6**, 428–42.
SAWYER, W. H. (1960). *Endocrinology*, **66**, 112–20.
SAWYER, W. H. (1961). *Meth. med. Res.* **9**, 210–19.
SAWYER, W. H. (1968). In *Neurohypophysial Hormones and Similar Polypeptides*, pp. 717–49. Ed. Berde, B. Berlin, Heidelberg, New York: Springer-Verlag.
VIZSOLYI, E. & PERKS, A. M. (1969). *Nature* **223**, 1169–71.
YAKOVLEVA, I. V. (1965). *Arkhiv. Anat. Gist. i Embryol.* **48**, 79–90.

VASOPRESSIN METABOLISM IN THE FOETUS AND NEWBORN

By W. R. SKOWSKY, R. A. BASHORE, F. G. SMITH AND D. A. FISHER

The Endocrine Research Laboratories, UCLA-Harbour General Hospital, Torrance, California, The Gwynne Hazen Cherry Renal Research Laboratory, Department of Pediatrics, UCLA Medical Center, Los Angeles, California, and the Departments of Obstetrics and Gynecology and Pediatrics, of the School of Medicine of the University of California, at Los Angeles

INTRODUCTION

There is considerable information available regarding foetal water and electrolyte content and maternal–foetal water exchange, and many scholarly reviews have appeared dealing with renal function and the control of water balance in the newborn. Little is known, however, about the secretion of vasopressin or its role in the foetus. One of the major difficulties has been the lack of a sensitive and specific vasopressin assay procedure requiring only small quantities of serum in which the deleterious effects of vasopressinase can be inhibited. We have recently developed such methodology and the present paper presents results of our preliminary studies of foetal vasopressin metabolism in the monkey and sheep. To present this data in perspective we begin with a short review of current knowledge of vasopressin physiology in the newborn.

VASOPRESSIN PHYSIOLOGY IN THE NEWBORN

The primary functions of vasopressin are the homeostatic control of body water volume and the maintenance of isotonicity of body fluids by regulation of free water excretion by the kidney. To these ends vasopressin secretion is regulated by two systems of receptors: the osmoreceptors located in man at or near the supraoptic and paraventricular nuclei in the anterior hypothalamus, and a series of volume receptors, the most important being the left atrial stretch receptors and the carotid sinus baroreceptors. The work of Verney (1947) indirectly established not only that the rate of vasopressin secretion was increased by intracarotid hyperosmotic infusions but that this stimulus must be sustained; a 5-minute carotid artery infusion of 5.5% hyperosmotic solution or a 15–25-minute infusion of a 2% solution is needed to increase vasopressin secretion. It is believed that the change in cell volume induced by the extracellular solute is the effective stimulus.

Peters (1935) in the early 1930s suggested that the 'fullness' of the intravascular space is sensed by the organism, contraction leading to an antidiuresis and expansion causing a diuresis. Much subsequent work has firmly established the existence of thoracic volume receptors and it has been suggested recently that these might be divided into high and low pressure systems. The former include the arterial system, primarily the aortic arch baroreceptors and the carotid receptors; the latter include the right heart, left atrium, and the pulmonary vessels. Share (1967) postulated that relatively small changes in vascular volume are first detected within the low pressure system since small changes in volume would produce large changes in pressure leading to alterations in stretch receptor impulses. For example, a small increase in central blood volume evokes an increased rate of firing of the stretch receptors; these stimuli are transmitted via the vagus nerves to the hypothalamus where they inhibit vasopressin secretion. Recently Moore (1971) has suggested that in the sheep neither the osmoreceptor nor the volume receptor system is dominant over the other, but act in concert to maintain the volume and isotonicity of body fluids. However, his suggestion was based on bioassay of AVP values which are subject to marked variability in the physiological range. Robertson & Mahr (1972), using a sensitive radioimmunoassay for AVP, have re-examined this and demonstrated that hypovolaemia in the absence of changes in serum osmolality do not affect serum AVP values. It seems likely, therefore, that AVP secretion is primarily regulated by the osmoreceptor system and that volume depletion serves as a secondary stimulus for AVP release.

Vasopressin antidiuresis is mediated at the renal tubule via a plasma membrane receptor–adenyl cyclase complex and the generation of cyclic AMP; and this mechanism is modulated importantly by adrenergic receptors. Sandler *et al.* (1961) observed that newborn infants kept at an ambient temperature of 75 °F excreted larger urine volumes than infants at a temperature of 85 °F. Fisher (1967) documented that this 'cold diuresis' was, in fact, due to an increase in free water clearance, presumably secondary to a reduction in renal tubular water reabsorption. In subsequent studies he showed that norepinephrine infused at a rate of 600 μg/h would block the antidiuresis of simultaneously infused vasopressin (10–20 mU/h) in water-loaded young adult subjects, and suggested that norepinephrine inhibition of vasopressin antidiuresis might be the mechanism of 'cold diuresis' in the newborn and the adult subjected to relatively low environmental temperatures (Fisher, 1968).

The interplay between adrenergic receptors and vasopressin antidiuresis was further studied by Klein *et al.* (1971) and Levi *et al.* (1971) who showed that the norepinephrine blockade of vasopressin antidiuresis can be blocked with phentolamine, an alpha adrenergic receptor blocker. In addition, isoproterenol (a beta receptor stimulator) will evoke a vasopressin-like antidiuresis, even in rats with congenital diabetes insipidus; moreover, propranolol, a beta adrenergic receptor blocker, will block this antidiuresis, but will not inhibit vasopressin antidiuresis. These studies suggested the existence of a complex system of receptor sites on the renal tubular plasma membrane; one receptor for vasopressin and separate alpha and beta adrenergic receptors which modulate in some way the vasopressin effect. Schrier *et al.* (1972) have recently reported that isoproterenol does not produce an antidiuretic effect in hypophysectomized dogs or when infused directly into the renal artery, and these authors feel that the beta adrenergic antidiuretic hormone-like effect is mediated centrally via stimulation of vasopressin release. However, there are presently no studies measuring serum AVP levels following isoproterenol infusion. Thus, although it is clear that alpha receptor inhibition of vasopressin antidiuresis is mediated at the level of the kidney, the mechanism of beta receptor antidiuresis is not entirely clear.

The embryological development of this intricate system of peripheral and central receptors, vasopressin synthesis, storage and release mechanisms and renal vasopressin receptor-response systems has not been clearly defined. Information available in the newborn has suggested that these systems are functional. The neonate characteristically demonstrates a low maximal renal tubular excretory capacity with a high glomerular filtration fraction, a low urine solute excretory rate, a low urine flow rate, and delayed excretion of a water load. However, these observations do not imply an immature or hypofunctioning hypothalamic-neurohypophyseal system. The low rate of solute excretion and low urine volumes probably reflect a low solute diet and low rate of glomerular filtration. The studies of Fisher *et al.* (1963) suggest that both the osmoreceptor and volume receptor systems are functional at birth. In infants 2½ to 40 hours of age, the infusion of 1 to 2.5 ml/kg of isotonic dextran increases free water clearance without altering glomerular filtration rate or plasma osmolality. Similarly the infusion of hypertonic saline to increase plasma osmolality 3–4% will decrease free water clearance without changing glomerular filtration rate. After an oral water load, the newborn can dilute his urine to the range of 50 mOsm/l in spite of the fact that the excretion of the total water load is delayed (Smith, 1959). The delayed excretion merely reflects the low rate of glomerular filtration.

The kidney at birth is not capable of concentrating tubular urine to the extent seen in adults (1000–1400 mOsm/l) but can effectively concentrate to 550–600 mOsm/l. It has been suggested that this limitation might be due to an immature hypothalamic-neurohypophyseal system and a reduced capacity to secrete vasopressin. Demant *et al.* (1958) and Hradcova & Heller (1962) could not detect measurable plasma vasopressin levels in infants under four months of age. And Janovsky *et al.* (1965) found no plasma vasopressin in hydrated infants up to $5\frac{1}{2}$ months of age but consistently detected low levels in hydrated infants more than 5 months of age. Caution must be exercised in the interpretation of these results, however. Vasopressin was measured by bioassay, which is fraught with inherent methodological errors when dealing with low levels of vasopressin (i.e. less than 2 μIU/ml). In addition the presence of circulating vasopressinase in the neonatal circulation during the first weeks of life may lead to false negative bioassay values.

The responses in the newborn to hypertonic saline or dextran infusions suggest not only that the neonatal pituitary is capable of vasopressin secretion in response to osmotic or volume stimuli but that the kidney is capable of responding to endogenous vasopressin. Ames (1953), moreover, has reported the presence of urinary vasopressin (> 1.25 mU/ml) in 23 infants between birth and 3 days of age after fasting. And Heller and Zaimis have shown that the pituitary of the newborn contains 350–400 mU of vasopressin at birth (Heller & Zaimis, 1949), an amount adequate to achieve prolonged antidiuresis since exogenous doses of 1–2 mU vasopressin have been observed to evoke maximal urine osmolality (550–600 mOsm/l) (Barnett & Vesterdal, 1953; Heller, 1944). Thus, it seems likely that the low maximal urine osmolality in the newborn reflects an inherent limitation in renal concentrating capacity, rather than reduced vasopressin secretion or responsiveness. This limitation might be due to a decreased ability to maintain a hypertonic medullary interstitium and hence a reduced effectiveness of the counter-current concentrating mechanism.

Using a sensitive bioassay procedure which inhibits the action of vasopressinase, Hoppenstein *et al.* (1968) found high plasma vasopressin levels at birth in infants after normal vaginal delivery (mean = 35 μIU/ml) and relatively low levels in infants born by Caesarean section (mean = 0.83 μIU/ml). Maternal plasma vasopressin levels obtained simultaneously were also low (mean = 2 μIU/ml). These authors interpreted the results as indicating a foetal vasopressin response to the stress of vaginal delivery. Further studies in the newborn indicated that AVP blood levels declined during the first 22 hours after birth and that infants 3 days to 3 months of age responded to surgical stress with significant increases in AVP plasma levels; however, a direct relationship of age to AVP response was not absolutely established.

Recently Chard *et al.* (1971) measured plasma AVP in cord blood using a sensitive radioimmunoassay. The levels of AVP in 12 specimens obtained at the time of vaginal delivery averaged 550 μU/ml for arterial cord plasma and 80 μU/ml for venous cord plasma. Lower levels were found in four infants delivered by Caesarean section: 52 μU/ml in arterial cord plasma and 21 μU/ml in venous cord plasma. These authors also suggested that the elevated AVP levels reflect increased vasopressin secretion from the newborn posterior pituitary gland, possibly secondary to the stress of delivery.

Thus, it seems clear that the newborn infant has the capacity to synthesize vasopressin; that the osmolar and volume stimuli controlling vasopressin release are operative at birth; and that the renal AVP receptor mechanisms, probably including alpha adrenergic receptors, are functional at birth.

FOETAL VASOPRESSIN METABOLISM

In contrast, data regarding foetal vasopressin physiology are meagre. Vasopressin has been clearly isolated from the foetal pituitary of the sheep (Vizsolyi & Perks, 1969). Moreover, Alexander and co-workers (1971) measured plasma AVP by bioassay in the sheep foetus as early as 107 days' gestation and state that the basal levels (less than 10–90 μU/ml) are not related to the age of the foetus. Furthermore in all animals studied, there was a marked rise in AVP in response to haemorrhage to a maximum of 1800 μU/ml in the 140-day foetus.

We recently examined the kinetics of AVP within the foetal animal using radioimmunoassay measurements in a system that inactivates the high levels of circulating vasopressinase. The foetal rhesus monkey was studied at 140–155 days' gestation (term = 165 days). The mean basal serum AVP level was 7.4 μU/ml. When 3% hypertonic saline was infused at a dose of 12 ml/kg or 10% dextran in saline at a dose of 10 ml/kg, all animals showed a marked rise in serum AVP levels 30 minutes after the infusion. The rise in AVP at 30 minutes ranged from 2 to 10 times baseline with a mean increment of 21 μU/ml (Table 1).

A pulse dose of radiolabelled AVP was injected into the monkey foetus and the plasma disappearance measured within the foetal–placental circulation. The mean half-time was 11.6 minutes and the extrapolated mean volume of distribution (VD) was 0.62 l. Blood production rate of AVP in the foetus was calculated as:

$$\text{PR (mU/h)} = \frac{0.693 \times \text{VD (l)}}{t_{\frac{1}{2}}\text{ (h)}} \times \text{AVP conc. (mU/l).}$$

Table 1. *Stimulation of foetal AVP ($\mu U/ml$) after hyperosmolar stimuli*

Foetus number	Control	Serum AVP concentration (μIU/ml): 3 % Saline infusion (12 ml/kg ≃ 5 ml) 15 min	30 min	45 min	10 % Dextran infusion (10 ml/kg ≃ 4 ml) 15 min	30 min	45 min
M-1	4.0	—	31.0	8.8	—	21.0	4.5
M-2	8.8	4.6	20.7	—	—	21.0	14.0
M-3	9.5	< 1.5	14.0	—	13.0	21.0	—
Mean	7.4	3.0	21.9	8.8	13.0	21.0	9.2

The mean value was 17.5 mU/h. When this value is related to foetal weight the PR is 42.2 mU/kg·h; and when it is related to combined foetal–placental weight, PR is 29.5 mU/kg·h (Table 2).

Table 2. *Blood half-life, volume of distribution, serum concentration and production rate of arginine vasopressin in foetal rhesus monkeys*

Monkey	$t_{\frac{1}{2}}$ (min)	VD (l)	AVP (μIU/ml)	AVP production rate mIU/h	mIU/kg·h
M-1	11.4	1.10	4.8	19.3	42.8
M-2	9.6	0.50	8.8	19.1	47.8
M-3	13.6	0.47	9.5	13.7	36.3
M-4	11.7	0.43	11.8	18.0	40.1
Mean	11.6	0.62	8.7	17.7	42.2
SEM	0.8	0.16	1.5	1.1	1.9

Similar studies were conducted in sheep of 120–130 days' gestation, measuring turnover of ^{125}I-labelled AVP in both the maternal and foetal circulations (Table 3). The mean half-time in the foetus was 6.3 minutes with an extrapolated mean volume of distribution of 0.55 l. The disappearance of labelled AVP within the maternal circulation did not follow a single exponential curve and meaningful plasma half-life values could not be obtained. However, the $t_{\frac{1}{2}}$ of blood AVP in the foetus was statistically less than in the non-pregnant ewe ($t_{\frac{1}{2}}$ = 19.4 minutes). The blood PR of AVP was calculated in maternal sheep by first estimating metabolic clearance rates (MCR) of labelled hormone as one/(area beneath the plasma disappearance curves). Then, blood PR was calculated as:

$$\text{PR (mU/h)} = \text{MCR (l/h)} \times \text{AVP conc. (mU/l)}.$$

Table 3. *Blood half-life, metabolic clearance rate, serum concentration, volume of distribution and production rate of arginine vasopressin in maternal and foetal sheep*

Sheep no.	Maternal $t_{\frac{1}{2}}$ (min)	Maternal MCR (l/min)	Maternal AVP conc. (μIU/ml)	Maternal AVP production rate (mIU/h)	Maternal AVP production rate (mIU/kg·h)	Foetal $t_{\frac{1}{2}}$ (min)	Foetal VD (l)	Foetal AVP conc. (μIU/ml)	Foetal AVP production rate (mIU/h)	Foetal AVP production rate (mIU/kg·h)
V-1	—	—	—	—	—	7.6	0.50	52.8	144	35.5
V-2	0.5–6.7	0.17	1.9	19.8	0.33	—	—	—	—	—
V-3	1.5–5.4	0.24	0.8	11.7	0.20	5.5	0.69	25.6	133	34.6
V-4	1.0–5.3	0.38	1.8	41.3	0.60	6.1	0.62	58.0	245	60.3
V-5	1.2–4.9	0.23	0.4	5.6	0.09	6.5	0.38	34.6	84	30.8
V-6	1.1–10.8	0.20	1.4	16.6	0.29	6.0	0.56	74.7	290	71.4
Mean	1.02–6.62	0.25	1.3	19.0	0.30	6.3	0.55	49.1	179	46.5
SEM	0.16–1.08	0.04	0.3	6.1	0.09	0.4	0.05	8.7	38	8.1

Estimated AVP production rate on a weight basis in the foetus was 46.5 mU/kg·h, more than one-hundred-fold greater than in the mother (0.30 mU/kg·h) (Table 3). This high rate of vasopressin production in the foetal sheep is remarkably similar to that estimated in the foetal monkey. And this high foetal production rate is reflected in the high mean serum AVP values: 49.1 μU/ml in the foetal sheep as compared to 1.26 μU/ml in the mother. Thus, it appears that the foetal hypothalamic posterior pituitary axis is functional and autonomous, maintaining a high rate of AVP secretion during the last trimester of pregnancy. Moreover, the system responds to appropriate hyperosmotic stimuli.

The significance of these elevated levels of AVP within the foetal circulation is not yet clear. Such high levels might induce pharmacological pressor effects within the umbilical circulation. Indeed, Carter *et al.* (1969) have demonstrated that injections of 100–500 mU vasopressin into the femoral artery of the foetal rabbit will cause marked constriction of the uterine artery with decreased uterine blood flow and diminished filling of the placental sinuses. This shunting of blood away from the placenta was found to lead to foetal death if infusion of 2–8 mU/min were maintained for a one-hour period. The possible benefit of a significant but sublethal pressor effect in the foetus is obscure.

Chard *et al.* (1971) have raised the possibility that the elevated levels of foetal vasopressin may play a role in stimulating labour. Certainly, pharmacological levels of vasopressin do possess myometrial contractile properties. However, our recent studies have failed to demonstrate any placental transfer of vasopressin. It has been demonstrated recently (Fylling, 1971) that vasopressin has a marked stimulating effect on progesterone levels during early pregnancy; perhaps the elevation of progesterone within the foetal–placental unit may contribute to the onset of labour.

In summary, it has been demonstrated that the foetal posterior pituitary, like the foetal anterior pituitary, is functional during late gestation and at the time of delivery, and the level of function exceeds that of the mother. The secretion rate of vasopressin in the foetus is significantly greater than the corresponding maternal rate and the blood AVP levels in the foetus are quite high. The significance of these elevated levels of peptide, however, and particularly their contribution to the maintenance of the homeostatic milieu of the unborn foetus, remain to be determined.

This research is supported by United States Public Health Service Grants HD-06335, HD-04460 and HD-04684 from the National Institute of Child Health and Human Development of the National Institutes of Health, Bethesda, Maryland, USA.

REFERENCES

ALEXANDER, D. P., BRITTON, H. G., FORSLING, M. L., NIXON, D. A. & RADCLIFF, J. G. (1971). *J. Physiol.* **213**, 31P.

AMES, R. G. (1953). *Pediatrics* **12**, 272.

BARNETT, H. L. & VESTERDAL, J. (1953). *J. Ped.* **42**, 99.

CARTER, A. M., GOTHLIN, J. & BENGTSSON, L. P. (1969). *Acta Obstet. Gynae. Scand.* **48**, Suppl. 3, 132.

CHARD, T., HUDSON, C. N., EDWARDS, C. R. W. & BOYD, N. R. H. (1971). *Nature* **234**, 352.

DEMANT, F., NEUBAUER, E., SRSEN, S. & TISCHLER, V. (1958). *Cesk. Fysiol.* **7**, 286.

FISHER, D. A., PYLE, H. R., PORTER, J. C., BEARD, A. G. & PANOS, T. C. (1963). *Amer. J. Dis. Child* **106**, 137.

FISHER, D. A. (1967). *Pediatrics* **40**, 636.

FISHER, D. A. (1968). *J. Clin. Invest.* **47**, 540.

FYLLING, P. (1971). *Acta Endocrinol.* **66**, 273.

HELLER, H. (1944). *J. Physiol.* **102**, 429.

HELLER, H. & ZAIMIS, E. J. (1949). *J. Physiol.* **109**, 162.

HOPPENSTEIN, J. M., MILTENBERGER, F. W. & MORAN, W. H. (1968). *Surg. Gynec. Obstet.* **90**, 966.

HRADCOVA, L. & HELLER, J. (1962). *Paediat. Acta* **17**, 531.

JANOVSKY, M., MARTINEK, J. & STANINCOVA, V. (1965). *Acta Paediat. Scand.* **54**, 543.

KLEIN, L. A., LIBERMAN, B., LAKS, M. & KLEEMAN, C. R. (1971). *Amer. J. Physiol.* **221**, 1657.

LEVI, J., GRINBLAT, J. & KLEEMAN, C. R. (1971). *Am. J. Physiol.* **221**, 1728.

MOORE, W. W. (1971). *Fed. Proc.* **30**, 1387.

PETERS, J. P. (1935). *Body Water*, p. 405. Springfield, Ill.: Charles C. Thomas Publishing Company.

ROBERTSON, G. & MAHR, E. (1972). Abstract, 64th Am. Soc. for Clin. Invest., Atlantic City, N.J., 1972.

SANDLER, M., RUTHVEN, C. R. J., NORMAND, I. C. S. & MOORE, R. E. (1961). *Lancet* **1**, 485.

SCHRIER, R. W., LIEBERMAN, R., UFFERMAN, R. C. (1972). *J. clin. Invest.* **51**, 97.

SHARE, L. (1967). *Am. J. Med.* **42**, 701.

SMITH, C. A. (1959). *The Physiology of the Newborn Infant*, 3rd edition, p. 320. Springfield, Ill.: Charles C. Thomas Publishing Company.

VERNEY, E. B. (1947). *Proc. R. Soc. Med.* **B135**, 25.

VIZSOLYI, E. & PERKS, A. M. (1969). *Nature* **233**, 1169.

HUMORAL REGULATION OF ERYTHROPOIESIS IN THE FOETUS

By ESMAIL D. ZANJANI, ANTHONY S. GIDARI, EDWARD N. PETERSON, ALBERT S. GORDON AND LOUIS R. WASSERMAN

Department of Physiology, Mount Sinai School of Medicine of the City University of New York, NY, Department of Obstetrics and Gynecology, University of Pittsburgh School of Medicine, Pittsburgh, Pa., and Department of Biology, Graduate School of Arts and Science, New York University, New York, NY

INTRODUCTION

In mammalian species, erythropoiesis in the adult is regulated by the hormone erythropoietin (Ep). Thus hypoxia, the fundamental erythropoietic stimulus, has been shown to exert its influence through Ep (Krantz & Jacobson, 1970). The production of Ep is controlled by the relative availability of oxygen to the tissue(s) concerned with its synthesis. The primary organ serving this function is the kidney (Gordon & Zanjani, 1970), although other sites may exist also (Schooley & Mahlmann, 1972).

In general, subjection of adult animals to phlebotomy, hypobaric hypoxia, or to treatment with haemolytic or histotoxic agents results in an increased production of Ep. Bilateral nephrectomy effectively suppresses the response of the animal to these stimuli (Krantz & Jacobson, 1970). Suppression of Ep production and erythropoiesis, on the other hand, ensues following the induction of polycythemia (e.g. by hypertransfusion), starvation or subjection to a hyperoxic environment.

A different pattern of response has been observed in neonatal animals. In this regard, nephrectomy or starvation does not affect red cell production in the 10-day-old rat (Lucarelli, Porcellini, Carnevali, Carmena & Stohlman, 1968). Moreover, hypoxia is an ineffective erythropoietic stimulus in rats less than 35 days old. It should be noted, however, that the erythropoietic response of the neonate appears to depend upon the stage of hematopoietic development of the animal. Thus while starvation is ineffective in the 10-day-old rat, the newborn guinea pig responds to starvation with suppressed erythropoiesis (Lucarelli *et al.* 1968).

Studies concerning the mechanisms controlling red cell production in the mammalian foetus have been sparse mainly because of the relatively small size and inaccessibility of the foetus in most laboratory animals.

In 1969, we reported our success in evoking the appearance of Ep in the plasma of foetal lambs with experimentally-induced haemolytic anaemia (Zanjani, Horger, Gordon, Cantor & Hutchinson, 1969). More recently, we demonstrated the suppression of erythropoiesis in hypertransfused foetal goats, and the lack of effect on Ep production by nephrectomy in these foetuses. In addition, we have examined the role of Ep in foetal red cell production and the influence of maternal anaemia and/or nephrectomy on erythropoiesis in foetal goats. These studies are described in this report.

METHODS

All experiments were performed in pregnant goats at approximately 110–120 days of gestation. Direct access to the foetal circulation was achieved by surgically implanting a catheter in the femoral vein of the foetus. In some studies, maternal and foetal nephrectomies were performed at the time of catheter emplacement. Details of the general surgical and cannulation procedures have been published (Zanjani *et al.* 1969).

Anaemia was induced in the foetuses by removing 45–50 ml of blood from each foetus. This volume of blood was equivalent to approximately 1.5–2 per cent of the foetal body weight. The withdrawal of this quantity of blood resulted in a drop in foetal hematocrit from a mean of 39 per cent to 20 per cent. Maternal anaemia was induced by removing quantities of blood corresponding to 2 per cent of the maternal body weight. In cases where nephrectomized animals were employed, bleeding was initiated 6 hours after surgery. Foetuses were rendered polycythemic by intravenous injections, on two successive days, of 40 ml of packed red blood cells obtained from freshly drawn maternal blood.

Erythropoietin, prepared from anaemic human urine, was given intravenously for two successive days (100 IU/day/foetus) beginning 24 hours after the second transfusion. To gauge the effects of polycythemia, anaemia and human Ep on erythropoiesis, 10 μCi $^{59}FeCl_3$ was administered intravenously 72 hours after the second injection of blood in the control group and 24 hours after the second dose of Ep in the treated, transfused foetuses. Anaemic foetal goats were injected with radio-iron 56 hours after bleeding. The per cent red blood cell-^{59}Fe incorporation was calculated 48 hours later. The blood volume in normal foetuses was assumed to be 5 per cent of the body weight; in the anaemic and transfused foetuses it was estimated to be 4 and 6 per cent of the body weight respectively.

Plasma samples were obtained from foetal and maternal goats at the intervals shown in the Tables. The volume of blood drawn from each foetus at each sampling period was 7–9 ml; the plasma was rapidly harvested and the cells, suspended in saline, were injected back into the foetus. Erythropoietin levels in plasmas were determined in transfusion-induced polycythemic mice as described by Erslev & Kazal (1968).

RESULTS

Table 1 demonstrates that bleeding significantly enhanced erythropoiesis in foetal goats. This was evident from the increased rate of red blood cell–radio-iron uptake and the reticulocytosis in the anaemic foetus as compared to normal foetuses. That the erythropoietic response was probably due to an increased production of Ep by the anaemic foetus was shown by the fact that while Ep was not present in detectable quantities in plasmas of normal or transfused foetuses, significant amounts of the hormone were found in the anaemic foetal plasma samples (Table 1). The importance of Ep in foetal erythropoiesis was further demonstrated by the finding that the administration of human urinary Ep to foetuses with suppressed erythropoiesis (due to the elevation of the circulating red cell mass by hypertransfusion) resulted in a marked rise in the red blood cell–radio-iron incorporation values (Table 1). This was accompanied by an increase in peripheral reticulocyte and bone marrow erythroblast levels.

Table 1. *Effect of anaemia, polycythaemia and erythropoietin on erythropoiesis in goat foetuses* in utero*

Type of foetus	Hct (%)	IU Ep activity/ml plasma†	Reticulocyte (%)	48-Hour RBC-^{59}Fe uptake	Cells/mg marrow $\times 10^{-5}$ Erythroblast	Small lymphocyte
Normal	39.4 ± 0.5	ND‡	5.9 ± 0.4	27.8 ± 2.9	1.8	1.3
Anaemic	20.5 ± 0.3	1.1	9.2 ± 0.8	36.2 ± 3.4	—	—
Polycythaemic	51.8 ± 1.6	ND	0.8 ± 0.05	6.1 ± 0.9	0.45	2.9
Polycythaemic given 200 IU Ep	52.0 ± 1.5	—	10.7 ± 0.8	29.6 ± 3.2	1.2	3.1

* Results from 4–5 foetuses in each group are presented here. The values represent mean $\pm$ 1 SEM.

† International Units of erythropoietin.

‡ Non-detectable activity.

The data presented in Table 2 again indicate that appreciable quantities of Ep were present in the plasmas of bled foetuses. In addition, increased levels of Ep were seen when bled foetuses of nephrectomized mothers were examined. No Ep was detected in plasmas obtained from normal or from nephrectomized foetuses which were not subjected to bleeding; in addi-

Table 2. *Effect of maternal and foetal nephrectomy on erythropoietin production by anaemic goat foetuses**

Type of mother	Type of foetus	IU Ep activity/ml plasma† Maternal	Foetal
Normal	Normal	ND‡	ND
Normal	Anaemic	ND	1.6
Normal	Nephrectomized	ND	ND
Normal	Nephrectomized, anaemic§	ND	1.4
Nephrectomized	Anaemic	ND	1.4
Nephrectomized	Nephrectomized, anaemic	ND	1.3

* Results from 3 animals are included in each group.
† International Units of erythropoietin determined in the transfusion-induced polycythaemic mouse (see Erslev & Kazal, 1968).
‡ Non-detectable activity.
§ Anaemia was induced 6 hours after nephrectomy.

tion, simultaneously-obtained maternal plasma samples did not exhibit elevated Ep levels (Table 2). Table 2 also shows that the response of bilaterally nephrectomized foetuses to bleeding did not differ from that of intact foetuses. Moreover, the absence of maternal kidneys did not affect this response.

Table 3. *Effect of bleeding of normal and nephrectomized pregnant goats on erythropoietin production in the mother and the foetus**

	IU Ep activity/ml plasma† Maternal: hours after maternal blood loss								
Treatment of the mother	0	3	5	7	9	10	12	15	24
Phlebotomy alone	←———			ND‡			———→	0.1	1.6
Nephrectomy followed by phlebotomy§	←———				ND				———→

	Foetal: hours after maternal blood loss						
Treatment of the mother	0	3	5	7	9	10	18
Phlebotomy alone	ND	0.5	1.0	1.4	1.0	0.9	1.3
Nephrectomy followed by phlebotomy§	ND	0.4	1.2	—	1.1	—	1.2

* Results from 4 animals are included in each group.
† International Units of erythropoietin determined in the transfusion-induced polycythaemic mouse (see Erslev & Kazal, 1968).
‡ Non-detectable activity.
§ Phlebotomy was initiated 6 hours after nephrectomy.

Bleeding was also a potent stimulus of Ep production in maternal goats (Table 3). However, the rise in maternal plasma Ep levels did not become evident until 15 hours after the induction of the anaemia. Thereafter, maternal plasma samples contained significant quantities of Ep (Table 3). Bilaterally nephrectomized pregnant goats, however, failed to respond to bleeding. Thus no Ep was present in the plasmas of nephrectomized, bled mothers at any of the time intervals studied (Table 3). Determinations of Ep levels in foetal plasma samples, on the other hand, revealed that considerable quantities of Ep were present in foetal circulation within 3 hours after maternal bleeding. The same pattern of response was evident when foetuses of nephrectomized, bled mothers were studied. Once again samples of foetal plasma obtained at 3 hours after the mother was bled contained significant quantities of Ep (Table 3).

DISCUSSION

The results presented here provide further support for our contention (Zanjani & Gordon, 1971) that the foetus is capable of producing Ep in response to hypoxic stimuli. Moreover, while the maternal erythropoietic status plays an important role in foetal erythropoiesis, this influence is not exerted directly but rather indirectly by reinforcing existing mechanisms in the foetus which control red cell formation. That Ep produced by the foetus has a primary influence on foetal erythropoiesis is shown by the fact that red blood cell formation is stimulated in the anaemic foetuses. In addition, polycythemic foetuses exhibited a decreased rate of erythropoiesis; administration of Ep to these foetuses resulted in the resumption of red cell production. The suppression of erythropoiesis in the foetus by transfusion-induced polycythemia suggests that the oxygen supply–demand state of the foetus is, as in the adult, the fundamental stimulus for Ep production.

Unlike the adult, however, Ep production in the foetus is not affected by nephrectomy, suggesting an extra-renal origin of Ep in the foetus. Extra-renal production of Ep in some nephrectomized, adult mammals (e.g. rat and man) has been reported (Schooley & Mahlmann, 1972; Naets & Wittek, 1968). The physiological role of extra-renal Ep in the intact animal, however, has not been ascertained. Additionally, the site(s) of production of extra-renal Ep has not yet been established. Although the liver (Burke & Morse, 1962) and the spleen (de Franciscus, De Bella & Cifaldi, 1965) have been implicated as extra-renal site(s) of Ep formation, these possibilities remain to be investigated in foetal goats.

It is clear, however, that the elevated levels of Ep in stimulated, nephrectomized foetuses cannot be attributed to the mother. Thus Ep production in nephrectomized, bled foetuses proceeds unhindered even in the absence of maternal kidneys. In this regard, previous experiments

(Zanjani & Gordon, 1971) have ruled out the possibility of Ep transfer across the placenta in foetal goats.

The independent regulation of foetal erythropoiesis was initially suggested by the experiments of Jacobson and his co-workers who observed that transfusion of pregnant mice resulted in an inhibition of maternal but not foetal erythropoiesis (Jacobson, Goldwasser, Gurney, Fried & Plzak, 1959). More recently, Matoth & Zaizov (1971) did not detect any enhancement of foetal erythropoiesis when red cell production in pregnant rats was stimulated by injections of Ep. On the other hand, increased production of red blood cells was observed in foetuses of rats rendered hypoxic by exposure to lowered atmospheric pressure or haemorrhage (Matoth & Zaizov, 1971). Similarly, Kaiser, Cummings, Reynolds & Marbarger (1958) observed that exposure of pregnant ewes to hypobaric hypoxia resulted in an increase in red cell numbers in the foetus. Matoth & Zaizov (1971) attributed the enhanced erythropoietic response of the foetus to the passage of maternal Ep to the foetus.

Although the studies reported here confirm the observations of these investigators, the results provide support for an alternative explanation. Thus, while the induction of anaemia in pregnant goats led to elevated levels of Ep in the foetus, this increase first became evident at a time when no increase in maternal Ep production had occurred. Furthermore, the results suggest that Ep production in pregnant goats was completely inhibited by nephrectomy. However, the production of Ep by the foetuses of renoprival, anaemic goats was similar to that observed in foetuses of anaemic but otherwise intact mothers.

It appears, therefore, that the increase in the rate of foetal erythrocyte formation in response to maternal anaemia is due to an increased availability of Ep from sources within the foetus. In this regard, the foetus appears to be more sensitive to reductions in maternal RBC levels than the mother. This may be related to the fact that the oxygen dissociation curve from foetal blood is normally shifted to the left of that for the adult (Behrman, 1968) suggesting that the foetus (at the period of gestation studied) is normally on the brink of hypoxia. The superimposition of maternal anaemia would therefore exacerbate the oxygen deficiency in the foetus resulting in the prompt production of higher amounts of Ep by the foetus.

SUMMARY

1. Bleeding of foetal goats resulted in stimulation of erythropoiesis in these foetuses. This increase was preceded by a rise in the production of erythropoietin (Ep). Erythropoietin production in response to bleeding was not affected when foetuses were nephrectomized prior to induction of anaemia.

2. Red cell production in the foetus was suppressed following induction of polycythemia by transfusion with maternal red blood cells. Administration of human urinary Ep to transfused foetuses, however, resulted in marked stimulation of foetal erythropoiesis.

3. Bleeding of pregnant goats resulted in increased production of Ep in both the mother and the foetus. The rise in foetal Ep, however, occurred several hours before the appearance of Ep in maternal circulation. Bilateral nephrectomy inhibited the response of the mother to bleeding; foetuses of nephrectomized, bled goats, however, continued to produce the hormone.

4. The data suggest that erythropoiesis in the foetus is regulated, as in the adult, by Ep. Moreover, while the maternal erythropoietic status plays an important role in foetal erythropoiesis, this influence is exerted not directly, but indirectly by reinforcing existing mechanisms in the foetus involved in the production of Ep.

This work was supported in part by Grant 1R01-AM 15525 from the National Institute of Arthritis and Metabolic Diseases and Grant 5R01-HE 03357-14 from the National Heart Institute of the National Institutes of Health, USPHS, and by the Albert A. List, Frederick Machlin, and Anna Ruth Lowenberg Funds.

REFERENCES

BEHRMAN, R. D. (1968). *J. Appl. Physiol.* **25**, 224–9.

BURKE, W. T. & MORSE, B. S. (1962). In *Erythropoiesis*, eds. Jacobson, L. O. & Doyle, M., pp. 111–19. Grune and Stratton, New York.

DE FRANCISCUS, P., DE BELLA, G., CIFALDI, S. (1965). *Science* **150**, 1831–3.

ERSLEV, A. J. & KAZAL, L. A. (1968). *Proc. Soc. Exp. Biol. Med.* **129**, 845–9.

GORDON, A. S. & ZANJANI, E. D. (1970). In *Regulation of Hematopoiesis*, ed. Gordon, A. S., pp. 413–57. Appleton-Century Crofts, New York.

JACOBSON, L. O., GOLDWASSER, E., GURNEY, C. W., FRIED, W. & PLZAK, L. (1959). *Ann. N.Y. Acad. Sci.* **77**, 551–73.

KAISER, I. H., CUMMINGS, J. N., REYNOLDS, S. R. M. & MARBARGER, J. P. (1958). *J. Appl. Physiol.* **13**, 171–8.

KRANTZ, S. B. & JACOBSON, L. O. (1970). *Erythropoietin and the Regulation of Erythropoiesis*, University of Chicago Press, Chicago, Ill.

LUCARELLI, G., PORCELLINI, A., CARNEVALI, C., CARMENA, A. & STOHLMAN, F., Jr (1968). *Ann. N.Y. Acad. Sci.* **149**, 544–59.

MATOTH, Y. & ZAIZOV, R. (1971). *Israel J. Med. Sci.* **7**, 839–42.

NAETS, J. P. & WITTEK, M. (1968). *Blood* **31**, 249–51.
SCHOOLEY, J. C. & MAHLMANN, L. J. (1972). *Blood* **39**, 31–8.
ZANJANI, E. D. & GORDON, A. S. (1971). *Israel J. Med. Sci.* **7**, 850–6.
ZANJANI, E. D., HORGER, E. O., III, GORDON, A. S., CANTOR, L. N. & HUTCHINSON, D. L. (1969). *J. Lab. Clin. Med.* **74**, 782–8.

GLUCAGON IN THE RAT FOETUS

By J. GIRARD, R. ASSAN* AND A. JOST
Laboratoire de Physiologie comparée, Université Paris VI,
9, quai Saint-Bernard, Paris 5e

Recent studies using electron microscopic techniques and radioimmunoassays have permitted the identification of A_2 cells (Orci *et al.* 1969; Perrier *et al.* 1969; Perrier, 1970) and the detection of glucagon (Orci *et al.* 1969) in the pancreas of the rat foetus on and after 18.5 days of gestation, but nothing is known about the secretion of glucagon in the foetal rat.

We measured the glucagon content of pancreas and plasma of 18.5- and 21.5-day-old rat foetuses and we studied some factors affecting the glucagon secretion in 21.5-day-old foetuses.

MATERIAL AND METHODS

Pregnant Sherman Rats (CNRS) were fed with laboratory rat chow (UAR) *ad libitum.* Females were caged with a male for one night and recognized pregnant by abdominal palpation fourteen days later.

Under pentobarbital anaesthesia of the pregnant rats (Nembutal 30 mg/kg i.p.) the foetuses were successively exteriorized from the uterus, the placenta and cord remaining *in situ*, and foetal blood was collected at the level of the armpit artery in polyethylene microtubes containing anti-protease (Iniprol 2000 UIP) cooled in ice.

The plasmas were rapidly separated by centrifugation at 4 °C and stored at −20 °C.

The pancreases of the foetuses were dissected with the aid of a magnifier, weighed, and glucagon was extracted by the method of Kenny (1955).

GLUCAGON ASSAY

Glucagon was determined by radioimmunoassay according to Assan *et al.* (1971): the plasma glucagon was measured with a specific antiserum K47 (Heding, 1971) which does not cross-react with the gut glucagon-like fraction and permits a measure of pancreatic glucagon in plasma (PG). The glucagon was determined in pancreatic extracts with an antiserum, PVP 8 (Assan *et al.* 1965), which cross-reacts with the gut glucagon-like fraction and which gives a measure of total glucagon-like immunoreactivity (GLI).

* Hôtel Dieu, Place du Parvis Notre-Dame, Paris 4e, France.

PLACENTAL TRANSFER

Under pentobarbital anaesthesia, a catheter was inserted into the maternal carotid artery for the collection of the blood samples and the abdomen was opened. ^{131}I-glucagon (1 μCi) was injected into the maternal saphenous vein. Maternal and foetal blood was collected simultaneously after 2–30 minutes. Chromato-electrophoresis of 100 μl plasma was run on Whatman 3 MC strips in Veronal 0.1 M (pH = 8.6) during 1 hour at 750 V. Under these conditions, intact ^{131}I-glucagon remained at the origin and the damage products and iodide migrated towards the anode. The strips were dried at 120 °C and their radioactivity measured in an autogamma chromato-scanner (Nuclear Chicago).

EXPERIMENTAL VARIATIONS OF GLUCAGON SECRETION ON DAY 21.5

Pentobarbital anaesthetized pregnant rats were perfused through the saphenous vein with pig insulin (NOVO) 800 mU/ml or 30% glucose solution. At zero time 1 ml of the solutions was injected, followed by a perfusion at the rate of 1.2 ml/hour. After one hour, foetal blood was collected and glucagon (PG) was assayed.

In some experiments, three or four foetuses in one uterine horn were injected s.c. with L-norepinephrine (10 μg/10 μl), or L(+)-arginine (6 mg/100 μl) or zinc-protamine insulin (1 U/25 μl). In the opposite uterine horn an equal number of control foetuses received the same volume of NaCl 9‰. Foetal blood was collected 30 minutes after norepinephrine, 30 minutes after arginine and 12 hours after insulin.

Blood glucose was measured with the glucose-oxidase method (Huggett & Nixon, 1957).

RESULTS

Glucagon was present in the plasma of 18.5- and 19.5-day-old foetuses at a concentration of 230 pg/ml. Its concentration increased to 490 pg/ml on day 20.5 and decreased again on day 21.5 (290 pg/ml) when the foetal glycaemia increased (Fig. 1).

The pancreatic glucagon concentration increased from 2.4 ng/mg on day 18.5 to 3.1 ng/mg on day 19.5, it remained the same on day 20.5 (2.9 ng/mg) and increased on day 21.5 (3.7 ng/mg) (Fig. 1).

These results suggest that the increase in plasma glucagon on day 20.5 results from an increased release by the foetal pancreas; this is supported by the absence of placental transfer of glucagon from the mother to the

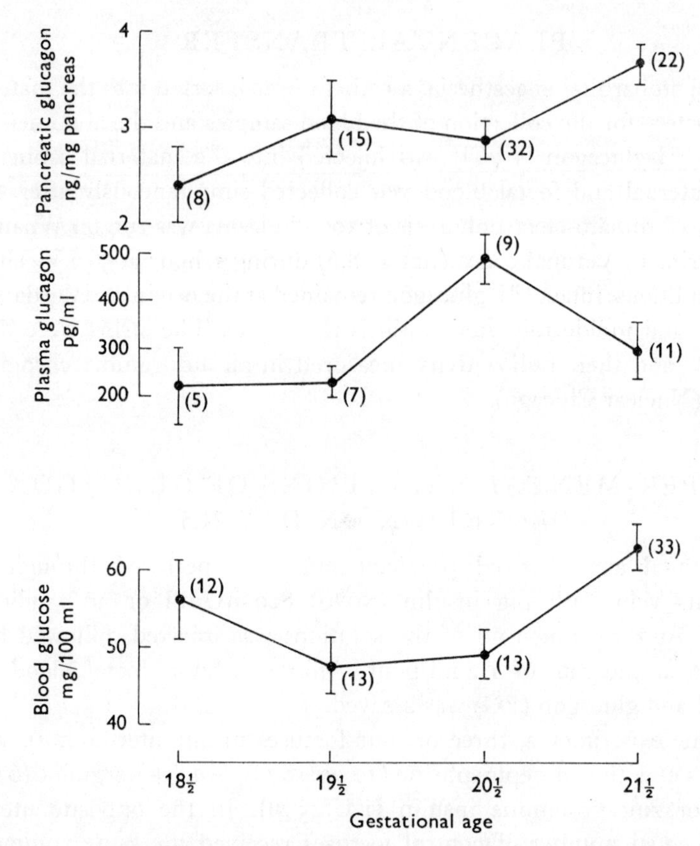

Fig. 1. Blood glucose, plasma and pancreatic glucagon in the rat foetus at the end of gestation. Means ± SEM, number of cases in brackets.

foetus. Labelled glucagon injected into the mother cannot be detected in the foetal blood for the following 30 minutes. The foetal origin of plasma glucagon is thus substantiated (Fig. 2).

Some factors known to affect glucagon secretion in adult animals were studied (Table 1).

(1) During acute foetal hyperglycaemia the plasma glucagon did not decrease. Prolonged hyperglycaemia has not yet been produced.

(2) During acute foetal hypoglycaemia (insulin infused to the mother one hour before) no statistically significant increase of plasma glucagon was obtained. After prolonged foetal hypoglycaemia (fasting mothers or insulin given to the foetus 12 hours before) a significant increase in plasma glucagon was seen.

(3) Norepinephrine acted as a fast and potent stimulus foı glucagon secretion (30 minutes).

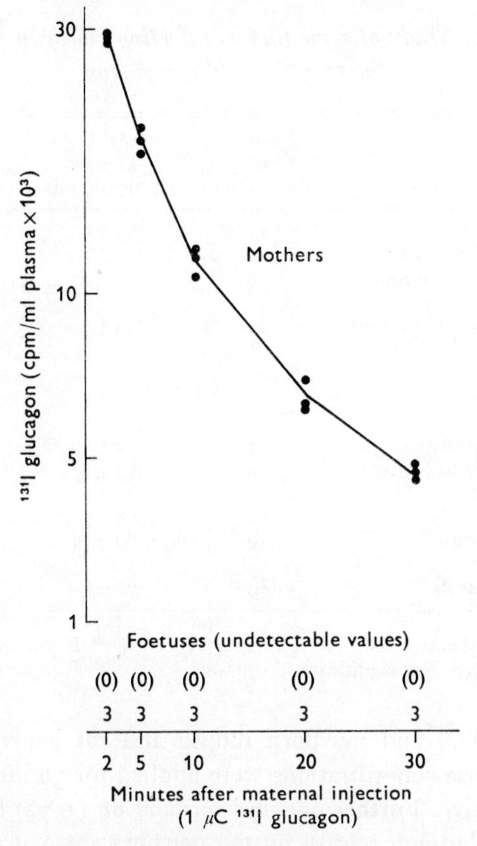

Fig. 2. Absence of placental transfer of glucagon from mother to foetus when labelled glucagon was injected into the mother.

(4) Arginine was the most potent agent for glucagon secretion in the foetus.

DISCUSSION

This study demonstrates that glucagon is secreted by the foetal rat pancreas *in vivo* at the end of gestation.

An increase in glucagon secretion seems to occur on day 20.5 at the time when ultrastructurally A_2 cells show signs of complete maturation (Perrier, 1970).

During the day before birth, the secretion of glucagon can be increased by arginine and norepinephrine. Even large variations in blood glucose do not appear to be stimuli for glucagon secretion *in vivo*, if they are applied for a short period of time (1 hour). Lernmark & Wenngren (1972) and Edwards *et al.* (1972) recently observed no variations in glucagon

Table 1. *Study of some factors affecting glucagon secretion in 21.5-day-old rat foetus*

Experimental conditions	Number of observations	Foetal blood glucose (mg/100 ml)	Foetal plasma glucagon (pg/ml)
No treatment	11	63 ± 9	285 ± 40
Mother fasted for 4 days	14	41 ± 3	380 ± 41*
Glucose perfusion to mother (1 hour)	9	150 ± 20	280 ± 25 NS
Insulin perfusion to mother (1 hour)	12	32 ± 7	330 ± 40 NS
Insulin to foetus (12 hours)	7	25 ± 5	400 ± 32
Littermate controls	7	54 ± 10	200 ± 30†
Norepinephrine to foetus (30 minutes)	19	75 ± 6	440 ± 25
Littermate controls	19	68 ± 7	180 ± 15†
Arginine to foetus (30 minutes)	19	80 ± 10	650 ± 50
Littermate controls	19	76 ± 6	210 ± 24†

Means ± SEM; NS Not significant; * $P < 0.05$; † $P < 0.01$

release from foetal and newborn mouse and rat pancreas when large changes in glucose concentrations were applied for 30 minutes; however, arginine was active. Furthermore Edwards *et al.* (1972) found that octanoate inhibits glucagon release by the newborn rat pancreas *in vitro* (30 minutes).

However, our results clearly suggest that prolonged hypoglycaemia increases glucagon secretion in the foetus. This was first suggested by Jacquot (1971), who found that liver glucose-6-phosphatase activity was increased in foetuses of starved mothers; it is known that the activity of the enzyme is increased by glucagon given to rat foetuses (Greengard & Dewey, 1967).

Dawkins (1963) and Yeung & Oliver (1968) speculated that the transient postnatal hypoglycaemia increases the glucagon secretion which in turn stimulates glycogenolysis and gluconeogenesis. The predicted increase in glucagon at birth has been verified (Girard, Bal & Assan, 1971, 1972; Blasquez *et al.* 1972); however, the rapidity of the response and the insensitivity of A_2 cells to acute glucose variations seem to rule out hypoglycaemia as the physiological stimulus for glucagon release in the newborn rat. The presence of nervous structures at the contact of endocrine cells in the pancreas (Perrier, 1970) and the effectiveness of norepinephrine as a glucagon-releasing stimulus in the 21.5-day-old rat foetus, suggest that the sympathetic nervous system could play a role in glucagon secretion at birth.

REFERENCES

ASSAN, R., ROSSELIN, G., DROUET, J., DOLAIS, J. & TCHOBROUTSKY, G. (1965). *Lancet* **2**, 91.

ASSAN, R., TCHOBROUTSKY, G. & DEROT, M. (1971). *Horm. Metab. Res.* Suppl. 3, 82.

BLASQUEZ, E., SUGASE, T., BLASQUEZ, M. & FOA, P. P. (1972). *Proc. 4th Intern. Congress Endocrinol. Washington.* Abst. no. 196. Excerpta Medica Foundation, Amsterdam ICS, no. 526.

DAWKINS, M. J. R. (1963). *Ann. N.Y. Acad. Sci.* **111**, 203.

EDWARDS, J. C., ASPLUND, K. & LUNDQUIST, G. (1972). Personal communication.

GIRARD, J., BAL, D. & ASSAN, R. (1971). *Diabetologia* **7**, 481.

GIRARD, J., BAL, D. & ASSAN, R. (1972). *Horm. Metab. Res.* **4**, 168.

GREENGARD, O. & DEWEY, M. K. (1967). *J. Biol. Chem.* **242**, 2986.

HEDING, L. G. (1971). *Diabetologia*, **7**, 10.

HUGGETT, A. ST G. & NIXON, D. A. (1957). *Lancet* **2**, 368.

JACQUOT, R. (1971). In *Hormones in Development*, eds. M. Hamburgh & E. J. W. Barrington. Appleton-Century-Crofts, New York.

KENNY, A. J. (1955). *J. Clin. Endocr.* **15**, 1089.

LERNMARK, A. & WENNGREN, B. I. (1972). Personal communication.

ORCI, L., LAMBERT, A. E., ROUILLER, C., RENOLD, A. E. & SAMOLS, E. (1969). *Horm. Metab. Res.* **1**, 108.

PERRIER, H. (1970). *Diabetologia* **6**, 605.

PERRIER, H., PORTE, A. & JACQUOT, R. (1969). *C. r. Acad. Sci., Paris* **269**, 841.

YEUNG, D. & OLIVER, I. T. (1968). *Biochem. J.* **108**, 325.

ONTOGENESIS OF GROWTH HORMONE, INSULIN, PROLACTIN AND GONADOTROPIN SECRETION IN THE HUMAN FOETUS

By MELVIN M. GRUMBACH AND SELNA L. KAPLAN
Department of Pediatrics, University of California
San Francisco, San Francisco, California 94122

The influence of the foetal endocrine glands on the morphogenesis and functional maturation of the foetus is well established; studies in experimental animals and of structural and functional developmental defects in man provide firm evidence of the participation of the foetal endocrine system in these processes (see reviews by Jost, 1966; Jost & Picon, 1970; Hamburgh & Barrington, 1971). Hormones in the foetal circulation can arise from maternal and foetal endocrine glands and the placenta. The role of the maternal and foetal hormonal milieu on development and regulatory and homeostatic mechanisms in the human foetus is poorly understood. The purpose of this presentation is to discuss ongoing studies in our laboratory on one aspect of the development of the foetal endocrine system – the chronology of the synthesis and secretion by the human foetus of certain pituitary hormones (growth hormone, prolactin, FSH and LH) and insulin. Significant transfer across the placenta has not been demonstrated for any of these polypeptide hormones (Gitlin, Kumate & Morales, 1965).

The nervous system is a major regulator of anterior pituitary function; this control is mediated through the hypothalamus and its neurohumoral chemotransmitters (Harris, 1970). Hypothalamic hypophysiotropic hormones synthesized by neurosecretory neurons in the basal hypothalamus (hypophysiotropic area) are released from nerve endings in the median eminence into the hypophyseal portal circulation and transmitted to the anterior pituitary. We shall attempt to correlate the evolution of the hypothalamus and its chemotransmitter pathway with the regulation of the foetal anterior pituitary gland by the hypothalamic releasing and inhibiting factors.

The pituitary gland of the human foetus has the capacity to synthesize and store protein hormones in the first trimester. Acidophils have been identified by the 9th week of gestation by histochemical (Pearse, 1953; Falin, 1961; Pasteels, 1967), immunochemical (Ellis, Beck & Currie, 1966) and electron microscopic techniques (Dubois, 1968). Foetal pituitary tissue grown in culture synthesizes and secretes growth hormone, gonadotropins, thyrotropin and prolactin (Pasteels, Brauman & Brauman, 1963; Solomon, Grant, Burr, Kaplan & Grumbach, 1969; Gitlin & Biasucci, 1969). We have utilized immunochemical techniques to detect and quantify the pituitary content and serum concentration of foetal HGH (Kaplan & Grumbach, 1962, 1967, 1971; Kaplan, Grumbach & Shepard, 1972), FSH and LH (Kaplan, Grumbach & Shepard, 1969). FSH and LH, prolactin (HPr), and insulin were measured by radioimmunoassays utilizing the double antibody method as described elsewhere (Morgan & Lazarow, 1963; Youlton, Kaplan & Grumbach, 1969; Burr, Sizonenko, Kaplan & Grumbach, 1970; Jacobs, Maziz & Daughaday, 1972). For the HPr assay, guinea pig anti-ovine prolactin, ^{131}I-human prolactin, and a human prolactin standard provided by Dr U. J. Lewis, were used. Pituitary glands and blood were obtained from aborted foetuses (spontaneous or therapeutic). Gestational age was estimated from crown-rump measurements and fertilization (ovulation) age (Shepard, 1969).

We recognize the limitations in the measurement of the hormone content of an endocrine gland and not its flux, in the interpretation of the serum concentration of a hormone without knowledge of possible differences between the foetus and the child in metabolic clearance rate, and in the assessment of foetal age, and the potential or unrecognized artifacts related to the state of the foetuses studied.

GROWTH HORMONE

No differences in the immunological activity of HGH were detected in foetal pituitary glands or sera when compared with purified HGH or that extracted from fresh adult pituitary glands. The HGH in the foetal pituitary had the same characteristics as that in the adult pituitary when examined by immunoelectrophoresis, disc gel electrophoresis, and starch gel electrophoresis (Kaplan, Grumbach & Shepard, 1972).

Immunoreactive GH was detected in the pituitary gland as early as 68 days of gestation, the youngest foetus studied. The mean HGH concentration (μg/mg pituitary gland) and content (μg) in 117 foetal pituitary glands from 68 days' gestation, to term are shown in Figs. 1 and 2 and contrasted with values in children 1 month to 1 year of age. The GH content of the human pituitary glands rose from 0.44 ± 0.20 (SEM) μg at 10 to 14 weeks of gestation to 577.7 ± 90 at 30 to 34 weeks, and 675.2 ± 112.3 at 35 to 40 weeks. Significant differences were obtained for each 4-week interval with the exception of the 30 to 34 weeks' and the 35 to 40 weeks' gestation groups. The HGH concentration increased between 10 to 14 weeks' and 25 to 29

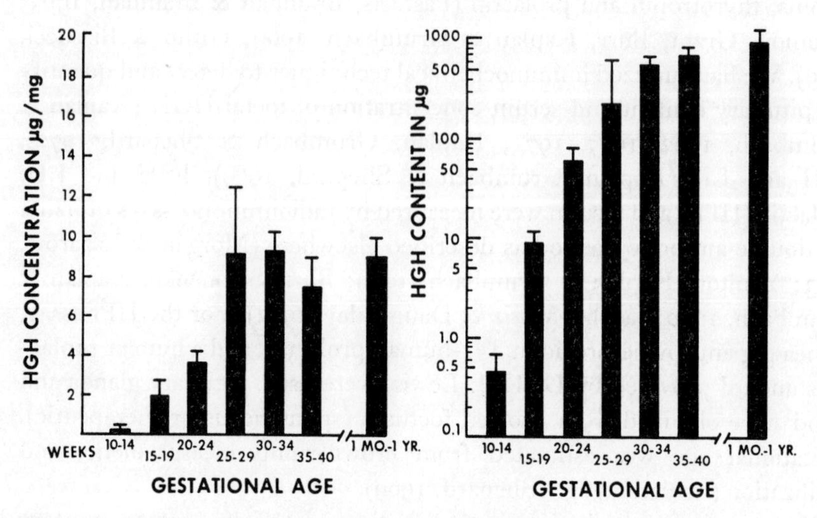

Fig. 1. The concentration (mean ± SEM) of HGH (µg/mg) is plotted on the left panel and the content (mean ± SEM) of HGH in µg in the right panel at 5 week intervals during gestation and for post-natal ages 1 month to 1 year.

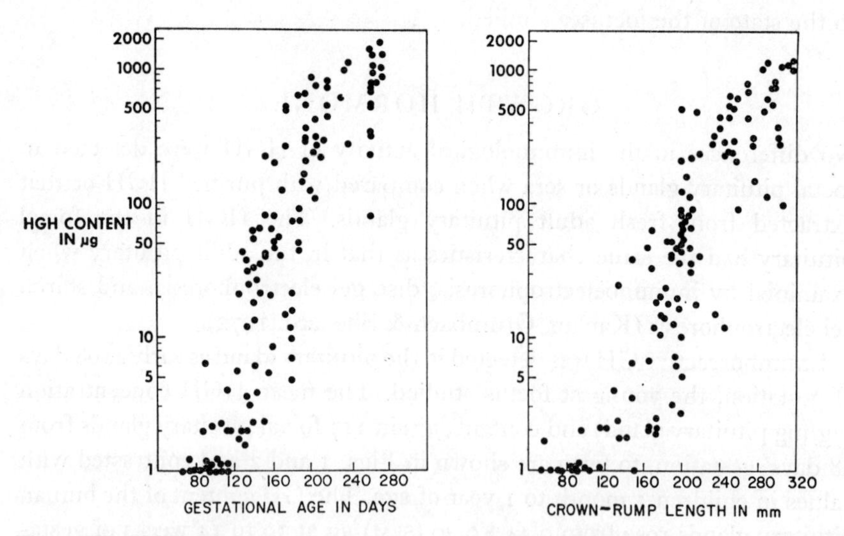

Fig. 2. The HGH content (µg) of human foetal pituitary glands is plotted on a log scale on the ordinate. On the left panel, gestational age in days is indicated on the abscissa and on the right panel, crown–rump length in mm.

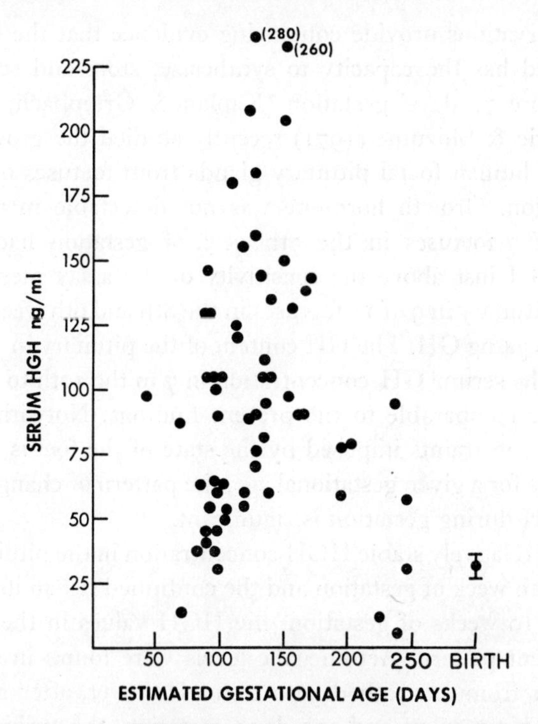

Fig. 3. The concentration of HGH (ng/ml) in foetal serum is plotted on the ordinate against gestational age in days on the abscissa.

weeks' gestation but plateaued thereafter. The HGH content and concentration in the pituitary glands of infants 1 to 12 months old were comparable to the values in foetuses from age 25 to 29 weeks' gestation in concentration and from age 30 to 34 weeks in content.

The concentration of serum GH was studied in 62 foetuses from 70 to 245 days' gestation (Fig. 3). The youngest foetus had a serum concentration of 14.5 μg/ml, with a sharp increase and some peak values greater than 175 ng/ml by 20 to 24 weeks, with a gradual decrease toward term. The mean GH concentration at 10 to 14 weeks was 65.2 ± 9.6 ng/ml; at 15 to 19 weeks 114.9 ± 12.5; at 20 to 24 weeks 119.3 ± 19.8; at 25 to 29 weeks 72.0 ± 6.5; and at 30 to 40 weeks 26.5 ± 11.5. The mean GH concentration in umbilical venous serum at delivery was 33.5 ± 4.2 ng/ml. There was a significant increase between 10- to 14-week and 15- to 19-week foetuses, no difference in mean values from 15 to 19 weeks through 25 to 29 weeks, and a significant decrease between 25 to 29 weeks' gestation and term.

No correlation between the serum GH concentration and content of the pituitary was found when paired serum and pituitary specimens were compared in 17 foetuses ($r = 0.11$, $p > 0.5$).

These observations provide convincing evidence that the human foetal pituitary gland has the capacity to synthesize, store and secrete growth hormone before 70 days' gestation (Kaplan & Grumbach, 1962, 1967). Matsuzaki, Irie & Shizume (1971) recently studied the growth hormone content of 47 human foetal pituitary glands from foetuses of age 6 to 16 weeks' gestation. Growth hormone was not detectable in the 9 6-week foetuses; 2 of 7 foetuses in the 7th week of gestation had measurable amounts of GH just above the sensitivity of the assay method (0.1 ng), whereas the pituitary in 9 of 11 foetuses in the 8th and 9th week of gestation contained 2 to 34 ng GH. The GH content of the pituitary in 10 to 20 week foetuses and the serum GH concentration in 7 in the 10th to 16th week of gestation were comparable to the present findings. Notwithstanding the interpretative constraints imposed by the state of the foetus and the wide range of values for a given gestational age, the pattern of change in pituitary and serum GH during gestation is significant.

Despite the relatively stable HGH concentration in the pituitary between the 25th to 40th week of gestation and the continued rise in its GH content until the last 10 weeks of gestation, the HGH values in the foetal serum show a different pattern. Acromegalic levels were found in all but a few of the foetuses from the 70th day to term. However, after a rise of GH in the serum between 70 and 100 days' gestation, the strikingly elevated serum levels of GH between 100 and 200 days (with the suggestion of a peak at about 150 days) decrease in the last 10 weeks of gestation. Even though the mean concentration in umbilical venous serum at term is elevated, it is lower than in 15- to 29-week foetuses. This pattern of change in the HGH concentration in serum can be correlated with the differentiation and development of the pituitary gland, of the hypothalamic monoaminergic network, the hypothalamic neurosecretory neurons and median eminence, of the hypophysial portal system, and of the neurophysiological maturation of the central nervous system. A scheme of these interrelationships is shown in Fig. 4; the details are given elsewhere (Kaplan *et al.* 1972).

The morphological details of the gross and cellular differentiation of the foetal adenohypophysis have been studied more extensively, but here too there are notable gaps in our knowledge. Relatively little information is available on the developmental morphology and ontogeny of the other components of the hypothalamic-hypophyseal complex and its restricted vascular communication system in man or even in experimental animals (Raiha & Hjelt, 1957; Rinne, Kivaio & Talanti, 1962; Humphrey, 1964; Dreyfus-Brisoe, 1966; Bergstrom, 1968). Hyyppa (1972) identified the median eminence in an 8-week-old foetus and stated that it resembled that

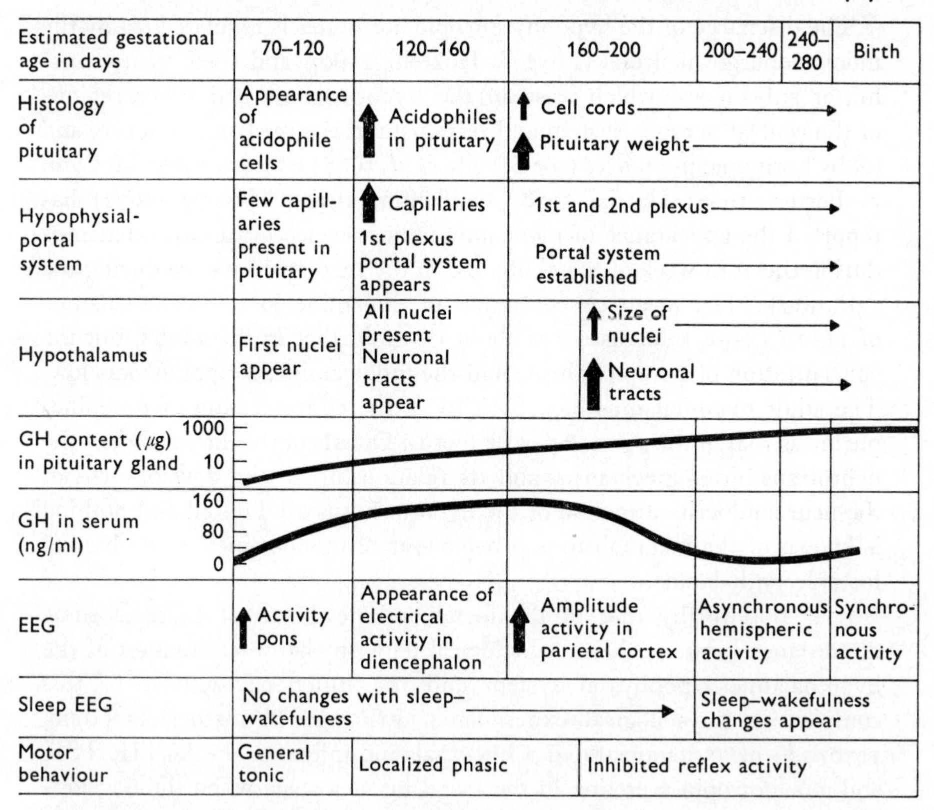

Fig. 4. The ontogeny of growth hormone secretion by the human foetus as correlated with histological changes in the pituitary, and the development of the portal system and central nervous system.

of the adult by the 11th week. The hypothalamic-hypophysiotropic releasing and inhibiting factors are synthesized and secreted by peptidergic neurosecretory neurons which terminate in the median eminence. Wurtman has called these cells neuroendocrine transducers. Two of these factors – thyrotropin-releasing factor (TRF), and luteinizing hormone and follicle stimulating hormone-releasing factor (LRF) – have been characterized and synthesized (Burgus *et al.* 1969; Nair *et al.* 1970; Matsuo *et al.* 1971; Amoss *et al.* 1971). Reichlin and his associates (Mitnick & Reichlin, 1971; Reichlin, 1972) recently reported that the biosynthesis of TRF and LRF by fragments of the ventral hypothalamus of the rat is an enzymatic process which does not involve the ribosomes. The hypophysiotropic hormones are transmitted to the effector site in the anterior pituitary via a primary capillary network in the median eminence, the portal veins and a secondary capillary network in the adenohypophysis (see Harris, 1970; Guillemin, 1971).

The discharge of the hypophysiotropic hormones is regulated (*a*) by the monoaminergic neurons (Fuxe & Hokfelt, 1969), and their neurotransmitter substances, which arise in the hypothalamus and other regions of the central nervous system and terminate in the median eminence, and (*b*) by hormonal influences (see Schally *et al.* 1968; Everett, 1969; McCann & Porter, 1969; Harris, 1970; Guillemin, 1971). Hyyppa (1972) has reported the appearance of monoamine fluorescence in the hypothalamus during the 10th week of foetal life and in the median eminence during the 13th week. The mean concentration of dopamine in the hypothalamus of 11- to 15-week foetuses was about two-fold that of the adults, but the concentration of norepinephrine and the indoleamine, serotonin, was low. The adult hypothalamus has about twice the concentration of norepinephrine as that of the 10- to 24-week foetus. Details of the maturation of the neurotransmitter mechanism and its relationship to the development of the neuroendocrine function of the hypothalamus are limited and nothing is known of the metabolism and behaviour of monoamines in the human foetal hypothalamus.

It is noteworthy that in the foetal and newborn rat there is some discordance between the morphological data on the differentiation of the hypothalamic-hypophyseal system and the functional activity of this complex in physiological experiments. Jost and his associates (1966, 1970*a*, *b*) have demonstrated a hypothalamic influence on ACTH, TSH and gonadotropin secretion in the foetal rat at a stage when the development of the hypophyseal-portal system is incomplete. Recent work by Halasz, Kosaras & Lengvari (1972) suggests, however, that this discrepancy may be resolved by more detailed morphological studies; they demonstrated the presence of an hypophyseal-portal system in the 18-day rat foetus. Studies of the ultrastructure of the median eminence (Kobayashi *et al.* 1968; Smith, 1970; Halasz *et al.* 1972) suggest that hypophysiotropic hormones are synthesized prior to differentiation of granules in the foetal pituitary cells (Yoshimura *et al.* 1970). But it is not known whether the foetal hypophysiotropic hormones have a significant effect on the differentiation of the foetal adenohypophysis. Further, there are limitations in translating results obtained in lower experimental animals to primates. For example, the structural maturation of the foetal hypothalamic-hypophyseal system occurs relatively much earlier in gestation in man than in the rat. Jost (1953, 1966) has pointed this out for many aspects of the foetal endocrine system.

We have interpreted the sequence of development of the neuroendocrine regulation of GH secretion (Fig. 5) by the adenohypophysis in the following manner: (1) the capacity of the foetal pituitary gland to synthesize and

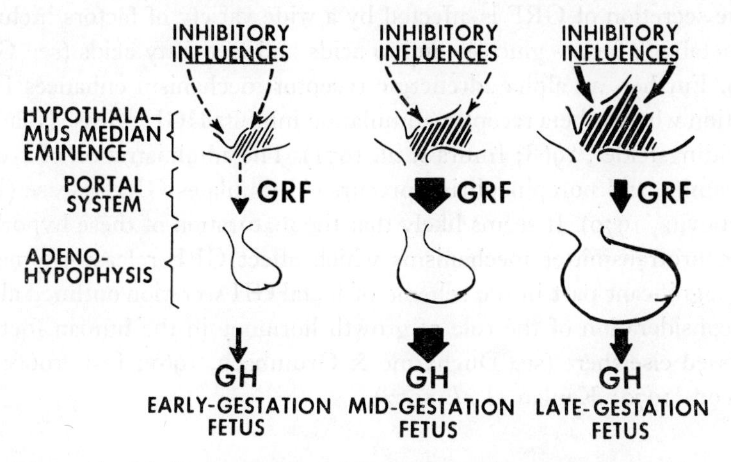

Fig. 5. Graphical representation of development of regulatory mechanisms for control of GH secretion in the human foetus.

secrete growth hormone occurs early in foetal development – synthesis of immunoreactive growth hormone appears to begin during the 7th to 8th week of gestation (Ellis *et al.* 1966) and by the 10th week secretion of growth hormone is detectable (Kaplan *et al.* 1972). These events coincide with the appearance of, and increase in, pituitary acidophils (Falin, 1961). Even though morphological studies suggest that the hypophyseal-portal system does not provide circulatory linkage between the median eminence and adenohypophysis until later in development (Niemineva, 1949; Räiha & Hjelt, 1957), it is uncertain whether the secretion of growth hormone at this stage occurs autonomously or is, at least in part, regulated by hypothalamic growth hormone releasing factor (GRF). (2) By midgestation, with the further maturation of the median eminence, and its hypophysiotropic area, the portal system, and continued growth of the adenohypophysis, we suggest that relatively unrestrained tonic release of GRF results in intense stimulation of growth hormone secretion by the pituitary acidophils. (3) In late gestation, especially during the last 10 weeks, neural inhibitory influences become operative – possibly related in part to the maturation of the hypothalamic monoaminergic neuronal network and correlating with the development of neurophysiological functions (Fig. 5). (4) Further maturation of regulatory mechanisms which affect GH secretion occurs during the first few postnatal weeks. The acquisition of sleep-mediated growth hormone release, which may be used as an index of the attainment of a fully functional mechanism, occurs in infants over 3 months of age (Shaywitz *et al.* 1971; Vigneri & D'Agata, 1971).

The secretion of GRF is affected by a wide variety of factors including the metabolic fuels – glucose, amino acids and free fatty acids (see Glick, 1969). Further, an alpha adrenergic receptor mechanism enhances HGH secretion whereas beta receptor stimulation inhibits HGH release (Blackard & Heidingsfelder, 1968; Imura *et al.* 1971). The administration of L-dopa, a dopamine and norepinephrine precursor, stimulates GRF release (Boyd & Lebovitz, 1970). It seems likely that the maturation of these hypothalamic neurotransmitter mechanisms which affect GRF release postnatally has a significant part in the scheme of foetal GH secretion outlined above.

A consideration of the role of growth hormone in the human foetus is discussed elsewhere (see Ducharme & Grumbach, 1961; Jost, 1966; Jost & Picon, 1970; Kaplan *et al.* 1972).

Studies in anencephaly

Anencephalic or apituitary foetuses characteristically have hypoplastic adrenal glands with a small foetal zone (Bernirschke, 1956). This is explicable as a consequence of foetal ACTH deficiency in the anencephalic foetus owing to the hypothalamic-hypophyseal defect (see Jost & Picon, 1970). Less well understood is the effect of anencephaly on the secretion of other pituitary hormones.

Growth hormone levels were determined in specimens obtained from the umbilical vein or foetal heart in nine anencephalic infants (Fig. 6). The mean concentration of GH was 7.0 ng/ml with a range of 1.0 to 17.0 ng/ml, which is significantly lower than the 33.5 ± 4.3 ng/ml present in umbilical vein specimens from normal infants at term. Stimulation with arginine infusion, insulin-induced hypoglycaemia and glucose infusion were performed in three liveborn male anencephalic infants (Fig. 7). The two infants (H and M) with very low basal values (1.0, 1.5 ng/ml) had no significant rise in plasma GH in response to any of the stimuli used. In one infant (L) the GH concentration of plasma increased from a fasting level of 8.4 ng/ml to 20 and 11.4 ng/ml, respectively, in response to hypoglycaemia and arginine infusion. In this infant (L), the maximum insulin response to arginine was 32 μU/ml. In the non-responder (infant H), the peak insulin concentration following arginine infusion was low (7 μU/ml). The insulin response to oral glucose was blunted (12 μU/ml) and was associated with a diabetic glucose tolerance curve.

Grunt & Reynolds (1970) demonstrated in an anencephalic infant that the insulin response to an intravenous glucose load and the glucose disposal rate were comparable to those in normal newborns. However, the peak growth hormone concentration in that infant was 10 ng/ml.

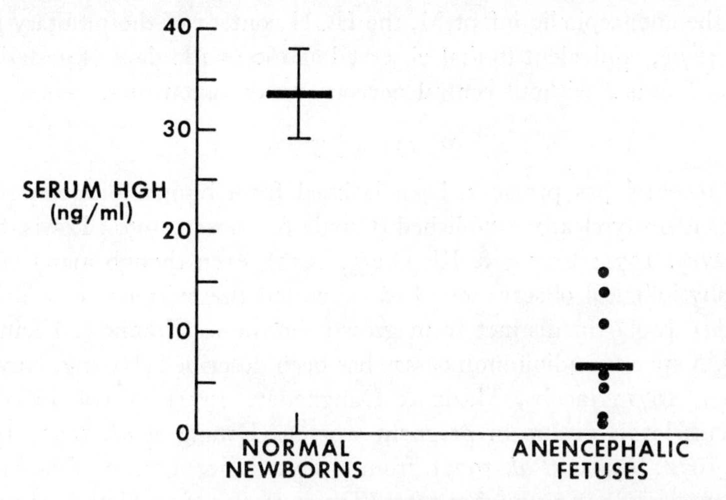

Fig. 6. Comparison of the mean level of growth hormone in normal newborns (33·5 ± 4·3 ng/ml) and in anencephalic foetuses (7 ng/ml).

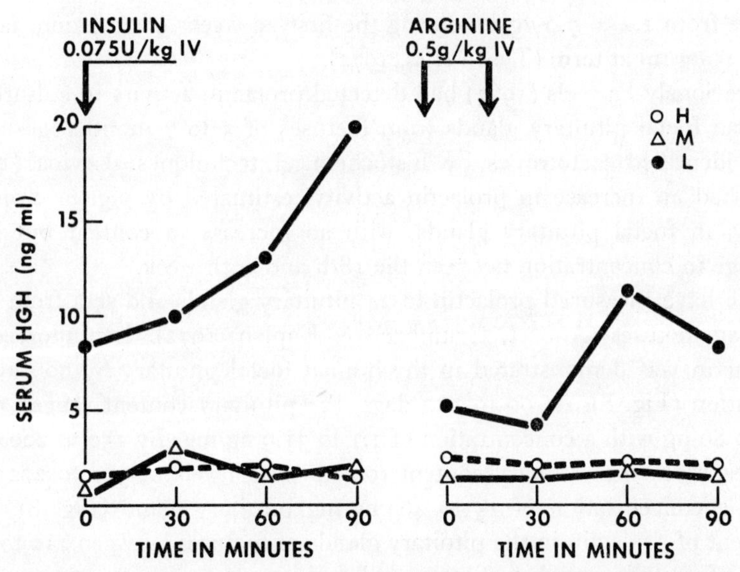

Fig. 7. The serum growth hormone response to arginine infusion and insulin-induced hypoglycaemia in 3 anencephalic infants. Only 1 of 3 infants showed a significant rise in GH.

These limited data suggest that the hypothalamic-hypophysial defect in anencephaly is variable, but clearly in some affected foetuses there is a severe disturbance in the hypothalamic regulation of growth hormone secretion. Furthermore, the insulin response to provocative stimuli may depend, in part, on the presence of circulating growth hormone.

In the anencephalic infant M, the HGH content of the pituitary gland was 140 μg, equivalent to that observed at 160 to 180 days of gestation in human foetuses without central nervous system anomalies.

PROLACTIN

Only recently has prolactin been isolated from human pituitary glands and its identity clearly established (Guyda & Friesen, 1971; Lewis, Singh & Seavey, 1971; Frantz & Kleinberg, 1972), even though many clinical and physiological observations had suggested the existence of a primate pituitary prolactin distinct from growth hormone (Frantz & Kleinberg, 1972). A specific radioimmunoassay has been described (Hwang, Guyda & Friesen, 1971; Jacobs, Maziz & Daughaday, 1972). A rise in plasma prolactin levels occurs in pregnant women (Hwang *et al.* 1971; Jacobs *et al.* 1972; Tyson *et al.* 1972), from a mean concentration of 24.8 ng/ml at 10 weeks to 207.3 ng/ml at term (Tyson *et al.* 1972). Umbilical venous plasma from full term infants has a mean concentration of 235.1 ng/ml. Of interest are the high concentrations of prolactin in amniotic fluid which range from 1.2 to 7.0 μg/ml during the first 20 weeks of gestation, falling to 0.35 μg/ml at term (Tyson *et al.* 1972).

Previously Pasteels (1963) had detected prolactin activity in cultures of human foetal pituitary glands from foetuses of 5 to 7 months' gestation and identified lactotropes by histochemical techniques. Levina (1968) reported an increase in prolactin activity, estimated by pigeon crop sac assay, in foetal pituitary glands, with an increase in content but little change in concentration between the 18th and 40th week.

We have measured prolactin in 24 pituitary glands and sera from nine human foetuses (Aubert, Grumbach & Kaplan, 1972). Immunoreactive prolactin was demonstrated in the human foetal pituitary by 90 days of gestation (Fig. 8). At 90 to 110 days, the pituitary content ranged from 28 to 86 ng with a concentration of 1.1 to 31.9 ng/mg. By 150 to 200 days of gestation, the pituitary content rose to peak levels of 560 to 2225 ng, with a concentration of 15 to 463 ng/mg pituitary gland (Fig. 8). The content of prolactin in the pituitary glands of foetuses between 210 to 280 days of gestation was in a comparable range.

Throughout gestation, there was a significant correlation of the content of prolactin with the content of growth hormone in the human foetal pituitary gland ($r = 0.5$, $p < 0.01$). The growth hormone content at all stages of gestation was 20- to 100-fold higher.

Prolactin was present in significant concentrations in human foetal sera at 28 weeks of gestation. A five-fold increase in serum prolactin was seen by 35 weeks of gestation, with comparable levels in umbilical vein samples

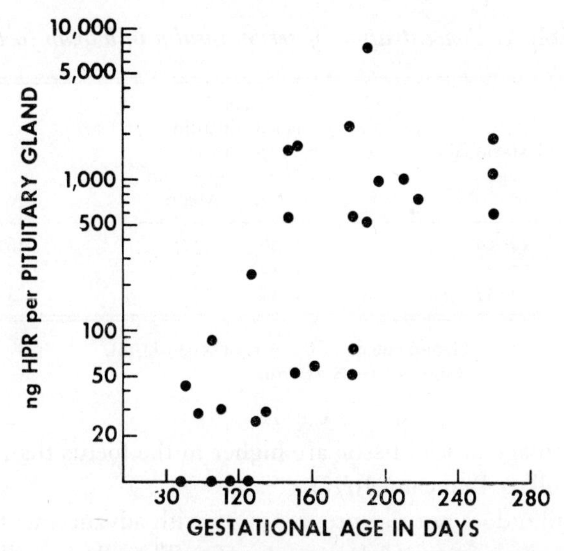

Fig. 8. The content of prolactin in human foetal pituitary glands is plotted in nanograms on the ordinate against gestational age in days on abscissa.

at term. The levels of serum prolactin in the foetus during late gestation are similar to those observed in pregnant women at term.

Moger & Geschwind (1971) first found significant prolactin concentrations in sheep foetal plasma after 122 days, with the exception of one 72-day foetus which had a barely detectable level. This correlated with the age at which lactotropes were first identified tinctorially in the foetal pituitary gland (Stokes & Boda, 1968), although localization by immunofluorescent staining was found by the 88th day. Similar observations were reported by Alexander *et al.* (1972) who first detected significant levels in foetuses older than 136 days.

INSULIN

Serum immunoreactive insulin was determined in 42 foetuses aged 84 to 245 days by a radioimmunoassay which does not discriminate between insulin and proinsulin. The mean concentration was 8.0 ± 1.5 μU/ml with a range of 1 to 28 μU/ml (Table 1); in three, the concentration was less than 1 μU/ml. These findings are consistent with histological localization of foetal insulin in the pancreatic islets by 12 weeks of gestation (Grillo & Shima, 1966). The content of insulin in the human foetal pancreas increases steadily from 12 to 24 weeks (Rastogi, Letarte & Fraser, 1970), with a more marked increment in insulin content by 34 to 40 weeks of gestation (Steinke & Driscoll, 1965). Both the insulin concentration and the

Table 1. *Concentration of serum insulin in human foetus*

Gestational age (weeks)	Foetal serum insulin (μU/ml) Range	Mean	N
11–19	1–28	5.4	26
20–30	1–10	3.7	10
34–37	2–17	11.3	3

Group mean = 8.0 ± 1.5 (SEM) μU/ml.
Range = 1–28 μU/ml.

relative percentage of islet tissue are higher in the foetus than in the adult (Wellman, Volk & Brancato, 1971).

Serum insulin does not rise progressively with advancement in development of the human foetus (Adam *et al.* 1969; Thorell, 1970; Kaplan *et al.* 1972). Regulation of carbohydrate metabolism has been studied extensively in the simian foetus by Chez and associates (Mintz, Chez & Horger, 1969; Chez, Mintz, Horger & Hutchinson, 1970; Chez, Mintz & Hutchinson, 1971), and in the ovine foetus by Bassett & Thorburn (1971). Acute administration of glucose or arginine has a limited effect on insulin secretion in the simian and ovine foetus, but glucagon or glucose plus theophylline significantly increase insulin release in the simian foetus (Chez *et al.* 1971).

FSH AND LH

Very little is known about the ontogenesis of human foetal pituitary gonadotropins or of their activity, if any, in the human foetus. Pearse (1953) identified cells with the characteristics of gonadotropes in foetal pituitary glands as early as the 8th week and Gitlin & Biasucci (1969) demonstrated FSH release by foetal pituitary glands in organ culture.

In the rabbit, Jost (1953, 1966), in an elegant series of experiments, has clearly demonstrated that decapitation of male rabbit foetuses prior to, or during, the critical period of sex differentiation impairs foetal testicular function and results in female differentiation or incomplete masculinization of the external genitalia. This does not occur if gonadotropins are injected into the foetus at the time of hypophysectomy. Such a relationship has not been established in man although, as in the rabbit and other placental mammals, the human foetal testis plays a crucial role in male sex differentiation, and absence of a foetal testis and its hormones early in foetal life results in development of a female genital tract (see Van Wyk & Grumbach,

1968). These dichotomous observations on the effect of the foetal pituitary gland on sex differentiation of the lower genital tract in the rabbit and man may be attributable to the secretion of chorionic gonadotropin (which has potent LH activity) by human, but not the rabbit, placenta. HCG is secreted mainly unidirectionally into the maternal circulation; the large concentration gradient between maternal and foetal blood at term (Geiger, Kaiser & Franchimont, 1971) suggests that only a small proportion of the daily secretion of this placental hormone is transferred to the foetus. However, during the critical period of differentiation of the urogenital sinus and external genitalia the concentration of HCG is at its peak in maternal blood, and physiologically significant levels may be present in the foetus.

The normal differentiation of the genital tract in the male anencephalic or apituitary foetus has been explained by the action of HCG transmitted to the foetus on Leydig cell function (Grumbach & Barr, 1958; Jost, 1966; Bearn, 1968). Nevertheless, anencephalic foetuses often exhibit hypoplasia of the ovaries or testes (Ch'in, 1938; Zondek & Zondek, 1965; Bearn, 1968) and, in the male, impaired growth and development (but not differentiation) of the external genitalia (Bearn, 1968).

In an attempt to clarify the ontogeny of pituitary FSH and LH, we measured these glycoprotein hormones in foetal pituitary extracts and serum by radioimmunoassay (Kaplan, Grumbach & Shepard, 1969). LER-869 was used for the human FSH standard and LER-960 for the human LH standard. The rabbit anti-human FSH serum which we utilized has a high degree of specificity for human FSH (Sizonenko *et al.* 1970); however, the anti-HCG serum does not distinguish HCG from LH and reacts equally with both hormones (Burr *et al.* 1970). In view of this limitation the foetal serum values are expressed as 'LH-HCG'. Recently, antisera to the β subunit of HCG have been generated which, according to Vaitukaitis, Braunstein & Ross (1972), discriminate between HCG and human LH. We plan to use this antiserum to the β subunit of HCG in future studies on foetal serum.

Fig. 9 shows the relationship between the content of FSH and LH in foetal pituitary glands and gestational age. Immunoreactive FSH and LH were detected as early as 68 days of gestation, the youngest foetus studied. The FSH content was 4.3 ng and the LH content, 28.5 ng. Six of seven foetuses with gestational ages of 68 to 80 days had measurable pituitary LH (1.8 to 105 ng) and three of four foetuses of the same gestational age had 1 to 4 ng of FSH in the pituitary. Although there was great variation in the amount of FSH and LH in the pituitary of foetuses of comparable gestational age, the mean content of LH rose from 82.0 ng at 10 to 14

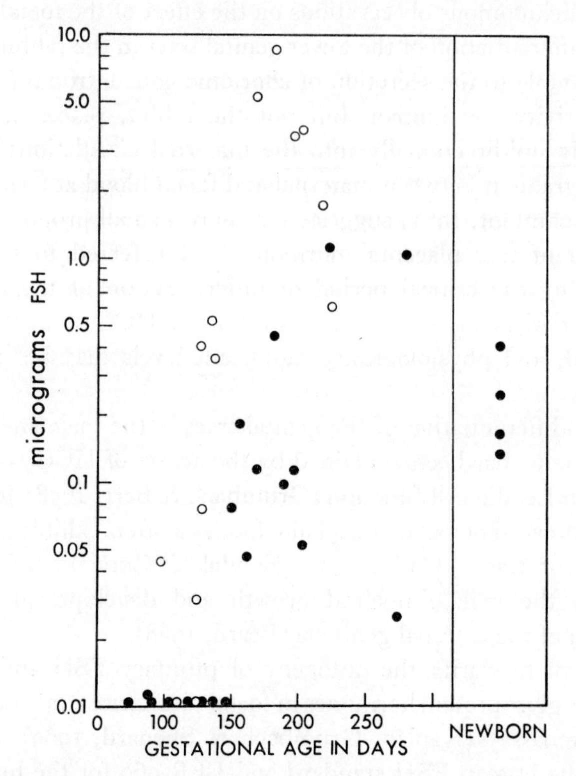

Fig. 9 (*a*) The total content of FSH in μg in the pituitary gland of human foetuses is plotted on log scale on ordinate against gestational age in days on abscissa. Open circles indicate female and closed circles male foetuses.

weeks' to 1906.1 ng at 25 to 29 weeks' gestation. The mean FSH content increased from 9.1 ng to 1532.0 ng during the same period. The mean value for LH in the pituitary between 150 to 200 days was about two-fold greater than that in newborn infants, whereas the mean FSH content was about five-fold greater in the pituitary gland of 150- to 200-day foetuses than in newborn infants. This difference is even more striking when the amount of FSH and LH per mg wet pituitary weight is compared (Fig. 10).

During the first four years after birth, the pituitary FSH content increases to 1.7 μg, about ten times the amount at term, and the amount of LH to 5.2 μg, about five-fold the mean value at term.

In contrast to the pattern of GH content of the pituitary during gestation, which steadily increases during foetal life and remains high postnatally, the FSH and LH content increases during gestation, reaches peak values at about 200 days, then decreases to term.

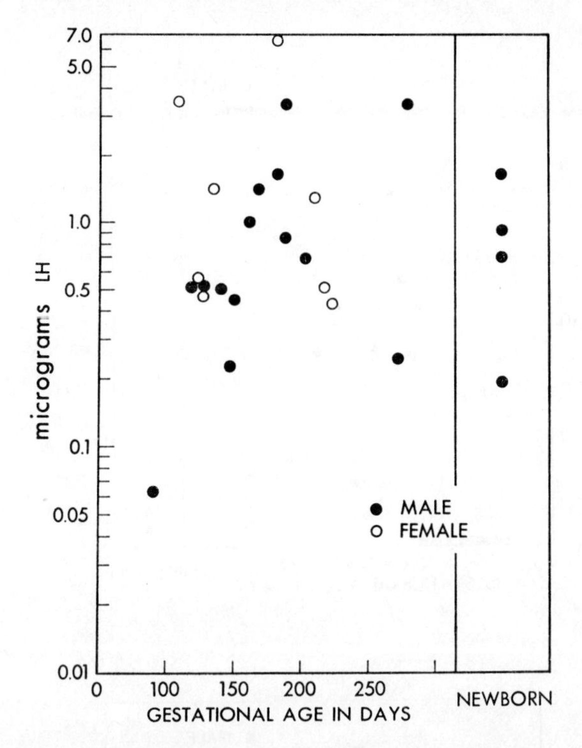

Fig. 9 (*b*) The content of LH in μg in the human foetal pituitary gland is plotted on ordinate against gestational age in days on abscissa. The content of LH in the pituitary gland of newborns is plotted on the right for comparison. Open circles indicate females and closed circles males.

Of interest are the sex differences in pituitary gonadotropin values. The pituitary of female foetuses between 24 and 29 weeks' gestation contains significantly more FSH and LH than male foetuses of the same gestational age. There is suggestive evidence of a higher LH/FSH ratio in male foetuses during the same period but the sample size is small. Levina (1968) has suggested that the amount of pituitary FSH (determined by bioassay) and of immunoreactive LH is greater in female than male foetuses after 4 to 5 months of gestation.

Studies in anencephaly

The pituitary gland from an anencephalic foetus born at term contained only about 2 per cent of the content of FSH and LH as the glands of newborn infants. Serum FSH was not detectable in any of the anencephalic infants studied whereas serum 'LH-HCG' was within the range found in normal newborn infants.

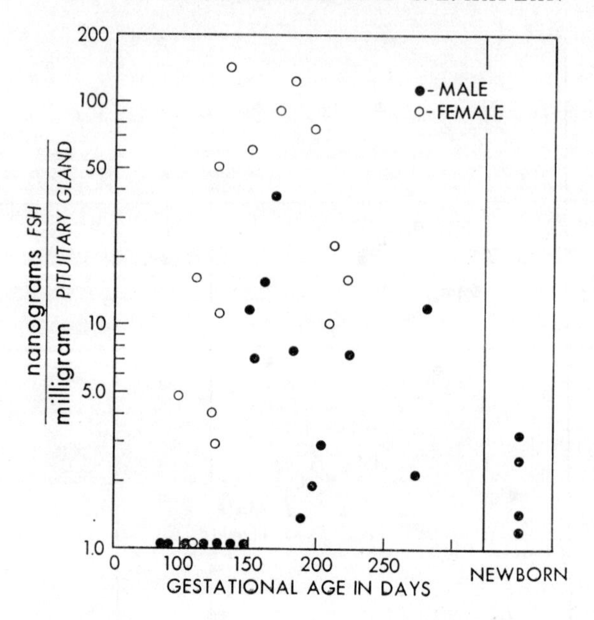

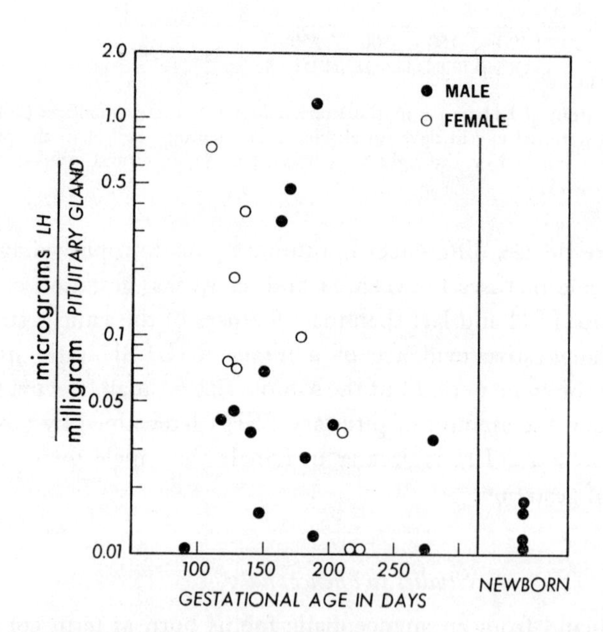

Fig. 10 (*a*) The concentration of FSH (ng/mg) in the human foetal pituitary gland is plotted on the ordinate on a log scale against gestational age in days on the abscissa. Open circles indicate female and closed circles male foetuses.

(*b*) The concentration of LH (μg/mg) in the human foetal pituitary gland is plotted on the ordinate on a log scale against gestational age in days on the abscissa. The concentration of LH in the pituitary gland of the newborn is indicated on the right panel.

These data demonstrate the capacity of the foetal pituitary gland to synthesize and store gonadotropin by day 68. Studies of the serum concentration of gonadotropin indicate that the foetal anterior hypophysis also secretes these hormones. Fig. 11 illustrates the pattern of change in concentration of serum FSH during gestation. At 84 days' gestation, the earliest foetal serum examined, the serum level of FSH was 11.0 ng/ml. The overall trend was one of increasing serum FSH concentration to about 150 days. Most of the values exceeded 15 ng/ml; the highest concentration was 50 ng/ml. These levels are comparable to those found in castrate patients and post-menopausal women. In contrast, the mean concentration of serum FSH in adult males is 3.1 ± 0.49 ng/ml (August, Grumbach & Kaplan, 1972). Beginning at about 150 days' gestation and extending to term, there is a striking fall in serum levels to term; at term, umbilical venous serum contains less than 1 ng/ml. Although these data suggest that serum values of FSH between 110 and 140 days' gestation are higher in females than in males, the larger number of specimens from female foetuses gave a skewed sex distribution.

The data relating the immunoreactive 'LH-HCG' levels in foetal sera to gestational age are shown in Fig. 11. The two earliest foetuses studied were an 83-day male and an 84-day female who had values of > 100 and 195 ng/ml, respectively (equivalent to > 1 to 2 IU/ml HCG). Again a wide range of values was found. The highest values clustered between 84 and 140 days, after which the 'LH-HCG' levels tended to fall to < 20 ng/ml.

The high serum values coincide approximately with the foetal period which exhibits the sharp increase in pituitary LH content. According to Albert & Berkson (1951), serum HCG levels in the mother, estimated by bioassay, peak at about 46 days after conception (60 days after the last menstrual period) and fall sharply by about the 65th day to a concentration which remains relatively stable throughout the remainder of pregnancy. Especially pertinent to our data is the pattern of serum immunoreactive HCG reported by Varma, Larraga & Selenkow (1971). The primary HCG peak (mean 163 IU/ml) occurred between 8 to 10 weeks of pregnancy (6 to 8 weeks post-conception), followed by a nadir (mean 12 IU/ml) at 17 to 19 weeks and a gradual rise in the mean concentration to a secondary peak (63 IU/ml) at 35 to 37 weeks of pregnancy. It is of interest that the nadir of the maternal serum HCG curve at 105 to 120 days occurs during an interval of continued high concentration of serum LH-HCG in the foetus. This leads us to surmise that foetal LH is secreted by the adenohypophysis and contributes significantly to the serum 'total' immunoreactive LH-HCG.

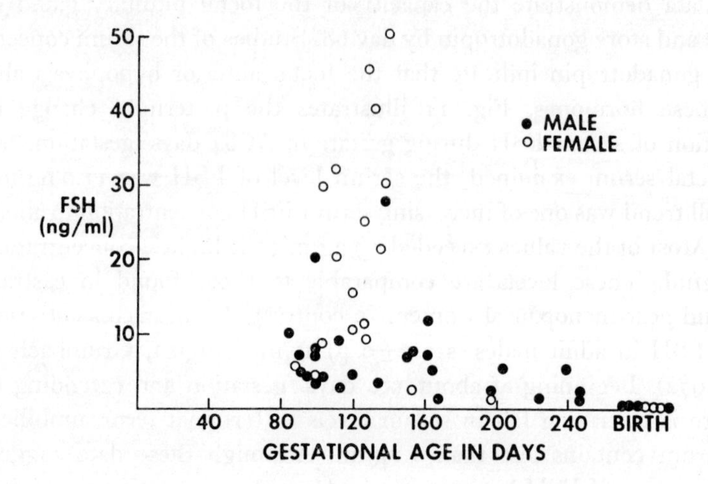

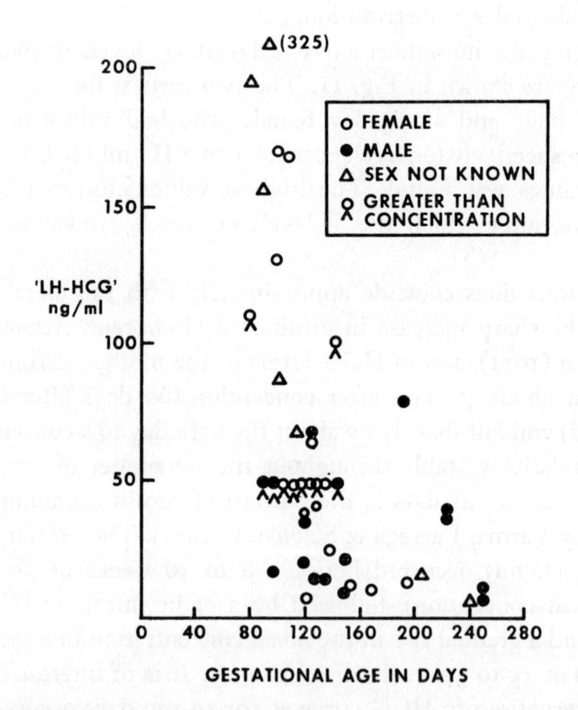

Fig. 11. (*a*) The concentration of serum FSH in ng/ml in human foetuses is plotted on the ordinate against gestational age in days on the abscissa. Serum FSH tends to be higher in females than males. Peak concentrations are noted between 100 to 140 days with a rapid decrease by 160 days.

(*b*) The concentration of serum 'LH-HCG' in ng/ml in the human foetus is plotted on the ordinate against gestational age in days.

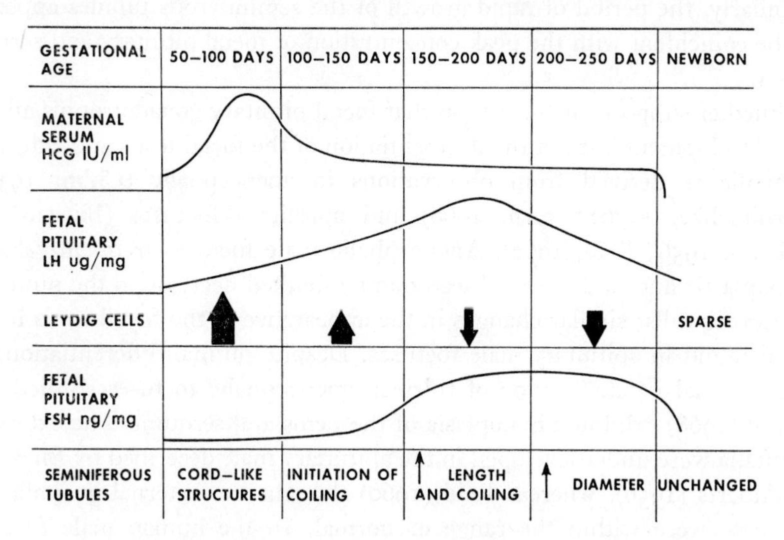

Fig. 12. LH and FSH content of the human foetal pituitary is correlated with developmental changes in the human foetal testis.

The data on foetal gonadotropins can be interpreted in relation to foetal sex development and to hypothalamic regulation of FSH and LH secretion in the foetus.

Foetal sex development

We have correlated the chronology of morphological changes in the foetal gonads and lower genital tract during gestation with the maternal concentration of serum HCG and the content of foetal pituitary FSH and LH (Fig. 12). The foetal testis differentiates from the indifferent gonad during the 6th week (Gillman, 1948). The gonad which is coded to become an ovary does not differentiate until later.

Leydig cells appear in the foetal testis after the 8th week and increase greatly in number between the 10th and 18th week; after the 18th to 20th week there is a rapid decline in size and number (Gillman, 1948; Van Wagenen & Simpson, 1965; Pelliniemi & Niemi, 1969). Very recently a sharp peak in the concentration of serum and testicular testosterone in male foetuses has been described at 12 weeks (70 to 80 mm C-R) followed by a gradual decline to low levels by 24 weeks (Murphy & Diez D'Aux, 1972; Reyes, Faiman & Winter, 1972). These changes correlate well with the high concentration of foetal serum LH-HCG and the relatively high content of pituitary LH. The regression of Leydig cells, and the fall in serum and testicular concentration of testosterone correlate with the decrease in serum LH-HCG levels after 140 days of gestation.

Similarly, the period of rapid growth of the seminiferous tubules appears to be coincident with the peak concentration of foetal pituitary and serum FSH.

Further support for the notion that foetal pituitary gonadotropins affect the development but not the differentiation of the foetal testis and external genitalia is derived from observations in anencephalic (Ch'in, 1938; Bernirschke, 1956; Bearn, 1968) and apituitary foetuses (Blizzard & Alberts, 1956; Reid, 1960). Anencephalic male foetuses frequently have hypoplastic and undescended testes and a marked decrease in the number of Leydig cells; similar changes in the appearance of the foetal testis have been found in apituitary male foetuses. Despite normal differentiation of the external genitalia, nine of thirteen anencephalic foetuses studied by Bearn (1968) exhibited hypoplasia of the penis and scrotum. The external genitalia were underdeveloped in the apituitary male described by Blizzard & Alberts (1956), whereas Reid (1960) thought the external genitalia in his case were within the range of normal. In the human male foetus, neither the pituitary nor the hypothalamus plays a critical role in the differentiation of the testes or the male genital tract.

After completion of male differentiation of the external genitalia, which occurs during the 4th month, the foetal hypothalamic-pituitary gonadotropin mechanism appears to influence testicular function and the growth of the external genitalia. By this period in gestation, it seems that chorionic gonadotropin (which has declined in the maternal circulation) is generally insufficient to maintain normal foetal Leydig cell function to ensure normal growth of the external genitalia in the absence of adequate secretion of FSH and LH by the foetal adenohypophysis. The strikingly low content of FSH and LH we found in the pituitary gland of the anencephalic foetus is consistent with this hypothesis. Hence, the foetal hypothalamic-pituitary complex, by virtue of its effect on the foetal testis, plays a role in the normal development of the male external genitalia.

During the 11th to 12th week, germ cells in the primordial ovary begin to enter meiotic prophase, which characterizes the transition of oogonia into oocytes; the formation of primordial follicles reaches a maximum during the 20th to 25th week of gestation (Van Wagenen & Simpson, 1965; Falin, 1969) and it is during this period that some of the primordial follicles take on the appearance of primary follicles. After the 6th to 7th month of gestation, further follicular growth occurs and an occasional Graffian follicle is present. Ross (1972) has recently reveiwed this subject. Fig. 13 correlates the morphological development of the ovary with the content of foetal pituitary FSH and LH. The stage of intense transformation of oogonia into primordial follicles coincides with the peak concentra-

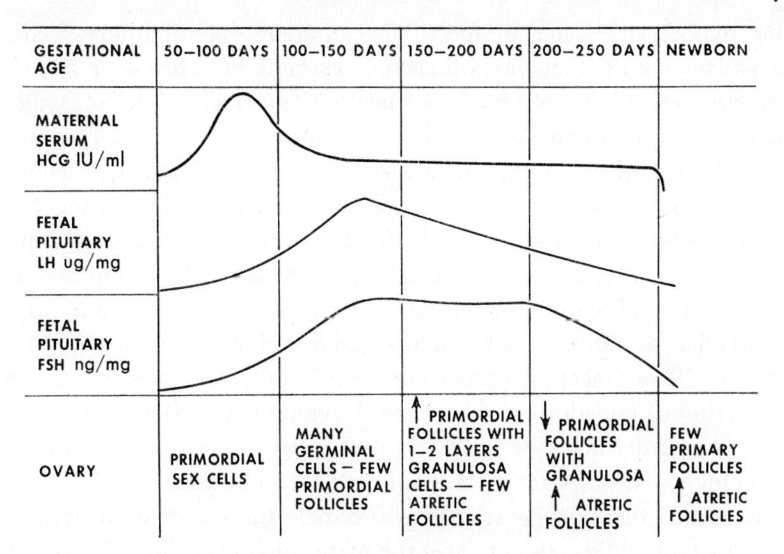

Fig. 13. LH and FSH content of the human foetal pituitary is correlated with developmental changes in the human foetal ovary.

tion of serum FSH in the foetus and a relatively high content of foetal pituitary FSH and LH.

According to Ch'in (1938) the female anencephalic foetus usually has abnormally small ovaries and hypoplastic primordial follicles. These findings support the contention that foetal pituitary gonadotropins exert an effect on the late stages of histogenesis of the foetal ovary.

HYPOTHALAMIC REGULATION OF FOETAL FSH AND LH

Two aspects of the control of the secretion of foetal gonadotropins will be considered: the development of the negative or inhibitory hypothalamic gonadal feedback mechanism and the apparent sex difference in the pituitary content of FSH and LH.

The pattern of change in the concentration of pituitary FSH and LH and of serum FSH and possibly LH during foetal life again suggests a sequence of increasing synthesis and secretion in which peak serum values reach castrate levels at between 14 and 24 weeks followed by a sharp decline which persists through the remainder of gestation (serum FSH is less than 1 ng/ml in umbilical cord serum at term). The high serum levels may be interpreted as a consequence of autonomous secretion of FSH and LH or of the relatively unrestrained secretion of the hypothalamic releasing factor, LRF. Indirect but not compelling evidence favours the latter hypothesis. Later in foetal development the inhibitory feedback mechanism matures and hypothalamic restraint becomes operative. The increasing sensitivity of the hypothalamic 'gonadostat' to sex steroids circulating in the foetus leads to suppression of LRF release. This results in greatly diminished synthesis and secretion of FSH and LH. Hence it would seem that the increased sensitivity of the hypothalamus to the inhibitory effects of sex steroids described in the pre-pubertal rats (Ramirez & McCann, 1963; Smith & Davidson, 1968) and which appears to be characteristic of the prepubertal state in man (Kulin, Grumbach & Kaplan, 1969; Kelch, Grumbach & Kaplan, 1972) first develops prenatally, in late foetal life. This concept departs from the generally held view that the onset of secretion of gonadotropins and the development of the hypothalamic-pituitary-gonadal negative feedback mechanism occurs during childhood.

The higher mean pituitary content and concentrations of FSH and LH in the female foetus suggest that in the male the secretion of androgens by the foetal Leydig cells enhances the maturation of the negative feedback mechanism in the male foetus. In contrast to the foetal testis, sex steroid output by the foetal ovary is negligible.

We wish to acknowledge the co-operation of the late Dr Dorothy Andersen and Drs Cynthia Barrett, William Blanc, Delbert Fisher, Arnold Klopper, Abraham Rudolph and Lotte Strauss in the collection of pituitaries and blood specimens from foetuses. Angeles Jardiolin and Inger Holst provided technical assistance. We thank the National Pituitary Agency for the Wilhelmi HGH and Dr Mary Root for the purified human and ovine insulin.

This work was supported in part by grants from the National Institute of Child Health and Human Development and the National Institute of Arthritis and Metabolic Diseases, NIH, USPHS.

REFERENCES

Adam, P. A., Teramo, K., Räiha, N., Gitlin, D. & Schwartz, R. (1969). *Diabetes* **18**, 409.

Albert, A. & Berkson, J. (1951). *J. clin. Endocrinol.* **11**, 805.

Alexander, D. P., Britton, H. G., Buttle, H. L. & Nixon, D. A. (1972). *Res. vet. Sci.* **13**, 188.

Amoss, M., Burgus, R., Blackwell, R., Vale, W., Fellows, R. & Guillemin, R. (1971). *Biochem. Biophys. Res. Commun.* **44**, 205.

Aubert, M. L., Grumbach, M. M. & Kaplan, S. L. In preparation 1972.

August, G. P., Grumbach, M. M. & Kaplan, S. L. (1972). *J. clin. Endocrinol.* **34**, 319.

Bassett, J. M. & Thorburn, G. D. (1971). *J. Endocrinol.* **50**, 59.

Bearn, J. G. (1968). *Acta Ped. Acad. Sc. Hung.* **9** (2), 1159.

Bergstrom, R. M. (1968). In *Ontogenesis of the Brain*, eds. L. Jilek & S. Trojan. Universitas Carolina, Prague, Czechoslovakia.

Bernirschke, K. (1956). *Obstet. Gynec.* **8**, 412.

Blackard, W. G. & Heidingsfelder, S. A. (1968). *J. clin. Invest.* **47**, 1407.

Blizzard, R. M. & Alberts, M. (1956). *J. Ped.* **48**, 782.

Boyd, A. E., iii & Lebovitz, H. E. (1970). *New Eng. J. Med.* **283**, 1425.
Burgus, R., Dunn, T. F., Desiderio, D. & Guillemin, R. (1969). *C. r. Acad. Sci. Paris* **269**, 1870.
Burr, I. M., Sizonenko, P. C., Kaplan, S. L. & Grumbach, M. M. (1970). *Ped. Res.* **4**, 25.
Chez, R. A., Mintz, D. H., Horger, E. D., iii & Hutchinson, D. L. (1970). *J. Clin. Invest.* **49**, 1517.
Chez, R. A., Mintz, D. H. & Hutchinson, D. L. (1971). *Metabolism* **20**, 805.
Ch'in, K. Y. (1938). *Chin. Med. J.* (Suppl.) **2**, 63.
Dreyfus-Briscoe, C. (1966). In *Human Development*, ed. F. Falkner. W. B. Saunders, Philadelphia.
Dubois, P. (1968). *C. r. Soc. Biol.* **162**, 689.
Ducharme, J. R. & Grumbach, M. M. (1961). *J. clin. Invest.* **40**, 243.
Ellis, S. T., Beck, J. S. & Currie, A. R. (1966). *J. Path. Bact.* **92**, 179.
Everett, J. W. (1969). *Ann. Rev. Physiol.* **31**, 383.
Falin, L. I. (1961). *Acta Anat.* **44**, 188.
Falin, L. I. (1969). *Acta Anat.* **72**, 192.
Frantz, A. & Kleinberg, D. (1972). *Rec. Prog. Horm. Res.* **28**. In press.
Fuxe, K. & Hokfelt, T. (1969). In *Frontiers of Neuroendocrinology*, p. 47, eds. W. F. Ganong & L. Martini. Oxford University Press.
Geiger, W., Kaiser, R. & Franchimont, P. (1971). *Acta Endoc.* **68**, 169.
Gillman, J. (1948). *Cont. Embryol. Carnegie Inst., Wash.* **32**, 81.
Gitlin, D. & Biasucci, A. (1969). *J. clin. Endocrinol.* **29**, 926.
Gitlin, D., Kumate, J. & Morales, C. (1965). *J. clin. Endocrinol.* **25**, 1599.
Glick, S. M. (1969). In *Frontiers of Neuroendocrinology*, p. 141, eds. W. F. Ganong & L. Martini. Oxford University Press.
Grillo, T. A. E. & Shima, K. (1966). *J. Endocr.* **36**, 151.
Grumbach, M. M. (1962). In *Immunoassay of Hormones*, Ciba Found. Coll. on Endocrinology **14**, 63.
Grumbach, M. M. & Barr, M. L. (1958). *Rec. Prog. Horm. Res.* **14**, 255.
Grunt, J. A. & Reynolds, D. W. (1970). *J. Ped.* **76**, 112.
Guillemin, R. (1971). *Adv. Met. Dis.* **5**, 1.
Guyda, H. J. & Friesen, H. G. (1971). *Biochem. Biophys. Res. Commun.* **42**, 1068.
Halasz, B., Kasaras, B. & Lengvari, I. (1972). In *Brain–Endocrine Interaction*, p. 27, eds. K. M. Knigge, D. E. Scott & A. Weindl. S. Karger, Basel.
Hamburgh, M. & Barrington, E. J. W. (1971). *Hormones in Development*. Appleton-Century-Crofts, New York.
Harris, G. W. (1970). In *Hypophysiotrophic Hormones of the Hypothalamus*, p. 1, ed. J. Meites. Williams & Wilkins Co., Baltimore.
Humphrey, T. (1964). *Prog. Brain Res.* **4**, 93.
Hwang, P., Guyda, H. & Friesen, H. (1971). *Proc. nat. Acad. Sci. USA* **68**, 1902.
Hwang, P., Guyda, H. & Friesen, H. (1972). *J. biol. Chem.* **247**, 1955.
Hyyppa, M. (1972). *Neuroendocrinology* **9**, 257.
Imura, H., Kato, Y., Ikeda, M., Morimoto, M., Yawata, M. & Fukase, M. (1971). *J. clin. Endocrinol.* **50**, 1069.
Jacobs, L. S., Mariz, I. K. & Daughaday, W. H. (1972). *J. clin. Endocrinol.* **34**, 484.
Jost, A. (1953). *Rec. Prog. Horm. Res.* **8**, 379.
Jost, A. (1966). In *The Pituitary Gland*, p. 299, eds. G. W. Harris & B. T. Donovan. University of California Press, Berkeley & Los Angeles.
Jost, A., Dupouy, J. P. & Geloso-Meyer, A. (1970). In *The Hypothalamus*, eds. L. Martini, M. Motta & F. Fraschini. Academic Press, New York.
Jost, A. & Picon, L. (1970). *Adv. Metab. Dis.* **4**, 123.

Kaplan, S. L. & Grumbach, M. M. (1962). *Am. J. Dis. Child.* **104**, 528.
Kaplan, S. L. & Grumbach, M. M. (1967). *Ped. Res.* **1**, 308.
Kaplan, S. L. & Grumbach, M. M. (1971). *Excerpta Medica* **236**, 382.
Kaplan, S. L., Grumbach, M. M. & Shepard, T. H. (1969). *Ped. Res.* **3**, 512.
Kaplan, S. L., Grumbach, M. M. & Shepard, T. H. (1972). *J. clin. Invest.* **51**.
Kelch, R. P., Grumbach, M. M. & Kaplan, S. L. (1972). In *Gonadotropins*, p. 524, eds. B. B. Saxena, C. G. Deling & H. M. Gandy. John Wiley & Sons, Inc., New York.
Kobayashi, T., Kobayashi, L., Yamamoto, K., Kaibara, M. & Ajika, K. (1968). *Endocrinol. Jap.* **15**, 337.
Kulin, H. E., Grumbach, M. M. & Kaplan, S. L. (1969). *Science* **166**, 1012.
Levina, S. E. (1968). *Gen. comp. End.* **11**, 151.
Lewis, U. J., Singh, R. N. P. & Seavey, B. K. (1971). *Biochem. Biophys. Res. Comm.* **44**, 1169.
McCann, S. M. & Porter, J. C. (1969). *Physiol. Rev.* **49**, 240.
Matsuo, H., Baba, Y., Nair, R. M. G., Arimura, A. & Schally, A. V. (1971). *Biochem. Biophys. Res. Comm.* **43**, 1334.
Matsuzaki, F., Irie, M. & Shizume, K. (1971). *J. clin. Endocrinol.* **33**, 908.
Mintz, D. H., Chez, R. A. & Horger, E. D., III (1969). *J. clin. Invest.* **48**, 176.
Mitnick, M. & Reichlin, S. (1971). *Science* **172**, 1241.
Moger, W. H. & Geschwind, J. J. (1971). *Experientia* **27**, 1479.
Morgan, C. R. & Lazarow, A. (1963). *Diabetes* **12**, 115.
Motta, M., Piva, F. & Martini, L. (1970). In *The Hypothalamus*, p. 463, eds. L. Martini, M. Motta & F. Fraschini. Academic Press, New York.
Murphy, B. E. P. & Diez D'Aux, R. (1972). In *Abstracts 4th Int. Congress of Endocrinol. Int. Congress Series* No. 256, 76.
Nair, R. M. G., Barrett, J. F., Bowers, C. Y. & Schally, A. V. (1970). *Biochem.* **9**, 1103.
Niemineva, K. (1949). *Acta Ped.* **38**, 366.
Pasteels, J. L. (1967). *Ann. Endocr.* **28**, 117.
Pasteels, J. L., Brauman, H. & Brauman, J. (1963). *C. r. Acad. Sci.* **256**, 2031.
Pearse, A. G. E. (1953). *J. Path. Bact.* **65**, 355.
Pelliniemi, L. J. & Niemi, M. (1969). *Z. Zellforsch.* **99**, 507.
Raiha, N. & Hjelt, L. (1957). *Acta Paediatrica* **46**, 610.
Ramirez, D. V. & McCann, S. M. (1963). *Endocrinology* **72**, 452.
Rastogi, G. K., Letarte, J. & Fraser, T. R. (1970). *Diabetologia* **6**, 445.
Reichlin, S. (1972). In *Abstracts 4th Int. Congress of Endocrinology. Int. Congress Series* No. 256.
Reid, J. D. (1960). *J. Ped.* **56**, 658.
Reyes, F., Faiman, C. & Winter, J. S. D. (1972). In *Abstracts 4th Int. Congress of Endocrinology. Int. Congress Series* No. 256.
Rinne, V. K., Kivalo, E. & Talanti, S. (1962). *Biol. Neonat.* **4**, 351.
Ross, G. T. (1972). In *Abstracts 4th Int. Congress of Endocrinology*. Int. Congress *Series* No· 256.
Schally, A. V., Arimura, A., Bowers, C. Y., Kastin, A. J., Sawano, S. & Redding, T. W. (1968). *Rec. Prog. Horm. Res.* **24**, 497.
Scharrer, B. (1970). In *The Neurosciences Second Study Program.* Schmitt, ed. Rockefeller University Press, New York.
Shaywitz, B. A., Finkelstein, J., Hellman, L. & Weitzman, E. D. (1971). *Pediatrics* **48**, 103.
Shepard, T. H. (1969). In *Endocrine and Genetic Dissorders of Childhood*, ed. L. I. Gardner. W. B. Saunders & Co., Philadelphia.

Sizonenko, P. C., Burr, I. M., Kaplan, S. L. & Grumbach, M. M. (1970). *Ped. Res.* **4**, 36.

Smith, E. R. & Davidson, J. M. (1968). *Endocrinology* **82**, 100.

Smith, G. C. (1970). *J. Anat.* **106**, 200.

Solomon, I. L., Grant, D. B., Burr, I. M., Kaplan, S. L. & Grumbach, M. M. (1969). *Proc. Soc. Exptl Biol. Med.* **132**, 505.

Steinke, J. & Driscoll, S. G. (1965). *Diabetes* **14**, 573.

Stokes, H. & Boda, J. M. (1968). *Endocrinology* **83**, 1362.

Thorell, J. I. (1970). *Acta Endocrinologica* **63**, 134.

Tyson, J. E., Hwang, P., Guyda, H. & Friesen, H. G. (1972). *Am. J. Obstet. Gynec.* **113**, 14.

Vaitukaitis, J. L., Braunstein, G. D. & Ross, G. T. (1972). *Am. J. Obstet. Gynec.* In press.

Van Wagenen, G. & Simpson, M. E. (1965). In *Embryology of the Ovary and Testis in Homo Sapiens and Macaca Mulatta*. Yale University Press, New Haven.

Van Wyk, J. J. & Grumbach, M. M. (1968). In *Textbook of Endocrinology*, ed. R. W. Williams. W. B. Saunders & Co., Philadelphia.

Varma, K., Larrage, L. & Selenkow, H. A. (1971). *Obstet. Gynec.* **37**, 10.

Vigneri, R. & D'Agata, R. (1971). *J. clin. Endocrinol.* **33**, 561.

Wellman, F. F., Volk, B. W. & Brancato, P. (1971). *Lab. Invest.* **25**, 97.

Yoshimura, F., Harumiya, K. & Kiyama, H. (1970). *Arch. Histol. J.* **31**, 333.

Youlton, R., Kaplan, S. L. & Grumbach, M. M. (1969). *Pediatrics* **43**, 989.

Zondek, L. H. & Zondek, T. (1965). *Biol. Neonat.* **8**, 329.

THYROID FUNCTION IN THE FOETAL LAMB

By G. D. THORBURN and P. S. HOPKINS

CSIRO, Division of Animal Physiology, Ian Clunies Ross Animal Research Laboratory, P.O. Box 239, Blacktown, NSW 2148, Australia

INTRODUCTION

Earlier studies on the function of the foetal thyroid have been limited to observations on laboratory animals (see review in Jost, 1971). In rats and rabbits it has been difficult to demonstrate any marked effect of thyroid deficiency in foetal life. Jost (1971) concluded from his studies in the rabbit that thyroid hormone is either not necessary for prenatal growth or that minimal amounts are adequate to meet the requirements. However, in both these species the foetal thyroid starts functioning relatively late in gestation; hence the young, which are born in an immature state, may leave the uterus before reaching the stage of development for which thyroid hormone is required. In many other species, including the sheep and man, the foetal thyroid starts functioning early in gestation and the young are born in a more mature state; in these species the foetus may be more susceptible to thyroid hormone deficiency.

In this review we are mainly concerned with recent studies on thyroid function in the foetal lamb. Initially, we establish that marked developmental defects following thyroidectomy of the foetus. Subsequently, we have been able to determine that the foetal lamb relies on its own pituitary–thyroid axis to supply the thyroid hormones that are essential for normal development. In this regard we have found that placental transfer of thyroid hormones is a relatively unimportant process. In the next section we discuss the factors responsible for the initiation and subsequent development of foetal thyroid function. Finally, we present data on the normal levels of thyroid hormones and thyrotrophin in the foetal plasma and the possible mechanisms controlling these hormone levels.

THE EFFECTS OF FOETAL THYROIDECTOMY

The initial suggestion that the foetal thyroid gland is important for normal development came from the study of Lascelles & Setchell (1959) who treated ewes with methylthiouracil during the last 3 to 4 months of gestation. In the lambs, they observed delayed osseous growth and maturity, abnormal plasma lipid patterns and clinical signs of goitre. Evaluation of their data is compromised to some extent by the fact that the ewes were also rendered hypothyroid by the treatment and that the degree of foetal hypothyroidism was difficult to assess.

In an endeavour to overcome these problems we have surgically excised the foetal thyroid gland (Hopkins & Thorburn, 1972), and thereby avoided interfering with maternal thyroid function. Most thyroidectomies were performed between 81 to 96 days' gestation. Eight of the ewes were allowed to lamb spontaneously; Caesarean sections were performed on the remaining 13 at 144 days' gestation. In only one foetus had there been obvious regrowth of thyroid tissue and in most of the foetuses thyroxine (T_4) was not detectable in cord plasma.

The following abnormalities have been attributed to a deficiency of thyroid hormone.

(i) *Length of gestation and brain development*

Prolonged gestation was a common feature in the ewes allowed to go to term; six of these animals lambed following a pregnancy period of 153–158 days (normal pregnancy = 150 days). These results are in agreement with earlier reports of prolonged pregnancy in ewes ingesting goitrogens (Sinclair & Andrews, 1954; Lascelles & Setchell, 1959). This phenomenon appears to stem from the effects of foetal rather than maternal hypothyroidism, since in the present study it was noted in euthyroid ewes.

In striking contrast with human athyroid neonates, no thyroidectomized lamb survived for longer than 30 h after birth. Indeed, all lambs that were thyroidectomized relatively early in gestation failed to establish spontaneous respiration and died soon after birth. Those lambs thyroidectomized later in gestation survived for periods up to 30 h. All attempts to induce these animals to stand and suck were unsuccessful. The rectal temperature of these lambs fell to approximately 33 °C within an hour of birth and remained subnormal until death.

These results support the concept that T_4 deficiency is associated with retarded development of the nervous system (Eayrs, 1960, 1964, 1971). Unfortunately, we were not able to make any other assessments of brain development in these lambs, although we might anticipate quite marked changes. In the rat, the period of ontogeny from birth to about 21 days of age is critical for the functional development of the CNS and it is during this period that thyroid hormones have been shown to exert their most striking effects on the morphological development of the brain (Eayrs, 1960; Hamburgh, Lynn & Weiss, 1964; Geel & Timiras, 1971). It will be important to determine a corresponding period for the lamb; this appears likely to occur during foetal life. Further discussion on the important role of thyroid hormone in brain development is outside the scope of this review but the reader is referred to a number of excellent papers on this subject in a recent book edited by Hamburgh & Barrington (1971).

(ii) *Growth*

In contrast to the results with rats and rabbits, thyroidectomy of the foetal lamb retarded growth (Hopkins & Thorburn, 1972). An assessment of the degree of dwarfing can be obtained by comparing the mean body weights of the athyroid foetuses (2.35 ± 0.09 kg, $n = 12$) and the controls (3.5 ± 0.07 kg, $n = 8$). However, the dwarfing of the thyroidectomized foetuses was not uniform; they had shorter limbs than the controls whereas the trunk measurements were relatively less affected by the treatment.

By comparison, the body weights of control and hypothyroid rats were similar until 10 to 12 days of post-natal age, after which time growth rate declined in hypothyroid animals (Geel & Timiras, 1971). Daily injections of T_4 from day 6 onwards prevented any reduction in body growth.

(iii) *Osseous growth and maturation*

It is known that thyroid hormone is required for the normal foetal development of the skeleton in man; congenital aplasia of the thyroid leads to a delayed ossification of those bones which begin to ossify shortly before birth (Wilkins, 1957). In our experiments (Hopkins & Thorburn, 1972) thyroidectomy of the foetal lamb clearly affected both maturation and growth of the appendicular skeleton. Similar findings were reported by Lascelles & Setchell (1959) in goitrogen-induced foetal hypothyroidism. Retarded longitudinal growth of the long bones was most pronounced in the metacarpals and metatarsals, though all bones were shorter than in the control lambs. Lack of ossification was apparent in the epiphysial centres of most of the long bones and also in many of the short bones. The degree of osseous maturity seen in the athyroid foetuses at Caesarean section (144 days) depended on the gestational age of the foetus at the time of surgery. For example, the appendicular skeleton of the foetus thyroidectomized at day 81 of gestation was very immature (Plate 1) while the osseous maturity of an animal thyroidectomized at day 92 of gestation was more advanced. Regional lamellation of bone tissue was noted in many of the athyroid foetuses, the proximal regions of the long bones, the distal femur and the bodies of many of the short bones (e.g. tarsals) were most commonly affected. Lamellated bone tissue in lambs born to ewes fed thiouracil during pregnancy was previously attributed to a set-back to the ewe caused by the administration of the drug (Lascelles & Setchell, 1959). However, the presence of such a condition in one thyroidectomized foetus and not in a sham-operated twin, suggests that the foetal hypothyroidism *per se* causes this condition.

(iv) *Development of wool follicles*

Perhaps the most striking effect of thyroidectomy in the foetal lamb was the interference with the maturation of the wool follicles (Hopkins & Thorburn, 1972). Two types of follicle develop in the skin at different gestational ages. The first-formed, primary follicles, are initiated as downgrowths from the epidermis at about day 50 of gestation. The secondary follicles commence downgrowth at about day 86 and undergo a process of branching between days 100–150 of gestation (Hardy & Lyne, 1956). The effects of foetal thyroidectomy at day 81 of gestation (i.e. before initiation of secondary follicles) on fibre growth at 144 days of gestation are well illustrated by the foetus in Plate 2. Sections of skin contained fibres, but the tips of the fibres had not reached the skin surface. Formation of secondary follicles had commenced normally but, in contrast to the primary follicles, none had developed to the stage of fibre formation. The ratio of the number of 'potential' secondary follicles (i.e. containing no fibre) to the number of primary follicles was however normal. Further observations (Chapman, Hopkins & Thorburn, unpublished observations) showed that maturation of primary follicles was retarded less by thyroidectomy later in gestation, but formation of fibres by secondary follicles was inhibited by thyroidectomy at all ages up to 96 days of gestation. The development of the sebaceous and sweat glands was retarded by thyroidectomy, the retardation tending to be greater the earlier thyroidectomy was performed. Keratinization of the epidermis was also dependent on the age at which thyroidectomy was performed, being inhibited by thyroidectomy prior to 90 days of gestation.

Our studies have shown therefore that thyroid hormones are not essential for the initial phase of follicle development or for the attainment of the genetically-determined ratio of secondary to primary follicles. However, the critical phase in the formation of fibres by the secondary follicles, involving the differentiation of cells in the follicle bulbs into the three layers of the inner root sheath, the fibre cuticle, the fibre cortex and the medulla (if present), is completely dependent on thyroid hormones. Likewise the conversion of the periderm to a keratinizing epidermis is also dependent on the presence of such hormones. These processes may be analogous to the metamorphosis of amphibians and of insects.

Table 1. *Fatty acid composition of foetal perirenal fat taken from control and athyroid foetuses at 144 days gestation*

Animal no.	Treatment	Fatty acid* 14:0	16:0	16:1	18:0	18:1	18:2	18:3
1352	Thyriodectomized	2.1	31.9	4.3	15.5	44.6	Tr.	
1351	Thyroidectomized	1.6	25.2	3.7	15.3	51.6	Tr.	
1356	Thyroidectomized	2.1	24.9	2.8	19.6	51.2	Tr.	
1345	Thyroidectomized	1.7	26.4	3.3	21.5	46.1	Tr.	
	Mean	1.9	27.1	3.3	18.0	48.4	Tr.	
	SE	0.13	1.63	0.26	1.54	1.78		
1359	Control	0.9	18.1	3.3	15.5	61.3	Tr.	
P218	Control	1.1	19.3	2.8	13.8	62.8	Tr.	
1823	Control	1.0	19.7	2.5	13.8	59.6	Tr.	
1783	Control	0.9	20.0	2.3	12.8	64.0	Tr.	
1758	Control	1.5	20.8	2.6	18.2	56.2	Tr.	
	Mean	1.1	19.6	2.7	14.8	60.8	Tr.	
	SE	0.11	0.44	0.17	0.95	1.36		

Tr. = trace.
* Values are expressed as % by weight of the total fatty acids.

(v) *Foetal lipids*

Thyroidectomy of the rabbit foetus increased the total lipid content of the body (Jost & Picon, 1958) and the plasma cholesterol levels of the foetus (Bearn, 1963). Plasma cholesterol levels were also elevated after thyroidectomy in the foetal lamb (Hopkins & Thorburn, 1972), although not to the same extent as has been reported in new-born lambs made hypothyroid by the administration of methylthiouracil to pregnant ewes (Lascelles & Setchell, 1959). In addition, preliminary studies have shown that there was an altered distribution of fatty acids in the adipose tissue of the thyroidectomized foetuses (Table 1; Cook, Scott, Hopkins & Thorburn, unpublished observations).

THE AUTONOMY OF FOETAL THYROID FUNCTION

Clearly thyroid hormone is needed for normal foetal development. It is important to establish whether the conceptus obtains this hormone from the maternal circulation. Several workers have shown that thyroidectomized animals can conceive and bear normal young (see review, Osorio & Myant, 1960). Spielman, Petersen, Fitch & Pomeroy (1945) showed that thyroidectomized cows inseminated with semen from thyroidectomized bulls gave birth to normal calves. There are also several reports of untreated myxoedematous women who became pregnant and bore normal children (Hodges, Hamilton & Keettal, 1952). In the sheep Falconer (1965) reported a high neonatal mortality rate of lambs born to athyroid ewes, although the likely reasons for this finding were not discussed. We have found that maternal thyroidectomy does not interfere with the normal growth and maturation of the ovine foetus up to 100 days' gestation (Hopkins, 1972).

We conclude that maternal thyroid hormone is not essential for normal foetal development though further studies of the post-natal viability of lambs born to athyroid ewes are indicated. Since the foetal thyroid gland of the sheep, cow and man does not commence to function until about the end of the first third of pregnancy (see below), it would seem that thyroid hormone is not essential for early foetal development. Later in gestation the foetal thyroid gland is apparently able to provide adequate amounts of hormone to meet the tissue requirements.

PLACENTAL TRANSFER OF THYROID HORMONES AND THYROTROPHIN

Although, under normal circumstances, the foetus apparently does not require thyroid hormone from the maternal circulation, it is still important to establish to what extent placental transfer may occur. Even limited transfer could be of value to a foetus with deficient thyroid function and a transfer in the opposite direction might ameliorate symptoms of thyroid deficiency in a hypothyroid mother. In this section, the recent work on the permeability of the ovine placenta to thyroid hormone and thyrotrophin (TSH) during the last third of pregnancy is discussed.

This question has been approached in two ways: firstly, by the use of hormones labelled with radioactive isotopes, and secondly, by the measurement of hormone levels in thyroidectomized foetuses, following administration of hormone to the ewe. Unfortunately most of the radioisotope experiments (Burrow, Anderson & Quilligan, 1969; Burrow & Anderson, 1970; Comline, Nathanielsz & Silver, 1970; Robin, Fang, Selenkow, Piasecki, Rauschecker & Jackson, 1970) have been published only in abstract form which makes critical analysis difficult. The general approach has been to implant catheters into the foetus and ewe, and then to inject tracer doses of $^{131}I\text{-}T_4$ and $^{125}I\text{-}T_4$ into the maternal and foetal circulations. Serial blood samples were drawn from each for periods up to 96 hours. In some studies, potassium iodide or perchlorate was administered to the ewe to limit the trapping of labelled iodide by the foetal and maternal thyroid glands. Burrow & Anderson (1970) and Robin *et al.* (1970)

concluded that a small bi-directional transfer of T_4 does take place in the sheep, but the direction is predominantly foetal–maternal. However, Comline *et al.* (1970) concluded that whereas passage of T_4 from mother to foetus can occur in small amounts, only negligible amounts pass in the reverse direction. In contrast to the above results Dussault, Hobel & Fisher (1971) failed to observe any placental transfer of labelled T_4 in either direction. In all these studies the plasma concentrations of T_4 in the maternal and foetal circulations were presumably normal, and since the concentration in the foetus is higher than the mother (see below), the concentration gradient would favour net foetal–maternal transfer if simple diffusion was the operative mechanism.

In an alternative approach (Hopkins & Thorburn, 1971), two foetuses (110 days' gestation) were thyroidectomized and catheters were inserted into the foetus and ewe. Despite a maternal T_4 concentration of 4–6 μg/100 ml, this hormone was not detected in the foetal plasma at any time during the remaining six weeks of gestation. The placental gradient was increased on a number of occasions by the administration of T_4 to the ewe without producing any measurable levels of T_4 in the foetal plasma (Fig. 1). Similarly, in an intact foetus (138 days' gestation) there was no measurable transfer of T_4 from ewe to foetus despite extremely high maternal T_4 levels.

Published data on the transfer of triiodothyronine (T_3) across the ovine placenta is limited at the present time. Dussault, Hobel, DiStefano, Erenberg & Fisher (1972) found that the ovine placenta is slightly more permeable to T_3 than T_4, a result consistent with their earlier observations in man (Dussault, Row, Lickrish & Volpe, 1969).

In our own laboratory, we have administered labelled T_3 to a ewe bearing a thyroidectomized foetus (Hopkins, 1972). The amounts of radioactivity in plasma samples collected from both the ewe and foetus during the following 24 h, before and after dialysis, are shown in Table 2. Whereas nearly all the radioactivity in the maternal plasma was of a non-dialysable form (i.e. T_3), the labelled material in the foetal plasma was virtually all dialysable. The amount of dialysable radioactivity detected in the foetus increased steadily during the period 2–24 h after the administration of ^{125}I-T_3 to the ewe. We concluded, therefore, that very little, if any, T_3 of maternal origin gained access to the foetal circulation.

There is general agreement that TSH does not cross the placenta in animals and man (for references see Dussault *et al.* 1969) and our own observations have confirmed this conclusion. We have found no measurable levels of TSH in the plasma of foetuses subjected to complete hypophysectomy (Fig. 2, see below), which indicates maternal TSH does not cross the placenta.

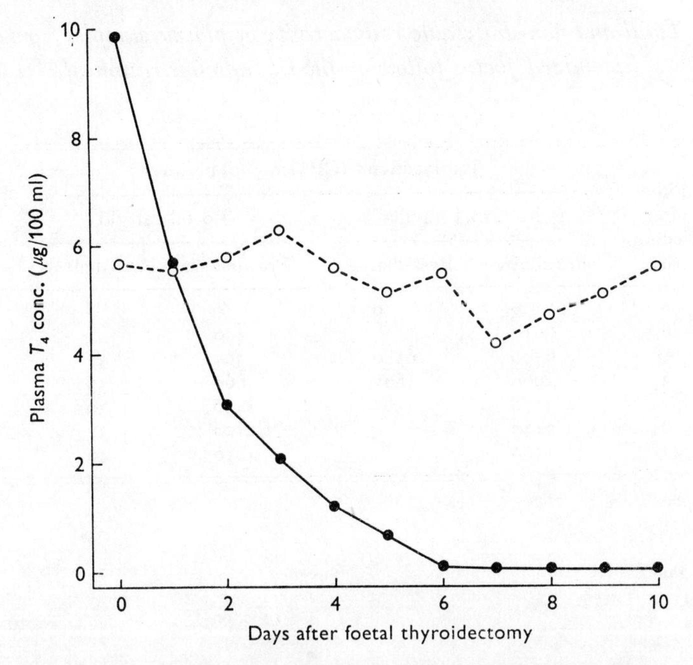

Fig. 1 (*a*). The effect of foetal thyroidectomy on the concentrations of T_4 in foetal and maternal plasma. Mean of 2. Foetal carotid ●——●. Maternal jugular ○---○.

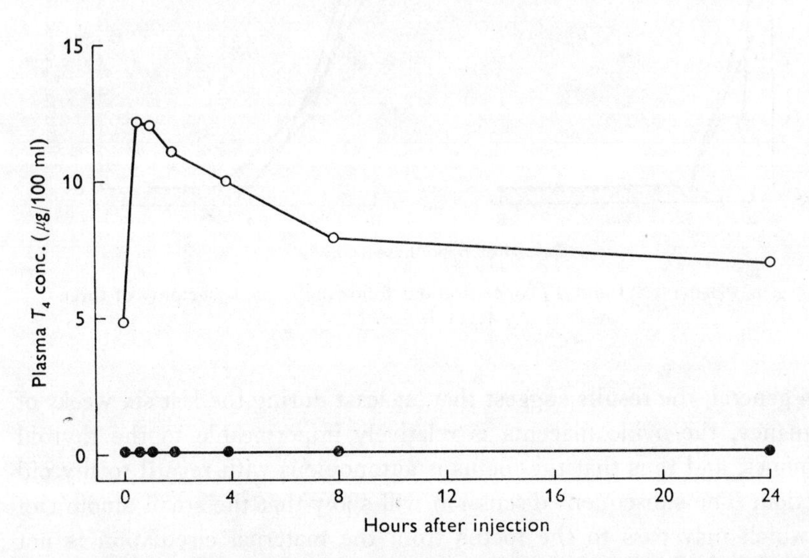

Fig. 1 (*b*). Plasma T_4 concentrations of an athyroid foetal sheep and its dam during the 24 h after i.v. T_4 administration (500 μg) to the ewe. Foetal carotid ●——●. Maternal jugular ○——○.

Table 2. *Total and non-dialysable radioactivity of plasma samples from a ewe and a thyroidectomized foetus following the i.v. administration of ^{125}I-T_3 to the ewe*

Time after injection (h)	Radioactivity (CPM/0.5 ml plasma)			
	Maternal jugular		Foetal carotid	
	Pre-dialysis	Post-dialysis	Pre-dialysis	Post-dialysis
0	0	0	0	0
0.5	9010		450	
1	6790	6580	560	31
2	4960	4805	860	33
4	3770	3565	1430	30
6	3140		2120	37
24	1360		4010	40

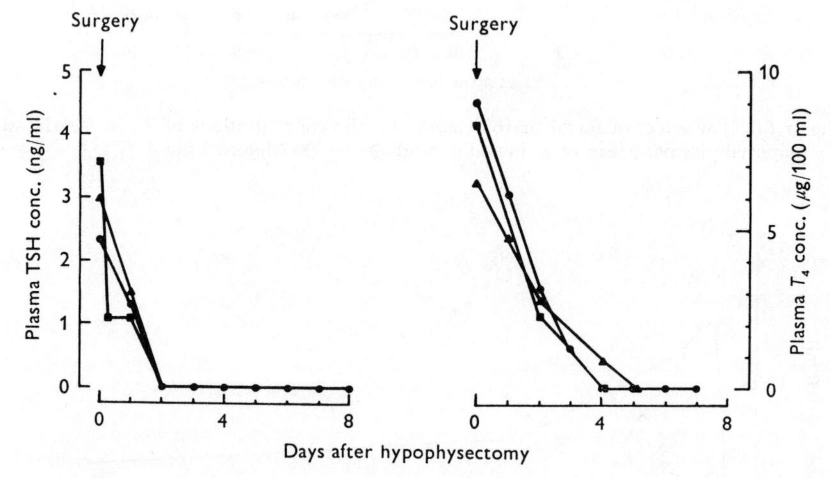

Fig. 2. Plasma TSH and T_4 concentrations following hypophysectomy of three foetal lambs.

In general, the results suggest that, at least during the last six weeks of pregnancy, the ovine placenta is relatively impermeable to the thyroid hormones, and thus that the foetus is autonomous with regard to thyroid function. The subsequent discussion will show that the small amount of T_3 which may pass to the foetus from the maternal circulation is not sufficient to prevent either the large increase in TSH or the major developmental defects which we have observed after foetal thyroidectomy.

THE DEVELOPMENT OF FOETAL THYROID FUNCTION

It is apparent that the foetal lamb does not depend on maternal thyroid hormones and that the foetal thyroid can provide adequate amounts of hormone for normal development. In this section, we examine the stage of development at which the foetal thyroid commences to function (i.e. secrete T_4) and the stimulus for the initiation of thyroid function.

Anatomical development

The first signs of organization of the thyroid cells into follicles were apparent by 50 days' gestation (Barnes, Warner, Marks & Bustad, 1957; Hopkins, Schneider & Thorburn, manuscript in preparation) and during the next ten days numerous follicles, many containing colloid were observed (Plate 3*a* and *b*). During the next eight weeks of foetal life, the number, size and colloid content of the follicles increased progressively so that by about 115 days' gestation, the foetal thyroid exhibited most of the histological features of the adult gland (Plate 3*c* and *d*). In the rat, by comparison, development of the foetal thyroid gland is not only more rapid, but occurs much later in gestation with comparable changes occurring between days 17 and 20 of gestation (Noumura, 1959; Feldman, Vasequez & Kurtz, 1961).

Limited information is available on the development of the pituitary of the foetal lamb. In our studies (Hopkins, Schneider & Thorburn, manuscript in preparation), an undifferentiated pituitary was present in the youngest foetus examined (45 days' gestation). By day 55 of gestation the anterior and posterior parts of the gland were almost in apposition, and the pars intermedia and hypophysial cleft were clearly apparent in the intervening space. Cellular orientation was also pronounced with many cells of the anterior pituitary forming the cord-like appearance characteristic of the tissue.

Functional development

The first signs of thyroid function in the foetal lamb have also been observed at about day 50 of gestation. At this time, the foetal thyroid starts to accumulate iodine (Barnes *et al.* 1957) and detectable levels of T_4 (0.2 μg/100 ml) and TSH (3.6 ng/ml) are present in the foetal plasma (Table 3). During the next 50–60 days, there is a steady increase in the levels of T_4 and free thyroxine (FT_4) in the plasma, with concentrations reaching 8.6 μg/100 ml and 3.9 ng/100 ml respectively by 115 days' gestation (Table 3). The concentration of TSH was relatively unchanged during this time (Hopkins, Wallace & Thorburn, 1973).

Table 3. *Mean plasma TSH, T_4 and FT_4 concentrations of ewes and their foetuses. Figures in parenthesis indicate number of animals*

Stage of gestation (days)	Maternal jugular			Foetal carotid		
	TSH ng/ml	T_4 μg/100 ml	FT_4 ng/100 ml	TSH ng/ml	T_4 μg/100 ml	FT_4 ng/100 ml
50	2.8 (2)			3.5 (2)	0.2 (2)	
55				3.5 (2)	0.3 (2)	
60		7.6 (2)	1.6 (2)	3.8 (2)	1.0 (2)	1.1 (2)
65		7.3 (2)	1.9 (2)		1.9 (2)	1.1 (2)
75		7.6 (2)	1.7 (2)		4.9 (2)	1.8 (2)
95		7.1 (2)	2.0 (2)		8.6 (2)	3.4 (2)
115	2.2 (4)	8.1 (6)	1.8 (2)	3.0 (4)	8.6 (6)	3.9 (2)
120	2.1 (4)	7.8 (6)	2.1 (2)	3.0 (4)	9.0 (6)	4.1 (2)
125	2.4 (4)	7.5 (6)	1.9 (2)	2.7 (4)	9.4 (6)	4.1 (2)
130	2.6 (4)	7.7 (6)	2.0 (2)	2.9 (4)	9.7 (6)	4.0 (2)
135	2.5 (4)	6.5 (6)	1.5 (2)	2.6 (4)	11.2 (2)	4.2 (2)
140	2.5 (4)	6.5 (6)	1.6 (2)	2.9 (4)	10.2 (6)	4.4 (2)
145	2.6 (4)	6.6 (6)	1.5 (2)	2.8 (4)	9.6 (6)	2.6 (2)
150	3.0 (4)	7.0 (6)	1.7 (2)	2.7 (4)	6.1 (6)	2.0 (2)

The progressive increase in the concentrations of FT_4 and T_4 in the foetal plasma between 50 and 110 days' gestation probably reflects an increase in the rate of T_4 secretion by the gland; these changes in hormone concentration correlate well with the progressive histological development of the gland during this time. Observations in the human foetuses (Greenberg, Czernichow, Reba, Tyson & Blizzard, 1970) suggest that the plasma thyroxine-binding globulin concentration may also increase during this period. However, since there is a simultaneous increase in the FT_4 concentration, it would seem that the rate of T_4 secretion was increasing more rapidly than the concentration of binding proteins.

Despite the relatively low levels of T_4 in the foetal circulation at 60–70 days' gestation, a relatively large proportion of this hormone is in the free form (Table 3); the FT_4 level at 60 days (1.1 ng/100 ml) was similar to the concentration in maternal plasma (1.8 ng/100 ml). Significant levels of T_4 in the foetal circulation from the end of the first third of gestation suggest that thyroid hormone may start to exert its action on target tissues at that time. It is not surprising, therefore, that the ossification centres of the limbs (Lascelles, 1959; Hopkins, Schneider & Thorburn, manuscript in preparation) and the development of the primary wool follicles (Hardy & Lyne, 1956) of the foetal lamb are first seen about this time.

From the available evidence, it is difficult to determine whether foetal pituitary stimulation is essential for the initiation of either the histological development (formation of follicles and colloid) or functional capacity of the thyroid gland. This question, however, has been studied in some detail in the rat. Geloso (1967, 1971) found that foetal rats hypophysectomized on day 16.5 of gestation (iodine accumulation by thyroid gland commences at day 17.5) exhibited only a slight capacity to trap iodide by the end of gestation. The author concluded that some development of the gland may occur in the absence of the foetal pituitary, but that the functional capacity of the thyroid remained well below normal. However, it is possible that thc functional changes observed by Geloso (1967) and Noumura (1959) were caused by small amounts of either maternal TSH or a placental thyrotrophin. The extent to which the findings may apply to the foetal lamb cannot be assessed, but they do suggest that only minimal histological and functional development of the thyroid gland can occur in the absence of thyroid stimulating hormone (TSH).

THE CONTROL OF FOETAL THYROID FUNCTION

Plasma levels and secretion rates of T_4 and T_3

We have described the increase in T_4 which occurs between 50 and 100 days' gestation. The use of chronically cannulated foetuses has permitted us to examine the day-to-day regulation of T_4 levels in the ovine foetus *in utero* (Hopkins, Schneider & Thorburn, manuscript in preparation). Between 100 and 135 days of gestation, the mean T_4 concentration in foetal plasma increased gradually from a level of 8.6 μg/100 ml to 11.2 μg/100 ml (Table 3). During this time, the T_4 levels in the foetal plasma were significantly higher than the corresponding maternal values. During the last 10 days of pregnancy, the foetal levels decreased to reach maternal levels at term. Day-to-day variation in the plasma T_4 concentration of an individual foetus was relatively small, whereas the variation between animals was marked. The foetal and maternal T_4 values reported by Dussault *et al.* (1971) were generally lower than our values. Their results also showed a marked variation between individual foetuses in the T_4 values (range 4.5–10.8 μg/100 ml). In our animals, this variation between animals did not appear to be related to variations in plasma TSH concentrations.

The concentration of FT_4 in the foetal plasma was also significantly higher than the maternal concentration during the period from 115 to 140 days' gestation. The mean foetal level was 4.2 ng/100 ml compared with the mean maternal level of 1.8 ng/100 ml (Table 3). The foetal FT_4 levels also decreased to maternal levels during the last 10 days of pregnancy.

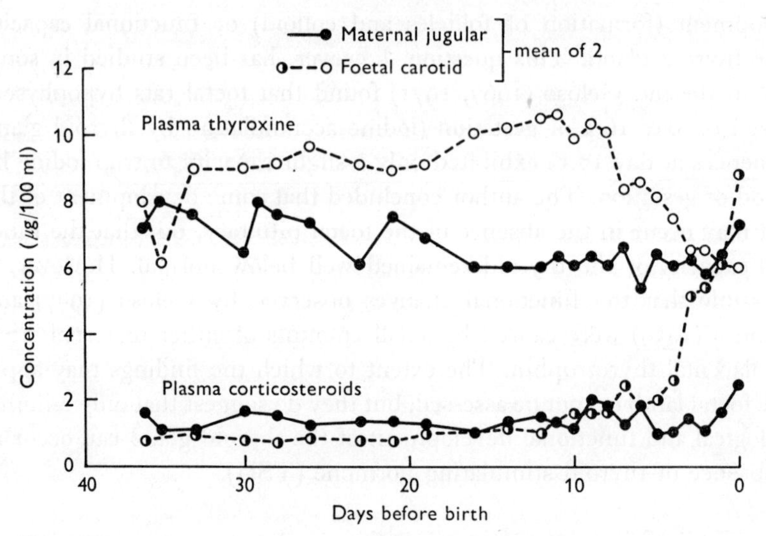

Fig. 3. Mean plasma thyroxine and corticosteroid concentrations of 2 ewes and their foetuses during the last 37 days of pregnancy.

These relatively high concentrations of T_4 and FT_4 in foetal plasma are consistent with the high secretion rates of T_4 which have been found in foetal lambs (Dussault *et al.* 1971). Using an isotope dilution technique, they reported a mean T_4 secretion rate for foetal lambs (105–145 days' gestation) of 104 μg/day (41 μg/kg/day) compared with a corresponding maternal value of 333 μg/day (5.5 μg/kg/day). The mean maternal values for the T_4 half time ($t_{\frac{1}{2}}$) (26 hours), serum T_4 concentrations (5.5 μg/100 ml), T_4 volume of distribution (9.3 litres or 15 % of body weight) that they reported are in general agreement with earlier observations in pregnant sheep (Annison & Lewis, 1959; Freinkel & Lewis, 1957). Corresponding mean foetal values were 17 hours, 7.4 μg/100 ml, and 1.36 litres. Their results suggest that the T_4 secretion rate (per unit body weight) in the foetal lamb during the last third of gestation greatly exceeds the maternal rate (8 times).

We have studied the disappearance of endogenous T_4 in foetal lambs following thyroidectomy; in this situation T_4 disappeared from the plasma with a $t_{\frac{1}{2}}$ of approximately 40 h. These values are considerably longer than those measured by Dussault and colleagues using labelled hormones (mean 17 h). The animals used in each study were of similar gestational age, and the difference is not easy to reconcile.

Recently, similar studies on the second thyroid hormone, T_3, have been reported (Dussault, Hobel, DiStefano & Fisher, 1971; Dussault, Hobel, DiStefano, Erenberg & Fisher, 1972). Initially they reported serum T_3

concentrations of 88 and 66 ng/100 ml in the mother and foetus respectively. In their subsequent paper, using a radioimmunoassay for T_3, they were unable to detect any T_3 in 4 out of the 6 foetuses but failed to comment on this discrepancy. No firm conclusions can be drawn from these results at the present time.

It should be noted that the volumes of distribution, particularly for T_3 (mean 5.1 l) reported by Dussault *et al.* (1971, 1972) are extremely large considering the weight of the foetuses (mean 2.0 kg). In cases, such as the foetus, where there is an extremely rapid peripheral metabolism of thyroid hormones, there are numerous difficulties in assessing accurately the distribution and turnover of the labelled hormone (cf. Ingbar & Freinkel, 1955; Surks & Oppenheimer, 1969; Woeber, Sobel, Ingbar & Sterling, 1970). Indeed, under such circumstances, one might question the validity of analysing the data according to a single compartmental model. Thus, the secretion rates calculated from these kinetic data, particularly in the case of T_3, should be accepted with some reservation until validated by some independent means.

Plasma concentration of thyrotrophin in the foetal lamb

During the last 20 days of pregnancy foetal plasma TSH concentrations (mean 3.2 ng/ml SE $\pm$ 0.2 ng/ml) were slightly higher than the corresponding maternal values (2.6 ng/ml SE $\pm$ 0.1 ng/ml) (Hopkins, Wallace & Thorburn, 1973; Table 3). The mean plasma TSH concentrations of 5 ewes and their foetuses during the last 15–35 days of foetal life are shown in Fig. 4. During the last 10 days of pregnancy a decrease in the plasma TSH concentration was apparent in only 2 of the 5 foetuses.

Hormonal changes following foetal thyroidectomy

Following thyroidectomy, T_4 virtually disappeared from the circulation of adult sheep and foetal lambs in 4–6 days, but no appreciable change in plasma TSH concentration was seen during the first 5 days after surgery (Fig. 5; Hopkins, Wallace & Thorburn, 1973). After 5–7 days, the TSH concentrations started to increase, and in the adult animals this increase continued for another 10–15 days. In the foetus, however, the TSH concentrations continued to increase. In one foetus, thyroidectomized 56 days previously, the TSH concentration was in excess of 200 ng/ml.

The similarity in the pattern of hormonal changes following thyroidectomy in the adult animal and foetal lamb indicates that a functional feedback mechanism already exists in the foetal lamb at this stage of gestation (i.e. 110 days). The higher levels of TSH reached in foetal plasma after

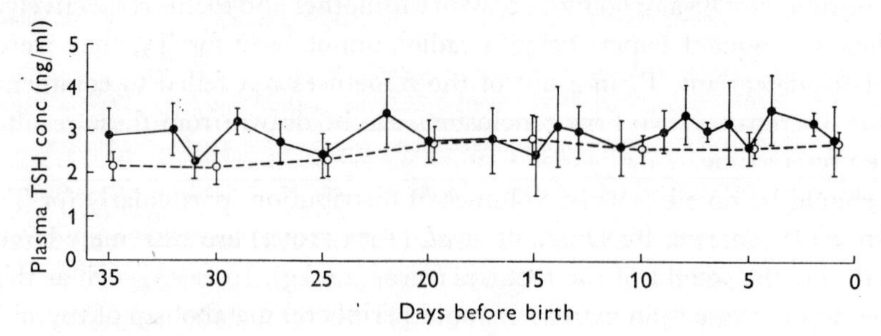

Fig. 4. Mean plasma TSH concentrations of 5 ewes (○----○) and their foetuses (●——●) during the last 5 weeks of pregnancy. Vertical bars represent SE.

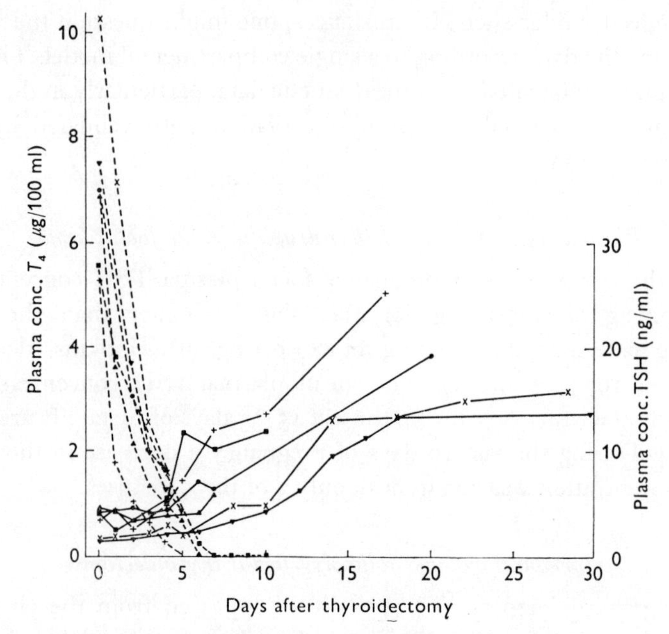

Fig. 5. Plasma TSH (——) and T_4 (----) concentrations following thyroidectomy of adult (● ▲ ▼) and foetal (× +) sheep.

thyroidectomy indicate that the foetal pituitary may have greater ability to secrete TSH. It is apparent that the response of the pituitary to decreasing T_4 levels is slow and it may be that the mechanisms responsible for increasing TSH secretion are relatively slow (Lemarchand-Beraud, 1969).

The progressive increase in the foetal TSH levels after thyroidectomy provides further evidence that the placental transfer of thyroid hormone from mother to foetus is extremely limited.

Hormonal changes following foetal hypophysectomy

Using the technique of Liggins, Kennedy & Holm (1967) we have performed hypophysectomies on foetal lambs during the last third of gestation (Hopkins, Schneider & Thorburn, manuscript in preparation). After hypophysectomy, the plasma TSH concentration decreased to zero within 48 hours and the plasma T_4 concentration decreased to zero over 4–5 days (Fig. 2). Clearly, the ability of the thyroid to secrete T_4 is entirely dependent upon trophic stimulation by the foetal pituitary. We can conclude that TSH (as measured by our immunoassay) is derived from the foetal pituitary and that no material of placental origin, able to crossreact with our antiserum, is present in the plasma of the foetal lamb. Furthermore, these results provide further support for the conclusion that neither TSH nor T_4 cross the ovine placenta.

In contrast, the results of Geloso (1967, 1971) suggest that thyroid function in the foetal rat is not completely under pituitary control. Geloso (1971) has suggested that maternal TSH may cross the placenta but the amounts are too small to permit the further increase in the thyroidal iodine/plasma iodine ratio that is normally seen. Nevertheless, the amount of TSH is apparently adequate to prevent the expected decrease in the T/P ratio after hypophysectomy (Taurog, Tong & Chaikoff, 1957).

Hormonal changes following section of the pituitary stalk

We have investigated the hypothalamic control of pituitary function by sectioning the pituitary stalk of foetal lambs. This operation was performed by inserting a small metal spatula through a mid-line fenestration of the frontal bones, and then passing it caudally to section the optic chiasma and pituitary stalk close to the dorsal surface of the sphenoid bone (Dr G. C. Liggins, personal communication).

The changes in the plasma TSH and T_4 concentration in foetal lambs after stalk section are shown in Fig. 6 (Hopkins, Schneider & Thorburn, manuscript in preparation). In general, TSH concentrations decreased initially but then recovered, the net effect being an overall decrease from pre-operative levels. The changes in T_4 concentration were more variable. In some foetuses the T_4 concentration decreased rapidly, reaching low concentrations within 5 days whereas in others, the T_4 concentration decreased to about 50% of the initial values and in some cases remained at this level for the next 30 days.

The relatively small change in TSH concentration following foetal stalk section contrasts with the findings that growth hormone (GH) levels

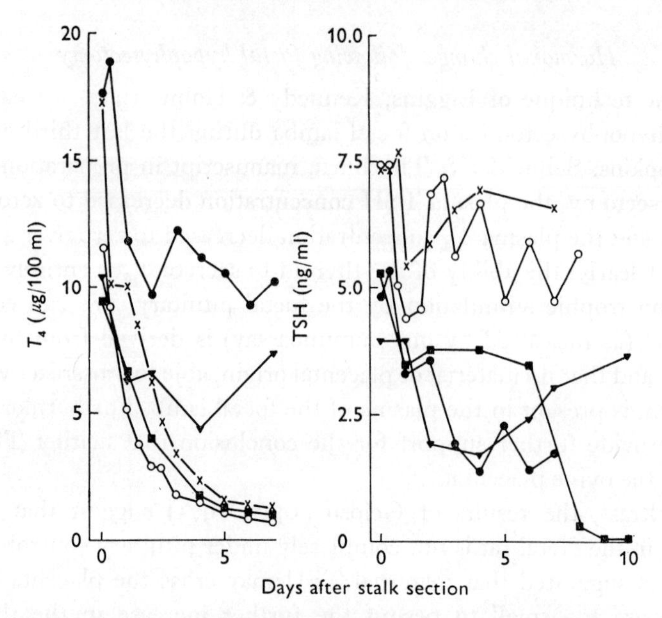

Fig. 6. Plasma T_4 and TSH concentrations following pituitary stalk-section of 5 foetal lambs.

decreased dramatically after stalk section (Wallace, Stacy & Thorburn, manuscript in preparation). There appeared to be no direct relationship between the changes in the plasma concentrations of TSH and T_4 following stalk section; indeed, in those foetuses in which the T_4 concentration decreased to very low levels, the TSH concentrations were maintained at relatively normal levels. A direct comparison of the present results with the earlier classical observations of Brown-Grant, Harris & Reichlin (1957) on the stalk-sectioned rabbit is not possible because of the different indices of thyroid activity used. However, our results are in general agreement with their conclusion that the thyroid activity of effectively stalk-sectioned rabbits was intermediate between that of normal and hypophysectomized animals (Brown-Grant *et al.* 1957; Harris, 1964).

The results in the foetal lamb are in agreement with earlier observations on the foetal rat (Jost & Geloso, 1967; Jost, 1971). Jost (1971) reported that the thyroid glands of encephalectomized ('stalk-sectioned') rat foetuses looked normal, and responded to maternally administered propylthiouracil with characteristic hyperplasia, although the blood T_4 levels were reduced. The normal histological appearance of the thyroid glands of encephalectomized foetal rats suggests that their TSH levels are relatively normal despite the reduced T_4 levels. Furthermore, the hyperplastic response in these foetuses following maternal propylthiouracil administration indicates that the pituitary can increase the TSH concentration in response to a further lowering of T_4 concentration and hence that a feedback mechanism is operative even in the absence of hypothalamic control.

The feedback mechanism

The effects of foetal thyroidectomy (Hopkins *et al.* 1973) clearly demonstrate the ability of the foetal pituitary to secrete massive amounts of TSH when T_4 is absent from the foetal plasma. The ability of the foetal thyroid to respond to this trophic stimulation is shown by the presence of goitres in new-born lambs born to ewes grazing on iodine-deficient or goitrogenic pastures (for references see George, Farleigh & Harris, 1966) or to ewes given methylthiouracil during pregnancy (Lascelles & Setchell, 1959). On the other hand, if foetal T_4 levels are increased by the administration of exogenous T_4, TSH levels in the foetus are depressed (Hopkins, 1972).

It is therefore apparent that a feedback control operates in the ovine foetus, at least during the last third of pregnancy. Nevertheless, there are features of the feedback mechanism of the foetus which differ from those of the adult. In particular, although the plasma TSH concentrations in foetal and adult sheep are similar (3.2 ng/ml and 2.6 ng/ml; Table 3), the plasma FT_4 concentrations of the foetus are more than double those of the mother ($\simeq$ 4.2 ng/100 ml compared with $\simeq$ 1.8 ng/100 ml). Moreover, the T_4 secretion by the thyroid gland is maintained at a higher level in the foetus than in the mother despite similar levels of TSH in their plasma. The question of how the feedback mechanism operates in the foetus cannot be answered on the evidence presently available. However, it is possible that the same mechanism exists in both the foetus and ewe but that it is set at a higher level in the foetus.

REFERENCES

ANNISON, E. F. & LEWIS, D. (1959). *J. agric. Sci., Camb.* **52**, 79–86.

BARNES, C. M., WARNER, D. E., MARKS, S. & BUSTAD, L. K. (1957). *Endocrinology* **60**, 325–8.

BEARN, J. G. & PILKINGTON, T. R. E. (1963). *Nature* **198**, 1005–6.

BROWN-GRANT, K., HARRIS, G. W. & REICHLIN, S. (1957). *J. Physiol., Lond.* **136**, 364–79.

BURROW, G. N., ANDERSON, G. G. & QUILLIGAN, E. J. (1969). *Proc. 51st Meeting, The Endocrine Society*, p. 126.

BURROW, G. N. & ANDERSON, G. G. (1970). *Clin. Res.* **18**, 356.

COMLINE, R. S., NATHANIELSZ, P. W. & SILVER, M. (1970). *J. Physiol., Lond.* **207**, 3–4P.

DUSSAULT, J., ROW, V. V., LICKRISH, G. & VOLPE, R. (1969). *J. clin. Endocr. Metab.* **29**, 595–603.

Dussault, J. H., Hobel, C. J. & Fisher, D. A. (1971). *Endocrinology* **88**, 47–51.
Dussault, J. H., Hobel, C. J., DiStefano, J. J. & Fisher, D. A. (1971). *Proc. 53rd Meeting, The Endocrine Society*, p. 186.
Dussault, J. H., Hobel, C. J., DiStefano, J. J., Erenberg, A. & Fisher, D. A. (1972). *Endocrinology* **90**, 1301–8.
Eayrs, J. T. (1960). *Brit. med. Bull.* **16**, 122–7.
Eayrs, J. T. (1964). In *Brain–Thyroid Relationships*, pp. 60–71. Eds. M. P. Cameron & M. O. O'Connor. Boston: Little, Brown and Co.
Eayrs, J. T. (1971). In *Hormones in Development*, pp. 345–56. Eds. M. Hamburgh & E. J. W. Barrington. New York: Meredith Corp.
Falconer, I. R. (1965). *Nature* **205**, 703.
Feldman, J. D., Vasquez, J. J. & Kurtz, S. M. (1961). *J. biophys. biochem. Cytol.* **11**, 365–83.
Freinkel, N. & Lewis, D. (1957). *J. Physiol., Lond.* **135**, 288–300.
Geel, S. E. & Timiras, P. S. (1971). In *Hormones in Development*, pp. 391–401. Eds. M. Hamburgh & E. J. W. Barrington. New York: Meredith Corp.
Geloso, J. P. (1967). *Ann. Endocr., Paris* **28** (Suppl. 1), 1–79.
Geloso, J. P. (1971). In *Hormones in Development*, pp. 793–9. Eds. M. Hamburgh & E. J. W. Barrington. New York: Meredith Corp.
George, J. M., Farleigh, E. A. & Harris, A. N. A. (1966). *Aust. vet. J.* **42**, 1–4.
Greenberg, A. H., Czernichow, P., Reba, R. C., Tyson, J. & Blizzard, R. M. (1970). *J. clin. Invest.* **49**, 1790–803.
Hamburgh, M., Lynn, E. & Weiss, E. P. (1964). *Anat. Rec.* **150**, 147–62.
Hamburgh, M. & Barrington, E. J. W. (Eds.) (1971). *Hormones in Development.* New York: Meredith Corp.
Hardy, M. H. & Lyne, A. G. (1956). *Aust. J. biol. Sci.* **9**, 423–41.
Harris, G. W. (1964). Ciba Foundation Study Group No. 18. Eds. M. D. Cameron & M. O'Connor. London: Churchill Ltd.
Hodges, R. E., Hamilton, H. E. & Keettal, W. C. (1952). *Archs. intern. Med.* **90**, 863.
Hopkins, P. S. & Thorburn, G. D. (1971). *J. Endocr.* **49**, 549–50.
Hopkins, P. S. (1972). Ph.D. Thesis, Macquarie University, Sydney, Australia.
Hopkins, P. S. & Thorburn, G. D. (1972). *J. Endocr.* (In press.)
Hopkins, P. S., Wallace, A. L. C. & Thorburn, G. D. (1973). *J. Endocr.* (In press.)
Ingbar, S. H. & Freinkel, N. (1955). *J. clin. Invest.* **34**, 808–19.
Jost, A. & Picon, L. O. (1958). *C. r. hebd. Séanc. Acad. Sci.* (*Paris*) **246**, 1281–3.
Jost, A. & Geloso, A. (1967). *C. r. hebd. Séanc. Acad. Sci.* (*Paris*) **265**, 625–7.
Jost, A. (1971). In *Hormones in Development*, pp. 1–18. Eds. M. Hamburgh & E. J. W. Barrington. New York: Meredith Corp.
Lascelles, A. K. (1959). *Aust. J. Zool.* **7**, 79–86.
Lascelles, A. K. & Setchell, B. P. (1959). *Aust. J. biol. Sci.* **12**, 445–65.
Lemarchand-Beraud, T., Scazziga, B. R. & Vannotti, A. (1969). *Acta endocr., Copenh.* **62**, 593–606.
Liggins, G. C., Kennedy, P. C. & Holm, L. W. (1967). *Am. J. Obstet. Gynec.* **98**, 1080–6.
Noumura, T. (1959). *Jap. J. Zool.* **12**, 301–18.
Osorio, C. & Myant, N. B. (1960). *Brit. med. Bull.* **16**, 159–64.
Robin, N. I., Fang, V. S., Selenkow, H. A., Piasecki, G. J., Rauschecker, H. F. J. & Jackson, B. T. (1970). *Clin. Res.* **18**, 370.
Sinclair, D. P. & Andrews, E. D. (1954). *New Zealand vet. J.* **2**, 72–9.
Speilman, A. A., Petersen, W. E., Fitch, J. B. & Pomeroy, B. S. (1945). *J. Dairy Sci.* **28**, 329–37.

SURKS, M. I. & OPPENHEIMER, J. H. (1969). *J. clin. Invest.* **48**, 685–95.
TAUROG, A., TONG, W. & CHAIKOFF, I. L. (1957). In *Colloquia on Endocrinology*, Ciba Foundation, pp. 59–78. Eds. G. E. W. Wolstenholme & E. C. P. Millar. London: J. & A. Churchill Ltd.
WILKINS, L. (1957). *The Diagnosis and Treatment of Endocrine Disorders in Childhood and Adolescence*, 2nd ed. Springfield, Illinois: Charles C. Thomas.
WOEBER, K. A., SOBEL, R. J., INGBAR, S. H. & STERLING, K. (1970). *J. clin. Invest.* **49**, 643–9.

THYROID HORMONE METABOLISM IN THE FOETUS

By A. ERENBERG and D. A. FISHER

Department of Pediatrics, Harbor General Hospital, and the School of Medicine of the University of California at Los Angeles

INTRODUCTION

Thyroid physiology has been extensively investigated in both the child and adult and the role of thyroid hormones in postnatal growth and metabolism is reasonably clear. The role of thyroid hormones in foetal growth and metabolism, however, has been more difficult to define. The use of animal models of different species, extrapolation of data from these models to the human foetus, and the influence of non-thyroid variables in these experiments have contributed to this problem. In addition, the influence of placental hormones and placental transfer of maternal thyroid hormones must be taken into consideration. The present paper reviews the now considerable data regarding thyroid function in the human mother and foetus *in utero* to which we and others have contributed, as well as our more recent studies in the sheep model. The former studies, of necessity, consist largely of static measurements of hormone and protein concentrations whereas kinetic data are available in the sheep.

FOETAL–MATERNAL THYROID FUNCTION IN MAN

The thyroid gland is derived embryologically from the entoderm of the foregut, with its point of origin demarcated in later life as the foramen caecum. Development of the human foetal thyroid gland can be divided into three general periods, the pre-colloid, the beginning colloid, and the follicular growth phases (Shepard, 1968). The pre-colloid period, from 47 to 72 days' gestation, and the beginning colloid stage, from 73 to 80 days' gestation, are characterized by rapid thyroid growth and gradual increase in weight of the foetal gland relative to body weight. After this time thyroid gland growth slows to parallel that of body growth and relative weight remains constant at about 0.05% of the body weight (Shepard, Andersen & Andersen, 1964). Thyroglobulin synthesis has been studied by incubating foetal thyroid gland cells *in vitro*, with ^{14}C-labelled amino acids (Gitlin & Biassucci, 1969). Such studies indicate that the foetal thyroid gland contains thyroglobulin as early as 29 days post-conception, and this amount, as measured by radioautography, increases with advancing gestational age. Colloid appears histologically during the second phase of development, first as central intrafollicular cellular accumulations and later as extracellular colloid pools derived from the intracellular accumulations of several concentric cells (Shepard, 1968). The follicular stage also is characterized by changes in the glycogen content, size and configuration of the thyroid cells.

Studies designed to determine the time of onset of function of the foetal thyroid gland have been conducted by Hodges (Hodges *et al.* 1955) and Evans (Evans *et al.* 1967) who administered ^{131}I to mothers prior to therapeutic abortions. The results show that the foetal thyroid gland begins to accumulate radioiodine by 10 to 12 weeks' gestation. Although absolute radioiodine uptake is highest at term, when the results are expressed as per cent uptake per gram of thyroid tissue, the highest concentration, 5 % per gram tissue, is found at 20 to 24 weeks' gestation (Evans *et al.* 1967). Radioautographic studies of organ cultures of human foetal thyroid glands of 45 to 115 days' gestation conducted after incubation with radioiodine show definite organic binding of ^{125}I and synthesis of iodothyronines at about 75 days' gestation (Shepard, 1967).

The onset of TSH production by the human foetal pituitary has been determined by studies of pituitary cell cultures incubated with ^{14}C-labelled amino acids (Gitlin & Biassucci, 1969) and by measurement of immunoreactive foetal pituitary and serum TSH at the time of therapeutic abortion (Fisher *et al.* 1970; Greenberg *et al.* 1970; Fukuchi *et al.* 1970). Pituitary cell cultures from foetuses of 4 to 18 weeks' gestation demonstrated TSH synthesis by 14 weeks. Foetal pituitary TSH is detectable in foetuses as young as 12 weeks and serum TSH by 9–10 weeks. There appears to be a rather abrupt increase in pituitary TSH content at about 18 to 22 weeks of gestation (Fisher *et al.* 1970), and a gradual increase thereafter to a plateau at about 32 weeks when the content is about 10 mIU/gland. When expressed as μU TSH per gram foetal body weight the maximum foetal pituitary TSH content is reached at about 17 weeks' gestation (Fukuchi *et al.* 1970).

Foetal serum TSH levels are low before 18 weeks (mean 2.4 μU/ml), but between 18 and 22 weeks an abrupt increase is observed, to a mean of 9.6 μU/ml, and these relatively high concentrations persist until the time of delivery (Fisher *et al.* 1969, 1970; Czernichow *et al.* 1971). Paired maternal and foetal serum cord blood samples obtained after 20 weeks' gestation consistently reveal a foetal-to-maternal serum TSH gradient;

higher foetal than maternal values are found in all cases, whether delivered vaginally or by Caesarean section.

The role of human chorionic thyrotropin (HCT) in foetal thyroid physiology is unknown. Though Hennen, Pierce & Freychet (1969) reported HCT to be detectable in maternal serum, especially during the first trimester, others have not been able to measure HCT activity in pregnancy sera (Hershman & Starnes, 1971); foetal serum HCT levels have not been investigated. It is unlikely that the foetal serum immunoreactive TSH measured in the foetus is due to cross-reactivity of HCT in the TSH radioimmunoassay since this cross-reaction is of a low order and very high levels of HCT would be necessary to account for levels of TSH reported at this time (Fisher *et al.* 1969).

TBG in foetal serum measured as T4 binding capacity increases from 2 μg% at 7 weeks' gestation to about 25 μg% at term (Greenberg *et al.* 1970). Maternal TBG concentrations increase rapidly during the first 8 weeks of pregnancy and then more gradually to values of about 45 μg% T4 binding capacity at term (Erenberg, Hobel & Fisher, 1973). These increases are presumably due to stimulation by progressively increasing oestrogen concentrations. Thyroxine (T4) is found in all foetal serum samples measured at 7 weeks' gestation and beyond and values range from a mean of 2.6 μg% at 11 to 18 weeks to a mean of 11.2 μg% at term (Fisher *et al.* 1969). The rise in foetal serum T4 generally parallels the rise in serum TBG concentration. Maternal serum total T4 levels are elevated by 8 weeks, and also tend to parallel the gradual increase in TBG observed between 8 weeks and term (Erenberg *et al.* 1973).

Foetal serum free thyroxine concentration (FT4) increases progressively throughout gestation from a mean of 1.85 ng% between 11 and 18 weeks, to 2.49 ng% at 23–24 weeks. At term, the mean value of 2.9 ng% actually exceeds the maternal level of 2.3 ng% (Fisher *et al.* 1969). Thus the serum FT4 gradient is maternal–foetal early in gestation and may be foetal–maternal at term.

We recently have measured total triiodothyronine (T3) in maternal serum throughout pregnancy and found that levels are significantly lower during the first 20 weeks than during the latter half of gestation, mean levels being 170 ng% and 208 ng% respectively (Erenberg *et al.* 1973). There is also a high correlation between the total serum T3 concentration and TBG levels throughout gestation. In contrast to maternal serum, we have observed that T3 levels in cord blood are low, with a mean of 50 ng%. And preliminary data indicate that serum T3 is not measureable ($<$ 18 ng%) in foetal serum samples of 12 to 32 weeks' gestation. Mean per cent dialysable T3 values measured in paired maternal–foetal samples at delivery are higher in cord blood than in maternal blood (cord blood 0.295; maternal blood 0.238) (Dussault *et al.* 1969); however, the absolute serum free T3 concentration is higher in maternal than in cord blood.

Thus, it appears that the early development of the human foetal thyroid gland and the synthesis of thyroglobulin are not TSH dependent. Iodothyronine synthesis, in contrast, is TSH dependent and appears by 7 to 8 weeks' gestation. Between 7–8 and 18 weeks, foetal TSH synthesis and secretion are minimal and the levels of foetal serum TSH and T4 are quite low. Near midgestation foetal pituitary TSH content and foetal serum TSH are observed to increase and there is a parallel increase in foetal thyroidal radioiodine concentration. These events correlate with development of the pituitary portal vascular system and suggest that the foetal hypothalamic-pituitary system has functionally matured and/or that thyrotropin releasing hormone (TRH) production is augmented. As a consequence, TSH synthesis and release are stimulated and there is a progressive increase in foetal serum free thyroxine concentration. These events occur independently of the changes in maternal thyroid function parameters for there is no correlation between maternal and foetal serum T4, FT4, T3, FT3 or TSH concentrations at any time during gestation. Moreover, TSH does not cross the placental barrier and placental transfer of T4 and T3 are minimal at best. Thus the foetal hypothalamic-pituitary-thyroid axis in the euthyroid foetus is autonomous of the maternal system.

FOETAL–MATERNAL THYROID FUNCTION IN SHEEP

In order to better understand thyroid hormone kinetics in the foetal–placental–maternal unit, we have investigated the kinetics and placental transfer of iodothyronines in the sheep model. These studies were conducted first in the euthyroid foetus, and then in the thyroidectomized foetus (Dussault *et al.* 1971, 1972; Fisher *et al.* 1972; Erenberg *et al.* 1972*a*, *b*).

Thyroid metabolism in euthyroid foetal sheep

Studies in euthyroid foetal sheep were conducted as follows: one- to four-year-old date bred ewes, obtained from a local source, were maintained at environmental temperatures of 57 to 85 °F, and given free access to alfalfa and water. Under spinal anaesthesia a uterotomy was performed on ewes of 90 to 145 days' gestation at which time an indwelling exteriorized arterial catheter was inserted into the foetus, and a jugular vein catheter into the mother for serial blood sampling. The normal duration of gestation in the sheep is 150 days. After a period of several days to assure recovery from the surgical procedure, tracer doses of ^{125}I-labelled T4 or T3 and ^{131}I-labelled T4 or T3 were injected into the mother and foetus, respectively, and serial blood samples collected for up to 96 hours. 400 mg of potassium percholorate was administered orally to the mother every 12 hours during the period of blood sampling to prevent secretion of endogenous labelled iodothyronines. Alkali-washed butanol extracts of 250 μl of each serum sample were prepared in duplicate, and counted for ^{125}I and ^{131}I activities. Serum T4 concentration was measured by the Murphy-Pattee method and serum T3 concentration by radioimmunoassay in 250 μl of whole serum. Dialysable T4 and T3 were measured using a modification of the method of Sterling & Brenner (Sterling, 1966). TSH was measured using a heterologous radioimmunoassay system developed in our laboratories. The maximal binding capacity of serum thyroid hormone binding proteins was measured using reverse flow electrophoresis in a Beckman system. To identify the form of radioactivity crossing the placenta, unwashed butanol extracts of serum, supernatant solutions from PBI precipitates of serum, and dialysates from maternal and foetal serum were chromatographed on Whatman No. 3 paper using butanol–acetic acid and tertiary amyl alcohol–ammonia solvent systems (Dussault *et al.* 1972).

Thyroxine secretion was investigated in six ewes of 105 to 145 days' gestation (Dussault *et al.* 1971): triiodothyronine secretion in six preparations between 98 and 125 days' gestation (Dussault *et al.* 1972). Fig. 1 shows a composite semi-logarithmic plot of the disappearance of ^{131}I-T4 from foetal sera and ^{125}I-T4 from maternal sera. Values were recorded as per cent of injected dose per litre of plasma and plotted as mean and SEM. From this data, the mean T4 volume of distribution (VD) was estimated by extrapolating the final single exponential disappearance curves to zero time and calculating the apparent volume of hormone distribution by a simple dilution formula. The mean $t_{\frac{1}{2}}$ values were determined by inspection and fractional K values (fractional clearance of extrathyroidal hormone pools) calculated for the foetuses and mothers; mean fractional T4 clearance rates were calculated from the volume of distribution and the K values.

Fig. 2 summarizes the T3 kinetic data in the mothers and foetuses.

Table 1 lists the VD, $t_{\frac{1}{2}}$, serum concentrations, and calculated turnover (secretion) rates for T4 in the maternal–foetal pairs. Table 2 shows the T3 data. The mean foetal serum T4 and T3 levels were 7.4 μg % and < 18 ng % respectively; maternal values were 5.5 μg % and 79 ng %. Foetal serum T3 was detectable above the lower limits of the radioimmunoassay (18 ng %) in only 2 of 10 fetuses. The apparent mean distribution volumes for T4 and T3 in the mothers were 9.33 and 28.8 l

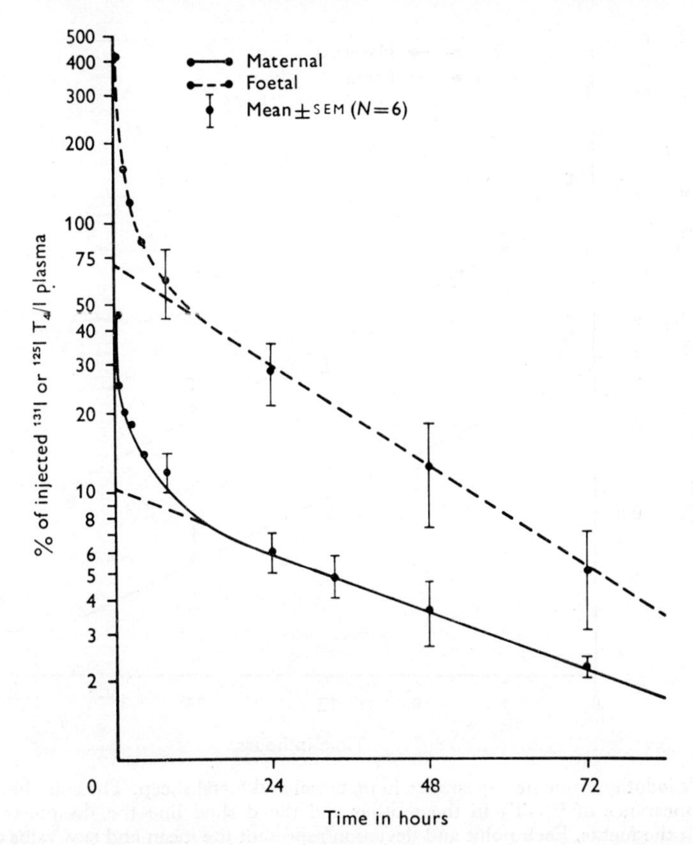

Fig. 1. Thyroxine disappearance in maternal and foetal sheep. The solid line shows the disappearance of ^{125}I-T4 in the mother and the dashed line the disappearance of ^{131}I-T4 in the foetus. Each point and deviation represent the mean and SEM values for six animals. From Dussault, Hobel & Fisher (1971).

respectively. Mean foetal VD values were 1.36 l for T4 and 5.1 l for T3. The foetal values were relatively larger because they represent the combined foetal and placental distribution volumes. The mean maternal and foetal $t_{\frac{1}{2}}$ values for T4 were 1.08 days and 0.72 days respectively; for T3 the respective values were 7.0 hours and 5.5 hours. The mean T4 turnover (secretion) rates were 333 μg/day for the mothers and 104 μg/day for the foetuses, or 5.5 and 41 μg/kg/day, respectively. Mean values for maternal and foetal T3 turnover (secretion) were 67 and < 2.8 μg/day, or 1.43 and < 1.45 μg/kg/day, respectively.

Table 3 lists the total serum thyroxine, free thyroxine and maximal T4 binding capacity of TBG (TBG cap.) in these animals; Table 4 shows total serum T3, free T3 and TBG cap. (Fisher *et al.* 1972). The mean per cent dialysable T4 was significantly higher in foetal blood than in maternal

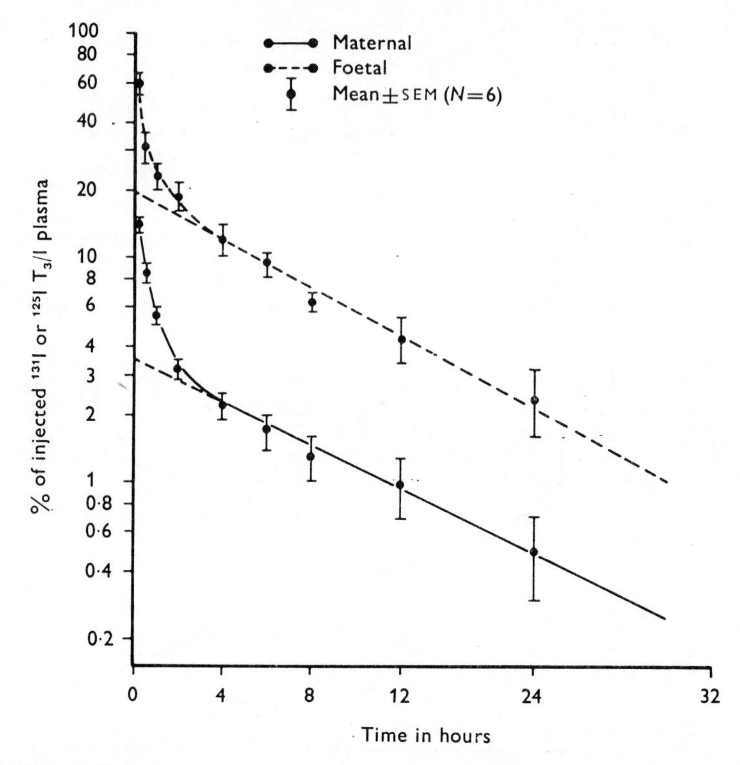

Fig. 2. Triiodothyronine disappearance in maternal and foetal sheep. The solid line shows the disappearance of ^{125}I-T3 in the mother and the dashed line the disappearance of ^{131}I-T3 in the foetus. Each point and deviation represent the mean and SEM values for six animals. From Dussault *et al.* (1972).

blood (0.066% *v.* 0.044%, $p < 0.01$), as was the absolute FT4 (5.1 ng% *v.* 2.4 ng%, $p < 0.01$). The mean per cent dialysable T3 also was higher in foetal than in maternal sera (0.51 *v.* 0.26%, $p < 0.001$); but the mean free T3 concentration was lower in the foetus (< 90 pcg% *v.* 176 pcg%). The mean binding capacity of foetal TBG in the two studies was similar (6.9 *v.* 8.8 ug%). This also was true for maternal sera (17.0 *v.* 16.1 μg%). In sheep, there is no T4 binding pre-albumin, and there is no increase in maternal TBG concentration during pregnancy. This presumably is due to the fact that placental oestrogen production in the sheep is minimal.

We recently have reported a second T4 binding protein in foetal sheep serum and have referred to this as T4 binding foetal protein (TBFP) (Fisher & Lamb, 1972). This protein migrates anodal to alpha-1-globulin and behind albumin, and may represent 'fetuin', an alpha-1-glycoprotein present in foetal and newborn serum of several species. Purified 'fetuin' will bind T4 *in vitro*, but the binding affinity is less than that of TBG. The

Table 1. *Volume of distribution, half-life, serum concentration and secretion rate of thyroxine in maternal and foetal sheep*

	Maternal					Foetal				
				T4 S*					T4 S*	
	VD (l)	$t_{\frac{1}{2}}$ (days)	T4 Conc. (μg/100 ml)	(μg/day)	(μg/kg·day)	VD (l)	$t_{\frac{1}{2}}$ (days)	T4 Conc. (μg/100 ml)	(μg/day)	(μg/kg·day)
Thyroidectomized foetus										
Mean	12.6	1.42	8.5	533	11.9	0.97	0.99	< 0.7	< 4.95	< 3.23
SEM	0.8	0.10	0.44	46	2.1	0.11	0.11	—	< 0.40	< 0.60
Normal foetus										
Mean	9.33	1.08	5.5	333	5.5	1.36	0.72	7.4	104	40.8
SEM	1.86	0.04	0.63	51	1.0	0.14	0.1	0.87	19	5.5

* T4 turnover.

Table 2. *Volume of distribution, half-life, serum concentration and secretion rate of triiodothyronine in maternal and foetal sheep*

	Maternal					Foetal				
				T3 S*					T3 S*	
	VD (l)	$t_{\frac{1}{2}}$ (h)	T3 Conc. (ng/100 ml)	(μg/day)	(μg/kg·day)	VD (l)	$t_{\frac{1}{2}}$ (h)	T3 Conc. (ng/100 ml)	(μg/day)	(μg/kg·day)
Thyroidectomized foetus										
Mean	26.7	6.7	94	62.2	1.27	7.5	10.0	< 18	< 2.3	< 1.05
SEM	4.5	0.5	5.3	13.0	0.26	1.1	0.3	—	< 0.39	< 0.19
Normal foetus										
Mean	28.8	7.0	79	67	1.43	5.1	5.5	< 18	< 2.8	< 1.45
SEM	3.6	1.5	13.0	16.4	0.36	0.85	0.6	< 2.7	< 0.60	< 0.18

* T3 turnover.

Table 3. *Serum thyroxine, free thyroxine and TBG capacity in maternal and foetal sheep*

	Maternal				Foetal			
	T4 (μg/100 ml)	%FT4	AFT4* (ng/100 ml)	TBG† (μg/100 ml)	T4 (μg/100 ml)	%FT4	AFT4* (ng/100 ml)	TBG† (μg/100 ml)
Mean	5.36	0.044	2.4	17.0	7.5	0.066	5.1	6.9
SEM	0.75	0.004	0.5	2.3	1.1	0.005	1.1	0.7

* Free thyroxine concentration.
† T4 binding capacity.

Table 4. *Serum triiodothyronine, free triiodothyronine and TBG capacity in maternal and foetal sheep*

	Maternal				Foetal			
	T3 (ng/100 ml)	%FT3	AFT3* (pcg/100 ml)	TBG† (μg/100 ml)	T3 (ng/100 ml)	%FT3	AFT3* (pcg/100 ml)	TBG† (μg/100 ml)
Mean	74	0.263	176.0	16.1	< 18	0.506	< 90	8.8
SEM	13.2	0.046	34.4	2.05	< 2.7	0.029	< 14	0.7

* Free triiodothyronine concentration.
† T4 binding capacity.

maximal T4 binding capacity, however, is quite high (> 600 μg %). Fetuin also binds T3 *in vivo* and *in vitro*. The significance of fetuin in foetal thyroid hormone metabolism is not clear, but maximal T4 binding in foetal sheep serum is observed in the inter-alpha globulin (TBG) area so that TBFP appears to play a secondary role to TBG in foetal T4 transport.

There was no placental transfer of butanol-extractable radioiodine-labelled T4 either in the maternal-to-foetal (M–F) or foetal-to-maternal (F–M) directions (Dussault *et al.* 1971). Significant placental transfer of butanol-extractable labelled T3 was observed in both the M–F and F–M directions as shown in Fig. 3 (Dussault *et al.* 1972). The fractional transfer rate of labelled T3 was greater in the F–M than M–F direction; the concentration of labelled foetal T3 in maternal blood at two to four hours was ten times greater than the concentration of labelled maternal T3 in foetal blood. Absolute net hormone transfer, however, was minimal; using a simple two-compartment model, we calculated that net transfer of unlabelled T3 was less than 1 μg/day in the maternal-to-foetal direction.

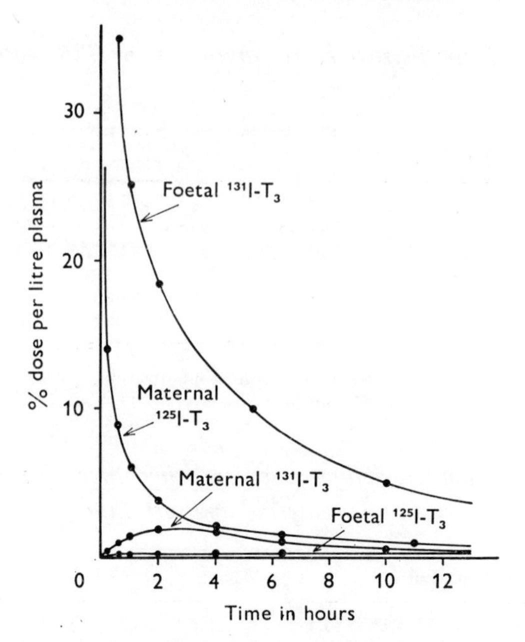

Fig. 3. Arithmetic plots of the disappearance of labelled T3 from maternal and foetal sera and the appearance of foetal labelled T3 in maternal serum and maternal labelled T3 in foetal serum. The data represent mean values of six experiments. From Dussault *et al.* (1972).

These observations of minimal iodothyronine transfer across the placenta are in agreement with most earlier studies in the rat, guinea pig, monkey and man (see Review, Dussault *et al.* 1971).

Fig. 4 shows the results of the chromatography in the BAA solvent system of unwashed butanol extracts of maternal and foetal serum and of serum dialysates; the extracted and measured radioactivity was, in fact, T3 (Dussault *et al.* 1972). In addition, however, there was a substance which appeared in the aqueous dialysate of foetal and maternal sera, moving just ahead of iodine, and not present in the butanol extracts. The chromatographic and solubility characteristics of this material suggest that it may be the T3 sulphate conjugate previously identified in foetal and maternal blood of monkeys by Schultz *et al.* 1965. However, sulphatase hydrolysis was not conducted.

To assess whether TSH crossed the placenta, tracer doses of ^{125}I-labelled and ^{131}I-labelled bovine TSH (BTSH) were injected into six mother and foetal pairs respectively, and serial blood samples drawn over the next 120 minutes (Dussault *et al.* 1973 in preparation). After precipitation with excess antibody, the serum samples were counted for both ^{125}I and ^{131}I activities. These data indicated that there was no transfer of the purified

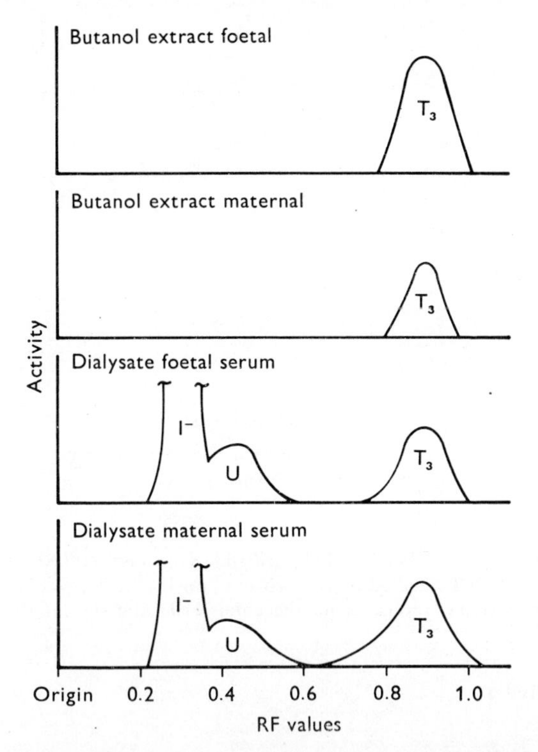

Fig. 4. Drawings of butanol–acetic acid (BAA) chromatograms of foetal and maternal serum 2–4 hours after administration of ^{131}I-T3 to the foetal sheep. The upper two panels show the T3 peaks in butanol extracts of maternal serum. The lower two panels show the iodide and unidentified peaks of radioactivity in dialysates of maternal and foetal blood at 2–4 hours (see text for details). From Dussault *et al.* (1972).

BTSH in either direction (Figs. 5 and 6). The mean volumes of distribution of BTSH were 3.2 l and 0.375 l in the mother and foetus, respectively; the mean disappearance times ($t_{\frac{1}{2}}$) were 33.6 min in the mother, and 46.8 min in the foetus.

In summary, iodothyronine turnover in maternal sheep is relatively greater than in man, although the ratio of T4 to T3 turnover is similar (Oddie *et al.* 1971): 333 μg T4/day and 67 μg T3/day = 5.0/1 in sheep *v.* 87 μg T4/day and 23 μg T3/day = 3.8/1 in man. And it is clear that during the last trimester of pregnancy in the sheep, thyroxine turnover in the foetus is nearly 8 times that of its mother (41 μg/kg/day *v.* 5 μg/kg/day). Moreover, this high foetal T4 secretion rate is associated with relatively higher levels of total T4, per cent dialysable FT4 and absolute free T4 values in foetal serum than in maternal serum. These results are similar to our earlier observations of relative maternal and foetal T4 and FT4 levels in human foetal sera near term.

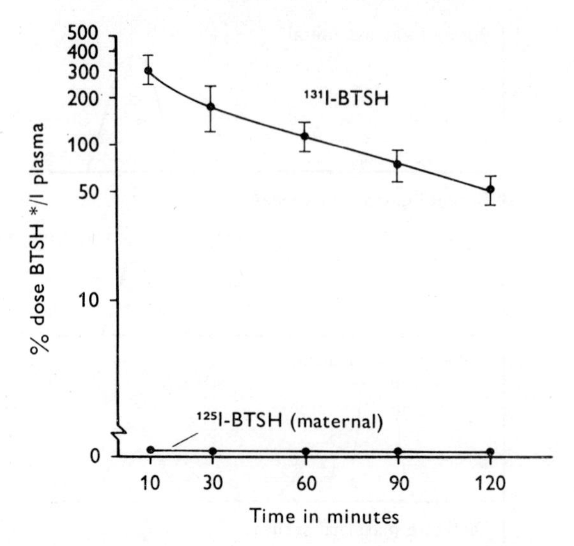

Fig. 5. The disappearance of ^{131}I-labelled purified bovine TSH (BTSH) from foetal serum and the appearance of ^{125}I-labelled BTSH (given to the mother) in foetal serum. The lower line, showing no activity, indicates no placental TSH transfer in the maternal–foetal direction.

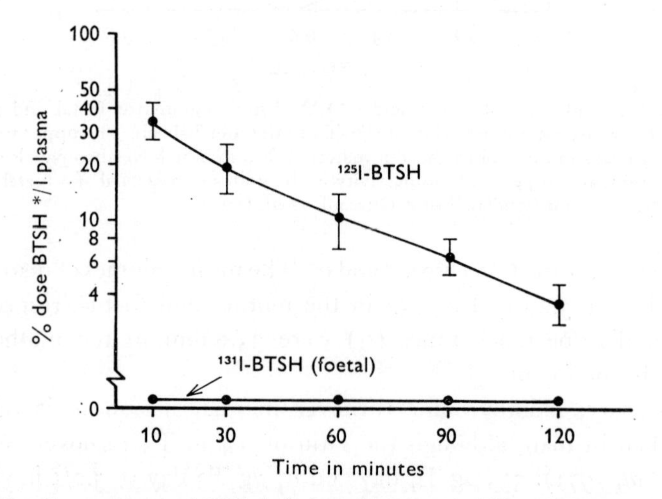

Fig. 6. The disappearance of ^{125}I-labelled purified bovine TSH (BTSH) from maternal serum and the appearance of ^{131}I-labelled BTSH (given to the foetus) in maternal serum. The lower line, showing no activity, indicates no placental TSH transfer in the foetal–maternal direction.

Serum T3 concentration is low in foetal sheep as in the human foetus and the T4/T3 turnover ratio is quite high. The serum T4/T3 concentration ratio in the maternal sheep was 74/1, while that in foetal sheep was > 417/1 (Dussault *et al.* 1972). Foetal and maternal T4/T3 turnover ratios were > 31/1 and 4.6/1, respectively. Possible explanations for these observations could be (*a*) a high T4/T3 ratio in the foetal thyroid gland; (*b*) a low rate of extrathyroidal conversion of T4 to T3 in peripheral tissues; or (*c*) preferential T4 excretion in conjugated form via the biliary system in the foetus. Further experiments to explore these possibilities were conducted in the thyroidectomized foetuses.

Thus, we conclude that in the sheep as in man the foetal hypothalamic-pituitary-thyroid axis is autonomous of the maternal system. Placental TSH and iodothyronine transfer are minimal; there are marked and opposite gradients of free T4 and free T3 concentrations across the placenta; and the foetal thyroid is functioning at a much higher level than is the maternal gland. Moreover, T3 metabolism appears to be qualitatively different in the mother and foetus.

Thyroid metabolism in hypothyroid foetal sheep

It is well known that the athyrotic human cretin at birth shows no evidence of growth retardation, although osseous development and central nervous system maturation may be delayed. Thus, it is possible (*a*) that placental thyronine transfer is increased when the foetus is athyrotic; (*b*) that the iodothyronines are not necessary for normal foetal growth; or (*c*) both factors might be involved. To investigate the first possibility we conducted studies of foetal and maternal thyroid hormone metabolism in sheep after foetal thyroidectomy (Erenberg, 1972*a*, *b*); studies of the effect of foetal thyroidectomy on foetal growth are in progress.

Under spinal anaesthesia, a uterotomy was performed through a flank incision. The foetal neck was isolated, exposed, and a thyroidectomy performed. Indwelling exteriorized foetal carotid artery catheters were placed at this time, or at a repeat uterotomy performed 25 to 30 days post-thyroidectomy. The foetus received tracer doses of either ^{125}I-T4, ^{125}I-T3 or ^{125}I-BSTH and the mother received tracer doses of the respective ^{131}I-labelled hormone. All samples were prepared and counted as in the euthyroid studies except that serum T4 levels were measured in 25 μl serum samples by radioimmunoassay. Nineteen to 43 days post-thyroidectomy, the preparations were sacrificed and an autopsy performed on each foetus. Pre-thyroidectomy, the mean foetal serum T4 value was 12 ug % and this level fell by post-operative day 3 to a mean of 1.7 μg %. Foetal serum T4 values measured 6 weeks post-thyroidectomy were < 0.7 μg %, the lower limit of our assay. The mean maternal serum T4 value was 8.5 μg % and did not change significantly throughout the study period. Foetal serum T3 values were undetectable in all samples taken during these studies. Pre-thyroidectomy, the mean maternal serum T3 value was 79 ng %, and this level did not change significantly throughout the study period.

T4 kinetics and placental transfer were studied in five preparations in which foetal thyroidectomy was performed between 90 and 125 days' gestation. Two animals were injected with labelled T3 4 days post-thyroidectomy and two animals were injected 32 to 35 days post-thyroidectomy.

Table 1 summarizes the T4 kinetic and concentration data, and Table 2 summarizes similar data for the T3 studies. The maternal and foetal T4 VD, and $t_{\frac{1}{2}}$, were similar to values in the euthyroid animals. The T4 secretion rate for athyrotic foetuses was < 4 μg/kg/day, far less than that for the euthyroid foetuses. The disappearance rate of labelled T3 from serum was slower in the thyroidectomized than in the euthyroid foetuses; the metabolic clearance rates, however, were nearly similar, mean values being 15.7 l/day and 12.6 l/day in euthyroid and hypothyroid foetuses, respectively. Since T3 concentrations were very low, T3 turnover could not be quantified.

Placental transfer of butanol-extractable labelled T4 and T3 activities occurred in both directions. The mean fractional rate of T4 transfer was greater in the foetal-to-maternal direction than in the maternal-to-foetal direction; the maximum concentration, at 24 to 36 hours, of foetal T4 label in maternal serum was twice the maximum concentration of maternal T4 label in foetal serum. Calculated net placental T4 transfer, however, was less than 1 μg/day in the maternal-to-foetal direction.

The mean fractional rate of T3 transfer was greater in the foetal-to-maternal direction; the maximum concentration of foetal T3 label in maternal serum at 1 to 2 hours was 1–5 times the maximum concentration of maternal T3 label in foetal serum at 12 hours. Calculated net T3 transfer was less than 0.7 μg/day in the maternal-to-foetal direction. Again chromatography of the serum extracts revealed that the measured radioactivity crossed the placenta primarily as T4 or T3. A compound with chromatographic characteristics similar to the T3 sulphate conjugate also was found in the aqueous dialysate of foetal and maternal sera.

Foetal serum TSH levels, 20 to 29 days post-thyroidectomy, were 300 to 1500 μU/ml. There was no transfer of labelled BTSH in either direction across the placenta. It also appears that the disappearance time of labelled BTSH in the athyrotic foetus was prolonged relative to that in the euthyroid foetus.

In summary, following thyroidectomy, the foetus rapidly becomes hypothyroid, as evidenced by foetal serum T4 levels of < 0.7 μg %, serum T3 levels of < 18 ng %, and serum TSH levels of 300 to 1500 μU/ml.

The high foetal-to-maternal TSH gradient and high maternal-to-foetal T4 and T3 gradients across the placenta after foetal thyroidectomy further substantiate the conclusion that the foetal pituitary–thyroid system functions autonomously of the maternal system. The sheep placenta is impermeable to BTSH, and relatively impermeable to both thyroxine and triiodothyronine, so that only minimal quantities of maternal hormone are transferred to the foetus whether euthyroid or hypothyroid. The estimate of about 1.7 μg/day of maternal iodothyronine transferred across the placenta to the thyroidectomized foetus would supply less than 2% of the total daily iodothyronine turnover in the euthyroid foetus.

In the euthyroid foetal sheep preparation, foetal thyroid function was characterized by a high T4/T3 turnover ratio. Since in the thyroidectomized foetus serum levels of T4 and T3 were below the limits of our radioimmunoassay systems and since iodothyronine transfer in the maternal-to-foetal direction was negligible, the athyrotic foetus was used to study extra-thyroidal conversion of T4 to T3. 25 μg of synthetic thyroxine (Synthroid) were injected into the carotid artery of three athyrotic foetuses, and serial blood samples were collected over a period of 48 hours for measurements of serum T4 and T3 concentrations by radioimmunoassay. Fig. 7 shows mean concentrations of both maternal and foetal serum T3 and T4 in these animals. The mean pre-injection values were < 0.7 μg% and < 18 ng% in each foetus. Within 10 minutes after injection, there was a marked rise in both foetal serum T4 and T3 concentrations, with a gradual fall over the subsequent 48 hours. There was no change in maternal serum T4 and T3 levels.

The rapid initial increase in serum T3 probably is due to T3 contamination of the injected T4. This would disappear, however, with a $t_{\frac{1}{2}}$ of about 10 hours and largely disappear within 24 hours. And by 6–12 hours the labelled T3 concentration had plateaued at a level of 0.23% to 0.30% of the T4 level (Fig. 8). This value is less than the value of 0.57% observed in thyroidectomized adult sheep (Fisher, Chopra & Dussault, 1972*b*) but clearly indicates a significant monodeiodination of T4 to T3. Using the formula derived from Schwartz *et al.* (1971), the fractional T4-to-T3 conversion rates can be calculated to be 0.017 and 0.030 respectively for thyroidectomized foetal and adult sheep, suggesting that the foetal sheep may have a lower rate of T4-to-T3 conversion than the adult sheep.

To investigate the second possible explanation for the low foetal serum T3 levels, the T4 and T3 content of thyroid gland homogenate hydrolysates were measured in seven foetal and five maternal thyroid glands. The mean T4 content of the foetal and maternal glands was 13.6 ± 4.6 and 141.1 ± 28.0 (SEM) μg/g wet weight of thyroid tissue, respectively; mean T3

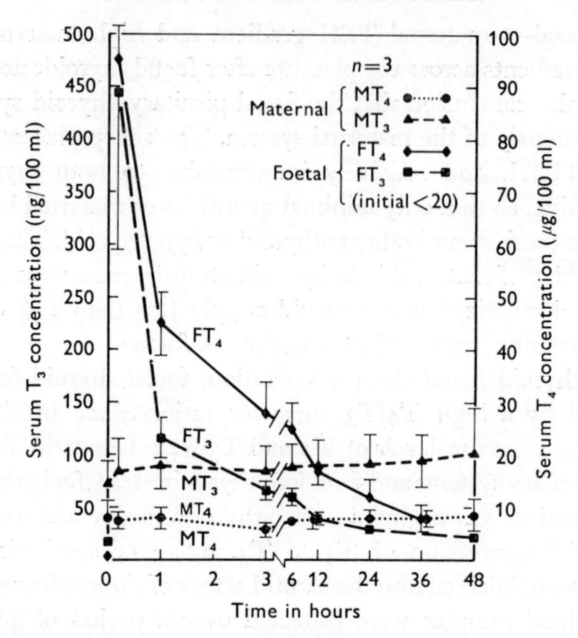

Fig. 7. T4 and T3 concentrations in maternal and foetal sera after administration of synthetic Na-l-thyroxine, intravascularly, to the thyroidectomized sheep foetus. Initial foetal serum T4 and T3 concentrations were unmeasurable. After T4 administration T4 and T3 concentrations both increased to high levels and gradually decreased during the subsequent 48 hours. Maternal values remained essentially unchanged. Each point represents mean and SEM values for three animals.

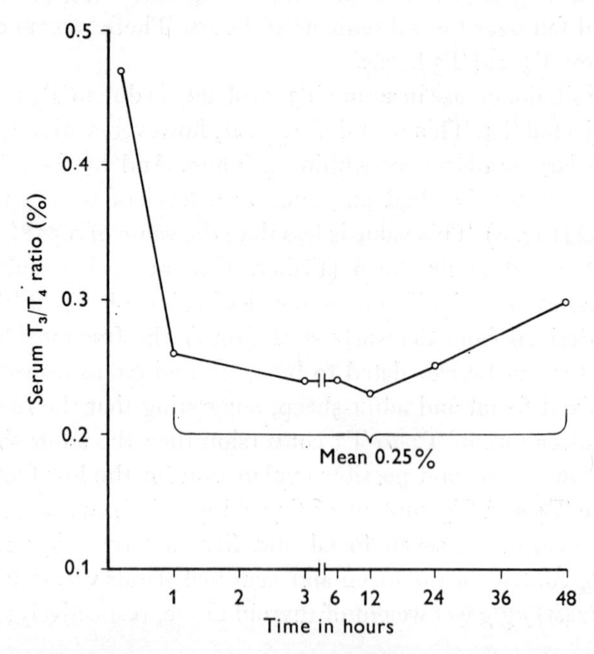

Fig. 8. Serum T3/T4 ratios (expressed as per cent) in foetal sera after administration of synthetic Na-l-thyroxine to the thyroidectomized sheep foetus. The concentration data from which these values were derived are shown in Fig. 7.

content of foetal and maternal glands was 0.27 ± 0.05 and 3.8 ± 0.09 (SEM) μg/g wet weight of thyroid respectively. Mean thyroid gland T4/T3 ratios were 67 ± 25 and 41 ± 4 (SEM) in the foetal and maternal glands, respectively, and these values were not statistically different. Thus, the low foetal serum T3 levels and the high T4/T3 turnover ratio in the foetus do not seem to be accountable on the basis of decreased T3 secretion or decreased T4-to-T3 conversion although both of these factors may contribute. We have not yet investigated biliary T4 excretion in the foetus.

Supported by United States Public Health Service Grant HD-4270 from the National Institute of Child Health and Human Development of the National Institutes of Health, Bethesda, Maryland, USA.

REFERENCES

CZERNICHOW, P., GREENBERG, A. H., TYSON, J. & BLIZZARD, R. M. (1971). *Pediat. Res.* **5**, 53.

DUSSAULT, J. H., ROW, V. V., LICKRISH, G. & VOLPE, R. (1969). *J. Clin. Endocr. Metab.* **29**, 595.

DUSSAULT, J. H., HOBEL, C. J. & FISHER, D. A. (1971). *Endocrinology* **88**, 47.

DUSSAULT, J. H., HOBEL, C. J., DISTEFANO, J. J., III, ERENBERG, A. & FISHER, D. A. *Endocrinology* **90**, 1301.

DUSSAULT, J. H., HOBEL, C. J. & FISHER, D. A. (1973). In preparation.

ERENBERG, A., OMORI, K., OH, W. & FISHER, D. A. (1972*a*). *Clin. Res.* **20**, 279.

ERENBERG, A., OMORI, K., OH, W. & FISHER, D. A. (1972*b*). *Pediat. Res.* **6**, 350.

ERENBERG, A., HOBEL, C. J. & FISHER, D. A. (1973). In preparation.

EVANS, T. C., KRETZCHMER, R. M., HODGES, R. E. & SONG, C. W. (1967). *J. Nuclear Med.* **8**, 157.

FISHER, D. A., ODELL, W. D., HOBEL, C. J. & GARZA, R. (1969). *Pediatrics* **44**, 526.

FISHER, D. A., HOBEL, C. J., GARZA, R. & PIERCE, C. A. (1970). *Pediatrics* **46**, 208.

FISHER, D. A., DUSSAULT, J. H., ERENBERG, A. & LAM, R. W. (1972). Submitted, *Pediat. Res.*

FISHER, D. A. & LAM, R. W. (1972). *Nature.* (In press.)

FISHER, D. A., CHOPRA, I. J. & DUSSAULT, J. H. (1972). *Endocrinology.* (In press.)

FISHER, W. D., VOORHESS, M. L. & GARDNER, L. I. (1963). *J. Pediat.* **62**, 132.

FUKUCHI, M., INOUE, T., ABE, H. & KUMAHARA, Y. (1970). *J. Clin. Endocr. Metab.* **31**, 565.

GITLIN, D. & BIASSUCCI, A. (1969). *J. Clin. Endocr. Metab.* **29**, 926.

GREENBERG, A. H., CZERNICHOW, P., REBA, R. C., TYSON, J. & BLIZZARD, R. M. (1970). *J. Clin. Invest.* **49**, 1790.

HENNEN, G., PIERCE, J. A. & FREYCHET, P. (1969). *J. Clin. Endocr. Metab.* **29**, 581.

HERSHMAN, J. M. & STARNES, W. R. (1971). *Prog. Endocr. Soc. Mtng*, p. A-131.

HODGES, R. E., EVANS, T. C., BRADBURY, J. T. & KEETEL, W. C. (1955). *J. Clin. Endocr. Metab.* **15**, 661.

MYANT, N. B. (1958). *Clin. Sci.* **17**, 75.

ODDIE, T. H., FISHER, D. A., DUSSAULT, J. H. & THOMPSON, C. S. (1971). *J. Clin. Endocr. Metab.* **33**, 653.

SCHULTZ, M. A., FORSANDER, J. B., CHEZ, R. A. & HUTCHINSON, D. L. (1965). *Pediatrics* **35**, 743.

SCHWARTZ, H. L., SHAPIRO, H. C., SURKS, M. I. & OPPENHEIMER, J. H. (1971). *J. Clin. Invest.* **50**, 2033.
SHEPARD, T. H., ANDERSEN, H. H. & ANDERSEN, H. (1964). *Anat. Record* **148**, 123.
SHEPARD, T. H. (1967). *J. Clin. Endocr. Metab.* **27**, 945.
SHEPARD, T. H. (1968). *Gen. Comp. Endocr.* **10**, 174.
STERLING, K. & BRENNER, M. A. (1966). *J. Clin. Invest.* **45**, 153.

RECENT STUDIES ON THE SEXUAL DIFFERENTIATION OF THE BRAIN

By K. BROWN-GRANT
Medical Research Council Neuroendocrinology Unit,
Department of Human Anatomy,
South Parks Road, Oxford OX1 3QX

1. INTRODUCTION

Goodman (1934) showed that cyclic ovulation occurred in ovarian grafts placed in the eye of an ovariectomized rat but that grafts in the eyes of male rats castrated as adults did not develop corpora lutea. Pfeiffer (1936) established that this functional difference was dependent upon the presence of the testis during the post-natal period. Early castration resulted in the female pattern of function in a genetic male when an ovary was grafted and transplantation of a testis in the neonatal period to a genetic female suppressed the ability to sustain cyclic ovulation. The presence or absence of ovaries during the neonatal period was found not to influence these changes. Suppression of receptivity in females receiving a testis graft was also observed. The imposition by the testis of male characteristics on a system otherwise destined to be female is fully in line with current views on the mechanism of sexual differentiation of the internal and external genitalia (Jost, 1972). Pfeiffer suggested that it was the functional capacity of the anterior pituitary that was altered but the evidence is now overwhelming that it is the brain that undergoes sexual differentiation during the neonatal period in the rat under the influence of an agent secreted by the testis (Harris, 1970).

In the twenty years following the publication of Pfeiffer's paper occasional reports appeared that indicated that the effects of a testis graft in suppressing ovulation could be duplicated by the repeated administration of gonadal steroids to female rats and mice during the neonatal period. The demonstration by Barraclough and Leathem for mice and by Barraclough for rats (see Barraclough, 1967) that a single injection of testosterone propionate in the early post-natal period was sufficient to produce the syndrome of anovulation, persistent vaginal cornification, and sterility, was a major contribution that came at a time of rapid development in the then new field of neuroendocrinology (Harris, 1955). The ease and reliability with which anovulatory, androgenized female rats and mice can be prepared in large numbers has allowed many aspects of the masculinization of the brain to be studied and the work has been reviewed in detail in recent publications (Flerko, 1971; Gorski, 1971; Goy, 1970; Harris, 1970). These will be cited as sources of references to the primary literature on particular aspects of this subject. The topics discussed in detail in this paper are those where it appears that current studies are yielding results that are likely, over the next few years, to stimulate further work which may lead to a better understanding of both the general nature and the detailed mechanism of the striking and permanent effects produced by transient changes in hormonal environment at a critical stage of development on the subsequent functional capacity of the brain.

2. NATURE OF THE ACTIVE HORMONAL AGENT

Unless otherwise indicated the studies described in this paper have been performed on rats. There is no doubt that the natural masculinizing agent comes from the testis nor that a single injection of testosterone propionate (TP) or repeated injections of free testosterone (T) during the neonatal period can produce the anovulatory syndrome. The current view of the mechanism of action of testosterone is that reduction of the double bond in Ring A to yield 5α-dihydrotestosterone (DHT) may be an essential step and that the reduction product DHT rather than T itself is the active agent at the level of the genome (Ohno, 1971). DHT is a potent androgen when administered systemically as judged by its effects on seminal vesicles and prostate gland but there are now several reports (cited by Davidson & Levine, 1972) that DHT or its 17-β propionate derivative, either alone or in combination with 5α-androstane-3α,17β-diol, were not effective in producing the anovulatory syndrome when injected into neonatal female rats. This finding raises some interesting questions. It is possible that at least some of the androgen 'receptors' in the brain differ from those in, for instance, the seminal vesicle. DHT did not maintain or restore male mating behaviour in the castrated male rat though there is evidence that DHT can act on other neurones forming part of the negative feedback system for the control of LH secretion and thought to be located in the medio-basal hypothalamus (Davidson & Levine, 1972). Further evidence that the brain 'receptors' for androgen may differ from those of peripheral tissues is provided by the finding that the synthetic steroids cyproterone or cyproterone acetate (CA), which are potent anti-androgens with respect to simulation of seminal vesicle growth by T or DHT, are themselves weak androgens with respect to male mating behaviour in the rat (Davidson & Bloch, 1969).

An alternative explanation for the failure of DHT to produce the anovulatory syndrome in female rats may be that unlike T it is not a potential oestrogen precursor. It has been known for some time that oestrogen as well as androgen can 'androgenize' the female rat with respect to ovulation and may be more effective (Flerko, 1971). This has been confirmed in recent studies on rats of the Wistar strain. Following the injection of 0.3, 1.0, 3.0 or 10.0 μg of oestradiol benzoate (OB) in oil s.c. on Day 4 of post-natal life, 18, 42, 64 and 87% respectively of female rats were anovulatory at 150 days of age. The corresponding figures after the injection of 3.0, 10.0, 30.0 or 100.0 μg of TP were 20, 69, 94 and 96%. The ability of hypothalamic or limbic system tissue, but not cerebral cortex, to convert androgen to oestrogen *in vitro* has recently been reported (Naftolin, Ryan & Petro, 1972) and preliminary results indicating that an anti-oestrogen (MER 25) can protect the neonatal female rat against the masculinizing action of TP have been reported (P. C. McDonald & C. Doughty, *J. Endocr.* (1972), **55**, 455.

Against the view that the active agent acting on the brain may be oestrogen, possibly generated *in situ*, are the findings that cyproterone acetate (CA) will protect the neonatal female rat from masculinization by TP and to some degree protect the genetic male rat from the effects of his own testicular secretion (see Gorski, 1971). These actions of CA need not necessarily be due to its anti-androgenic potency, however. As noted above CA can act as an androgen with respect to male mating behaviour in the rat, implying, presumably, some binding to a 'receptor' site on some neurones. Possibly it could also bind to neurones in the neonate and, not itself being aromatizable, interfere with the transformation of testosterone to oestrogen. CA is also a more potent progestational agent than progesterone itself. Progesterone can protect against masculinization by TP to some extent (Gorski, 1971) and the protective action of CA may be related to its progestational rather than its anti-androgenic properties. Finally there is some evidence that CA may be a weak anti-oestrogen in certain tests (Neri, Monahan, Meyer, Afonso & Tabachnik, 1967) and its protective action could be related to this.

The possibility that, paradoxically, masculinization of the brain may be brought about by oestrogen raises an intriguing question as to the state of the brain in genetic males with the syndrome of testicular feminization. The basis of this disorder, which has been described in rats, mice, bulls and man, is a genetically determined inability of the tissues to respond to T or DHT (Ohno, 1971). If an ovary were transplanted into a genetic male mouse or rat with this disorder after castration as an adult, would cyclic ovulation occur or not? If sexual differentiation of the brain is due to oestrogen, then ovulation should not occur provided adequate amounts of aromatizable androgen had been produced during the neonatal period. Behavioural studies after castration and hormone replacement do not appear to have been carried out on either rats or mice with this syndrome, but would also be of great interest (see Section 4).

3. THE FUNCTIONAL BASIS OF THE ANOVULATORY STATE

The absence of ovulation in the female rat treated with steroids neonatally and in the male castrated when adult and given an ovarian graft could, theoretically, be due either to an abnormal basal level of gonadotrophin secretion resulting in an inadequate or excessive secretion of steroid hormones by the ovary or to the failure of a relatively normal output of steroid hormones from the ovary to trigger the ovulatory surge of gonadotrophin secretion. The evidence in favour of the first possibility is neither extensive nor convincing. Despite earlier reports to the contrary, based on measurements by bioassay, basal plasma levels of LH and FSH in the androgenized female determined by immunoassay were close to those found in normal females at the dioestrous stage of the cycle and plasma oestradiol concentrations, though above the values seen at the metoestrous stage of the cycle, were below peak values observed on the morning of pro-oestrus (Johnson, 1971; Naftolin, Brown-Grant & Corker, 1972). Moreover if exogenous gonadotrophin was used to stimulate ovarian function in androgenized females before puberty, the output of oestrogen was increased as in the normal rat but neither the spontaneous nor the progesterone-induced ovulation seen in normal rats occurred in the androgenized animals (Brown-Grant, Quinn & Zarrow, 1964).

A defective capacity to respond to normally adequate stimulation by ovarian steroids with an abrupt increase in gonadotrophin secretion seems more likely to be the explanation. Such a facilitatory action, sometimes loosely referred to as a positive feedback action, of oestrogen (or possibly oestrogen plus progesterone in the rat) is now generally accepted as the likely mechanism for the triggering of the ovulatory surge of gonadotrophins in species that ovulate spontaneously (Schwartz & McCormack, 1972). Possible abnormalities of the facilitatory actions of steroids were looked for in experiments on gonadectomized rats based on the procedures developed by Taleisnik, Caligaris & Astrada (1969).

Normal male and female rats and female rats that were anovulatory following the injection of either 1.25 mg of TP or 250 μg of OB on Day 4 of post-natal life were gonadectomized as adults and tested 1 to 2 months later. Plasma LH concentrations were determined by immunoassay before

Table 1. *Plasma radioimmunoassayable LH concentrations (ng/ml NIH-S13-LH) in normal male and female Wistar rats and females treated with TP or OB on Day 4 of post-natal life. Values are means ± SEM and figures in parentheses indicate numbers of animals*

Type of animal	Intact	Gonadectomized (1–2 months)	Post-OB (72 h)	Post-OB + OB (5 h)	Post-OB + P (5 h)
Normal female	2 ± 1 (18)	33 ± 3 (22)	16 ± 2 (16)	205 ± 36 (7)	215 ± 27 (8)
Normal male	2 ± 1 (14)	54 ± 3 (6)	21 ± 5 (6)	8 ± 2 (7)	4 ± 1 (6)
TP-treated female	2 ± 1 (8)	16 ± 5 (6)	9 ± 1 (6)	4 ± 1 (6)	7 ± 1 (9)
OB-treated female	1 ± 1 (11)	7 ± 3 (11)	5 ± 1 (6)	6 ± 2 (5)	4 ± 1 (6)

treatment, 72 h after the i.m. injection of 20 μg of OB in oil and 5 h after a second i.m. injection of either 20 μg OB or 2.5 mg of progesterone (P) at 12.00, 72 h after the initial priming injection of oestrogen. The results were clear cut (Table 1); the only animals showing an increase in plasma LH concentration after the second injection of steroid were the normal females. Males and previously anovulatory females showed no change or a decrease following the second injection. The increase in plasma LH following ovariectomy was somewhat reduced in the androgen-treated females and greatly reduced in the oestrogen-treated females. All types of rats showed a decrease in plasma LH after the first injection of steroid. Using a different schedule of steroid treatment Neill (1972) has also shown that the facilitatory action of oestrogen on LH secretion was absent in males and neonatally androgen-treated females but was present in normal females and males castrated early in the neonatal period. He also demonstrated that similar differences existed in the occurrence of an increase in prolactin secretion following oestrogen administration.

In summary, the female rat treated with either androgen or oestrogen neonatally is rendered anovulatory and loses the capacity to show the facilitatory action of oestrogen or oestrogen plus progesterone on gonadotrophin secretion. Conversely, the male castrated at birth but not the male castrated when adult can sustain cyclic ovulation and can show these positive responses. It seems likely that the defects in these facilitatory responses are a major factor in the genesis of the anovulatory state.

4. ALTERATIONS IN BEHAVIOUR

On the basis of experiments performed on guinea pigs, Phoenix, Goy, Gerall & Young (1959) proposed that androgenic hormones had an organizational action on the brain during development with respect to the type and quality of sexual behaviour that the animal would show after treatment with exogenous hormones when adult. Rather similar results to theirs have been reported by many workers who have studied both homo- and heterotypical behaviour in rats, mice and hamsters treated with steroids or castrated during the neonatal period. Either androgens or oestrogen are reported to reduce or suppress the capacity to show the female type of behaviour and to enhance the capacity to display male behaviour. In contrast to the effects of hormone treatment or neonatal castration on the pattern of gonadotrophin secretion, the effects on sexual behaviour are neither easy to assess nor clear cut and quite divergent views have been expressed about the heuristic value of the concept of organization of the brain when applied to sexual behaviour (Beach, 1971; Goy, 1970; Whalen, 1968).

Male behaviour in the laboratory rodent is studied quantitatively by measuring both the frequency and the latency of mounting, penile intromission and ejaculation when the test animal is placed with a receptive female (Davidson & Bloch, 1969). The interpretation of changes in male behaviour induced by neonatal steroid administration or castration is complicated by the fact that unless the phallus is exposed to androgen during the neonatal period, growth in response to androgen administered when adult will not be normal. This difficult and as yet unresolved problem is discussed by Beach (1971) and by Davidson & Levine (1972).

In the female rodent a crude index of sexual behaviour can be obtained by housing the females with males overnight and examining the vaginal smear for the presence of sperm the next morning. A more precise method is to observe directly the behaviour of the test animal when placed with an active male. When mounted a receptive animal stops, arches its back and elevates the head and pelvis. The tail is either elevated or moved to one side. This is the *lordosis response*, and a generally accepted measure of female behaviour in the rat is the Lordosis Quotient (LQ), the number of mounts resulting in a lordosis response (which may or may not be associated with penile intromission and possibly ejaculation) divided by the total number of mounts occurring during the test and multiplied by 100 (range of values 0–100). Both methods have been used to study the capacity for female mating behaviour in normal and experimental rats. The results have provided evidence against the view that the ability to display lordosis can be suppressed in the female and is suppressed in the normal male rat but at the same time they have provided support for the view that exposure to either androgen or oestrogen in the neonatal period does alter the responsivity of the brain to the hormonal stimuli that induce female mating behaviour and also modifies the effects of mating on gonadotrophin secretion.

Table 2. *Incidence of mating in anovulatory rats and of ovulation in rats that mated for the Wistar strain treated with different doses of TP or OB on Day 4 of post-natal life and tested at 100–150 days of age*

Dose of OB (μg)	3	10	30	100	250		
Numbers mating	14/14	11/13	10/11	10/26	4/27		
%	100	85	91	38	15		
Numbers ovulating	12/14	4/11	4/10	0/10	0/4		
%	86	36	40	0	0		
Dose of TP (μg)	3	10	30	100	250	1250	2500
Numbers mating	5/5	26/26	28/30	44/45	27/28	35/35	8/8
%	100	100	93	98	96	100	100
Numbers ovulating	2/5	4/20	5/28	5/42	1/27	0/35	0/8
%	40	20	18	12	4	0	0

In the first series of experiments intact female rats that had received various doses of either OB or TP on Day 4 of postnatal life and were anovulatory at 100 to 150 days of age were caged with males overnight. The occurrence of mating was confirmed by the presence of sperm in the vaginal smear the next morning and the occurrence of ovulation by examining the oviduct for newly shed eggs (Table 2). Under these conditions a very high percentage of all TP-treated rats mated, even those treated with high doses. Many rats that had received 100 or 250 μg of OB neonatally failed to mate. However these animals had very small ovaries and uteri and the vaginal smears were comparable to those obtained at the dioestrous stage of the normal cycle, i.e. these rats were showing the persistent dioestrous syndrome (Flerko, 1971; Gorski, 1971). When adequate exogenous oestrogen was provided such animals also became receptive (Table 3).

The retention by the rats treated with low doses of steroid of the ability to ovulate following mating despite the suppression of 'spontaneous' ovulation is an interesting observation. It raises some intriguing questions as to the ontogenetic development of a similar functional state in animals such as the rabbit, cat and ferret that are normally 'induced' ovulators.

A decreasing proportion of the rats that mated ovulated as a result of mating as the dose of steroid administered on Day 4 was increased (Table

Table 3. *Lordosis quotients (LQ) for groups of gonadectomized rats tested as adults after treatment with oestrogen alone or oestrogen plus progesterone. Values are group means ± SEM though this is not a very appropriate measure of variation. LQ without treatment was always between 0 and 4*

Type of rat	No.	LQ after oestrogen alone	LQ before progesterone	LQ 4–6 h after progesterone
Wistar, adult when gonadectomized				
Normal female	7	76 ± 8	24 ± 7	97 ± 1
Normal male	7	86 ± 4	30 ± 4	4 ± 2
Female, TP Day 4	8	79 ± 11	48 ± 12	28 ± 9
Normal female	8	80 ± 5	60 ± 7	99 ± 1
Female, OB Day 4	6	68 ± 8	28 ± 5	14 ± 3
Normal female	6	79 ± 8	41 ± 4	98 ± 2
Lister males				
Castrated Day 1	5	32 ± 10	*81 ± 6	100 ± 0
Castrated Day 7	5	11 ± 3	44 ± 6	66 ± 9
Castrated Day 60	8	10 ± 4	41 ± 7	40 ± 4

* Separate experiment.

2). It seems probable that the failure of these animals to ovulate was due to a failure to release LH in response to the stimulus of mating. Eight intact rats that had been treated with 100 μg of TP on Day 4 of life and were not ovulating were mated under observation at 150 days of age (mean LQ 91 ± 5) and blood was taken for determination of LH concentration 40 min later. Each rat received 20–25 penile intromissions over a period of 15 min, sufficient stimulation to ensure ovulation in more than 90% of rats of this strain in persistent oestrus as a result of exposure to constant light. Plasma LH concentration (ng/ml NIH-S13-LH) was 1.6 ± 0.1 before mating and 2.5 ± 0.5 40 min after mating. Corresponding values for rats exposed to constant light were 1.5 ± 0.2 (mean of 7) and 47.5 ± 13.7 (mean of 10) (K. Brown-Grant, J. M. Davidson & Fenella Greig, unpublished).

The high proportion of rats mating in these experiments was unexpected. More detailed tests confirmed the high degree of spontaneous receptivity in neonatally androgen-treated rats of this strain (Wistar rats, obtained from Scientific Products Farm, Margate, Kent) and extended the observations to other strains. Some of the results obtained are shown in Table 3. Gonadectomized animals received 50 μg OB i.m. daily for 2 or 3 days before being tested. After a lapse of several days they were retested and if the LQ was then low received 1.25 mg of progesterone i.m. and were retested 4–6 h later. It is clear that high levels of receptivity could be induced by treatment with large doses of oestrogen alone in normal females as expected but also in males of the Wistar strain castrated as adults or females rendered anovulatory by the injection of 1.25 mg TP or 250 μg OB on Day 4 and ovariectomized as adults. What was equally striking was that facilitation of the lordosis response by progesterone is seen only in the normal females. Males of a second strain (Hooded Lister, also obtained from Scientific Products Farm) showed a lower response to oestrogen but illustrated that facilitation by progesterone was retained in males castrated in the neonatal period (Table 3).

The high levels of receptivity shown by females that had received large doses of steroids in the neonatal period were not in agreement with the majority of previous reports though retention of receptivity despite suppression of ovulation by low doses of steroid is well known (Gorski, 1971). These results are perhaps less surprising when the fact that males of this strain castrated as adults also showed a high level of receptivity is kept in mind. This last finding is in agreement with recent studies of Davidson & Levine, as is the lack of facilitation by progesterone in the male castrated as adult and the retention of this characteristic in the neonatally castrated male. The absence of progesterone facilitation in the androgenized female confirms the findings of Clemens, Shrine & Gorski (see Davidson & Levine, 1972). The reasons for the disagreements between different groups of workers as to the degree of receptivity observed in males and in females treated with steroids in the neonatal period are not obvious. Strain differences may be important but do not provide a complete explanation. For instance Wistar and commercially available Lister rats given 1.25 mg TP on Day 4 and anovulatory when tested intact at 90–100 days of age had a mean LQ of 89 ± 3 (mean of 12) and 79 ± 5 (mean of 9) respectively. In contrast Lister rats from the closed colony used in the original studies of Harris & Levine (1965) treated and tested in the same way had a mean LQ of only 24 ± 9 (mean of 15) though after treatment with high doses of oestrogen for several days (50–100 μg OB) the mean LQ was 70 ± 11.

The conditions of testing may be of critical importance. The work reported here is based on tests made with two experienced and active males for each test animal. The tests were run for two consecutive periods of 10 min and the test animals, generally an experimental and a control, were exchanged between the two test cages in use simultaneously after the first 10 min. A high rate of mounting was ensured by changing any sluggish males and immediately replacing any who ejaculated; generally 50–100 mounts were observed over the 20 min test period. It appears from the literature that in many cases the LQ on which the assessment of receptivity is made is calculated from results of 5–10 mounts or from an unspecified but probably small number of mounts achieved by a single male over a 5–10 min test period. Such procedures may be adequate when the test animal is a normal female but may be misleading when experimental animals are being tested, particularly when the *capacity* to display lordosis is the essential point under investigation. Abnormal females and males have frequently been observed to show no or few lordosis responses over the first few minutes of testing or the first 10–20 mounts and yet subsequently to respond in this manner to 90–100% of mounts. Selected records for individual androgen-treated rats and mean values for randomly selected normal female rats of the same strain are shown in Table 4. Lister refers to rats of the strain used by Harris & Levine (1965). It is obvious that the LQ based on a few mounts gives a very different picture of the receptivity of the androgen-treated female from that based on prolonged testing. Under very different experimental conditions a progressive increase in receptivity with prolonged testing has also been observed by Clemens, Hiroi & Gorski (Gorski, 1971).

Table 4. *Rapid and progressive increases in the LQ may occur during the course of a prolonged test of receptivity in neonatally androgen-treated female rats* (1.25 *mg TP on Day* 4)

Type of rat		LQ for first 6 mounts	LQ for first 10 mounts	LQ for 0–10 min	LQ for 10–20 min	Overall LQ
Intact Wistar female, TP Day 4		33	60	73	100	81
		17	40	63	90	73
		0	20	24	66	50
		17	20	29	43	33
		13	20	37	58	45
	Mean of 5	16±5	32±8	45±10	71±10	56±9
Normal Wistar female, ovariectomized and treated with oestrogen	Mean of 5	54±16	53±17	49±16	46±21	47±17
Intact Lister TP Day 4, treated with oestrogen		0	0	61	100	76
		0	10	22	95	61
		17	50	76	100	78
		0	30	56	55	56
		0	30	70	100	81
	Mean of 5	3±3	24±9	57±9	90±9	70±5
Normal Lister, intact treated with oestrogen	Mean of 5	75±20	75±19	78±14	74±9	76±11

The 'response probability' for the lordosis response in the rat under particular experimental conditions was clearly altered by neonatal steroid treatment or neonatal castration, particularly with respect to the test situation where progesterone was administered to an animal already under the influence of a low level of oestrogen. Beach (1971) argues forcibly that interpretation of experimental results in this field in terms of altered 'response probability' is the most likely to lead to a better understanding of the nature of the changes produced. His article is strongly recommended to anyone interested in the important problem of early hormonal influences on later sexual behaviour.

The ease with which large numbers of 'masculinized' rats, mice and hamsters can be prepared has enabled studies to be performed on many aspects of behaviour apart from mating that are known to differ between the sexes in order to determine whether the differences are related to early hormonal effects. In many cases the exhibition of these patterns of behaviour, unlike mating, is not dependent on the gonadal hormone levels at the time of testing, and as Goy (1970) points out this excludes altered sensitivity to gonadal hormones as an explanation for the differences observed. References to studies in this area are given by Davidson & Levine (1972) and Gorski (1971).

5. DIRECT STUDIES OF THE EFFECTS OF NEONATAL STEROID ADMINISTRATION ON THE BRAIN

The available evidence strongly suggests that the pre-optic area (POA) is the region of the brain where the integration of the neural and hormonal inputs that together initiate the ovulatory surge of gonadotrophins occurs and that the POA is also concerned in mating behaviour (Raisman & Field, 1971*a*). It is not surprising, then, that this area has received particular attention in studies in which the distribution of radioactive steroids, the effects of steroid administration on RNA and protein synthesis, and possible morphological differences have been studied.

No evidence of preferential uptake and retention of testosterone by specific regions of the brain after administration to neonatal rats has been obtained and preferential uptake of labelled oestradiol does not develop until about Days 15–20 of post-natal life (Gorski, 1971). Possible long-term changes in the oestrogen 'receptor' content of the brain after neonatal steroid administration have been sought but the evidence is conflicting (Flerko, 1971).

Gorski (1971) describes experiments in which various agents known to interfere with protein or RNA synthesis have been tested for their ability to prevent masculinization of the brain by exogenous androgen. Changes in brain metabolism following neonatal steroid administration have not been extensively studied by direct methods. The work that has been carried out appears to have been greatly influenced by the studies of Gorski and his associates in which a variety of 'blocking' agents were used and which suggested that irreversible changes in the brain were produced within a few hours by the administration of TP. It should be noted, however, that some of the earlier results have proved difficult to reproduce (see Gorski, 1971). In addition the findings of Alklint & Norgren (1970) have suggested that exposure to androgen for days, rather than hours, was necessary and it has recently been shown that a small amount of TP was much more effective in blocking ovulation when given in divided doses over a period of days than when given as a single injection (P. J. Sheridan, M. X. Zarrow & V. H. Denenberg, personal communication, 1972).

Clayton, Kogura & Kraemer (1970) reported that testosterone depressed RNA synthesis in all brain areas studied except the medial amygdala and the medial pre-optic area 3–4 h after injection. It should be noted that the treatment they gave (50 μg of free testosterone in propylene glycol) would probably not have been sufficient to produce the anovulatory syndrome. In Oxford, H. K. Darrah, B. Liddiard, P. C. B. MacKinnon & A. W. Rogers (unpublished) have also used autoradiographic methods to examine possible differences between male and female rats and female rats given either 1.25 mg TP or 60 μg OB in oil with respect to protein synthesis in different areas of the brain 10–12 h after steroid administration on Day 4 of post-natal life. They compared grain densities over the medial pre-optic area, medial and lateral portions of the amygdala with those over control areas (putamen, ventrolateral nucleus of the thalamus and an area of neocortex). No significant differences were found in the medial pre-optic area or lateral amygdala between males and females or between control females and females injected with androgen or oestrogen. However the ratios of grain density over the medial amygdala to grain densities over the control areas were slightly higher in males than in females and were significantly higher in androgen-treated (but not oestrogen-treated) females than in control females (see p. 545). No differences between males, females and treated females were seen in any area in rats examined at 30 days of age.

The suggestions of early changes in the medial amygdala are of interest as the corticomedial nucleus projects to the medial pre-optic area via the stria terminalis. The sensitivity of the POA to electrical or electrochemical stimulation of LH release may be abnormal in androgen-treated female rats when adult and the normal male differs from the female. When LH release following chemical or electrochemical stimulation of the medial amygdala is examined the differences are more striking, normal females responding but males and androgenized females failing to respond (see Arai (1971*a*) for references). These findings could be taken to indicate that the efferent

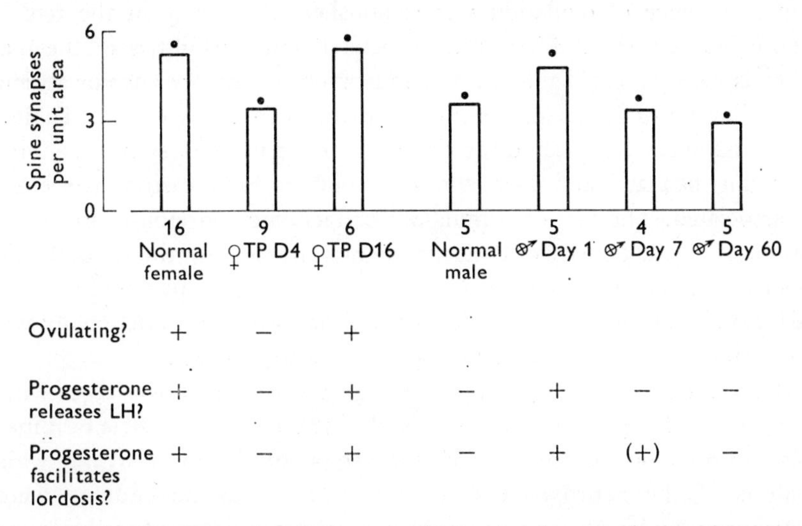

Fig. 1. The density of normal synapses on dendritic spines (expressed as number per unit area) in the region of the medial pre-optic area of Lister rats defined by the presence of degenerating synapses after section of the stria terminalis (G. Raisman & Pauline Field, unpublished). The dots indicate SEM. Details of the treatments and of the functional tests are given in the text.

pathway from the POA is intact and potentially functional in both male and female rats and androgenized females but that the pattern of afferent connections was qualitatively or quantitatively altered in males and androgenized females in such a way that the responses to direct or indirect stimulation differed from those seen in normal females. A possible morphological basis for such a functional difference has recently been described.

The first stage was a quantitative analysis at the electron microscopic level of the pattern of synaptic endings in the neuropil of the areas of the MPOA and of the ventromedial nucleus of the hypothalamus that contained degenerating synapses after section of the stria terminalis (Raisman & Field, 1971*a*) which was extended to include a comparison between male and female rats of the Wistar strain. A significant difference was established between the sexes in the ratio of non-degenerating synapses on dendritic spines and to those on dendritic shafts in the MPOA but not in the ventromedial nucleus, the female MPOA showing a relative excess of endings on dendritic spines (Raisman & Field, 1971*b*). These findings have been confirmed in rats of the Lister strain (G. Raisman & P. Field, unpublished) and extended to include females treated with 1.25 mg TP on Day 4 and males castrated on Day 1, Day 7 or Day 60 of post-natal life. Some preliminary results are shown in Fig. 1 where the index used is density of synapses on dendritic spines per unit area. The occurrence or non-occurrence of ovulation was established at autopsy in the females whose brains were studied and the behavioural capabilities of the male animals were tested (Table 2). The indications of the presence or absence of facilitation of LH release are extrapolations from the data in Table 1. The functional changes and the changes in the pattern of synaptic endings following neonatal androgen administration or early castration seem to be associated. The loss of stimulatory or facilitatory responses to progesterone when the density of dendritic spine synapses and the ratio of spine to shaft synapses is reduced is consistent with the current hypothesis that this ultrastructural feature of the neuropil is associated functionally with the excitatory process (Diamond, Gray & Yasargil, 1970).

It should be emphasized that although the presence of a *degenerating* synapse in the electron microscopic field after section of the stria terminalis was the operational definition of the region of the POA within which analysis of the neuropil was conducted, the synaptic endings whose distribution was analysed and found to be sexually differentiated were not degenerating, i.e. do not belong to neurones whose cell bodies are in the corticomedial nucleus of the amygdala and whose axons travel in the stria terminalis to the POA. The origin of the fibres whose synapses show this sexually differentiated distribution is not known at present but is clearly of great interest, as are the projections of the neurones receiving these afferents.

6. SUMMARY OF THE FINDINGS IN RODENTS AND POSSIBLE IMPLICATIONS FOR OTHER SPECIES

An anovulatory state dependent upon alterations in brain function can be produced consistently in rats, mice and hamsters by the early post-natal administration of certain androgens or oestrogens or by pre-natal exposure to androgens in the guinea-pig. Where it has been tested, early castration of the genetic male prevents the normal loss of the capacity to secrete gonadotrophins cyclically when an ovary is transplanted. In the rat the functional defect appears to be the inability of oestrogen or oestrogen plus progesterone to facilitate gonadotrophin secretion. Associated with these endocrine abnormalities are alterations in the probability that homo- or heterotypical mating behaviour will be exhibited under different conditions of hormonal stimulation and under different conditions of testing. Changes in many types of non-sexual but sexually dimorphic behaviour have also been observed to follow experimental manipulation of gonadal steroid hormone levels in the early post-natal period. In the rat certain recent findings suggest that, paradoxically, the active agent at the cellular level may be oestrogen rather than androgen though in the natural process of sexual differentiation the testis is the source of the hormone responsible. In general there seems to be only a limited period during development when these effects can be easily and reproducibly obtained. However recent studies on rats showing a delayed onset of anovulation following the administration of low doses of androgen in the neonatal period suggest that oestrogen may still contribute to the development of the anovulatory state in the adult (Arai, 1971*b*). This possibility should be investigated in other species. The biochemical basis of the masculinizing action of steroid on the brain is unknown, as is the precise site which is affected. Recent neuroanatomical studies have shown that a sexual dimorphism normally exists in the pattern of synaptic endings in a restricted portion of the medial pre-optic area in the rat. Experimental procedures that alter the pattern of gonadotrophin secretion and mating behaviour also alter the pattern of synaptic endings in this area in the appropriate direction, suggesting a functional importance for these morphological differences.

The other spontaneously ovulating mammalian species widely available for studies are the domestic and farm animals (dog, sheep, goat, cow, pig), the sub-human primates and man. All these are species with extended gestation periods and all have placentae with the necessary enzymes to form oestrogens from androgens (Ryan, 1969). The timing of testis development and of the differentiation of the genital tract and external genitalia suggests that the period of sexual differentiation of the brain, if it occurs, will be pre-natal except perhaps in the dog.

Though naturally occurring abnormalities of sexual development in domestic and farm animals are well documented (Short, 1970) little experimental work seems to have been performed with the objective of establishing the normal neuroendocrine and behavioural differences between the sexes and attempting to influence them by pre-natal hormone treatment, other than the behavioural studies of Beach on dogs (see Davidson & Levine, 1972). If one characteristic of the masculinized brain is an inability to mediate a facilitatory action of oestrogen on gonadotrophin secretion (see Section 3) then the ram, intact or castrate, could be tested very easily; facilitation of LH release by oestrogen in the ewe has been demonstrated by several groups (see Schwartz & McCormack, 1972).

Experimental studies on the Rhesus monkey are discussed by Goy (1970) and Goy & Phoenix (1971). Repeated administration of TP to pregnant monkeys resulted in female offspring with masculinized external genitalia. These animals showed a striking delay in the onset of menstruation but subsequently had cycles of normal length and in later cycles measurements of plasma progesterone in the luteal phase showed that at least 5 out of 9 animals were still ovulating at the age of 4–5 years. It will be of great interest to learn whether these animals later become anovulatory. If they do, then they may perhaps be like the rats with the syndrome of delayed anovulation, still ripening follicles and secreting oestrogen.

Tests of the ability of oestrogen to stimulate LH release at this stage in these animals would be of great value; the effects of oestrogen on plasma LH levels in gonadectomized normal male and female monkeys have already been shown to differ (Yamaji, Dierschke, Hotchkiss, Bhattacharya, Surve & Knobil, 1971). On the other hand it may be that if they become anovulatory these monkeys will be found to be in a state analogous to the menopause. Normally the monkey (like all mammals other than man) does not show this phenomenon (Krohn, 1955), which in the human is generally considered to be related to exhaustion of the ovarian stock of oocytes. It may be that the treatment which androgen has reduced the oocyte stock in these monkeys as it has been reported to do in the mouse (Peters, Sørensen, Byskov, Pedersen & Krarup, 1970).

Behavioural studies of the androgenized monkeys during the prepubertal period have shown a clear diversion of the patterns of play and of the development of mounting behaviour in the male direction. It may be difficult to determine whether female sexual behaviour as adults has been affected because of the anatomical abnormalities and also because of the marked differences in patterns of behaviour between different male–female pairs and the influence of the hormonal state of the female on the performance of the male (Michael, 1971).

The situation in the human appears to be similar in some respects to that in the Rhesus monkey. Masculinization of the external genitalia was observed at birth in girl babies unwittingly exposed to exogenous progestational steroids with androgenic properties administered to the mother during pregnancy, or exposed to endogenous androgen secreted in abnormal amounts by their own adrenal cortices in cases of congenital adrenal hyperplasia (Federman, 1968). These latter cases do not menstruate if untreated but if adequately treated and reared as girls after surgical correction of any anatomical abnormality, they ovulate regularly and can become pregnant. Odell & Swerdloff (1968) showed that progesterone facilitation of LH and FSH release after oestrogen priming occurred in the majority of women tested but not in intact or castrated men. This response is probably not the basis of the ovulatory surge of gonadotrophins in women (see Schwartz & McCormack, 1972) so that it may have been abolished or modified in these cases of female pseudohermaphroditism, and this possibility could be tested.

Patterns of sexual behaviour in the human are so influenced by gender assignment and social conditioning that it is difficult to assess any possible effects of pre-natal exposure to androgens in genetic females or the effective lack of exposure in genetic males with the syndrome of testicular feminization. Money (1971) has reviewed the available data, much of it collected by his own group. Briefly, adult sexual behaviour in androgenized females correctly assigned and treated shows no clear abnormality; attitudes towards the maternal role may be rather negative in contrast to the possibly unmasculinized genetic males with testicular feminization who show a completely female attitude to both sex and child rearing. Two features of the androgen-exposed females are of interest; the first is a tendency towards 'tomboyishness' in their pre-pubertal play which is very reminiscent of the findings in androgenized monkeys, and the second is an apparently well-validated increase in intelligence quotient when compared to a closely matched control group of normal women.

It seems highly probable that the concept, developed mainly from work on rodents, of an action of an agent of testicular origin on the brain at a critical stage of development to alter the ability to respond to certain steroid hormone inputs with an increase in gonadotrophin secretion and to alter the reactivity of the neural systems serving both sexual and some forms of non-sexual behaviour, will be found to be valid for all mammals. Present evidence suggests, however, that the manifestations of these changes in non-rodent species may be more subtle and more difficult to quantify.

Two recently published notes (Anon., 1972; Fortier, 1972) have emphasized the dominant role played by the late G. W. Harris in the development of neuroendocrinology in all its aspects. His own early contributions to the problem of sexual differentiation of the brain and its experimental control are reviewed in his Upjohn Lecture (Harris, 1964) and a definitive account of this and later work was in preparation at the time of his death in November 1971. He had also begun and until shortly before his death was actively engaged on work in Oxford on the effects on the offspring of androgen and anti-androgen treatment of pregnant Rhesus monkeys. At the conclusion of his posthumously published Dale Lecture (*J. Endocr.* **53**, i–xxiii) he asked 'And also, what benefits will this work and knowledge confer on human welfare?' Hopefully, the benefit from current studies of the type described here will be a better understanding of the actions of hormones on the brain during development and their effects on subsequent function and of the possible role of such effects in the genesis of aberrant behaviour in man.

I would like to express my thanks to colleagues both in the Unit and the Department and elsewhere who not only gave me access to unpublished work but also gave their valuable time to discuss it with me. Dr Geoffrey Raisman and Dr Andy Rogers read this paper at an early stage of preparation and made many helpful suggestions.

REFERENCES

ALKLINT, T. & NORGREN, A. (1970). *Acta Physiol. Scand.* **80**, 5A (Abstr.).

ANON. (1972). *J. Endocr.* **53**, 185–6.

ARAI, Y. (1971*a*). *Endocrinol. Japonica* **18**, 211–14.

ARAI, Y. (1971*b*). *Experientia* **27**, 463–4.

BARRACLOUGH, C. A. (1967). Chap. 19, pp. 62–100. In *Neuroendocrinology*, Vol. 2. Eds. L. Martini & W. F. Ganong. New York and London: Academic Press.

BEACH, F. A. (1971). Pp. 249–96. In *The Biopsychology of Development*. Eds. E. Tobach, L. R. Aronson & E. Shaw. New York and London: Academic Press.

BROWN-GRANT, K., QUINN, D. L. & ZARROW, M. X. (1964). *Endocrinology* **74**, 811–13.

CLAYTON, R. B., KOGURA, J. & KRAEMER, G. C. (1970). *Nature* **226**, 810–12.

DAVIDSON, J. M. & BLOCH, G. (1969). *Biol. Reprod., Supplement 1*, 67–92.

DAVIDSON, J. M. & LEVINE, S. (1972). *Ann. Rev. Physiol.* **34**, 375–408.

DIAMOND, J., GRAY, E. G. & YASARGIL, G. M. (1970). Pp. 213–22. In *Excitatory Synaptic Mechanism*. Eds. J. Jansen & P. Andersen. Oslo: Scandinavian University Books.

FEDERMAN, D. D. (1968). *Abnormal Sexual Development*. Philadelphia: W. B. Saunders Company.

FLERKO, B. (1971). Pp. 41–80. In *Current Topics in Experimental Endocrinology*. Eds. L. Martini & V. H. T. James. New York and London: Academic Press.

FORTIER, C. (1972). *Endocrinology* **90**, 851–4.

GOODMAN, L. (1934). *Anat. Rec.* **59**, 223–51.

GORSKI, R. A. (1971). Chap. 9, pp. 237–90. In *Frontiers in Neuroendocrinology, 1971*. Eds. L. Martini & W. F. Ganong. New York and London: Oxford University Press.

GOY, R. W. (1970). *Phil. Trans. R. Soc. B.* **259**, 149–62.

GOY, R. W. & PHOENIX, C. H. (1971). Chap. 19, pp. 193–202. In *Steroid Hormones and Brain Function*. Eds. C. H. Sawyer & R. A. Gorski. Los Angeles: University of California Press. U.C.L.A. Forum Med. Sci. No. 15.

HARRIS, G. W. (1955). *Neural Control of the Pituitary Gland*. London: Arnold.

HARRIS, G. W. (1964). *Endocrinology* **75**, 627–48.

HARRIS, G. W. (1970). *Phil. Trans. R. Soc. Lond. B.* **259**, 165–77.

HARRIS, G. W. & LEVINE, S. (1965). *J. Physiol.* **181**, 379–400.

JOHNSON, D. C. (1971). *Proc. Soc. Exp. Biol. Med.* **138**, 140–4.

JOST, A. (1972). *Johns Hopkins Med. J.* **130**, 38–53.

KROHN, P. L. (1955). Pp. 141–61. In *Ciba Foundation Colloquia on Ageing*, Vol. 1. Eds. G. W. Wolstenholme & M. Cameron. London: Churchill.

MICHAEL, R. P. (1971). Chap. 12, pp. 359–98. In *Frontiers in Neuroendocrinology, 1971*. Eds. L. Martini and W. F. Ganong. New York and London: Oxford University Press.

MONEY, J. (1971). Chap. 30, pp. 325–38. In *Steroid Hormones and Brain Function*. Eds. C. H. Sawyer & R. A. Gorski. Los Angeles: University of California Press.

NAFTOLIN, F., BROWN-GRANT, K. & CORKER, C. S. (1972). *J. Endocr.* **53**, 17–30.

NAFTOLIN, F., RYAN, K. J. & PETRO, Z. (1972). *Endocrinology* **90**, 295–8.

NEILL, J. D. (1972). *Endocrinology* **90**, 1154–9.
NERI, R. O., MONAHAN, M. D., MEYER, J. G., AFONSO, B. A. & TABACHNIK, I. A. (1967). *Europ. J. Pharmacol.* **1**, 438–44.
ODELL, W. D. & SWERDLOFF, R. S. (1968). *Proc. Nat. Acad. Sci. U.S.A.* **61**, 529–36.
OHNO, S. (1971). *Nature* **234**, 134–7.
PETERS, H., SØRENSEN, I. N., BYSKOV, A. G., PEDERSEN, T. & KRARUP, T. (1970). Pp. 351–61. In *Gonadotrophins and Ovarian Development*. Ed. W. R. Butt., A. C. Crooke & M. Ryle. London: Livingstone.
PFEIFFER, C. A. (1936). *Am. J. Anat.* **58**, 195–225.
PHOENIX, C. H., GOY, R. W., GERALL, A. A. & YOUNG, W. C. (1959). *Endocrinology* **65**, 369–82.
RAISMAN, G. & FIELD, P. M. (1971*a*). Chap. 1, pp. 3–44. In *Frontiers in Neuroendocrinology, 1971*. Eds. L. Martini & W. F. Ganong. New York and London: Oxford University Press.
RAISMAN, G. & FIELD, P. M. (1971*b*). *Science* **173**, 731–3.
RYAN, K. J. (1969). *Excerpta Medica Int. Congress Series* **183**, 120–31.
SCHWARTZ, N. B. & MCCORMACK, C. E. (1972). *Ann. Rev. Physiol.* **34**, 425–72.
SHORT, R. V. (1970). *Phil. Trans. R. Soc. B.* **259**, 141–8.
TALEISNIK, S., CALIGARIS, L. & ASTRADA, J. J. (1969). *J. Endocr.* **44**, 313–21.
WHALEN, R. E. (1968). Pp. 303–40. In *Perspectives in Reproduction and Sexual Behaviour*. Ed. M. Diamond. Bloomington: Indiana University Press.
YAMAJI, T., DIERSCHKE, D. J., HOTCHKISS, J., BHATTACHARYA, A. N., SURVE, A. H. & KNOBIL, E. (1971). *Endocrinology* **89**, 1034–41.

Note added in proof (*see p.* 538). The level of significance was $0.05 > P > 0.02$. Additional experiments have to confirm this difference in the cortico-medial amygdala and the original findings must be accepted as being due to chance (A. W. Rogers, personal communication).

THE ADRENAL CORTEX AND INTESTINAL ABSORPTION OF MACROMOLECULES BY THE NEW-BORN ANIMAL

By R. N. HARDY, V. G. DANIELS, K. W. MALINOWSKA AND P. W. NATHANIELSZ

The Physiological Laboratory, Cambridge CB2 3EG

The new-born animal is dependent for its protection against infection upon the passive immunity it obtains from maternal antibodies. There are, however, considerable species differences in the time at which maternal antibodies gain access to the circulation of the young animal. In certain species, including man, the transfer occurs *in utero*, in others, such as the ox, sheep, pig and horse, little prenatal transfer takes place and the antibodies acquired are those found in the colostrum which subsequently pass unchanged across the villous epithelial cells of the small intestine during the first few hours after birth. A third group of species, exemplified by the rat, obtains significant amounts of antibody *in utero*, but also absorbs antibodies from colostrum and milk for a period of weeks after birth. The existence of a prolonged period of postnatal transfer in the rat permits the study of the intestinal changes which result in the loss of the ability to absorb antibodies and other macromolecules (closure).

The work to be described comprises experiments performed in the rat in an attempt to determine the part played by the adrenal gland in the control of closure. The extent to which the results of this work may be applicable in other species will also be discussed.

The first indication that the adrenal cortex might be implicated in closure came from the observation of Halliday (1959) and Clark (1959) who demonstrated that large doses of adrenal steroids could cause precocious closure in rats; antibody absorption from the intestine ceased 14 days after birth following steroid injection at 10 days in comparison with normal closure which occurs 21 days after birth. We have recently extended these observations using the intestinal uptake of [^{125}I]polyvinyl pyrrolidone (PVP) K60 (mean mol. wt 160,000) as a more precise quantitative index of macromolecular transport (Fig. 1) (Clarke & Hardy, 1969).

It can be seen from Fig. 2*a* that cortisone acetate, given by a single subcutaneous injection as early as either 5 or 12 days after birth, caused a progressive decline in PVP uptake over the following 4–6 days. This decline in PVP uptake can be related to histological changes in the terminal intes-

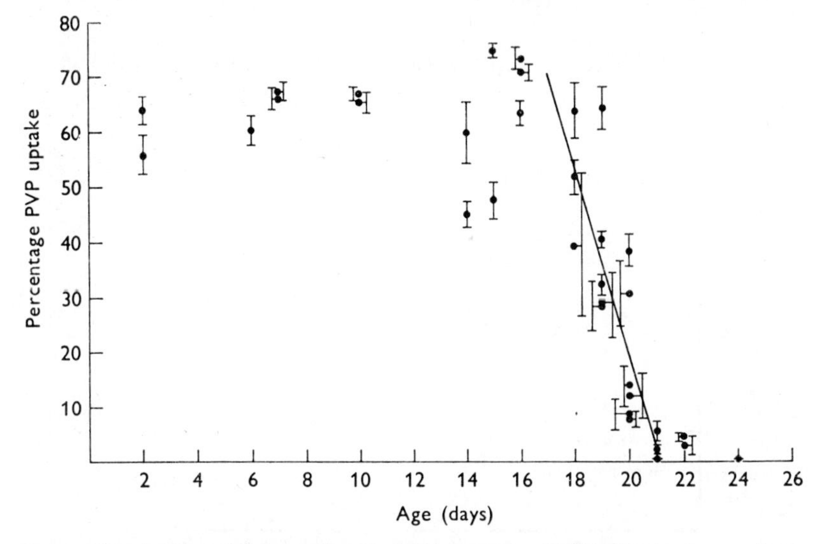

Fig. 1. The changes with increasing age in the uptake of (^{125}I)PVP by the small intestine. Each point represents the mean ± SEM, the average number of rats per point being four. The straight line is the best fit by the method of least squares for results from animals 18–21 days old inclusive. Ordinate: per cent PVP uptake. Abscissa: age in days. (From Clarke & Hardy, 1969.)

tine closely comparable with those seen during normal closure. Thus, the vacuolated villous epithelial cells, characteristically associated with the ability to absorb macromolecules, are progressively displaced from the villus by more mature non-vacuolated cells emerging from the crypts. The decline in PVP uptake thus represents the decrease in the number of vacuolated cells and, therefore, the time-course of the closure process can be related to the villous epithelial cell turnover-time (Clarke & Hardy, 1969).

Fig. 2*b* shows the results of injecting corticosterone, the principal glucocorticoid in the rat. Once again there is a decrease in PVP uptake, although the effect is transient, moreover, unlike that of cortisone acetate, it is not associated with histological changes in the villous epithelial cells, all of which remain vacuolated.

It can be concluded from these experiments that both steroids depress macromolecular uptake, but that the mechanism of their action differs. Cortisone acetate induces the production of non-vacuolated cells by the crypts of Lieberkühn, whereas corticosterone depresses uptake by the cells, without obvious change in their histological appearance.

The results of steroid injection raise the question as to whether natural closure is related to the endogenous adrenal steroid production. In an attempt to answer this question, plasma glucocorticosteroid concentrations

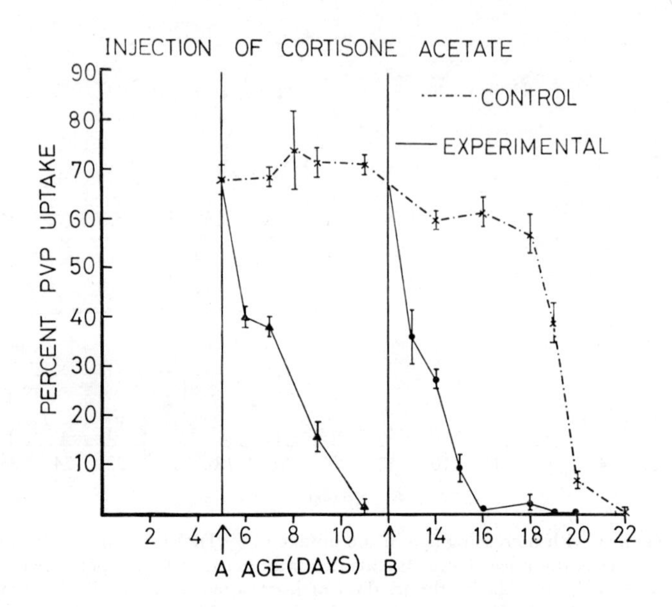

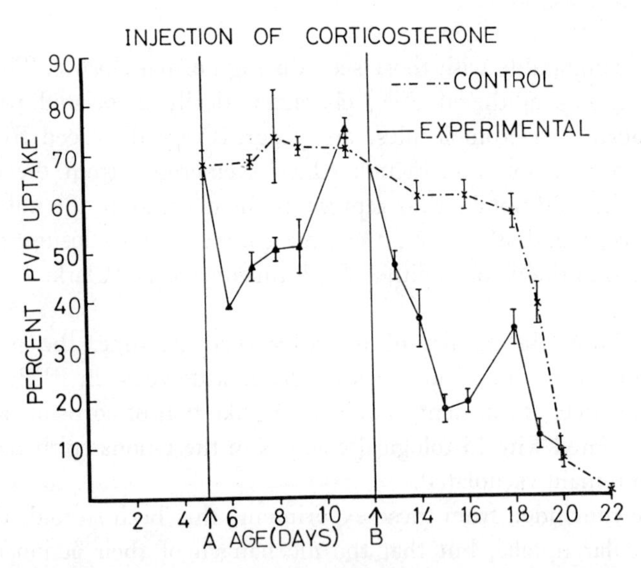

Fig. 2. The effect of cortisone acetate and corticosterone administration on intestinal uptake of PVP. Control PVP uptake (×); uptake after steroid injection, 2.5 mg corticosterone or 5 mg cortisone acetate 5 days after birth (▲); 5 mg corticosterone or 5 mg cortisone acetate 12 days after birth (●). Vertical lines represent ±SEM, where this exceeds the dimensions of the plotted point. Ordinate: per cent PVP uptake. Abscissa: age after birth (days).

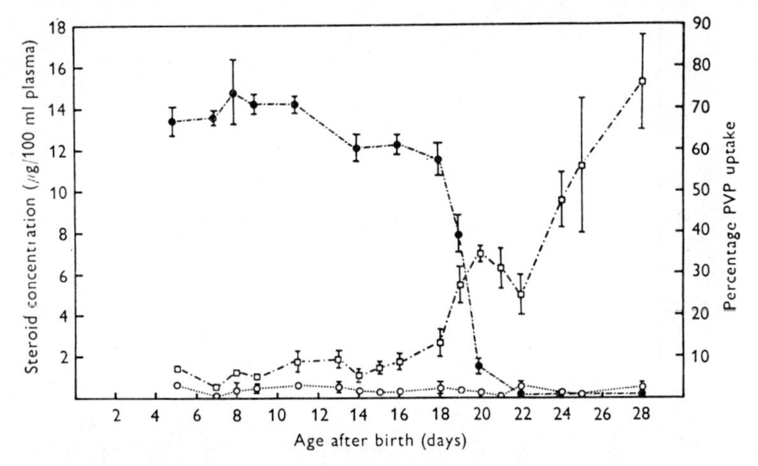

Fig. 3. Relationship between plasma adrenocortical steroid concentrations and polyvinyl pyrrolidone (PVP) uptake in young rats. The points represent mean values for at least four rats. The vertical lines indicate ±SEM where this exceeds the dimensions of the plotted point. ● = PVP uptake; □ = corticosterone; ○ = cortisol. (From Daniels, Hardy, Malinowska & Nathanielsz, 1972.)

were measured by competitive protein binding in young rats 5–28 days of age. It can be seen from Fig. 3 that although the cortisol concentrations remained low throughout this period, there was a marked increase in corticosterone concentrations coincident with closure (Daniels, Hardy, Malinowska & Nathanielsz, 1972).

The two preceding lines of evidence suggested that the adrenal cortex may be causally related to closure. It remained, however, to establish whether the function of the adrenal cortex was an absolute prerequisite for closure. This was determined by examining the effect of bilateral adrenalectomy on day 18 on PVP uptake during the normal closure period (Daniels & Hardy, 1971, 1972). It can be seen from Fig. 4 that this procedure delayed closure by about 4 days, but it did not prevent it completely. In fact, histologically the delayed closure following bilateral adrenalectomy closely resembled the normal closure process.

It seems then that the adrenal cortex is necessary for closure to occur at the normal time and that natural closure is associated with an increase in the plasma concentration of the principal glucocorticosteroid. However, closure still takes place, albeit somewhat delayed, even after removal of both adrenal glands when the plasma adrenal glucocorticosteroid concentrations are extremely low. The stimulus for closure following adrenalectomy remains unclear but is currently under investigation.

The extent to which the results obtained in the rat are applicable to other species has been the subject of some preliminary experiments

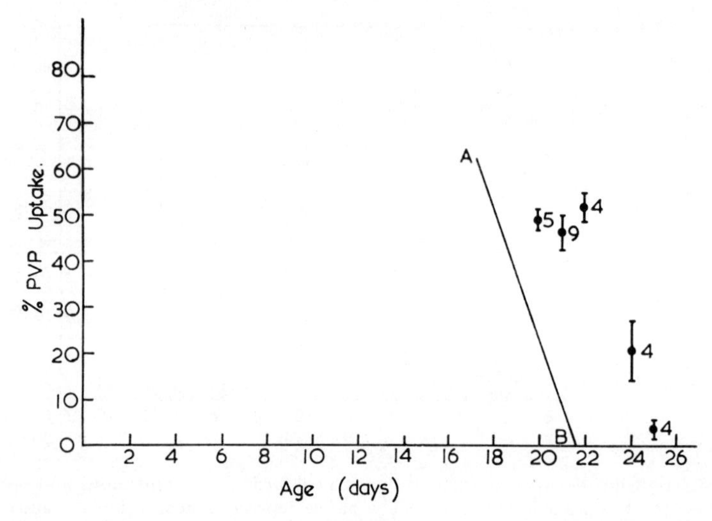

Fig. 4. Effect of adrenalectomy on the uptake of PVP by the small intestine of the young rat. A–B, Regression line calculated by the method of least squares for sham-adrenalectomized animals aged 18–22 days old inclusive. ●, Mean and SEM for groups of animals adrenalectomized at 18 days (figures indicate number of animals). (From Daniels & Hardy, 1972.)

(Malinowska, Hardy & Nathanielsz, 1972). In the guinea-pig, closure occurs within 24–48 hours of birth and this may be associated with the extremely high concentrations of cortisol in the plasma at this time. In the rabbit, closure occurs 18–22 days after birth, at which time there is an increase in plasma cortisol concentrations.

From the results at present available, it seems that the adrenal cortex in certain species may play a part in controlling the phase of intestinal development which terminates the absorption of macromolecules.

REFERENCES

Clark, S. L. (1959). *J. Biophys. Biochem. Cytol.* **5**, 41–9.

Clarke, R. M. & Hardy, R. N. (1969). *J. Physiol.* **204**, 113–25.

Daniels, V. G. & Hardy, R. N. (1971). *Proc. Int. Union of Physiological Sciences*, Vol. ix, 130. Munich: German Physiological Society.

Daniels, V. G. & Hardy, R. N. (1972). *Experientia* **28**, 272.

Daniels, V. G., Hardy, R. N., Malinowska, K. W. & Nathanielsz, P. W. (1972). *J. Endocr.* **52**, 405–6.

Halliday, R. (1959). *J. Endocr.* **18**, 56–66.

Malinowska, K. W., Hardy, R. N. & Nathanielsz, P. W. (1972). *J. Endocr.* **55**, 397–404.

SESSION 6

PARTURITION

HORMONAL CONTROL OF PREGNANCY AND PARTURITION; A COMPARATIVE ANALYSIS

BY R. W. ASH,* J. R. G. CHALLIS,* F. A. HARRISON,* R. B. HEAP,* D. V. ILLINGWORTH,* J. S. PERRY* AND N. L. POYSER†

*ARC Institute of Animal Physiology, Babraham, Cambridge and
†Department of Pharmacology, University of Edinburgh,
1 George Square, Edinburgh

Hormonal control systems concerned with the maintenance of pregnancy and the induction of parturition differ widely among mammalian species. In some species the foetus plays an important role in the induction of parturition, whereas in others its presence has little, if any, effect, and the hormone pattern of pregnancy resembles that of the non-pregnant or pseudopregnant condition. In the ferret, for example, the duration of pregnancy and pseudopregnancy is similar (42 days); the secretory activity of the corpus luteum is also closely similar in both conditions. Peripheral progesterone levels reach maximum values of about 30 ng ml^{-1} 10–30 days after mating and then decline gradually to less than 10 ng ml^{-1} before term (Hammond & Heap, unpublished observations). In rats and mice (and monkeys), it has long been recognized that after the removal of foetuses, some placentae will survive and be delivered at the normal time, though delivery may be prolonged or abnormal. In contrast, the ovaries have an important role in parturition in the rat. If they are removed 48 h before term, preceding the time of rapid increase in ovarian oestrogen secretion (Yoshinaga, Hawkins & Stocker, 1969), normal parturition is arrested, although it can be maintained by a single injection of oestradiol propionate (10 μg) given 24 h after ovariectomy (Csapo, 1969). Foetal corticosterone secretion also increases during late gestation in the rat (Kamoun, 1970). At the present time, however, the evidence that the foetus plays an active role in the onset of parturition in these species is inconclusive.

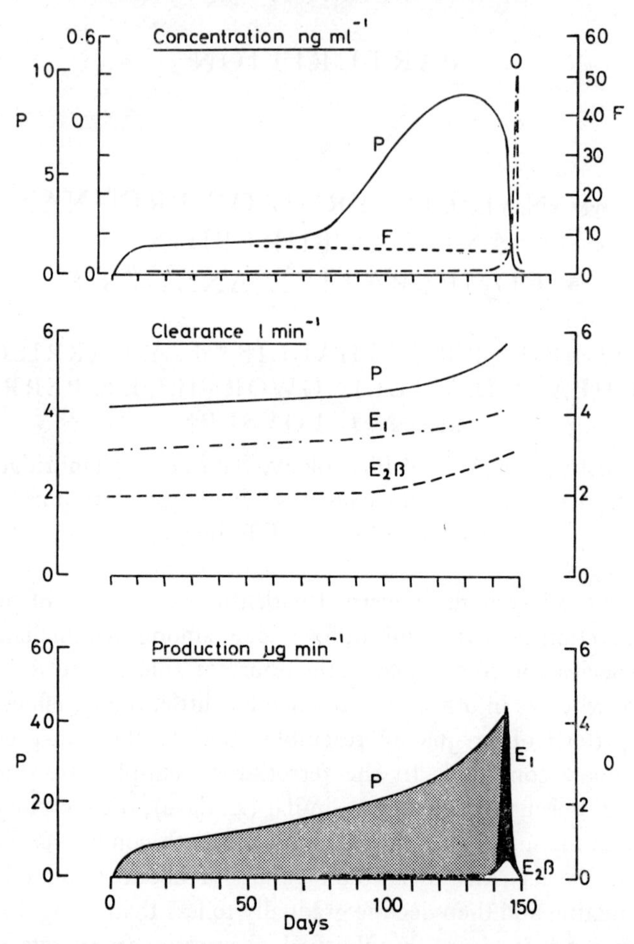

Fig. 1. Progesterone and oestrogens in pregnant sheep. P, progesterone (Bassett, Oxborrow, Smith & Thorburn, 1969; Bedford, Harrison & Heap, 1972); O, total unconjugated oestrogens (Challis, 1971); F, cortisol (Paterson & Harrison, 1967); E_1, oestrone; $E_2\beta$, oestradiol-17β (Challis, Harrison & Heap, 1972). All values relate to measurements in jugular vein blood.

In the sheep, on the other hand, there is evidence that parturition is influenced by a system that involves the foetal pituitary and foetal adrenals (Liggins, 1969; Liggins, Grieves, Kendall & Knox, 1972). The ewe somehow recognizes the presence of a conceptus in the uterus even before attachment, and the suppression of the oestrous cycle is related to the early neutralization of a luteolytic mechanism and the continued secretion of progesterone, first by the corpus luteum and later by the placenta. Parturition is preceded by complex hormonal interactions that include an

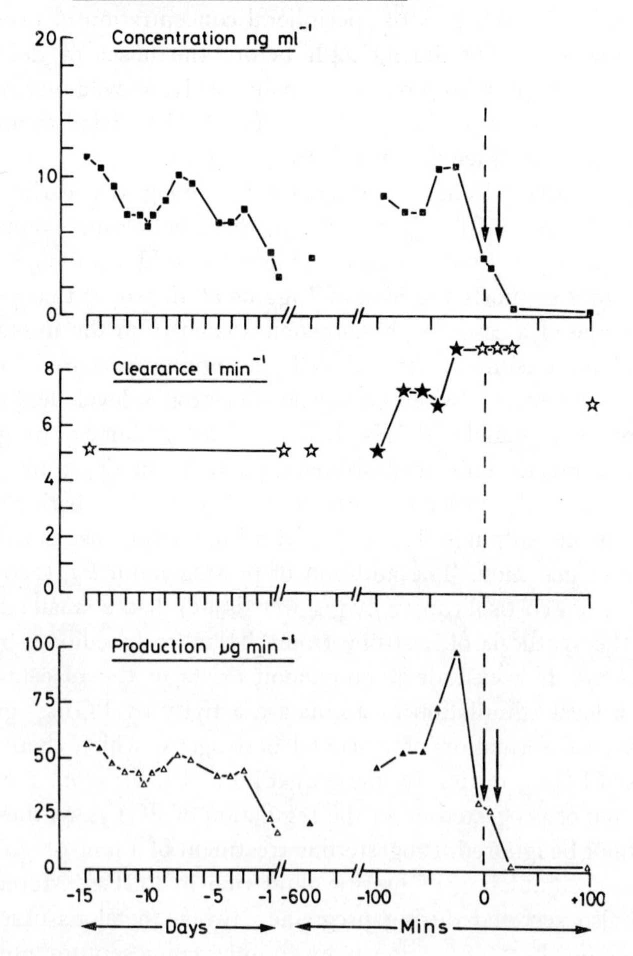

Fig. 2. Kinetics of progesterone metabolism in a pregnant sheep at term. Progesterone concentration in jugular vein blood measured by competitive protein binding and radioimmunoassay. At 225 min before delivery of the first lamb (t = 0), [^{3}H]progesterone was infused at a constant rate to measure metabolic clearance rate in jugular vein blood. Production rate was calculated as described elsewhere (Bedford, Harrison & Heap, 1972). Open symbols refer to values previously published (Bedford *et al.* 1972).

increased foetal secretion of corticosteroids; a decreasing peripheral level of progesterone; a sharp rise in the production of oestradiol-17β and oestrone (Fig. 1), and in the uterine secretion of prostaglandin $F_{2\alpha}$, $PGF_{2\alpha}$ (Challis, Harrison, Heap, Horton & Poyser, 1972). There is a marked rise in absolute uterine blood flow (see Bedford, Challis, Harrison & Heap, 1972), presumably related to the sensitivity of the capillary bed to oestrogens (Greiss & Anderson, 1970; Huckabee, Crenshaw, Curet, Mann & Barron,

1970). The onset of normal parturition is not always associated with progesterone withdrawal. Whereas the peripheral concentration of progesterone may decrease gradually during 24 h before the onset of delivery, the production rate of progesterone may remain steady, or even increase during the critical time before delivery. These changes are related to an increase in progesterone clearance rate at this time (Fig. 2).

Although foetal adrenal hypertrophy is among the earlier changes associated with parturition in the sheep, these endocrine events may be initiated even earlier. Evidence concerning the steroid metabolizing activity of placentomes supports the idea of Liggins *et al.* (1972) that parturition is the outcome of a cascade phenomenon arising from the interactions of several endocrine components. Minced placentomes prepared from sheep within 36 h of delivery (when endogenous oestrogen levels had started to increase) were incubated with ^{3}H-labelled androstenedione or progesterone. There was a greatly enhanced aromatase activity in tissue obtained immediately before the onset of parturition (Fig. 3). Incubation with ^{3}H-labelled oestrone sulphate demonstrated a high sulphatase activity at the same time of gestation. The addition of prostaglandin $F_{2\alpha}$ (200 ng/g wet weight), but not cortisol (up to 5 μg/g wet tissue), had a small stimulatory effect on the synthesis of oestrone from [^{3}H]androstenedione, but not of oestradiol-17β. If a cascade phenomenon exists in the placenta, it may comprise a local stimulation of aromatase activity by $PGF_{2\alpha}$, giving rise to an increased formation of placental oestrogens, which then stimulate an increased $PGF_{2\alpha}$ output by the gravid horn (Challis *et al.* 1972). However, the role of progesterone in the regulation of $PGF_{2\alpha}$ synthesis by the uterus cannot be ignored. Progesterone treatment of a non-pregnant sheep caused the accumulation of uterine fluid rich in $PGF_{2\alpha}$. Moreover, this fluid was also secreted during pregnancy; when foetal tissues were excluded from one horn of the uterus by an intervening septum, uterine fluid containing $PGF_{2\alpha}$ was found in this non-gravid horn (Amoroso, Harrison, Heap & Poyser, 1972).

Further support for the view that the increased production of oestrogens before parturition may be partially derived from Δ^4-3-keto-steroid precursors arises from the identification in whole blood of [^{3}H]oestrone and [^{3}H]oestradiol-17β formed from [^{3}H]progesterone continuously infused during the last four hours of pregnancy. In one sheep 28 min before delivery, the conversion ratio of progesterone (production rate, 98 μg/min^{-1}) to oestrone was 4.18 %. From a mean metabolic clearance rate of 3.868 l min^{-1} (Challis, Harrison & Heap, 1972) a transfer constant (Horton & Tait, 1966) of progesterone to oestrone was calculated, 1.79 %, so that the amount of oestrone formed from progesterone was probably about 1.8 μg

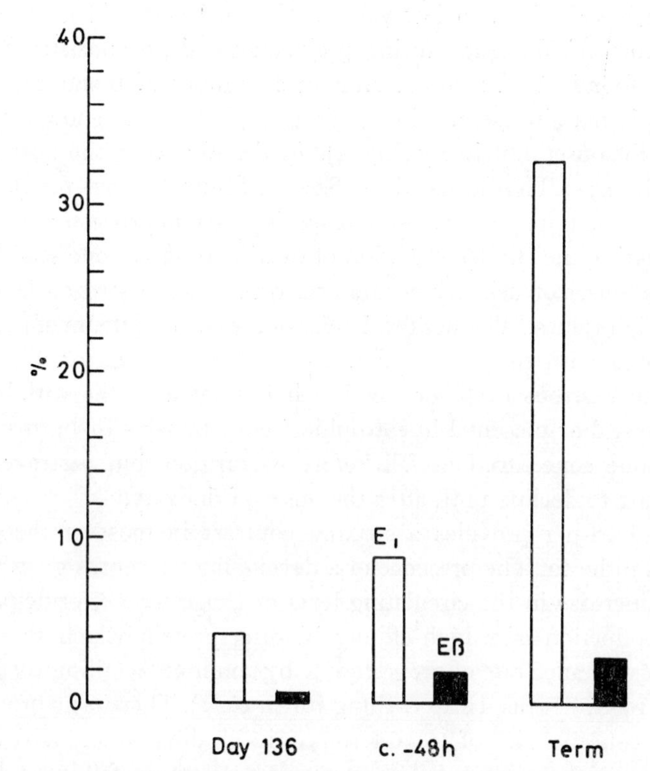

Fig. 3. The percentage conversion of [³H]androstenedione (about 1 μCi, 40 ng) to oestrone (open column) and oestradiol-17β (black column) by minced placentomes. Tissue was removed from three pregnant sheep on day 136, about 48 h before parturition, and at term during second-stage labour. Incubations were carried out for 1 h in a gas phase of 95 % oxygen, 5 % CO_2, without added co-factors.

min^{-1}. This compares with an estimated oestrone production rate of about 28 μg min^{-1} (endogenous level, 7.3 ng/ml blood) and indicates that a small proportion of the oestrone produced immediately before parturition is derived from progesterone.

In the pig, the presence of a developing conceptus is recognized from an early stage (about day 12) and pregnancy depends on the ovarian secretion of progesterone throughout, and on the neutralization of a uterine luteolytic factor. In pregnant sows, plasma progesterone levels remain relatively constant (about 10 ng ml^{-1}) until the day of parturition, and decline rapidly 3 to 6 hours before the birth of the first piglet. Plasma oestrogen (total unconjugated) concentration rises gradually in late pregnancy and reaches very high values (2–4 ng ml^{-1}) at parturition. Oestrone is a major component. Oestrogen levels decrease during delivery and by the time all the placentae have been expelled the concentration of unconjugated oestrogen has declined to low values. Maternal corticoid concentrations (mainly cortisol) fluctuate during pregnancy and no consistent increase has been found at parturition. Among a number of blood components studied, plasma glucose and inorganic phosphate levels show a transient rise after the onset of delivery (Fig. 4). In the pig, as in the goat (Challis & Linzell, 1971; Thorburn, Nicol, Bassett, Shutt & Cox, 1972), another species in which pregnancy maintenance depends on progesterone secreted by the ovaries, the sudden cessation of ovarian progesterone secretion just before the onset of delivery occurs at a time when oestrogen levels have reached high values. Whether the demise of the corpus luteum of pregnancy is associated with the release of a uterine luteolytic factor, like that of the normal oestrous cycle, or whether it is related to the withdrawal of a hypophyseal or placental luteotrophic factor, remains to be discovered. Progesterone concentrations fall before parturition, but oestrogen levels do not start to decline until after the onset of delivery.

The guinea-pig provides a striking contrast in most of the features discussed hitherto. The presence of a developing conceptus gives rise to a 100-fold increase in the circulating level of progesterone, principally due to the production of a high-affinity binding protein which reduces the metabolic clearance rate of progesterone by about 90% (Illingworth, Heap & Perry, 1970; Challis, Heap & Illingworth, 1971). There is no pronounced fall in the concentration of progesterone or its binding protein at parturition. Nor is there a sharp increase in oestrogens but rather a fall in total unconjugated oestrogens. The role of the foetus in parturition is also ambiguous. Foetal adrenal weights increase during pregnancy but only in proportion to body weight. Dexamethasone given as a single or repeated dose from day 57 of pregnancy fails to induce parturition when injected into the dam (up to 8 mg day^{-1} for at least 7 days) or foetus (up to 100 μg day^{-1} for at least 7 days). In further contrast to the sheep, in which maternal hypophysectomy does not affect the time of parturition (Denamur & Martinet, 1961), hypophysectomized pregnant guinea-pigs fail to deliver their young except when given oxytocin. Thus, in guinea-pigs hypophysectomized as early as the first day after a single fertile mating, live foetuses are carried to term. In eight of nine hypophysectomized pregnant guinea-pigs, parturient activity failed to develop (Table 1), though live foetuses were delivered by Caesarian section when the animals showed signs of distress.

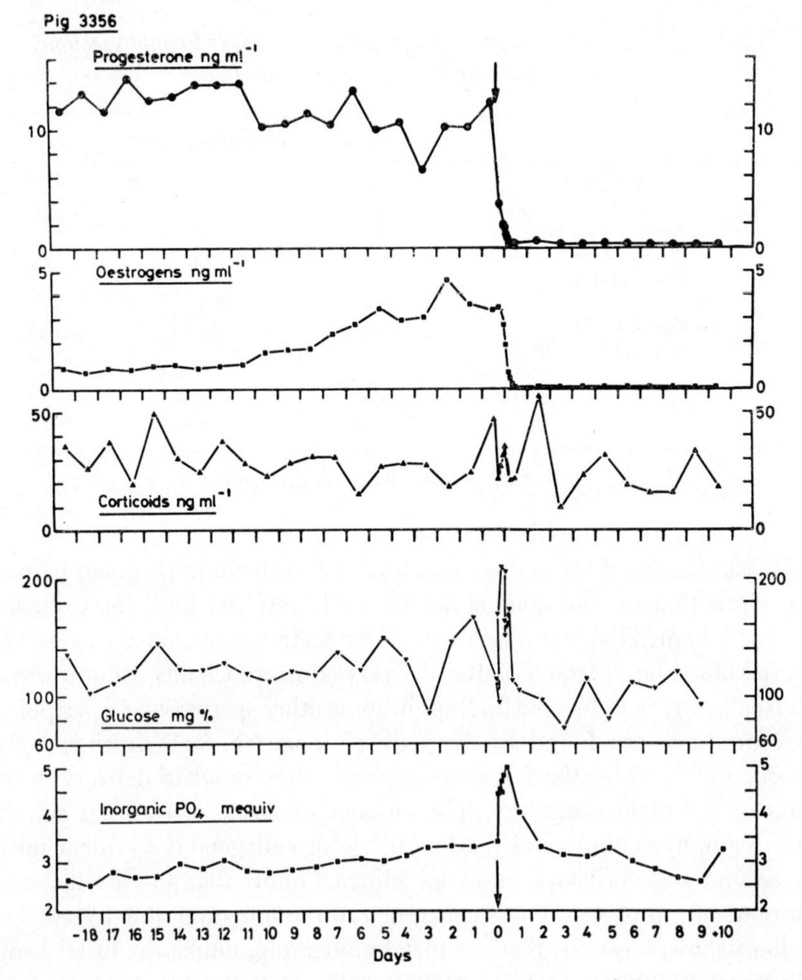

Fig. 4. Plasma concentrations of some blood constituents in a sow (5th pregnancy) during pregnancy and parturition (14 piglets born). An indwelling ear vein catheter (Ash, Banks, Broad & Heap, 1972) was established. Progesterone and corticoids were measured by competitive protein-binding techniques (Thornburn, Bassett & Smith, 1969; Bassett & Hinks, 1969), total unconjugated oestrogens by radioimmunoassay (Challis, Heap & Illingworth, 1971), glucose by the glucose oxidase method (Huggett & Nixon, 1957) and inorganic phosphate by the method of King & Wootton (1956). The arrow denotes the day of farrowing.

In the ninth animal, spontaneous delivery occurred on day 70 of pregnancy. In this animal, completeness of hypophysectomy was not confirmed with certainty, though maternal adrenals were small, ovarian follicular activity had ceased, and no pituitary remnant was seen at autopsy. Pelvic relaxation was noted in all animals, though complete relaxation occurred only in the case of spontaneous delivery.

Table 1. *Hypophysectomy in pregnant guinea-pigs; hypophysectomy performed up to 4 days after mating*

	Intact	Hypophysectomy[a]
No. animals	9	9
No. with live foetuses	9	4
Normal delivery	9	(1)
Placental wt (g)	7.2	7.0
No. foetuses	3.9 ±0.5	2.6 ±0.4
Foetal wt (g)	102±3	97±6
Adrenal wt (mg)		
maternal	313±23	129±9*
foetal	11.8 ±0.6	21.9* ±2.3

[a] Autopsy on Day 63–71 of pregnancy.
* $P < 0.001$.

Measurements of the endogenous levels of oxytocin in pregnant guinea-pigs show that the hormone is not normally released until the expulsive stage of labour. High concentrations of oxytocin are sustained in maternal arterial blood for at least 1 h after delivery (Burton, Challis, Illingworth & McNeilly, 1972), unlike the findings in some other species where peripheral concentrations are found to fluctuate (Fitzpatrick & Walmsley, 1965; Chard, 1972). Thus, the factors precipitating the onset of delivery in the guinea-pig remain a mystery. The absence of normal parturient activity after hypophysectomy, and the low levels of endogenous oxytocin up to the second stage of labour in intact animals, imply that another maternal hormone of pituitary origin is concerned in the initiation of delivery. The findings show, however, that the foetal guinea-pig, unlike the foetal lamb, fails to trigger the onset of parturition if the maternal pituitary has been removed.

This baffling array of differences precludes any overall explanation of the regulation of parturition in those species considered. Whereas the hypothesis proposed by Liggins (1968, 1969) concerning the onset of parturition in sheep draws attention to the important role of the foetus, the evidence at present available in other species is not yet sufficient to suggest that the foetus triggers the onset of parturition in a wide variety of animals. Despite the convincing evidence provided by Liggins and his colleagues, it should be noted that Lanman & Schaffer (1968) found that after foetal decapitation in sheep 57–85 days pregnant, delivery occurred at 144, 147 and 168 days in three sheep. The total adrenal weight in each foetus was only 150–191 mg, appreciably less than that found in normal foetuses. Although

Progesterone in pregnancy

Species	Concentration	Clearance	Production	Binding
Woman	↑↑↑	–	↑↑↑	n.d.
Sheep	↑	–	↑	n.d.
Guinea-pig	↑↑↑	↓↓↓	↑	↑↑↑

Fig. 5. The progesterone requirements of pregnancy. Dash denotes no marked change; n.d. indicates not detectable. Plasma concentration, metabolic clearance rate, production rate and concentration of progesterone-binding protein in woman (Short, 1961; Little, Tait, Tait & Erlenmeyer, 1966; Little & Billiar, 1968), sheep (Bassett, Oxborrow, Smith & Thorburn, 1969; Bedford, Harrison & Heap, 1972) and guinea-pig (Heap, 1969; Challis, Heap & Illingworth, 1971; Illingworth, Heap & Perry, 1970).

the foetuses grew, albeit slowly in two cases, and delivery occurred in the absence of a functional foetal pituitary (or of hypertrophy of the foetal adrenals), it is unclear whether delivery was associated with some pathological condition of the surviving foetuses.

It is worth recalling the constancy of gestation length within a species, and its variability among species. Thus in the ferret, although pregnancy and pseudopregnancy are similar in duration, it seems unlikely that gestation length is determined solely by the inherent life-span of the corpus luteum – a relatively imprecise mechanism. A remarkable feature of parturition in eutherian mammals is the finding that the delivery of young, a process of high survival value, apparently embraces a wide divergence of hormonal control mechanisms. A search for some common denominator in different species must have regard not only for variants of foetal endocrine changes that are associated with the onset of parturition, but also for those pronounced maternal changes that occur in response to what has been termed 'the maternal recognition of pregnancy' (Short, 1969). It has long been understood that the maintenance of pregnancy is dependent on progesterone, for when progesterone secretion is interrupted, either experimentally or accidentally, gestation is rapidly terminated. Yet even the progesterone requirements of pregnancy and foetal well-being may be met by one, or more, of several means; by a continued production of ovarian progesterone at a rate comparable with that of the luteal phase of the oestrous cycle (as in the pig); by an increased production of placental progesterone (as in the sheep and woman); or by a marked decrease in progesterone clearance from blood accompanied by a modest rise in ovarian and placental production, a 'progesterone-conserving' mechanism (as in the guinea-pig; see Fig. 5). Thus it appears that the maternal recognition of pregnancy is accompanied by changes in the endocrine pattern of the mother by which gestation is maintained, while the induction of parturition is preceded by a complex series of hormonal events in which the foetus may play a part. Whether the control of pregnancy and parturition can be adequately explained in terms of hormonal regulation, or whether the variety of endocrine changes is related in some way to other phenomena such as the maternal tolerance of a developing homograft within the uterus and its subsequent rejection at parturition, are fascinating questions for the future.

We gratefully acknowledge the skilled assistance of Miss Grace Needham, Mrs N. Ackland, Mrs S. Broad, Mr G. Bailes, Mr P. Banks, Mr A. Henville and Mr G. Jenkin. DVI and JRGC acknowledge financial support from the Lalor Foundation.

REFERENCES

AMOROSO, E. C., HARRISON, F. A., HEAP, R. B. & POYSER, N. L. (1972). *J. Endocr.* (In press.)

ASH, R. W., BANKS, P., BROAD, S. & HEAP, R. B. (1972). *J. Physiol., Lond.* **226**, 40–1 *P*.

BASSETT, J. M. & HINKS, N. T. (1969). *J. Endocr.* **44**, 387–403.

BASSETT, J. M., OXBORROW, T. J., SMITH, I. D. & THORBURN, G. D. (1969). *J. Endocr.* **45**, 449–57.

BEDFORD, C. A., CHALLIS, J. R. G., HARRISON, F. A. & HEAP, R. B. (1972). *J. Reprod. Fert. Suppl.* **16**, 1–23.

BEDFORD, C. A., HARRISON, F. A. & HEAP, R. B. (1972). *J. Endocr.* **55**, 105–18.

BURTON, A. M., CHALLIS, J. R. G., ILLINGWORTH, D. V. & MCNEILLY, A. S. (1972). *J. Physiol., Lond.* **226**, 94–5 *P*.

CHALLIS, J. R. G. (1971). *Nature, Lond.* **229**, 208.

CHALLIS, J. R. G., HARRISON, F. A. & HEAP, R. B. (1972). (In *Endocrinology of Pregnancy and Parturition*, ed. C. G. Pierrepoint. Cardiff: Alpha Omega Alpha.

CHALLIS, J. R. G., HARRISON, F. A., HEAP, R. B., HORTON, E. W. & POYSER, N. L. (1972). *J. Reprod. Fert.* (In press.)

CHALLIS, J. R. G., HEAP, R. B. & ILLINGWORTH, D. V. (1971). *J. Endocr.* **51**, 333–45.

CHALLIS, J. R. G. & LINZELL, J. L. (1971). *J. Reprod. Fert.* **26**, 401–4.

CHARD, T. (1972). *J. Reprod. Fert. Suppl.* **16**, 121–38.

CSAPO, A. (1969). In *Progesterone: its regulatory effect on the myometrium.* Ciba Foundation Study Group No. 34, pp. 13–42. London: Churchill.

DENAMUR, R. & MARTINET, J. (1961). *Annls Endocr.* **22**, 755–9.

FITZPATRICK, R. J. & WALMSLEY, C. F. (1965). In *Advances in Oxytocin Research*, ed. J. H. M. Pinkerton, pp. 51–71. London and New York: Pergamon Press.

GREISS, F. C. & ANDERSON, S. G. (1970). *Am. J. Obstet. Gynec.* **106**, 30–8.
HEAP, R. B. (1969). *J. Reprod. Fert.* **18**, 546–8.
HORTON, R. & TAIT, J. F. (1966). *J. clin. Invest.* **45**, 301–13.
HUCKABEE, W. E., CRENSHAW, C., CURET, L. B., MANN, L. & BARRON, D. H. (1970). *Q. Jl exp. Physiol.* **55**, 16–24.
HUGGETT, A. ST G. & NIXON, D. A. (1957). *Lancet* **2**, 368–70.
ILLINGWORTH, D. V., HEAP, R. B. & PERRY, J. S. (1970). *J. Endocr.* **48**, 409–17.
KAMOUN, A. (1970). *J. Physiol., Paris* **62**, 5–32.
KING, E. J. & WOOTTON, I. D. P. (1956). In *Microanalysis in Medical Biochemistry*, 3rd ed., p. 77. London: Churchill.
LANMAN, J. T. & SCHAFFER, A. (1968). *Fert. Steril.* **19**, 598–605.
LIGGINS, G. C. (1968). *J. Endocr.* **42**, 323–30.
LIGGINS, G. C. (1969). In *Foetal Autonomy*. Ciba Foundation Symposium, eds. G. E. W. Wolstenholme & M. O'Connor, pp. 218–31. London: Churchill.
LIGGINS, G. C., GRIEVES, S. A., KENDALL, J. Z. & KNOX, B. S. (1972). *J. Reprod. Fert. Suppl.* **16**, 85–103.
LITTLE, B. & BILLIAR, R. B. (1968). *Excerpta med. Fdn 3rd Intern. Congr. of Endocr., Mexico*, pp. 871–9.
LITTLE, B., TAIT, J. F., TAIT, S. A. S. & ERLENMEYER, F. (1966). *J. clin. Invest.* **45**, 901–12.
PATERSON, J. Y. F. & HARRISON, F. A. (1967). *J. Endocr.* **37**, 269–77.
SHORT, R. V. (1961). In *Hormones in Blood*, ed. C. H. Gray & A. L. Bacharach, pp. 379–437. London and New York: Academic Press.
SHORT, R. V. (1969). In *Foetal Autonomy*. Ciba Foundation Symposium, eds. G. E. W. Wolstenholme & M. O'Connor, pp. 2–26. London: Churchill.
THORBURN, G. D., BASSETT, J. M. & SMITH, I. D. (1969). *J. Endocr.* **45**, 459–69.
THORBURN, G. D., NICOL, D. H., BASSETT, J. M., SHUTT, D. A. & COX, R. I. (1972). *J. Reprod. Fert. Suppl.* **16**, 61–84.
YOSHINAGA, K., HAWKINS, R. A. & STOCKER, J. F. (1969). *Endocrinology* **85**, 103–12.

Note added in proof. In recent experiments it was found that parturition occurred normally in hypophysectomized guinea-pigs which received injections of cortisone acetate daily from day 55.

FOETAL PARTICIPATION IN THE PHYSIOLOGICAL CONTROLLING MECHANISMS OF PARTURITION

By G. C. LIGGINS

Postgraduate School of Obstetrics and Gynaecology,
University of Auckland, Auckland, New Zealand

The reproductive physiologist interested in the problems of maintenance of pregnancy and the mechanisms of initiation of labour was understandably led by his own and others' experiments in rabbits and similar laboratory rodents to the conclusion that the mother is responsible for determining the duration of pregnancy and the time of its termination. It was found not only that the hormones upon which the maintenance of pregnancy depends are elaborated in the maternal ovaries in these species but also that the continued function of the ovaries is dependent upon hormones secreted by the maternal adenohypophysis. And finally, the successful completion, if not initiation, of labour, is accomplished with the aid of oxytocin released from the maternal neurohypophysis.

The first clear statement that the concept of maternal control of pregnancy might not apply equally well to other mammalian species came from a British obstetrician, Malpas (1933), who explained his observations of prolonged human pregnancies in association with foetal anencephaly by proposing that the abnormal foetuses failed to make their usual contribution to the mechanisms initiating labour. In addition, Malpas concluded that 'the foetal adrenal, pituitary or nervous system, perhaps in combination, are the tissues possibly concerned in the actual excitation of the neuromuscular expulsive mechanisms'. This novel idea was brought into sharper focus by the classic work of Kennedy, Kendrick & Stormont (1957) in which they demonstrated the absence of anterior pituitary tissue in foetal calves delivered alive by Caesarian section many weeks beyond term in the course of a genetically determined syndrome of prolonged gestation. Their work was elaborated upon by Holm (1967) who studied another naturally occurring form of prolonged gestation in cattle and found evidence of foetal adrenal hypofunction as well as a cytological abnormality of the anterior pituitary gland.

Recent work in certain other species, particularly sheep, has confirmed the important part that the foetus plays in the endocrine regulation of myometrial function. Such species synthesize steroid hormones in the trophoblast of the placenta and can usually dispense with their ovaries in the latter part of pregnancy. A foetal influence on placental biosynthetic processes is easy to appreciate and for a time the disparity between mammals that are placenta-dependent and those that are corpus luteum-dependent was resolved by postulating two distinct classes of mammals: those needing functioning ovaries throughout pregnancy and in which control of parturition is a maternal function, and those elaborating 'ovarian' hormones in the placenta and in which control of parturition is a foetal function.

In the light of discoveries of the last few years, a classification of this sort is no longer tenable. There is evidence to show that the conceptus of corpus luteum-dependent species exercises an important influence on the maternal ovaries although in concert with hormones of the maternal anterior pituitary. Thus, rather than distinctive classes, there may be a spectrum of degrees to which the foeto-placental unit participates in the control of myometrial function, ranging from almost complete, as in sheep, to less complete but nevertheless essential, as in rabbits and goats.

In this paper, the evidence supporting a major role of the foetal lamb in the mechanisms controlling parturition is reviewed and is compared briefly with the evidence supporting a similar foetal role in other species.

HYPOTHALAMO-PITUITARY-ADRENAL SYSTEM

Function in the foetal lamb

Prolonged gestation associated with congenital anomalies of the hypothalamus and anterior pituitary provide strong circumstantial evidence that the foetal lamb participates in the regulation of myometrial function. Two such syndromes occurring under natural conditions in sheep share a common aetiology, both being the result of ingestion by the dam of a teratogenic agent. In one, the agent is a steroidal alkaloid, cyclopamine, contained in the foliage of skunk cabbage (Binns, James & Shupe, 1964) and the anomalies take the form of a cyclopian deformity of the head which includes dislocation of the pituitary from the hypothalamus. In the other, the teratogenic agent, although not yet identified, is contained in a shrub, *Salsola tuberculata*. The oversized, postmature lambs have no anatomical anomalies but the function of the hypothalamus and pituitary is disordered (Basson, Morgenthal, Bilbrough, Marais, Kruger & van der Merwe, 1969).

These naturally-occurring syndromes of prolonged gestation were reproduced experimentally in sheep by surgically ablating the foetal

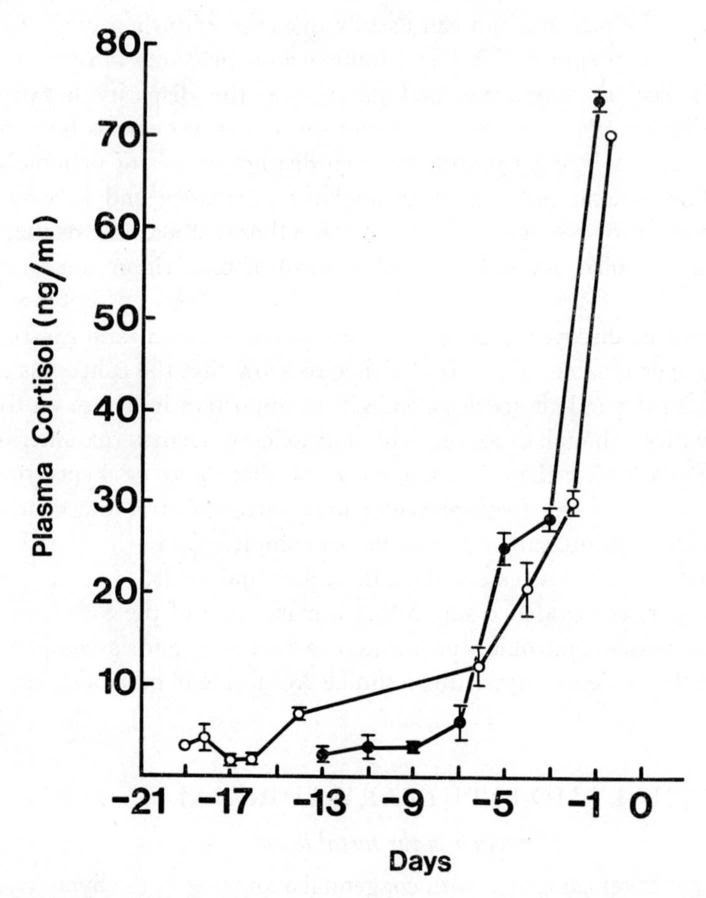

Fig. 1. Concentration of cortisol in the plasma of two foetal lambs from the time of implantation of catheters in the carotid arteries until spontaneous parturition at term. Day 0 = day of parturition.

pituitary (Liggins, Kennedy & Holm, 1967; Comline, Silver & Silver, 1970; Bosc, 1972), by dividing the foetal pituitary stalk (Liggins, 1969) or by removing both foetal adrenals (Drost & Holm, 1968; Liggins, 1969). The results of these procedures left no doubt about either the importance of the foetal contribution to the initiation of parturition or the identity of the endocrine organs concerned in mediating the foetal influence. However, some uncertainty remained as to whether the foetal lamb actually triggered parturition or whether it merely ensured the environment necessary for a trigger of maternal origin to operate effectively. Support for the more active, former alternative was found by Liggins (1968) who showed that stimulation of the foetal (but not maternal) adrenals with corticotrophin (ACTH) caused premature delivery. He also found that the effect of adrenal activity on parturition was mediated by a glucocorticoid, probably cortisol, rather than by a mineralocorticoid (Liggins, 1969).

The pattern of plasma cortisol levels in the foetus suits it ideally to serve as a trigger (Fig. 1). During the last 7–10 days of gestation there is a slow rise in concentration until about 24 hours before parturition when there is a rapid and many-fold increase in levels (Bassett & Thorburn, 1969; Comline, Nathanielsz, Paisey & Silver, 1970). If the terminal rise in corticosteroid levels is prevented by foetal hypophysectomy or adrenalectomy prolonged gestation occurs. On the other hand, if a sharp rise in corticosteroid levels is stimulated prematurely by ACTH, delivery follows within a few days even though the pregnancy is as much as 60 days from term.

Serial measurements of metabolic clearance rates and production rates of cortisol (Comline *et al.* 1970) and of secretion rates (Fairclough & Liggins, 1972) have established that the rise in cortisol levels is the result of an increase in the rate of secretion. The stimulus to the adrenal cortex is presumably from ACTH but this has not yet been confirmed by assays of plasma concentrations of ACTH. The effects of hypophysectomy and section of the pituitary stalk strongly support this idea, yet there is evidence that changes within the adrenal cortex itself are partly responsible. Relative inactivity of the enzyme 11β-hydroxylase was found to be present in immature adrenocortical tissue incubated with suitable substrates by Anderson, Pierrepoint, Griffiths & Turnbull (1972); with advancing maturity, activity of both the enzyme and the ratio of cortisol:11-deoxycortisol increased. Furthermore, although the ability of the adrenal cortex to respond acutely to ACTH administered before day 130 is low, the responsiveness rises sharply near term (Fairclough & Liggins, 1972). And finally, the adrenal mass increases rapidly in the last week of pregnancy (Comline & Silver, 1961). The combined effects of these three adrenocortical changes is to enhance the rate of secretion of cortisol in response to a steady stimulus by ACTH and to cause a dramatically rapid increase in rate of response to rising levels of ACTH.

Nathanielsz, Comline, Silver & Paisey (1972) discussed the possible alterations in foetal hypothalamic function that might lead to increased release of ACTH and they concluded that the factors triggering hypothalamic activity are unknown. There is some evidence from changes in cortisol levels consequent upon the central effects of administered pyrogen or lowered pH (Nathanielsz *et al.* 1972) or hypoglycaemia (Liggins, 1973) that hypothalamic sensitivity increases with maturity. However, the nature of any physiological stimuli is uncertain. Thorburn, Nicol, Bassett, Shutt & Cox (1972) proposed that maturation of the hypothalamic thermoreceptors may be the determining factor because they found that the foetus was hyperthyroid compared with its mother until 7–10 days before term when the foetal thyroxine concentration decreased to maternal levels. They interpreted this as the onset of 'awareness' by the foetus of its hot environment and suggested that the response to the heat stress is suppression of release of thyrotrophin releasing factor and stimulation of corticotrophin releasing factor.

Clearly, it is important to understand the factors responsible for determining the pattern of activity of the foetal adrenal cortex since they constitute the triggering mechanism of parturition. The speculation of Nathanielsz *et al.* (1972) that the initial stimulus is from maturing tissues such as lungs and central nervous system that are essential to extrauterine survival has teleological merit.

Functions in the foetuses of other mammalian species

Limitations on experimentation in women necessarily lead to difficulties in establishing with any certainty a role for the human foetus in the initiation of parturition. Nevertheless, there is circumstantial evidence supporting the proposal that the human foetus plays an important part in determining the time of its birth. Syndromes of prolonged gestation similar to that seen in sheep have been described in association with foetal anomalies including anencephaly (Anderson, Laurence & Turnbull, 1969) and adrenal hypoplasia (O'Donohue & Holland, 1968). Measurements of cortisol levels in the human foetus are not practicable but urinary assays of corticosteroid sulphates which are thought to be of foetal origin suggest a marked increase in corticosteroid secretion in the later weeks of pregnancy (Klein, Kertesz, Chan & Giroud, 1971). The foetal adrenal cortex is the main source of the precursors of placental oestrogen production and it is known that these precursors, principally dehydroepiandrosterone sulphate, are secreted in rapidly increasing amounts in late pregnancy.

Cattle have many similarities to sheep in relation to foetal control of labour. Syndromes of prolonged pregnancy have been described in genetically determined malformations of the foetal hypothalamus and pituitary by Kennedy *et al.* (1957). Continuous infusion of ACTH into foetal calves as in foetal lambs, causes labour within two or three days (R. A. S. Welch, personal communication). Glucocorticoids administered in large doses to the mother within six weeks of term stimulate parturition, probably because the placenta is sufficiently permeable to steroids to permit entry of the glucocorticoid into the foetal circulation (Adams & Wagner, 1969). Naturally-occurring abortion in cattle is commonly due to foetal infections and when the aborted foetuses are delivered alive it is usual to find hypertrophy of the foetal adrenals (Kennedy, 1971).

There is convincing evidence of a foetal influence on labour in goats despite the fact that maintenance of pregnancy in this species depends on ovarian rather than placental function. The cause of a naturally-occurring syndrome of habitual abortion occurring in in-bred Angora goats was identified by van Rensberg (1963) as foetal adrenal hyperplasia. Experimental stimulation of the foetal adrenal cortex by ACTH likewise causes premature delivery (Thorburn *et al.* 1972).

Experiments in rhesus monkeys suggest that the foetal pituitary-adrenal system is involved to some extent in the control of parturition. Chez, Hutchinson, Salazar & Mintz (1970) hypophysectomized eleven foetal monkeys and five survived the operation and continued to grow. Four of the five foetuses were delivered more than $2\frac{1}{2}$ weeks beyond term whereas pregnancy was not extended by sham operations or by maternal hypophysectomy. Mueller-Heubach, Myers & Adamsons (1972) determined the duration of pregnancy in eight monkeys after foetal adrenalectomy. In only one foetus was extirpation complete; in this monkey, delivery occurred 12 days beyond the mean gestation period. The remaining foetuses were delivered before or at term and hyperplastic remnants of cortical tissue were found at autopsy. These results suggest that mechanisms other than foetal adrenal activity can eventually initiate labour in this species.

Little information is available concerning small mammals. Ablation of pituitaries or adrenals of all foetuses of litters in rabbits has been attempted, so far apparently unsuccessfully. Injection of dexamethasone into amniotic sacs of foetal rabbits is followed by premature delivery (Kendall & Liggins, 1972). The conceptus in rats apparently contributes to the control of ovarian function since although abortion occurs after dislocation of all placentae of a litter, pregnancy continues to term when one placenta remains intact (Weist, Kidwell & Balogh, 1969). A surge of adrenocortical activity occurs in foetal rats shortly before term (Levine & Treiman, 1969) but it is not known whether this is related in any way to parturition.

Finally, there is evidence in a wide variety of mammals ranging from those with extremely short gestation periods such as marsupials to those with long gestation periods such as horses, that foetal genotype is a major factor determining variability of gestation length.

ADRENOCORTICAL-PLACENTAL INTERACTIONS

Function in the foetal lamb

The endocrine changes occurring in the placenta of the foetal lamb in association with elevated levels of circulating corticosteroids have been studied both at term in normal animals and earlier in gestation in experi-

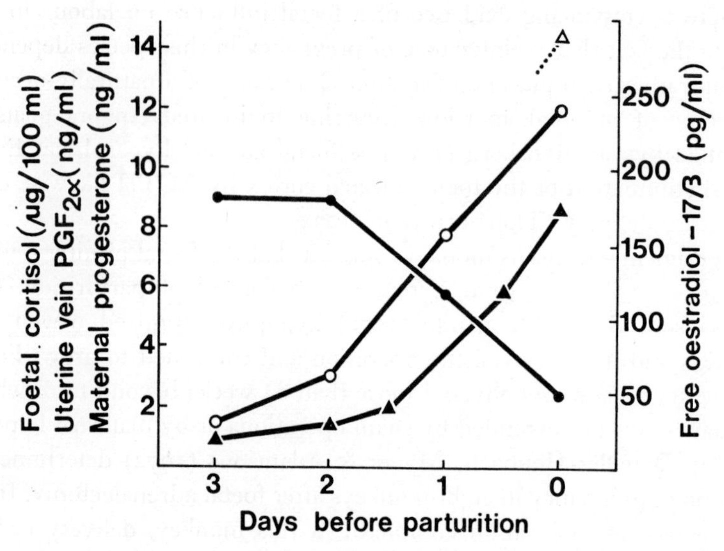

Fig. 2. Endocrine changes preceding premature parturition induced by a foetal infusion of ACTH. Progesterone (●) and oestradiol-17β (○) in maternal jugular vein samples; cortisol (▲) in foetal carotid arterial samples; $PGF_{2\alpha}$ (△) in uterine vein samples. Reproduced, by permission, from Liggins *et al.* (1972).

mental preparations infused intravascularly with ACTH or corticosteroids. The close approximation of the patterns of hormone concentrations after experimental adrenal stimulation to those observed at term supports the contention that the latter changes are a direct consequence of increased adrenocortical activity.

Three major endocrine events herald parturition (Fig. 2). In the peripheral plasma of the ewe, a sharp fall in the concentration of progesterone begins a few days before parturition and by the time uterine contractions appear, levels reach those in non-pregnant animals during the luteal phase (Bassett & Thorburn, 1969; Fylling, 1970; Stabenfeldt, Osburn & Ewing, 1970). The concentration of progesterone in uterine vein blood, normally 3–4 times higher than in peripheral plasma, drops to low levels, confirming the cause of falling concentrations as cessation of placental production of progesterone. These changes in progesterone are duplicated by infusing either ACTH or dexamethasone into the foetus (Liggins, Grieves, Kendall & Knox, 1972).

Within 24 hours of parturition a surge of unconjugated oestrogens appears in peripheral blood (Challis, 1970; Thorburn *et al.* 1972). The levels in uterine vein blood are the same or lower than in peripheral blood, an observation that led Thorburn *et al.* (1972) to suggest that the oestrogen surge resulted from increased release from the placenta of conjugated oestrogen which was subsequently hydrolysed in maternal tissues. The mechanism of the oestrogen surge remains obscure. It can be induced by infusing ACTH into the foetus and might be explained by increased production of an oestrogen precursor such as androstenedione in the adrenal cortex. But the surge is also inducible by dexamethasone (Liggins *et al.* 1972) which probably suppresses foetal adrenal androgen production. The foetus contains a large store of oestrogens in the form of circulating sulphoconjugates which might escape through the placenta if placental permeability were suddenly to alter. However, we found no change in the placental resistance to the passage of cortisol within 24 hours of labour and it seems unlikely that altered permeability is specific to steroid conjugates.

The third event observed before parturition is a peak of prostaglandin $F_{2\alpha}$ ($PGF_{2\alpha}$) in uterine vein blood coinciding with the surge of unconjugated oestrogen (Liggins & Grieves, 1971; Thorburn *et al.* 1972). The magnitude of the peak is the same whether the onset of parturition is spontaneous at term or is induced by ACTH or dexamethasone (Liggins *et al.* 1972). The source of $PGF_{2\alpha}$ is probably the maternal placenta and the myometrium since these tissues contain a high concentration of $PGF_{2\alpha}$ at the time when $PGF_{2\alpha}$ is present in uterine vein blood. The stimulus to $PGF_{2\alpha}$ synthesis is unlikely to be a direct action of foetal cortisol since neither maternal placenta nor myometrium is in direct contact with the foetal circulation. Moreover, labour is not stimulated by massive doses of maternally-administered corticosteroids until within about 10 days of term although foetally-administered corticosteroids are effective much earlier. The precise temporal relationship between the levels of unconjugated oestrogen and $PGF_{2\alpha}$ suggests that oestrogen is the stimulus to synthesis. This is further supported by observations of the effects of single doses of stilboestrol on levels of $PGF_{2\alpha}$; within 24 hours of an injection of stilboestrol (20 mg) the concentration of $PGF_{2\alpha}$ in uterine vein blood, maternal placenta and myometrium rises to levels similar to those at the onset of parturition (Liggins, 1973).

Function in the foetuses of other species

Of the three endocrine events preceding parturition in sheep, only one (the rise in $PGF_{2\alpha}$) occurs in women. No change occurs in the plasma concentration of either progesterone or of oestrogens until labour is well advanced.

Karim (1968) found $PGF_{2\alpha}$ in the maternal circulation of women in labour although it was not detectable before labour started. In addition, Karim & Devlin (1967) found the advent of labour to be associated with a rapid rise in concentration of prostaglandins in amniotic fluid. The stimulus to release of prostaglandins in women in unknown. The absence of changes in oestrogen and progesterone is incompatible with their being responsible but it is possible that release is stimulated by dissociation of the foetal membranes and the maternal decidua. The decidua is the site of high concentrations of prostaglandins and mechanical separation of the membranes from the uterine wall or any form of physical damage to the membranes is an effective means of inducing labour.

The human foetal adrenal cortex has important interactions with the placenta. It is the main site of synthesis of the precursor, dehydroepiandrosterone sulphate, utilised by the trophoblast in the synthesis of oestrogen. The foetal adrenal cortex thus has the potential for influencing systems in the maternal compartment and it was shown by Turnbull, Anderson & Wilson (1967) that a correlation existed between the duration of pregnancy and the ratio of urinary oestriol:oestrone determined at the 34th week. There is no evidence that foetal corticosteroids directly influence placental function in women. Indeed, the effect on oestrogen metabolism of treatment with corticosteroids is opposite to that in sheep in that oestrogen production is markedly reduced because of suppression of precursor formation by the foetal cortex.

Preparturitional endocrine changes in cattle are comparable to those in sheep. Plasma progesterone levels fall sharply about 24 hours before labour (Donaldson, Bassett & Thorburn, 1970) and urinary oestrogens rise (Osinga, 1970). There are no reports of prostaglandin assays. Glucocorticoids given to the cow in large doses induce labour in the last six weeks of pregnancy and progesterone and oestrogen levels alter as they do before spontaneous parturition (Jochle, 1971). It is likely that the greater susceptibility of cattle than sheep to maternally-administered corticosteroids reflects earlier changes in placental permeability and entry of steroid into the foetal compartment.

The results of stimulation with ACTH of the adrenals of the foetal goat are of great interest in connection with the possibility of foetal influences on ovarian function. Thorburn *et al.* (1972) observed that infusion of ACTH into foetal goats causes premature parturition preceded by falling progesterone levels and rising levels of unconjugated oestrogens and $PGF_{2\alpha}$. The principal source of progesterone in pregnant goats is ovarian; thus, the effect of foetal cortisol activity is ultimately reflected in changes in ovarian function. The 'messenger' between conceptus and corpus luteum is uncertain but it could be either $PGF_{2\alpha}$, which has luteolytic properties in several species, or an unidentified luteotrophic agent.

Proof is lacking of an adrenocortical-placental interaction in corpus luteum-dependent species other than the goat. Premature parturition in rabbits given intra-amniotic injections of corticosteroids is preceded by premature luteolysis and falling levels of circulating progesterone (Kendall & Liggins, 1972), suggesting as in the goat that a 'messenger' derived from the conceptus influences corpus luteum function. The behaviour of oestrogens and prostaglandins at the time of luteolysis has not been studied.

PLACENTAL-MYOMETRIAL INTERACTIONS

Functions in the sheep

The physiological roles of each of the three hormonal changes preceding parturition have been investigated in sheep by observing the effects on myometrial activity of administration of each hormone in turn. Progesterone given to the ewe in daily doses of 80 mg fails to prolong pregnancy (Bengtsson & Schofield, 1963) although this dose exceeds the normal secretion rate of progesterone in late pregnancy. Doses of progesterone up to 150 mg daily also fail to prevent parturition induced by glucocorticoids infused into the foetal lamb (Liggins *et al.* 1972), but labour may be protracted. Assays of progesterone concentrations in plasma and myometrium demonstrate that such doses prevent the fall in concentration that usually occur before the onset of labour. These findings suggest that withdrawal of progesterone, although aiding the progress of labour, is inessential to the initiating mechanism and they cast some doubt on the validity of the concept of the 'progesterone block', at least in sheep.

Oestrogen in the form of a single injection of stilboestrol (20 mg) given to pregnant sheep in late pregnancy causes increased uterine activity and sometimes parturition (Hindson, Schofield & Turner, 1967). The onset of uterine activity is preceded by a marked increase in the sensitivity of the myometrium to oxytocin (Fig. 3) but progesterone levels are unaltered. The interpretation of these findings is difficult because of present uncertainties about the acute actions of oestrogen on uterine smooth muscle. *In vitro*, oestrogens have inhibitory effects on uterine contractility in various species, whereas *in vivo* they are said to have oxytocic effects or to increase responsiveness to oxytocin. This paradox may be resolved in relation to the uterus of the pregnant sheep by the finding that oestrogen stimulates the synthesis of $PGF_{2\alpha}$ in myometrium and placenta and causes its release in uterine vein blood (Liggins, 1973).

Continuous intra-aortic infusion of $PGF_{2\alpha}$ at a similar rate (10 μg/min) to that at which $PGF_{2\alpha}$ drains from the uterus in venous blood preceding

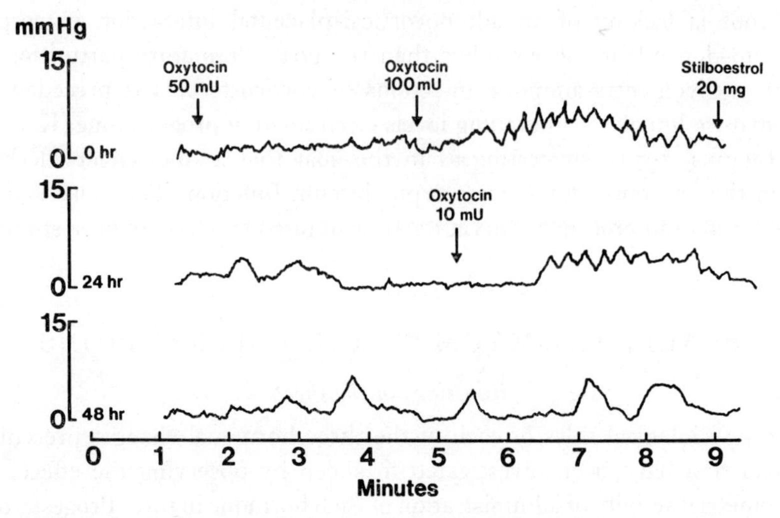

Fig. 3. Effect of a single maternal injection of stilboestrol (20 mg) on the uterine response to intravenous injections of oxytocin. The response shown in each instance is to the threshold dose. Uterine activity recorded by means of an intra-amniotic catheter and strain-gauge transducer.

labour causes an increase in myometrial sensitivity to oxytocin and the onset of uterine contractions without an alteration in the secretion rate of progesterone (Fig. 4). These effects are identical to those caused by the administration of oestrogen and since oestrogen, in addition, stimulates synthesis of $PGF_{2\alpha}$ (Fig. 5), it is possible that the action of oestrogen on uterine contractility is mediated by $PGF_{2\alpha}$. If this is so, the main determinant of uterine contractility is $PGF_{2\alpha}$, the short-term effects of oestrogen and progesterone perhaps being largely dependent on their respective actions on the rate of synthesis of $PGF_{2\alpha}$.

Functions in other species

The actions of oestrogen and progesterone on pregnant human myometrium *in vivo*, apart from those relating to uterine growth and vascularity, are uncertain. Labour occurs in the presence of unchanged concentrations of both hormones and administration of massive doses of either oestrogen or progesterone does not modify the course of labour or the time of its onset. On the other hand, $PGF_{2\alpha}$ and PGE_2 have potent oxytocic properties and are widely used for induction of either mid-trimester abortion or labour at term (Karim & Sharma, 1970). However, nothing is known as yet about the physiological mechanisms controlling the release of prostaglandins at the onset of labour in women. One might speculate, however, that the controlling mechanism is foetal in origin.

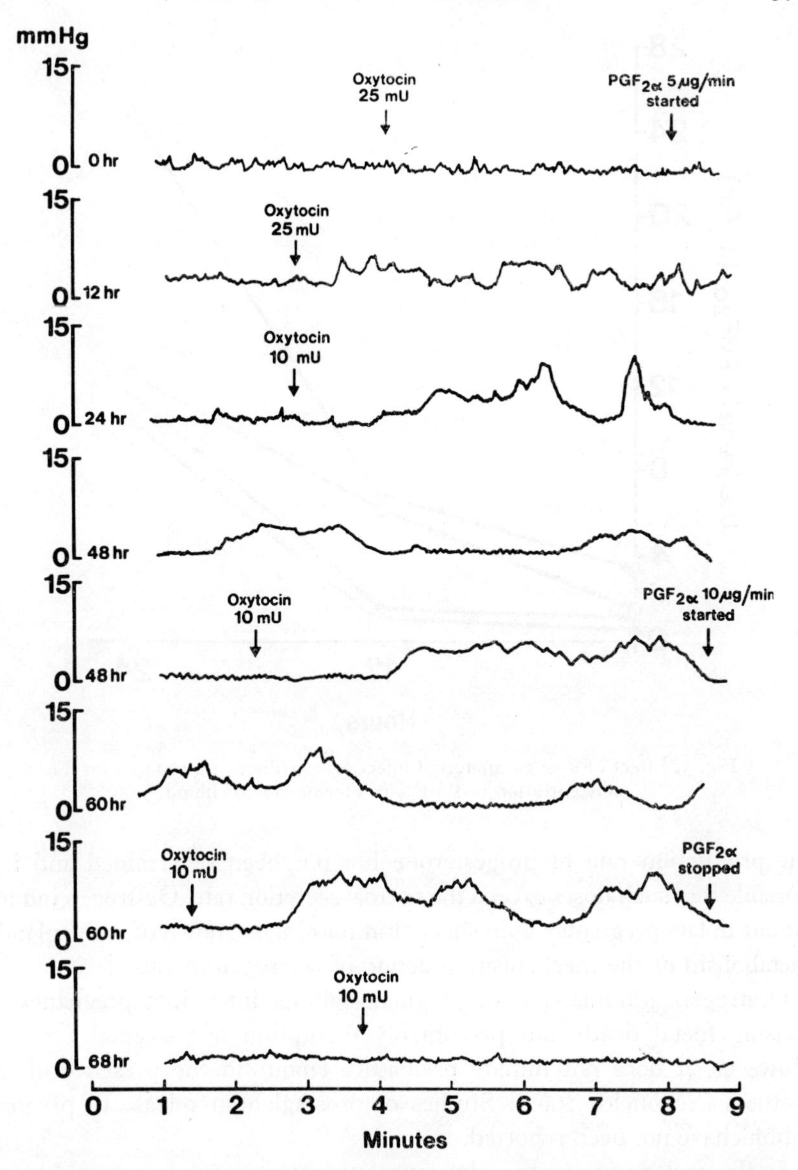

Fig. 4. Effect of a continuous intra-aortic infusion of $PGF_{2\alpha}$ at a rate of 5–10 μg/min on uterine sensitivity to oxytocin.

Cattle show similarities to sheep in the response of the pregnant uterus to steroid hormones. Administration of progesterone prolongs pregnancy and also blocks the action of corticosteroids in inducing premature delivery (Jochle, 1971). The uterus appears relatively sensitive to progesterone in that these effects are observed with doses of only 100 mg/24 h. However,

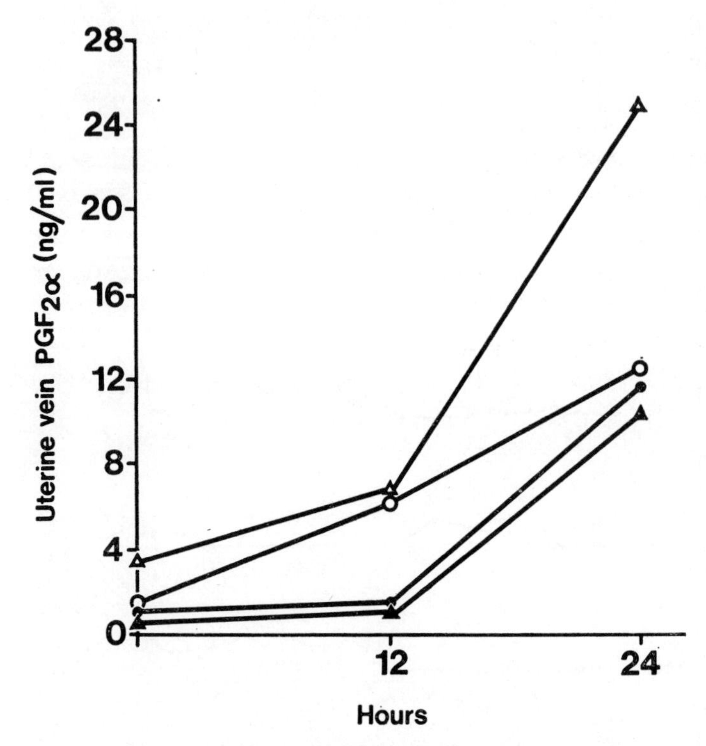

Fig. 5. Effect of a single maternal injection of stilboestrol (20 mg) on the concentration of $PGF_{2\alpha}$ in uterine venous blood.

the production rate of progesterone has not been determined and it is possible that such doses exceed the normal secretion rate. Oestrogen induces labour in late pregnancy as in sheep, but nothing is known of prostaglandin metabolism or the mechanism of action of oestrogen in cattle.

Oestrogen administered to pregnant rabbits interrupts pregnancy by causing foetal death and presumably disruption of placental function. However, it does not initiate premature labour in the presence of live foetuses (Schofield, 1962). Studies of prostaglandin release in pregnant rabbits have not been reported.

In the goat, progesterone administration prolongs pregnancy and blocks the inducing action of foetal infusions of ACTH. No information is available about the action of exogenous oestrogen on release of prostaglandins or on uterine mobility. The close temporal relationship of peaks of oestrogen and $PGF_{2\alpha}$ before parturition as found by Thorburn *et al.* (1972) suggest that similar interactions are present in the goat as are found in the ewe.

Pregnancy in rats is prolonged by progesterone administration and interrupted by oestrogen and prostaglandins, raising the possibility that in this species too, the interrelationships found to exist in sheep and goats may apply equally well.

MATERNAL-MYOMETRIAL INTERACTIONS

Despite the emphasis in this paper on the role of the foetus, the importance of maternal factors influencing parturition should not be overlooked. Observations of a circumstantial nature point to marked, although relatively brief, maternal control of labour. Examples such as characteristic day: night ratios of births and delayed labour in the presence of adverse environmental circumstances are known in a variety of species and are usually attributed to neural influences on the myometrium.

Oxytocin released from the maternal posterior pituitary is commonly relegated to a supporting role in the late stages of labour because it is only then that its concentration in the blood may rise. Indeed, in women, doubt has been cast recently on the likelihood of oxytocin playing any part in labour; a sensitive radioimmunoassay failed to detect oxytocin in most women even during the second stage of labour (Chard, Hudson, Edmonds & Boyd, 1971). Nevertheless, the possibility remains that low levels of circulating oxytocin may share with other endocrine factors in providing the complex environment that allows the uterine smooth muscle cells to begin contracting rhythmically in a manner that dilates the cervix and expels the conceptus.

Maternal disorders of various sorts may hasten the onset of labour and cause premature delivery. But in contrast to the foetus, the mother is not known to be the cause of any syndromes of prolonged pregnancy nor does ablation of any maternal organ cause failure of labour. Perhaps these facts place correctly in perspective the relative contributions of mother and foetus to the physiological mechanism controlling parturition.

POSSIBLE PATHWAY FROM FOETAL HYPOTHALAMUS TO MYOMETRIUM IN SHEEP

'Broadly speaking, labour is a change of environment induced by the foetus when all its tissues have reached a certain degree of harmonious development, or to use an obstetrical term, at the period of optimum viability' (Malpas, 1933). The nature of the signals from developing tissues that communicate their maturation to the foetal endocrine system remains to be elucidated but it is clear that the foetal hypothalamus serves as the means of their expression. Through the pituitary, the hypothalamus

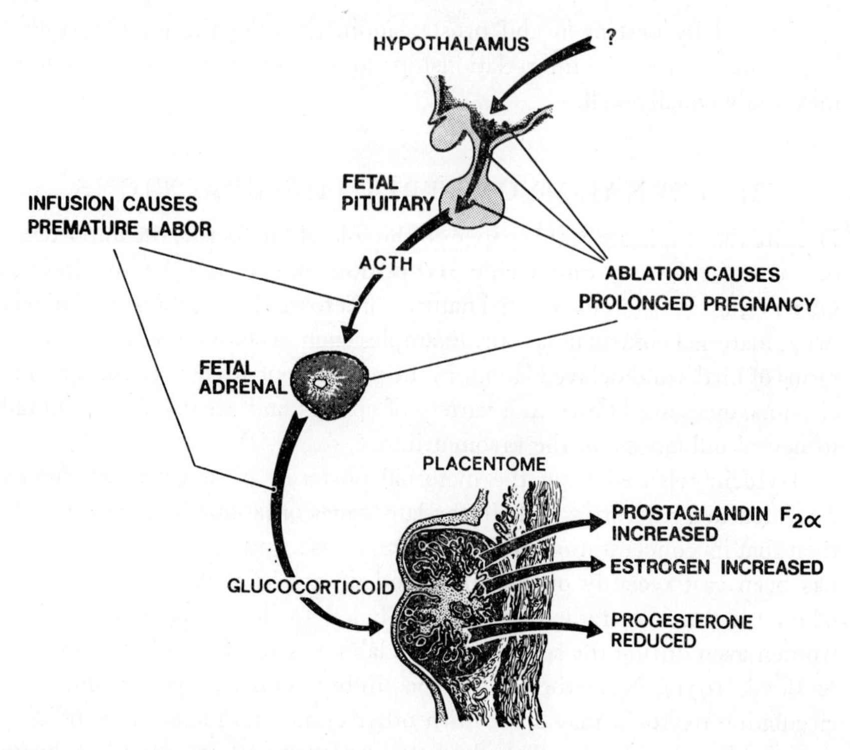

Fig. 6. Pathway by which the foetal endocrine system influences myometrial contractility. Experimental methods for manipulating the pathway and their effects are illustrated.

stimulates increased secretion of cortisol from the foetal adrenal. Cortisol, in turn, acts on the placenta, causing changes in the secretion of progesterone, oestrogen and $PGF_{2\alpha}$ (Fig. 6). It is suggested that the changes in myometrial contractility that lead to parturition are mainly induced by the action of $PGF_{2\alpha}$ and that the major function of oestrogen and progesterone is in regulating the release of $PGF_{2\alpha}$. Maternal endocrine and neural influences can only briefly alter the sequence of events initiated by the foetus.

CONCLUSIONS

There is now good evidence, both circumstantial and experimental, to show that the conceptuses of several species are largely responsible for determining the endocrine changes that initiate labour. In the two species most thoroughly investigated from this point of view, the sheep and the goat, it has been established that the foetal adrenal, by means of its secretion of cortisol, regulates the production of progesterone, oestrogens and prostaglandins. It seems likely that the adrenals of human and bovine foetuses also are involved in the mechanisms controlling parturition. In the human, adrenal action is possibly mediated by the secretion of oestrogen precursors rather than by corticosteroids. In other species, the applicability of the concept of foetal participation in the control of parturition remains to be investigated.

The recent discovery that prostaglandins are liberated at the onset of labour in humans, sheep and goats may require revision of traditional notions regarding the control of myometrial function by progesterone and oestrogen.

I am indebted to Mr R. J. Fairclough, Miss Susan Grieves, Miss June Kendall and Mr B. S. Knox who have shared with me the work described in this paper. I also wish to thank Mr A. Mekkelholt and Mr H. Wattam for their meticulous care of experimental animals and Dr J. T. France for advice on biochemical matters. The financial support of The Wellcome Trust is gratefully acknowledged.

REFERENCES

ADAMS, W. M. & WAGNER, W. C. (1969). *Biol. Reprod.* **3**, 223.

ANDERSON, A. B. M., LAURENCE, K. M. & TURNBULL, A. C. (1969). *J. Obstet. Gynaec. Br. Commonw.* **76**, 199.

ANDERSON, A. B. M., PIERREPOINT, C. G., GRIFFITHS, K. & TURNBULL, A. C. (1972). *J. Reprod. Fert.* Suppl. 16, 25.

BASSETT, J. M. & THORBURN, G. D. (1969). *J. Endocr.* **43**, 449.

BASSON, P. A., MORGENTHAL, J. C., BILBROUGH, R. B., MARAIS, J. L., KRUGER, S. P. & VAN DER MERWE, J. L. DE B. (1969). *Onderstepoort J. Vet. Res.* **36**, 59.

BENGTSSON, L. PH. & SCHOFIELD, B. M. (1963). *J. Reprod. Fert.* **5**, 423.

BINNS, W., JAMES, L. F. & SHUPE, J. L. (1964). *Ann. N.Y. Acad. Sci.* **111**, 571.

BOSC, M. J. (1972). *C. r. hebd. Séanc. Acad. Sci., Paris* **274**, 93.

CHALLIS, J. R. G. (1970). *Nature, London* **229**, 203.

CHARD, T., HUDSON, C. N., EDWARDS, C. R. W. & BOYD, N. R. H. (1971). *Nature, London* **234**, 352.

CHEZ, R. A., HUTCHINSON, D. L., SALAZAR, H. & MINTZ, D. H. (1970). *Am. J. Obstet. Gynec.* **108**, 643.

COMLINE, R. S., NATHANIELSZ, P. W., PAISEY, R. B. & SILVER, M. (1970). *J. Physiol.* **210**, 141P.

COMLINE, R. S. & SILVER, M. (1961). *J. Physiol.* **156**, 424.

COMLINE, R. S., SILVER, M. & SILVER, I. A. (1970). *Nature, Lond.* **225**, 789.

DONALDSON, L. E., BASSETT, J. M. & THORBURN, G. D. (1970). *J. Endocr.* **48**, 599.

DROST, M. & HOLM, L. W. (1968). *J. Endocr.* **40**, 293.

FAIRCLOUGH, R. J. & LIGGINS, G. C. (1972). *N.Z. Med. J.* (In press.)

FYLLING, P. (1970). *Acta Endocr. Copnh.* **65**, 273.

HINDSON, J. C., SCHOFIELD, B. M. & TURNER, C. B. (1967). *Res. vet. Sci.* **11**, 159.

HOLM, L. W. (1967). *Adv. vet. Sci.* **11**, 159.

JOCHLE, W. (1971). *Folia Vet. Latina* **1**, 229.

KARIM, S. M. M. (1968). *Br. med. J.* **4**, 618.

KARIM, S. M. M. & DEVLIN, J. (1967). *J. Obstet. Gynaec. Br. Commonw.* **74**, 230.

KARIM, S. M. M. & SHARMA, S. D. (1971). *J. Obstet. Gynaec. Br. Commonw.* **78**, 294.

KENDALL, J. Z. & LIGGINS, G. C. (1972). *J. Endocr.* **29**, 409.
KENNEDY, P. C. (1971). *Fed. Proc.* **30**, 110.
KENNEDY, P. C., KENDRICK, J. W. & STORMONT, C. (1957). *Cornell Vet.* **47**, 160.
KLEIN, G. P., KERTESZ, J. P., CHAN, S. K. & GIROUD, C. J. P. (1971). *J. clin. Endocr.* **32**, 333.
LEVINE, S. & TREIMAN, L. J. (1969). *Foetal Autonomy*. Ciba Foundation Symposium, eds. G. E. W. Wolstenholme & M. O'Connor, p. 271. London: Churchill.
LIGGINS, G. C. (1968). *J. Endocr.* **42**, 323.
LIGGINS, G. C. (1969). *Foetal Autonomy*. Ciba Foundation Symposium, eds. G. E. W. Wolstenholme & M. O'Connor, p. 218. London: Churchill.
LIGGINS, G. C. (1973). *Recent Prog. Horm. Res.* (In press.)
LIGGINS, G. C. & GRIEVES, S. A. (1971). *Nature, Lond.* **232**, 629.
LIGGINS, G. C., GRIEVES, S. A., KENDALL, J. Z. & KNOX, B. S. (1972). *J. Reprod. Fert.* Suppl. 16, 85.
LIGGINS, G. C., KENNEDY, P. C. & HOLM, L. W. (1967). *Am. J. Obstet. Gynec.* **98**, 1080.
MALPAS, P. (1933). *J. Obstet. Gynaec. Br. Commonw.* **40**, 1046.
MUELLER-HEUBACH, E., MYERS, R. E. & ADAMSONS, K. (1972). *Am. J. Obstet. Gynec.* **112**, 221.
NATHANIELSZ, P. W., COMLINE, R. S., SILVER, M. & PAISEY, R. B. (1972). *J. Reprod. Fert.* Suppl. 16, 39.
O'DONOHUE, N. W. & HOLLAND, P. D. J. (1968). *Arch. Dis. Child.* **43**, 717.
OSINGA, A. (1970). *Oestrogen Excretion by the Pregnant Bovine and its Relation with Some Characters of Gestation and Parturition.* Wageningen: Veenman & Zonen.
SCHOFIELD, B. M. (1962). *J. Endocr.* **25**, 95.
STABENFELDT, G. H., OSBURN, B. E. & EWING, L. C. (1970). *Am. J. Phys.* **218**, 571.
THORBURN, G. D., NICOL, D. H., BASSETT, J. M., SHUTT, D. A. & COX, R. I. (1972). *J. Reprod. Fert.* Suppl. 16, 61.
TURNBULL, A. C., ANDERSON, A. B. M. & WILSON, G. R. (1967). *Lancet* **2**, 627.
VAN RENSBERG, S. J. (1963). *S. Afr. med. J.* **37**, 1114.
WEIST, W. G., KIDWELL, W. R. & BALOGH, K. (1969). *Endocrinology* **82**, 844.

THE ROLE OF THE POSTERIOR PITUITARIES OF MOTHER AND FOETUS IN SPONTANEOUS PARTURITION

By T. CHARD

Departments of Obstetrics and Gynaecology, and Chemical Pathology, St Bartholomew's Hospital, London E.C.1.

Of the many factors which can influence the contractility of the uterus at term, none can be considered as pre-eminent in the initiation and maintenance of spontaneous labour. The posterior pituitary and its hormones are merely one of these factors.

The evidence for the involvement of the posterior pituitary hormones, and in particular oxytocin, is both direct and indirect (see Chard, 1972). The direct evidence comes from measurement of circulating oxytocin levels, a procedure fraught with methodological problems. The indirect evidence arises from observations on the uterine-stimulating effects of oxytocin, on the effects of hypophysectomy on the outcome of pregnancy, and on the inhibitory effect of alcohol on spontaneous uterine contractions (see Caldeyro-Barcia, 1971; Fuchs *et al.* 1971).

The recent development of highly specific and sensitive radioimmunoassays for oxytocin, vasopressin, and neurophysin (e.g. Chard *et al.* 1970; Edwards *et al.* 1972; Martin *et al.* 1972) has permitted a fresh approach using direct measurement of circulating levels. The results of these studies will be presented here.

MATERNAL OXYTOCIN IN ANIMAL PARTURITION

Radioimmunoassay (Chard *et al.* 1970; McNeilly *et al.* 1971) has confirmed earlier studies using bioassays in showing a massive release of oxytocin during the expulsive phase of labour, but relatively little during the earlier stages. The increase in oxytocin is accompanied by a simultaneous release of neurophysin (McNeilly *et al.* 1972). It should be noted that these observations, in common with those of previous studies, were performed on jugular venous plasma, the direct drainage of the pituitary gland in animal species. The levels found are likely to be an overestimate of those reaching the uterus.

MATERNAL OXYTOCIN IN HUMAN PARTURITION

The radioimmunoassay studies (Chard *et al.* 1971; Gibbens *et al.* 1972) disagree with earlier results based on biological assays. The latter have indicated oxytocin levels, in peripheral plasma, of 100 μU/ml or more. By contrast, immunoassays suggest that the levels at this site rarely exceed 10 μU/ml, and that the release occurs in a series of 'spurts'. The frequency of the spurts increases as labour progresses, reaching a peak in the second stage (Fig. 1). The discrepancy between these and previous investigations can probably be attributed to methodological factors, and also, as in the excellent studies of Coch *et al.* (1965), to a difference in the site of sampling.

The low circulating levels of oxytocin are well reflected by the absence of any rise in urine oxytocin excretion during labour, despite the fact that an increase can be produced by infusion of hormone at a rate of 1 mU/min (Boyd & Chard, 1973). Our own studies have also failed to show a rise in circulating neurophysin, though Legros & Franchimont (1972) found a steady increase throughout pregnancy.

FOETAL OXYTOCIN IN ANIMAL PARTURITION

In the guinea-pig, there is a striking increase in the hormone content of the foetal posterior pituitary at or around the time of delivery (Burton & Forsling, 1972). In the cow, the levels of neurophysin I are higher in the foetus than the mother at the time of delivery (Robinson *et al.* 1972); release of neurophysin I is thought to reflect release of oxytocin.

FOETAL OXYTOCIN IN HUMAN PARTURITION

High levels of oxytocin and vasopressin can be found in cord blood at the time of delivery (Chard *et al.* 1971); in the case of vasopressin, this confirms the earlier results of Hoppenstein *et al.* (1968). The levels are highest in the umbilical artery, indicating the foetal pituitary as the origin. Maximal levels are found at the time of vaginal delivery, and the lowest levels when the patient is not in labour (Fig. 2). The release is therefore associated with the process of labour. High levels of neurophysin in neonatal plasma have been reported by Legros & Franchimont (1972).

DISCUSSION

Activation of the posterior pituitaries of both mother and foetus during labour is in little doubt. However, there are many unanswered questions,

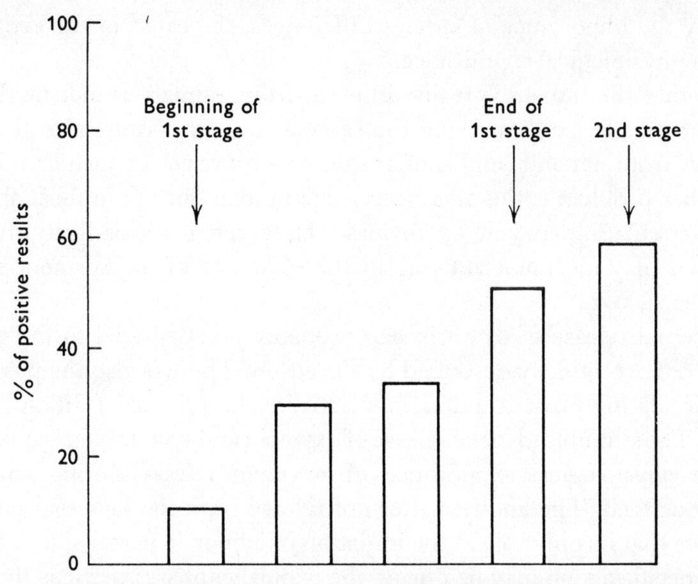

Fig. 1. The frequency of positive results (oxytocin level greater than 1 μU/ml) in serial samples collected from women during spontaneous labour.

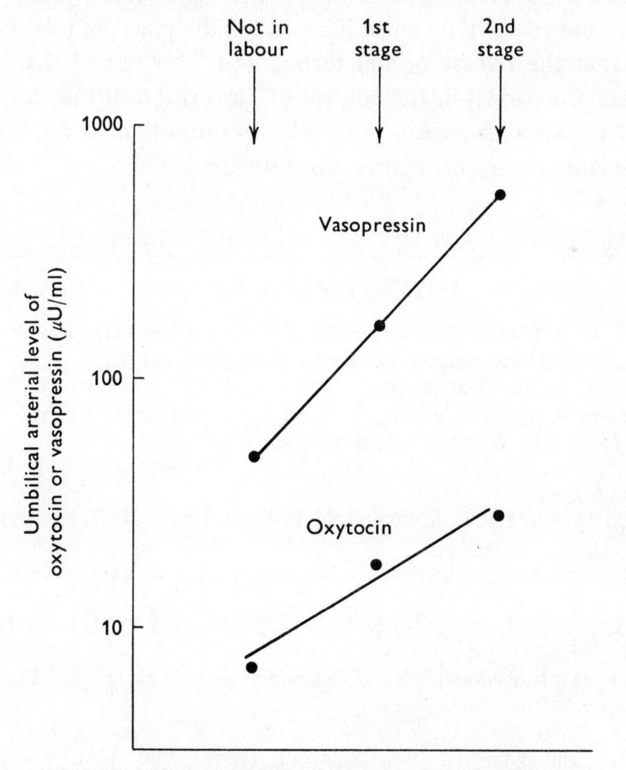

Fig. 2. Umbilical arterial levels of oxytocin and vasopressin in the human foetus at the time of delivery.

notably the importance of species differences, the cause of the activation, and its physiological significance.

Whether the human is really different from animals cannot be decided without parallel studies, using the same assay, on peripheral and central samples from patients, and similar samples from species such as the goat. A further problem is the presence, in the human but not in most animals, of the circulating enzyme oxytocinase. However, it seems likely that this does not play an important role in the clearance of the hormone *in vivo* (see Chard, 1972).

Maternal release of oxytocin can probably be attributed to the type of spinal reflex originally described by Ferguson. There is also the possibility that circulating prostaglandins may activate the pituitary (Gillespie *et al.* 1972). The stimulus to foetal release of oxytocin and vasopressin is unknown.

The physiological significance of oxytocin release during labour is poorly defined. The known pattern of release from the maternal pituitary suggests that its only role in the initiation of labour is permissive, while its most specific action may be during the expulsive phase; perhaps the function at this stage is to ensure full retraction of the uterus once it is emptied of its contents. Oxytocin released into the foetal circulation is well-placed to influence uterine activity since it can cross the placenta relatively easily. Knowing that the release occurs throughout labour, and that the foetal mechanisms are critical in the control of parturition, it may be suggested that foetal release of oxytocin is at least as important as the maternal in the maintenance of uterine contractions during labour.

REFERENCES

BOYD, N. R. H. & CHARD, T. (1973). *Am. J. Obstet. Gynec.* (In press.)
BURTON, A. M. & FORSLING, M. L. (1972). *J. Physiol.* **221**, 6P.
CALDEYRO-BARCIA, R., MELANDER, S. & COCH, J. A. (1971). In *Endocrinology of Pregnancy*, p. 235. Ed. Fuchs, F. & Klopper, A. Harper & Row: New York.
CHARD, T. (1972). *J. Reprod. Fert.*, Suppl. 16, 121.
CHARD, T., BOYD, N. R. H., FORSLING, M. L., MCNEILLY, A. S. & LANDON, J. (1970). *J. Endocr.* **48**, 223.
CHARD, T., HUDSON, C. N., EDWARDS, C. R. W. & BOYD, N. R. H. (1971). *Nature* **234**, 352.
COCH, J. A., BROVETTO, J., CABOT, H. M., FIELITZ, C. A. & CALDEYRO-BARCIA, R. (1965). *Am. J. Obstet. Gynec.* **91**, 10.
EDWARDS, C. R. W., CHARD, T., KITAU, M. J., FORSLING, M. L. & LANDON, J. (1972). *J. Endocr.* **52**, 279.
FUCHS, F. (1971). In *Endocrinology of Pregnancy*, p. 306. Ed. Fuchs, F. & Klopper, A. Harper & Row: New York.
GIBBENS, D., BOYD, N. R. H. & CHARD, T. (1972). *J. Endocr.* **53**, LIV.
GILLESPIE, A., BRUMMER, H. C. & CHARD, T. (1972). *Brit. Med. J.* **1**, 543.

HOPPENSTEIN, J. M., MILTENBERGER, F. W. & MORAN, W. H. (1968). *Surgery Gynec. Obstet.* **127**, 966.

LEGROS, J. J. & FRANCHIMONT, P. (1972). *Clin. Endocr.* **1**, 99.

McNEILLY, A. S., FORSLING, M. L. & CHARD, T. (1971). In *Radioimmunoassay Methods*, p. 556. Ed. Kirkham, K. & Hunter, W. M. Churchill Livingstone: Edinburgh & London.

McNEILLY, A. S., MARTIN, M. J., CHARD, T. & HART, I. C. (1972). *J. Endocr.* **52**, 213.

MARTIN, M. J., CHARD, T. & LANDON, J. (1972). *J. Endocr.* **52**, 481.

ROBINSON, A. G., ZIMMERMAN, E. A. & FRANTZ, A. G. (1972). *Metabolism* **20**, 1148.

IONIC CURRENTS IN A PREGNANT MYOMETRIUM

By C. Y. KAO
Department of Pharmacology,
State University of New York,
Downstate Medical Center, Brooklyn, N.Y. 11203

Some steps involved in the complex process of parturition are concerned with peripheral mechanisms that control and effect excitation and contraction of the uterine smooth muscle. Up until recently, nearly all studies on excitation phenomena in the uterine muscle have been descriptive observations on spike discharges of myometrial cells. Several years ago, Anderson (1969) succeeded in adapting a double sucrose-gap technique to make a voltage-clamp study on an oestrogen-dominated non-pregnant myometrium. This communication is concerned with a voltage-clamp study of ionic currents underlying some excitatory phenomena in myometrial cells from a pregnant uterus close to term. It is trivial, but perhaps necessary, to restate that although individual cells in a pregnant myometrium are probably the largest mammalian smooth muscle cells readily available for experimentation, they measure only about 20 μm in their largest diameters, and taper towards the ends for a total length of about 200 μm. With such sizes, it is difficult to employ certain techniques which are commonplace in work on nerves, skeletal and cardiac muscle fibres. An important advantage of a double sucrose-gap technique is that external electrodes are used; and if sufficient care is taken, it is possible to make observations on the same group of cells for as long as 45 minutes during which several changes of solutions can be made. Briefly, the technique consists of bathing isolated strips of myometrium from pregnant rat uterus in five pools of solutions, three pools of Krebs-bicarbonate solution (or some electrolyte solution of a different composition) separated from one another by two cuffs of isotonic sucrose solution. The non-electrolyte sucrose, in replacing extracellular ions in the segments of the myometrium it bathes, increases the extracellular resistance so much that the main paths for the flow of electric current between the segments bathed in the electrolyte solutions are intracellular. The portion of the myometrial preparation in the central electrolyte pool (the 'node') is the important region. For a successful voltage clamp experiment, the 'node' should be less than 100 μm, a length in which reasonable uniformity could be expected from a known space constant of

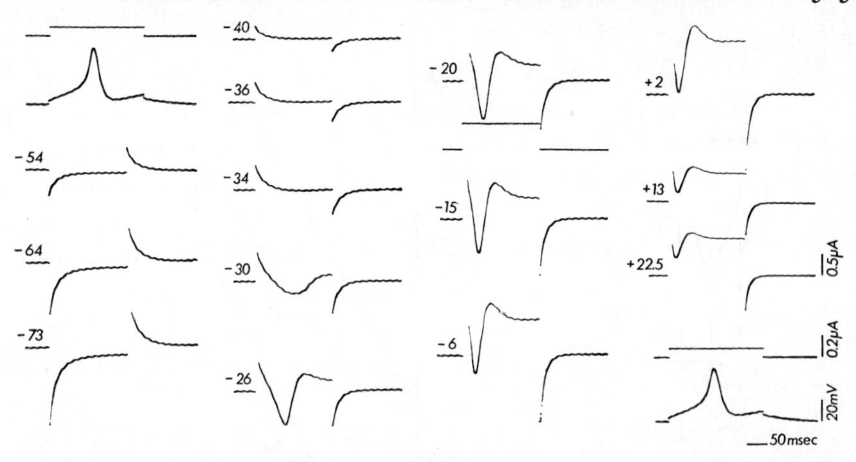

Fig. 1. Records of currents due to step voltage changes imposed in a voltage-clamp series. Myometrium from a 21-day pregnant rat uterus. Frames at top left and bottom right corners are constant current spike responses before and after voltage clamp series. Resting potential in this preparation was −44 mV which was also the holding potential in voltage clamp. Numbers attached to left of traces indicate voltage levels to which step changes were made. Actual voltage traces have been omitted for clarity, except for −20 mV. Inward current is maximum at −20 mV, and squareness of voltage step indicates quality of voltage control. Except for +13 and +22.5 mV which have lower current amplifications, all other traces have same current calibration as marked at right lower corner frame. Note slowness of inward current, and inactivating delayed outward current.

1.8 to 2 mm for this preparation. The observations made by means of this technique cannot be obtained by any other means, and although the information obtained is not perfect, it is useful in providing some understanding of the ionic currents underlying some of the excitatory phenomena in the myometrium.

The main conclusions about ionic currents in the myometrium is that they are, in principle, very similar to those occurring in some better-studied excitable tissues, such as the squid giant axon (Hodgkin, Huxley & Katz, 1952), and the frog skeletal muscle (e.g. Adrian, Chandler & Hodgkin, 1970). There are important differences in details; and the following discussion will emphasize these differences with the assumption that the similarities are understood. Fig. 1 shows some current records obtained during a voltage-clamp experiment, and Fig. 2 the current–voltage relation for the initial current and for the late current.

The ionic nature of the initial current is not completely resolved, because of some peculiarities of the myometrial preparation. Myometrial cells probably contain a good deal of intracellular Na^+ (see Kao, 1967) which could leak into the extracellular phase whenever $[Na^+]_0$ is reduced. Therefore, in experiments testing the sodium hypothesis in the myometrium, $[Na^+]_0$ is somewhat uncertain. The initial current in the myometrium is

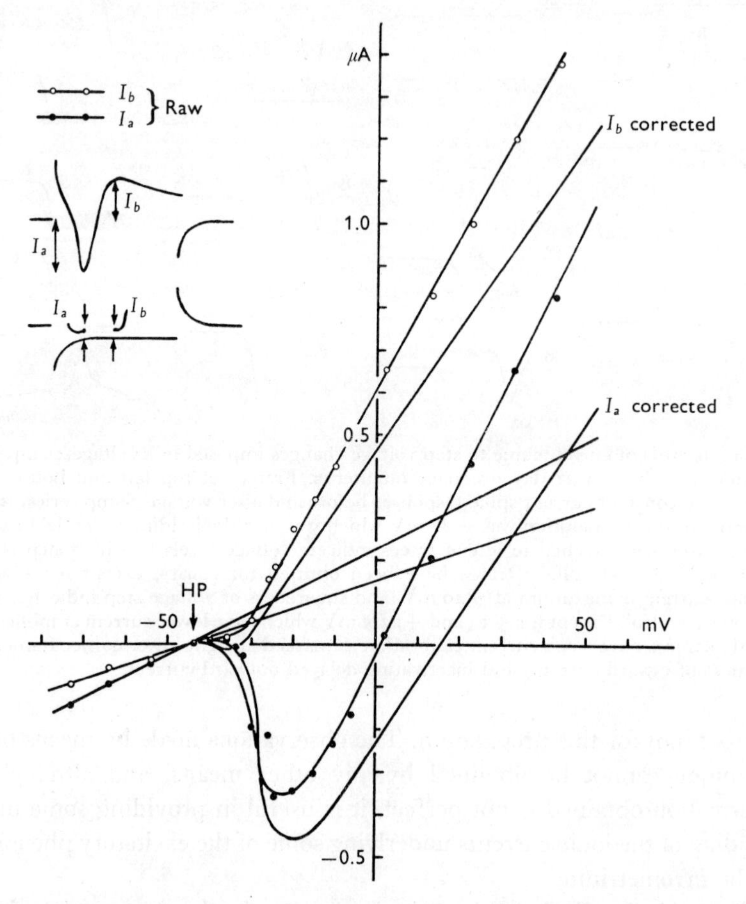

Fig. 2. Current–voltage relations in voltage-clamp condition. Inset shows how early current (I_a) and late current (I_b, now identified as potassium current, see text) were measured on depolarizing and hyperpolarizing voltage steps. Lines joining circles (hollow and filled) are experimental points. Continuous lines are currents after correction for leakage currents. HP = holding potential, which is same as resting potential. Note currents are total currents. Equilibrium potential for early current (E_a) is +16 mV.

also insensitive to tetrodotoxin (Anderson, 1969) which has been used with great effectiveness in many other tissues (see Kao, 1966). Nevertheless, when $[Na^+]_0$ is reduced, the magnitude of the initial inward current is reduced, and the equilibrium potential (E_a) shifts towards the resting potential (Anderson, 1969; Kao, McCullough & Davidson, 1970). More quantitatively, when $[Na^+]_0$ was reduced from 143 mM to 71.5 mM, the mean shift in E_a in three preparations agreed with a theoretical shift of a sodium equilibrium potential (E_{Na}) to within 4.4 mV. However, there are aberrant observations in these experiments which preclude a firm conclusion on the ionic nature of the initial current: the shift in E_a is not in good agreement with theory if large reductions of $[Na^+]_0$ were made. Even in very low $[Na^+]_0$ some inward current persisted, which could be abolished upon removal of external Ca^{2+} (Kao *et al.* 1970; Kao, 1971). A reasonable conclusion to draw at this time is that Na^+ is probably the main carrier of the initial current in the rat myometrium, and that under extreme conditions some other ion can also carry the current.

The nature of the late current has been resolved by experiments of the following type: a myometrial preparation was depolarized by a solution containing 149 mM K^+, all of the external Na^+ and most of the Cl^- having been replaced by K_2SO_4. In this solution $[K^+]_i$ and $[K^+]_0$ are much closer to each other than in Krebs solution in which $[K^+]_0 = 5.9$ mM. When such a depolarized preparation was clamped back to the original resting potential that it had when in Krebs solution, and then electrically depolarized, the late current which had always been outwards when in Krebs solution became inwards within a limited voltage range (Kao *et al.* 1971). This changed direction of current flow is explained on the basis that the elevated $[K^+]_0$ permitted an inward movement of K^+ under the appropriate conditions (see, e.g., Frankenhauser, 1962).

Many details of the ionic currents remain to be investigated, but the little that is known has been encouraging in the use of the voltage-clamp technique to study the reactions of the pregnant myometrium to some neurohormones. For instance, it is known in classical pharmacology that the pregnant rat myometrium relaxes upon the action of adrenaline, a phenomenon which is closely related to a hyperpolarization of the myometrial cells. There are several suggestions on the possible mechanism of this hyperpolarization, such as an increase in the potassium conductance of the myometrial membrane, a decrease in the sodium conductance, and an increase in the rate of an electrogenic sodium pump. When the adrenaline hyperpolarization in the rat myometrium was investigated with a voltage-clamp technique (Kao *et al.* 1971), it was readily evident that neither the conductance associated with initial current (sodium?) nor that with the late current (potassium) was substantially altered. The equilibrium potential of the initial current was also unchanged. The important change was a genuine increase in the potassium equilibrium potential, an observation which led to the conclusion that a change in the electromotive force of the potassium cell was more important than any possible changes in conductances. This conclusion must in turn call for a serious reappraisal of the role of metabolic (perhaps extra-membrane or other indirect) effects of adrenaline in bringing about readily observable changes in electrophysiological properties (Kao, 1971).

Another neurohormone of particular interest to myometrial physiologists is oxytocin. All that I can say at this time about the action of oxytocin on the pregnant rat myometrium is that under voltage-clamp conditions, nothing dramatically different happens on the various ionic conductances, the maximum inward current, and the inactivation of the initial current. With such a disappointing outcome, I suspect that the direction to look in a search for an explanation of the electrophysiological actions of oxytocin may be in the pacemaker activity of the myometrium. In the few instances in our experience in which spontaneous activities occurred under voltage-clamp conditions, the recurrent large inward currents associated with spike discharges have been preceded by a slowly increasing inward current. When this slow current became smaller or changed to outward, spontaneous discharges tended to cease. There is no evidence at this time to state whether the slow inward current (possibly responsible for pacemaker depolarization discussed in Kao, 1967) is due to an increased inward leak of Na^+ or a decreased outward leak of K^+. However, because of the finding that in the rat myometrium the potassium equilibrium potential is very close to the resting potential (at which the preparations were usually held), it would seem that the driving force for an outward leak of K^+ would be rather small. Perhaps, pacemaker activity in the myometrium is more dependent on inward leakage of Na^+, and perhaps it is at this point that oxytocin exerts its primary action. Obviously, these comments are nothing more than speculations for use as possible leads to further investigations; hopefully, the voltage-clamp technique could provide some useful answers.

Previously unpublished work in this communication was supported in part by a grant from the National Institute of Child Health and Human Development (HD 378). Participation in the Barcroft Centenary Symposium is supported in part by this grant and a grant from the State University of New York Downstate Medical Center.

REFERENCES

Adrian, R. H., Chandler, W. K. & Hodgkin, A. L. (1970). *J. Physiol.* (*London*) **208**, 607–44.

Anderson, N. C. (1969). *J. Gen. Physiol.* **54**, 145–65.

Frankenhauser, B. (1962). *J. Physiol.* (*London*) **160**, 40–5.

Hodgkin, A. L., Huxley, A. F. & Katz, B. (1952). *J. Physiol.* (*London*) **116**, 424–48.

Kao, C. Y. (1966). *Pharm. Rev.* **18**, 997–1049.

Kao, C. Y. (1967). In *Cellular Biology of the Uterus*, ed. R. M. Wynn. Appleton-Century-Crofts, New York.

Kao, C. Y. (1971). In *Researches in Physiology*, Liber Memoralis for Chandler McC. Brooks; ed. F. F. Kao, M. Vassalle & K. Koizumi. Aulo Gaggi, Bologna.

Kao, C. Y., McCullough, J. R. & Davidson, H. L. (1970). *Pharmacologist* **12**, 300.

Kao, C. Y., McCullough, J. R. & Davidson, H. L. (1971). *Fed. Proc.* **30**, 384.

DOES THE FOETAL HYPOPHYSEAL–ADRENAL SYSTEM PARTICIPATE IN DELIVERY IN RATS AND IN RABBITS?

By A. JOST

Laboratoire de Physiologie comparée, Université Paris VI,
9, Quai Saint-Bernard, Paris 5e

The discovery that in sheep functional integrity of the foetal pituitary and adrenal glands is necessary for normal delivery has aroused much interest. In sheep several lines of evidence are at hand (see paper by G. C. Liggins): (1) experimental destruction of the foetal pituitary or ablation of the foetal adrenals prevents parturition; (2) infusion of ACTH or of a corticosteroid into the foetus induces premature delivery; (3) the foetal plasma cortisol concentration increases sharply during the days preceding parturition; (4) premature parturition can be produced by injecting large amounts of corticosteroids into the pregnant ewe (Adams & Wagner, 1969).

The foetal share – if any – in timing parturition in rats or in rabbits remains unknown. Professor K. W. Cross in his letter of invitation to this symposium urged me to discuss the problem or to present any recent work concerning the foetal pituitary–adrenal status at birth in small laboratory animals.

It should first be noticed that the large number of young in these animal species renders surgery on every foetus difficult and hazardous in normal-sized litters because of the injury to the uterus. Too small litters are somewhat abnormal.

Rat

In this species the placentas survive after surgical removal of the foetuses, and are delivered at expected term; according to Selye, Collip & Thomson (1935) 'the length of pregnancy was not considerably altered by the removal of the embryos'. This was confirmed by Kirsch (1938).

Large doses of cortisone acetate given to pregnant rats during the second half of gestation decrease the size of the foetal adrenals, and delay and disturb delivery (Courrier, Colonge & Baclesse, 1951). Large doses of corticosteroids (up to 4 mg cortisone), given subcutaneously to two to four foetuses per litter, depressed the adrenals, but did not hasten delivery (the rats were sacrificed after 21.5 days of pregnancy) (Jost, Jacquot & Cohen, 1955).

The sharp increase in foetal plasma cortisol which precedes delivery in

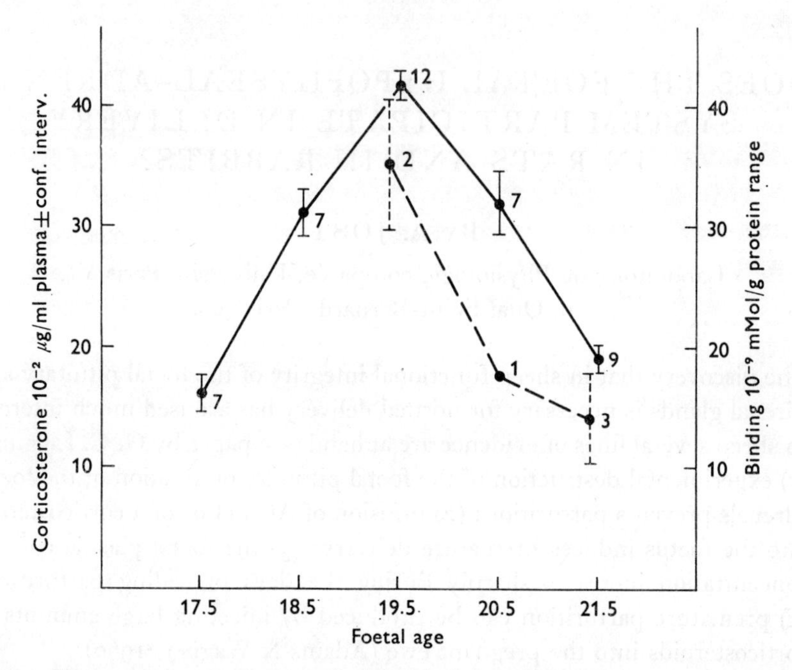

Fig. 1. Plasma corticosterone determined by A. Cohen (heavy line) and corticosterone binding capacity of serum proteins measured by E. Nunez and colleagues (interrupted line). Proteins were determined by the Lowry method. Number of determinations indicated near experimental points. (Conf. Interv. = Confidence Interval.)

sheep and in goats is not observed in rats. Cohen (1972) recently determined the foetal plasma corticosterone concentration (fluorimetric method). She found an increase in plasma corticosterone between days 17.5 and 19.5, followed by a definite decrease until day 21.5 (less than 12 hours before expected delivery) (Fig. 1).

The day before birth growth of the foetal adrenals is practically stopped and the pituitary cortico-stimulating activity is very reduced (Cohen, 1963). This seems to result from a feedback effect since the adrenals hypertrophy in foetuses submitted to metopirone (Dupouy, 1971 and unpublished data). Increased negative feedback accompanied by decreased total corticosterone might result from a change in the corticosterone binding capacity of the foetal plasma. Nunez, Savu, Engelmann, Benassayag, Crepy & Jayle (1971) had found that the plasma binding capacity for corticosterone was higher in the 19-day-old foetus than in the newborn. They studied 19- to 21-day-old foetuses. The binding of corticosterone to proteins was measured by Sephadex G100 gel filtration at 4 °C, after incubation of serum with 1-2-^{3}H corticosterone for 30 min at 4 °C. It was verified that the differences in binding capacity according to age could not be accounted for by changes in plasma proteins. The preliminary results (Fig. 1) indicate that the binding capacity of the foetal plasma decreases when the total corticosterone also declines. These observations give no estimate of the unbound corticosterone, but they do not rule out the possibility that it is unchanged or even slightly increased; this might contribute to the depressed pituitary cortico-stimulating activity before birth. Therefore, it still remains impossible to correlate variations of plasma cortiscosterone in the rat foetus and delivery.

Rabbits

The effects of corticosteroids injected into the pregnant doe on the outcome of pregnancy depend upon the stage of pregnancy and on the dose (Table 1). Courrier & Colonge (1951) and Robson & Sharaf (1952) observed foetal death *in utero* when large doses of cortisone acetate (20 or 25 mg/day) were given from day 12 to 15 onwards. Denamur (1966, and unpubl.) injected cortisol for 5 days: between days 19 and 24 only the highest dose (15 mg/day) produced abortion (empty uterus at the end of treatment); between days 24 and 29, a dose of 7.5 mg/day had the same effect; the dose of 3.75 mg/day induced premature delivery during the night preceding day 29. Adams & Wagner (1969) also observed discharge of the foetuses within three days after a single dose of 0.5 to 1.2 dexamethasone on day 27. Kendall & Liggins' (1972) data are somewhat at variance; they obtained premature delivery only when the single dose of dexamethasone was given on day 25 and only with the highest dosage (4 mg/day). Similarly, intra-amniotic injection of dexamethasone induced premature delivery only if made on day 25. (It should be noted that hormones introduced in the amniotic cavity are probably readily released into the uterine lumen.) In many experiments I gave large doses of ACTH or corticosteroids subcutaneously to two or three decapitated foetuses per litter between days 21 and 26 and recovered the young by Caesarian section on day 28 and sometimes 29 (Jost & Jacquot, 1955; Jost, 1961). As a whole, injection of large daily doses of corticosteroid into pregnant rabbits seems to result in intra-uterine foetal death and resorption at mid pregnancy and in discharge of the foetuses during the last ten days; a similar difference is observed when rabbits are castrated at varying stages of pregnancy (Hammond, 1925; Klein, 1932).

In order to verify if a spurt of foetal pituitary hormones might play a role in delivery, I made experiments in which only decapitated foetuses were left *in utero* after day 23. The number of young was reduced either by extracting all the foetuses but two to four on day 23, or by ligating on day 2 the tube on the side where the ovary showed the largest number of

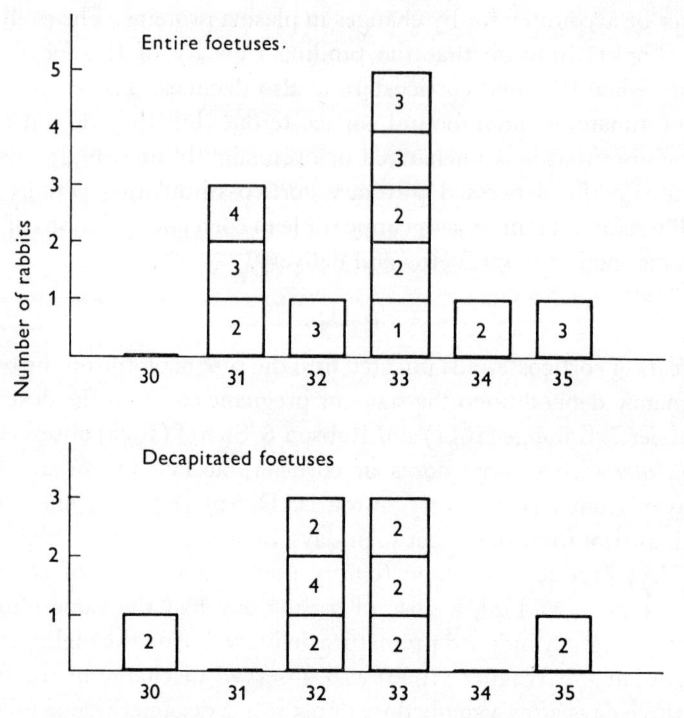

Fig. 2. Rabbits: date of delivery. Number of foetuses shown in each square.

ovulations; on day 23, the foetuses were decapitated (1 to 4 foetuses). The results were somewhat deceptive: in 41 experiments a rather high proportion of decapitates died and were rejected macerated (14 animals); others were never recovered perhaps because they were immediately eaten at birth (these animals were found empty at palpation on day 34 or 35) (19 animals). The date of delivery could be ascertained in only 8 animals (recovery of the young) and compared with delivery in 11 controls with a similarly limited number of foetuses (Fig. 2). In both the controls and in the experimental animals there is a wide variability in the date of delivery but no prolonged intrauterine retention of the pituitaryless foetuses was obtained.

CONCLUSIONS

The data summarized in this paper are still too incomplete to be very conclusive. The experiments on rabbits do not suggest any very important participation of the foetal hypophyseal–adrenal system in parturition.

Table 1. *Effects of corticosteroids and of castration on outcome of pregnancy in the rabbit*

Period of gestation	Treatment	Result (Number of animals)	Reference
From day 12 or 15 on	Cortisone acet. (20–25 mg/day)	Resorption of conceptuses	(1), (2)
Days 19–24	Cortisol: 15 mg/day	Abortion (5/6)	(3)
	,, 7.5 or 3.75 mg	Rare abortion (2/12)	(3)
Days 24–29	Cortisol: 7.5 mg	Abortion (7/7)	(3)
	,, 3.75 mg	Premature delivery on day 28 (4/5)	(3)
On day 27	Dexamethasone 0.5 to 1.2 mg	Premature delivery within 3 days (11/11)	(4)
On days 22, 26 or 27	Dexamethasone 1 to 4 mg	Normal delivery on day 31 (23)	(5)
On day 25	Dexamethasone 1 to 2 mg	Delivery on day 33 (7)	(5)
	,, 4 mg	Premature delivery on day 29 (3)	(5)
Before day 20	Castration	Resorption of conceptuses	
Between days 20 and 27	Castration	Premature discharge	(6), (7)

(1) Courrier & Colonge (1951); (2) Robson & Sharaf (1952); (3) R. Denamur, Thesis (Sci.), Paris (1966) and unpublished data; (4) Adams & Wagner (1969); (5) Kendall & Liggins (1972); (6) Hammond (1925); (7) Klein (1932).

REFERENCES

ADAMS, W. M. & WAGNER, W. C. (1969). *J. Amer. Vet. Med. Assoc.* **154**, 1396–7.

COHEN, A. (1963). *Arch. Anat. microsc. Morphol. expérim.* **52**, 277–407.

COHEN, A. (1972). *Horm. Metab. Res.* (In press.)

COURRIER, R. & COLONGE, A. (1951). *C. r. Acad. Sci. (Paris)* **232**, 1164–6.

COURRIER, R., COLONGE, A. & BACLESSE, M. (1951). *C. r. Acad. Sci. (Paris)* **233**, 333–6.

DENAMUR, R. (1966). Thesis Sci., Fac. Sci. Paris.

DUPOUY, J. P. (1971). *C. r. Acad. Sci. (Paris)* **273**, 962–5.

HAMMOND, J. (1925). *Reproduction in the Rabbit.* Oliver and Boyd, Edinburgh and London.

JOST, A. (1961). *The Harvey Lectures* **55**, 201–26.

JOST, A. & JACQUOT, R. (1955). *Ann. Endocrinologie (Paris)* **16**, 849–72.

JOST, A., JACQUOT, R. & COHEN, A. (1955). *C. r. Soc. Biol.* **149**, 1319–22.

KENDALL, J. Z. & LIGGINS, G. C. (1972). *J. Reprod. Fert.* **29**, 409–13.

KIRSCH, R. E. (1938). *Am. J. Physiol.* **122**, 86–93.

KLEIN, M. (1932). *C. r. Soc. Biol.* **109**, 932–4.

NUNEZ, E., SAVU, L., ENGELMANN, F., BENASSAYAG, C., CREPY, O. & JAYLE, M. F. (1971). *C. r. Acad. Sci. (Paris)* **273**, 242–5.

ROBSON, J. M. & SHARAF, A. A. (1952). *J. Physiol., London* **116**, 236–43.

SELYE, H., COLLIP, J. B. & THOMSON, D. L. (1935). *Endocrinology* **19**, 151–9.

HORMONAL FACTORS IN PARTURITION IN THE RABBIT

By P. W. NATHANIELSZ, MARGARET ABEL and G. W. SMITH
The Physiological Laboratory, Cambridge CB2 3EG

In recent years much information has accumulated concerning the mechanisms which initiate and support labour in the pregnant sheep (for reviews see Suppl. 16, *J. Reprod. Fertil.* 1972). The ability to obtain sequential blood samples from, and to administer various metabolites and hormones continuously into, the foetal and maternal vasculature make the sheep preparation an extremely powerful method for investigating the pregnant state and the events occurring during parturition. It seemed to us that the rabbit and other smaller species had been the victims of neglect due to the relative difficulty of performing similar chronic experiments. This omission appeared unsatisfactory especially when the investigations on the progress of parturition in the rabbit had been the subject of so much attention in the 1960s (see Schofield, 1968). These earlier studies were extremely important in delineating the possible roles of progesterone and oxytocin in labour. We, therefore, devised an experimental preparation which could be infused continuously via the aorta and sampled whenever required from the inferior vena cava. A pressure-sensitive radiotransducer was placed in the uterus to record uterine contractions in the unrestrained animal (Nathanielsz & Abel, 1972).

METHODS

105 rabbits from the Laboratory Animal House, predominantly New Zealand Whites, were used in this investigation. Normal delivery for unoperated animals in the colony is 32 days ± 26.2 h (mean ± SEM, $n = 6$). Catheterisation of the maternal aorta and inferior vena cava under Nembutal anaesthesia and the constant infusion system have been described previously (Nathanielsz & Abel, 1972). All rabbits were operated on at Day 21 (D21) of gestation. Prostaglandin F2α (PGF2α) and cortisol sodium succinate solutions were made up in sterile physiological saline containing heparin (20 units/ml). Progesterone was administered intramuscularly (i.m.).

RESULTS

Induction of parturition by infusion of cortisol

(*a*) *At different doses of infusion continued throughout the experiment until delivery occurs*

Eleven rabbits were catheterised on D21 of pregnancy and infused continuously thereafter with saline through the maternal aortic catheter. Delivery occurred 185 ± 26.7 h (mean ± SEM) later. Live litters were delivered in nine out of the eleven experiments. Thus although infusion with saline alone does appear to result in slightly earlier parturition than in unoperated animals (see above) this difference is not significant. The fact that live litters are usually obtained also suggests that the saline-infused animals represent an adequate control group.

Fig. 1 shows the effect of different rates of intravascular infusion of cortisol into the maternal circulation on the induction of delivery. Infusion of 1.24 mg cortisol/h and 0.31 mg cortisol/h resulted in delivery in a significantly shorter time than when saline was infused into controls of the same gestational age ($p < 0.01$).

It is apparent that the highest dose rates, 0.31–1.24 mg cortisol/h, had very similar effects. Parturition occurred about 72 hours later with very great reproducibility. The means ± SEM for these two groups (eleven animals in all) were 72.9 ± 6.3 h and 72.2 ± 3.4 h respectively. Parturition was accompanied by many of the signs of normal delivery such as extensive plucking of fur, nest building, and marked lactation. This reproducibility of delivery provided an accurate baseline against which to compare different experimental regimes. When comparing other experimental situations which either precipitate or delay the onset of parturition it should be noted that on these doses of cortisol infused into the mother (0.31–1.24 mg/h), no delivery occurred earlier than 56 h and none later than 82 h.

(*b*) *At constant dose rate* (1.24 *mg/h*) *for different lengths of time*

From the results shown in Fig. 2 it is apparent that infusion of cortisol for a minimum of 6 h was adequate to produce parturition even if the infusion was then discontinued. With shorter periods of infusion premature delivery did not always occur. This is a clear demonstration of an all-or-none effect – if the cortisol infusion to the mother has triggered some secondary mechanisms, parturition will occur despite the termination of the cortisol infusion. Thus in Fig. 2 ten animals can be considered to have had a dose exceeding the threshold whilst four had only subthreshold doses.

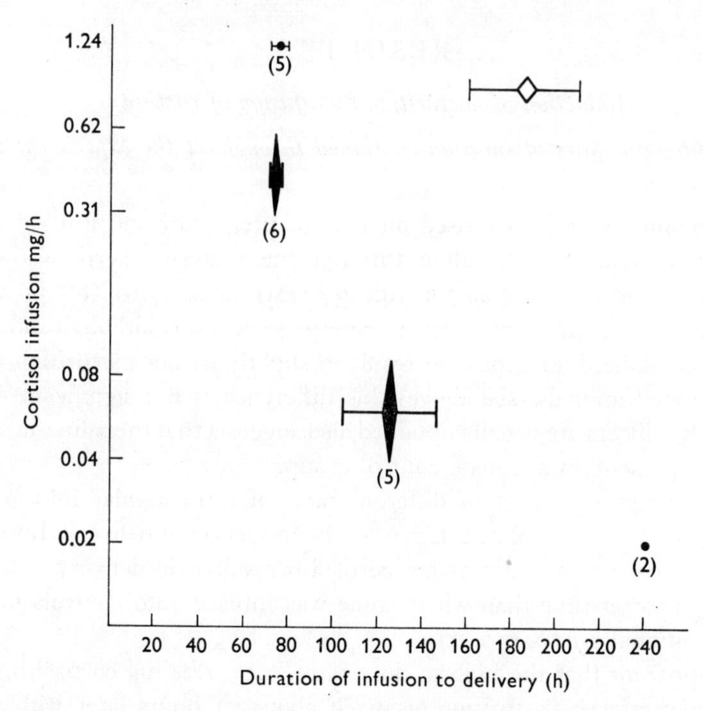

Fig. 1. Initiation of parturition in the pregnant rabbit with the administration of cortisol by continuous infusion from D21 into the maternal vasculature. The open diamond represents the control group which was not infused with cortisol. Continuous saline was infused in the control group. Horizontal bars represent SEM.

(*c*) *Induction of parturition by the administration of cortisol to the foetus*

In two experiments 0.6 mg cortisol/h was infused into each of two amniotic sacs in two D21 pregnant rabbits. Delivery occurred after 43.5 and 67 h. The total dose of cortisol was thus 1.2 mg cortisol/h. When only one sac was infused at 0.6 mg cortisol/h in two separate rabbits, delivery occurred in 74.5 and 144 h.

(*d*) *Inhibition of induction of parturition with maternally administered cortisol by administration of exogenous progesterone to mother*

10.0 mg progesterone administered i.m. on D21–D27, 5.0 mg on D28 and 2.5 mg on D29 to six rabbits infused with 0.31 mg cortisol/h via the maternal aorta, prevented delivery before D30 when all foetuses were still alive (Fig. 3). In one animal also shown in Fig. 3, 1.0 mg progesterone daily also delayed parturition on this does of cortisol. Progesterone also blocked premature parturition at the highest dose of cortisol used – 1.24 mg/h in one rabbit. When viewing the significance of these eight proges-

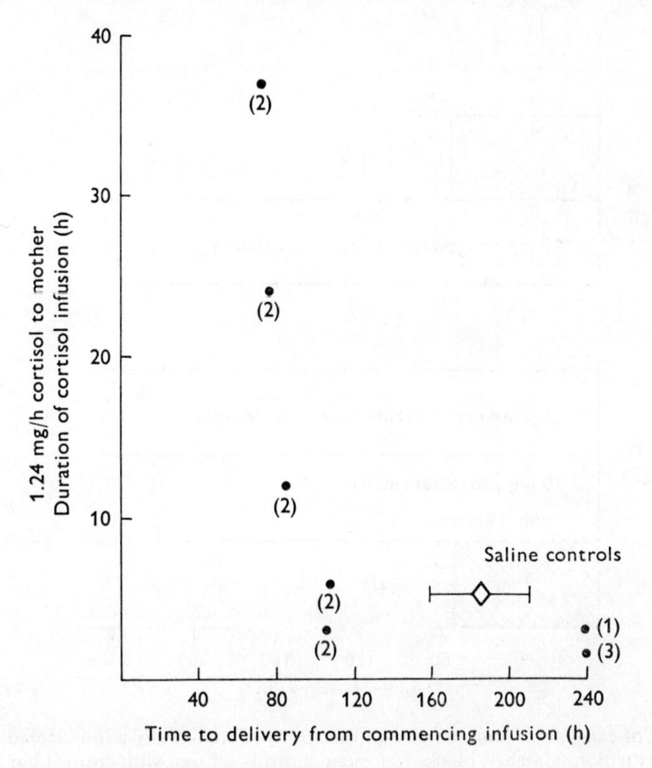

Fig. 2. Effect of different periods of infusion of cortisol (1.24 mg/h) into the aorta of pregnant rabbits from D21, on the time to delivery. The open diamond represents the control group which was not infused with cortisol. Horizontal bars represent SEM.

terone-treated animals, none of which had been allowed to deliver, it should be remembered that all rabbits infused with cortisol at these rates would have delivered by 82 h (i.e. on D25).

Induction of parturition by intra-aortic infusions of prostaglandin F2α

Since PGF2α has been shown to be luteolytic in several species (Duncan & Pharris, 1970) and the results above demonstrate that the induction of parturition by cortisol can be inhibited by progesterone, the effect of PGF2α administered via the maternal aorta was investigated. The importance of the route of administration of PGF2α cannot be overstressed. Much conflicting data is present in the literature as to whether PGF2α is luteolytic or not in different species. In view of the multitude of routes of administration and pharmacological doses used, it appeared to us that a controlled quantitated approach with a period of continuous infusion was required.

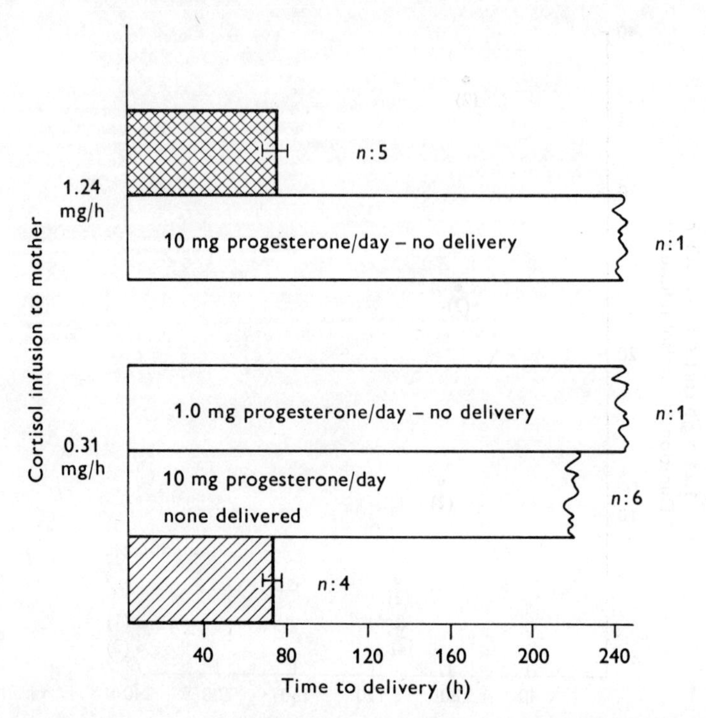

Fig. 3. Effect of exogenous progesterone on the ability of maternally administered cortisol to initiate parturition. Hatched blocks represent controls infused with cortisol but with no exogenous progesterone. Indented lines at termination of progesterone-treated experiments indicate that delivery had not occurred and Caesarian section was performed.

Fig. 4 demonstrates the effect of continuous infusions of PGF2α into the maternal aorta over a wide dose range. When 2.25–75 μg PGF2α was infused per hour, delivery occurred in a significantly shorter interval than with the highest dose of cortisol ($p < 0.01$). When similar concentrations of PGF2α were infused into the inferior vena cava, parturition still occurred earlier than in the saline controls, but using this route the time interval to delivery was more variable. Side-effects of the infusion also appeared more frequently with intravenous infusion.

In twelve animals PGF2α was infused into the maternal aorta for differing time intervals from D21. The infusion was then changed to saline until the animal delivered. At three different dose rates, 0.15, 0.6 and 3.0 μg PGF2α/h, there was a minimum duration for the exogenous PGF2α to have the effect of precipitating delivery. At 0.15 μg PGF2α/h, 8, 10 and 12 h infusions produced delivery after a mean of 38 h. When PGF2α was infused at this dose rate for only 4 or 6 h, delivery had not occurred by D31 when the live foetuses were delivered by Caesarian section.

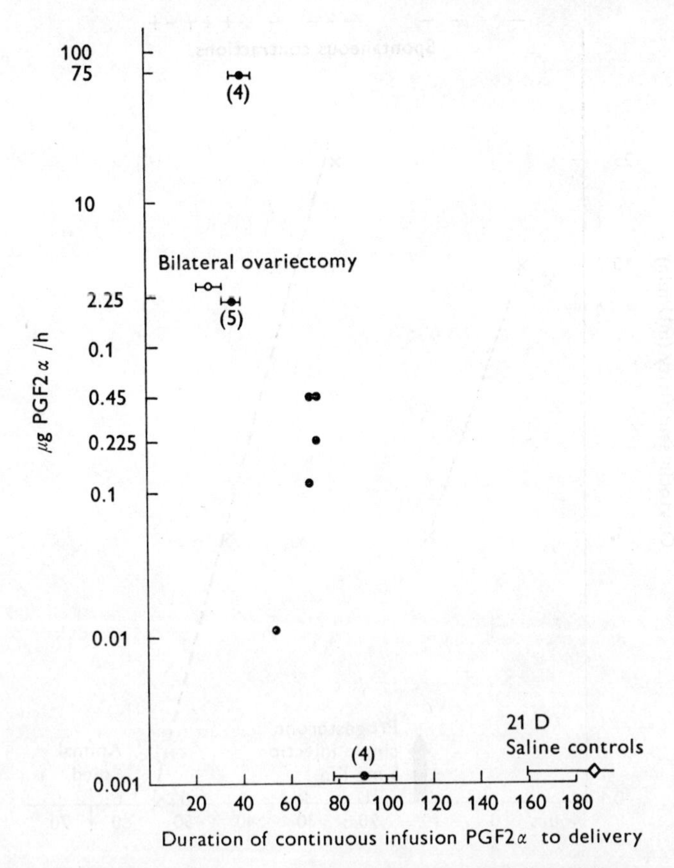

Fig. 4. Effect of different rates of continuous maternally administered intra-aortic PGF2α on time interval to delivery. Horizontal bars represent SEM. Open circle represents animals bilaterally ovariectomised on D21 and the open diamond represents the saline-infused controls. *Note that neither of these two last groups were infused with PGF2α at any time.*

Recordings of uterine contractions during PGF2α infusions

In animals on infusions of PGF2α from 11.25 ng to 75 μg/h, uterine contractions were never observed before 15 h. After this time, contractions slowly increased in frequency and amplitude until delivery was accomplished. In four animals the sensitivity to intra-aortic oxytocin increased throughout the period of PGF2α infusion. This effect is shown in one animal in Fig. 5. In this animal the increase in oxytocin sensitivity was reversed by maternally administered progesterone.

Effect of progesterone on PGF2α-induced labour

Four animals were infused continuously with 2.25 μg PGF2α/h and given 2.0 mg progesterone i.m. per day. Caesarian sections were performed on

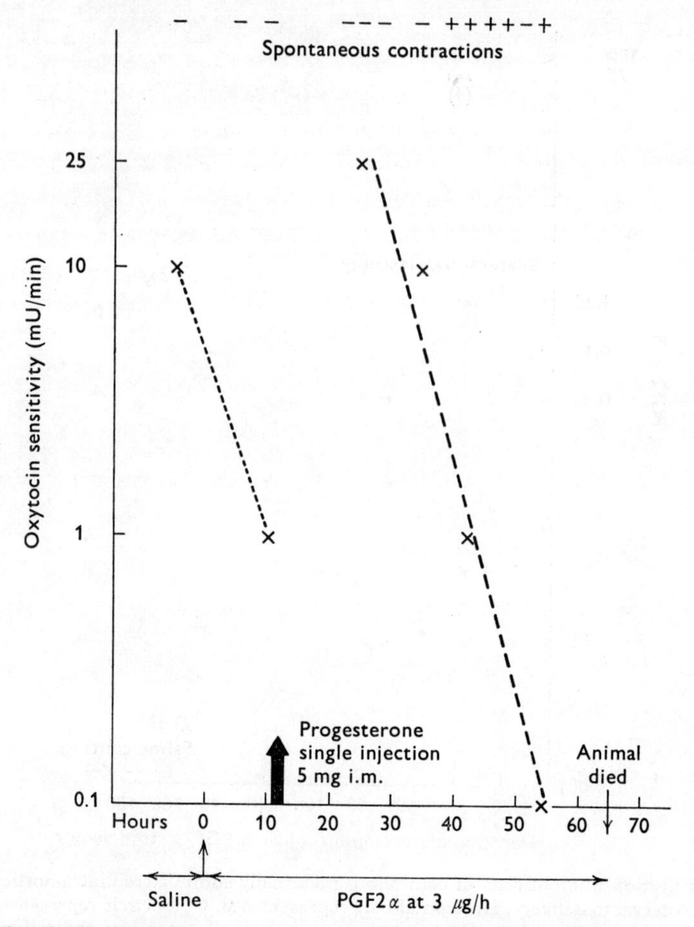

Fig. 5. Development of spontaneous contractions and oxytocin sensitivity to continuous intra-aortic PGF2α infusion at 3 μg/h. Oxytocin sensitivity was tested by observing the lowest continuous intra-aortic infusion rate at which uterine contractions could be obtained within 5 min.

three animals at 140, 120 and 109 h. All foetuses were still present *in utero* but in distinction to the animals treated with cortisol and progesterone, all foetuses were dead. The third animal was weaned off progesterone and delivered one dead and three live foetuses 264 h after the PGF2α infusion was started. The difference between this progesterone-treated group and the animals infused with 2.25 μg PGF2α/h, but not injected with progesterone, is statistically significant ($p < 0.01$).

Effect of infusion of Arachidonic acid

Since prostaglandin synthesis depends on the availability of essential fatty acids, as a result of discussions with Dr Peter Ramwell, we decided to investigate the effect of infusion of arachidonic acid via the maternal aorta. Seven animals were infused with arachidonic acid in 5 % autologous plasma. Two animals infused with 100 μg arachidonic acid/h delivered in 46 and 40 h. One animal infused at 50 μg/h delivered after 24 h. Three animals infused with 10 μg arachidonic acid/h delivered after 25, 35 and 38 h. One animal infused at 1 μg/h delivered one foetus after 47 h and the rest of the litter in the next 23 h. Indeed delivery in several of the animals in this experimental group was a prolonged process with considerable time intervals between delivery of individual foetuses. A similar disco-ordinate type of delivery is seen after ovariectomy.

DISCUSSION

Although glucocorticoids have been shown to play a role in parturition in monotocous species, little evidence has been presented to date suggesting that similar mechanisms may be operating in polytocous species. Kendall & Liggins (1972) and Nathanielsz & Abel (1972) were able to precipitate parturition in the rabbit by maternal administration of glucocorticoids. The results presented here demonstrate that this effect has a minimum latency of about 72 h even if adequate doses of cortisol are administered for only 6 h. This observation strongly suggests that cortisol is acting through some intermediary mechanism.

The latency to delivery is shorter when cortisol is administered from the foetal side of the placenta and in this respect the pregnant rabbit resembles the pregnant sheep. Since the minimal duration of maternally administered cortisol is of the order of 6 h it is impossible to make quantitative comparisons of the efficacy of cortisol from the maternal and foetal side unless the threshold dose (from both directions) is discovered experimentally. It appears that 7.44 mg cortisol is adequate when administered from the maternal side.

Intra-aortic infusion of PGF2α induced parturition with a significantly shorter latency than after cortisol administration. Using an estimate of aortic blood flow it would appear that PGF2α is active when the circulating molar concentration is about 4×10^{-12} M at the ovary. Progesterone could inhibit PGF2α-induced labour and reverse the increase in oxytocin sensitivity which occurs during PGF2α infusion.

All these results would be explicable on the basis of the following sequential series of events: exogenously administered cortisol leading to PGF2α production leading to luteolysis. It should be noted that there is only a slightly longer latency to parturition after PGF2α infusion at high dose rates than after ovariectomy. In these experiments PGF2α may be considered as performing a chemical ovariectomy and it is not surprising therefore that exogenous progesterone is found to inhibit the induction of premature labour by PGF2α. Kendall & Liggins (1972) demonstrated that plasma progesterone levels fall when cortisol is infused into the doe, and this may be achieved by the stimulation of prostaglandin production by cortisol. Arachidonic acid, a prostaglandin precursor, will also induce parturition but with a longer latency than PGF2α.

Newborn rabbits' plasma contains high concentrations of cortisol rather than corticosterone, which is the major steroid in adult rabbits (Malinowska, Hardy & Nathanielsz, 1972). It may yet be demonstrated that the rabbit foetus plays a role in the initiation of parturition, as does the sheep foetus. Indirect evidence suggests that the foetal adrenal is active before birth in this species. Foetal plasma cortisol levels in the rabbit rise slightly over the last few days of intrauterine life (Malinowska, Hardy & Nathanielsz, unpublished observations) and the rise is accompanied by a slight fall in plasma corticosterone. Taken together, these changes result in a high cortisol: corticosterone ratio. There is some evidence for differential function of cortisol and corticosterone in adult rabbits and recently it has been reported that the nuclear receptors in the developing foetal rabbit lung will bind cortisol preferentially to corticosterone (Giannopoulos, Mulay & Solomon, 1972). The significance of the qualitative difference in adult and foetal adrenocortical secretion in relation to the roles of the foetal and maternal adrenals in the progress of parturition in this species merits further study.

Our thanks are due to Mr D. Clarke and Mr L. Bancroft for their care of the experimental animals. Mr B. Secker and Mr A. Cattell were responsible for the construction of the constant infusion system. Gestone was kindly donated by Dr F. P. Diggins of Paines and Byrne Ltd. PGF2α was the gift of Dr John E. Pike, The Upjohn Co., Kalamazoo. This work has been assisted by an MRC Grant No. G/969/601/8 and ARC Grant No. 9/34. One of us, GWS, would like to acknowledge assistance given by the H. E. Durham Fund, King's College, Cambridge.

REFERENCES

DUNCAN, G. W. & PHARRIS, B. B. (1970). *Fed. Proc.* **29**, 1232–9.

GIANNOPOULOS, G., MULAY, S. & SOLOMON, S. (1972). *Biochem. Biophys. Res. Comm.* **47**, 411–18.

KENDALL, J. Z. & LIGGINS, G. C. (1972). *J. Reprod. Fert.* **29**, 409–11.

MALINOWSKA, K. W., HARDY, R. N. & NATHANIELSZ, P. W. (1972). *J. Endocr.* **55**, 397–404.

NATHANIELSZ, P. W. & ABEL, M. (1972). *J. Endocr.* **55**, 617–18.

SCHOFIELD, B. M. (1968). *Adv. Reprod. Phys.* **3**, 9–32.

FOETAL HYPOTHALAMIC AND PITUITARY LESIONS, THE ADRENAL GLANDS AND ABORTION IN THE GUINEA PIG

By B. T. DONOVAN and M. J. PEDDIE
Department of Physiology, Institute of Psychiatry,
De Crespigny Park, London SE5 8AF

In the course of a wide-ranging study of the development of the neural control of endocrine function in the guinea-pig, electrolytic lesions have been made in the hypothalamus, median eminence and pituitary gland of foetal guinea-pigs. Depending upon the location of the lesions, differential effects upon adrenal growth and the induction of abortion have been observed and are the subject of the present contribution.

Fifty-four pregnant females were used, with lesions placed in the foetuses on days 40 or 50 of pregnancy, which was dated from the time of ovulation. The lesions were made under barbiturate anaesthesia, through a laparotomy opening in the abdomen, by inserting a platinum-tipped electrode through the uterine wall and the skull vault of the foetus to the base of the brain and passing 3–5 mA d.c. for 30–50 seconds. The brain electrode was positive and an indifferent electrode was held against the uterus near the foetal head. Loss of amniotic fluid was prevented by ligature of the puncture point in the uterus as the electrode was withdrawn. From one to four foetuses were operated on in any one female, depending upon the size of the litter. To minimize operative trauma lesions were placed in those foetuses approached most readily. When possible, at autopsy the heads of the foetuses were perfused with formol-saline; and the adrenal glands and other organs were removed, weighed and fixed. Serial sections of the heads and endocrine organs were prepared.

Lesions were placed in foetuses carried by 38 females at 40 days. Seven females aborted between 47 and 56 days, 20 females were killed at 50 days and 11 females were killed at 60 days. Of 16 females operated upon at 50 days, eight aborted between 52 and 56 days and eight were killed at 60 days. When abortion occurred it took place between 47 and 56 days, despite the placement of lesions in some cases at 40 days and in others at 50 days. Abortion was not restricted to the operated foetuses within the uterus; all foetuses were evacuated. The occurrence of abortion was not related to the number of foetuses in the uterus, while the presence of more undisturbed than operated foetuses did not prevent premature delivery.

All aborted foetuses had suffered midline baso-medial hypothalamic damage which partially involved the median eminence region. Damage to peripheral or extrahypothalamic structures, major stalk-median eminence lesions, or destruction of the pituitary gland were not associated with abortion. Four foetuses in four pregnant females had sustained baso-medial hypothalamic damage and were not aborted, but the lesions were placed at 40 days and the animals were killed at 50 days. Midline baso-medial hypothalamic damage was lacking in those animals which did not abort, and which were killed at 60 days after making lesions at 40 or 50 days. Lesions of the pituitary stalk or pituitary, or located outside the hypothalamus, were evident at 60 days.

Analysis of the changes in adrenal activity is complicated by the different sizes of the adrenal glands between foetuses of the same weight and age but with different mothers, and by an apparent sex difference within litters. The glands of the females are larger than those of males of equivalent body weight, although female foetuses tend to be smaller than their male siblings. It is thus most appropriate to compare the adrenal weights of operated foetuses with those of control foetuses of the same sex and gestation – an ideal seldom attained. However, it appears that destruction of the pituitary gland at 40 days of foetal age depresses adrenal growth, as determined at 50 days. The zona fasciculata was atrophic. Thus the mean adrenal weight of four foetuses with the pituitary destroyed at 40 days, autopsied at 50 days, was 5.59 ± 0.05 mg as compared with 8.98 ± 0.05 mg for 12 littermate controls. This difference is significant at the 0.01 % level. In four foetuses at 60 days with the pituitary destroyed at 40 days, the adrenal weights were 10.94 ± 0.37 mg, compared with 17.5 ± 0.03 for seven littermate controls; this difference is significant at the 0.05 % level. Damage to the stalk median eminence region, to the hypothalamus, or elsewhere in the brain did not alter adrenal gland weight. Because of growth of the organ, it was not practicable to destroy the pituitary gland completely in 50-day foetuses, but severe damage to the median eminence or peripheral hypothalamic damage or lesions elsewhere in the brain in five foetuses did not affect adrenal gland growth by 60 days.

The possibility that adrenal gland weight may have been altered in the aborted foetuses was examined, but marked deviations from normal were infrequent. Nevertheless, in one case in which unlesioned control littermate foetuses of the same sex were available, hypertrophy of the glands of the lesioned foetuses seemed to have occurred: control females: 15.95 and 15.3 mg; lesioned females: 29.7 and 20.6 mg.

Although the case remains to be proven, there seems little doubt that damage to the hypothalamus can cause abortion in the guinea-pig. Abortion is unlikely to be a non-specific consequence of abdominal interference, or of operations upon the uterus, because it seldom follows the injection of fluids into the foetus after a comparable surgical exposure of the uterus. Of seven pregnant females so treated at 40–55 days of pregnancy, none aborted. Further, abortion was associated with damage to the midline baso-medial hypothalamus, and not elsewhere, and occurred after a considerable lapse of time post-operatively. Abortion is unlikely to have been due to foetal death since the 15 litters examined very soon after expulsion from the uterus could not have been dead more than 3–4 hours, and frequently the foetuses were still vital.

The processes underlying the abortion are far from clear. It is remarkable that damage to the hypothalamus of one or more foetuses in a litter can cause the abortion of all, for the changes induced in the operated animals must be transmitted to the uterus as a whole. Despite the information available for the sheep (Liggins *et al.* 1972) it is not known whether the pituitary and adrenal glands are concerned with parturition in the guinea pig. A trophic influence does not appear to be exerted by the hypothalamus over the pituitary gland during foetal life, for hypothalamic damage at 40 or 50 days did not depress adrenal growth as did pituitary damage, although it is feasible that an inhibitory influence may be operative. Hypothalamic damage could then favour ACTH release and adrenal growth (observed in one litter of the present series) and the rise in adrenal activity could be great enough to precipitate delivery. However, destruction of the pituitary stalk did not cause adrenal enlargement while the administration of 1–4 i.u. ACTH to five foetuses in four pregnant females has failed to cause abortion. Dexamethasone (0.25 mg) has also been administered to foetuses at 40 days but in this dose proved highly toxic.

We wish to thank Maureen Harrison for her excellent technical help and the Medical Research Council for financial support.

REFERENCE

LIGGINS, G. C., GRIEVES, S. A., KENDALL, J. Z. & KNOX, B. S. (1972). *J. Reprod. Fert.* Suppl. 16, 85–104.

PARTURITION IN THE LARGER HERBIVORES

By R. S. COMLINE, MARIAN SILVER, P. W. NATHANIELSZ and L. W. HALL

The Physiological Laboratory and the
Department of Veterinary Clinical Studies,
Cambridge CB2 3EG

There is now good evidence that normal parturition in the sheep is closely linked to and probably dependent upon an increased secretion of cortisol from the foetal adrenal cortex in the last 48–72 h of gestation. The characteristic, abrupt increase in foetal adrenal weight, together with the rise in plasma cortisol values, is eliminated by electro-coagulation of the foetal pituitary or by stalk section and pregnancy is thereby prolonged. Conversely, premature labour can be induced in this species by administration of cortical hormones to the foetus (for references see Liggins, Grieves, Kendall & Knox, 1972; Nathanielsz, Comline, Silver & Paisey, 1972).

This relationship between the foetal hypothalamo–pituitary–adrenal system and parturition has not, however, been so precisely defined in other species. In primates adrenalectomy in the monkey (Mueller-Heubach, Myers & Adamsons, 1972) and some forms of anencephaly in the human (Anderson, Laurence & Turnbull, 1969) may result in prolonged gestation, while in cattle genetical abnormalities of the cranium and central nervous system of the foetus are associated with extended pregnancy (Holm, Parker & Galligan, 1961). In this species also, premature labour has been reported after large single doses of dexamethazone administered to the mother towards the end of pregnancy (Adams, 1969). However, there is no evidence for an abrupt foetal adrenal hypertrophy in the calf similar to that which occurs just before birth in the lamb (Fig. 1), so that the role of the foetal adrenal cortex in the initiation of parturition in the cow seems somewhat equivocal.

The present experiments were carried out to provide sequential measurements of plasma cortisol levels in the conscious foetal calf during the latter part of gestation and throughout parturition. In addition, some preliminary observations were made on cortisol levels in the foetal foal, since in this species there is some evidence for foetal adrenal hypertrophy immediately before birth (Fig. 1).

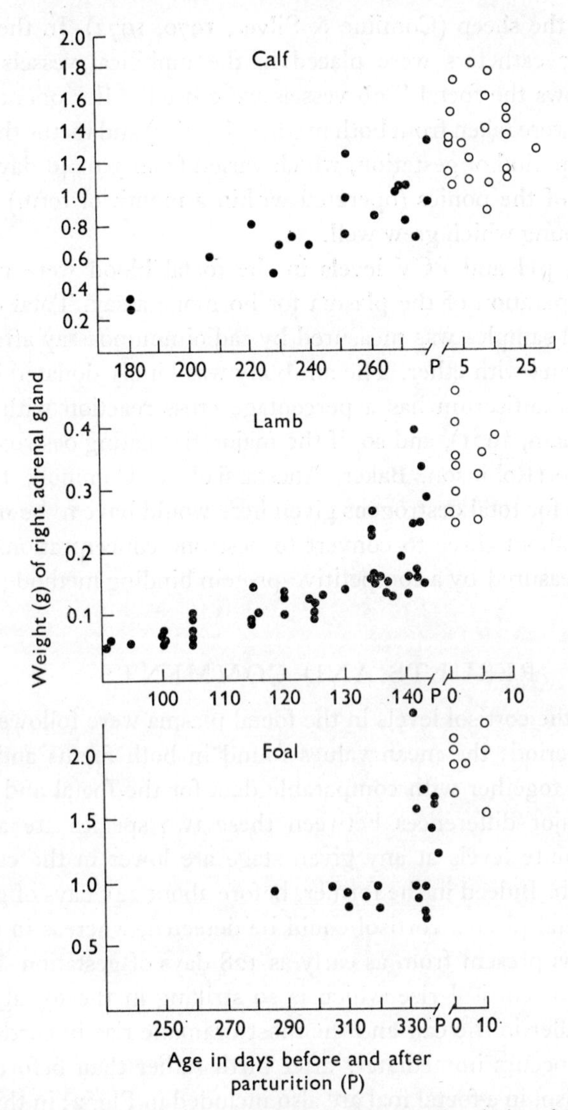

Fig. 1. Changes in foetal adrenal weight during gestation (●) and after birth (○) in the calf, lamb and foal (from Comline & Silver, 1971).

METHODS

Jersey cows or small ponies (200–350 kg body weight) of known gestational age were used. In these large animals the success of the preparation was dependent upon the maintenance of the foetal pH and of placental exchange during the operation, and attention to strict aseptic precautions thereafter. The catheters were inserted by techniques modified from those reported previously for the sheep (Comline & Silver, 1970, 1972). In the majority of preparations catheters were placed in the umbilical vessels while in three of the cows the foetal limb vessels were used. After operation daily blood samples were taken from both mother (jugular) and foetus throughout the remaining period of gestation, which varied from 3 to 40 days. All the cows and two of the ponies (operated within a month of term) delivered live, healthy young which grew well.

Blood gases, pH and PCV levels in the foetal blood were monitored daily before separation of the plasma for hormone assay. Total oestrogen in the maternal samples was measured by radioimmunoassay after extraction of the plasma with ether. The antibody was kindly donated by Dr G. Abraham. This antiserum has a percentage cross reaction with oestrone of 35% (Abraham, 1971), and so, if the major circulating oestrogen in the cow is oestrone (Robinson, Baker, Anastassiads & Common, 1970), the absolute values for total oestrogens given here would have to be multiplied by a factor of about three to convert to oestrone concentrations. Plasma cortisol was measured by a competitive-protein binding method (Murphy, 1967).

RESULTS AND COMMENTS

In seven cows the cortisol levels in the foetal plasma were followed during the perinatal period; the mean values found in both foetus and calf are given in Fig. 2 together with comparable data for the foetal and neonatal lamb. Two major differences between these two species are apparent; firstly the absolute levels at any given stage are lower in the calf foetus than in the lamb. Indeed in the former, before about 245 days of gestation, virtually no foetal plasma cortisol could be detected, whereas in the lamb 1–2 μg/100 ml is present from as early as 128 days of gestation. Secondly the preparturient cortisol rise which is so striking in the foetal sheep is very much smaller in the calf and the most dramatic rise in cortisol levels in this species occurs immediately after birth rather than before. Values for plasma cortisol in a foetal foal are also included in Fig. 2; in this animal there was no apparent change before parturition but a large rise in cortisol immediately afterwards. Resting plasma cortisol levels in the foetal foal appear to be higher than those of either foetal lamb or calf as indicated by the values in Fig. 2 and results from other foals sampled earlier in gestation.

The observations on the foal gave no evidence for a foetal adrenal component in parturition in the mare but further data are obviously needed in this species. The findings on the calf foetus suggested that, either the rise in plasma foetal cortisol required to trigger parturition is very small in the cow, or some other factor or factors are involved in the

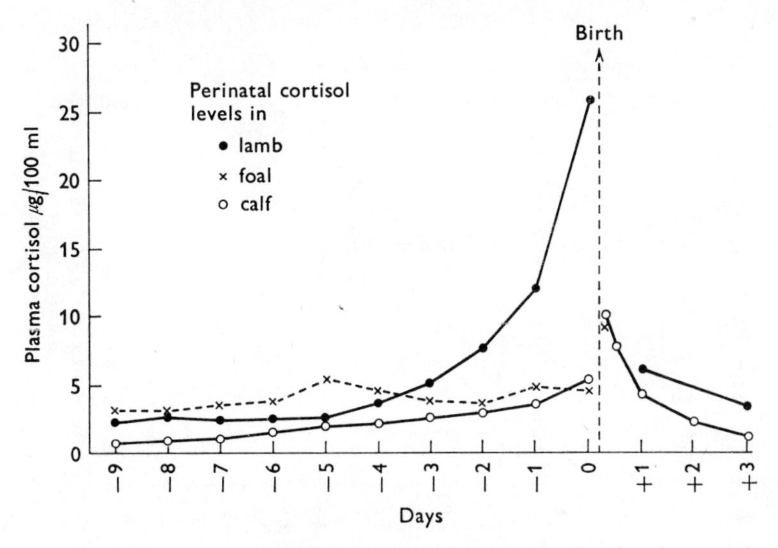

Fig. 2. Comparison of the mean plasma cortisol levels in the foetal and neonatal calf (○) and lamb (●); (×) values for a foal.

mechanism. In order to test this hypothesis further we investigated the effects of exogenously administered cortisol or Synacthen given to the foetus at 245–250 days, i.e. before any likelihood of spontaneous delivery. Cortisol (100 mg four times daily) raised the foetal levels far above any seen in the untreated animal (Fig. 3*c*), yet delivery did not occur until six days later. This dose of cortisol was equivalent to a level of dexamethasone which, in the sheep foetus, resulted in parturition two days later (Liggins, 1969). A 5–6-day time lag was also observed when Synacthen was given to the foetus. The foetal plasma cortisol values found during Synacthen administration (0.5 mg/day given as four doses) are shown in Fig. 3*b*; the levels are comparable with those observed in an animal which delivered spontaneously (Fig. 3*a*). These preliminary results show that it is possible to induce parturition in the cow with either cortisol or ACTH given to the foetus, but that 5–6 days seem to be required to trigger the mechanism, irrespective of the actual foetal cortisol level produced.

A feature associated with parturition in the chronically catheterised calf foetus which had not been observed in the sheep under similar conditions was the non-delivery of the placenta even after 24 h from birth. This may have been related to the premature delivery of all the calves; the mean gestational age at birth was 264 days whereas term in the cow is 281 days. However, premature delivery in the sheep is not normally associated with retained placenta and the foetal plasma cortisol levels found in the lamb under these circumstances are no different from those at or near term.

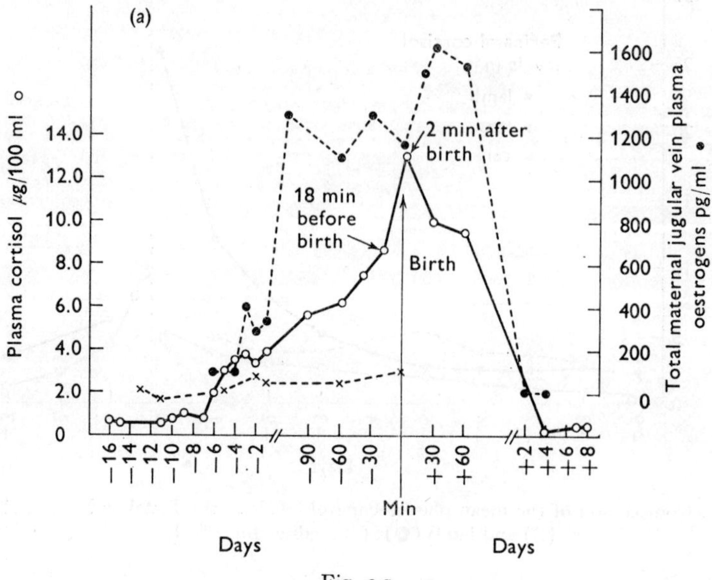
(a)
Plasma cortisol μg/100 ml ○
Total maternal jugular vein plasma oestrogens pg/ml •
2 min after birth
18 min before birth
Birth
Min
Days
Days

Fig. 3a.

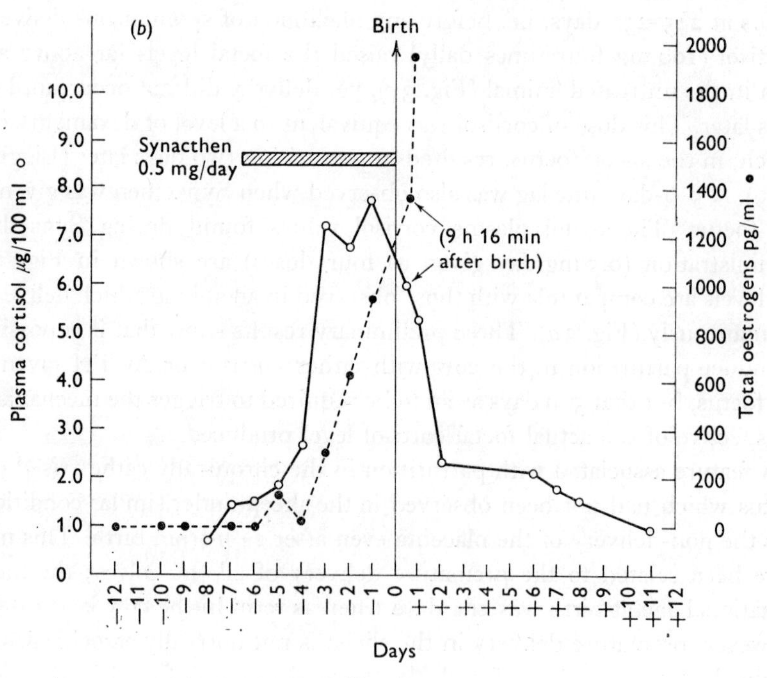
(b)
Birth
Synacthen 0.5 mg/day
(9 h 16 min after birth)
Plasma cortisol μg/100 ml ○
Total oestrogens pg/ml •
Days

Fig. 3b.

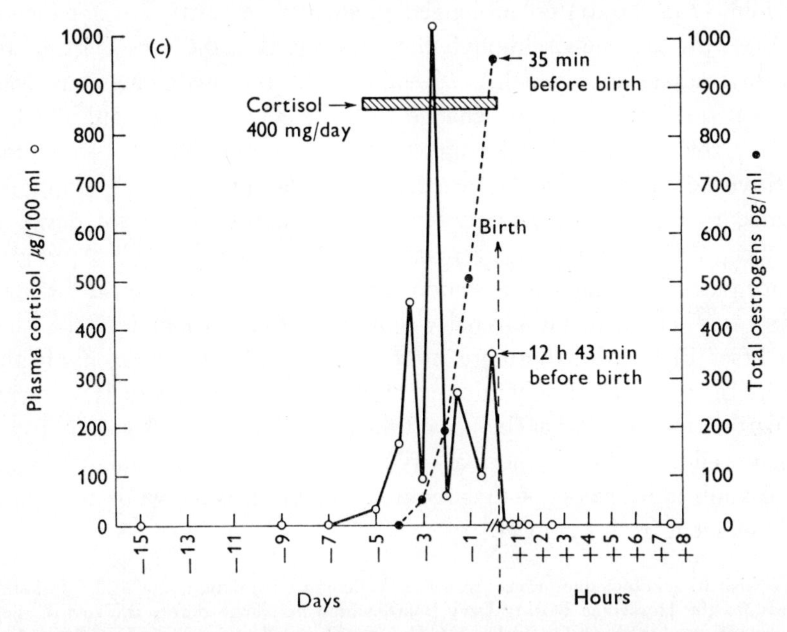

Fig. 3*a*-*c*. Perinatal foetal plasma cortisol (○), maternal plasma cortisol (×) and maternal plasma oestrogen (●) values during (*a*) spontaneous delivery; (*b*) parturition induced by the administration of Synacthen to the foetus; and (*c*) induction by the injection of cortisol into the foetus; for doses of Synacthen and cortisol see text. Duration of treatment is shown by the shaded bar.

Retained placenta was never associated with parturition in the mare at any stage of gestation.

In the sheep foetus the rise in plasma cortisol precedes all hormonal changes seen in the mother before delivery, i.e. the rise in oestrogen, the fall in progesterone and the rise in PGF2α. In the present investigations maternal plasma oestrogen levels were also measured in some experiments. In all the spontaneous deliveries examined there was a rise in maternal oestrogen which followed the increase in the foetal plasma cortisol; an example is given in Fig. 3*a*. The oestrogen levels were also measured in the two animals in which parturition was induced artificially. In both, the maternal oestrogen levels increased after the cortisol (Fig. 3*c*) or Synacthen (Fig. 3*b*) treatment had begun and the foetal plasma cortisol levels had risen. In this respect, therefore, the changes in the cow are similar to those found in the sheep before parturition.

The comparatively small rise in prenatal plasma cortisol in the calf and the lack of any adrenal hypertrophy at this time raises the problem of whether the secretion of the foetal adrenal cortex is the primary stimulus to parturition in the cow. Our findings, together with the indirect evidence of Holm *et al.* (1961) on prolonged pregnancy in cattle, are not incompatible with this view, although they suggest that the sensitivity of the placental tissue, upon which the foetal cortisol presumably acts, may vary between species; it may also change with the stage of gestation. Thus, in the cow, the critical period during which the placental tissue becomes more sensitive to glucocorticoids, whether of foetal or maternal origin (cf. Adams, 1969), seems to be between 260 days and term (~ 281 days). In the sheep relatively more exogenous cortical hormone may be required to induce parturition early in gestation than later on. Even at term there are wide variations in foetal adrenal weight in individual lambs of the same breed (see Fig. 1). Furthermore, after incomplete hypophysectomy in this species the presence of only minimal amounts of cortical tissue is necessary for parturition to occur at the usual time (Comline, Silver & Silver, 1970). It may well be that under normal circumstances cortisol production by the foetal lamb is in excess of that required for its delivery while the calf is more efficient in this respect.

We wish to acknowledge grants from the Wellcome Foundation, the Milk Marketing Board and the Horserace Betting Levy Board which helped to defray the cost of these experiments, and to thank Dr Burley of Ciba Laboratories Ltd for a supply of Synacthen. We also wish to thank Mr R. Proudfoot for his careful observation and care of the animals, and Miss Helen Shiers, Mr P. Hughes and many other laboratory staff for their technical assistance.

REFERENCES

Abraham, G. E. (1971). *Methods of Hormone Analysis*, Ed. Breuer & Kruskemper.
Adams, W. M. (1969). *J. Am. vet. med. Ass.* **154**, 261–5.
Anderson, A. B. M., Laurence, K. M. & Turnbull, A. C. (1969). *J. Obstet. Gynaec. Br. Commonw.* **76**, 196–9.
Comline, R. S. & Silver, M. (1970). *J. Physiol.* **209**, 567–86.
Comline, R. S. & Silver, M. (1971). *J. Physiol.* **216**, 659–82.
Comline, R. S. & Silver, M. (1972). *J. Physiol.* **222**, 233–56.
Comline, R. S., Silver. M. & Silver, I. A. (1970). *Nature, Lond.* **225**, 739–40.
Holm, L. W., Parker, H. R. & Galligan, S. J. (1961). *Am. J. Obstet. Gynec.* **81**, 1000–8.
Liggins, G. C. (1969). *J. Endocr.* **45**, 515–23.
Liggins, G. C., Grieves, S. A., Kendall, J. Z. & Knox, B. S. (1972). *J. Reprod. Fert.* Suppl. 16, 85–103.
Mueller-Heubach, E., Myers, R. E. & Adamsons, R. (1972). *Am. J. Obstet. Gynec.* **112**, 221–6.
Murphy, B. E. P. (1967). *J. clin. Endocr. Metab.* **27**, 973–90.
Nathanielsz, P. W., Comline, R. S., Silver, M. & Paisey, R. B. (1972). *J. Reprod. Fert.* Suppl. 16, 39–59.
Robinson, R., Baker, R. D., Anastassiads, P. A. & Common, R. M. (1970). *J. Dairy Sci.* **53**, 1592–5.

SESSION 7

SURFACTANT COLLOQUIUM

PREVENTION OF RESPIRATORY DISTRESS SYNDROME BY ANTEPARTUM CORTICOSTEROID THERAPY

By G. C. LIGGINS and R. N. HOWIE

Postgraduate School of Obstetrics and Gynaecology,
University of Auckland, Auckland, New Zealand

The stimulus to this clinical trial came from our gross observations of the behaviour of the lungs of foetal lambs born after a few days' treatment with either corticotrophin or a glucocorticoid (see page 564 in these proceedings). Although our interest was primarily in the physiology of initiation of parturition, the viability of extremely premature lambs born as a consequence of experimental triggering of the parturitional mechanisms was too striking to overlook (Liggins, 1969*a*, *b*). Partially inflated lungs were found at necropsy in lambs of gestational ages shown by previous workers (Brumley, Chernick, Hodson, Normand, Fenner & Avery, 1967) to be associated with a complete lack of pulmonary alveolar stability. This raised the possibility that glucocorticoids accelerated functional maturation of foetal lungs as postulated by Buckingham, McNary, Sommers & Rothschild (1968) by analogy with the effects of corticosteroids on the developing intestine. Confirmation of accelerated pulmonary maturation by corticosteroids was rapidly forthcoming from deLemos, Shermata, Knelson, Kotas & Avery (1970) and Kotas & Avery (1971), who showed that the effect occurred not only in the foetal lamb but also in foetal rabbits. With the knowledge that the phenomenon was not peculiar to the ovine species and that the human foetus could be subjected to treatment with corticosteroids simply and apparently safely, a controlled investigation of the effects of maternally-administered corticosteroids on the incidence of the respiratory-distress syndrome (RDS) in premature infants was started in December 1969.

Two groups of patients were admitted to the trial. The first group consisted of women admitted in threatened or established premature labour at 24–36 weeks of pregnancy. Whenever possible delivery was delayed for 24–72 hours by intravenous infusions of ethanol or, more recently, of salbutamol (Liggins & Vaughan, 1972). The second group included women in whom premature delivery before 36 weeks was planned because of an obstetrical complication. Patients were randomly assigned to a treatment or control group. Ampoules containing the agents for injection were arranged in random order and serially numbered so that ward staff were unable to identify the two groups. Patients in the treatment group were given two injections at an interval of 24 hours of a mixture of betamethasone acetate 6 mg and betamethasone phosphate 6 mg. The control group were given cortisone acetate 6 mg as a placebo having identical appearance and approximately one-seventieth of the potency of the dose of betamethasone.

Treatment was considered complete for the purposes of analysis as soon as the second injection was given, but efforts were made to delay delivery for an arbitrarily defined period of at least 72 hours after the first injection. Amniotic fluid samples were obtained by amniocentesis from a number of women. Lecithin–sphingomyelin (L/S) ratios were determined by thin-layer chromatography (Gluck, Kulovich, Borer, Brenner, Anderson & Spellacy, 1971).

A diagnosis of RDS was made only when both clinical and radiological signs were present as follows: clinical signs of grunting respirations and chest retraction during the first three hours and persisting beyond the first six hours after delivery, and radiological signs of fine generalised granularity of lung fields with air bronchogram.

A detailed account of the results of the trial is published elsewhere (Liggins & Howie, 1972). A summary of the major findings is presented in Tables 1 and 2 which show that both the perinatal mortality and the incidence of RDS were significantly lower in the treated group than in the controls. However, the difference in incidence of RDS is confined to infants delivered at less than 32 weeks' gestation. Infants delivered within 24 hours of the first injection of betamethasone derived no apparent benefit from treatment, those delivered from 24–48 hours were possibly less likely to develop RDS and the full effects of treatment appeared after 48 hours (Table 3). There were no deaths from RDS or intraventricular haemorrhage in betamethasone-treated infants who delivered at least 24 hours after the first injection. An excess of antepartum foetal deaths occurred in the group of patients with severe hypertensive syndromes who were treated electively before planned premature delivery. The difference between treated and control groups is not significant but until the trial is extended to establish whether this distribution has occurred by chance the possibility remains that corticosteroid treatment may adversely affect placental function when severe damage is already present.

Table 1. *Composition and outcome of treated and control groups of infants of unplanned deliveries occurring at least 24 hours after entry to trial*

	Betamethasone-treated group		Control group		
Mothers	89		74		
Infants	94		78		
Sex: male	52 (55.3 %)		48 (61.5 %)		
female	42 (44.7 %)		30 (38.5 %)		
Gestational age (days after LMP*, mean ± SD*)					
(*a*) at entry to trial	222 ± 21		225 ± 20		
(*b*) at delivery	249 ± 31		244 ± 29		
Birth weight (g, mean ± SD)	2350 ± 810		2280 ± 780		
Infant outcome	No.	%	No.	%	*P*
Foetal deaths, antepartum	—	0.0	—	0.0	
Foetal deaths, intrapartum	1	1.1	3	3.8	> 0.05
Early neonatal deaths	3	3.2	11	14.1	< 0.02
Perinatal deaths	4	4.3	14	17.9	< 0.01
Survived 7 days	90	95.7	64	82.1	
All infants	94	100.0	78	100.0	

* LMP = first day of last normal menstrual period. SD = standard deviation.

Table 2. *Incidence of RDS according to gestational age at delivery in liveborn infants of unplanned deliveries at least 24 hours after entry to trial*

	Betamethasone-treated group			Control group			
	No. of	RDS		No. of	RDS		
	infants	No.	%	infants	No.	%	*P*
Gestational age at delivery:							
26 and under 32 weeks	17	2	11.8	23	16	69.6	0.02
32 and under 37 weeks	43	2	4.7	29	2	6.9	
37 weeks and over	33	0	0.0	23	0	0.0	
All liveborn infants	93	4	4.3	75	18	24.0	0.002

The L/S ratio was determined in amniotic fluid samples taken before and after betamethasone or placebo treatment in 20 patients. Small increases were sometimes found in L/S ratios of second samples taken 1–6 days after the first, but in this respect there was no difference between betamethasone and control groups. Consequently, the L/S ratio after corticosteroid treatment did not correlate with the occurrence of RDS (Table 4).

Table 3. *Occurrence of RDS in liveborn infants related to entry–delivery interval*

Entry–delivery interval	Betamethasone-treated group No.	RDS	%RDS	Control group No.	RDS	%RDS	P
Under 24 hours	29	7	24.1	22	7	31.8	NS
24 and under 48 hours	20	2	10.0	19	7	36.8	NS
2 and under 7 days	28	1	3.6	24	8	33.3	0.03
7 days and over	45	1	2.2	32	3	9.4	NS
All live births	122	11	9.0	97	25	25.8	0.003
All infants born alive over 24 hours after entry to trial	93	4	4.3	75	18	24.0	0.002

Table 4. *Lecithin–spingomyelin ratio and occurrence of RDS in* 16 *patients delivering within six days of entry into trial*

	Lecithin–sphingomyelin ratio		
	1	1–1.5	1.5
Betamethasone-treated	6 (1*)	2 (0)	2 (0)
Control	2 (2†)	1 (1*)	3 (0)

Figures in parentheses = no. with RDS.
* = survived; † = both died.

Six anencephalic foetuses, all with low liquor L/S ratios, were injected intramuscularly with betamethasone alcohol 4 mg and betamethasone phosphate 1 mg. Forty-eight hours later the L/S ratio was unchanged in five and increased in one.

We conclude that maternal administration of large doses of corticosteroids for more than 24 hours before delivery at less than 32 weeks of pregnancy helps to protect the newborn from RDS. The liquor lecithin/sphingomyelin ratio does not reflect changes in pulmonary maturation induced by corticosteroids.

REFERENCES

BRUMLEY, G. W., CHERNICK, V., HODSON, W. A., NORMAND, C., FENNER, A. & AVERY, M. E. (1967). *J. clin. Invest.* **46**, 863.

BUCKINGHAM, S., MCNARY, W. F., SOMMERS, S. C. & ROTHSCHILD, J. (1968). *Fed. Proc.* **27**, 328.

DELEMOS, R. A., SHERMETA, D. W., KNELSON, J. H., KOTAS, R. V. & AVERY, M. E. (1970), *Am. Rev. res. Dis.* **102**, 459.

Gluck, L., Kulovich, M. V., Borer, R. C., Brenner, P. H., Anderson, G. G. & Spellacy, W. N. (1971). *Am. J. Obstet. Gynec.* **109**, 440.

Kotas, R. V. & Avery, M. A. (1971). *J. appl. Physiol.* **30**, 358.

Liggins, G. C. (1969*a*). *Foetal Autonomy*. Ciba Foundation Symposium. Eds. G. E. W. Wolstenholme & M. O'Connor. Churchill, London, pp. 142 and 218.

Liggins, G. C. (1969*b*). *J. Endocr.* **45**, 515.

Liggins, G. C. & Howie, R. N. (1972). *Pediatrics*. (In press.)

Liggins, G. C. & Vaughan, G. S. (1972). *J. Obstet. Gynec. Br. Commonw.* (In press.)

PULMONARY SURFACTANT AND ITS ASSAY

BY JOHN A. CLEMENTS* AND RICHARD J. KING
Cardiovascular Research Institute,
University of California at San Francisco

In some ways, the most cataclysmic event in life is the first breath, the transition from an aquatic to an atmospheric existence, the expansion of the lungs with air. It is a truism by now to say that this transformation is made possible by pulmonary surfactant, a magical substance that lines the alveoli, stabilizes them against collapse, and allows them to remain fluid-free. The phrase 'pulmonary surfactant' is tossed off so casually these days that one tends to forget that its existence was hypothetical just a few years ago and in Sir Joseph Barcroft's time had not even been postulated. Its nature is still not precisely known. Our purpose in this paper, therefore, is to give a progress report on the analysis of the surfactant and to show how new analytical data can make certain physiological studies more meaningful, especially those that concern development of the foetal lungs. We shall concentrate on recent work in our own laboratories.

Since analysis of the surfactant requires that it be removed from the lungs, our first task has been to establish strict criteria, deduced from the behaviour of the alveolar surfaces and the structure of the lungs, that must be met by a candidate surfactant (King & Clements, 1972). These are physical properties such as surface film collapse pressure and density, film collapse rate, adsorption rate, film compressibility, and thermal dependence of these parameters; chemical composition consistent with phylogenetic information; and stoichiometry.

Using ultracentrifugation and density gradients we have been able to prepare fractions of the surfactant from endobronchial lavage fluid that closely approach theoretical purity and meet such criteria for quality. These materials are reproducible in composition, have equilibrium buoyant density of 1.089 (SD 0.002), give values of surface tension *in vitro* below 10 dynes/cm at 37 °C, and maintain their composition and properties through eight additional ultracentrifugations. They contain 74% phospholipids, among which saturated lecithin comprises 56%. Dipalmitoyl

* Career Investigator of the American Heart Association.
This work was supported by Grants HL-6285, HL-5251, and HL-14201 from the National Heart and Lung Institute.

lecithin (DPL) is the predominant molecule. This is the backbone of the surfactant, from a physicochemical point of view; it makes the principal contributions to the mass and to the stability of surface films of the surfactant. There are consistent small amounts of other phospholipids (7%), cholesterol (8%), and triglycerides (3%). Of especial interest is the invariable appearance of proteins to the extent of 10% of these materials, and it is these components of the surfactant fractions which most engage our attention in this paper.

To harvest the proteins we lyophilize the surfactant and extract the lipids from it at −15 °C in ether–ethanol (3:1) according to the method of Scanu *et al.* (1969). The precipitate contains 100% of the proteins and 11·6% of the lipids, which remain bound to the proteins under these conditions, but which can be removed with a stronger solvent mixture, a phenomenon also observed by Reiss (1971). The precipitate is dried and suspended in 0.05 M borate, pH 9.0, containing 0.1% sodium dodecyl sulphate, by sonication for 10 minutes at 80 watts. The suspension is clarified by centrifugation at 89,000 × **g** for two hours, and the water-clear supernatant fluid is used for characterization of the protein by chromatography on Sephadex G-150 with the same solvent. Two fractions appear, the first in the void volume (210 ml) and the second at about half the bed volume (620 ml) of the column. Of the recovered proteins 76% are in fraction 1 and 24% in fraction 2. We have identified these components by electrophoresis on polyacrylamide gels and by immunochemical analysis, using appropriate protein standards. Fraction 1 contains three identifiable proteins, serum albumin, immunoglobulin G, and a non-serum protein. Except for traces of albumin and IgG, fraction 2 consists of the same non-serum protein. From densitometry and staining of the polyacrylamide gels we calculate that this protein accounts for about two-thirds of the total surfactant protein. We estimate its molecular weight at 10,000 to 11,000 daltons by electrophoresis in polyacrylamide gels containing sodium dodecyl sulphate, with and without 2-mercaptoethanol.

In previous work (Klass, King & Clements, 1971) we obtained antiserum against surfactant and showed that it is specific for an antigen localized at the alveolar surface and in certain parenchymal cells of the lung, but does not react with plasma and other tissues. Using this antiserum we find (Fig. 1) that the protein of fraction 2, the 10,000 dalton component in polyacrylamide electrophoresis, and the intact surfactant contain identical antigens. Needless to say, this result greatly intensifies our interest in the protein of fraction 2.

On analysis it contains 16.6 mg of organic phosphorus, 13.5 mg of hexose, and 5.6 mg of sialic acid per 100 mg of protein. Extraction of lipids

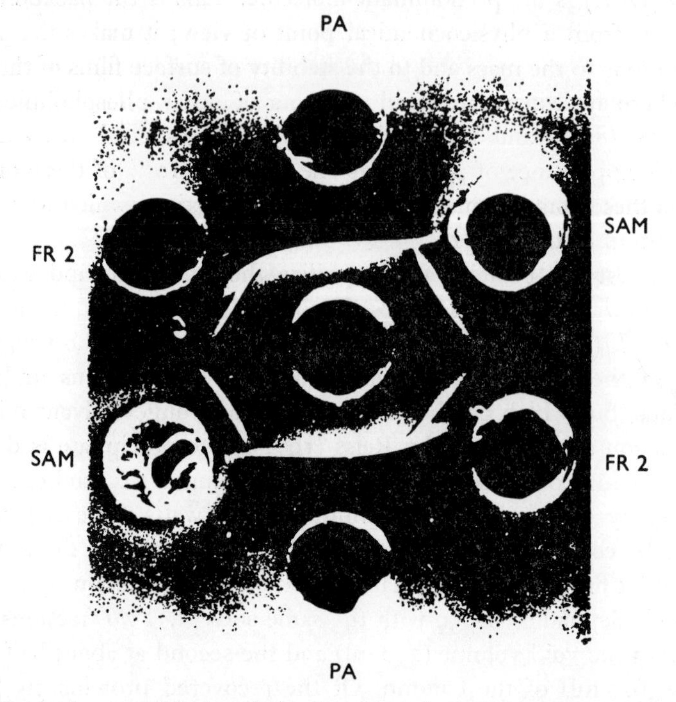

Fig. 1. Test by immunodiffusion of identity of proteins in intact surface active material (SAM), protein of fraction 2 (FR 2), and the 10,000 dalton antigen isolated by polyacrylamide electrophoresis. Antiserum was placed in the central well.

from the protein with chloroform–methanol (2:1) and two-dimensional thin-layer chromatography yield 56% lecithin, 32% phosphatidylglycerol, 5% phosphatidylserine, and 7% neutral lipids. Acid hydrolysis of the residual protein and ion-exchange chromatography of the liberated amino acids reveal a startling pattern (Table 1) in which hydrophobic amino acids outnumber hydrophilic ones two-to-one and acidic amino acids outnumber basic ones by a factor of 2. These facts demonstrate that the protein is new and unique among mammalian peptides and well-suited to a role as the apoprotein of pulmonary surfactant.

Its uniqueness and antigenicity also suit it for immunoassay of the surfactant, as Fig. 2 demonstrates. This method, which utilizes a quantitative precipitin reaction, is specific, sensitive and reproducible. We have found that addition of lipids or serum proteins does not interfere with the assay. It can be expected, therefore, to give reliable results when applied to unfractionated tissue and body fluids.

We have already used it to demonstrate some interesting facts. For example, our intact surfactant is purified 50-fold with respect to whole

Table 1. *Amino acid composition of fraction 2 protein*

	*Residues/mole protein		*Residues/mole protein
His	1	Try	2
Thr	2	Tyr	3
Arg	2	1/2 Cys	8
Ser	5	Gly	11
Lys	4	Ala	7
Glu	7	Phe	4
Asp	6	Pro	7
		Met	2
		Ilu	3
		Leu	8
		Val	4
Total polar+ = 27		Total non-polar+ = 59	

* 24-hour hydrolysis in 6 N HCl, 110 °C. Nearest integer to the mean is calculated from 4 determinations. Minimum molecular weight is 10,700 daltons.

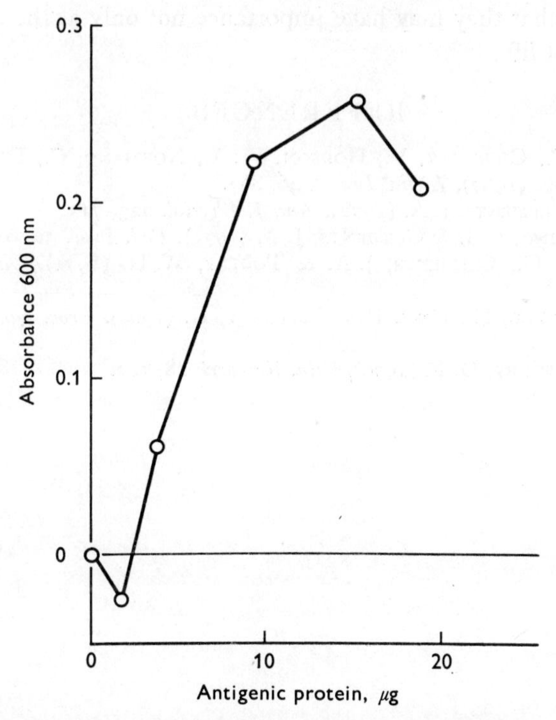

Fig. 2. Assay of surfactant antigen by a quantitative precipitin test. The quantity of precipitate, measured by absorbance at 600 nm, is related to the amount of antigenic protein added by a characteristic equivalence curve.

lung tissue. By comparison it is purified only 20-fold when saturated lecithin is chosen as the basis for calculation. This discrepancy implies, as others have suggested (Young & Tierney, 1970), that much of the saturated lecithin in the lung is not directly related to the surfactant, even though it is surface active. If true, this idea would enable us to explain how saturated lecithin (Brumley *et al.* 1967) and 'surface active material' (Platzker, Clements & Tooley, 1971) appear in the foetal lung in advance of surfactant.

We have also used immunoassay to demonstrate the appearance of surfactant apoprotein in the fluid within the foetal lung at a time when serum proteins are not detectable in it, and we have shown that similar apoproteins exist in surfactants from dog, lamb and man. It should therefore soon be possible to use specific antibodies for developmental studies of surfactant in amniotic and foetal lung fluids, for isolation of surfactant in metabolic investigations, and for assay in small biopsy specimens.

Our experiments unfortunately do not prove that the apoprotein of surfactant has important physiological roles. Its remarkable hydrophobicity and its ability to bind phospholipids suggest that in time such roles may be found and that they may have importance not only at the first breath but throughout life.

REFERENCES

BRUMLEY, G. W., CHERNICK, V., HODSON, W. A., NORMAND, C., FENNER, A. & AVERY, M. E. (1967). *J. clin. Invest.* **46**, 863.

KING, R. J. & CLEMENTS, J. A. (1972). *Am. J. Physiol.* **223**, 715.

KLASS, D. J., KING, R. J. & CLEMENTS, J. A. (1971). *Fed. Proc.* **30**, 619.

PLATZKER, A. C. G., CLEMENTS, J. A. & TOOLEY, W. H. (1971). *Clin. Research* **19**, 232.

SCANU, A., POLLARD, H., HIRZ, R. & KOTHARY, K. (1969). *Proc. nat. Acad. Sci.* **62**, 171.

YOUNG, S. & TIERNEY, D. F. (1970). *Clin. Research* **18**, 192.

THE PULMONARY SURFACTANT IN FOETAL AND NEONATAL LUNGS

BY MARY ELLEN AVERY
Department of Pediatrics, Faculty of Medicine,
McGill University, Montreal

HISTORICAL BACKGROUND

Interest in the forces operating at interfaces was prominent among physical chemists during the late 19th and early 20th century. Wilhelmy described his method of recording the forces of surface tension exerted on a thin, partially submerged metal plate in 1863. Perhaps the earliest published observation on the phenomenon of a change in surface tension with contamination of a clear water surface was communicated in the form of a letter to Lord Rayleigh by Miss Agnes Pockels. She wrote

> My Lord, – Will you kindly excuse my venturing to trouble you with a German letter on a scientific subject? Having heard of the fruitful researches carried on by you last year on the hitherto little understood properties of water surfaces, I thought it might interest you to know of my own observations on the subject. For various reasons I am not in a position to publish them in scientific periodicals, and I, therefore, adopt this means of communicating to you the most important of them.

Lord Rayleigh submitted the letter to *Nature* in 1891 with an accompanying note,

> I shall be obliged if you can find space for the accompanying translation of an interesting letter which I have received from a German lady, who with very homely appliances has arrived at valuable results respecting the behaviour of contaminated water surfaces.

Miss Pockels then proceeded to describe her rectangular tin trough, with which she could vary surface area and record the changes in surface tension. She noted that the surface tension of a perfectly clean water surface remained constant, but the addition of camphor or flour lowered surface tension to a definite value, and sugar and soda interacted in lowering surface tension. She further observed that the surface layer of water can take up more of a soluble substance than the internal liquid, and when that happens, surface currents are established on contraction of the film.

The systematic studies of Langmuir, published in 1917, included a description of his trough to measure the differences in surface pressure between surfaces separated by a horizontal float and moveable barrier. The dependence of surface tension on surface area was well known to the physical chemists of that era, and thoroughly reviewed by N. K. Adam in his classic text *The Physics and Chemistry of Surfaces*, first published in 1930.

During this same era there was less evidence of awareness of the biological applications of the new knowledge in physical chemistry. However, one premedical student became interested in the lively topic of surface forces, and in 1895, Joseph Barcroft presented his first communication entitled 'The properties of the surface of liquids', later published in summary form in *Reports of the Belfast Natural History Society*. As far as I can tell, that was Sir Joseph's last publication on that topic. I am sure he would be among the most interested in seeing how in the past 15–20 years biologists have rediscovered surface forces, and thought through their significance, particularly with respect to the lung.

If one thinks of the lung as an emulsion of air in liquid, or as an organ with an immense internal surface area (70 m^2 in the adult) with a gas–liquid interface, sharply curved with air spaces arranged in parallel, it is immediately obvious that the forces of surface tension should be considered. When one further reflects that these forces tend to reduce the area of a surface, it follows that surface tension should increase the elastic recoil of the lung; conversely, if the lung is to remain aerated, it might be useful if these forces were reduced by the presence of something capable of lowering surface tension, a surfactant. That such is the case is now a familiar story, first deduced by Pattle (1955) on inspection of bubbles expressed from the lung, and by Clements who demonstrated the capability of materials expressed from the lung to change surface tension with surface area (Clements, 1957).

Since these now classic observations were made, their ramifications on our understanding of aspects of structure and function of lung have been pursued by workers in a number of fields, and summarized in several reviews (Clements, 1962; Avery & Said, 1965; Pattle, 1965; Scarpelli, 1968). It is the purpose of this review to focus on the implications of the surfactant story as it bears on our understanding of the foetal and neonatal lung.

ONTOGENY

Interest in the ontogeny of the surfactant dates from Pattle's first extensive article in 1958, when he noted that bubbles expressed from lungs of foetal rabbits and guinea pigs were stable only late in foetal life. He noted that spontaneous respiration occurred only in animals of sufficient gestational age to have a normal lung lining film (1958, 1961).

HUMANS

Meanwhile Avery & Mead (1959) were investigating the time of appearance of the surfactant in human foetal lungs. They were aware of the studies of Gruenwald, reported first in 1947 and amplified in 1955, on the unusual expansion patterns of the lungs of premature infants and those with hyaline membrane disease. Gruenwald's observation that such lungs could be expanded evenly with kerosene, but only unevenly with air, led him to assume 'that a higher interfacial (surface) tension accounts not only for the higher pressure necessary to introduce air, but also for the fact that air accumulates in the lungs in the fewest and largest possible bubbles, thus having the smallest possible surface' (1955). Avery & Mead (1959) used the Wilhelmy balance technique as described by Clements to study the lungs of stillborn and liveborn infants obtained at autopsy. They found that most infants under 1.1 kg birth weight, and those of any weight with atelectasis and hyaline membranes, lacked sufficient surface active material to lower surface tension significantly on the film balance. Later the correlation of surface tension measurements of lung extracts with abnormal pressure–volume characteristics was established on neonatal lungs (Gruenwald *et al.* 1962; Reynolds *et al.* 1968). A reduction in lung lipid phosphorus likewise correlated with abnormal pressure–volume relationships (Rufer, 1968).

Later more extensive study of surface properties of human lungs has established a variable time of appearance of stable bubbles, on the whole present in foetuses over 700 g weight (Kumode, 1968), stable pressure–volume relationships in some foetuses of 300–500 g (Gruenwald, 1962) and demonstration of a reduction in lung phospholipids among a group of infants from 26 to 36 weeks' gestational age who had hyaline membrane disease at autopsy (Brumley, 1967) (Fig. 1).

Morphological studies of foetal lungs confirm the variability in the human with respect to time of appearance of type II alveolar cells, containing osmiophilic inclusions (Spear *et al.* 1969).

The relationship of surfactant deficiency to hyaline membrane disease was pursued by Gandy *et al.* (1970). In morphological studies they found osmiophilic inclusions as early as twenty weeks in the human, and nearly always present by twenty-four weeks, except in infants with hyaline membrane disease. Numerous inclusions usually indicated normal surfactant. Boughton *et al.* (1970) confirmed the low lecithin content of lungs of infants with inadequate surfactant. In a series of 95 autopsy specimens of lung, when lung lecithin content was greater than 8 g/100 g of lung, surfactant was normal. In 73 of these lungs the lecithin content was less

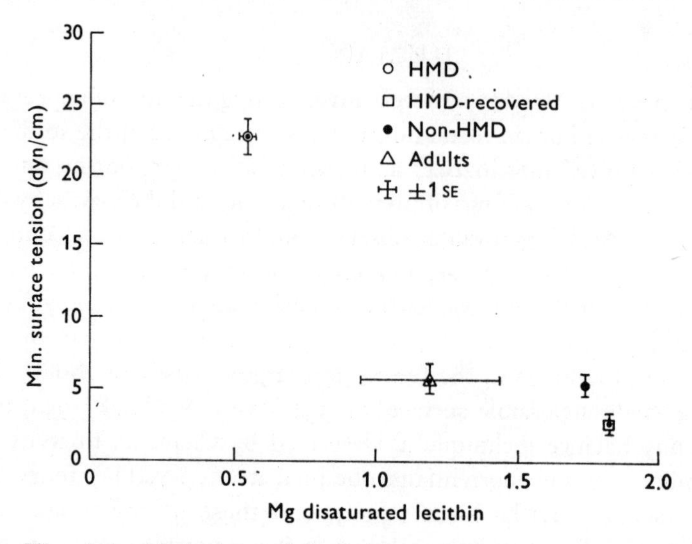

Fig. 1. The relationship between minimal surface tension on compression of a surface film from lung extracts with the calculated minimal amount of disaturated lecithin extractable from lungs obtained at autopsy. (Data of Brumley *et al.* 1967.)

than 8 g/100 g, and all of those with lack of surfactant were in this group. However, 31 infants with low lecithin had demonstrable surfactant. They explain these findings by suggesting that lecithin content measures the reserve from which surfactant can come. Surfactant utilization after birth could account for the few infants born with minimal respiratory distress who subsequently die from the disorder.

Studies of the pathways of synthesis of saturated lecithins have been reported in the human by Gluck *et al.* (1972) on pharyngeal aspirates and mucus.

BIOSYNTHETIC PATHWAYS, ACTIVE IN LUNG

(1) CDP-choline + D-α,β-diglyceride $\rightarrow$ lecithin.

(2) *S*-adenosyl-L-methionine $\rightarrow$ CH_3 + phosphatidyl ethanolamine (PE) $\rightarrow$ phosphatidyl methyl ethanolamine (PME) + CH_3 $\rightarrow$ phosphatidyl dimethyl ethanolamine (PDME) + CH_3 $\rightarrow$ lecithin.

Foetal lung of 18 and 20 weeks showed slight incorporation by the cytidine diphosphate choline (CDP-choline) pathway, and no evidence of methylated derivatives of phosphatidyl ethanolamine, i.e. phosphatidyl dimethyl ethanolamine (PDME). By 22–24 weeks' gestation, methylation of phosphatidyl ethanolamine was evident at birth, since aspirates from all newborn and premature infants contained PDME. An increase in lecithin content in lungs of humans at about 36 weeks' gestation is accompanied by greater activity of the CDP choline pathway. Gluck comments that the human differs from rabbits and sheep in having greater evidence of synthesis of lecithin by the methylation pathway until such time as the CDP choline becomes more active. The methylation pathway produces principally α-palmitic/β-myristic acid lecithin whereas the CDP choline pathway produces the more stable α-palmitic/β-palmitic acid lecithin according to Gluck *et al.* (1972).

LAMBS

The ontogeny of the surfactant has been carefully studied in both lambs and rabbits. Interest in the lamb was stimulated by the wealth of studies available on neonatal adaptions in lambs, many of which appear in Sir Joseph Barcroft's classic volume, *Researches on Pre-natal Life*, and Geoffrey Dawes' *Foetal and Neonatal Physiology*. It seemed useful to examine changes in the pulmonary surfactant in lungs of foetal and neonatal lambs to permit correlations with other studies (Orzalesi *et al.* 1965; Brumley *et al.* 1967; Adams *et al.* 1971). In general, lambs born before 120 days' gestation have lungs that are 'immature' in the sense that they are incapable of supporting life after birth. Before that time the terminal air spaces are lined by cuboidal epithelium, and alveolar cells have not differentiated to the extent of forming osmiophilic inclusions. In the study of Orzalesi *et al.* the percentage of cells containing inclusions increased from 3 per cent at 121 days to 12 per cent at term (147 days). When more than 4 per cent of the cells contained inclusions, surface activity of lung extracts increased. In general, the lungs had detectable surfactant by 126 days; later Howatt *et al.* (1965) noted that the more cephalad lobes had surfactant a few days earlier than the more caudad ones. The increases in phospholipid content, and especially disaturated phosphatidyl choline correlated with stable pressure–volume behaviour. Brumley *et al.* (1967) noted that the foetus near term had concentrations of phospholipids in excess of those found after birth, as well as those in mature animals (Figs. 2 and 3). The possibility was raised that a 'reservoir' of surfactant was present before birth, and partially depleted with air breathing. The kinetics of synthesis and turnover of lung phospholipids before and after birth deserve further study. Meanwhile, some quantification is known from the very recent studies of Platzker *et al.* (1972). They cannulated the tracheae of foetal lambs *in utero*, attached a thin-walled latex bag, and harvested the tracheal

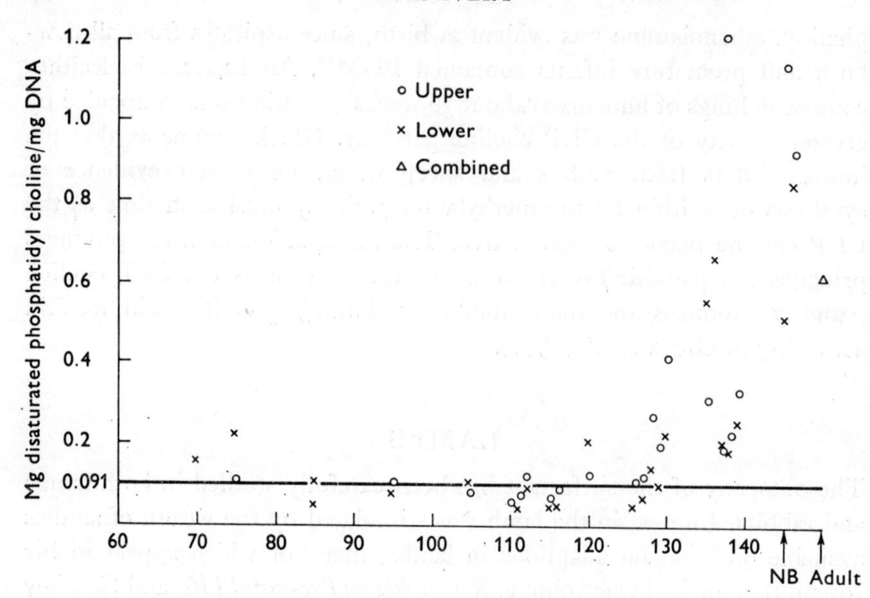

Fig. 2. Comparisons of calculated minimum disaturated lecithins with gestational age in the foetal and newborn lamb and ewe. Note the greater concentrations in upper (cephalad) compared to lower (caudad) lobes. (Data of Brumley *et al.* 1967.)

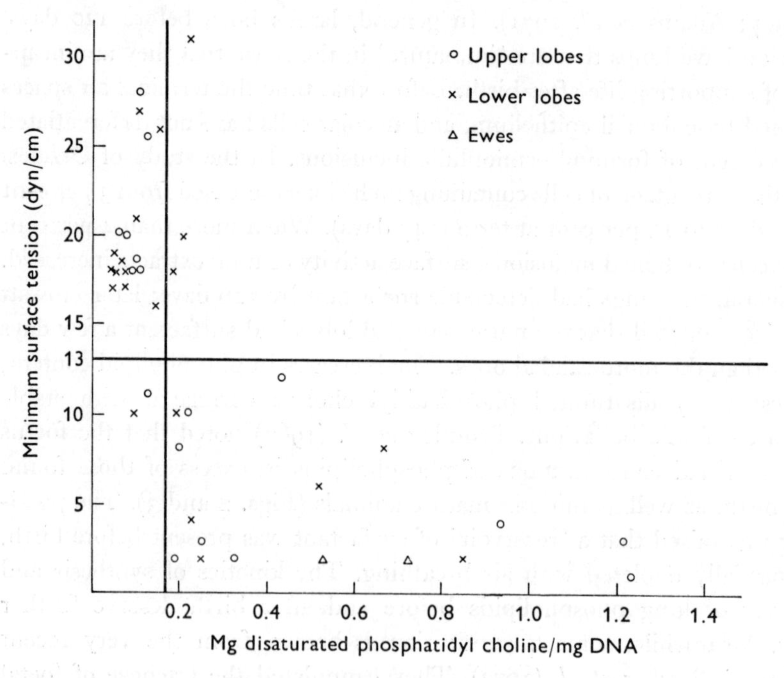

Fig. 3. The correlation between concentration of lecithin and minimal surface tension in lungs of lambs. (Data of Brumley *et al.* 1967.)

effluent. Measurements of surface active material in the effluent showed it first detectable at 120–122 days. Concentrations were 5.2 mg/ml at 126 days, and 15.8 mg/ml at 132 days.

The question of effects on postnatal lung function of surfactant deficiency was studied systematically by Adams *et al.* (1971). Twin lambs were delivered prematurely from 105 to 132 days gestation; one was sacrified at birth, the other allowed to breathe. Ten of the fourteen lambs allowed to breathe developed respiratory distress, and all ten had levels of disaturated phosphatidyl choline below 10 mg/g dry weight of lung. Since non-breathing control twins of those with distress had similarly low tissue levels, Adams *et al.* postulated the prenatal origin of the respiratory distress in their lambs related to immaturity of the lung rather than postnatal consumption or inactivation of disaturated phosphatidyl choline. Support for a prenatal deficiency in humans is apparent from inability of lung extracts from some stillborn infants to lower surface tension (Avery & Mead, 1959). Support for possible postnatal utilization of surfactant in excess of the capacity to synthesize it is apparent from the studies of Boughton *et al.* (1970) in humans. They noted that surfactant deficiency was not found in stillbirths over 30 weeks' gestational age. Some infants seem to have only mild respiratory distress for some minutes or even hours, but proceed to die with lung devoid of demonstrable surfactant.

Studies on synthesis of phospholipids by lung from labelled precursors have demonstrated *in vitro* and *in vivo* the rapid uptake of the precursors and their incorporation into phospholipids by the alveolar type II cells (Felts, 1964; Buckingham *et al.* 1966). Chida & Adams (1967) show in the foetal lamb an increased uptake of palmitate-1-C^{14} into lecithin by lung slices from 120 days to term and in a few days after birth. Adult ewes showed a lesser incorporation of palmitate into lecithin, expressed per mg protein, compared with the late foetal and neonatal lambs.

RABBITS

The rabbit is the other animal in which extensive studies of the ontogeny of the surfactant have been reported. One of the reasons for studying this animal was the ease of timing pregnancies precisely, the short gestation (31 days) and year around availability in most laboratories. The large litter size of 6–15 per litter was an advantage for study of within-litter variability when gestational age was a constant.

During the last third of intrauterine life the foetus grows rapidly, from a weight of 12 g at 25 days to 48 g at 31 days. The lung of the 19-day rabbit foetus is glandular, with columnar epithelium lining the potential airway. By day 23, the epithelium becomes cuboidal, and by day 24 primitive alveolar septae are evident. From day 24 to 27 there is a marked increase in osmiophilic inclusions, with differentiation of the type II cell complete by day 28. The epithelial cells are rich in glycogen until days 25–27 when the glycogen content decreases, and disappears after day 28 (Kikkawa *et al.* 1968).

The pulmonary surfactant is present in low concentration on day 26, but readily demonstrable on a surface balance by day 28. Stable bubbles, and stable pressure–volume curves become manifest at the same time, between days 28 and 29 (Humphreys & Strang, 1967; Kotas & Avery, 1971).

A marked increase in lecithin and sphingomyelin occurs on day 29 when stable pressure–volume relationships are achieved (Kikkawa *et al.* 1968). Extensive biochemical studies on the developing rabbit lung by Gluck and colleagues (1967) establish the large concentrations of saturated lipids in the foetal lung with a 3–5-fold increase of total alveolar lecithin after one hour of air breathing. The amount of acetone-precipitated, alveolar wash lecithin increased 20–30-fold with breathing, suggestive of a postnatal release from tissue reserves into the airway, as well as possible increased synthesis. The principal pathways for the *de novo* biosynthesis of lecithin in the foetal rabbit lung appear to be: (1) the incorporation of CDP-choline + D-α,β-diglyceride and (2) the triple methylation of phosphatidyl ethanolamine. The CDP-choline pathway peaked between 25 and 27 days then declined toward term, and the methylation pathway became more prominent at 28 days. After birth 90 per cent of the *de novo* synthesis, however, was by the incorporation of choline (Gluck *et al.* 1967).

Synthesis of dipalmitoyl lecithin in the adult rabbit lung appears possible through several pathways, including the conversion of lysophosphatidyl choline to phosphatidyl choline by direct acylation (Elsbach, 1966) or the reaction of two molecules of lysophosphatidyl choline to form phosphatidyl choline and glycerolphosphoryl choline (Wolfe *et al.* 1970). Phosphatidyl choline in the adult rabbit lung is also formed by the reaction of diglyceride with CDP-choline (Kennedy pathway). It remains unclear as to the relative contribution of these various pathways to total surfactant production.

INDUCTION OF SURFACTANT

Against the background of the descriptive studies of the appearance of the surfactant in foetal lungs, and as an incidental observation made during studies on the initiation of parturition in the lamb, Liggins noted greater aeration in lungs of lambs after dexamethasone infusion than would be expected. He suggested the possibility of steroid induction of the surfactant (1969). Knowing of Liggins' initial observation by way of a personal communication at a combined meeting of the New Zealand Obstetrical and Paediatric Societies in 1968, Avery and others pursued the possibility that accelerated appearance of the pulmonary surfactant could be achieved by induction of the capacity of type II cells to synthesize it by steroids. The first suggestion of such a possibility was made by Buckingham in 1968, by analogy with changes demonstrated by Moog (1953) in the foetal intestinal epithelium after steroid treatment. Accelerated lung maturation in the lamb was reported by deLemos *et al.* (1969, 1970). They infused hydrocortisone into one lamb from each of the seven sets of twins, and used the other as a control. In each cortisol-treated twin the lung was 'more mature' than in the control. Further evidence of acceleration of surfactant protection by glucocorticoids in the lamb was reported by Platzker *et al.* (1972). Surfactant production could be detected in dexamethasone-treated lambs as early as 108 days of gestation. At 126 days' gestation the tracheal flux of surface-active material increased 370 per cent after two days of treatment.

In order to see if the ability to accelerate synthesis of the surfactant was unique to the lamb, or might occur in other species, Kotas & Avery conducted systematic studies on 228 foetal rabbits from 24 days of gestation to term (31 days) (1971). The trends in time were for a greater distensibility of the lung when inflated with air post-mortem, and greater stability on deflation, indicative of the presence of the surfactant. One foetus in each of 15 litters was given a single dose of a long acting glucocorticoid, 9α-fluoroprednisolone, on day 24. After sacrifice on day 26 or 27, the lungs of the treated foetuses resembled those of a 29–30-day foetus. Wang *et al.* (1971) studied these same lungs morphologically and demonstrated an increase in attenuation of the alveolar cells with more osmiophilic inclusions per cell in the steroid-treated rabbits. Motoyama *et al.* (1971) confirmed these findings with the injection of 0.5 to 1 mg of hydrocortisone sodium succinate into the foetal rabbit, and further noted that the injected foetuses on delivery were more active and breathed better than their littermate controls. The enhanced survival of prematurely delivered rabbits after cortisol injection 48 hours earlier was further studied by Taeusch *et al.* (1972). They found that survival correlated with the degree of maturity of the lung, and was significantly increased after cortisol. An effect on the foetus adjacent to the treated one was also evident, as if some cortisol could diffuse to the neighbour. The longest survival was after direct injection, next longest among the neighbours, and shortest among the more distant littermates, suggesting a dose–response relationship (Fig. 4). Farrell &

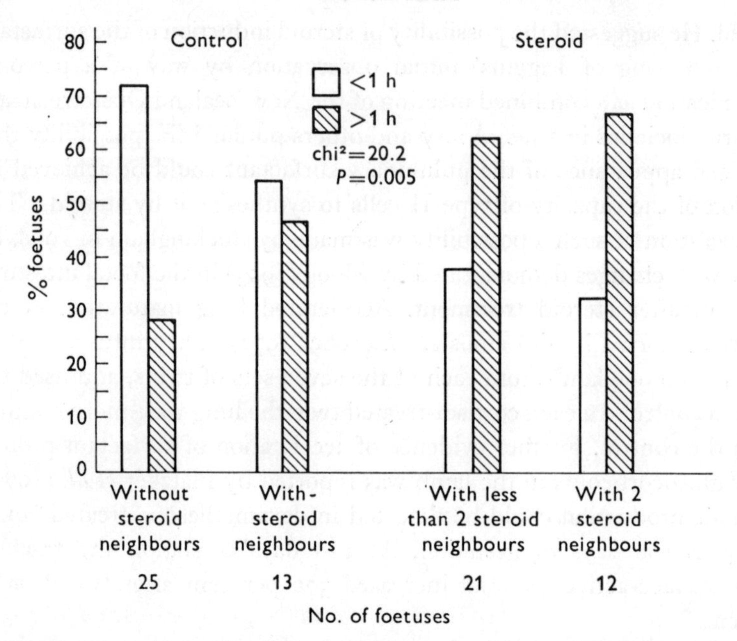

Fig. 4. The percentage of newborn rabbits who survived over one hour was greatest in those receiving steroid when others in the litter also received steroid (columns on right). Survivors under one hour (black columns) were greatest when least exposed to steroid (columns on left). These findings suggest a dose–response relationship.

Zachman (1972) repeated the studies of Kotas & Avery (1971) with a 9α-fluoroprednisolone acetate injected into foetal rabbits at 23–24 days, and measured the lecithin content of the lung parenchyma a few days later. The treated group had 96 mg/g dry weight of lung, compared to 70 mg/g in the controls. They also examined four enzymes in the two major biosynthetic pathways. The only significant change was in the level of phosphorylcholine-glyceride transferase, which was elevated 45 per cent in the steroid-treated group, suggesting acceleration of the choline incorporation pathway.

Further evidence for the role of the intact pituitary–adrenal axis for the maturation of the foetal lung comes from experimental ablation of the hypophysis or adrenals. Foetal hypophysectomy in goats led to a delay in lung maturation (deLemos *et al.* 1971); and in decapitated foetal rats a similar observation was reported by Blackburn *et al.* (1972). They found enlarged lungs with more immature cells in the rats decapitated on the 16th day of gestation. Lung lecithin at term was decreased in the decapitated foetuses.

The mechanism by which cortisol accelerates lung maturation is not established. However, recently it has been suggested that steroid hormones

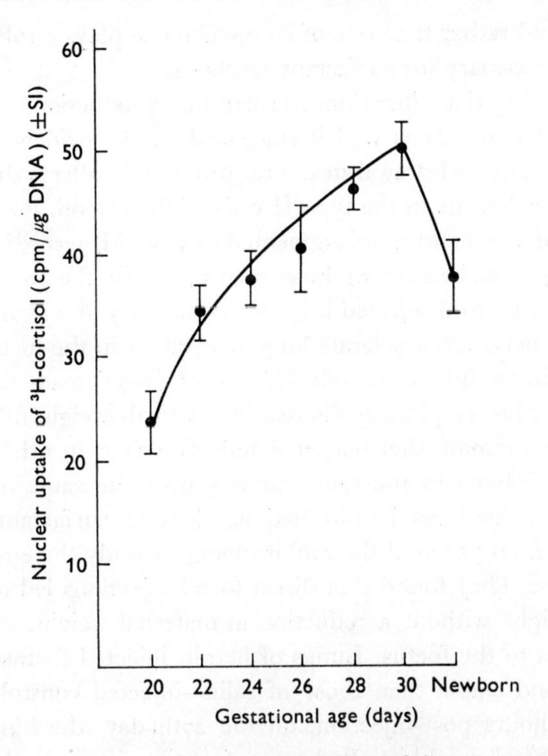

Fig. 5. Changes in uptake of ^{3}H-cortisol (per unit DNA) by lung nuclei of the developing rabbit foetus. Each point represents the mean of 5–6 determinations. For each determination, lungs from all foetuses of the same litter (5–10 foetuses) were pooled, the tissue was minced and a one-gram aliquot of the minces was incubated in 10 ml of Eagle's Hela medium containing 1×10^{-7} M ^{3}H-cortisol for 2 h at 37 °C. The nuclear uptake of ^{3}H-cortisol was then measured as described by Giannopoulos *et al.* (1972).

require specific receptor molecules in their target tissue in order for steroids to affect it. Ballard & Ballard (1972) examined foetal rabbit tissues for the presence of glucocorticoid receptors using ^{3}H-dexamethasone. Binding sites were found in the soluble cytoplasmic fraction from all foetal tissues studied, but was in highest concentration in the lung. The concentration was constant during the last 12 days of gestation. Nuclear binding sites for tritiated cortisol were demonstrated in foetal lung by Giannopoulos *et al.* (1972) (Fig. 5). They further showed the specifity of the sites for compounds with glucocorticoid activity, and the failure to bind progesterone or testosterone. The uptake of tritiated cortisol increased from day 20 and reached a maximum at 28–30 days of gestation. The uptake was somewhat lower in newborn rabbits. Thus cortisol-binding sites are present in both nuclei and cytoplasm of foetal lung cells, suggesting that this compound rather than one of its metabolites plays a role in induction of enzymes necessary for surfactant synthesis.

The possibility that alterations in surfactant production can be achieved with other compounds as well is suggested by the evidence that thyroid administration or withdrawal in the rat profoundly affects the numbers of osmiophilic inclusions in the type II cells of the alveoli. Both storage and production of surfactant were augmented by thyroid, according to Redding *et al.* (1972). Acceleration of lung maturation in the foetal rabbit was achieved with thyroid injected into the foetuses by Wu *et al.* (1971).

Heroin likewise can accelerate lung maturation in the foetal rabbit, and presumably in the human as well. Glass *et al.* (1971) made the remarkable observation of less respiratory distress in low birth-weight infants who had withdrawal symptoms after maternal addiction than in other equally low birth-weight infants in the same nursery over the same interval. They suggested that perhaps heroin had accelerated surfactant production. Taeusch *et al.* (1972) used the rabbit model to study the effects of heroin in the foetus. They found that direct foetal injections led to a reduction in foetal weight without a reduction in maternal weight, suggestive of a general stress to the foetus. Lungs of heroin-injected foetuses were more distensible and stable than those of saline-injected control foetuses. At sacrifice 72 hours post-injection, on the 27th day, the lungs resembled those of 29–30-day rabbits. In human foetuses, it is possible that withdrawal from heroin could increase foetal adrenal cortical output and thus accelerate lung maturation, since an excess excretion of corticoid metabolites has been found in adults during withdrawal from addiction (Eisenman *et al.* 1961). Alternatively, heroin may act directly on the foetal lung to induce maturation.

Acceleration of surfactant production would surely be useful to prevent or ameliorate respiratory distress in infants. Conservation of existing stores would also be constructive. Clements (1970) noted the evidence that extremes of lung volume and rapid ventilation tended to deplete the lungs of surfactant; low volume because it promotes atelectasis and requires the formation of a new surface film with reinflation; high volume because the return to a functional residual volume could promote film collapse and removal of surfactant. The application of positive end-expiratory pressure to avoid collapse and allow a longer interval for gas exchange has produced substantial immediate benefit to the infants and the promise of lower mortality (Gregory *et al.* 1971).

PRENATAL DETECTION

The possibility of prenatal detection of the surfactant from studies on amniotic fluid was first suggested by Gluck *et al.* (1971). They demonstrated changes in lecithin and sphingomyelin concentrations throughout the last weeks of gestation, with a sharp increase in lecithin from 34 to a peak at 36 weeks. Sphingomyelin tended to peak at 30–32 weeks, and fell after 35 weeks of gestation. Gluck *et al.* suggested that the ratio of lecithin to sphingomyelin would be a useful way to obviate problems with differing amounts of amniotic fluid, and should be predictive of the respiratory distress syndrome. When the ratio of lecithin to sphingomyelin exceeds 2.0, the infant is unlikely to have respiratory distress; between 1.5 and 1.9, the distress will be mild; and a ratio of 1.0 to 1.49 is predictive of immaturity of the lung and moderate to severe respiratory distress. Whitfield *et al.* (1972) agree that the determination is predictive of respiratory distress in the infant. Others have questioned the validity of these predictors, and suggest that the concentration of lecithin in the amniotic fluid is a better indicator of the likelihood of respiratory distress than is the lecithin–sphingomyelin ratio (Nelson, 1969; Bhagwanani *et al.* 1972).

A more rapid and more readily available test was described by Clements *et al.* (1972). Their method depends on the ability of the pulmonary surfactant to generate stable bubbles in the presence of ethanol. By their method, the surfactant becomes detectable in amniotic fluid at about 33 weeks, although variable from 25 weeks to term. Amniotic fluid is mixed with an equal volume of 95 per cent ethanol, shaken vigorously for 15 seconds and allowed to stand for 15 minutes. Observation of the tube would allow small stable bubbles to be detected indicating the presence of surfactant. The test is made semiquantitative by preparing serial dilutions of amniotic fluid.

These studies are of significance since they permit prediction of which foetuses may be capable of breathing normally after birth. Occasionally delivery can be delayed until lung maturity is established. Unfortunately, not all births can be postponed to the ideal moment, so means of conserving pre-existing surfactant or supporting infants without it deserve continued study.

What would be the logical extensions of these findings on the developmental aspects of the pulmonary surfactant? Surely the further exploration of the effects, beneficial or adverse, of treatment of the foetus with cortisol before or shortly after delivery deserves study. Other areas of endeavour might be the study of the kinetics of surfactant synthesis and degradation in the foetus, and the factors that affect it. How important is a critical oxygen tension, hydrogen ion concentration or blood flow? How much benefit can be gained from continuous positive airway pressure and why? What are the regulators of surfactant metabolism? What are the crucial substrates for synthesis? Investigators in many parts of the world are concerned with these issues.

REFERENCES

Adam, N. K. (1930). *Physics and Chemistry of Surfaces*. Oxford University Press.

Adams, F. H. *et al.* (1965). *J. Pediat.* **66**, 357.

Adams, F. H., Fujiwara, T. & Latta, H. (1971). *Biol. Neonate* **17**, 198.

Avery, M. E. & Mead, J. (1959). *Am. J. Dis. Child.* **97**, 517.

Avery, M. E. & Said, S. (1965). *Med.* **44**, 503.

Ballard, P. L. & Ballard, R. A. (1972). *Pediat. Res.* **6**, 338.

Barcroft, J. (1895). *Rep. Belfast nat. Hist. Soc.*, p. 24.

Barcroft, J. (1947). *Researches on Pre-Natal Life*. Charles Thomas, Springfield, Ill.

Bhagwanani, S. G., Fahmy, D. & Turnbull, A. C. (1972). *Lancet* **1**, 159.

Blackburn, W. R. *et al.* (1972). *Lab. Invest.* **26**, 306.

Boughton, K., Gandy, G. & Gairdner, D. (1970). *Arch. Dis. Child.* **45**, 311.

Brumley, G. W., Hodson, W. A. & Avery, M. E. (1967). *Pediatrics* **40**, 13.

Brumley, G. W. *et al.* (1967). *J. clin. Invest.* **46**, 863.

Buckingham, S. *et al.* (1966). *Am. J. Path.* **48**, 1027.

Buckingham, S. *et al.* (1968). *Fed. Proc.* **27**, 328.

Chida, N. & Adams, F. H. (1967). *Pediat. Res.* **1**, 364.

Clements, J. A. (1957). *Proc. Soc. exp. Biol. Med.* **95**, 170.

Clements, J. A. (1962). *Physiologist* **5**, 11.

Clements, J. A. (1970). *Am. resp. Dis.* **101**, 984.

Clements, J. A. *et al.* (1972). *New Eng. J. Med.* **286**, 1077.

Dawes, G. (1968). *Foetal and Neonatal Physiology*. Year Book Medical Publishers Inc., Chicago.

deLemos, R. A. *et al.* (1969). *Pediat. Res.* **3**, 505.

deLemos, R. A. *et al.* (1970). *Am. Rev. resp. Dis.* **102**, 459.

deLemos, R. A., Diserens, W. & Halki, J. (1971). *Combined Abst. Am. Ped. Soc. & Soc. for Ped. Res.* 272.

Eisenman, A. (1961). *J. Pharm. exp. Ther.* **132**, 226.

Elsbach, P. (1966). *Biochem. biophys. Acta* **125**, 510.

Farrell, P. M. & Zachman, R. D. (1972). *Ped. Res.* **6**, 337.

Felts, J. M. (1964). *Health Physics* **10**, 973.

Gandy, G., Jacobson, W. & Gairdner, D. (1970). *Arch. Dis. Child.* **45**, 289.

Giannopoulos, G., Mulay, S. & Solomon, S. (1972). *Biochem. biophys. Res. Comm.* **47**, 411.

Glass, L. *et al.* (1971). *Lancet* **2**, 685.

Gluck, L., Sribney, M. & Kulovich, M. (1967). *Pediat. Res.* **1**, 247.

Gluck, L. *et al.* (1967). *Pediat. Res.* **1**, 237.

Gluck, L. *et al.* (1971). *Am. J. Obstet. Gynec.* **108**, 440.

Gluck, L. *et al.* (1972). *Pediat. Res.* **6**, 81.

Gregory, G. A. *et al.* (1971). *New Eng. J. Med.* **284**, 1333.

Gruenwald, P. (1947). *Am. J. Obstet. Gynec.* **53**, 996.

Gruenwald, P. (1955). *Bull. Margaret Hague Maternity Hosp.* **8**, 100.

Gruenwald, P. (1960). *Lancet* **1**, 230.

GRUENWALD, P. *et al.* (1962). *Proc. Soc. exp. Biol. Med.* **109**, 369.
HOWATT, W. F. *et al.* (1965). *Clin. Sci.* **29**, 239.
HUMPHREYS, P. W. & STRANG, L. B. (1967). *J. Physiol., London* **192**, 53.
KIKKAWA, Y., MOTOYAMA, E. K. & GLUCK, L. (1968). *Am. J. Path.* **52**, 177.
KOTAS, R. V. & AVERY, M. E. (1971). *J. appl. Physiol.* **30**, 358.
KUMODE, S. (1968). *Acta paediat. Jap.* **10**, 51.
LANGMUIR, I. (1917). *J. Am. chem. Soc.* **39**, 1848.
LIGGINS, G. C. (1969). *J. Endocr.* **45**, 515.
MOOG, F. (1953). *J. Exp. Zool.* **124**, 329.
MOTOYAMA, E. K. *et al.* (1971). *Pediatrics* **48**, 547.
NELSON, G. H. (1969). *Am. J. Obstet. Gynec.* **105**, 1072.
ORZALESI, M. M. *et al.* (1965). *Pediatrics* **35**, 373.
PATTLE, R. E. (1961). *J. Pathol. Bacteriol.* **82**, 333.
PATTLE, R. E. (1955). *Nature* **175**, 1125.
PATTLE, R. E. (1958). *Proc. R. Soc. London (Series B)* **148**, 217.
PATTLE, R. E. (1965). *Physiol. Rev.* **45**, 48.
PLATZKER, A. C. G. *et al.* (1972). *Pediat. Res.* **6**, 406.
POCKELS, A. (1891). *Nature* **43**, 437.
REDDING, R. A., DOUGLAS, W. H. J. & STEIN, M. (1972). *Science* **175**, 994.
REYNOLDS, E. O. R. *et al.* (1968). *Pediatrics* **42**, 758.
RUFER, R. (1968). *Respiration* **25**, 441.
SCARPELLI, E. M. (1968). *The Surfactant System of the Lung*. Lea and Febiger, Philadelphia.
SPEAR, G. S. *et al.* (1969). *Biol. Neonat.* **14**, 344.
TAEUSCH, H. W. Jr, *et al.* (1972). *Pediat. Res.* **6**, 75.
TAEUSCH, H. W., Jr, HEITNER, M. & AVERY, M. E. (1972). *Am. Rev. resp. Dis.* **105**, 971.
WANG, N. S. *et al.* (1971). *J. appl. Physiol.* **30**, 362.
WHITFIELD, C. R. *et al.* (1972). *Br. Med. J.* **2**, 85.
WILHELMY, L. (1863). *Ann. Physiol.* **119**, 177.
WOLFE, B. M. J. *et al.* (1970). *Can. J. Biochem.* **48**, 170.
WU, B. *et al.* (1971). *Physiologist* **14**, 253.

SURFACTANT PRODUCTION AND FOETAL MATURATION

By LOUIS GLUCK and MARIE V. KULOVICH
University of California, San Diego School of Medicine,
La Jolla, California 92037, USA

The studies in our laboratory on the biosynthesis of surface active lecithin in human foetal lung and the maturation of human foetal lung as reflected by the surface active phospholipids in amniotic fluid showed, when compared, dramatic differences in biochemical development of surfactant between primates and other mammalia. Generalizations based on data gathered from sub-primate species may be misleading.

SURFACTANT IN LUNG OF THE RABBIT FOETUS

Concentrations of surface active lecithin in homogenates of foetal lungs of rabbits (and also in sheep) rise during late gestation. The surface active lecithin, however, is almost entirely intracellular until very late in gestation (Gluck, Motoyama, Smits & Kulovich, 1967; Gluck, Sribney & Kulovich, 1967; Kikkawa, Motoyama & Gluck, 1968). If intact, non-breathed lungs of rabbit foetuses delivered by Caesarean section are lavaged gently with saline and the 'alveolar wash' examined, surface active lecithin is not found until day 28 of a 31-day gestation when a minute amount is detectable (Gluck, Motoyama, Smits & Kulovich, 1967; Kikkawa, Motoyama & Gluck, 1968).

Prior to breathing, even at term, the surface active alveolar lecithin constitutes only about 11 per cent of the total alveolar lecithin fraction (Gluck, Motoyama, Smits & Kulovich, 1967). By day 29, at 90 per cent of gestation, there is adequate lecithin production in the foetal lung and at delivery foetuses have sufficient alveolar stability to survive with no respiratory difficulties.

Once the delivered foetus begins to breathe there is a great outpouring of surface active lecithin from the cells while at the same time a fantastic rate of synthesis of surface active lecithin occurs. In the term foetus, by one hour of life the concentration of surface active lecithin in alveolar wash has increased *30-fold* over the concentration found in the term non-breathed lung (Kikkawa, Motoyama & Gluck, 1968). The proportion of total lecithin in alveolar wash which is surface active rises from 11 to 45 per cent by 1 hour (Gluck, Motoyama, Smits & Kulovich, 1967). However, even more incredible is the rate of synthesis of surface active lecithin in alveolar wash; radioactive precursors injected into the 29-day foetus at birth (by Caesarean section) show *100 per cent* labelled surface active lecithin in alveolar wash at one hour of life (Gluck, Scribney & Kulovich, 1967)!

The fatty acid ester structure of the surface active alveolar lecithin in the rabbit (or sheep) newborn is primarily dipalmitoyl lecithin, with palmitic acid on both the α and β carbons (Gluck, Landowne & Kulovich, 1970); this is the end product of the reaction in the enzymatic biosynthesis Pathway I, associated with the enzyme *phosphocholine* (*cytidyl*)-*transferase* : I. Cytidine-diphospho-choline + D-α,β-diglyceride → lecithin. This enzymatic pathway does not become fully active until about day 29 when it responds to birth by extremely rapid synthesis of surface active lecithin (Gluck, Motoyama, Smits & Kulovich, 1967).

SURFACTANT IN LUNG OF THE HUMAN FOETUS

Although there are gross similarities in surfactant metabolism between rabbit and human foetuses, the differences are profound. Pathway I in the human becomes mature (significantly active) at about 35 weeks' gestation (Gluck, Kulovich, Eidelman, Cordero & Khazin, 1972; Gluck, Kulovich, Borer, Brenner, Anderson & Spellacy, 1971). In measurements on a limited number of stillborn foetuses, the concentrations of intracellular surface active lecithin rose during gestation to very high levels, similar to those in the rabbit. However, surface active lecithin in alveolar wash is present as early as about 20–22 weeks (Gluck, Kulovich, Eidelman, Cordero & Khazin, 1972). By 22–24 weeks surface active alveolar wash lecithin is present in easily detectable quantities. The concentrations, both in lung tissue and in alveolar wash, continue to increase (Gluck, Kulovich & Brody, 1966; Gluck, Kulovich, Eidelman, Cordero & Khazin, 1972).

The predominant surface active lecithin prior to biochemical maturity at about 35 weeks' gestation is α-palmitic/β-myristic, or palmitoylmyristoyl lecithin, and is formed by Pathway II (Gluck, Sribney & Kulovich, 1967; Gluck, Landowne & Kulovich, 1970; Gluck, Kulovich, Eidelman, Cordero & Khazin, 1972), catalysed by a transmethylation enzyme:

II. Phosphatidyl ethanolamine + 3 CH_3 (from *S*-adenosyl-L-methionine) → lecithin.

At about 35 weeks the phosphocholine cytidyl transferase pathway (I) becomes mature and there is an increase in β-carbon palmitic acid on surface active lecithin (Gluck, Kulovich, Eidelman, Cordero & Khazin, 1972). At term and certainly by 18 hours of life, there are about equal concentrations of surface active dipalmitoyl and palmitoylmyristoyl lecithins (Gluck, Kulovich, Eidelman, Cordero & Khazin, 1972).

AMNIOTIC FLUID LECITHINS

Comparisons of the surface active phospholipids in amniotic fluid, lecithin and sphingomyelin between humans and rabbits (as well as sheep) show that extremely little phospholipids enter amniotic fluid in the rabbit where even at term concentrations are very low.

In the human pulmonary phospholipids are secreted into the amniotic fluid from about 20–22 weeks (Gluck, Kulovich, Borer, Brenner, Anderson & Spellacy, 1971). Initially concentrations of lecithin and sphingomyelin are low but rise as pregnancy progresses to relatively high concentrations.

The relationships of lecithin and sphingomyelin to each other are diagnostic of the state of maturity of the foetal lung and form the basis of a precise diagnostic test, the lecithin/sphingomyelin (L/S) ratio (Gluck, Kulovich, Borer, Brenner, Anderson & Spellacy, 1971). Concentrations of sphingomyelin early in gestation exceed those of lecithin until about week 30 of gestation when the two are equal, after which lecithin concentration exceeds that of sphingomyelin. At 35 weeks' gestation, a sharp rise in lecithin concentration heralds biochemical maturity of foetal lung (Gluck, Kulovich, Borer, Brenner, Anderson & Spellacy, 1971). When compared by reflectance densitometry, an L/S ratio of 2.0 or more marks the mature lung when there will be no RDS following birth.

INFLUENCES IN DEVELOPMENT BY MATERNAL DISEASE

The timetables for development of lung in the human foetus and the appearance of the surface active phospholipids are orderly for pregnancies that are normal and uncomplicated. A series of maternal conditions have been evaluated, however, which have a profound influence on the timing of surfactant maturation. Most of these conditions produce high risk pregnancy and may be associated with very high foetal and neonatal mortality and morbidity (Gluck & Kulovich, 1972; Gluck, Kulovich & Gould, 1972).

One group of pregnancies with abnormally small placentae, placental infarctions, fibrinous degeneration, and small umbilical cords, are associated with marked *acceleration of maturation* of foetal lung function (plus liver and CNS functions as well). L/S ratios in this group mature prior to the 33rd week of gestation and as early as 27–28 weeks. The maternal conditions include hypertensive disease ('chronic' toxaemia, renal disease, cardiovascular disease), sickle-C disease, some heroin and morphine addicts, diabetes mellitus classes D, E, F, and placental diseases (retroplacental bleeding, insufficiency, circumvallate placenta) (Gluck & Kulovich, 1972).

Another group of maternal conditions produces a *delay in biochemical maturation of lung*. Diabetes mellitus, classes A, B, C, is the prime example. Here maturation does not occur until week $36\frac{1}{2}$ to $37\frac{1}{2}$. Certain cases of hydrops foetalis, non-hypertensive chronic glomerulonephritis, and the smaller of a pair of non-parasitic monochronial twins also are associated with delays in the time of surfactant maturation (Gluck & Kulovich, 1972).

These findings suggest that it is of paramount importance to identify the mechanisms controlling the onset of enzymatic action for synthesis of surface active lecithin for both the fundamental biological information and the clinical significance. Theoretically this information could provide a clinical means to overcome the morbid aspects of prematurity.

Studies are supported by research grants HD-01299, HD-00989, FR-05358, HD-04143, HD-04380, and HE-14169 from the National Institutes of Health; Maternal and Child Health funds of the California State Department of Health; and the Gerber Products Company.

REFERENCES

Gluck, L. & Kulovich, M. V. (1972). *Am. J. Obst. Gynec.* (In press.)
Gluck, L., Kulovich, M. V., Borer, R. C., Jr, Brenner, P. H., Anderson, G. G. & Spellacy, W. N. (1971). *Am. J. Obst. Gynec.* **109**, 440.
Gluck, L., Kulovich, M. V. & Brody, S. J. (1966). *J. Lipid Res.* **7**, 570.
Gluck, L., Kulovich, M. V., Eidelman, A. I., Cordero, L. & Khazin, A. F. (1972). *Pediat. Res.* **6**, 81.
Gluck, L., Kulovich, M. V. & Gould, J. B. (1972). *Pediat. Res.* **6**, 409.
Gluck, L., Landowne, R. A. & Kulovich, M. V. (1970). *Pediat. Res.* **4**, 352.
Gluck, L., Motoyama, E. K., Smits, H. L. & Kulovich, M. V. (1967). *Pediat. Res.* **1**, 237.
Gluck, L., Sribney, M. & Kulovich, M. V. (1967). *Pediat Res.* **1**, 247.
Kikkawa, Y., Motoyama, E. K. & Gluck, L. (1968). *Am. J. Pathol.* **52**, 177.